Advances in Spectros
and Sensing

NATO Science Series

A Series presenting the results of scientific meetings supported under the NATO Science Programme.

The Series is published by IOS Press, Amsterdam, and Springer in conjunction with the NATO Public Diplomacy Division.

Sub-Series

I. Life and Behavioural Sciences	IOS Press
II. Mathematics, Physics and Chemistry	Springer
III. Computer and Systems Science	IOS Press
IV. Earth and Environmental Sciences	Springer

The NATO Science Series continues the series of books published formerly as the NATO ASI Series.

The NATO Science Programme offers support for collaboration in civil science between scientists of countries of the Euro-Atlantic Partnership Council. The types of scientific meeting generally supported are "Advanced Study Institutes" and "Advanced Research Workshops", and the NATO Science Series collects together the results of these meetings. The meetings are co-organized by scientists from NATO countries and scientists from NATO's Partner countries – countries of the CIS and Central and Eastern Europe.

Advanced Study Institutes are high-level tutorial courses offering in-depth study of latest advances in a field.
Advanced Research Workshops are expert meetings aimed at critical assessment of a field, and identification of directions for future action.

As a consequence of the restructuring of the NATO Science Programme in 1999, the NATO Science Series was re-organized to the four sub-series noted above. Please consult the following web sites for information on previous volumes published in the Series.

http://www.nato.int/science
http://www.springer.com
http://www.iospress.nl

Series II: Mathematics, Physics and Chemistry – Vol. 231

Advances in Spectroscopy for Lasers and Sensing

edited by

Baldassare Di Bartolo

Boston College, Chestnut Hiil, MA,
U.S.A.

and

Ottavio Forte

Boston College, Chestnut Hill, MA,
U.S.A.

Published in cooperation with NATO Public Diplomacy Division

Proceedings of the NATO Advanced Study Institute on
New Developments in Optics and Related Fields
Erice, Sicily, Italy
6-21 June, 2005

A C.I.P. Catalogue record for this book is available from the Library of Congress.

ISBN-10 1-4020-4788-6 (PB)
ISBN-13 978-1-4020-4788-6 (PB)
ISBN-10 1-4020-4787-8 (HB)
ISBN-13 978-1-4020-4787-9 (HB)
ISBN-10 1-4020-4789-4 (e-book)
ISBN-13 978-1-4020-4789-3 (e-book)

Published by Springer,
P.O. Box 17, 3300 AA Dordrecht, The Netherlands.

www.springer.com

Printed on acid-free paper

Printed in the Netherlands.

CONTENTS

PREFACE

This volume presents the Proceedings of the Institute "New Development in Optics and Related Fields," held in Erice, Sicily, Italy, from the 6th to the 21st of June, 2005. This meeting was organized by the International School of Atomic and Molecular Spectroscopy of the "Ettore Majorana" Center for Scientific Culture.

The purpose of this Institute was to provide a comprehensive and coherent treatment of the new techniques and contemporary developments in optics and related fields. Several lectures of the course addressed directly the technologies required for the detection and identification of chemical and biological threats; other lectures considered the possible applications of new techniques and materials to the detection and identification of such threats.

Each lecturer developed a coherent section of the program starting at a somewhat fundamental level and ultimately reaching the frontier of knowledge in the field in a systematic and didactic fashion. The formal lectures were complemented and illustrated by additional seminars and discussions. The course was addressed to workers in spectroscopy-related fields from universities, laboratories and industries. Senior scientists were encouraged to participate. The Institute provided the participants with an opportunity to present their research work in the form of short seminars or posters.

The secretary of the course was Ottavio Forte.

The participants came from 18 different countries: Belarus, Bulgaria, France, Georgia, Germany, Italy, Lithuania, Poland, Portugal, Romania, Russia, Spain, Switzerland, The Netherlands, Turkey, United Kingdom, Ukraine, and the Unites States of America. There were 30 formal lectures, three interdisciplinary lectures, 10 seminars and 21 posters. In addition this year there were six "This is My Country" presentations in which the presenter introduced his country to all participants. Two round-table discussions were held. The first round-table discussion took place during the first week of the school in order to evaluate the work done and consider suggestions and proposals regarding the organization, format and presentation of the lectures. The second round-table was held at the conclusion of the course so that the participants could comment on the work done during the entire meeting and discuss various proposals for the next course of the International School of Atomic and Molecular Spectroscopy.

I wish to acknowledge the sponsorship of the meeting by NATO, NASA, the ENEA Organization, Boston College, the Italian Ministry of University and Scientific Research and Technology, the USA National Science Foundation, and the Sicilian Regional Government.

I would like to thank the Co-Director of the Course, Academician Alexander Voitovich, the members of the organizing committee (Prof. Steve Arnold, Dr. Giuseppe Baldacchini, Dr. Norman Barnes, Prof. Claus Klingshirn, Prof. Eric Mazur, Prof. Ralph von Baltz, and Prof. Martin Wegener), the secretary of the course, Mr. Ottavio Forte, Prof. John Collins, and Prof. Gonul Ozen for their help in organizing and running the course.

I am looking forward to our activities at the Majorana Centre in years to come, including the next 2007 meeting of the International School of Atomic and Molecular Spectroscopy.

Baldassare (Rino) Di Bartolo
Director of the International School of
Atomic and Molecular Spectroscopy of
the "Ettore Majorana" Center

Beau mot que celui de chercheur, et si preférable à celui de savant! Il exprime la saine attitude de l'esprit devant la verité: le manque plus que l'avoir, le désir plus que la possession, l'appétit plus que la satieté.

Jean Rostand

LIST OF PAST INSTITUTES

Advanced Study Institutes Held at the "Ettore Majorana" Centre
in Erice, Sicily, Italy, organized by the International School of Atomic
and Molecular Spectroscopy:

1974 – Optical Properties of Ions in Solids
1975 – The Spectroscopy of the Excited State
1977 – Luminescence of Inorganic Solids
1979 – Radiation-less Processes
1981 – Collective Excitations in Solids
1983 – Energy Transfer Processes in Condensed Matter
1985 – Spectroscopy of Solid-State Laser Type Materials
1987 – Disordered Solids: Structures and Processes
1989 – Advances in Non-radiative Processes
1991 – Optical Properties of Excited State in Solids
1993 – Nonlinear Spectroscopy of Solids: Advances and Applications
1995 – Spectroscopy and Dynamics of Collective Excitations in Solids
1996 – Workshop on Luminescence Spectroscopy
1997 – Ultra-fast Dynamics of Quantum Systems: Physical Processes and Spectroscopic Techniques
1998 – Workshop on Advances in Solid State in Luminescence Spectroscopy
1999 – Advances in Energy Transfer Processes
2000 – Workshop on Advanced Topics in Luminescence Spectroscopy
2001 – Spectroscopy of Systems with Spatially Confined Structures
2002 – Workshop on the Status and Prospects of Luminescence Research
2003 – Frontiers of Optical Spectroscopy: Investigating Extreme Physical Conditions with Advanced Optical Techniques
2004 – Workshop on Advances in Luminescence Research

The Participants

"The Directors"

"The Workers"

SPECTROSCOPY OF PHOTONIC ATOMS: A MEANS FOR ULTRA-SENSITIVE SPECIFIC SENSING OF BIO-MOLECULES

STEPHEN ARNOLD AND OPHIR GAATHON
Microparticle Photophysics Laboratory(MP3L)
Polytechnic University, Brooklyn, N.Y. 11201
arnold@photon.poly.edu,sarnold935@aol.com,

Abstract

By combining bio-recognition using bio-nano-sensors with the transduction capability of resonant dielectric micro-cavities, we demonstrate the specific identification of protein and DNA. This is accomplished through the spectroscopy of a microcavity as molecular adsorption takes place on its functionalized surface. A surprising aspect is that this sensing paradigm is not only ultra-sensitive but also provides a measure of molecular weight and the thickness of molecular monolayers on the microcavity surface.

1. Introduction

The specific detection and measurement of the concentration of biological entities (e.g. protein, DNA, virus) is important since the appearance of specific bio-molecules can alter our lives for better and worse An optical approach to this task is what this paper is all about. It fits into a new area, Bio-Photonics, for which there is much current activity.[1]

One must first realize that most bio-molecules (e.g. protein, DNA) fall into the nanoscopic realm. For example, in our blood a particular protein, Human Serum Albumin (HSA) is prominent. It is about 3 nm in size, however with a molecular weight of 66,438 (also specified as 66.438 kDa) it contains a vast multitude of vibrational modes, principally associated with 20 amino acids. Since virtually all protein molecules contain these same amino acids (constructed in different sequences), the IR spectra from one protein to the other are virtually the same. Biology runs itself on these proteins. Each of the approx. 30,000 genes within our nuclear DNA encode for a separate protein. They are biologically distinguishable, but optically very similar. Faced with this difficulty an optical specialist has to rethink the means by which to identify a given protein. Some years ago our laboratory came to a painful conclusion. Molecular spectroscopy is an unlikely candidate!

B. Di Bartolo and O. Forte (eds.), Advances in Spectroscopy for Lasers and Sensing, 1–18.

To understand how to proceed we attempted to learn from nature. There are all sorts of ways in which nature can make light, but there is no light inside our bodies. Yet our bodies recognize tens of thousands of different proteins well enough to enable us to function. Biological interaction involves multiple unique chemical bonding. Each type of interaction involves complementary molecular structure from all bodies involved. In essence it is no different then the key to lock match (but on a much smaller scale). For example in a sensitized person, the allergic reaction to certain toxic proteins on the surface of a pollen grain causes a specific antibody to engulf the invading allergen like a lock covering a key. This highly specific physio-chemical recognition will not occur with other proteins. Another example is the hybridization that occurs between complimentary strands of DNA. If we mimic biology, we would also sense bio-molecules through theses "dark interactions", rather than looking for changes in the molecular spectrum. The trick is to distribute a variety of distinguishable locks on separate transducers and sensitively determine onto which lock the key attaches.

As a transducer we have chosen a resonant dielectric microcavity whose state functions resemble those of a hydrogen atom. This "photonic atom" will be found to shift the frequency of its resonant modes by the surface adsorption of a few bio-molecules.[2,3]

In what follows we first describe the photonic atom (PA) descriptively followed by some of its more detailed physics. Following this we discuss the sensitivity of the PA to molecular adsorption. Finally we describe specific detection schemes for protein and DNA without labels.

2. Photonic Atom

Photonic Atoms are dielectric microspheres that demonstrate high Q resonances associated with Whispering Gallery Modes (WGMs).

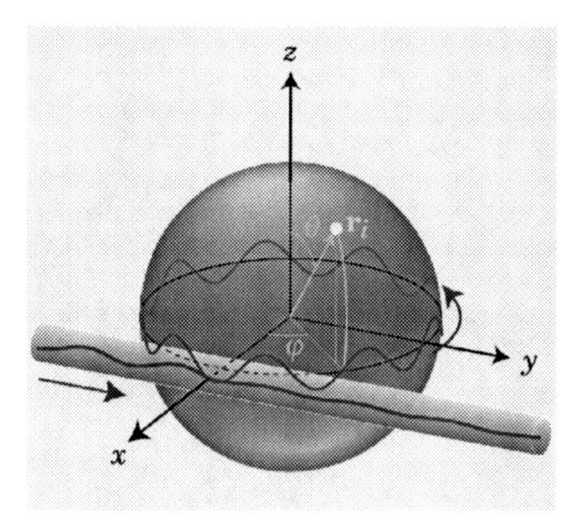

Fig. 1 Illustration of coupling of optical energy between a guided wave in an tapered optical fiber and a Photonic Atom mode in a microsphere.[4] The vector $\mathbf{r}_i$ points to a small nano-perturbation.

A high Q resonance is one with a very narrow linewidth in frequency δf; $\delta f = f/Q$. From a geometrical point of view light is confined within the microsphere by total internal reflection (TIR) against the inner surface. As with all systems that confine light by TIR an exponentially decaying evanescent field extends just beyond (~200 nm) the sphere radius (~100 μm). It is this field which allows the resonant mode to be influenced from the outside. Fig. 1 shows the wave point of view. When light circumnavigates the sphere and returns in phase the system is resonant. Here evanescent coupling to a guided wave in a tapered optical fiber (dia. 4 μm) stimulates the mode, leading to a dip in the transmission spectrum through the fiber.[5] The wave circumnavigating the sphere is reminiscent of a *de Broglie matter wave* circling the electron orbit in a Bohr atom. Indeed the detailed description of this atomic system is similar to the description of the photon orbit in the dielectric sphere (Fig. 1).

In hydrogen the electron orbits are distinguished by the parameter set ν, l, m, s representing the principle (radial) quantum number, the angular momentum quantum number, the azimuthal quantum number (z-projection of angular momentum), and the spin quantum number, respectively. In the photonic atom the first three are still used, but the spin quantum number is replaced with the polarization state of the orbit, which we will designate as P. As in the case of the spin, P has two values designated as TE for transverse electric and TM for transverse magnetic. The simple wave depicted in Fig. 1 is transverse electric with the electric field oscillating tangent to the surface. The electric field of a TM resonance oscillates in the plane of the orbit. One can specifically select TE or TM modes by polarizing light in the coupling fiber either tangent or perpendicular to the surface of the microsphere, respectively. We will designate a particular mode by P^{ν}_{lm}. The mode in Fig. 1 happens to be equatorial with its angular momentum directed along the z-axis. For such a mode l = m. The designation for this particular mode is $TE^{1}_{13,13}$.

Each photonic atom mode occurs at a specific optical frequency. However, since these frequencies are sensitive to the circumference of the orbit even a nanoscopic particle adsorbed on the surface can shift this frequency. This frequency shift provides the transduction principle for the Photonic Atom Bio-sensor.

3. Photonic Atom Physics: Heuristics

To understand the sensitivity of a Photonic Atom mode to perturbation we take a heuristic approach. Consider a mode in the equatorial plane (Fig. 2a) and imagine that we add a layer of identical material to the surface of the microsphere (Fig. 2b).
To maintain the same mode l must be invariant. This scales the whole problem up and as a consequence the wavelength within the mode is changed in proportion to the thickness of the layer. On this basis the fractional increase in wavelength $\delta\lambda/\lambda$ will be approximately equal to the fractional increase in radius t/a,

$$\frac{\delta\lambda}{\lambda} \approx \frac{t}{a}. \tag{1}$$

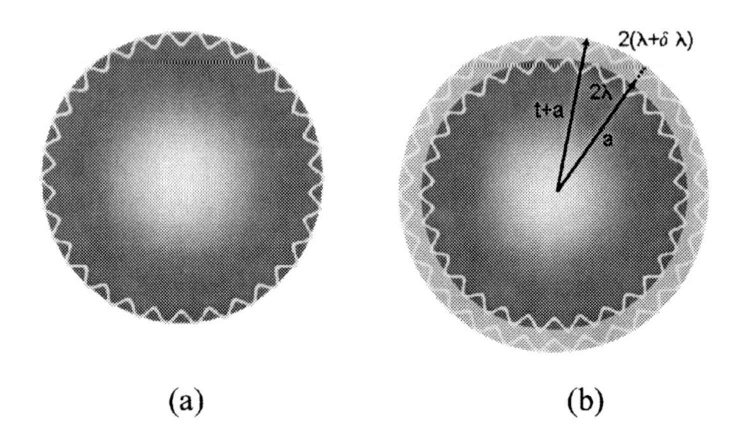

Fig. 2 (a) Photonic Atom Mode; (b) Anticipated wavelength change caused by the addition of a spherically symmetric layer.

Suppose now that the adsorbing material is 1.0 nanometer in thickness. For a sphere having a 100 μm radius, the fractional shift in wavelength according to Eqn. 1 would be $\sim 10^{-5}$. This is smaller than the resolution of a grating spectrometer, but is a "piece of cake" (i.e. easy) for the microsphere as a spectrometer. The reason is that the resonances of a microsphere are extremely narrow. Resonances with Q's of 10^7 (linewidth of 1 part in 10^7) are considered "broad".[6] But such a "broad" line would shift 100 times its linewidth for a 1.0 nm layer. For it to shift just one linewidth requires only a 10 picometer layer (i.e. one tenth the size of a hydrogen atom). So such a small perturbation should be easy to observe. In fact, it is not difficult to define a resonance position to $1/50^{th}$ of a linewidth, which allows for the observation of a small fraction of a monolayer.

One should doubt Eqn. 1. Although it is heuristically appealing, it is limited. After all, adsorbed bio-molecules most likely don't have the same dielectric properties as the glass we will ultimately use for our microcavity. In addition, we have clearly simplified the properties of the modes. Further understanding can only be obtained from a solution for $\delta\lambda/\lambda$ from Maxwell's equations.

4. Photonic Atom Physics 101

A meaningful attempt to obtain an exact solution to the problem of layer perturbation was described within the proceedings of the previous school.[7] That work has been extended. Herein we will describe the method of calculation in brief, and report the major result.

Our approach relies on a quantum analog of the electrodynamic problem. It is our belief that exposure to quantum mechanics is more ubiquitous to the sciences (e.g. Physics, Chemistry, and Biology) than electromagnetics. So where it is possible we reduce the vector time-harmonic electromagnetic problem to a scalar quantum analog. This is not particularly difficult in the case of the spherical dielectric cavity since as we have already pointed out there are similarities between this problem and the atomic problem.

As a first step we must solve the source free vector Helmholtz equation,

$$\nabla^2 \mathbf{E} + k^2 \mathbf{E} = 0. \tag{2}$$

We are particularly interested in modes for which the field and its derivative are continuous at the boundary, just as the wave function would be in quantum mechanics. Of course we would also like to reduce Eqn. 1 to a scalar equation. There is a simple choice. We construct the field as the angular momentum operator acting on a scalar function, $\mathbf{E} = \hat{L}\psi$. Since the angular momentum operator has only angular components **E** will be tangent at the microsphere surface. With this choice for the field the problem of solving the vector wave equation is easily reduced to the solution of a Schrödinger-like equation for $\psi_r = r\psi$.[8] The effective energy E_{eff} for this quantum analog is the square of the free space wave vector; $E_{eff} = k_0^2$, and the effective potential $V_{eff} = k_0^2(1-n^2) + l(l+1)/r^2$, where n is the radial refractive index profile, and l is the angular momentum quantum number of a particular mode. A layer perturbation corresponds to changing n^2 from the surface out to a thickness t by $\delta(n^2)$. The first order perturbation has an identical form as we are use to in quantum mechanics,

$$\delta E_{eff} \approx \langle \psi_r | \delta V_{eff} | \psi_r \rangle, \tag{3}$$

where ψ_r is constructed from the appropriate quasi-normalized functions.[9] After substituting for the major components in Eqn. 1 we find that the fractional perturbation in the effective energy is

$$\frac{\delta(k_0^2)}{k_0^2} = -2\frac{\delta\lambda}{\lambda} = -\left[\frac{2t}{a}\left(\frac{\delta(n^2)}{n_s^2 - n_m^2}\right)\right] \times \left[\frac{L}{t}\left(1 - e^{-t/L}\right)\right] \tag{4}$$

where n_s, and n_m are the refractive indices of the sphere (silica, 1.47), and its environment (water, 1.33), respectively, a is the sphere radius, and L is the evanescent field length for grazing incidence; $L = (\lambda/4\pi)(n_{eff}^2 - n_m^2)^{-1/2}$. If we allow $t/L \ll 1$ and the dielectric constant to be the same as the sphere, Eqn.2 reduces to $\delta\lambda/\lambda \approx t/a$. Aside from being consistent with our heuristic thinking (Eqn. 1), Eqn. 2 has an interesting structure which allows us to anticipate a number of measurement possibilities in relation to bio-layers.

5. Photonic Atom Sensor as a Nanoscopic Ruler [10]

First it should be possible to measure not only the thickness of a nanolayer, but also its excess dielectric constant, simply by obtaining the wavelength shifts of two separate resonance in the same sphere and at the same time. This is because Eqn. 2 has a particularly simple structure when one considers that the principle wavelength dependence is contained within the evanescent field length in the rightmost factor on the right hand side. By a judicious choice of the wavelength regions to be used, the leftmost factor on the right hand side can be considered relatively constant. Consequently, by taking a ratio of the fractional shift at one wavelength λ_1 to that at a longer wavelength λ_2, we arrive at a particularly simple expression that provides the design principle for this surface analysis technique. This ratio S is

$$S = \frac{\left(\frac{\delta\lambda}{\lambda}\right)_1}{\left(\frac{\delta\lambda}{\lambda}\right)_2} \approx \frac{L_1\left[1-e^{-t/L_1}\right]}{L_2\left[1-e^{-t/L_2}\right]}. \tag{5}$$

For an ultra thin layer (i.e t/L_1, $t/L_2 << 1$) S approaches 1, whereas for a thick layer (i.e t/L_1, $t/L_2 >> 1$) S approaches L_1 / L_2, which with n_{eff} taken constant is just λ_1/λ_2. For our experiments this ratio is (760nm/1310nm) = 0.58. For these chosen wavelengths S falls off in an approximate exponential fashion in between the two extreme cases with a characteristic length of t_c = 192 nm [i.e. $S \approx (L_1/L_2) + (1 - L_1/L_2)\exp(-t/t_c)$]. Measuring S therefore allows us to estimate t. With t in hand, Eqn. 2 gives the excess dielectric constant of the layer, $\delta(n^2)$.

We have performed wavelength multiplexing experiments while forming nano-layers on a silica microsphere surface. Light from two current tunable distributed feed-back (DFB) lasers with nominal wavelengths of 760 nm and 1310 nm was coupled to a single mode fiber (Nufern 780-HP)(Fig. 3). A portion of the fiber was tapered down to 3 μm diameter by acid erosion to facilitate coupling to WGMs of a silica microsphere approx. 350 μm in diameter.[11] The microsphere and fiber were contained within a temperature controlled 1ml cuvette containing buffer solution and a magnetic stirrer. Beyond this cuvette the fiber was led to an InGaAs detector. By scanning both lasers with a synchronous ramp we observe that the light from each independently stimulates WGMs in the microsphere and gives a distinct transmission spectrum with a superposition of resonant dips from each. By observing which resonances disappear as either laser is shut off, the resonances are easily associated with the 760 nm and 1310 nm region. In this way resonances can be identified and tracked.

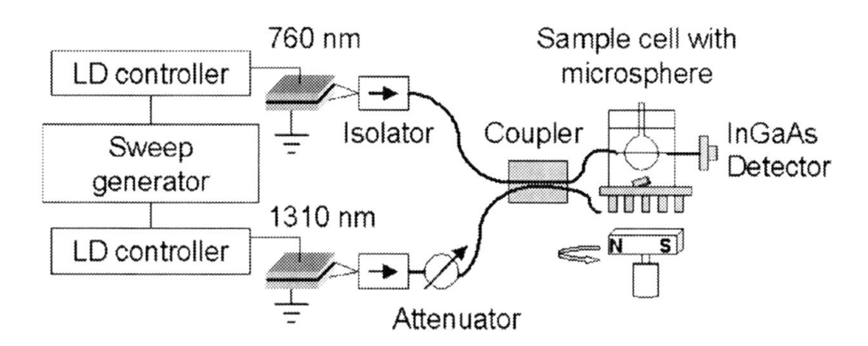

Fig. 3 Experimental setup for wavelength multiplexing of a micro-cavity.[10]

As a test of our perturbation theory we constructed two experiments at the extreme limits. First we built a monolayer of a protein (Bovine Serum Albumin, BSA) much thinner (3 nm) than the evanescent field length (~100 nm).[4,12] Fig. 4 shows a record of the data in real time. The individual curves are typical, and clearly show little difference in the overall shift. In a separate experiment, we added an infinite layer thickness perturbation by increasing the refractive index of the surrounding medium (water) by adding NaCl. In each case the shifts of resonances centered about two wavelengths, λ_1 = 760 nm and λ_2 = 1310 nm were measured. These shift are plotted against each other in Fig. 5. In a separate experiment, we added an infinite layer thickness perturbation by increasing the refractive index of the surrounding medium (water) by adding NaCl. In each case the shifts of resonances centered about two wavelengths, λ_1 = 760 nm and λ_2 = 1310 nm were measured. These shift are plotted against each other in Fig. 4. For BSA adsorption S was measured to be 1.04, which compares well with the $t/L \ll 1$ limit for Eqn. 3 of 1.00. For the NaCl experiment, S was measured to be 0.54 whereas the $t/L \gg 1$ limit of Eqn. 3 corresponds to S = 0.58. This good agreement tests the wavelength dependence of our shift equation and allows the technique to be used as a nanoscopic ruler.

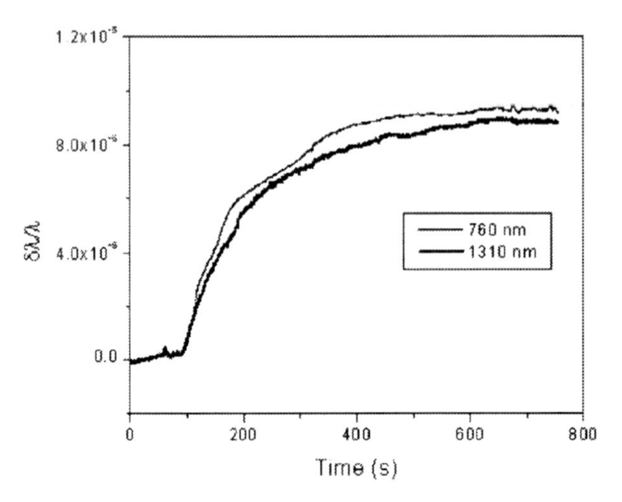

Fig. 4 Resonances shifts at two wavelength [λ_1 = 760 nm (thin) and λ_2 = 1310 nm (bold)] due to BSA adsorption.[10]

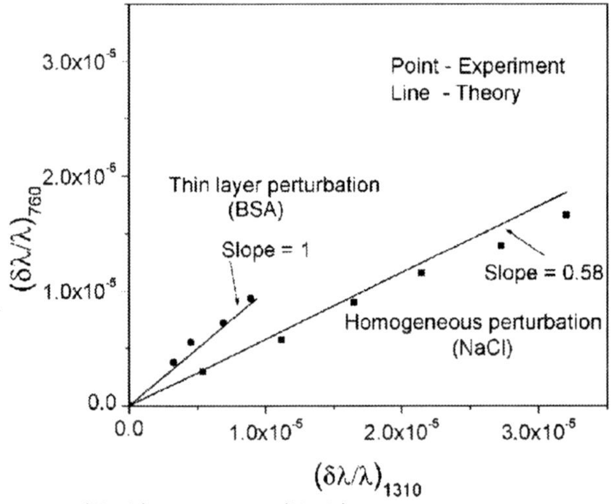

Fig. 5 The plot of $(\delta\lambda/\lambda)_{760\ nm}$ against $(\delta\lambda/\lambda)_{1310\ nm}$ for BSA layer and NaCl addition. The lines are the result of layer perturbation theory.[10]

In addition upon the determination of t using the multiplexing experiment and Eqn. 3, the result can be funneled back into Eqn. 2 for the determination of the $\delta(n^2)$

perturbation. For the NaCl solutions the relationship between concentration and the refractive index increment is well documented. By using these tables, the δn arrived at from salt shift measurements enabled us to back out the salt concentration. Concentrations measured in this way were found to be within 3% of the injected concentrations. The challenge is to use this approach to determine the thickness and δn of a dilute soft condensed layer of intermediate thickness.

Poly-L-lysine (PLL) is a polymer that sticks to silica at biological pH (7.2), and takes on an extreme positive charge in water. Consequently PLL is favored as a means for adsorbing bio-molecules with negative charge. However the physical properties of PLL are difficult to measure since it deposits in a thin layer with an extremely low contrast in a water environment. We used a PLL solution from Sigma (P8920, 0.1% w/v in water, the average molecular weigh 225,000 g/mol), which is commonly used in biology to treat glass slides. To generate a layer, 40 μl of the PLL solution was injected into 900 μl of PBS surrounding the microsphere. We observed a shift toward longer wavelength that saturated in the usual Langmuir fashion[13] for monolayer formation. However, the fractional shift at saturation ~2 x 10^{-6} was well below anything we had seen previously. The slope S based on the average of a number of experiments was 0.82. The slope fed back into Eqn. 3 gives a thickness of approximately 110 nm, which is reasonable considering the molecular structure of the polymer. After substituting this thickness into Eqn. 2, we determined the water excess increment in optical dielectric constant to be $\delta^2(n) = 0.0033$. Consequently $\delta n = 0.0012$, which is indeed small.

6. Protein Molecular Weight Sensitivity of a Photonic Atom Sensor[14]

Water-soluble protein molecules are 1-10 nm in size, considerable less than the length of the evanescent field. With this restricted dimension Eqn. 2 can be reduced to an approximate form

$$\frac{\delta\lambda}{\lambda} = \frac{t}{a}\left(\frac{\delta(n^2)}{n_s^2 - n_m^2}\right) \tag{6}$$

Eqn. 4 suggests that we can normalize data for equatorial radius variations from one microsphere to another by plotting the shift as (a·δλ/λ). We see from Eqn. 4 that this normalized shift is simply proportional to the thickness of the layer. This suggests that protein with differing molecular weight may give different normalized shifts. Based on this reasoning we carried out a number of experiments in which we adsorbed protein molecules with molecular weights ranging over more than two orders of magnitude. For each molecule a saturation shift was established by varying the concentration in solution until the saturated shift plateaus. Fig. 6 shows each of these so-called isotherms. Indeed the larger molecular weight species show the largest overall shifts.

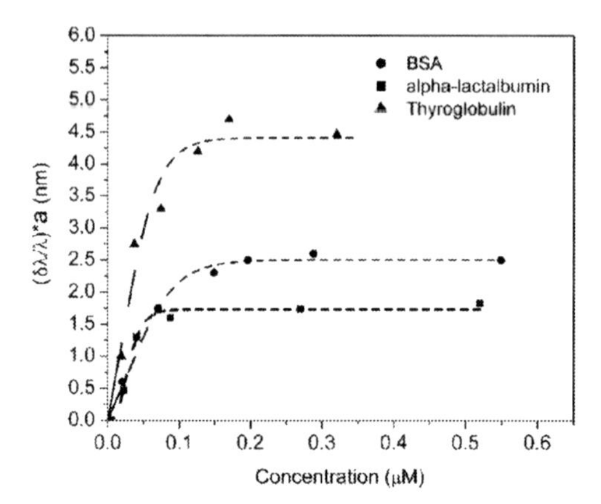

Fig. 6 Adsorption isotherms of protein: α-Lactalbumin, BSA and Thyroglobulin with molecular weight of 14.3 kDa, 66 kDa and 670 kDa respectively. The dotted lines are to guide your eyes.[14]

Fig. 7 is a compendium of the isotherms plateaus for various molecular weights MW. As one can see there is a definite power law behavior. The shift is proportional to $MW^{1/3}$. Considering that globular protein have almost identical densities and dielectric properties due to their common components (i.e. amino acids), $\delta(n^2)$ can be expected to

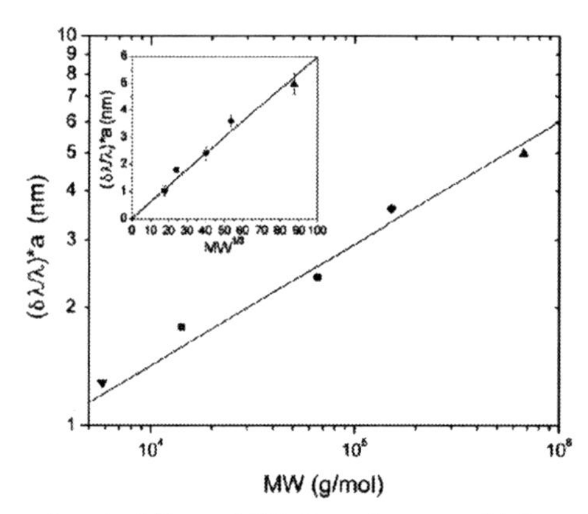

Fig. 7 Resonance wavelength shift vs. MW. Form lowest to highest: Insulin, α-Lactalbumin, BSA, γ-Globulin, and Thyroglobulin. The line was drawn manually to fit the data points. Inset: $(\delta\lambda/\lambda)$*a against MW $^{1/3}$.[14]

be nearly constant at saturation. Under these circumstances t dominates the shift. With the protein volume proportional to the molecular weight, the thickness would be expected to be proportional to the cube root of the molecular wt. The small scatter of the points around the $MW^{1/3}$ dependence is surprising considering that protein have different shapes from one to another. It may well be that protein deformation on adsorption plays a pivotal role in allowing this simple dependence.

7. Specific Detection of Protein [12]

Up until now we have considered protein adsorbed on a silica surface. There are a number of ways in which this can be facilitated. On means is to functionalize the silica surface with an agent such as 3-aminopropyltriethoxysilane (APTES). This compound reacts with the silanol groups on the silica surface and at biological pH leaves an exposed NH_3^+ rug. This is ideal for a protein that acquires a negative charge at this pH; the charge attraction holds the protein in place. Many protein acquire a net negative charge. So the attachment to the surface is not specific.

To make the attachment specific requires the lock to key relationship that we spoke about earlier. A graphic example can be illustrated in the case of BSA. By chemically attaching biotin (vitamin H) to BSA we generate a modified protein with one major suitor. This so-called biotinylated BSA attaches with one of the strongest known physiochemical bonds to a protein known as streptavidin. Fig. 8 pictorially shows the hypothetical configuration.

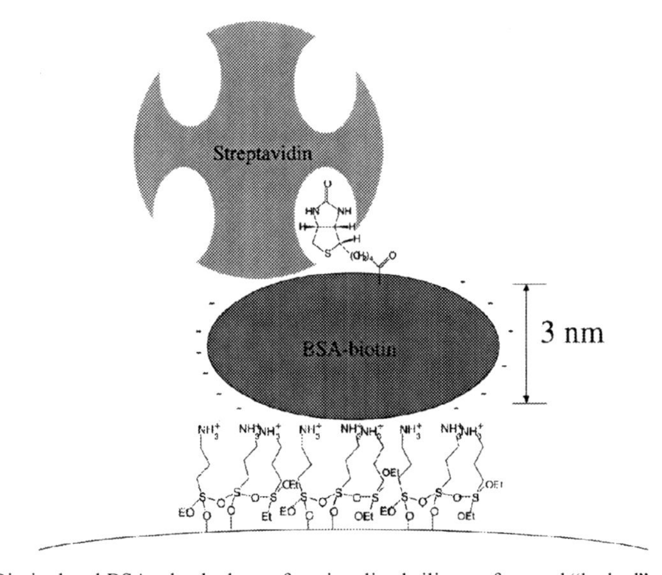

Fig. 8 Biotinylated BSA adsorbed on a functionalized silica surface and "locked" to Streptavidin.

Recall that a typical BSA adsorption experiment on a NH_3^+ rug is shown in Fig.4. This experiment was carried out at 1μM concentration. As the concentration is increased no additional shift is seen, indicating that the shift is associated with a full monolayer. Apparently the protein is more electrostatically interested in the surface than other BSA. The same effect occurs with BSA-biotin, however when streptavidin is introduced a pronounced increase in signal is registered (Fig. 9). Only streptavidin demonstrates this effect. In addition the increase in signal associated with the streptavidin is only slightly smaller than the BSA signal, consistent with its slightly smaller molecular weight (based on 1:1 stoichiometry for the binding event.).

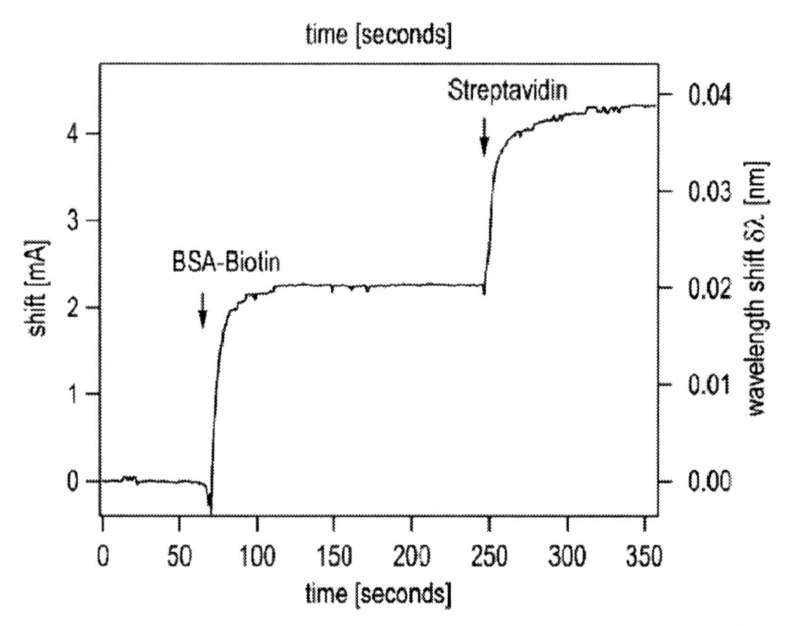

Fig. 9 Detection of streptavidin binding to previously surface immobilized BSA biotin.[12]

8. Discrimination of Single Nucleotide Polymorphisms in DNA[15]

In 1953 a revolution began in molecular biology with the discovery of the structure of DNA by Crick and Watson.[16] They provided a mechanism for genetic replication (information transfer) and opened the door for the interpretation of nucleotide sequences. With the full genome sequences for a number of species revealed over the past few years, we are now at the beginning of a revolution in genetic analysis.

A major interest is in detecting single defects in DNA (a.k.a. single nucleotide polymorphisms, SNP). SNPs typically occur in fewer than 1 in 1000 bases and are being widely used to better understand disease processes, thereby paving the way for genetic-based diagnostics and therapeutics. Our goal in this section is to understand how our bio-sensor can be used to reveal an SNP. The bio-nano-probe in this case is not a modified protein such as biotinylated BSA, but a complimentary single strand of DNA.

Complimentarity as it is applied to DNA has to do with associations between individual molecules on separate strands. The molecular units, known as bases, are distinguished by the letters A, G, C, and T. The rules are simple: A associates with T and G associates with C. To understand whether you have a particular stand of DNA a compliment to that strand can be immobilized near the microsphere surface. A shift in the resonance frequency signals the binding event (a.k.a. hybridization).

Suppose a strand of DNA (a.k.a. as an oligonucleotide, abbrev. oligo), which we will call the target, contains 11 bases that spell out the nonsensical word GATAGAGTCAG. The compliment is. CTATCTCAGTC. A change in the compliment in the 5th letter from the left from C to A is clearly a defect in relation to the target. We would like to design a way to use our photonic atom biosensor in order detect this defect. Since one molecular defect out of eleven bases is a small difference, the target could partially bind to the defective strand. A means for discriminating this imposter from the real thing has to be invented. The solution takes advantage of the frequency domain nature of our photonic atom sensor.

Since it would be extremely difficult to generate two high Q spheres which would have the same resonance spectra,[17] each sphere is distinguishable by its resonant frequency. So the resonant frequency marks the microsphere and locates it in the frequency domain. Attaching a particular word of DNA to a given sphere locates that DNA in frequency space. If we mark one sphere (call it S1) with many duplicates of CTATCTCAGTC and another S2 with duplicates of CTATATCAGTC, then a target GATAGAGTCAG should seek out the sphere S1 preferentially. The implementation of this idea fits beautifully into the fiber-sphere coupling scheme since the spectrum taken through the fiber should reveal each sphere separately, and two spheres sitting on the same fiber can be bathed simultaneously in a sample of stirred fluid containing the target (i.e. similar to the cell in Fig. 2, but with two spheres). Then it is only a matter of tracking the shift in the resonant dips from each as the target is adsorbed. This basic idea is depicted by the cartoon in Fig. 10.

Of course, there are many details to making the Photonic Atom SNP Detector a reality. For one thing one has to functionalize the silica surface of the microsphere to adsorb the oligos of DNA. To do this Frank Vollmer attached a biotinylated dextran polymer to the silica surfaces of S1 and S2, and allowed both to be reacted with streptavidin.[14] Since streptavidin has four sites for attaching biotin (Fig. 8) three sites remained for adding biotinylated-DNA. Of course, these DNA additions must be done separately in order that S1 is covered with CTATCTCAGTC and S2 with CTATATCAGTC. In addition there is a primary physical effect that must be verified.

The physical effect has to do with the independence with which the resonators can be stimulated. The approach to checking this is to first place one sphere in contact with the fiber and observe the spectrum. After this the other makes contact. Fig.11 shows the way in which the spectra are changed. So long as the surface of the spheres are separated by more than ~200 nm there is no apparent interaction. The resonances which appear after the first sphere touches the fiber are preserved upon contact with the second sphere. The second sphere, however, contributes its own resonances to the spectrum, as our frequency domain argument anticipated.

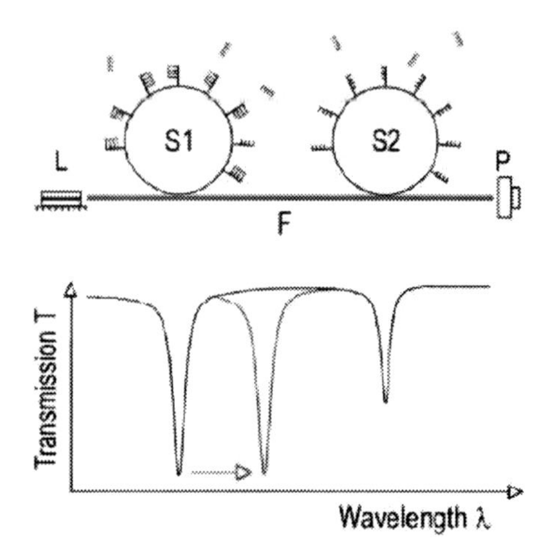

Fig. 10 Two microspheres touching the same fiber and having separate words of DNA are bathed in fluid containing a target. If the SNP sensing idea is correct then the sphere containing the compliment to the target should preferentially shift its resonance (cartoon: lower figure).

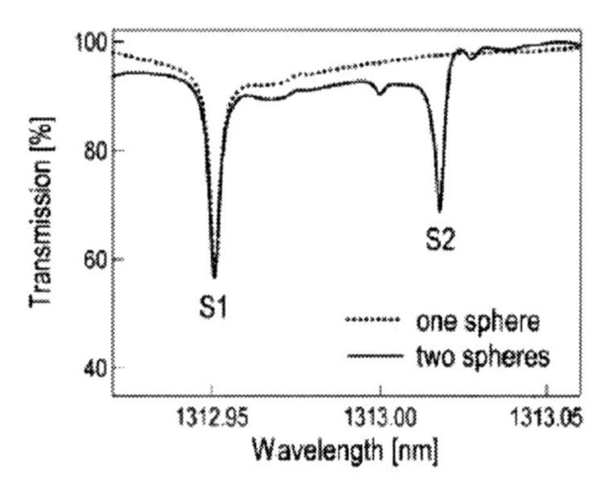

Fig. 11 The spectrum after one sphere contacts an eroded optical fiber about 4μm in diameter (solid line), and when both spheres are in contact (dotted line).[15]

The results of the SNP experiment are shown in Fig.12A. As the target is injected into the cell the frequencies of the resonances associated with microspheres S1 and S2 begin to change. Both show jagged fluctuations in the frequencies of approximately equal intensity. These fluctuations are apparently due to the pipetting process in which

10 μl of the target solution is injected into the 1 ml cell containing a buffer solution. Following these fluctuations the wavelength of the resonance associated with S1 shows a systematic red shift. Microsphere S2 shows a substantially smaller shift (less than one sixth).

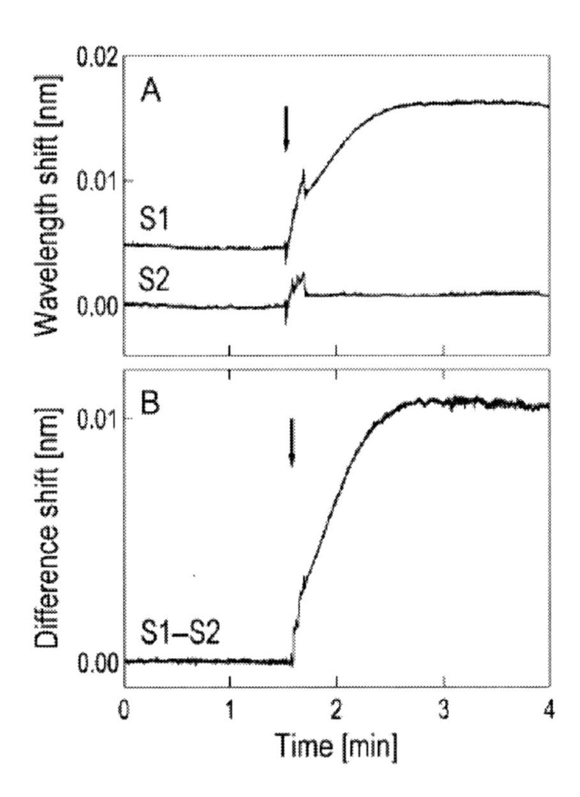

Fig. 12A Shift of resonances of spheres S1 and S2 following target injection. 12B Differential shift, $(\delta\lambda)_{S1}$ -$(\delta\lambda)_{S2}$.[15]

The original fluctuations may be considered as a common mode noise similar to the sort of noise that enters the inputs of a differential amplifier. If so a simple subtraction between the shifts associated with the S1 signal and the S2 signal should cancel this common mode noise. Fig. 12B shows this differential signal. Here we see the real time discrimination of one letter of DNA with a very large signal-to-noise ratio, 54!

9. Future Directions

By reducing the microsphere radius our sensitivity will increase[4] allowing us to see smaller and smaller fractions of a monolayer, until we may ultimately reach a single bio-particle. At this point it may be possible to count label-free molecules one at a time rather than wait for a full monolayer to form. Each step in increased wavelength will be directly proportional to the polarizability α and therefore to the molecular weight. It may be unnecessary for molecules to land on the surface. If they simply flow through the evanescent field of a mode, then the frequency shift will pulse, and molecular counting and analysis becomes possible. Of course, there are noise problems associated with this idea. The molecules are under intense Brownian forces. The Brownian noise interval will be severely reduced if the molecule is carried by a flow. Here we require a flow system, in the form of a micro-fluidic channel.

At the same time as the flow system is assembled our tapered and tenuous fiber waveguide must be replaced with something more robust. Instead of employing lithography and micro-manufacturing techniques[18] we favor supporting the tapered fiber with a substrate. In contrast to the aerogel supports demonstrated by the Mazur group for supporting nano-fibers,[19] we favor a less porous material with the refractive index of water. Recent work at the Polytechnic[20] and elsewhere on fluoro-polymers, have show that this moldable material is essentially non-porous, as robust as plexiglass (i.e. PMMA), and has the refractive index of water.

Fig. 13 shows the micro-fluidic concept,[21]

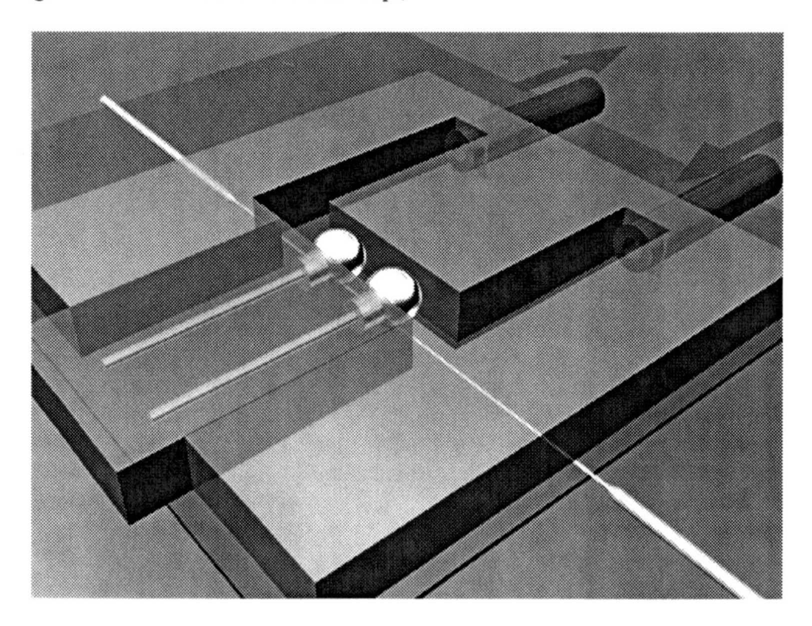

Fig. 13 Dual-sphere sensor consists of micro-fluidic channel with tapered fiber sandwiched between two layers of fluoro-polymer. Spheres are mounted on a polymer cassette while coupled to the fiber. Flow passes from right to left

Acknowledgements

Steve Arnold thanks Rino DiBartolo for his kind invitation to lecture at this school and for once again introducing him to the land of Rino's birth.

This was the third time Steve attempted to make a course out of his research. By learning from those who have done this in the past years, he gained a great deal of insight into the technique. Steve is particular indebted to Eric Mazur and Ralph von Baltz for their encouragement.

Without the prodding and encouragement from Ottavio Forte this manuscript would not have been completed on time. Bravo Ottavio!

Both of us would like to thank the Othmer Institute for Interdisciplinary Studies at the Polytechnic, and the NSF (BES Grant No. 0522668) for support.

References

1. P. N. Prasad, *Introduction to Biophotonics* (Wiley-Interscience, New Jersey, 2003).
2. A. Serpenguzel, S. Arnold, and G. Griffel, "Excitation of resonances of microsphere on an optical fiber", Opt. Lett. **20**, 654-656 (1995).
3. S. Arnold, "Microspheres, Photonic Atoms, and the Physics of Nothing", American Scientist **89**, 414-421 (2001).
4. S. Arnold, M. Khoshima, I. Teraoka, S. Holler, and F. Vollmer, "Shift of whispering gallery modes in microspheres by protein adsorption", Opt. Lett. **28**, 272(2003).
5. J. C. Knight, G. Cheung, F. Jacques, and T. A. Birks, "Phase-matched excitation of whispering-gallery-mode resonances by a fiber taper," Opt. Lett. **22**, 1129(1997).
6. M. L. Gorodetsky, A. A. Savchenkov, and V. S. Ilchenko, "Ultimate Q of optical microsphere resonators," Opt. Lett. **21**, 453(1996).
7. S. Arnold, M. Noto, and F. Vollmer, "Consequences of extreme photon confinement in micro-cavities: I. ultra-sensitive detection of perturbations by bio-molecules," in *Frontiers of optical spectroscopy: investigating extreme physical condition with advanced optical techniques,* Ed. Baldassare DiBartolo (Kluwer Publishers, 2005).
8. S. Arnold, and S. Holler, "Microparticle photophysics: Fluorescence microscopy and spectroscopy of a photonic atom," in *Cavity-enhanced spectroscopies,* R. D. van Zee, and J. P. Looney, eds. (Academic, San Diego, Calif., 2002), pp. 227-253.
9. E. S. C. Ching, P. T. Leung, and K. Young, "Optical processes in microcavities-the role of quasinormal modes," in *Optical process in microcavity,* R. K. Chang, and A. J. Campillo, eds. (World Scientific, Singapore, 1996), pp. 1-76.
10. M. Noto, F. Vollmer, D. Keng, I. Teraoka, and S. Arnold, "Nanolayer Characterization through Wavelength Multiplexing a Microsphere Resonator", Opt. Lett. **30**, 510-512(2005).
11. J. P Laine, B. E. Little, and H. A. Haus, "Etch-eroded fiber coupler for whispering-gallery-mode excitation in high-Q silica microspheres," IEEE Photon. Tech. Lett. **11**, 1429-1430 (1999).

12. F. Vollmer, D. Braun, A. Libchaber, M. Khoshsima, I. Teraoka, and S. Arnold, "Protein detection by optical shift of a resonant microcavity," Appl. Phys. Lett. **80**, 4057-4059 (2002).
13. I. N. Levine, *Physical Chemistry*, 4th Ed., (McGrew-Hill, INC., New York, 1995), p.366.
14. M. Noto, M. Khoshsima, D. Keng, I. Teraoka, V. Kolchenko and S. Arnold, "Molecular weight dependence of a whispering gallery mode biosensor" Appl. Phys. Lett. (in press, 2006).
15. F. Vollmer, S. Arnold, D. Braun, I. Teraoka, and A. Libchaber, "Multiplexed DNA Quantification by Spectroscopic Shift of Two Microsphere Cavities", Biophysical Journal **85**, 1974-1979(2003).
16. D. Watson and F. H. Crick, "Molecular structure of Nucleic Acids" *Nature* 171, 737-738(1953).
17. With a Q of 10^7, attempting to make two spheres spectrally the same over even 1 resonant width would require a tolerance in the equatorial radius of better then 1 part in 10^7 which for a 100μm radius is 0.01nm!
18. B. E Little, J. P Laine, D. R Lim, H. A Haus, and L. C Kimerling, and S. T. Chu, Opt.Lett.**25** 73(2000).
19. L. Tong, J. Lou, R. R. Gattass, S. He, X. Chen, L. Liu, and E. Mazur, "Assembly of Silica Nanowires on Silica Aerogels for Microphotonic Devices" Nano Lett.**5**, 259-262(2005).
20. F. Mikes, Y. Yang, I. Teraoka, T. Ishigure, Y. Koike, and Y. Okamoto,"Synthesis and Characterization of an Amorphous Perfluropolymer: Poly(perfluoro-2-methylene-4-methyl-1,3-dioxolane", Macromolecules **38**, 4237-4245(2005).
21. O. Gaathon, W. Jeck, I. Teraoka, and S. Arnold (in progress).

LASER SOURCES FOR HIGH RESOLUTION SENSING

GIUSEPPE BALDACCHINI
ENEA, Centro Ricerche Frascati,
00044 Frascati (Roma), ITALY
baldacchini@frascati.enea.it,

Abstract

The first laser has been invented in 1960, and since then the laser field exploded almost exponentially, covering a spectral range from less than 1 Å to more than 1 mm. Many of them have been used with outstanding results both in basic science, and in industrial and commercial applications, by changing for ever the same lifestyle of humankind. As far as spectroscopy is concerned, the laser light has started an unprecedented revolution and spectroscopic applications increased qualitatively and quantitatively in the last four decades as never before. Overall, the laser improved by several order of magnitude the resolution and sensitivity of well known classical spectroscopy, made possible the introduction of new spectroscopic methods, and allowed a wide diffusion of sophisticated but easily handled sensing devices for the environment, production processes, and medicine. However, this story is not yet concluded because the laser and spectroscopy fields are still on the move as a consequence of impressive technological advances.

1. Introduction

The laser is one of the most fantastic and versatile tool fabricated in recent times, although its story began long ago with the observation of natural optical phenomena, which have been always at the center of curiosity of mankind. However, the true nature of light started to emerge only in 19th century with the discovery of its variegated and conflicting properties, was completely understood in the frame of the new quantum mechanics at the beginning of the 20th century, and finally in 1954 the first amplification by stimulated emission was observed in ammonia with the realization of the first maser, microwave amplification by stimulated emission radiation. Soon after, in 1960 the first laser, light …., device was realized by using the well known red emission of the ruby crystal, and since then the laser field exploded almost exponentially, and thousands of different materials, in the state of solids, liquids, vapors, gases, plasmas, and elementary particles have lased up to now, as shown in Fig. 1. However, only a few of them became practical laser sources, which have been used with outstanding results both in basic science, and in industrial and commercial applications.

As far as spectroscopy is concerned, the laser light has started an unprecedented revolution because of its unique properties of monochromaticity, coherence, power, brightness and short-pulse regime, unrivaled by any other natural and artificial light source.

B. Di Bartolo and O. Forte (eds.), Advances in Spectroscopy for Lasers and Sensing, 19–32.

Spectroscopic applications increased qualitatively and quantitatively in the last four decades as never before, and they are still proceeding in parallel with the improving of the laser sources, which benefit also by a better spectroscopic knowledge of old and new materials [1]. Lately, the improvement of laser sources, and auxiliary optical and electronic equipments, is producing a growth of traditional laser spectroscopy with superior resolution and sensitivity. Moreover, the outstanding increase of the output power give rise to completely new spectroscopic effects ranging from atomic to nuclear physics, pulses as short as atto-second (10^{-18} s) are used to manipulate the coherence of quantum excite states, and the Terahertz region is being explored by the first laser sources.

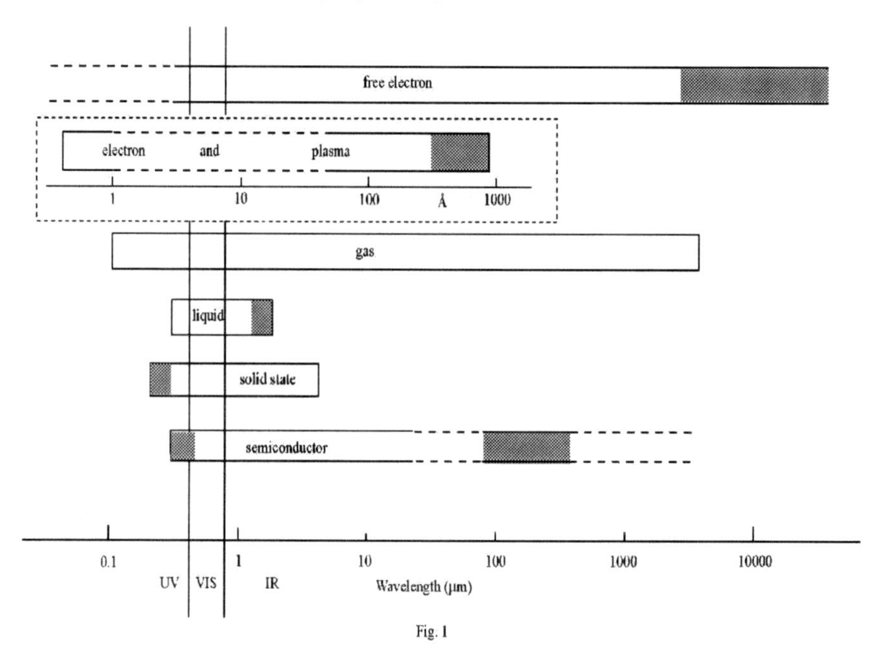

Figure 1. Approximate spectral ranges of laser sources according to the state of matter of the active material. The grey areas show the laser emissions added during the last 10 years, while the dashed lines refers to regions under development. The single spectral regions are not always filled with continuous bands, as it is the case of gases which emit in discrete lines.

2. Present Situation of Lasers

At moment laser technology has reached a high level of sophistication but, while there have been improvements and new lasers have been devised lately for a few types, for other ones there have not been marked advancements.

Among the latter, excimer, He-Cd, ion, copper vapor, dye, He-Ne, ruby, color center, carbon dioxide, far infrared gas, have reached a high degree of maturity, and they are considered a standard commodity in both industry and research.

A discussion apart deserve free electron (FEL) and x-ray lasers, which lately underwent processes of improvements both in theory and experiment, and there is a well founded expectation for betters sources. FELs offer several opportunities for applications in material science and biology but require longer times to be realized, while the situation is much more pregnant for x-ray lasers.

Anyway, the most impressive advancements so far have been accomplished by semiconductor diode (SDL) and solid state lasers (SSL). Not only classical diode junction lasers have been realized with superior qualities, but they have been extended to the blue and lately UV side of the spectrum, which opens a completely new world of applications in spectroscopy, sensing, and communication. New and old solid state materials, inorganic and organic compounds, are being explored and new laser emissions are discovered, by using both SDLs emitting in the blue as pumping sources and nonlinear optical properties. The emitted powers of organic lasers are not impressive, but their easy functioning and the possibility of miniaturization are two powerful drivers for their development.

As far as SDLs are concerned, in the last 10 years the SDL family improved the quality of the existing devices, increased the emitted power, and lately extended the wavelength range below 400 nm, i.e. in the much searched UV region [2]. When the GaN-based violet diode lasers were introduced in 1998, there were initially targeted for optical storage purposes, but it became immediately apparent that they could also be applied in bio-instrumentation, microscopy, graphics arts, and semiconductor industry, with great advantages with respect to the usual bulky ultraviolet sources. However, these new diode lasers had short lifetimes, low powers, environmental dependence, and poor beam quality, so that their use was considered problematic for practical uses. In time, not only the previous drawbacks were almost completely solved, but new materials pushed still in the UV their emission. Indeed, by adding Al to the GaN compounds the band gap moves up to 6 eV, by decreasing so the emission wavelength. Up to now, gallium aluminum nitride (GaAlN) devices are limited to 280 nm for light emitting diodes (LED) and 350 nm for SDLs. There are devices working at lower wavelengths, but they are still in the development stage for their output powers well below 1mW and lifetimes much less than hundreds of hours. Technical difficulties arise especially by the use of aluminum which is needed to increase the band gap, but does not allow to fabricate good electrical contacts which, on the contrary, can be easily made with indium, but at the price of decreasing the band gap. Figure 2 shows a schematic situation of the most common SDLs at disposal in 2004 to science and technology with satisfactory reliability. It also reports an indicative output power in cw regime of tunable diode lasers of the main semiconductor families, mostly commercial, as a function of wavelength. Above ~2.5 μm they emit only at low temperature, 200-4 K, and the power decreases with increasing wavelength, falling well below 1 mW for lead salt diode lasers above 10 μm which, notwithstanding the weak intensity, served so well and are still serving the scientific community of high resolution molecular spectroscopy. Below ~2.5 μm they work also at room temperature and the maximum power is emitted in the near infrared region, with the blue diodes very much improved and the ultraviolet ones surfacing above 1 mW. SDLs are also emitting in pulsed regime with larger tunable intervals with respect to cw regime, but never at very high peak powers because of irreversible damaging effects on the facets.

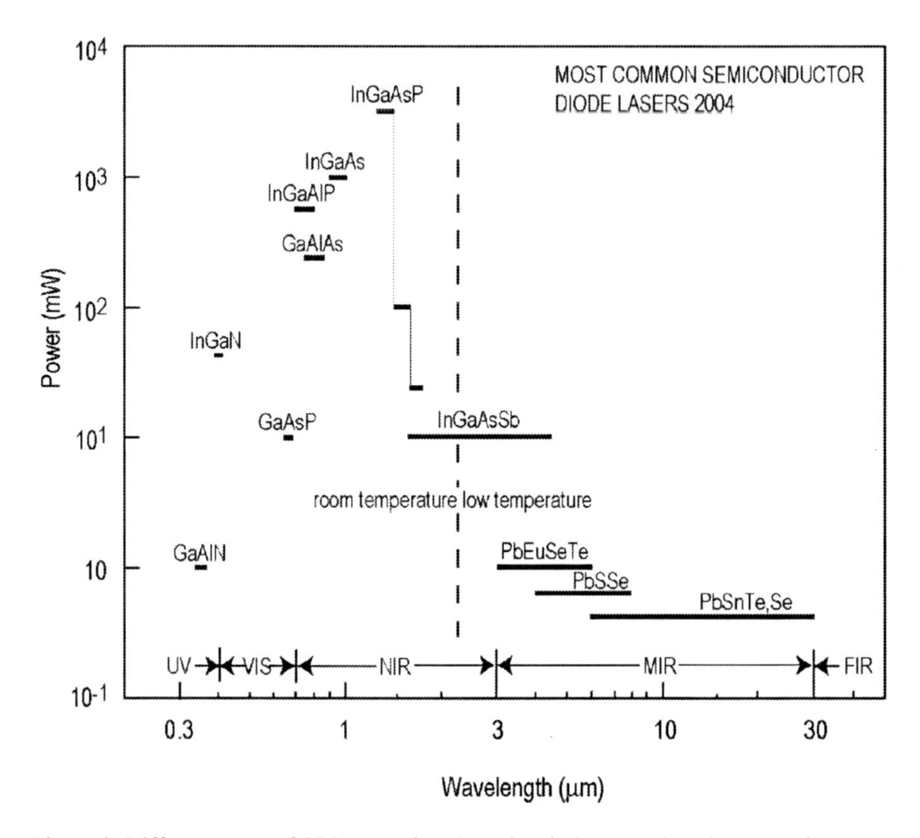

Figure 2. Different types of SDLs as a function of emission wavelength. Approximate cw power is also indicated together with the spectral range of tuning.

As far as SSLs are concerned, organic lasers are playing more and more an increasing role. Organic materials have been used as optical active media since the beginning of the laser era, because many of them were well known to be good light emitters. Indeed, dye lasers were invented very early in 1966. Also dielectric solids containing luminescent molecules and plastic materials have lased efficiently. However, the organic laser have an intrinsic problem arising from their thermal properties. Indeed, their thermal conductivity is rather poor so that it is impossible to use intense pumping for obtaining high power laser emissions. Moreover, also low power lasers have severe limits, because locally the temperature can grow to such high values to accelerate the degradation processes, so common in organic materials.

Notwithstanding the previous problems, organic materials have been widely used in optoelectronic devices, where they have been very successful, as in the OLED sector, Organic Light Emitting Devices, where high densities of energy are not required and one dimension has been squeezed to less than 1 μm [3]. One of the most successful organic

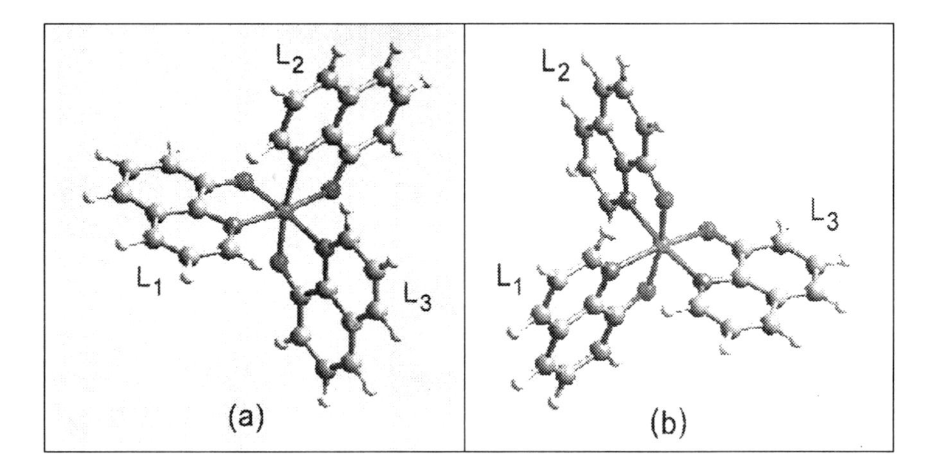

Figure 3. The molecule Alq_3, tris(8-hydroxyquinoline)aluminum ($C_{27}H_{18}AlN_3O_3$), is made out of three benzene ligands bound through nitrogen and oxygen with a central aluminum atom. It exists in two isomers, facial (a) and meridianal (b), being the latter the most frequent one, and when excited emits intense green light, but by association with other chemical compounds its emission covers the entire visible region.

material ever is the Alq_3, a metal-organic molecule shown in Fig. 3. Although its optical properties are very complex, as usually happens in organic compounds [4], it produces an intense greenish emission when excited by electrical current or by light in one of its absorption bands in the UV region of the spectrum, as shown in Fig. 4. It has been the first organic compound to be used successfully in a practical electroluminescent device, and it is still a material of choice for OLEDs although it is subjected to degradation phenomena [5], a subject still at the very center of intense research efforts [6]. In the laser field, there have been lately successful attempts to overcome the poor thermal conductivity by resorting to the thin film technology, as in the OLEDs. Indeed, it is well known that thin films of any kind of material can dissipate more easily consistent amount of heat into their substrates, usually made of good conducting materials. Anyway, thin films of organic luminescent compounds have lased in pulsed mode in the VIS and UV region of the spectrum [7,8], the latter with an energy of a few nJ, which is not a negligible figure. Particularly interesting for medical applications is the UV organic laser which can be scaled to micrometer sizes, together with a solid state pumping source.

Moving more toward higher photon energy, x-ray lasers have been the dream of a whole generation of researchers after the discovery of the first laser emission in 1960. But the very difficulties, which delayed the laser for six years after the first maser, were for the x-ray spectral region much worst. Indeed, the spontaneous emission was several order of magnitude bigger with respect to the visible spectrum, and reversing the population for amplification was not an easy task. It was very clear to everybody in the field that enormous amounts of energy should have been used to generate the right conditions for

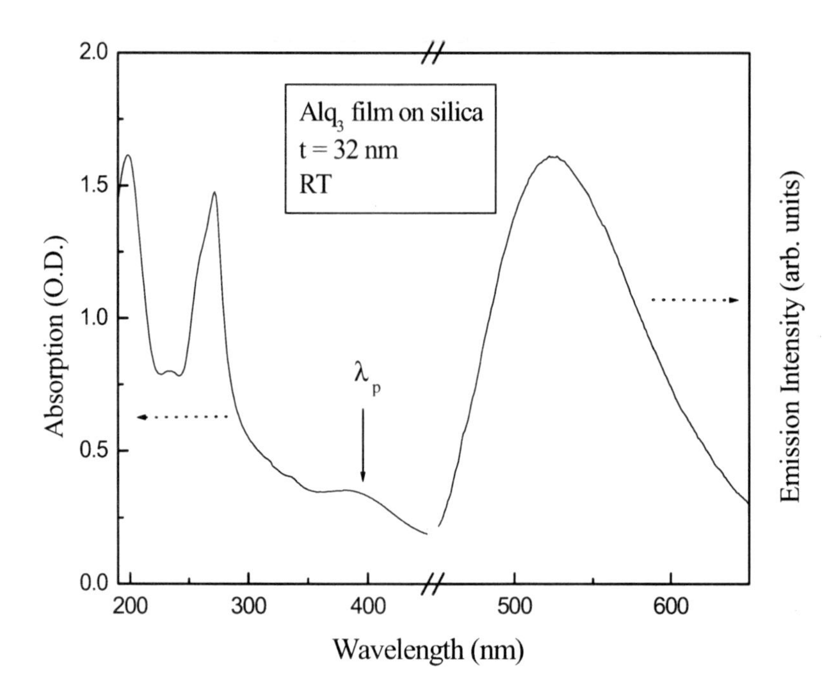

Figure 4. Absorption and emission spectra at room temperature of a 32 nm thick film of Alq_3 grown on a silica substrate kept at room temperature during evaporation. The emission has been excited at λ_p = 395 nm, as shown by the arrow.

lasing. Indeed, the first x-ray emission was produced via a small nuclear detonation in 1981. Since then the x-ray laser moved as in the following:
emission in plasma generated by pulsed lasers on solid targets, 1985,
table-top emission in a capillary discharge plasma, CDP, 1994,
generation via laser-electron Thomson scattering, 1996,
high harmonic generation, HHG, in noble gases by ultra short lasers, 1996,
imaging detectors for x-rays and LiF, 2002,
improved CDP and HHG x-ray lasers, 2002-2004,
x-ray spectroscopy developments, 1895-1981-2004,
proposal for production of x-ray pulses via interaction FEL+ fs laser, 2004.
Today x-ray lasers are commonly built in Table–Top version [9] and proposed by using x-ray FEL SASE, Stimulated Amplification of Spontaneous Emission, combined with femtosecond laser systems [10], but they are not so simple to be built and to operate.

However, their use is highly rewarding especially in exploring new technologies. Indeed, the story of their development and utilization is a further proof of the virtuous circle spectroscopy-laser-material-spectroscopy. Indeed, the LiF crystal proved to be the best image detector for soft x-rays [11], and allowed to observe details in the energy distribution of the laser beam as never before. Figure 5a is the image of an x-ray laser beam focused on a LiF film. The x rays produce defect centers inside the ionic crystallites composing the film which later on are excited with blue light, and as a consequence emit in the yellow-green spectrum. The image possesses a spatial resolution of better than 1 μm and can be observed indefinitely. In case the laser beam passes through a grid of square apertures bigger than the wavelength, the image reported in Fig. 5b is observed on the film. It is very similar to typical Fresnel and/or Fraunhofer diffraction patterns, which signify an x-ray beam of high spatial coherence. It is also possible to encode gratings in LiF, which opens the door of applications in the field of fotonic materials for optoelectronics [12]. So, a material known since long time for his role in color centers physics and dosimetry, resulted particularly useful also as imaging detector. Moreover, it also benefits from the x-ray laser for being transformed in active and passive optical material [13]. In particular, this wide band gap material has been used lately as an active material for a distributed-feedback laser emission at 710 nm [14].

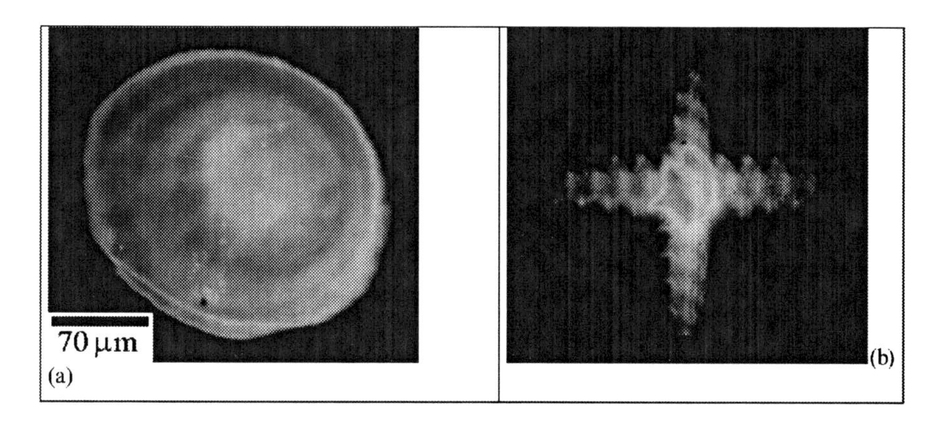

Figure 5. (a) Optical image of focusing the Ne-like Ar capillary discharge x-ray laser at λ=46.9 nm on a LiF film without filtering. The mexican-hat shape of the beam indicates a TEM_{00} mode. (b) Optical image of the same laser after passing through a metallic mesh (nickel) of 70 lpi (period 360 μm, thickness 60 μm) and being focused on a LiF crystal. The evident interference was at first not expected because the emission wavelength of the laser, λ=46.9 nm, is much smaller than the linear dimension of the square aperture, d=300 μm, of the metallic mesh and its wire, Φ=60 μm. The two images were observed with an optical microscope operating in fluorescence.

3. Recent Developments of Spectroscopic Sensors

As told in the previous section, SDLs emitting in the blue and violet region were especially sought for their mass application in optical storage, but lately they have been employed with success as inexpensive biosensors for biological agents and short distance communications [15]. It is well known, indeed, that the majority of organic compounds which are the constituent blocks of living organisms, have their absorption bands in the UV region of the spectrum, and so they can be easily "sensed" in absorption or emission regime by UV light sources, which can be used in loco coupled with sensitive detectors to the particular biological spectra. Communications may benefit by the "solar blind" spectrum, i.e. the solar light which does not reach the ground below about 350 nm, and by the existence of the Rayleigh scattering phenomenon, particularly intense in the UV region [16].

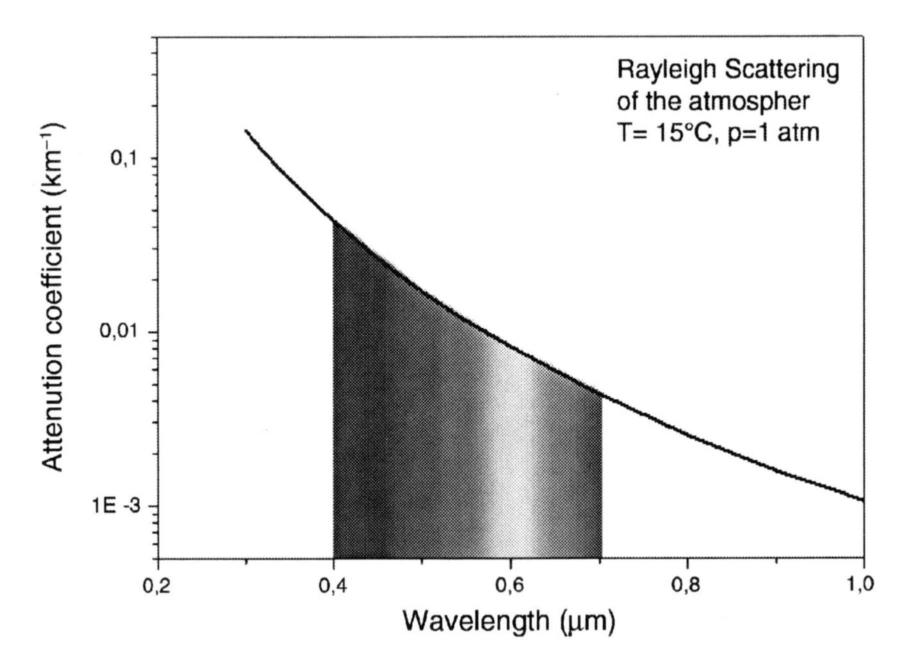

Figure 6. Molecular scattering coefficient $\alpha(\lambda)$, from the expression $I(\lambda) = I_0(\lambda) \exp [- \alpha(\lambda) L]$ commonly known as Beer-Lambert's or Bouguer's law, of the solar light from the atmosphere as a function of wavelength..

Figure 6 reports the optical attenuation coefficient of the Earth atmosphere as a function of wavelength, and it is easily observed the strong absorption below 500 nm, which is due to the well known λ^{-4} behaviour. In reality, the solar light is not absorbed but rather scattered by the atmosphere, which acquires the typical blue-sky color, as shown beautifully in Fig. 7. So, the UV source can be used at short distances without a line of sight, benefiting from the scattered light, strong absorption of atmosphere, and lacking of background sunlight. Indeed, the local scattering of the UV light insures a complete local coverage, the high atmosphere absorption avoids the light moving too far from the source and interferences with the solar radiation. Any UV luminous device with more than 1 mW of power can be used, but SDLs are simply better than LEDs, and so the continuous quest for them.

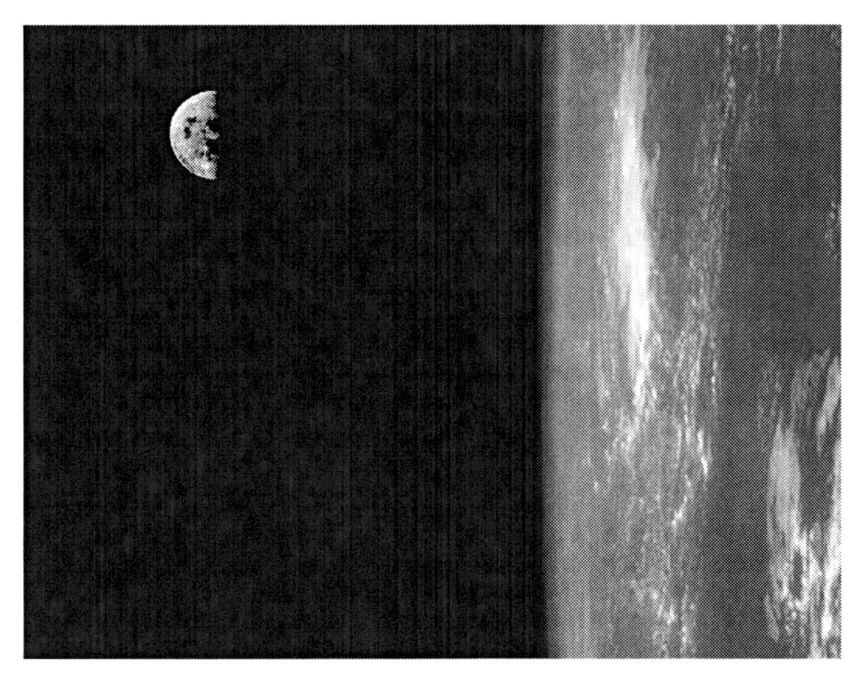

Figure 7. A space photo taken by a satellite showing part of the Earth on the right and the half Moon on the left. The blue color of the Earth is due to the Rayleigh scattering of the atmosphere, which is missing in the Moon because it is airless, like the deep space.

Another fast growing technology is the cavity ring-down spectroscopy, CRDS, which can provide ppb sensitivity in a dimension of a suitcase [17]. In practice, it measures the intensity decay of a laser beam while bouncing back and forth between two high reflecting mirrors. The main advantage of this optical gas sensor is that it measures the absolute optical losses inside an optical cavity, and so the absolute absorption of the gas under study. It is also selective in frequency, so that it recognizes the chosen molecular species whenever mixed with other gaseous components. Accordingly to the experts, this method will take over other older and well established systems like Fourier transform infrared, non dispersive infrared, and tunable diode laser absorption (TDLA). In practice, CDRS systems can be considered a sophisticated extension of the TDLA method coupled with multipass cells, which has been possible thanks to the improvements of the reflecting optics in recent years [18]. This technique was developed initially in the visible spectral region, but has been extended later on in the IR and UV region. It has been devised originally for pulsed laser sources, which is the logical tool for studying the decay of the laser intensity inside the optical cavity, but the method has also been extended to narrow-bandwidth cw lasers. In this case, spectroscopy is performed by measuring phase shifts of an intensity-modulated laser beam transmitted through a high-finesse cavity.

More recently, the well known frequency modulation techniques have been applied to CDRS and the sensitivity of the method increased by a factor 1000 [19]. The new method has been called NICE-OHMS (Noise-Immune Cavity-Enhanced Optical-Heterodyne Molecular Spectroscopy).

A completely different system, which is producing sub-Doppler spectra, is based on the ability to build nanometer gas cells, called also PILL-BOX CELLS. In practice these cells have a thickness varying from 100 nm up to 20 μm, while the transversal dimensions are macroscopic. The principle of functioning relays on the time of flight of the single molecules, which is of the order of ns and μs for longitudinal and transverse paths, respectively. As a consequence the longitudinal molecules have a higher absorption with respect to the other ones, and practically no-Doppler effect at all [20]. Resolution as high as 100 MHz have been obtained for the D1 line of Rb at 794 nm, which is better than $\Delta\nu/\nu = 3\ 10^{-9}$.

In the brief description of the four previous spectroscopic methods of sensing, we have not discussed at all about the laser sources, which however remain the pivotal tools of them all. Indeed, without the directionality, brilliance, monochromaticity, and pulsed regime of lasers, it would have proven impossible even to think at any of the previous methods.

4. Conclusions

It is usually accepted in modern times and countries that technological advancements are also stimulated by the quest for better and cheaper mass products and vice versa, so also a close look at the commercial side of the laser world has been accomplished [21].

Figure 8 displays the values of the global market of lasers during the last 15 years. After a low figure in 1993, due to a general recession, the world wide laser sales have markedly increased more than linearly until the year 2000, after which there has been an abrupt decreasing for the following two years, the great fall, with a consistent recovery in the last two years, also confirmed by the previsions for 2005. The 2001-2002 commercial debacle has been originated by the crumbling of the information technology which hits severely SDLs, and to a less extent or not at all the other families of lasers. The most common lasers of the latter ones have been reported in Fig. 9, which shows a marked decreasing of Excimer lasers in 2001-2003, as a consequence of their involvement in material processing for the information technology, and much less of Solid State, CO_2, and ION lasers. In particular SSLs, which are further broken down in four categories in Fig. 10, show a continuous positive trend especially in connection with SDL pumping and better technologies. Indeed, while the old but still appealing lamp pumped SSLs have slightly suffered by the great fall of the market, the laser pumped SSLs are becoming insignificant in the whole panorama, and the diode pumped SSLs show a steady increase, thanks to their high emission efficiencies and improved materials. So, the previous data confirm the close links existing among better technologies, innovative products, and market.

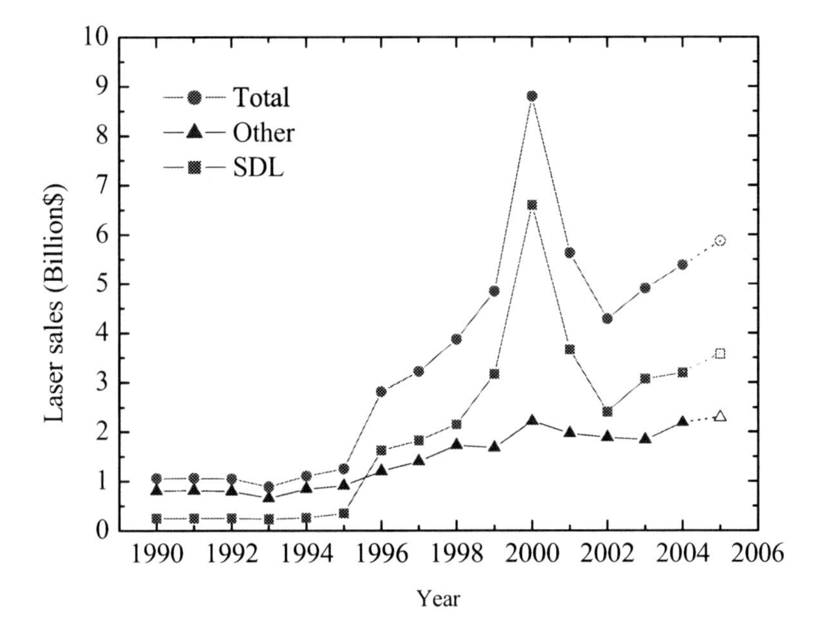

Figure 8. Worldwide commercial laser sales in the last 15 years, from Laser Focus World. Before 1995 only one curves was usually reported, but after that date a splitting between SDLs and all the other lasers has been deemed necessary, because of the abrupt surging of the former ones which represent today the most numerous family of all the lasers.

Moreover, it is a common opinion among the experts of the laser field, that further achievements should be expected from technological advancements in general and in material science in particular. Indeed, the fabrication of micro and nano optoelectronics devices, passive and active, is a result of our ability to assemble atoms and molecules in new artificial structures. This complex game is actually at its infancy, so that others promising results should be obtained especially from the development of new materials and processes.

As far as laser sensing is concerned, the accessible spectrum has been consistently increased in the last decade. In particular, the fast development of laser sources and auxiliary equipments, above all optics and electronics, prompted a wide diffusion of laser based sensing equipments with superior resolutions and sensitivity. Moreover, excitations in electrons, plasmas, gas, liquid and solid materials are being investigated systematically, especially in new regions of the spectrum and by using time resolved spectroscopy. As a consequence, new processes and luminescent materials with greater efficiency and stability may appear soon.

In this complex scenario, the global market takes advantage from the technological advancements, and also pushes them at the same time accordingly to the quest for better and cheaper products. As we have seen before, the total sales have been increasing in the last two years, and the previsions for the year 2005 are still for a further increase, which is a welcome situation after the great fall of the years 2001-2003.

In conclusion, although a few sectors show clear signs of maturity, the laser field looks well alive and a new golden age is awaiting ahead also in spectroscopy and laser sensing.

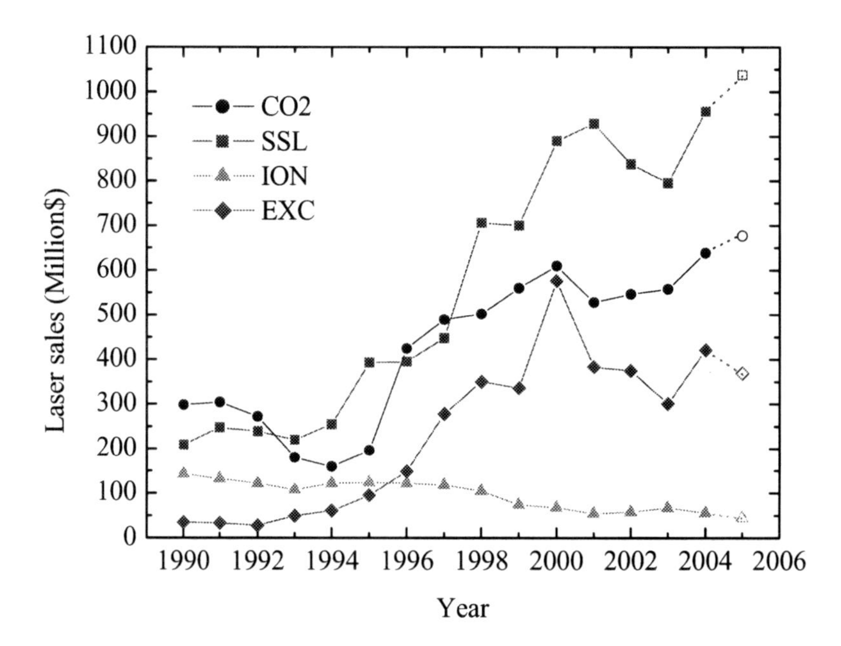

Figure 9. Worldwide commercial sales of the most common lasers different from SDLs in the last 15 years, from Laser Focus World. CO_2 gas laser with carbon dioxide, SSL solid sate laser, ION ionised gas laser, EXC excimer laser. Well before the 90s, ION lasers were more common than SSLs.

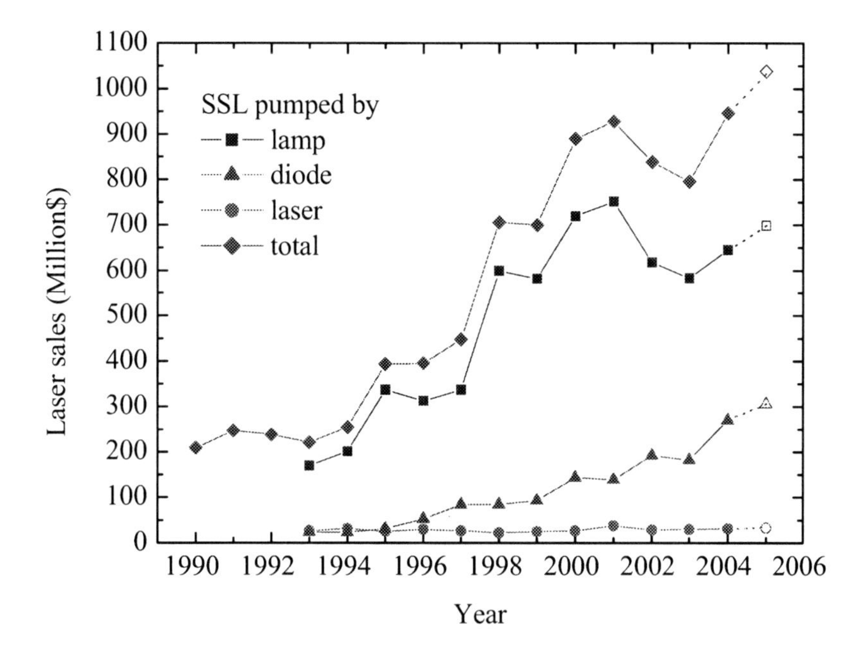

Figure 10. Worldwide commercial sales of SSLs split according to the different methods of optical pumping in the last 15 years, from Laser Focus World. Before the 90s the diode pumped category did not exist, while the laser pumped one was dominant.

5. Acknowledgements

The author is indebted with Lucilla Crescentini and Piero Chiacchiaretta for their precious help during the preparation of the manuscript.

6. References

[1] G. Baldacchini, *Lasers for Frontier Spectroscopy,* in Frontiers of Optical Spectroscopy, B. Di Bartolo and O. Forte, eds., Kluwer Academic Publishers, Dordrecht, 2005, p. 251-288.

[2] J. Hecht, Laser Focus World, February 2005, p. 95.

[3] G. Parthasarathy, J. Liu, and A.R. Duggal, *Organic Light Emitting Devices From Display to Lighting*, ECS Interface, Summer 2003, p. 42.

[4] G. Baldacchini, S. Gagliardi, R.M. Montereali, A. Pace, and R.B. Pode, Philosophical Magazine B 82, 669 (2002).

[5] F. Papadimitrakopoulos, X.-M. Zhang, and K.A. Higginson, IEEE J. Sel. Top. Quantum Electron. 4, 49 (1998).

[6] G. Baldacchini, T. Baldacchini, A. Pace, and R.B. Pode, Electrochem. Solid-State Lett. 8, J24 (2005).

[7] D. Schneider et al., Appl. Phys. B 77, 399 (2003).
[8] D. Schneider et al., Adv. Mater. 17, 31 (2005).
[9] G. Tomassetti et al., Eur. Phys. J.D 19, 73 (2002).
[10] E.L. Saldin, E.A. Schneidmiller, and M.V. Yurkov, Opt. Commun. 237, 153 (2004).
[11] G. Baldacchini et al., Appl. Phys. Lett. 80, 4810 (2002).
[12] G. Baldacchini et al., Rev. Sci. Instrum. 76, 113104 (2005).
[13] G. Baldacchini, J. Lumin. 100, 333 (2002).
[14] T. Kurobori et al., Phys. Stat. Sol. (c) 2, 637 (2005).
[15] J. Hecht, *Semiconductor Sources advance Deeper into the Ultraviolet*, Laser Focus World February 2005. p. 95.
[16] V.E. Zuev, *Laser-Light Transmission Through the Atmosphere*, in Laser Monitoring of the Atmosphere, E.D. Hinkley, ed., Springer-Verlag, Berlin, 1976, p. 29.
[17] J. Tirrel, *Cavity Ring-Down Spectroscopy Makes the Leaps out of the Laboratory*, Opto & Laser Europe, October 2004, p. 17.
[18] A. O'Keefe and D.A.G. Deacon, Rev. Sci. Instrum. 59, 2544 (1988).
[19] J. Ye, L.-S. Ma, and J.L. Hall, J. Opt. Soc. Am. B 15, 6 (1998).
[20] M. Ducloy, Europhys. News March/April 2004, p. 40.
[21] Laser Focus World, Issues January and February 2005, p. 83 and 69, respectively, for the latest data, and previous corresponding Issues for old data

RELAXATION AND DECOHERENCE: WHAT'S NEW?

RALPH v. BALTZ
Institut für Theorie der Kondensierten Materie
Universität Karlsruhe, D–76128 Karlsruhe, Germany *

1. Introduction

Relaxation and decoherence are omnipresent phenomena in macro–physics

- *Relaxation*: evolution of an initial state towards a steady state.
- *Decoherence*: decay of correlation $G(t,t') = \langle x(t)x(t')\rangle \to 0$ with time $|t-t'| \to \infty$. $x(t)$ can be any classical quantity or quantum observable which permits the linear superposition (i.e., x has a "phase").

Some examples are sketched in Fig. 1. For fields, like the electrical field $\mathcal{E}(\mathbf{r}, t)$ of a light wave, one may discuss relaxation and correlation for different space–time points which is called temporal and spatial coherence.

The microscopic origin of relaxation and decoherence is *irreversibility*, i.e., the coupling of a *system* to its environment by which a *pure state* is transformed into a *mixed state*. For instance, a two–level–system (like a spin 1/2 in a magnetic field or an atom under near resonant excitation) $\widehat{\rho}$ is a 2×2–matrix.

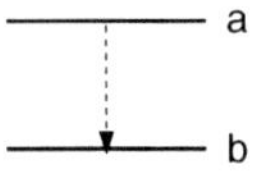

$$|a\rangle \;\to\; |b\rangle, \qquad \rho_{aa} \propto e^{-t/T_1}\,,$$
$$|a\rangle \;+\; |b\rangle, \qquad \rho_{ab} \propto e^{-t/T_2} \cos\left(\frac{E_a - E_b}{\hbar}\,t\right).$$

The diagonal elements $\rho_{aa} \geq 0$ and $\rho_{bb} = 1 - \rho_{aa} \geq 0$ give the populations of the levels, whereas the nondiagonal elements $\rho_{ba} = \rho^*_{ab}$ describe the coherent motion, i.e., an oscillating (electrical or magnetic) dipole moment. In simple cases relaxation and coherence can be characterized by two different relaxation times which are usually denoted by T_1 (diagonal elements) and T_2 (nondiagonal elements), $T_2 \leq 2T_1$.

---

* http://www.tkm.uni-karlsruhe.de/personal/baltz/

B. Di Bartolo and O. Forte (eds.), Advances in Spectroscopy for Lasers and Sensing, 33–62.

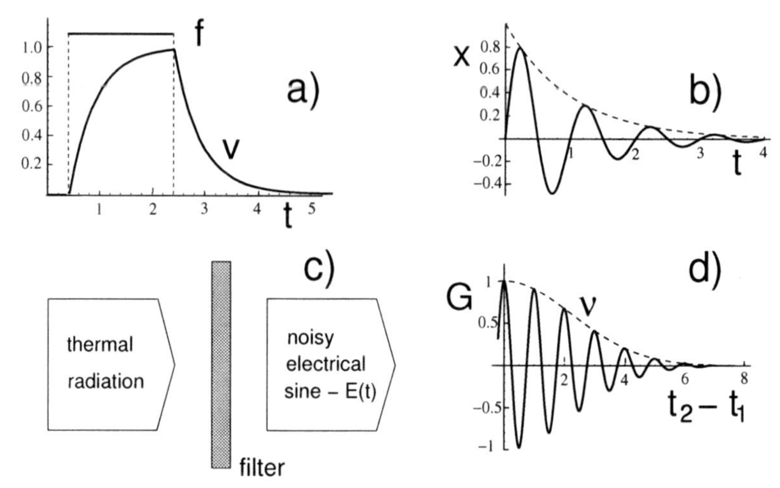

Figure 1. Upper panel: time dependence of two classical relaxing systems: (a) Particle in a viscous fluid subjected to an external force f(t), (b) harmonic oscillator. Lower Panel: (c) filtered thermal light, (d) correlation function ($\mathcal{V}$: fringe contrast in an interferometer).

This article provides an overview on typical relaxation and decoherence phenomena in classical and quantum systems, see also some previous Erice contributions[1]. In addition, various examples are given including optically generated exciton spin states in quantum dots which are currently of interest in spintronics[2]. Decoherence is also an important issue in connection with the appearance of a classical world in quantum theory and the quantum measurement problem.

2. Relaxation in Classical Systems

In the following we denote the relaxing quantity by "v" which may be also a dipole moment or position etc.

2.1. GENERAL PROPERTIES OF LINEAR RELAXING SYSTEMS

The prototype of relaxation is characterized by an exponential decay, $v(t) \propto \exp(-\gamma t)$, which is termed *Debye–relaxation.* γ is the relaxation rate and $\tau = 1/\gamma$ is the relaxation time. Including an external "force" $f(t)$ of arbitrary time–dependence, Debye–relaxation obeys a first order differential equation whose solution can be given in terms of its *Green–function* $\chi_{\rm D}(t)$

$$\dot{v}(t) \;+\; \gamma\, v(t) = f(t)\,, \tag{1}$$

$$v(t) \;=\; \int_{-\infty}^{\infty} \chi_{\rm D}(t-t')\, f(t')\, dt'\,, \tag{2}$$

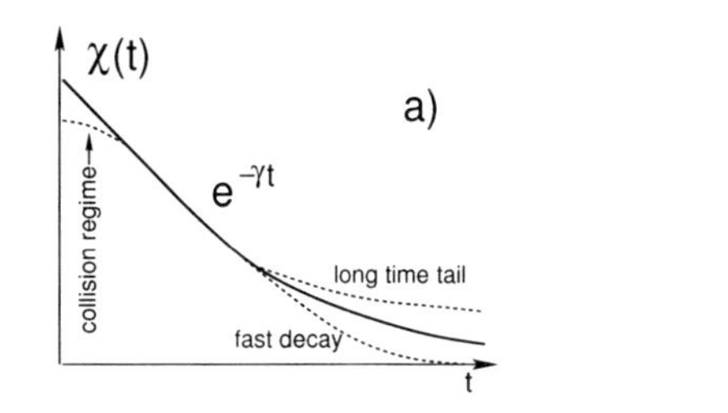

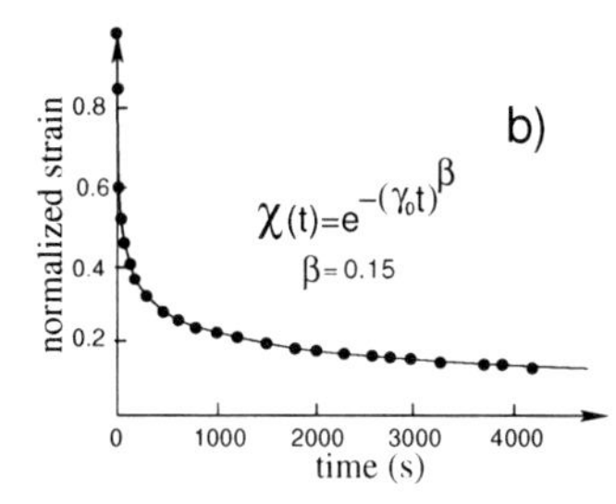

Figure 2. (a) Time dependence of typical (monotonous) relaxing processes. (b) Stretched exponential decay of strain recovery in polycarbonate[5].

$$\chi_{\mathrm{D}}(t) \;=\; e^{-\gamma t}\,\Theta(t)\,. \tag{3}$$

Instead of specifying an equation of motion for the relaxing quantity a *linear system* may be equally well characterized by an arbitrary causal function $\chi(t)$ ($\chi(t) \equiv 0$, $t < 0$) which – like a Green–function – gives the response with respect to a $\delta(t)$–pulse, see Fig. 2. As $\chi(t)$ fully describes the input–output relationship (2) this function is also called *system function.*

In frequency–space[1] the convolution integral (2) becomes a product

$$v(t) \;=\; \int_{-\infty}^{\infty} v(\omega)\, e^{-\mathrm{i}\omega\mathrm{t}}\, \frac{d\omega}{2\pi}\,, \qquad \text{etc.} \tag{4}$$

$$v(\omega) \;=\; \chi(\omega)\, f(\omega)\,. \tag{5}$$

Due to causality, the *dynamical susceptibility*[2]

$$\chi(\omega) = \int_0^{\infty} \chi(t)\, e^{\mathrm{i}\omega\mathrm{t}}\, dt = \int_0^{\infty} \chi(t)\, e^{-\omega_{\mathrm{i}} t}\, e^{\mathrm{i}\omega_{\mathrm{r}}\mathrm{t}}\, dt \tag{6}$$

is an analytic function in the upper part of the *complex ω–plane.* In $\mathrm{Im}\,\omega = \omega_{\mathrm{i}} < 0$, however, $\chi(\omega)$ must have singularities (apart from the trivial case $\chi(\omega)$ =const). In terms of singularities, Debye–relaxation provides the simplest case: it has a first order pole at $\omega = -\mathrm{i}\gamma$, see Fig. 3.

$$\chi_{\mathrm{D}}(\omega) = \frac{1}{\gamma - \mathrm{i}\omega} = \frac{\gamma}{\gamma^2 + \omega^2} + \mathrm{i}\frac{\omega}{\gamma^2 + \omega^2}\,. \tag{7}$$

The real part of the susceptibility, $\chi_1 = \mathrm{Re}\,\chi$, determines the cycle–averaged dissipated power whereas $\chi_2 = \mathrm{Im}\,\chi$ describes the flow of energy from the driving force to the system and vice versa ("dispersion").

[1] Fouriertransformation is with $\exp(-\mathrm{i}\omega\mathrm{t})$ so that it conforms with the quantum mechanical time–dependence $\exp(-\mathrm{iEt}/\hbar)$.

[2] The physical susceptibility may have in addition a prefactor depending on the relation between $f(t)$ and the physical driving "force", e.g., $E(t)$.

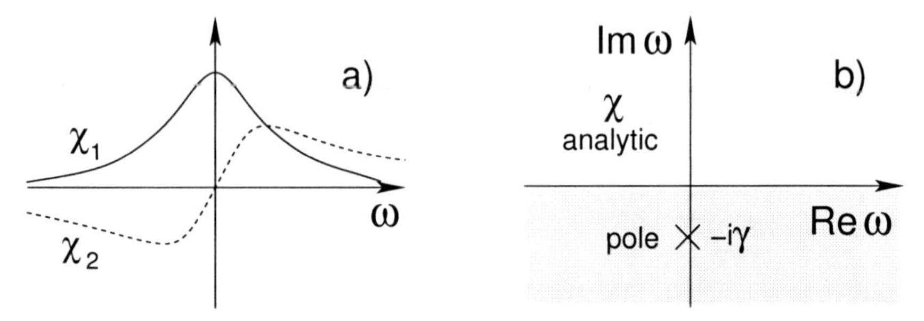

Figure 3. (a) Sketch of the real and imaginary parts of the Debye susceptibility. (b) Location of singularities of $\chi(\omega)$ in the complex ω–plane.

A characteristic property of Debye relaxation is the appearance of a semicircle when plotting the real/imaginary parts of $\chi(\omega)$ on the horizontal/vertical axis with ω as a parameter (Cole–Cole plot), see Fig. 4(a). Experimental data, however, are usually located below the semicircle. In many cases, the experimental results can be very well fitted by a "Cole-Cole function",

$$\chi_{CC}(\omega) = \frac{1}{1-(i\tau_0\omega)^{1-\alpha}}, \quad 0 \leq \alpha < 1. \tag{8}$$

Nevertheless, the benefit of such a fit is not obvious, see Fig. 4(b).

An often used generalization of (3) is a (statistical) distribution of relaxation rates in form of "parallel–relaxation"

$$\chi(t) = \int_0^\infty w(\gamma)\, e^{-\gamma t}\, d\gamma\, \Theta(t), \tag{9}$$

$$\int_0^\infty w(\gamma)\, d\gamma = 1, \quad w(\gamma) \geq 0. \tag{10}$$

$\chi(t)$ is the Laplace–transform of $w(\gamma)$.[3] Note, $\dot{\chi}(t=0) = -\int \gamma w(\gamma) d\gamma < 0$ is the negative first moment of $w(\gamma)$. Within this model, the Cole–Cole plot is always under the Debye–semicircle.

2.2. EXAMPLES

2.2.1. *Gaussian band of width Γ centered at γ_0*

$$w(\gamma) = \frac{1}{\sqrt{\pi}\Gamma} e^{-\frac{(\gamma-\gamma_0)^2}{\Gamma^2}}, \tag{11}$$

$$\chi(t) = \begin{cases} e^{-\gamma_0 t}, & \gamma_0 t \ll 1, \\ \frac{e^{-(\gamma_0/\Gamma)^2}}{\sqrt{\pi}\Gamma t}, & \gamma_0 t \gg 1. \end{cases} \tag{12}$$

[3] Equivalently, one may use a distribution function for the relaxation times $\tilde{w}(\tau)$.

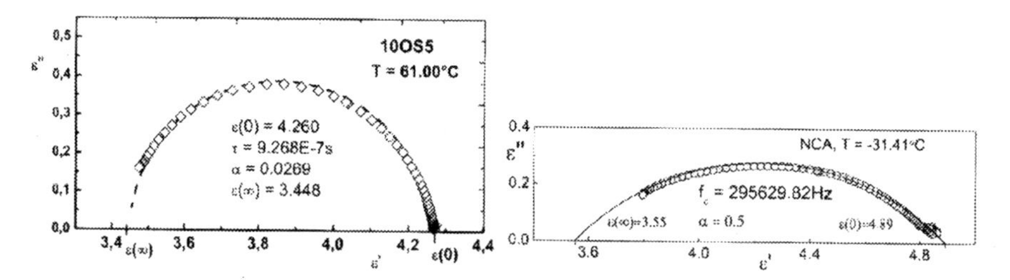

Figure 4. Relaxation of electrical dipoles in two different samples of liquid crystals, $\epsilon(\omega) = \epsilon(\infty) + \chi(\omega)$. (left) Almost pure Debye–type spectrum and (right) Cole-Cole relaxation. According to Haase and Wrobel[3].

2.2.2. *Power–law decay*

The following distribution (which has a broad maximum at $\gamma_m = n/\tau_0$) leads to a power–law decay

$$w(\gamma) = \frac{\tau_0^{n+1}}{n!}\,\gamma^n e^{-\tau_0\gamma}\,, \qquad \chi(t) = \left[\frac{\tau_0}{\tau_0 + t}\right]^{n+1}\,. \tag{13}$$

2.2.3. *Stretched exponential*

Relaxation in complex strongly interacting systems often follows a *stretched exponential* form[5], see also Fig. 2(b),

$$\chi(t) \;=\; e^{-(\gamma_0 t)^\beta}\,, \quad 0 < \beta \leq 1\,. \tag{14}$$

Kohlrausch[4] (p. 179) first suggested (14) to describe viscoelasticity but there are many examples (including transport and dielectric properties) which supports the stretched exponential as a universal law, see e.g., Refs. [6](a-c). As $\dot{\chi}(t=0) = -\infty$ the corresponding distribution function $w(\gamma)$ must either have a pathological tail $w(\gamma) \to 1/\gamma^{2-\delta}$ $(\delta > 0)$ or (14) does not hold for $t \to 0$, see Fig. 2. The long time behaviour, on the other hand, corresponds to the strong decrease of $w(\gamma)$ for small values of γ, i.e., large relaxation times.

An analytical example for the case $\beta = 1/2$ in terms of parallel relaxation can be found in tables of Laplace–transforms[7](Vol. 5, Ch. 2.2)

$$\chi(t) = e^{-\sqrt{\gamma_0 t}}\,\Theta(t)\,, \quad w(\gamma) = \sqrt{\frac{\gamma_0}{4\pi\gamma^3}}\,e^{-\frac{\gamma_0}{4\gamma}}\,. \tag{15}$$

$w(\gamma)$ has a broad maximum at $\gamma_m = \gamma_0/6$.

Very likely, the stretched exponential originates from many sequential dynamically correlated activation steps rather than from parallel relaxation[8].

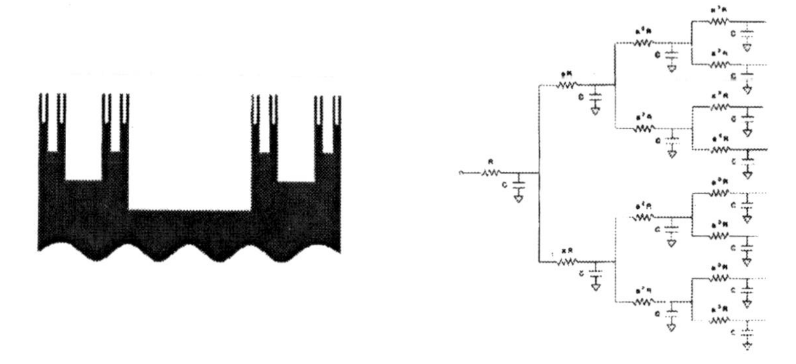

Figure 5. (left) Liu's Cantor block model of a rough interface between an electrolyte (black) and an electrode (white), two groves, each with four stages of branching are shown. (right) Equivalent circuit. According to Liu[11](a).

2.2.4. *Frequency dependent relaxation rate*

C–C–plots above the semicircle may arise in systems which either display

- "faster than exponential relaxation", e.g., $\chi(t) = (1-\gamma t)^2\Theta(t)\Theta(1-\gamma t)$,
- or have a frequency dependent relaxation rate with $\gamma(\omega) \to 0$ for $\omega \to \infty$ ("sequential relaxation")

$$\chi(\omega) = \frac{1}{-\mathrm{i}\omega + \gamma(\omega)}\,. \tag{16}$$

In the time domain, a frequency dependent $\gamma(\omega)$ leads to relaxation with a memory

$$\dot{x}(t) + \int_{-\infty}^{\infty} \gamma(t-t')\, x(t')\, dt' = f(t)\,, \tag{17}$$

where $\gamma(t)$ is the Fourier–transform of $\gamma(\omega)$. Causality requires that $\gamma(t) \equiv 0$ for $t < 0$ so that $\gamma(\omega)$ is an analytic function in $\mathrm{Im}\,\omega > 0$.

A simple approximation for $\gamma(\omega)$ is of "Drude" type

$$\gamma(\omega) = \gamma_{\mathrm{D}}\,\frac{\gamma_{\mathrm{c}}}{-\mathrm{i}\omega + \gamma_{\mathrm{c}}}\,. \tag{18}$$

$\gamma_{\mathrm{c}} \geq \gamma_{\mathrm{D}}$ may be interpreted as a "collision" rate. For this model $\chi(\omega)$ displays two poles in $\mathrm{Im}\,\omega < 0$ and a zero in $\mathrm{Im}\,\omega > 0$. For $\gamma_{\mathrm{D}} < \gamma_{\mathrm{c}} < 4\gamma_{\mathrm{D}}$, $\chi(t)$ even shows oscillatory behaviour. For experimental evidence see, e.g., measurements by Dressel et al.[9]. Finally, we mention that fractional kinetics can be also formulated by the new and fancy mathematics of "fractional derivatives"[10].

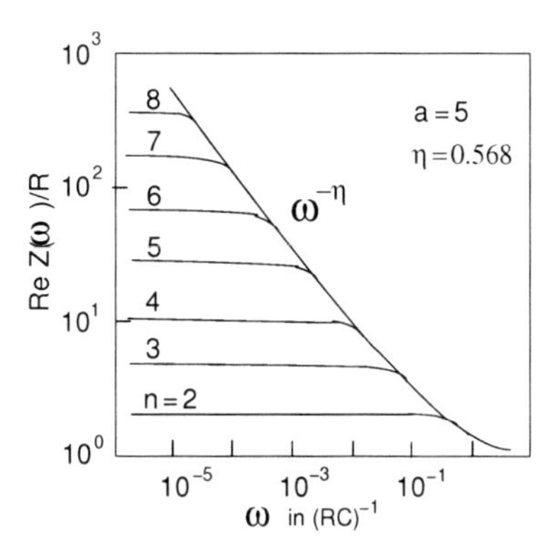

Figure 6. Frequency dependence of the real part of the impedance of the fractal network in Fig. 5 for finite and infinite stages. According to Liu[11](a).

2.3. IMPEDIANCE OF A ROUGH SURFACE

An interesting example for sequential relaxation is the *Cantor–RC–circuit* model[11] which is supposed to describe the impediance of a rough (fractal) metal surface[12], see Fig. 5. The impedance of the network is given by the inifinite continued fraction[11](b)

$$Z(\omega) \;=\; R+\frac{1}{j\omega C+\frac{2}{aR+\frac{1}{j\omega C+..}}} \tag{19}$$

which fulfills the exact scaling relation[4]

$$Z(\omega/a) \;=\; R+\frac{aZ(\omega)}{j\omega CZ(\omega)+2}\,. \tag{20}$$

For $\omega\to 0$ and $a>2$, Eq. (20) becomes $Z(\omega/a)=aZ(\omega)/2$ which implies $Z(\omega)\propto\omega^{-\eta}$ with $\eta=1-\ln(2)/\ln(a)=1-\bar{d}$, $\bar{d}$ is the fractal dimension of the surface, see Fig. 6.

For $a=1$ (yet unphysical) (20) becomes a quadratic equation for Z which can be solved analytically. Limiting cases are

$$\omega\to 0: \quad Z_{a\,=\,1}(\omega) \;=\; 2R\Big(1-jRC\omega\Big)\,, \tag{21}$$

$$\omega\to\infty: \quad Z_{a\,=\,1}(\omega) \;=\; R\Big(1+\frac{2}{(RC\omega)^2}\Big)-\frac{j}{2\omega C}\,. \tag{22}$$

[4] Electrotechnical convention: $j=-\mathrm{i}$, time dependence is by $\exp(+j\omega t)$.

3. Interaction of a Particle with a Bath

Many situations in nature can be adequately described by a system with one (or only a few) degrees of freedom ("particle") in contact with a rather complex environment modelled by a reservoire of harmonic oscillators or a "bath" of temperature T. In the classical limit the interaction with the bath is described by a stochastic force acting on the particle

$$F_{\text{bath}} = -\eta v + F_{\text{st}}\,, \tag{23}$$

where $-\eta v$ describes the slowly varying frictional contribution and F_{st} denotes a rapidly fluctuating force with zero mean $\overline{F_{\text{st}}(t)} = 0$. For a stationary Gaussian process, the statistical properties are fully characterized by its correlation function which in the case of an uncorrelated process reads

$$K_{\text{FF}}(t_1, t_2) = \overline{F_{\text{st}}(t+t_1)F_{\text{st}}(t+t_2)} = 2\eta k_{\text{B}}T\delta(t_2 - t_1)\,. \tag{24}$$

The overline denotes a time average and the constant $\eta(= M\gamma)$ is proportional to the viscosity.

The *Langevin equation*

$$M\ddot{q}(t) + \eta\dot{q}(t) + V'(q) = F_{\text{st}}(t) \tag{25}$$

describes, for example, a heavy Brownian particle of mass M immersed in a fluid of light particles and driven by an external force $F = -V'(q)$. Another example is Nyquist noise in a R–L circuit ($V \equiv 0$). For an overview see, e.g., Reif[13] (Sect. 15).

For a free particle (25) conforms with (1); the velocity–force and velocity–velocity correlation functions (in thermal equilibrium) follow from (2,3)

$$K_{\text{vF}}(t_1, t_2) = \overline{v(t+t_1)F_{\text{st}}(t+t_2)} = 2\gamma k_{\text{B}}T\,\chi(t_1 - t_2)\,, \tag{26}$$

$$K_{\text{vv}}(t_1, t_2) = \overline{v(t+t_1)v(t+t_2)} = \frac{k_{\text{B}}T}{M}\,e^{-\gamma|t_1 - t_2|}\,. \tag{27}$$

Eq. (27) includes the equipartition theorem $\overline{Mv^2(t)/2} = k_{\text{B}}T/2$. Under the influence of the stochastic force the particle describes *Brownian motion*[5]

$$q(t) = \int_0^t v(t')\,dt'\,, \tag{28}$$

$$\overline{q^2}(t) = \int_0^t \int_0^t K_{\text{vv}}(t', t'')\,dt'dt'' \to 2Dt\,, \quad (t \to \infty)\,, \tag{29}$$

where $D = \overline{v^2}/\gamma = k_{\text{B}}T/(M\gamma)$ is the diffusion coefficient.

[5] Note, one of Einstein's three seminal 1905–papers was on diffusion[14].

As an example how to eliminate the reservoire variables explicitly, we consider a classical particle of mass M and coordinate q, which is bilinearly coupled to a system of uncoupled harmonic oscillators

$$H = H_{\rm s} + H_{\rm res} + H_{\rm int}\,, \tag{30}$$

$$H_{\rm s} = \frac{p^2}{2M} + V(q), \tag{31}$$

$$H_{\rm res} = \sum_i \frac{p_i^2}{2m_i} + \frac{1}{2} m_i \omega_i^2 x_i^2\,, \tag{32}$$

$$H_{\rm int} = -q \sum_i c_i x_i + q^2 \sum_i \frac{c_i^2}{2m_i\omega_i^2}\,, \tag{33}$$

with suitable constants for c_i, ω_i, m_i. This model has been used by several authors and is nowadays known as the *Caldeira–Leggett model*[15]. For an elementary version see Ingold's review article in Ref.[16] (p. 213).

The equations of motion of the coupled system are:

$$M\ddot{q} + V'(q) + q\sum_i \frac{c_i^2}{m_i\omega_i^2} = \sum_i c_i x_i\,, \tag{34}$$

$$\ddot{x}_i + \omega_i^2 x_i = \frac{c_i}{m_i} q(t)\,. \tag{35}$$

As the reservoire represents a system of uncoupled oscillators (35) can be easily solved in terms of the appropriate Green–function

$$x_i(t) = x_i(0)\cos(\omega_i t) + \frac{p_i(0)}{m_i}\sin(\omega_i t) + \int_0^t \frac{c_i}{m_i\omega_i}\sin[\omega_i(t-t')]q(t')dt'\,. \tag{36}$$

Inserting (36) into (34) yields a closed equation for $q(t)$

$$M\ddot{q}(t) + \int_0^t M\gamma(t-t')\dot{q}(t')\,dt' + V'(q) = \xi(t)\,, \tag{37}$$

$$\gamma(t) = \frac{2}{\pi}\int_0^\infty \frac{J(\omega)}{\omega}\cos(\omega t)d\omega\,, \tag{38}$$

$$J(\omega) = \pi \sum_i \frac{c_i^2}{2Mm_i\omega_i}\delta(\omega-\omega_i)\,. \tag{39}$$

Comparison with (25) shows that the relaxation process acquired a memory described by $\gamma(t-t')$, i.e., a frequency dependent relaxation rate $\gamma(\omega)$, (non–Markovian process). $\xi(t)$ is the microscopic representation of the fluctuating force $F_{\rm st}(t)$ which depends on the initial conditions of the reservoire variables and is not explicitely stated here.

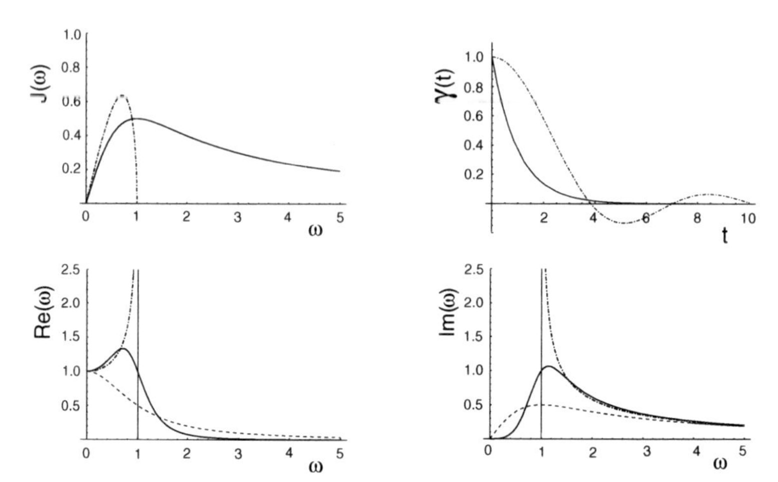

Figure 7. Upper panel: Spectral density of the reservoire oscillators (left) and memory functions (right). Solid lines: Caldeira–Leggett model with a Drude form, dashed lines: Drude–model, dashed dotted lines: Rubin model. (Dimensionless quantities, $M_0/M = 1$). Lower panel: Real and imaginary parts of the electrical conductivity $\sigma \propto \chi$.

For a finite number of reservoire oscillators the total system will always return to its initial state after a finite (Poincaré) recurrence time or may come arbitrarily close to it. For $N \to \infty$, however, the Poincaré time becomes infinite, simulating dissipative behavior. We therefore first take the limit $N \to \infty$ and consider the spectral density of reservoire modes $J(\omega)$ as a continous function. Frictional damping, $\gamma(t) = \gamma_0 \delta(t)$, is obtained for $J(\omega) \propto \omega$. A more realistic behavior would be a "Drude" function

$$J(\omega) = \gamma_0 \omega \frac{\gamma_{\mathrm{D}}^2}{\omega^2 + \gamma_{\mathrm{D}}^2}, \quad \gamma(t) = \gamma_0 \gamma_{\mathrm{D}} e^{-\gamma_{\mathrm{D}} t}, \tag{40}$$

which is linear for $\omega \to 0$ but goes smoothly to zero for $\omega \gg \gamma_{\mathrm{D}}$, see Fig. 7. Note, memory effects do not only lead to a frequency dependent scattering rate ($= \mathrm{Re}\,\gamma(\omega)$) but to a shift in the resonance frequency ($= \mathrm{Im}\,\gamma(\omega)$), too.

Another nontrivial yet exactly solvable model is obtained by a linear chain with one mass replaced by a particle of (arbitrary) mass M_0 (Rubin–model). The left and right semi–infinite wings of the chain serve as a "reservoire" to which the central particle is coupled.

As a result the damping kernels are

$$\gamma(t) = \frac{M_0}{M} \omega_L \frac{J_1(\omega_L t)}{t}, \tag{41}$$

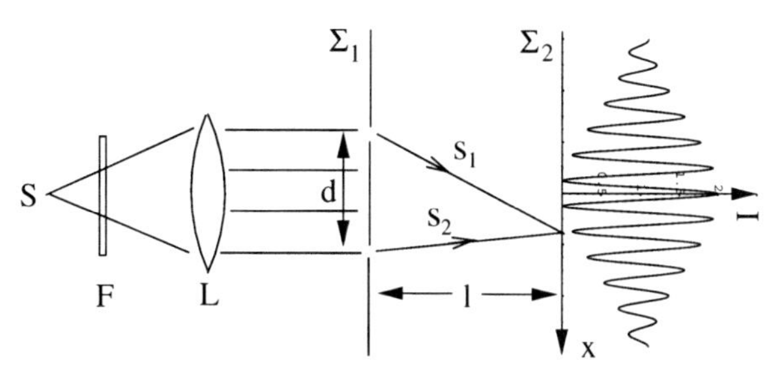

Figure 8. Arrangement of components for an idealized Young's interference experiment. Interferogram shown in the limit of infinitely small slits and $\lambda \ll d \ll l$. Gaussian spectral filter F.

$$\gamma(\omega) = \frac{M_0}{M}\begin{cases}\sqrt{\omega_L^2-\omega^2}+i\omega\,, & |\omega|<\omega_L\,,\\ i\frac{\omega_L^2\,sgn(\omega)}{\omega_L+\sqrt{\omega^2-\omega_L^2}}\,, & |\omega|>\omega_L\,,\end{cases} \tag{42}$$

$$J(\omega) = \omega\frac{M_0}{M}\sqrt{\omega_L^2-\omega^2}\,\Theta(\omega_L^2-\omega^2)\,. \tag{43}$$

$J_1(x)$ is a Bessel–function. For details see Fick and Sauermann book [17] (p. 255). In contrast to the "Drude case" $\gamma(t)$ shows oscillations and decays merely algebraically for large times, see Fig. 7.

$$\gamma(t) \to \frac{M_0}{M}\sqrt{\frac{2\omega_L}{\pi}}\frac{\sin[\omega_L t-\pi/4]}{t^{3/2}}\,. \tag{44}$$

These oscillations reflect the upper cut–off in $J(\omega)$ at the maximum phonon frequency ω_L.

4. Coherence in Classical Systems

The degree of coherence of a signal "$x(t)$" or a field $\mathcal{E}(\mathbf{r},t)$ etc., is measured in terms of correlation at different times (or space–time points)

$$G(t,t') = \langle x(t)x(t')\rangle. \tag{45}$$

For stationary, ergodic ensembles, time– and ensemble averages give identical results. Moreover, the following general properties hold[13]

- $G(t,t') = G(t-t') = G(|t-t'|)$,
- $|G(t-t')| \leq G(0)$,
- $G(t) = \int_{-\infty}^{\infty} |x(\omega)|^2\, e^{\mathrm{i}\omega t}\,\frac{d\omega}{2\pi}$, (Wiener–Khinchine theorem).

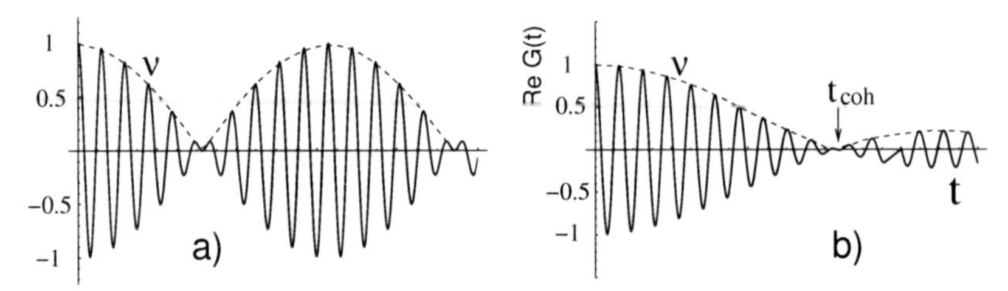

Figure 9. First order coherence functions as a function of time. (a) Two modes with $\omega_j = (1 \pm 0.05)\omega_0$ and (b) many uncorrelated modes with a "box" spectrum centered at ω_0 and full width $\Delta\omega = 0.1\omega_0$. $\mathcal{V}$ is the visibility of the fringes.

The prototype of a device to measure the optical coherence(–time) of light is the *Young double slit interference experiment*[6], see Fig. 8. In the following discussion, we shall ignore complications arising from the finite source diameter and consequent lack of perfect parallelism in the illuminating beam, diffraction effects at the pinholes (or slits), reduction of amplitude with distances s_1, s_2 etc., in order that attention be focused on the properties of the incident radiation rather than on details of the measuring device. Let $\mathcal{E}(\mathbf{r}, t)$ be the electrical field of the radiation at point $\mathbf{r}$ on the observation screen at time t. This field is a superposition of the incident field at the slits $\mathbf{r}_1, \mathbf{r}_2$ at earlier times $t_{1,2} = t - s_{1,2}/c$,

$$\mathcal{E}(\mathbf{r}, t) \;\propto\; \mathcal{E}^{\rm in}(\mathbf{r}_1, t_1) + \mathcal{E}^{\rm in}(\mathbf{r}_2, t_2)\,. \tag{46}$$

As a result, the (cycle averaged) light intensity $I \propto \overline{|\mathcal{E}(\mathbf{r}, t)|^2}$ on the screen can be expressed in terms of the correlation function $G(\mathbf{r}_2, t_2; \mathbf{r}_1, t_1)$

$$I(t) \;\propto\; G(1,1) + G(2,2) + 2{\rm Re}\, G(2,1), \tag{47}$$

$$G(\mathbf{r}_2, t_2; \mathbf{r}_1, t_1) \;=\; \langle \mathcal{E}^{(-)}(\mathbf{r}_2, t_2)\mathcal{E}^{(+)}(\mathbf{r}_1, t_1)\rangle\,. \tag{48}$$

$G(2,1)$ is short for $G(\mathbf{r}_2, t_2; \mathbf{r}_1, t_1)$ etc., and $\mathcal{E}^{(\pm)}(\mathbf{r}, t) \propto \exp(\mp {\rm i}\omega t)$ denote the positive/negative frequency components of the light wave. It is seen from Eq. (47) that the intensity on the second screen consists of three contributions: The first two terms represent the intensities caused by each of the pinholes in the absence of the other, whereas the third term gives rise to interference effects. [Note the difference between $G(2,1)$ and (45)].

For a superposition of many (uncorrelated) modes we obtain

$$\mathcal{E}(\mathbf{r}, t) \;=\; \sum_{\mathbf{k}} A_{\mathbf{k}} e^{i(\mathbf{k}\mathbf{r} - \omega_{\mathbf{k}} t)} + cc = \mathcal{E}^{(+)}(\mathbf{r}, t) + \mathcal{E}^{(-)}(\mathbf{r}, t)\,, \tag{49}$$

$$G(2,1) \;=\; \sum_{\mathbf{k}} |A_{\mathbf{k}}|^2 \; e^{{\rm i}[\mathbf{k}(\mathbf{r}_2 - \mathbf{r}_1) - \omega_{\mathbf{k}}(t_2 - t_1)]}\,. \tag{50}$$

[6] The Michelson interferometer and the stellar interferometer would be even better suited instruments to measure the spatial and temporal coherence independently[18].

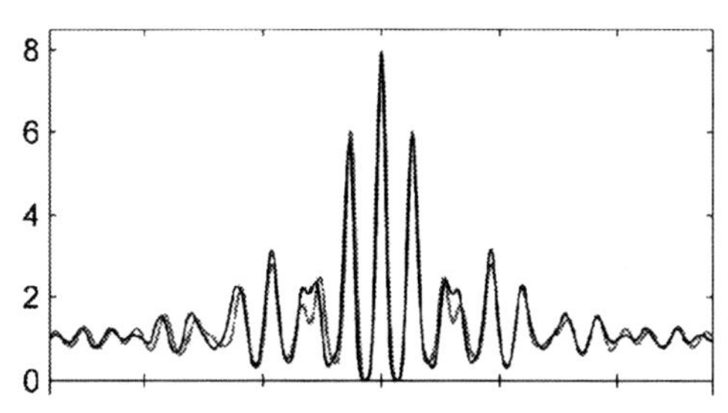

Figure 10. Experimental autocorrelation function of a 5fs laser oscillator at 82MHz and its spectral reconstruction ("background" = 1, time: in units of 10 fs). From Wegener[19].

In particular, for a gaussian spectral line centered at ω_0 (and $\mathbf{r}_2 = \mathbf{r}_1$), Eq. (50) becomes

$$I(\omega) = \exp\left[-\frac{(\omega-\omega_0)^2}{2(\Delta\omega)^2}\right], \tag{51}$$

$$G(t_2,t_1) = \int_{-\infty}^{\infty} I(\omega)\, e^{-i\omega(t_2-t_1)}\,\frac{d\omega}{2\pi}, \tag{52}$$

$$= e^{-i\omega_0(t_2-t_1)}\exp\left[-\frac{1}{2}[\Delta\omega(t_2-t_1)]^2\right]. \tag{53}$$

Eq.(52) gives an example of the famous Wiener-Khinchine theorem[13]: the correlation function is just the Fourier–transformed power spectrum $I(\omega_\mathbf{k}) \propto |A_\mathbf{k}|^2$. Filtering in frequency-space is intimately related to a correlation (or coherence)–time $\Delta t \propto 1/\Delta\omega$, see Figs. 8 and 9. [Similarly, spatial filtering, i.e., selecting $\mathbf{k}$–directions within $\Delta\mathbf{k}$ leads to spatial coherence.]

There is a deep relation between the fluctuations in thermal equilibrium (i.e. loss of coherence) and dissipation in a nonequilibrium state driven by an external force. This is the *fluctuation – dissipation theorem* which (for the example of Ch. 3) reads

$$\eta = \frac{1}{2k_\mathrm{B}T}\, K_\mathrm{FF}(\omega=0)\,. \tag{54}$$

Analogous to optics, coherence can be defined for any quantity which is additive and displays a phase or has a vector character, e.g., electrical and acoustic "signals", electromagnetic fields, wave–functions, etc. Coherence is intimately connected with reversibility, yet the opposite is not always true. At first sight, a process might appear as fully incoherent or random, nevertheless it may represent a highly correlated pure state which always implies complete coherence.

A beautiful example of an disguised coherent process is the spin– (or photon–) echo[20]. This phenomenon is related to a superposition of many

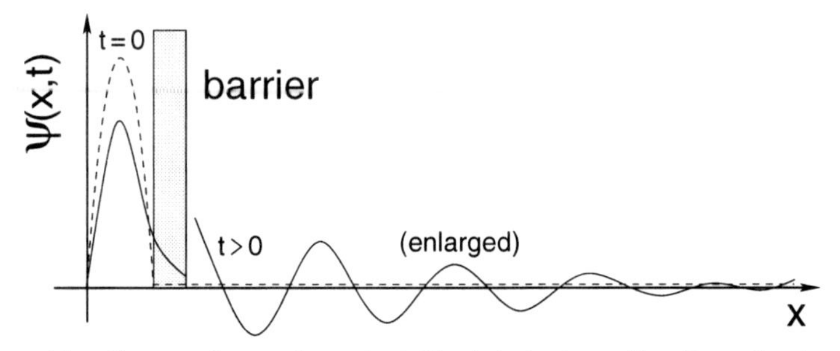

Figure 11. Gamow–decay of a metastable state by tunneling through a barrier.

sinusoidal field components with fixed (but random) frequencies. At $t = 0$ these components have zero phase differences and combine constructively to a nonzero total amplitude. Later, however, they develop large random phase differences and add up more or less to zero so that the signal resembles "noise". Nevertheless, there are fixed phase relations between the components at every time. By certain manipulations at time T a time–reversal operation can be realized which induces an echo at time $t = 2T$, which uncovers the hidden coherent nature of the state. Echo phenomena are always strong indications of hidden reversibility and coherence. For applications in semiconductor optics (photon–echo) see, Klingshirn[21], Haug and Koch[22], or previous Erice contributions[1](b). Another nice example is "weak localization" of conduction electrons in disordered materials[23].

A survey of second order coherence and the Hanbury–Brown Twiss effect can be found in Ref.[1](d).

5. Relaxation and Decoherence in Quantum Systems

5.1. DECAY OF A METASTABLE STATE

The exponential form of the radioactive decay $N(t) = N_0 \exp(-\lambda t)$ (or time dependence of the spontaneous emission from an excited atom) follows from a simple assumption which can be hardly weakened: $\dot{N} = -\lambda N$. Nevertheless, numerous authors have pointed out that the exponential decay law is only an approximation and deviations from purely exponential behaviour are, in fact, expected at very short and very long times, e.g., Refs.[24].

At short times the decay of any nonstationary state must be quadratic

$$|\Psi(t)\rangle = e^{-\mathrm{it}\widehat{\mathrm{H}}/\hbar}|\Psi(0)\rangle = \left[1 + (-\mathrm{it}\widehat{\mathrm{H}}/\hbar) + \frac{1}{2}(\ldots)^2 + \ldots\right]|\Psi(0)\rangle, \quad (55)$$

$$N(t) = |\langle\Psi(0)|\Psi(t)\rangle|^2 = 1 - (\Delta\widehat{H}/\hbar)^2 t^2 + \ldots. \quad (56)$$

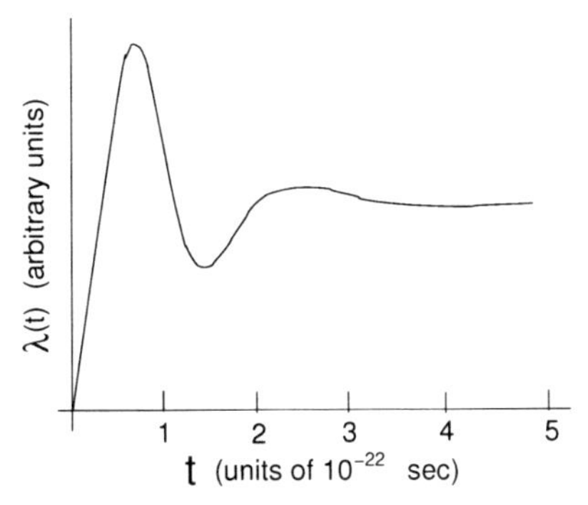

Figure 12. Calculated decay rate as a function of time. According to Avignone[26].

Here, $N(t)$ denotes the non–decay probability and $\Delta\widehat{H}$ is the energy–uncertainty in the initial state. Clearly, the transition rate $\lambda = -\dot{N} \propto t \to 0$ decreases linearly with $t \to 0$.[7]

On the other hand, at very long times the decay follows a power law $N \propto t^{-\alpha}$ which originates from branch cuts in the resolvent operator[24](a,b). Nevertheless, pure exponential decay can arise if the potential $V(x)$ decreases linearly at large distances, see, e.g., Ludviksson[24](c).

Up to date all experimental attempts to find these deviations failed[25]. Fore example, Norman et al.[25](d) have studied the β–decay of ^{60}Co at times $\leq 10^{-4}T_{1/2}$ ($T_{1/2}$ = 10.5min) and those of ^{56}Mn over the interval $0.3T_{1/2} \leq t \leq 45T_{1/2}$ ($T_{1/2}$ = 2.579h) to search for deviations from exponential decay but with a null result. Calculations by Avignone[26]

$$\lambda(t) = \frac{t}{\hbar^2} \int |\langle \Psi_f|\widehat{H}_{\text{int}}|\Psi_i\rangle|^2 \left[\frac{\sin(\frac{1}{2}\omega t)}{(\frac{1}{2}\omega t)}\right]^2 \rho(E_{\text{f}})\, dE_{\text{f}}\,, \tag{57}$$

demonstrate, however, that these experiments were $18 - 20$ orders of magnitude less time sensitive than required to detect pre–exponential decay, see Fig. 12. $\hbar\omega = [E_\gamma - (E_i - E_f)]$, $E_{\text{i,f}}$ are the nuclear eigenstate energies, $\Psi_{\text{i,f}}$ are the initial and final states of the total system "nucleus + radiation", and $\rho(E_{\text{f}})$ is the density of final states. For long times $[\ldots] \to 2\pi\delta(\omega)/t$ yielding Fermi's Golden rule and a constant value of λ for $t > 10^{-22}$s.

Another example is the decay of an excited atom in state a which decays into the ground state b by spontaneous emission of a photon. The state of combined system "atom + radiation field" is

$$|\Psi(t)\rangle = c_a(t)|a, 0_{\mathbf{k}}\rangle + \sum_{\mathbf{k}} c_{b,\mathbf{k}}(t)|b, 1_{\mathbf{k}}\rangle\,. \tag{58}$$

---

[7] This may also have far reaching consequences for the interpretation of the predicted proton–decay[25](c) ($T_{1/2} \approx 10^{15}$ times the age of the universe).

From the Schrödinger equation we get two differential equations for c_a, c_b which can be solved in the *Weisskopf–Wigner approximation*[27, 29](Ch. 6.3) which leads to pure exponential decay,

$$\dot{c}_a(t) = -\frac{\Gamma}{2} c_a(t), \quad |c_a(t)|^2 = \exp(-\Gamma t), \quad \Gamma = \frac{1}{4\pi\epsilon_0} \frac{4\omega_0^3 \mathbf{p}_{\mathrm{ab}}^2}{3\hbar c^3}. \tag{59}$$

$\mathbf{p}_{\mathrm{ab}}$ is the dipol–matrixelement and $\omega_0 = (E_a - E_b)/\hbar$.

In an infinite system (as assumed above) the decay of a metastable state into another (pure) state is irreversible, yet it is not related to dissipation or decoherence – the dynamics is fully described by the (reversible) Schrödinger equation. Here, irreversibilty stems from the boundary condition at infinity ("Sommerfeld's Austrahlungsbedingung"). Although there is little chance to reveal the quadratic onset in spontaneous decay of an atom in vacuum it may show up in a photonic crystal[30]. For induced transitions it became, indeed, already feasable, where it is called quantum–Zeno (or "watchdog") effect, see, e.g., Refs[31]. The decay of a metastable state has also received much attention for macroscopic tunneling processes[32].

5.2. DISSIPATIVE QUANTUM MECHANICS

On a microscopic level, dissipation, relaxation and decoherence are caused by the interaction of a system with its *environment* – dissipative systems are *open systems.* In contrast to classical (Newtonian) mechanics, however, in quantum systems dissipation cannot be included phenomenologically just by adding "friction terms" to the Schrödinger equation because (a) dissipative forces cannot be included in the Hamiltonian of the system, (b) irreversible processes transform a pure state to a mixed state which is described by a statistical operator $\widehat{\rho}$ rather than a wave function. Some useful properties are:

- $\widehat{\rho}$ is a hermitian, non–negative operator with $\mathrm{tr}\widehat{\rho} = 1$.
- For a pure state $\widehat{\rho} = |\Psi\rangle\langle\Psi|$ is a projector onto $|\Psi\rangle$.
- General mixed state: $\widehat{\rho} = \sum_\alpha p_\alpha |\phi_\alpha\rangle\langle\phi_\alpha|$ with $0 \leq p_\alpha \leq 1$, $\sum_\alpha p_\alpha = 1$.
- The dynamics of the total system is governed by the (reversible) *v. Neumann equation*

$$\frac{\partial}{\partial t}\widehat{\rho}(t) + \frac{\mathrm{i}}{\hbar}[\widehat{H}, \widehat{\rho}] = 0. \tag{60}$$

- To construct the density operator of the "reduced system" $\widehat{\rho}_{\mathrm{r}} = \mathrm{tr}_{\mathrm{res}}\widehat{\rho}$ one has to "trace–out" the reservoire variables which leads to irreversible behaviour governed by the *master equation*

$$\frac{\partial}{\partial t}\widehat{\rho}_{\mathrm{s}}(t) + \frac{\mathrm{i}}{\hbar}[\widehat{H}_{\mathrm{s}}, \widehat{\rho}_{\mathrm{s}}] = \widehat{C}(\widehat{\rho}_{\mathrm{s}}), \tag{61}$$

where $C(\widehat{\rho}_{\mathrm{s}})$ is the collision operator.

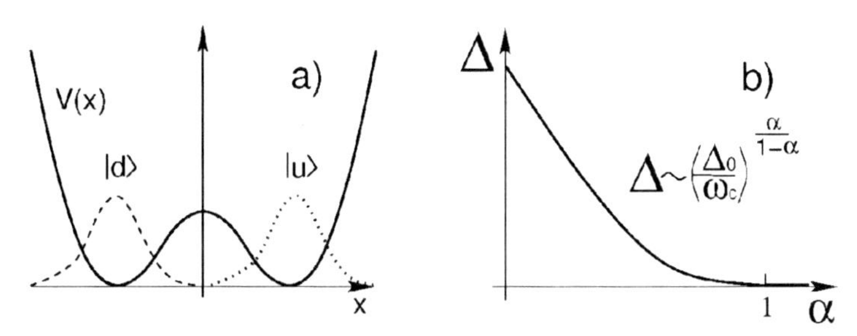

Figure 13. (a) Double well potential with localized base states $|d\rangle$ ("down"), $|u\rangle$ ("up"). (b) Renormalized level splitting as a function of damping.

A simple approximation for $\widehat{C}(\widehat{\rho}_{\rm r})$ is the *relaxation–time–approximation*

$$\widehat{C}(\widehat{\rho}_{\rm s}) = -\frac{1}{\tau}\left(\widehat{\rho}_{\rm s}(t) - \widehat{\rho}_{\rm s}^{\rm eq}\right), \tag{62}$$

where $\widehat{\rho}_{\rm s}^{\rm eq}$ denotes the statistical operator for the equilibrium state. In general, relaxation times τ for the diagonal and non–diagonal elements of $\widehat{\rho}_{\rm s}$ are different, the relaxation goes to a local equilibrium with a $(\mathbf{r}, t)$–dependent temperature and chemical potential[33], and memory effects may occur.

Finally we mention the *fluctuation–dissipation theorem* which establishes an important relation between the fluctuations in equilibrium of two observables $\widehat{A}$, $\widehat{B}$ and the dissipative part of the linear response of $\widehat{A}$ upon a perturbation $\widehat{H}_{\rm int} = -\widehat{B}b(t)$[13]. Nowadays dissipative quantum mechanics has become an important issue in the field of mesoscopic systems[16, 28] and quantum optics[29], see also contributions at previous Erice schools[1](a,c).

5.3. TWO LEVEL SYSTEMS (TLS)

We study a system with two base states $|u\rangle$, $|d\rangle$ with equal energies ϵ_0, e.g., a particle in a double–well potential, see Fig. 13.

$$|u\rangle = \begin{pmatrix} 1 \\ 0 \end{pmatrix}, \quad |d\rangle = \begin{pmatrix} 0 \\ 1 \end{pmatrix}, \tag{63}$$

In this base, the Hamiltonian and the (electrical) dipole operator read

$$\hat{H}_0 = \begin{pmatrix} \epsilon_0 & -\Delta \\ -\Delta & \epsilon_0 \end{pmatrix}, \quad \hat{D} = d_0 \begin{pmatrix} 1 & 0 \\ 0 & 1 \end{pmatrix}. \tag{64}$$

Here, $\pm\Delta$ is the tunneling splitting and $d_0 = ex_0$ is the magnitude of the (electrical) dipole moment. The eigenstates of $\widehat{H}_0$ are

$$|1\rangle = \frac{1}{\sqrt{2}}\begin{pmatrix} 1 \\ 1 \end{pmatrix}, \quad |2\rangle = \frac{1}{\sqrt{2}}\begin{pmatrix} 1 \\ -1 \end{pmatrix}. \tag{65}$$

In the basis of these energy–eigenstates $|1,2\rangle$, we have ($\epsilon_{1,2} = \epsilon_0 \mp \Delta$)

$$\hat{H}_0 = \begin{pmatrix} \epsilon_1 & 0 \\ 0 & \epsilon_2 \end{pmatrix}, \qquad \hat{D} = d_0 \begin{pmatrix} 0 & 1 \\ 1 & 0 \end{pmatrix}. \tag{66}$$

For the TLS, the density operator is a 2×2 matrix $\widehat{\rho} = \rho_{i,k}$. In particular, for a pure state $|\Psi\rangle = c_1|1\rangle + c_2|2\rangle$, we have

$$\hat{\rho}_{\text{pure}} = |\psi\rangle\langle\psi| = \begin{pmatrix} |c_1|^2 & c_1\, c_2^* \\ c_1^*\, c_2 & |c_2|^2 \end{pmatrix}. \tag{67}$$

The diagonal elements ρ_{11} and ρ_{22} yield the populations, whereas the off–diagonal elements describe the coherent motion of the dipole moment $d = \text{tr}(\widehat{\rho}\,\hat{D}) \propto \text{Re}\,\rho_{12}$.

We consider a TLS subjected to a time–dependent electrical field $\mathcal{E}(t)$

$$\hat{H} = \hat{H}_0 - \mathcal{E}(t)\,\hat{D}\,. \tag{68}$$

The v. Neumann–master equation (61) (in the basis of $\hat{H}_0$ eigenstates) reads

$$\dot{\rho}_{11}(t) + 2\omega_{\text{R}}(t)\text{Im}\,(\rho_{21}) = -\frac{1}{T_1}(\rho_{11} - \rho_{11}^{\text{eq}})\,, \tag{69}$$

$$\dot{\rho}_{22}(t) - 2\omega_{\text{R}}(t)\text{Im}\,(\rho_{21}) = -\frac{1}{T_1}(\rho_{22} - \rho_{22}^{\text{eq}})\,, \tag{70}$$

$$\left[\frac{d}{dt} + \text{i}\omega_0\right]\rho_{21}(t) - \text{i}\omega_{\text{R}}(\text{t})\,(\rho_{22} - \rho_{11}) = -\frac{1}{T_2}\rho_{21}\,. \tag{71}$$

$\omega_0 = (\epsilon_2 - \epsilon_1)/\hbar = 2\Delta/\hbar$ is the transition frequency, $\omega_{\text{R}}(t) = p_0\mathcal{E}(t)/\hbar$ denotes the (time dependent) Rabi–frequency and $\widehat{\rho}^{\text{eq}} = \exp(-\widehat{H}_0/k_{\text{B}}T)/Z$. Z is the partition function. The diagonal elements ρ_{11}, ρ_{22} give the population of the stationary states $|1\rangle$,$|2\rangle$, whereas the population of the "up" and "down" states and the dipole moment are determined by $\rho_{12} = \rho_{21}^*$

$$N_{\text{u}}(t) = \text{tr}(\widehat{\rho}(t)|u\rangle\langle u|) = \frac{1}{2} + \text{Re}\,\rho_{12}(t)\,, \tag{72}$$

$$d(t) = \text{tr}(\widehat{\rho}(t)\widehat{D}) = 2d_0\text{Re}\,\rho_{12}(t)\,. \tag{73}$$

Usually, Eqs. (69-71) are rewritten in terms of the inversion $I = \rho_{22} - \rho_{11}$ and (complex) dipole moment $\mathcal{P} = \rho_{21}$ or in terms of a pseudo spin vector $\boldsymbol{s}(t) = \text{tr}(\widehat{\rho}\widehat{\boldsymbol{\sigma}})$, where $\widehat{\boldsymbol{\sigma}}$ is the Pauli–spin–vector operator.

$$\frac{d\boldsymbol{s}(t)}{dt} = \boldsymbol{\Omega} \times \boldsymbol{s} + C(\boldsymbol{s})\,, \quad \boldsymbol{\Omega} = (-2\omega_{\text{R}}(t), 0, -\omega_0)\,. \tag{74}$$

The *Bloch equations* (74) permit a very suggestive physical interpretation: they describe the rotation of vector $\boldsymbol{s}$ around $\boldsymbol{\Omega}$. For applications in atomic

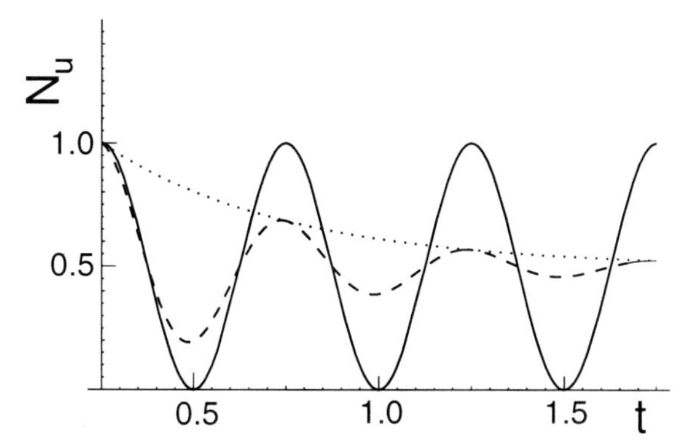

Figure 14. Dynamics of the population of the "up" state of the two–level–system $N_\mathrm{u}(t)$: Coherent (solid line), relaxing (dashed line), and fully incoherent motion (dotted line).

physics see, e.g., books by Allen and Eberly[20], semiconductor optics see Klingshirn[21], Haug andKoch[22], and previous Erice contributions[1](c).

As an illustration we discuss the following situations:

(a) At $t = 0$ the system is in the excited stationary state $|2\rangle$

$$\rho_{11}(t) = \rho_{11}^{\mathrm{eq}}\left(1 - e^{t/T_1}\right), \quad \rho_{21}(t) = 0\,. \tag{75}$$

(b) At $t = 0$ the system is in the "up" state

$$\rho_{11}(t) = \frac{1}{2}\,e^{-t/T_1} + \rho_{11}^{\mathrm{eq}}\left(1 - e^{t/T_1}\right), \quad \rho_{21}(t) = \frac{1}{2}\,e^{-\mathrm{i}\omega_0 \mathrm{t}}\,e^{-t/T_2}\,. \tag{76}$$

(c) For larger damping, one may neglect the coherent motion between states $|u\rangle$, $|d\rangle$ and set up rate equations for the diagonal elements $N_m = \rho_{mm}$

$$\dot{N}_m(t) \;=\; \sum_n \left(N_n \Gamma_{n\to m} - N_m \Gamma_{m\to n}\right), \tag{77}$$

$$\Gamma_{n\to m}\exp(-E_n/k_\mathrm{B}T) \;=\; \Gamma_{m\to n}\exp(-E_m/k_\mathrm{B}T)\,. \tag{78}$$

Eq. (78) is the *detailed balance* relation.

In our case $|u\rangle$, $|d\rangle$ have the same energy so that $\Gamma_{\mathrm{u}\to\mathrm{d}} = \Gamma_{\mathrm{d}\to\mathrm{u}} = \Gamma$. Using $N_\mathrm{u} + N_\mathrm{d} = 1$, Eqs. (77) can be solved easily, e.g., for $N_\mathrm{u}(0) = 1$, $N_\mathrm{d}(0) = 0$, we obtain

$$N_\mathrm{u}(t) = \frac{1}{2}(1 + e^{-2\Gamma t}), \quad N_\mathrm{d}(t) = \frac{1}{2}(1 - e^{-2\Gamma t})\,. \tag{79}$$

Some results of coherent, fully incoherent, and relaxation dynamics are displayed in Fig. 14.

The "state of the art" treatment of the dissipative TLS is layed out in the review article by Leggett et al.[34] and the book by Weiss[35]. The underlying model is

$$\widehat{H} = \frac{1}{2}\hbar\Delta\widehat{\sigma}_x + \frac{1}{2}q_0\widehat{\sigma}_z\sum_i c_i x_i + \widehat{H}_{\rm res}\,, \tag{80}$$

$$\alpha = \eta q_0^2/2\pi\hbar\,, \qquad J(\omega) = \eta\omega e^{-\omega/\omega_c}\,. \tag{81}$$

ω_c denotes a cut–off in the excitation spectum. For finite temperatures and small coupling to the bath $0 < \alpha < 1$ the levels become damped as well as the splitting ω_0 becomes smaller and eventually tends to zero at $\alpha = 1$, see Fig. 13(b). This conforms with the classical harmonic oscillator. [The path integral techniques as well as the explicit results are too involved to be discussed here.]

When the oscillation frequency ω_0 of the TLS is small compared with the Debye–frequency, there is a universal lower bound on the decoherence rate $\Gamma \ll \omega_0$ due to the atomic environment[36]

$$\langle x(t)x(0)\rangle = x_0^2 e^{-\Gamma t}\cos(\omega_0 t)\,, \quad \Gamma = \frac{M^2 x_0^2\omega_0^5}{2\pi\hbar\rho c_s^3}\coth(\frac{\hbar\omega_0}{2k_B T})\,. \tag{82}$$

ρ denotes the mass density, and c_s the speed of sound. For a NH_3 molecule ($M = 3\times10^{-23}$g, $\omega_0 = 10^{12}$/s, $x_0 = 2\times10^{-8}$cm, $\rho = 5$g/cm^3, $c_s = 10^5$cm/s) $\Gamma \approx 10^{10}$/s. For tunneling electrons the rate is much smaller.

Kinetic equations in the Markovian limit are derived in Refs.[44, 45].

5.3.1. *The neutral Kaon system*

Particle physics has become an interesting testing ground for fundamental questions in quantum physics, e.g. possible deviations from the quantum mechanical time evaluation have been studied in the neutral Kaon system[39]. These particles are produced by strong interactions in strangeness eigenstates ($S = \pm1$) and are termed K_0 $\bar{K}_0$ which are their respective antiparticles. As both particles decay (by weak forces) along the same channels (predominantly $\pi\pm$ or two neutral pions) there is an amplitude which couples these states. The "stationary" (CP–eigenstates) called K_s and K_l for "short" and "long" (or K_1, K_2). Decay of K_s and K_l (by weak forces), however, is very different and results from CP–violation. K_s decays predominantly into 2 pions with $\tau_s = 9\times10^{-10}$s whereas, the K_l decay is (to lowest order) into 3 pions with $\tau_l = 5\times10^{-8}$s, [39]. The oscillatory contribution in the decay rate corresponds to a small K_s/K_l level splitting of $\Delta m_{K0}/m_{K0} \approx 4\times10^{-18}$. For an overview see, *The Feynman Lectures on Physics*[37] (Vol. III, Ch. 11-5) and Källen's textbook[38].

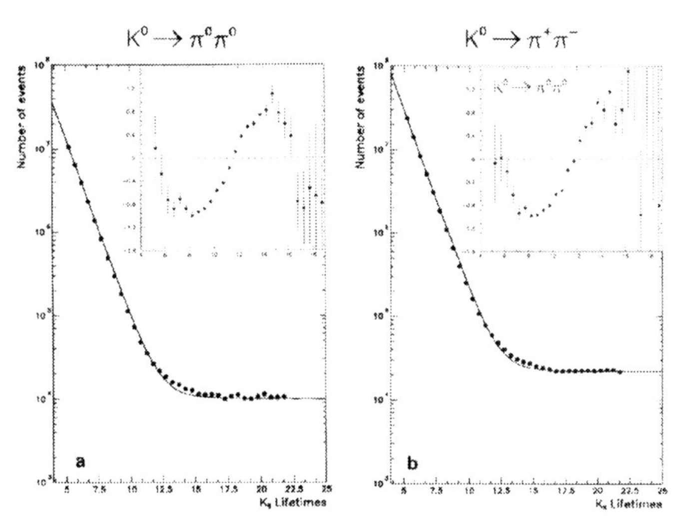

Figure 15. The rate of decay of Kaons to neutral and charged pions as a function of K_s life–time. Superimposed are the fitted lifetime distributions with the interference terms removed. The insets (interchanged?) show the interference terms extracted from the data. From Carosi et al.[39](a).

5.4. A PARTICLE IN A POTENTIAL

For the harmonic oscillator interacting with a reservoir Caldeira and Leggett [15] and Walls[40] provided an exact solution which shows a rich and intricate dependence on the parameters – too extensive to be discussed here. For weak coupling and high temperatures, however, the physics can be described by an equation of motion for the reduced density operator $\rho(x, x', t) = \langle x|\widehat{\rho}_{\mathrm{s}}|x'\rangle$ of the following form[41] (p. 57)

$$\begin{aligned} \frac{\partial}{\partial t}\rho(x, x'; t) \ &+ \ \frac{\mathrm{i}}{\hbar}\left(\widehat{H} - \widehat{H}'\right)\rho(x, x', t) \\ &= \ -\frac{\gamma}{2}(x - x')\left(\frac{\partial}{\partial x} - \frac{\partial}{\partial x'}\right)\rho(x, x'; t) \\ &\quad -\Lambda(x - x')^2\rho(x, x'; t)\,, \end{aligned} \tag{83}$$

$$\widehat{H} = -\frac{\hbar^2}{2m}\frac{\partial^2}{\partial x^2} + V(x)\,. \tag{84}$$

$\widehat{H}'$ is given by (84) by replacing x by x'. Eq. (83) is valid for a particle in an arbitrary potential. Relaxation rate γ and decoherence parameter Λ are considered as independent parameters. Scattering of the oscillator particle by a flux of particles from the environment enforces decoherence with a rate of

$$\Lambda = nv\sigma_{\mathrm{sc}}k^2\,. \tag{85}$$

TABLE I. Some values of the localization rate (1/s). From Joos in Ref.[41].

	free electron	10^{-3}cm dust particle	bowling ball
sunlight on earth	10	10^{20}	10^{28}
300K photons	1	10^{19}	10^{27}
3K cosmic radiation	10^{-10}	10^{6}	10^{17}
solar neutrinos	10^{-15}	10	10^{13}

nv is the flux and k the wave number of the incoming particles with cross section $\sigma_{\rm sc}$. For a recent paper on collisional decoherence see Ref.[43].

5.4.1. *Free particles*

First, we consider a superposition of two plane waves without coupling to the bath

$$\Psi(x,t=0) = \frac{1}{2}\left(e^{\mathrm{i}k_1\mathrm{x}} + e^{\mathrm{i}k_2\mathrm{x}}\right), \tag{86}$$

$$\rho_0(x,x';t) = \Psi(x,t)\Psi^*(x',t), \tag{87}$$

$$\rho_0(x,x;t) = 1 + \cos\left[(k_1-k_2)x + \frac{\hbar(k_1^2-k_2^2)}{2m}t\right]. \tag{88}$$

In the presence of the environment, the fringe contrast in the mean density will be reduced[40] (For notational simplicity, explicit results are for the particle density, i.e. the diagonal elements of $\widehat{\rho}$ only.)

$$\rho(x,x;t) = 1 + e^{-\eta}\cos\left[(k_1-k_2)x - \frac{1-e^{-\gamma t}}{\gamma}\frac{\hbar(k_1^2-k_2^2)}{2m}t\right], \tag{89}$$

$$\eta = \frac{2\hbar^2\Lambda}{m^2\gamma^3}\left[\gamma t/2 - \frac{3}{4} + e^{-\gamma t} - \frac{1}{4}e^{-2\gamma t}\right](k_1^2-k_2^2). \tag{90}$$

In the special case of negligible friction, $\gamma = 0$, the visibility of the interference fringes is strongly reduced, while the spatial structure remains unaffected.

The standard example of a gaussian wave packet with momentum $\hbar k_0$ and width a can also be tackled analytically, but the result is rather lengthy, hence we only state the result for the mean position and variance

$$\langle x\rangle = \frac{\hbar k_0}{m\gamma}(1-e^{-\gamma t}), \qquad (\Delta x)^2 = \frac{a^2}{2}\left(1 + \left[\frac{\hbar t}{ma^2}\right]^2\right) + \frac{\hbar^2\Lambda}{m^2}t^3. \tag{91}$$

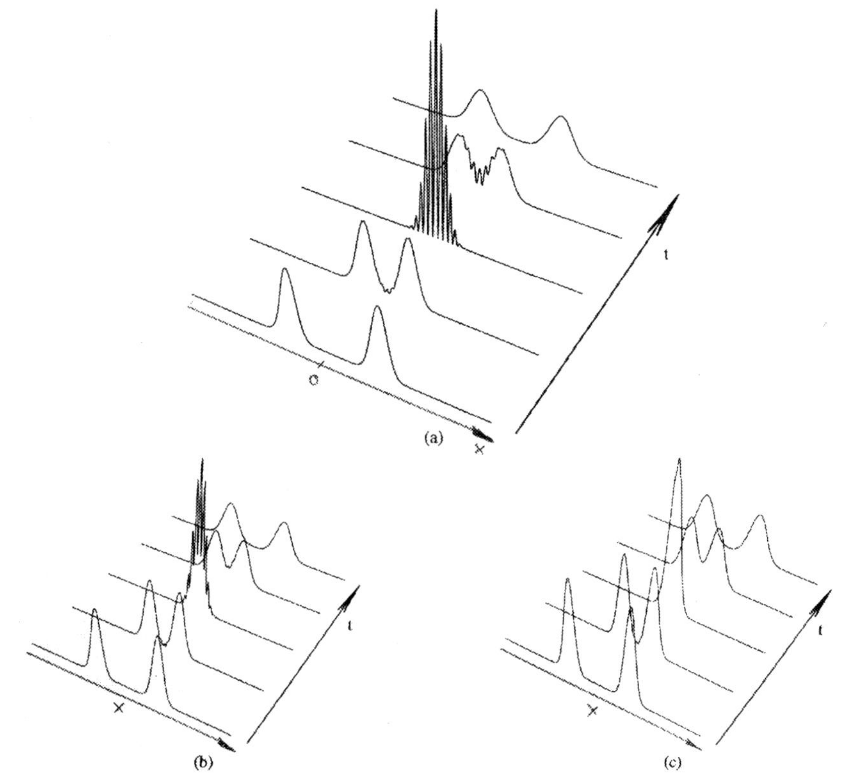

Figure 16. Harmonic oscillator which is initially in a superposition of two displaced ground state wave functions at $x = \pm x_0$. (a) pure quantum mechanics, (b) weak damping, (c) strong damping. Loss of coherence is much stronger than the damping of the oscillation amplitude. According to Joos' article in Ref.[41] (p. 35–135).

Note, $\langle x \rangle$ is independent of Λ whereas Δx is independent of γ. For details, see appendix A2 of Giulini's book[41].

5.4.2. *Harmonic oscillator*

When the two components of a wave function (or mixed state) do not overlap, a superposition of two spatially distinct wave packets can still be distinguished from a mixture, when the two wave packets are brought to interference as in the two–slit experiment. In general, the decoherence time is much smaller than the relaxation time $1/\gamma$

$$\tau_{\text{dec}} = \frac{1}{\gamma} \left(\frac{\lambda_{\text{dB}}}{\Delta x} \right)^2 . \tag{92}$$

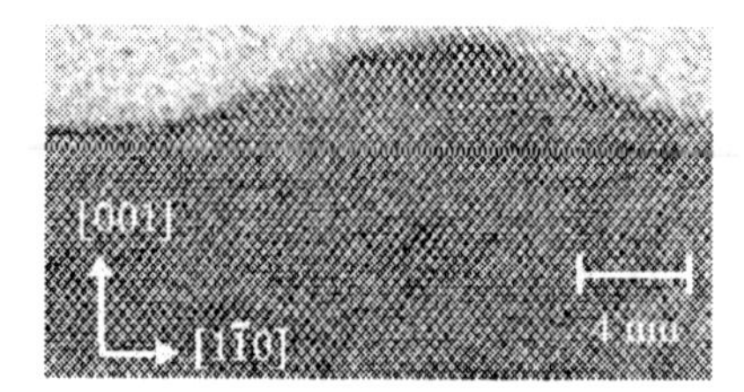

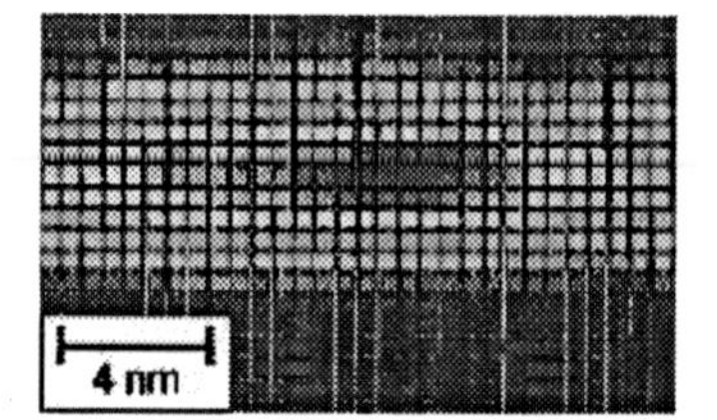

Figure 17. High resolution transmission electron micrographs of InAs/GaAs QDs (vertical and horizontal cross sections)[46].

$\lambda_{\rm dB} = \hbar/\sqrt{2mk_{\rm B}T}$ is the thermal de Broglie wave length and Δx is the width of the wave packet. For macroscopic parameters $m = 1$g, $\Delta x = 1$cm the decoherence time $\tau_{\rm dec}$ is smaller than the relaxation time γ by an enormous factor of $\sim 10^{40}$. This is the every day experience of the absence of interference phenomena in the macroscopic world. For the harmonic oscillator, this is easily seen from a superposition of two counterpropagating gaussian packets, see Fig. 16. A spectacular example is the interference of two Bose–Einstein condensates, see Wieman's Erice–article in Ref.[1](e).

6. Exciton Spins in Quantum Dots

The current interest in the manipulation of spin states in semiconductor nanostructures originates from the possible applications in quantum information processing[2]. Since most of the present concepts for the creation, storage and read–out of these states are based on (or involve) optical techniques one has to investigate the dynamics and relaxation of exciton (or trion) states rather than single carrier–spin states. Due to the discrete energy structure of quantum dots (QDs), inelastic relaxation processes are strongly suppressed with respect to quantum wells or bulk systems, e.g., $\tau_{\rm bulk} \approx 10$ps whereas $\tau_{\rm QD} \approx 20$ns.

Extensive experimental studies have identified the main features of the exciton fine structure in self–organized QDs by single–dot spectroscopy. Such QDs are usually strained and have an asymmetrical shape with a height smaller than the base size, see Fig. 17. The reduction of the QD symmetry lifts degeneracies among the exciton states and results, in particular, in a splitting of the exciton ground state. Thus, as a consequence of strain and confinement, the ground states of the QD heavy–hole (hh) and light–hole (lh) excitons are well-separated [$E_{\rm h\text{-}l} \approx 30 \ldots 60$meV] and the hh–exciton has the lowest energy, see Fig. 18(a-c). For an overview on optical properties of semiconductor quantum structures see Refs.[47](a,b).

The hh– and lh–exciton quartetts are characterized by the projections $J_z = \pm 1,\ \pm 2$ and $J_z = \pm 1,\ 0$ of the total angular momenta $J = 1, 2,$

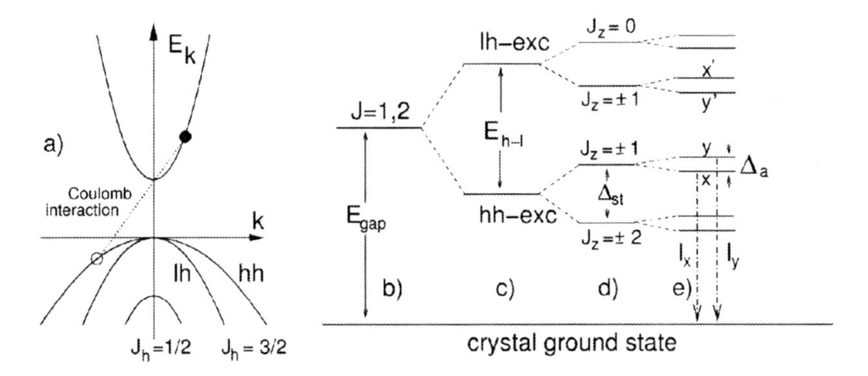

Figure 18. Sketch of bandstructure and exciton levels in III-V quantum dots ($J_h = 1/2$ split–off band and bulk heavy-light hole exciton splitting omitted).

respectively. The short-range exchange interaction splits the ground states of both hh– and lh–excitons into doubles [so–called singlet–triplet splitting, $\Delta_{st} \approx 0.2$meV, in CdSe: 1meV], see Fig. 18(d). The lateral anisotropy of a QD leads to a further splitting of the $|\pm 1\rangle$ levels (labeled by $|x\rangle$ and $|y\rangle$) with allowed dipole transitions to the crystal ground state which are linearly polarized along the two nonequivalent in–plane QD axes $[1, \pm 1, 0]$, see Fig. 18(e). This anisotropic splitting [$\Delta_a \approx 0.1 \ldots 0.2$meV] originates from the long–range exchange interaction in the elongated QDs.

Experimental studies on exciton–spin dynamics in QDs refer mostly to spin–coherence, i.e., they determine the transverse relaxation time T_2. Studies of the population relaxation of spin states (i.e. longitudinal relaxation time T_1) are rare since they require strict resonant excitation conditions. The only direct experimental studies on population relaxation between $|x\rangle$ and $|y\rangle$ under perfect resonant excitation were done by Marie's group[49] (Toulouse, France). To improve the signal to noise ratio, an ensemble of nearly identical InAs–QDs in a microresonator was used, see Fig. 19. Experimentally, these depopulation processes are analyzed by detecting a decay of the polarization degree $P = (I_x - I_y)/(I_x + I_y) = \exp(-t/\tau_{\text{pol}})$ of the luminescence upon excitation by a x– (or y–) polarized light pulse. (I_x, I_y denote the intensities of the x–, y–polarized luminescent radiation). The total intensity from both luminescent transitions to the crystal ground state is constant. Hence, relaxation is solely within the x–y doublett. The incomplete polarization degree at time $t = 0$ probably results from misalignement of the QDs in the microresonator. Despite the tiny x–y splitting of about 0.1–0.2meV, the relaxation is thermally activated with the LO–phonon energy of about 30meV. In conclusion, exciton–spin is totally frozen during the radiative lifetime $\tau_{\text{rad}} \approx 1$ns (even for high magnetic fields up to 8T). With decreasing size, increasing temperature, and large magnetic fields, however, the T_1–time becomes comparable to the exciton life–time. For details see Tsitsishvili[48](a,b).

Relaxation and decoherence on a single QD was studied by Henneberger's group[50] (Berlin, Germany) by analyzing time–resolved secondary emission. Because strict resonant excitation is faced with extreme stray light problems, LO–phonon assisted quasi–resonant excitation by tuning the laser source 28meV($= \hbar\omega_{LO}$) above the exciton ground state was used, see Fig. 20. The x/y doublet of the X_0 exciton with its radiative coupling to the crystal ground state represents a "V–type" system[29], where quantum beats in the spontaneous emission may occur upon coherent excitation (spectral of the laser pulse larger than the level splitting). In the present case the doublet consists of two linearly cross–polarized components so that interference is possible only by projecting the polarization on a common axis before detection. For excitation into the continuum (of the wetting layer) the PL decay is monotonous with a single time constant $1/\Gamma = 320$ps which is the anticipated radiative life–time. The beat period (which by chance is also 320ps) corresponds to a fine structure splitting of 13μV, not resolvable in the spectral domain. The fact that the beat amplitude decays with the same time–constant as the overall signal clearly demonstrates, no further relaxation and decoherence takes place once the exciton has reached the ground state doublet. However, a large 60% loss of the initial polarization degree of unknown origin was found.

7. Outlook

The transition between the microscopic and macroscopic worlds is a fundamental issue in quantum measurement theory[51]. In an ideal model of measurement, the coupling between a macroscopic apparatus ("meter") and a microscopic system ("atom") results in an entangled state of the "meter+atom" system. Besides the macroscopic variable of the display the meter supplies many uncontrolled variables which serve as a bath and irreversibly de–entangles the atom–meter state. For a recent overview on this field see contributions by Giulini[41] and Zurek[42, 52]. A nice overview on *Strange properties of Quantum Systems* has been given by Costa[1](d).

Brune et al.[53] created a mesoscopic superposition of radiation field states with classically distinct phases and, indeed, observed its progressive decoherence and subsequent transformation to a statistical mixture. The experiment involved Rydberg atoms interacting once at a time with a few photons coherent field in a high–Q cavity.

The interaction of the system with the environment leads to a discrete set of states, known as *pointer states* which remain robust, as their superposition with other states, and among themselves, is reduced by decoherence.

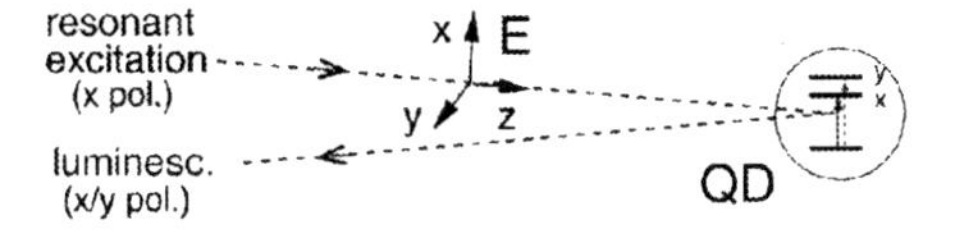

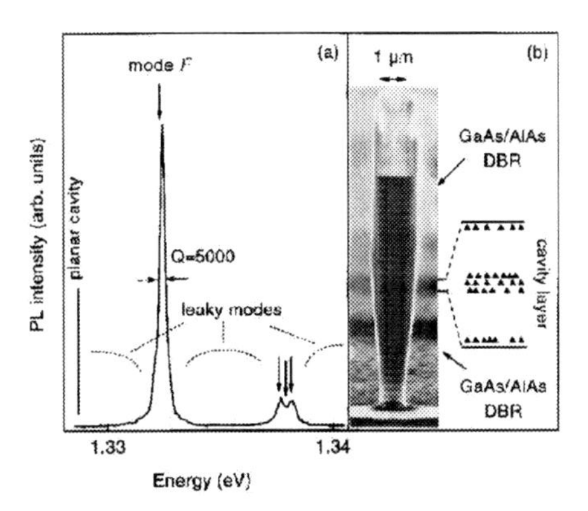

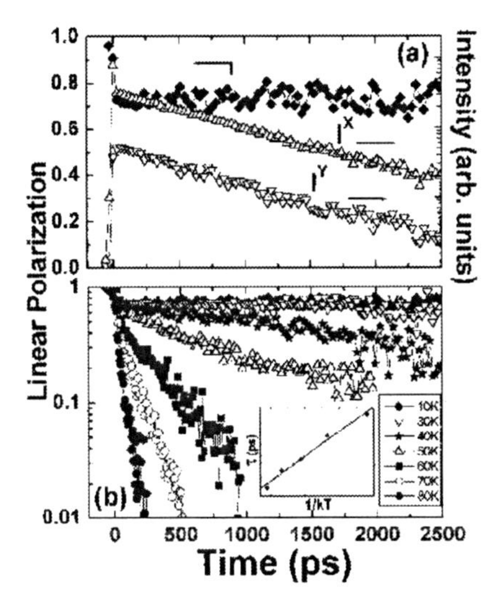

Figure 19. Experiment by Paillard et al.[49] on a system of many InAs/GasAs QDs. Upper panel: Setup, middle: microresonator structure, bottom: (a) Time dependence of the measured photoluminescence components copolarized I^x (Δ) and cross polarized I^y (∇) to the σ^x polarized excitation laser ($T = 10$K) and the corresponding linear polarization degree P_{lin} (♦). (b) Temperature dependence of the linear polarization dynamics. Inset: P_{lin} decay time as a function of $1/(k_B T)$.

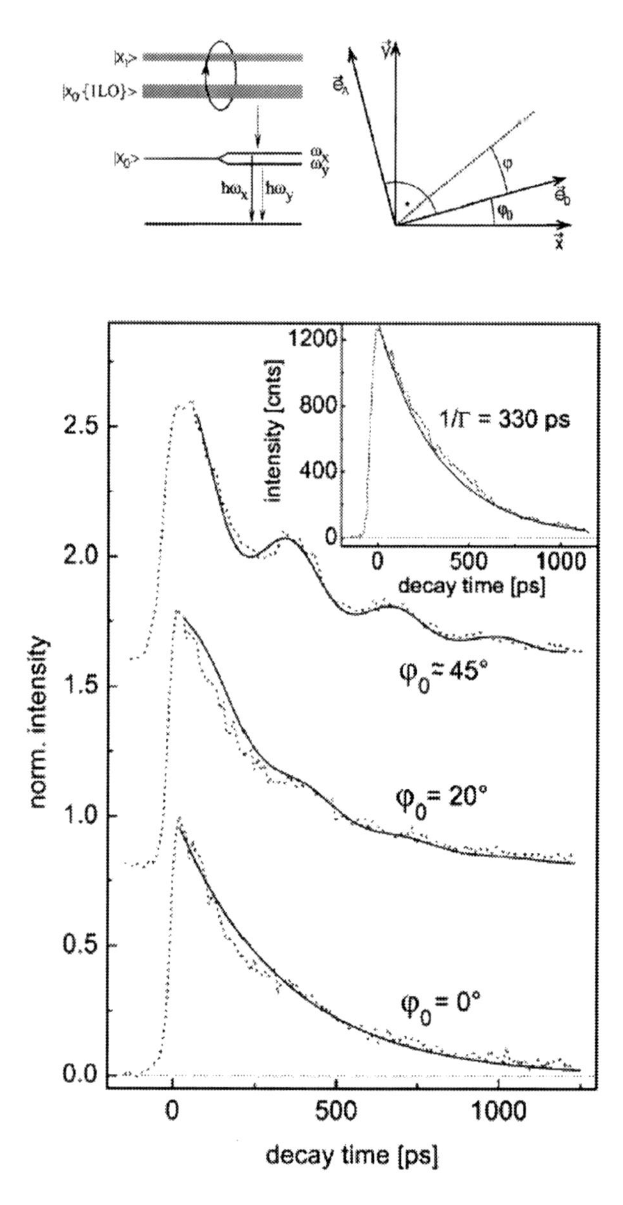

Figure 20. Experiment by Flissikowski et al.[50] on a single CdSe/ZnSe QD. Upper panel: Energy level scheme (left) and geometry (right). Cross alignment of the polarizers for excitation (e_0) and detection (e_A). Lower panel: Relaxation and quantum beats in the photoluminescence for different angles with respect to the fundamental QD axis. Inset: Excitation into the continuum for comparison.

8. Acknowledgements

Thanks to Prof. Rino Di Bartolo and his team, the staff of the Majorana Center, and all the participants who again provided a wonderful time in a stimulating atmosphere. This work was supported by the DFG Center for Functional Nanostructures CFN within project A2.

References

1. Proceedings of the International School of Atomic and Molecular Spectroscopy, Di Bartolo, B., et al. (Eds.), Majorana Center for Scientific Culture, Erice, Sicily:
   (a) (1985) *Energy Transfer Processes in Cond. Matter*, NATO ASI B 114, Plenum;
   (b) (1997) *Ultrafast Dynamics of Quantum Systems*, NATO ASI B 372, Plenum;
   (c) (2001) *Advances in Energy Transfer Processes*, World Scientific;
   (d) (2005) *Frontiers of Optical Spectroscopy*, Kluwer;
   (e) (2006) *New Developments in Optics and Related Fields*, Kluwer (to appear).
2. (a) Bonadeo, N. H. et al. (1998), Science, **282**, 1473;
   (b) Awschalom, D. D. et al. (Eds) (2002) *Semiconductor Spintronics and Quantum Computation*, Springer;
   (c) Žutić, I. et al. (2004), Rev. Mod. Phys., **76**, 323.
3. Haase, W. and Wròbel, S. (Eds.) (2003) *Relaxation Phenomena*, Springer.
4. Kohlrausch, R. (1854), Poggendorffs Annalen der Physik und Chemie **91**(1), 56.
5. Scher, H. et al. (1991), Phys. Today, Jan–issue, p. 26.
6. (a) Götze, W. and Sjögren, L. (1992), Rep. Progr. Phys. **55**, 241;
   (b) Phillips, J. C. (1996), Rep. Progr. Phys. **59**, 1133;
   (c) Jonscher, A. K. (1977), Nature **267**, 673.
7. Prudnikov, A. P. et al. (Eds) (1992) *Integrals and Series*, Gordon and Breach.
8. (a) Palmer, R. G. et al. (1984), Phys. Rev. Lett. **53**, 958;
   (b) Sturman B. and Podilov, E. (2003), Phys. Rev. Lett. **91**, 176602.
9. Dressel, M. et al. (1996), Phys. Rev. Lett. **77**, 398.
10. Sokolov, I. M. et al. (Nov. 2002), Physics Today, p. 48.
11. (a) Liu, S. H. (1985), Phys. Rev. Lett. **55**, 529 and (1986), PRL **56**, 268;
    (b) Kaplan, Th. et al. (1987), Phys. Rev. B**35**, 5379;
    (c) Dissado, L. A. and Hill, R. M. (1987), Phys. Rev. B**37**, 3434.
12. (a) Zabel, I. H. H. and Strout, D. (1992), Phys. Rev. B**46**, 8132;
    (b) Sidebottom, D. L. (1999), Phys. Rev. Lett. **83**, 983.
13. Reif, F. (1965) *Fundamentals of Statistical and Thermal Physics*, McGraw–Hill.
14. Einstein, A. (1905), Annalen der Physik (Leipzig), **17**, 549.
15. Caldeira, A. O. and Leggett, A. J. (1985), Phys. Rev. A **31**, 1059.
16. Dittrich, T. et al. (Eds) (1998) *Quantum Transport and Dissipation*, Wiley-VCH.
17. Fick, E. and Sauermann, G. (1990) *The Quantum Statistics of Dynamic Processes*, Solid State Sciences **86**, (Springer,).
18. Born, M. and Wolf, E. (1964) *Principles of Optics*, Pergamon.
19. Wegener, M., private communication.
20. Allen, L. and Eberly, H. (1975) *Optical Resonance and Two–Level Atoms*, Wiley.
21. Klingshirn, C. F. (2004) *Semiconductor Optics*, 3rd edition, Springer.
22. Haug, H. and Koch, S. W. (2004) *Quantum Theory of the Optical and Electronic Properties of Semiconductors*, World Scientific, 4th edition.

23. Bergmann, G. (1984), Phys. Rep. **107**, 1.
24. (a) Höhler, G. (1958), Z. Phys **152**, 546;
(b) Peres, A. (1980), Ann. Phys. (NY), **129**, 33;
(c) Ludviksson, A. (1987), J. Phys. A: Math. Gen. **20**, 4733.
25. (a) Wessner, J. M. et al. (1972), Phys. Rev. Lett. **29**, 1126;
(b) Horwitz, L. P. and Katznelson, E. (1983), Phys. Rev. Lett. **50**, 1184;
(c) Grotz, G. and Klapdor, H. V. (1984), Phys. Rev. C **30**, 2098;
(d) Norman, E. B. et al. (1988), Phys. Rev. Lett **60**, 2246;
(e) Koshino, K. and Shimizu, A. (2004), Phys. Rev. Lett. **92**, 030401.
26. Avignone, F. T. (1988), Phys. Rev. Lett. **61**, 2624.
27. Weisskopf, V. and Wigner, E. (1930), Z. Phys. **63**, 54.
28. Novotny, T. et al. (2003), Phys. Rev. Lett. **90**, 256801.
29. Scully, M. O. and Zubairy, M. S. (1997) *Quantum Optics*, Cambridge.
30. Busch, K. et al. (2000), Phys. Rev. E **62**, 4251.
31. (a) Itano, W. M. et al. (1990), Phys Rev. A **41**, 2295; (1991) ibid **43**, 5165, 5166;
(b) Kwiat, P. et al. (Nov. 1996), Scientific American, 52.
32. (a) Leggett, A. J. (1978), J. Physique C **6**, 1264;
(b) Schmid, A. (1986), Ann. Phys. (N.Y.), **170**, 333.
33. Etzkorn, H. et al. (1982), Z. Phys. B **48**, 109.
34. Leggett, A. J. et al. (1987), Rev. Mod. Phys. **59**, 1; (1995), ibid **67**, 725.
35. Weiss, U. (1993) *Quantum Dissipative Systems*, World Scientific.
36. Chudnovski, E. M. (2004), Phys. Rev. Lett. **92**, 120405; (2004), ibid. **93**, 158901 and 208901.
37. Feynman, R. P. et al. (1966) *The Feynman Lectures on Physics*, Addison Wesley.
38. Källén, G. (1964) *Elementary Particle Physics*, Addison–Wesley.
39. (a) Carosi, R. et al. (1990), Phys. Lett. B 237, 303;
(b) Alavi–Harati, A. et al. (2003), Phys. Rev. D **67**, 012005;
(c) Bertlmann, R. A. et al. (2003), Phys. Rev. A **68**, 012111.
40. Savage, C. M. and Walls, D. F. (1985), Phys. Rev. A **32**, 2316 and 3487.
41. Giulini, D. et al. (Eds) (1996) *Decoherence and the Appearance of a Classical World in Quantum Theory*, Springer.
42. Zurek, W. H. (2002) *Decoherence and the Transition from Quantum to Classical – Revisited*, Los Alamos Science, No. **27**, 2; (quant-phys/0306072).
43. Hornberger, K. and Sipe, J. E. (2003), Phys. Rev. A **68**, 012105.
44. Spohn, H. (1980), Rev. Mod. Phys. **53**, 569.
45. Kimura, G et al. (2001), Phys. Rev. A **63**, 022103.
46. Gerthsen, D. private communication.
47. (a) Kalt, H. *Optical Spectroscopy of Quantum Structures*, in Ref.[1](e);
(b) Lyssenko, V. *Coh. Spectr. of SC Nanostruct. and Microcavities*, in Ref.[1](e).
48. (a) Tsitsishvili, E. et al. (2002), Phys. Rev. B **66**, 161405(R);
(b) Tsitsishvili, E. et al. (2003), Phys. Rev. B **67**, 205330.
49. Paillard, M. et al. (2001), Phys. Rev. Lett. **86**, 1634.
50. Flissikowski, T. et al. (2001), Phys. Rev. Lett. **86**, 3172.
51. Wheeler, J. A. and Zurek, W. H. (Eds) (1983) *Quantum Theory and Measurement*, Princeton.
52. Zurek, W. H. (2003), Rev. Mod Phys. **75**, 715.
53. Brune, M. et al. (1996), Phys. Rev. Lett. **77**, 4887.

CHALLENGES FOR CURRENT SPECTROSCOPY: DETECTION OF SECURITY THREATS

NORMAN P. BARNES
NASA Langley Research Center
Hampton, VA 23681 U.S.A.
n.p.barnes@larc.nasa.gov,

1. Introduction

Finding better and less expensive method of detecting security threats is a huge global problem. Many countries have felt the effects of terrorist activities including: Russia, Spain, and Indonesia. Not only is terrorism world wide, it is a protracted activity that does not appear to be fading soon. The cost in human lives is horrendous, both in terms of the people lost and in terms of the grief of their families and friends. In addition to these incalculable losses are economic impacts that are created by the need of a security force and in lost time and inconvenience of the general public. In the United States alone, there are 660 million passengers annually. If each passenger is delayed by 10 minutes and is compensated at minimum wage rates, the economic impact is $575 million.

There are several classes of threats to consider including, but not limited to: chemical, biological, human, and radiological. Each class of threat can come in a variety of forms. Chemical threats could be phosgene, hydrogen cyanide, sarin, or mustard gas. Biological threats could be small pox, anthrax, or E-bola. Obviously, because of the multiplicity of threats, a detection system that can handle all or most threats is highly desirable. Human threats could be a suicide bomber or a hijacker. Radiological threats could be a nuclear bomb or even a conventional bomb encased in a radioactive material, or dirty bomb. Spectroscopy is currently positioned to aid in the detection of chemical threats and, with some sophistication, biological threats.

Threat detection devices must be sensitive and reliable in order to be useful. The required level of sensitivity depends on the chemical species being detected. The level of detection must be small compared with the level of lethality. This varies with the chemical species but is generally on the order of parts per million. The level of sensitivity of biological species is on the order of a particle per cubic meter. Detection devices must also be rapid; delays of even a few hours are unacceptable. Useful detection devices must have a low false alarm rate. The acceptable false alarm rate is application dependent. If the false alarm rate is too high, the operators will tend to ignore it or turn it off. Preferably, the device will not require contact and, in the best case, has a stand off range of kilometers.

B. Di Bartolo and O. Forte (eds.), Advances in Spectroscopy for Lasers and Sensing, 63–72.

Several spectroscopic techniques are available for the detection of threats. Perhaps the most straight forward technique is a measurement of the transmission of a probe beam tuned to an absorption feature of the species to be detected. A diminution in transmission is interpreted as the presence of the threat. Effects associated with the absorption of the probe beam, such as an increase in the temperature, can also be used. If the probe beam is absorbed, there is a possibility that the absorber will subsequently emit a photon. The wavelength of the emitted photon can correspond to the energy difference between a pair of energy levels characteristic of the absorber. With this technique, the wavelength of the probe beam must correspond to an absorption feature of the species to be detected. Detection of fluorescence from the particular absorber can indicate its presence. The emitted photon may also be wavelength shifted from the probe beam wavelength by an amount characteristic of a vibrational mode of the species to be detected. If the emitted wavelength is longer than the probe beam, the process is known as Raman scattering. With this technique, the probe beam need not correspond to an absorption feature of the species to be detected. Unfortunately, Raman scattering is a weak effect. If 2 probe beams are used, the emitted photon can be shifted to a wavelength shorter than the probe beam. This process is known as coherent anti resonant Raman scattering and can be mush stronger than Raman scattering.

All of the spectroscopic techniques described above depend on the availability of a probe source. The simplest probe source is a black body radiator, such as a tungsten filament lamp or even the sun. Black body sources produce a wide range of wavelengths so they can be used in wavelength scanning systems. However, they are isotropic emitters and usually lacking at ultraviolet wavelengths. Probes beams made using black body sources have to be collimated and filtered. Consequently, they are not very efficient. Lasers. On the other hand, they tend to generate more nearly collimated and monochromatic beams, nearly ideal for a probe. The wavelength range needed plays a major role in the selection of the laser. Excimer lasers and solid states laser coupled with nonlinear optical device can be employed in the ultraviolet. Carbon dioxide lasers, quantum cascade lasers, and solid state lasers coupled with parametric oscillators can be used in the mid infrared. Quantum cascade lasers and solid state lasers coupled with a difference frequency generator can be utilized in the far infrared or TeraHertz region. Each of these lasers has their own operational characteristics that must be taken into account in any system design.

Black body radiation curves can be calculated using Plank's radiation law. This yields the amount of power per unit area and per unit wavelength interval. Black body radiation curves for both a 300 and a 3000°K black body appear in Figure 1. It may be noted 300°K corresponds to a room temperature emitter while 3000°K corresponds to a tungsten filament. Even a 3000°K black body only produces roughly tens of microWatts of power around 10 μm for practical sizes and spectral bandwidths. While raising the temperature significantly increases short wavelength emission, it only increases long wavelength emission roughly linearly. On the other hand, the total emitted power increases as temperature to the fourth power with a concomitant increase in the power required.

2. Spectral Regions of Interest

The physics of the absorption process depends on the wavelength region. In the ultraviolet, absorption is associated with a change in the electronic energy levels. Besides the primary electronic transition, in molecules there can also be vibrational sublevels. However, the absorption features are so wide, individual vibrational features tend to coalesce into a wide bandwidth structure. In the mid infrared, absorption is associated with a change in only the vibrational levels. Besides the primary vibrational transition, in molecules there are also rotational sublevels. At low pressures, individual rotational transitions can be observed. However, at atmospheric pressures, pressure broadening dominates and individual rotational features tend to coalesce. In the far infrared, absorption is associated with rotational transitions. Pressure broadening is dominant at atmospheric pressures and features tend to be proportionally wide. To further illustrate these points, absorption spectra of trinitrotoluene in these regions can be considered.

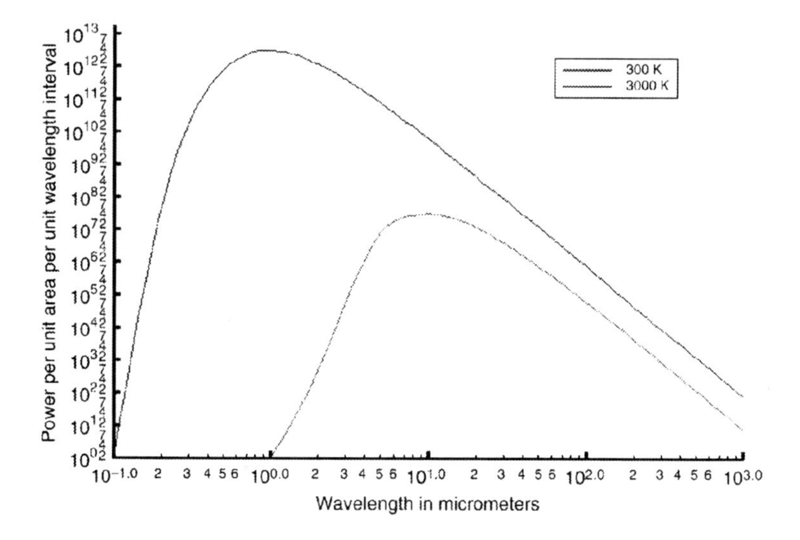

Figure 1. Black body emission versus wavelength.

Ultraviolet absorption of trinitrotoluene extends from 0.2 to 0.26 μm in a relative featureless band. Absorption peaks associated from the $X^2\Pi(v = 0)$ to the $A^2\Sigma^+(v = 1,2,3)$ vibrational transitions [1] are barely discernable. Lack of prominent absorption features makes unambiguous identification of molecules more difficult. Rather than using the presence or absence of an absorption feature to determine if a particular molecule is present, relative emission at some set of wavelengths must be measured and compared with a reference.

Absorption spectrum of trinitrotoluene in the mid infrared is dominated by 2 strong absorption peaks around 6.3 and 7.4 μm. Relatively strong and narrow absorption features spread over a wide range of wavelengths abets the identification process. These absorption features in trinitrotoluene are associated with the NO_2 complex. This particular complex is found in most materials that can be used for explosives and in few other compounds [2].

Absorption of trinitrotoluene in the far infrared [3] or TeraHertz region contains 2 peaks around 1.7 and 3.0 THz. Other explosive materials, such as Semtex, RDX, PETN, and HMX, have somewhat richer spectra. However, the features are not as rich nor as narrow as the features found in the mid infrared. The advantage of the TeraHertz wavelengths is their ability to penetrate most clothing materials. This makes concealment more difficult.

The ability to detect hazardous gasses depends on the vapor pressure of the gas. Vapor pressure in turn depends on the particular hazardous gas and the temperature. Vapor pressures of the several hazardous gasses vary can 6 orders of magnitude. Vapor pressure of a particular gas can vary more than an order of magnitude over a 50°C temperature range. Low vapor pressures make detection more difficult because of the paucity of absorbing molecules. On the other hand, if some chemical agent has a very low vapor pressure, its low vapor pressure may make it ineffective.

3. Optical Detection Schemes

An optical schematic of spectrophotometric detection scheme is shown in Figure 2. The probe beam is created by a light source, such as a tungsten lamp, that is collimated and propagated through a grating monochromator to select a narrow band of wavelengths. The system is made self calibrating by using a reflective chopper to split the beam into 2 probe beam paths, 1 goes through the sample and the other serves as a reference beam. Both beams are then temporally multiplexed on the same detector. The transmission is the electronic ratio of the time multiplexed signal divided by the reference.

This method represents the traditional spectroscopic approach to detection and identification. It has several advantages such as wavelength flexibility, being able to scan over a wide range of wavelengths. By using a single source and a single detector, it is self calibrating. It is also able to detect rather low levels of some gasses, dependent on the absorption cross section of the particular gas. It is limited by the need to bring the device to the gas sample that is the spectrometer is an in situ device. Methods of detecting even lower levels particular gas samples are needed.

The ring down method can achieve a much higher sensitivity than the simple spectrometer. A ring down device employs a laser source that is matched to a resonator. Matching implies both that the spatial mode of the laser is matched to the mode of the resonator and the wavelength of the laser is resonant in the resonator. When matched, the intensity in the resonator is much higher than the incident laser intensity. When the incident power is terminated, the intensity inside the resonator exponentially decays. The decay constant depends on the losses inside the resonator, including losses caused by absorption. The decay constant measured with and without the gas present are measured and compared. From this information, the absorption can be calculated.

A ring down device has a large advantage in sensitivity when compared with more conventional spectrometers. It is also insensitive to the power of the probe beam and the responsivity of the detector. This is a result of the measurement be determined by the decay constant rather than the measurement of the transmitted probe power. It does require that a single frequency laser probe with good beam quality be used. If the probe beam is scanned over wavelength, the scan must be stepwise over the resonant wavelengths of the ring down spectrometer. This method is an in situ measurement because it requires the sample to be injected into the device.

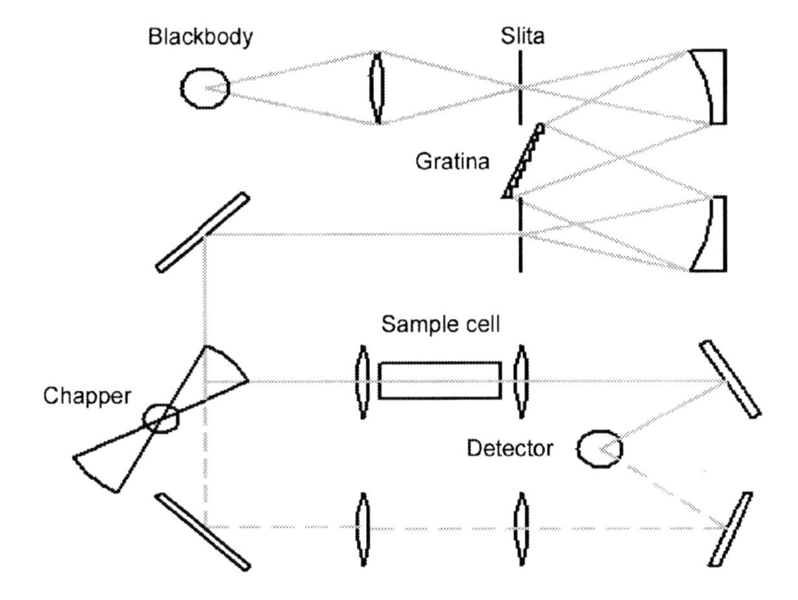

Figure 2. Optical schematic of spectrophotometer.

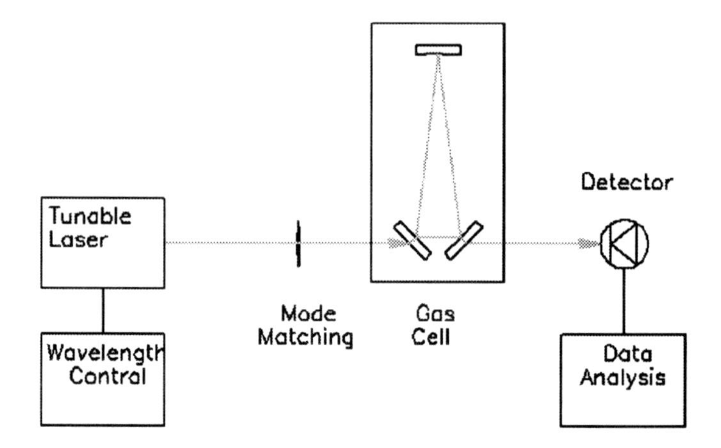

Figure 3. Optical schematic of a ring down spectrometer.

A photo acoustic detection method takes advantage of the heating of the sample by the pulsed probe beam. Heating causes a thermal expansion of the sample. Thermal expansion initiates a sound wave that can be detected by a sensitive microphone. Such a system can be made more sensitive by varying the pulse repetition frequency to match a resonance of the absorption cell, much like tuning an organ pipe. Typically, the microphone is placed at an antinode of the absorption cell for maximum sensitivity.

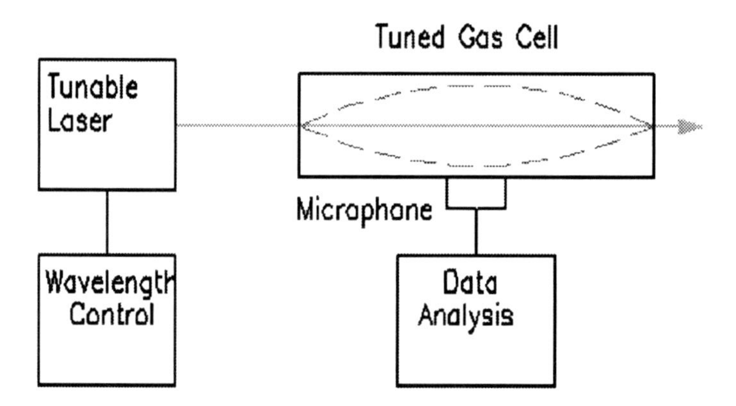

Figure 4. Optical schematic of photo acoustic spectrometer.

The sensitivity of the photo acoustic method is directly proportional to both the average laser power and the absorption coefficient and inversely proportional to the pulse repetition frequency to first order. The product of the first 2 factors is a measure of the heat absorbed and thus the thermal expansion of the gas. There are some practical limits on how lower the pulse repetition frequency can be. For example, if the pulse repetition frequency is too low, 1/f noise becomes the major factor in the signal to noise ratio and the sensitivity is degraded. Sensitivity can be increased by making the cell have a narrow acoustic resonance. Having a narrow acoustic resonance increases the sensitivity, however it slows down the wavelength tuning process. As with the previous methods, the sample must be injected into the device.

Unlike the previous schemes, a lidar can detect threats at a distance. Typically, a lidar produces laser pulses at 2 closely spaced wavelengths. The first pulse has a wavelength close to an absorption feature of a threat. As it propagates through the atmosphere, part of the pulse is reflected or scattered back toward the laser, primarily by aerosols. In addition, the pulse may impinge on a scattering surface or a retro reflector. Scattered radiation is then detected. A ratio of detected to transmitted energy is a measure of the scattering and collection efficiency of the detector. The second pulse is tuned to the absorption feature of the threat that attenuates it. However, the scattering and collection efficiency of the detector is essentially identical because the wavelengths are quite close. The ratio of the 2 signals is a measure of the absorption and thus the concentration level of the threat. With sensitive detectors, even the scattering from aerosols produces sufficient return to measure the concentration. Utilizing time of flight techniques, concentration can be measured as a function of range, which adds a powerful new dimension to the capability of a lidar.

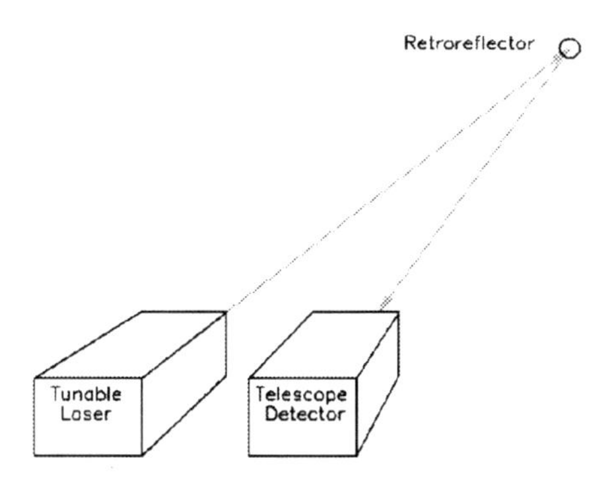

Figure 5. Optical schematic of lidar.

Threat detection can utilize fluorescence spectra as well as absorption spectra. In this case, a sample of the suspected threat is excited, usually by a short laser pulse. The laser must be tuned so that it is absorbed by the sample. After the laser pulse excites it, the sample may fluoresce, most often at longer wavelengths than the laser wavelength. The fluorescence spectrum can then be compared with a data base of known threats to identify it. Measuring the decay time constant of the fluorescence can also be used to help identify the threat. The possibility of identification makes this technique very appealing. However, not all materials emit fluorescence after absorbing the laser pulse. Fortunately, organic molecules tend to fluoresce. This fact is often used to detect them. A typical fluorescence system is shown in Figure 6.

If a material is excited with a laser pulse, it can also emit Raman shifted radiation as well as fluorescence. The laser need not be tuned to an absorption feature and the Raman shift is independent of the laser wavelength. The Raman scattered radiation is down shifted from the frequency of the pump laser by an amount that is characteristic of vibrational modes in the molecule. Therefore, the Raman shift can be employed to help identify the molecule. Unfortunately, Raman scattering is a nonlinear interaction and relatively weak. On the other hand, coherent anti Stokes scattering can be significantly stronger. While Stokes radiation has a longer wavelength than the laser that causes it, anti Stokes radiation is shorter than the pump laser. Anti Stokes radiation has the additional advantage of being more directional than spontaneous Raman scattering. To obtain anti Stokes radiation, a second laser that operates at the

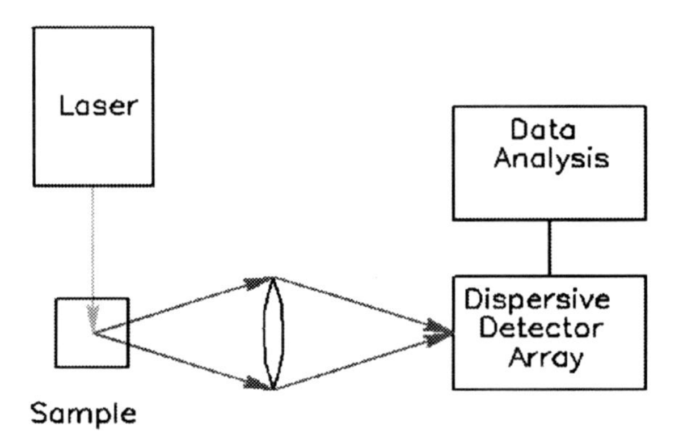

Figure 6. Optical schematic of fluorescence system.

Stokes wavelength is needed. The combined action of the pump and Stokes lasers produce radiation at the anti Stokes wavelength. The Raman and anti Stokes scattering processes are depicted in Figure 7.

TeraHertz radiation can be highly useful for threat detection because it is transmitted by many types of clothing. The transmission of TeraHertz radiation through several types of clothing materials is illustrated in Figure 8. The ability of TeraHertz radiation to penetrate clothing makes concealment of explosive devices more difficult. Because TeraHertz radiation is reflected by metallic objects, it is very useful for screening people who may carry explosive devices or weapons.

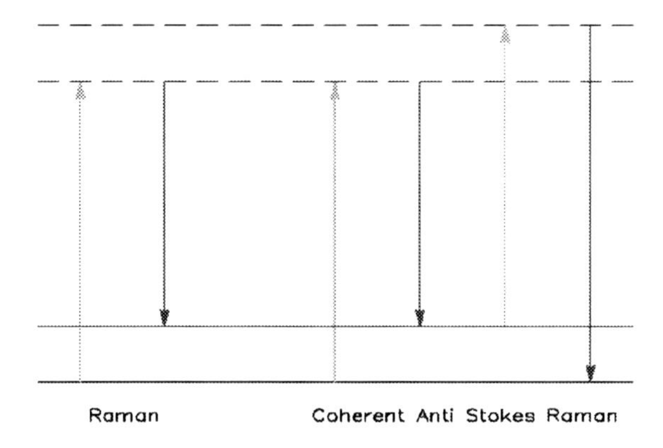

Figure 7. Energy level diagram of Stokes and anti Stokes processes.

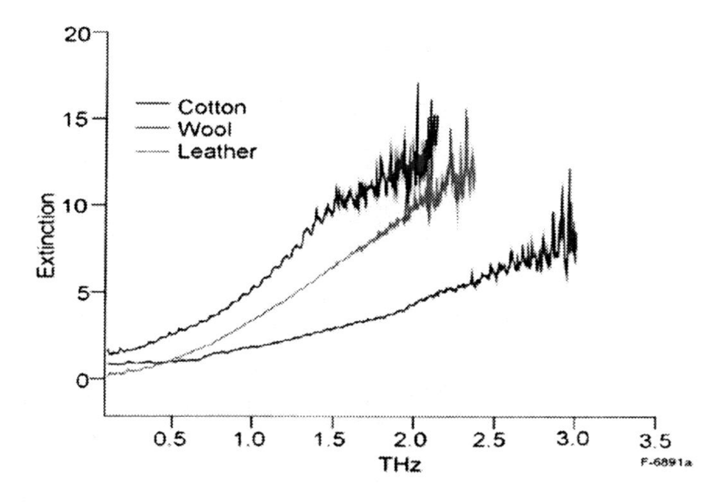

Figure 8. Transmission of clothing material versus frequency
Courtesy of Dr. David Cook, Physical Sciences Inc.

4. Summary

Threats to national security extract huge emotional and monetary resources from many nations. Furthermore, the threat is not confined to a single nation or even to a group of nations nor is it projected to disappear anytime in the near future. This situation provides an opportunity for spectroscopy and spectroscopic techniques to make a vital contribution.

Security threats take many forms including chemical, biological, human, and radiological. The most obvious threats that can be addressed by spectroscopic techniques are chemical threats. Chemical threats can be detected in several spectral regions including the ultraviolet, the mid infrared, and the far infrared or TeraHertz region. Ultraviolet absorption features are strong but broad, making identification more difficult. On the other hand, mid infrared absorption features can be rather narrow, aiding identification. Far infrared or TeraHertz systems have the advantage of being able to penetrate most clothing materials making concealment more difficult.

Several spectroscopic techniques are illustrated. The simplest method is the common spectrometer. Considerably more sensitive is a ring down spectrometer. A ring down spectrometer requires the source be resonant with the ring down spectrometer. A photo acoustic spectrometer relies on the heating of the gas in the cell by a probe beam when the probe beam wavelength corresponds to an absorption feature. Sensitivity of the photo acoustic spectrometer can be enhanced by adjusting the pulse repetition frequency to an acoustic resonance of the cell. Lidar can detect security threats at a distance. Lidar requires 2 laser wavelength, 1 of which is tuned to an absorption feature of the threat. Other detection techniques use fluorescence or Raman shifted radiation emitted by the excited molecules. The wavelength of the emitted radiation can be used for identification purposes.

5. References

1. Usachev, A.D., Miller, T.S., Singh, J.P, Yuen, F.Y., Jang, P.R., andMonts, D.L., (2001) "Optical Properties Of Gaseous 2,4,6, Trinitrotoluene In The Ultraviolet Region," *Appl. Spect.* **55** 125-129
2. Effenberger, F., "Explosive Vapor Detection Prevents Bombs On Board,"(1998) *OE Reports* **172,**
3. Private communication, Dr. David Cook, Physical Sciences Inc.

SOURCES FOR THREAT DETECTION

NORMAN P. BARNES
NASA Langley Research Center
Hampton, VA 23681 U.S.A.
n.p.barnes@larc.nasa.gov,

1. Introduction

The availability of laser sources at particular wavelengths enables several of the more sophisticated threat detection techniques. Methods for the detection of security threats concentrate on 3 spectral regions, the ultraviolet, mid infrared, and far infrared. In the ultraviolet region, electronic spectral features are of interest. In the mid infrared region, vibrational spectral features can be used. In the far infrared or TeraHertz region, rotational spectral features are of interest. Several different laser sources are available in each of these 3 spectral regions. Each spectral region will be discussed separately.

2. Ultraviolet

Because of their nature, electronic transitions in the ultraviolet tend to be characterized by broad spectral features. As a consequence, accurate wavelength tuning, to within a part in a million, and extremely narrow laser linewidths are not a major concern. For some detection methods, it may be sufficient for the laser to emit in the vicinity of some broad absorption features and rely on fluorescence or Raman shifted radiation to detect and identify the threat. Currently, there are 2 primary laser approaches for the ultraviolet; excimer lasers and harmonic generation and/or sum frequency generation of solid state lasers.

Excimers are molecules that exist only in the excited state. Noble gasses; such as Ar, Kr, and Xe; do not usually form molecules in the ground state because their outer electronic shell is filled. However, if an outer shell electron is promoted to the next higher electron shell, the noble gas can act in a manner similar to an alkali metal and form a molecule with a halogen; F, Cl, or Br. If the excited molecule is stimulated to emit a photon and transition to the ground state, the excimer dissociates almost instantaneously. From a laser physics aspect, this behavior is advantageous because it implies there is no population in the lower laser level. Thus, excimer lasers behave as 4 level lasers with their attendant efficiencies.

A list of existing excimer lasers with their wavelength and limiting efficiencies [1] are compiled in Table I. Excimer laser wavelengths are set by the nature of the electronic bond of the excimer. As is true for most electronic transitions, the spectral bandwidth of the transition is rather wide. The limiting efficiency is determined primarily by 2 factors, the ratio of the laser photon energy to the energy required to create the excimer and the extraction efficiency of the

B. Di Bartolo and O. Forte (eds.), Advances in Spectroscopy for Lasers and Sensing, 73–81.

laser. Although the excimer laser behaves like a 4 level laser, there still exists a nonsaturable absorption loss. Even if the gain of the laser is 10 times the loss, the extraction efficiency is only about 0.5. The limiting efficiency in Table I is the product of these 2 factors. Of course, other factors, such as the overlap between the pumped volume and the mode volume, will reduce the actual laser efficiency.

TABLE I. Wavelength and limiting efficiency of excimer lasers.

Molecule	Wavelength(μm)	Limiting Efficiency
ArF	0.193	0.161
ArCl	0.175	0.177
KrF	0.248	0.139
KrCl	0.222	0.127
XeF	0.483	0.086
XeCl	0.308	0.134
XeBr	0.282	0.147

Excimer lasers favor pulsed rather than continuous operation because of the problems associated with the discharge stability. Because of the short wavelength and the high level of excitation, excimer lasers are high gain devices. This complicates but does not rule out wavelength control. Even with wavelength control, the spectral bandwidth of the excimer laser will be limited by the Fourier transform limit.

Excimer lasers are powered by a high voltage, high current discharge. The lifetime of the upper level of excimer lasers is short and the electrical discharge tends to be unstable. These factors tend to favor a short duration electrical discharge, on the order of 10s of ns. A 10 kV, 10 kA pulse with a 10 ns duration delivers 1.0 J to the excimer laser. Switching these high voltage and high current discharges places a large burden on the switching device, typically a thyraton. Lifetime of these switching devices can also be a problem. Other lifetime concerns are associated with the corrosive nature of the excimer gas mixture in the laser and suitable optics that are transparent and laser induced damage resistant at ultraviolet wavelengths.

Solid state lasers when coupled with nonlinear optics can produce a line tunable ultraviolet source. The most common illustration of this is the frequency tripled and quadrupled Nd:YAG laser which produces an output at 0.355 and 0.266 μm. In reality, a 0.355 μm output is produced by frequency doubling the 1.064 μm Nd:YAG and the mixing the frequency doubled output with the residual 1.064 μm fundamental. Frequency quadrupling is actually second harmonic generation of the second harmonic output. This approach possesses several advantages including use of well developed Nd:YAG lasers, especially diode pumped Nd:YAG lasers. There are no toxic gasses to handle, electrical pulse lengths are longer, lower current, and lower voltage, and the number of optical components that are exposed to high power ultraviolet radiation is limited. However, the question of tunability should be addressed.

Although the Nd:YAG 1.064 μm transition is well known, somewhat less well known [2] is that Nd:YAG can operate efficiently on several other laser transitions as well. Work done at NASA Langley measured several important spectroscopic features and the laser performance on the stronger of these transitions. The linewidth, $\Delta\nu$, is a measure of the possible tuning range of each transition. Laser performance is characterized by a threshold, E_{ETH}, and a slope efficiency, σ_s. The former is the amount of electrical energy needed to begin laser action and the latter is the rate of increase in laser energy with increased pump energy above threshold. Threshold and slope efficiency data were taken with a flash lamp pumped Nd:YAG laser operating in a TEM_{00} transverse mode. It may be noted that almost all of the slope efficiencies are more than 0.01. Slope efficiencies for flash lamp pumped Nd:YAG lasers operating at 1.064 μm and multi transverse mode is on the order of 0.03. Thus the efficiency penalty for operation on other transitions need not be severe.

TABLE II. Parameters of Nd:YAG $^4F_{3/2}$ to $^4I_{11/2}$, near 1.064 μm, and $^4F_{3/2}$ to $^4I_{13/2}$, near 1.32 μm, laser transitions.

λ_L(μm)	$\sigma_e(10^{24}\ m^2)$	$\Delta\nu(cm^{-1})$	E_{ETH}(J)	σ_s
1.0520	9.62	5.7	4.93	0.0123
1.0614	20.62	4.8	2.60	0.0184
1.0641	27.74	6.0	1.99	0.0193
1.0737	15.15	5.2	3.62	0.0139
1.0779	6.26	9.9	7.63	0.0095
1.1120	3.97	12.1	10.28	0.0113
1.1160	3.95	14.4	11.48	0.0104
1.1225	3.97	11.4	11.26	0.0112
λ_L(μm)	$\sigma_e(10^{24}\ m^2)$	$\Delta\nu\ (cm^{-1})$	E_{ETH}(J)	σ_s
1.3187	8.92	4.5	6.64	0.0127
1.3336	4.03	3.7	11.82	0.0104
1.3381	9.57	5.0	6.53	0.0131
1.3564	7.35	6.6	8.42	0.0107

To expand the possible ultraviolet wavelengths, laser output at each of the transitions can be frequency tripled and quadrupled. Harmonics cover the spectral ranges from 0.440 to 0.452, 0.351 to 0.364, 0.330 to 0.339, and 0.263 to 0.281 μm. Unfortunately, there is a large gap in the spectral range from 0.281 to 0.330 μm. This gap can be alleviated if a laser operating on a $^4F_{3/2}$ to $^4I_{11/2}$ transition could be mixed with a laser operating on a $^4F_{3/2}$ to $^4I_{13/2}$ transition. If so, wavelengths in the range from 0.293 to 0.307 could be generated. The standard method of approaching this would be to use 2 lasers and synchronize their pulses to within a small fraction of the pulse length, typically a few nanoseconds.

NASA Langley developed a single laser [3] that synchronously generates Q-switched pulses from both the $^4F_{3/2}$ to $^4I_{11/2}$ transition and the $^4F_{3/2}$ to $^4I_{13/2}$ transition. Typically this does not occur as the transition with the highest round trip gain will deplete the population inversion before pulses operating on other transitions can extract a significant amount of energy. However, by carefully designing the resonator, a single laser can produce synchronous and collinear pulses. In this case, mixing becomes as simple as second harmonic generation. An optical schematic of the ultraviolet laser system appears in Figure 1. Because both pulses have the same output mirror and are produced in the same laser rod, they are collinear. Because both pulses have the same Q-switch, they are initiated at the same time. By careful resonator design, both pulses can have the same pulse evolution time interval. The inverse of the pulse evolution time interval plotted versus pump energy can be shown to be a collinear straight lines for both the 1.052 and 1.319 μm transitions.

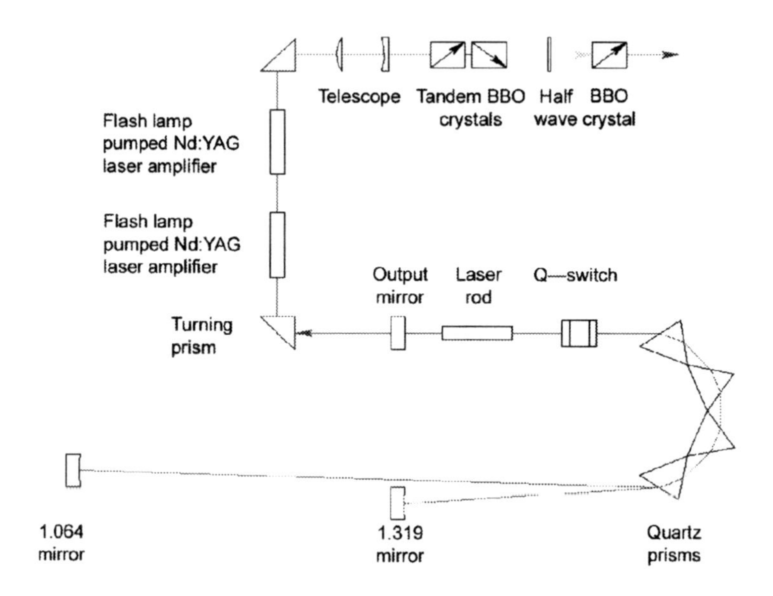

Figure 1. Optical schematic of STOP light laser.

With a single laser that produces both wavelengths for a sum frequency generation process, implementing this process becomes as simple as second harmonic generation. The Nd:YAG laser can designed to generate any pair of wavelengths from the $^4F_{3/2}$ to $^4I_{11/2}$ and the $^4F_{3/2}$ to $^4I_{13/2}$ transitions. With the laser depicted in Figure 1, the tandem BBO crystals can be rotated through a small angle to produce second harmonic of 1.319 μm, the sum frequency of 1.319 and 1.052 μm, or the second harmonic of 1.052 μm. The resulting wavelengths are at 0.660 μm, red, 0.585 μm, orange, or 0.526 μm, green. The red, orange, and green pulses are reminiscent of a STOP light, thus the name Synchronously Tunable Optical Pulses. Each of these sum frequency wavelengths can then be frequency doubled to generate ultraviolet radiation, in this case at 0.330, 0.293, or 0.263 μm.

3. Mid Infrared

There are at least good 3 choices for lasers that operate in the mid infrared region of the spectrum, nominally from 3.0 to 20 μm. This region is of particular interest because the spectral features are narrow compared with the wavelength. The existence of narrow and distinctive spectral features labels this region as the fingerprint region of the spectrum. Laser choices include CO_2, quantum cascade laser diodes, and solid state lasers coupled with nonlinear optics.

CO_2 lasers operate primarily in the 9.0 to 11.0 μm region. CO_2 is a linear molecule, a C atom flanked by 2 O atoms. The CO_2 molecule possesses 3 vibrational modes, the symmetric stretch, bending, and asymmetric stretch, shown in Figure 2. Laser operation can occur between the first excited asymmetric stretch mode, 001, and either the first excited symmetric stretch mode, 100, or the second excited bending mode, 020. From the 100 or 020 levels, the CO_2 molecule relaxes to the ground level through the 010 level. In normal isotopic CO_2, the 001 to 100 transition is around 10.6 μm and the 001 to 020 transition is around 9.6 μm.

Each vibational transition also has a series of rotational modes that are associated with it. These rotational modes have populations that are determined by Boltzmann statistics. Laser transitions are between individual rotational levels, denoted by J, allow a wide variety of CO_2 wavelengths to be produced. Quantum mechanical selection rules limit the transitions to change rotational quantum number by ±1, producing R and P branch transitions. A table of the available wavelengths [4] for J ranging from 2 to 60 appears in Table II. By using isotopic forms of CO_2, almost any wavelength from 9.1 to 12.3 μm can be generated.

CO_2 lasers are pressure broadened at atmospheric pressures to the point where individual J level transitions overlap. This provides for continuous tuning of the laser. On the other hand, if narrow linewidth operation is needed, a low pressure CO_2 device can be used if a suitable wavelength match can be found. In this case, the linewidth of an individual rotational transition is on the order of 1.0 GHz.

CO_2 lasers can be excited either by direct electrical discharge or by radio frequency excitation. Usually a N_2 molecule is excited which then transfers energy to the 001 level of CO_2. The former excitation method has the advantage of simplicity while the latter has the advantage of long life. Electrical excitation can be continuous or pulsed. When pulsed, electrical pulses are on the order of 1.0 μs and lead to short, gain switched, laser output pulses. Short pulses are suitable for lidar applications. Electrical to optical efficiencies can be quite high, in excess of 0.2. However, single transverse mode operation is needed, or if gas circulation is needed to cool the CO_2 gas mixture, the overall efficiency may drop to below 0.1.

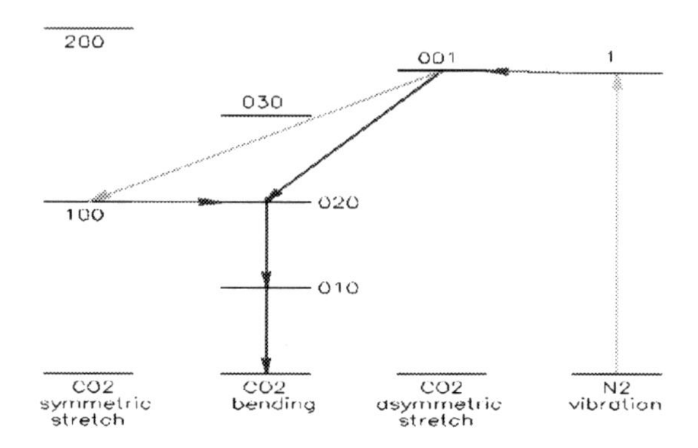

Figure 2. CO_2 energy level diagram.

TABLE III. CO_2 wavelengths available.

Isotope	Vibrational	Rotational	Wavelength Range (μm)
$^{12}C^{16}O_2$	001 to 100	P	10.42 to 11.07
		R	10.40 to 10.04
	001 to 020	P	9.41 to 9.95
		R	9.39 to 9.11
$^{13}C^{16}O_2$	001 to 100	P	10.97 to 11.66
		R	10.94 to 10.52
	001 to 020	P	9.94 to 10.44
		R	9.82 to 9.51
$^{12}C^{16}O_2$	001 to 100	P	11.57 to 12.31
		R	11.53 to 11.05
	001 to 020	P	10.19 to 10.84
		R	10.17 to 9.85

Quantum cascade lasers are diode lasers where a single electron has several chances to contribute to the laser output. They are unipolar devices that consist of many injectors and active regions, each consisting of several alternating thin layers of III-V materials, such as

$In_xGa_{1-x}As/Al_yIn_{1-y}$ As grown in InP or $GaAs/Ga_yAl_{1-y}As$ on GaAs substrates, Figure 3. Typically the layers are on the order of a nanometer thick. In the active region, several quantum wells are grown. The material and the depth of the quantum well determine its energy level. As a consequence, a wide variety of energy levels are available. There are usually 1 or 2 high lying energy levels and 2 low lying energy levels. Lasing occurs on a transition between the lowest of the high lying energy levels and the highest of the low lying energy levels. Usually the low lying energy levels are designed so they are separated by an energy corresponding to an optical phonon. This helps relax the lower laser level.

After making the transition, the electron enters the injector region to be transported to another active region. The injector region consists of many very thin alternating pairs of layers that are too thin to confine the electron. Instead a miniband and a minigap are formed. The miniband allows electrons to travel to the next active region where they enter the active region. Once in the active region, they are discouraged from leaving as the upper laser level corresponds to the minigap. A quantum cascade laser can have on the order of 25 pairs of injectors and active regions. Therefore, a single electron can generate up to 25 photons.

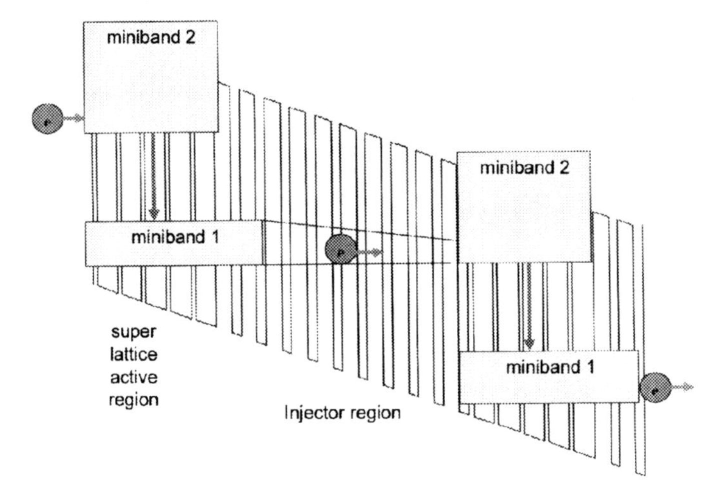

Figure 3. Quantum cascade laser band structure.

Short wavelength quantum cascade lasers can produce on the order of 1.0 W continuously at room temperature. Electrical to optical efficiency can be of quantum cascade lasers on the order of 0.01 at the present. However, an electrical to optical efficiency in excess of 0.04 has been claimed at room temperature. However, both average power output and laser efficiency benefit from cooling, even to liquid N_2 temperatures. As the laser wavelength becomes longer, cooling becomes more essential. Pulsed quantum cascade lasers do not suffer from heating as much. Thus, they tend to be more efficient.

With no line narrowing elements, the linewidth of a quantum cascade laser can be on the order of 10s of cm^{-1} wide. However, a distributed feed back structure can be grown on top of the quantum cascade laser structure. The distributed feed back structure can narrow the linewidth of the quantum cascade laser to a single mode. If the quantum cascade laser is pulsed, the wavelength is swept over frequency as the laser diode seeks an equilibrium temperature. A quantum cascade laser with a distributed feed back structure can sweep through a spectral interval on the order of 10 cm^{-1} as it warms up. This swept frequency effect can be used to advantage to perform a spectral scan over the absorption spectrum of a threat of interest. Although quantum cascade lasers can be fabricated at many mid infrared wavelengths, a single device is only temperature tunable over a relatively narrow spectral range, on the order of a few nanometers. The tuning rate with temperature is on the order of 0.5 nm/°K.

A single parametric oscillator can cover the entire mid infrared range. A parametric oscillator may be considered to be a photon splitter, that is, a pump photon is split into 2 photons. The sum of the energies of these 2 photons is equal to the pump photon energy; that is energy is conserved. For the interaction to be efficient, momentum must be conserved. Conservation of momentum is also referred to as phase matching. The momentum of a photon is proportional to the index of refraction. In normally dispersive crystals, the index of refraction decreases monotonically with increasing wavelength, discouraging phase matching. To achieve phase matching, nonlinear crystals where the refractive index depends on the polarization, that is birefringent crystals, are used. In birefringent crystals, 1 or both of the refractive indices depend on the direction of propagation. In this case, the generated wavelengths can be tuned by varying the direction of propagation. Even with birefringent crystals only a narrow spectral range of wavelengths can be approximately phase matched. With the proper choice of pump wavelength and nonlinear crystal, the entire 3.0 to 18.0 μm spectral range can be accessed. If a wide bandwidth source is satisfactory, the natural spectral bandwidth of the nonlinear interaction may produce a sufficiently narrow bandwidth. If a narrow spectral bandwidth source is desired, the parametric oscillator can be injection seeded [5].

4. Far Infrared

Laser sources similar to the laser sources that were applicable in the mid infrared are also applicable to the far infrared. Quantum cascade lasers can be readily designed for operation at far infrared wavelengths by making the width of the quantum well wider. However, the energy difference between the upper and lower laser levels becomes rather small for far infrared laser diodes. Far infrared quantum cascade lasers usually operate at extremely low temperatures, on the order of 10°K. At the present time their output power is low, on the order of 10 μW.

In principle, parametric oscillators can be made to operate in the far infrared however the gain is low. Parametric gain is proportional to $(\lambda_s\lambda_i)^{-1}$ where λ_s is the wavelength of the signal and λ_i is the wavelength of the idler, in this case the far infrared wavelength. Because the wavelength of the idler is so long, gain is low making threshold difficult to achieve. To circumvent this problem, difference frequency generation is employed. With difference frequency generation, strong laser sources are supplied at both the pump and the signal wavelengths. This ploy avoids having to achieve threshold as required by the parametric oscillator. However, it requires 2 lasers for operation.

Difference frequency generation can span the far infrared region of the spectrum. If a spectrally narrow source is needed, 2 narrow band lasers can be employed in the difference frequency process. On the other had, virtually the entire far infrared band can be generated by using a femtosecond pulsed laser in the difference frequency process. Of course, the generated spectral bandwidth will be limited by the phase matching condition. However, in some cases, variation of the phase matching condition is relatively insensitive to the far infrared wavelength.

Difference frequency efficiency is limited by the ratio of the pump wavelength to the far infrared wavelength. This limit is set by the fact that for each far infrared photon generated, a pump photon disappears. This limits the efficiency to approximately 0.01 or less. However, the difference frequency process does not require cooling which helps decrease the overall power consumption.

5. References

1. J.J. Ewing "Excimer Lasers At 30 Years," *Optics And Photonics News14* 26-31 (2003)
2. N.P Barnes, B.M. Walsh, and R.E. Davis, "Dispersive Tuning And Performance Of A Pulsed Nd:YAG Laser," Proceedings Of The Advanced Solid State Photonics Conference, San Antonio TX, (2003)
3. N.P. Barnes and B.M. Walsh, "Synchronous Tunable Optical Pulses," Advanced Solid State Photonics Conference, Santa Fe NM (2004)
4. C. Freed, L.C. Bradley, and R.G. O'Donnell, "Absolute Frequencies Of Lasing Transitions In Seven CO_2 Isotopes," *IEEE J. Quantum Elect. QE-16* 1195-1206 (1980)
5. N.P. Barnes, K.E. Murray, G.H. Watson, "Injection Seeded Optical Parametric Oscillator," Proceedings On Advanced Solid State Laser Conference, *13*, 356-360 (1992)

Yb^{3+}-DOPED CaF_2 FLUORIDE AS AN EXAMPLE OF OUR RESEARCH APPROACH IN SOLID-STATE LASER-TYPE CRYSTALS

GEORGES BOULON
Physical Chemistry of Luminescent Materials
Claude Bernard/Lyon1 University,
UMR 5620 CNRS, Bât.Kastler, La Doua, Villeurbanne, France
boulon@pcml.univ-lyon1.fr,

Abstract

$Ca_{1-X}Yb_XF_{2+X}$ crystals have been grown by three different methods: Czochralski and simple melting under CF_4 atmosphere and laser heated pedestal growth (LHPG) method under Ar atmosphere. Spectroscopic characterization (absorption, emission, Raman spectroscopy and decay curves) were carried out to identify Stark's levels of Yb^{3+} transitions in the different crystallographic sites of the cubic structure in $Ca_{1-X}Yb_XF_{2+X}$ crystals. Yb^{3+} concentration dependence of the experimental decay time was analyzed by using concentration gradient fibre in order to understand involved concentration quenching mechanisms. Under Yb^{3+} ion infrared pumping, self-trapping and up-conversion non-radiative energy transfer to unexpected rare earth impurities (Er^{3+}, Tm^{3+}) has been observed in visible region and interpreted by a limited diffusion process within the Yb^{3+} doping ion subsystem towards impurities. Main parameters useful for a theoretical approach of laser potentiality have been given as $\tau_{rad.} = 2.05$ ms, $N_0 = 7.47\times10^{21}$ cm^{-3} (32 mole%) and $N_m = 6.39 \times 10^{21}$ cm^{-3} (26.5 mole%). In addition high thermal conductivities measurements have been recorded confirming the good laser behaviour of Yb^{3+} -doped CaF2 as both CW diode pumped crystal laser source and tuneable laser source in the near IR range.

1. Introduction

Yb^{3+} is the most promising ion that can be used in a non-Nd^{3+} laser in the near-IR spectral range of 1030 nm, it means roughly in the same range of Nd^{3+} emission wavelength, under laser diode pumping with a smaller quantum defect than in Nd^{3+}-doped crystals (10% instead of 25%). The Yb^{3+} ion has several advantages compared with Nd^{3+} due to its very simple energy level scheme, constituting of only two levels. It makes possible to avoid up-conversion, excited state absorption and concentration quenching within a large concentration domain. Up to now, several oxides and fluorides were deeply investigated, and these spectroscopic properties as the Yb^{3+}-doped host materials leading to general methods of evaluation [1, 2, 3, 4]. Among such hosts, we are dealing with the main families of laser crystals as sesquioxides (Y2O3, Sc2O3, Lu2O3), garnets (YAG, GGG, LuAG), tungstates (KY(WO4)2, KGd(WO4)2) and fluorides $LiYF_4$, $LiLuF_4$, BaY_2F_8 and cubic KY_3F_{10} and CaF_2 as it was mentioned in the previous Erice's Book [5].

The present review has been focused on the very interesting Yb^{3+}-doped CaF_2 cubic fluoride host especially as an example of our research approach in solid-state type-laser crystals. Such un-doped crystals are now growing with large diameter mainly to be used as window materials for excimer lasers which are needed in the semiconductor technology as optical lithography. We think the growth of high quality CaF_2 single crystal, which are

B. Di Bartolo and O. Forte (eds.), Advances in Spectroscopy for Lasers and Sensing, 83–102.

characterized by a high value of the thermal conductivity (10 W m^{-1} K^{-1}), much higher than other fluoride laser crystals (5 W m^{-1} K^{-1} in $LiYF_4$) and as high as YAG (10.7 W m^{-1} K^{-1}), should also contribute to the development of compact diode-pumped Yb^{3+}-doped laser sources.

In this article, we are first investigating basic research based on the detailed spectroscopy of $Ca_{1-X}Yb_XF_{2+X}$ crystals which are grown by three different methods: Czochralski, simple melting under CF_4 atmosphere and laser heated pedestal growth (LHPG) technique under argon atmosphere. Following the last Erice's Book [5], we are indeed completing another article published on combinatorial chemistry to grow single crystals for the analysis of concentration quenching processes applied to Yb^{3+}-doped laser crystals. Analyzing quenching mechanisms is also useful to estimate potential development as optical materials, mainly as tuneable solid state laser source and ultra-short pulse laser source due to the broad band emission at room temperature. Then, we shall show our theoretical approach of the laser source optimization and we shall confirm how interesting such type of laser crystal is. In addition, as thermal properties of laser crystals have also to be optimized for any kind of application, we shall confirm high thermal conductivities measurements of $Ca_{1-X}Yb_XF_{2+X}$ crystals almost as high as Yb^{3+}-doped YAG.

2. Structure and Nature of Crystallographic Sites

CaF_2 is the typical fluorite-type crystal. It is known to form $Ca_{1-X}RE_XF_{2+X}$ type solid solution with trivalent rare-earth fluorides up to nearly 40 mole% of dopant concentration [6]. The structure of CaF_2 can be regarded as a simple cubic array of touching F^- ions with every second void occupied by Ca^{2+} ion. When Ca^{2+} ion is substituted by trivalent rare earth, charge balance compensating F^- ions should enter the fluorite structure. Up to now a lot of structural investigations have been carried out on $Ca_{1-X}RE_XF_{2+X}$ type materials [7-12] leading dominant defect structures. At low concentration, at least for dopant concentration less than 1 mole%, the additional F^- ion goes into a next nearest neighbour void of the trivalent rare earth ion giving isolated tetragonal (C_{4V}), trigonal (C_{3V}) or rhombic (C_{2V}) defects. These isolated defects were well studied previously by electron paramagnetic resonance and optical spectroscopy [7, 13-15]. However, for dopant concentration greater than 0.1 mole%, the anion clusters involving more than one dopant cation are formed [7]. These clusters were studied by X-ray diffraction, EXAFS (extended X-ray absorption fine structure) and neutron diffraction [7-12]. An interpretation of the crystallographic data has been already shown in [16]. There are a huge number of structural possibilities and each cuboctahedral-related cluster should have similar structure, hence its spectroscopic properties became an average of them. Then, it is hardly possible to separate them and to assign them exactly due to the spreading of the optical spectra like those of the disordered crystals.

3. Crystal Growths

$Ca_{1-X}Yb_XF_{2+X}$ crystals were prepared by two different methods as described below. It is important to compare the spectroscopic properties from the growth conditions because some properties may be changed depending the crystal qualities.

3.1. SYNTHESIS OF CRYSTALLINE SAMPLES BY SIMPLE MELTING FROM CZOCHRALSKI (CZ) MODIFIED AT TOHOKU UNIVERSITY

The main objective of this synthesis was to provide seed crystals and feed rods for Laser Heated Pedestal Growth (LHPG) technique to grow single crystalline fibres as described in the next section. $Ca_{1-X}Yb_XF_{2+X}$ (X = 0.005, 0.02, 0.05, 0.15 and 0.3) crystals were prepared by simply melting mixtures of commercially available powders of CaF_2 and YbF_3 with the purity of 4N. The furnace was driven with a 30 kW RF generator and a carbon crucible was used for melting materials. The furnace was vacuumed up to 10^{-4} Torr prior to synthesis to eliminate oxygen and/or water, and then CF_4 gas was slowly introduced. The melting was performed under CF_4 atmosphere. After completely melting materials, the furnace was slowly cooled down to room temperature (Figure 1).

Large crystalline ingots could be prepared very rapidly and efficiently by this method. Now, un-doped CaF_2 crystals of few inches diameter are industrially grown, but not high concentration trivalent rare earth doped crystals. The growth of trivalent rare earth doped crystals turn to be difficult since the occurrence of supercooling phenomena. Nevertheless, this method allowed preparing these crystals of few centimetre sizes only in few hours. Figure 1 shows some pieces of un-polished $Ca_{9.95}Yb_{0.05}F_{2.05}$ rough crystals to prove the sizes of pieces. Although the ingots are in fact disordered samples composed of some large single crystals, each part of ingot has very good transparency as a single crystal one. As the first objective of this synthesis, we could provide this ingot to seed crystals and feed rods for LHPG method as described in the next section. Moreover, this is worthwhile to study spectroscopic characterizations of such Yb^{3+}-doped samples to compare them with fibre samples grown by LHPG technique.

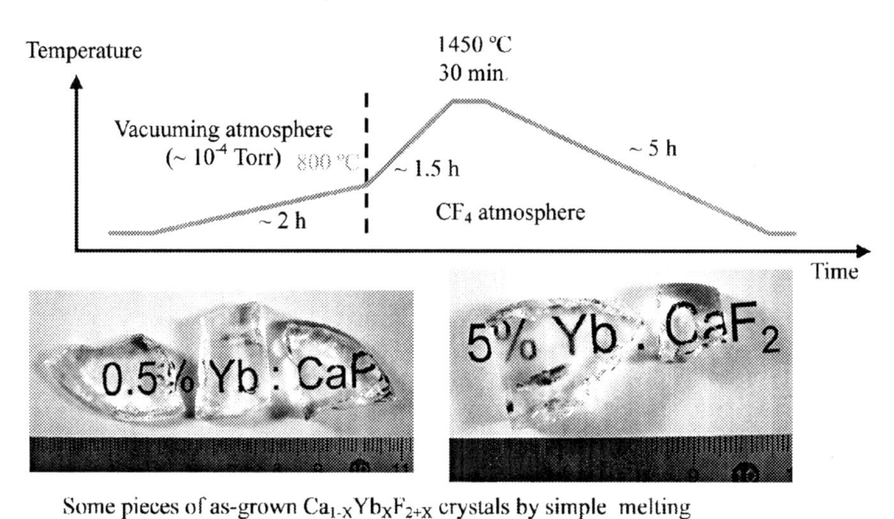

Some pieces of as-grown $Ca_{1-X}Yb_XF_{2+X}$ crystals by simple melting

Figure 1. Synthesis by simple melting process:diagram and pictures of samples.

3.2. CRYSTAL GROWTH BY THE LHPG METHOD AT THE UCBLYON1

The LHPG method [17] is one of the suitable methods for evaluating crystals as laser materials since it needs only few amounts of raw materials and the growth rate is much faster than other methods such as Czochralski and melting methods. And what is more, recently we have developed concentration gradient fibres to measure optical properties [18]. By the help of the special ceramic rod, which is composed of two different dopant concentrations and cut along a slant direction, it is possible to prepare original single-crystalline samples with variable concentration of dopant in an extremely wide range. Such fibres shown in Figure 2 are very useful for spectroscopic characterizations as shown recently [19-22]. As can be seen, transparent crystalline fibres with diameter about 1.0 mm were obtained.

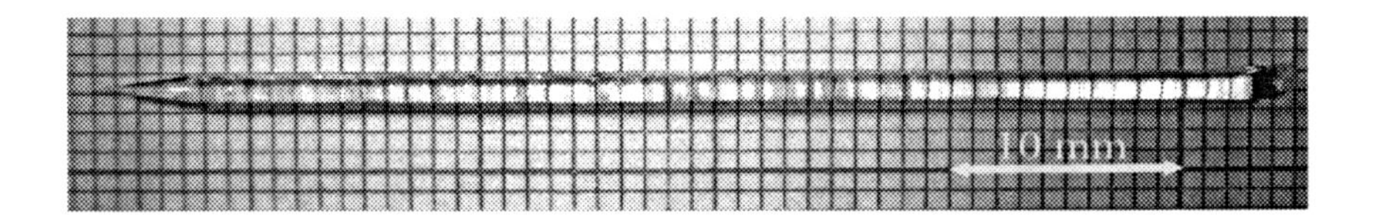

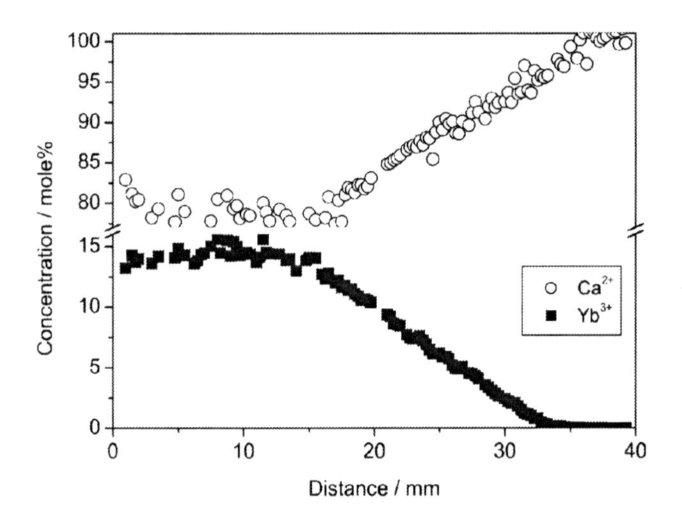

Figure 2. Concentration analysis data of 0-15 mole% Yb^{3+}: CaF_2 concentration gradient fibre analyzed by EPMA.

3.3. CZOCHRALSKI GROWTH AT TOHOKU UNIVERSITY

At the last stage of the optical characterizations, we have ordered crystals grown by the standard CZ technique to measure laser essays in excellent optical quality samples [23].

4. Assignment of Electronic and Vibronic Lines in Yb^{3+}-Doped Crystals

Although the Yb^{3+} ion has a simple electronic structure with only one excited state ($^2F_{5/2}$) above the ground state ($^2F_{7/2}$), the assignment of pure electronic lines is a rather difficult task due to a strong electron-phonon coupling.The degeneracy of the two multiplets is raised and seven Stark electronic levels are expected: four for the ground and three for the excited state, which have been labelled in Figure 3. Room temperature absorption and emission spectra of Yb^{3+}-doped CaF_2 crystals like in Figure 3 show clearly many more lines than can be expected for an electronic transition alone. The observation is the same in Y_2O_3 sesquioxide [19-21] and YAG garnet [22]. The reason is the strong electron-phonon coupling giving rise to both electronic and vibronic transitions.

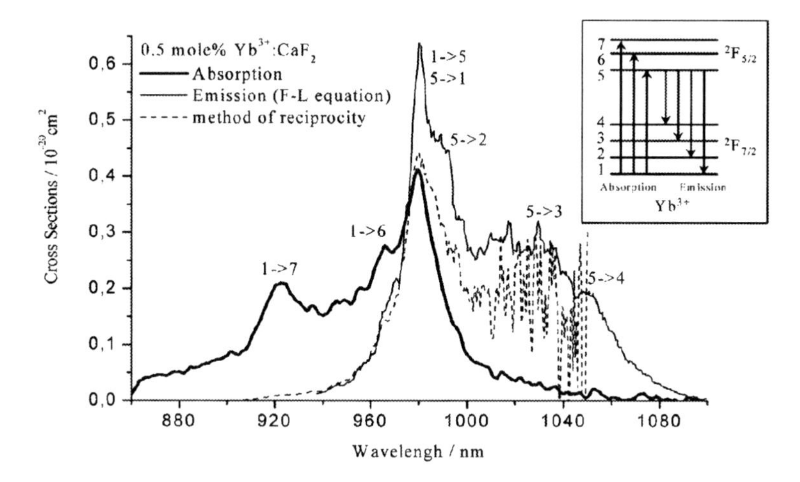

Figure 3. Absorption and emission spectra of 0.5% Yb^{3+}-doped CaF_2 fibre at room temperature. Cross-sections are shown in cm^2. The main electronic transitions have been assigned by using the energy diagram. Both Futchbauer-Ladenburg equation and method of reciprocity do not show a good agreement for the resonant 1-5 transition, as expected, due to the formation of clusters [16].

5. Yb^{3+} Concentration Quenching Processes

Gradient concentration fibre is one unique tool to make the correlation between Yb^{3+} concentration and lifetime measurements which have been measured in situ on the same sample in relation to the distance from the top of the crystallized rod. By using a reference made of a homogeneous single crystal fibre of well-defined composition, the two curves were correlated.

The lifetimes have been directly measured on each point of the concentration gradient fibre of 800 μm diameter, with a beam laser of about 500 μm size.Because the volume of materials excited is steady, radiative trapping due to geometrical effect can be supposed to be characterized by a constant value for each concentration, which should be weak due to the small size of the excited sample. Experimental lifetime values have been fitted to one exponential profile with an excellent agreement. Results are shown in Figure 4 as a function of Yb^{3+} concentration in CaF_2 [16] as it was done for two other samples Y_2O_3 [19-21], YAG [22].

5.1. EVALUATION OF THE RADIATIVE LIFETIME OF Yb^{3+}-DOPED CaF_2

The measurement of intrinsic lifetimes has received attention since a long time in the literature [24-27]. Especially, the determination of the Yb^{3+} radiative lifetime in crystals requires a lot of precaution. As an example, in YAG, measurements of the room temperature effective stimulated emission cross-section have ranged from 1.6×10^{-20} cm^2 to 2.03×10^{-20} cm^2 [22]. Depending of the concentration and size of samples, we have seen in the previous section that the self-trapping process is much or less involved. Gradient concentration fibre is a unique tool to make the correlation between Yb^{3+} concentration and decay time measurements which have been measured in situ on the same sample in relation to the distance from the top of the crystallized rod. By using a reference made of a homogeneous single crystal fibre of well-defined composition, the two curves were correlated. Because the volume of materials excited is steady, radiative trapping due to geometrical effect can be supposed to be a constant value for each concentration, and it should be weak due to the small size of the excited sample. The intrinsic lifetime can be read by following the concentration dependence to the lowest values in Figure 3. The value has been estimated 2.05 ± 0.01 ms.

The direct calculation of the spontaneous emission probability from the integrated absorption intensity [32] is adapted to confirm the radiative lifetime value. Radiative lifetime can be deduced from the absorption spectrum according to the formula (1):

$$1/\tau_{rad} = A_{if} = \frac{g_f}{g_i}\frac{8\pi n^2 c}{\lambda_0^{\,4}}\int \sigma_{fi}(\lambda)\, d\lambda \qquad (1)$$

Where the g's are the degeneracies of the initial and final states ($g_f = 4$ for $^2F_{7/2}$ and $g_i = 3$ for $^2F_{5/2}$), n is the refractive index ($n = 1.43$), c is the light velocity, λ_0 is the mean wavelength of the absorption peak (980 nm), $\sigma(\lambda)$ is the absorption cross-section at wavelength λ. The radiative lifetime was calculated as $\tau_{rad\ (theo.)} = 2.0$ ms by the equation (2) quite close to the experimental value of $\tau_{rad\ (exp.)} = 2.05$ ms. The lower value of $\tau_{rad\ (theo.)}$ with respect of the $\tau_{rad\ (exp.)}$ might be related to the over estimated value of the integrated absorption cross-section including all types of sites at room temperature and not only the highest population of the main square-antiprism sites.

5.2. SELF-TRAPPING AND SELF-QUENCHING PROCESSES

Identifying the origin of the concentration dependence of the observed experimental lifetime allows us understanding the excited state dynamics. An essay of the interpretations of, both, the radiative energy transfer and the non radiative energy transfer in Yb^{3+}-doped crystals, has already been published [16].The curves can be divided into two regimes in Figure 3:

(i) in the lowest concentration range (up to 1.3% in YAG, 5% in yttria and 10% here in CaF_2), the experimental lifetime increases as the dopant concentration increases. This regime is the indication of the radiative energy transfer (self-trapping) between 1 and 5 levels, by the 1↔5 resonant transitions leading to a lengthening of the fluorescence lifetime measured over the volume of the sample. Consequently, the decay times of a single isolated ion should be observed as seen in Figure 4 and dependence on size samples is effectively observed in Table 1. Strong spectral overlap between the fluorescence and absorption spectra enhances fluorescence re-absorption. The overlap of the 1↔5 resonant transition is clearly seen in the case of Yb^{3+}-doped CaF_2 in Figures 3 and 5.

Table 1. Experimental decay times increase depending on the sample size.

Samples	Decay times (ms)
Large piece (4 × 4 × 3 mm3) of simple melting	2.87
Small piece (1 × 1 × 1 mm3) of simple melting	2.07
LHPG fibre	2.10
Grained powders of simple melting	1.92

(ii) for the higher concentration range, the measured lifetime decreases when the doping rate increases up to 4% in YAG, 25% in yttria and 30% here in CaF_2. The second type of regime corresponds to the usual quenching process by a non-radiative energy transfer to defects and other impurities in the host (self-quenching). The decreased lifetime in the high concentration region can be, first, naturally assigned to an energy transfer through Yb^{3+} excited ion to unwanted traces of rare earth impurities (Er^{3+}, Tm^{3+}) which are located near by Yb^{3+} ions to explain the efficiency of the non radiative energy transfer observed in all laser crystals [16]. Another reason could be non radiative energy transfer from Yb^{3+} to unwanted OH^- impurities which are very often involved in the quenching mechanisms. However, in CaF_2 fluoride, OH- are excluded for chemical reason whereas in oxides OH^- have been shown to exist. The presence of OH^- groups in YAG has been characterized by an relatively high absorption coefficient of 3.5cm^{-1}. This difference might be the reason of the strongest maximum concentration value of the experimental decay time in fluorides (~10-12% in CaF2) than in oxides (~8% in Y2O3 and ~6% in YAG). Other physical reasons have been unsuccessfully searched in CaF_2 like: color centres, but we did not observe any new resonant absorption in the near IR with Yb^{3+} emission lines, or traces of Yb^{2+} ions from reduction of Yb^{3+} ions but we have shown that Yb^{2+} ions could only be involved under extrinsic gamma or X-Rays –irradiations [27].

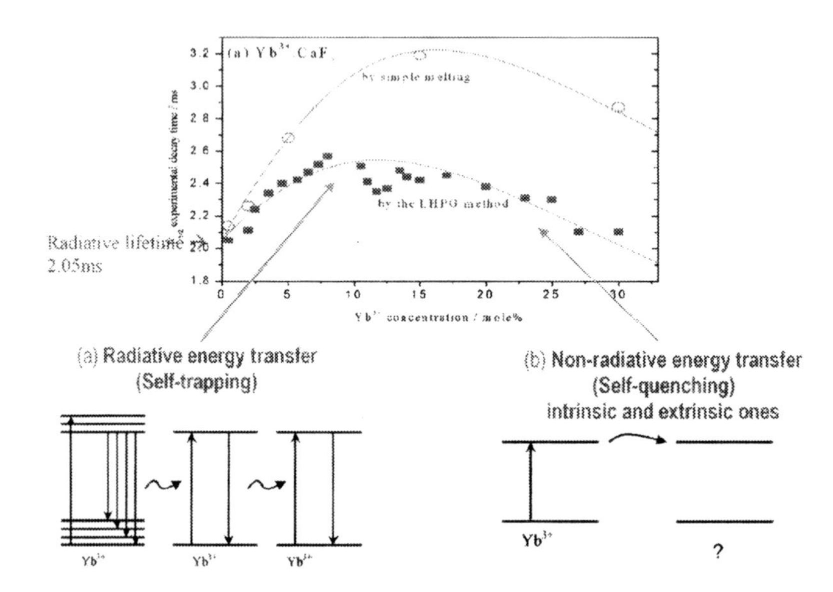

Figure 4. $^{2}F_{5/2}$ experimental decay time dependence of Yb^{3+}-doped CaF_2 samples grown either by Czchochralski simple melting (bulky) technique or by LHPG (fibre) technique.

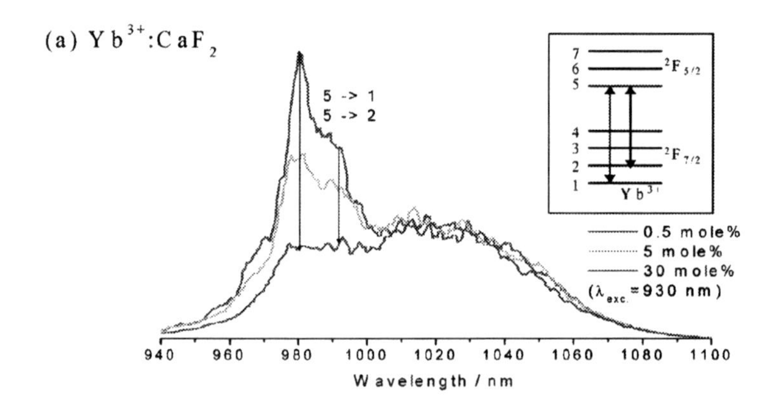

Figure 5. Emission spectra of different Yb^{3+} concentrations at room temperature. Strong re-absorption phenomenon of the resonant transitions is seen by 1-5 and 2-5 emission lines due to self-trapping effect when concentration is increasing from 0.5 moles%to 30 moles%.

5.3. MODEL TO INTERPRET RADIATION TRAPPING AND SELF-QUENCHING MECHANISMS

The combined case of self-trapping and self-quenching in the lifetime analysis of Yb, Er, and Ho doped Y_2O_3 single crystal fibre has been studied [28-29]. It has been shown that self-quenching, for a rather large doping range, is well described by a *limited diffusion* process within the doping ion subsystem towards impurities analogous to the doping ions themselves. *Fast diffusion* towards intrinsic non-radiative centers cannot explain the observed results. Assuming an electric dipole-dipole interaction (s = 6) between ions, the self-quenching behaviour can be simply described by:

$$\tau(N) = \tau(rad)/[1+(9/2\pi)(N/N_0)^2] \tag{2}$$

Where:
- τ(rad) is the measured radiative lifetime at weak concentration,
-*N* the ion doping concentration,
-*N*o is the doping concentration corresponding to the critical distance R_0 for which the non-radiative energy transfer is as probable as photon emission:

$$R_0 = (3/4\pi N_0)^{1/3}$$

In case of photon trapping be present, Eq. (1) has to be multiply by Eq. (2):

$$\tau\, trapping = \tau(rad)(1+\sigma Nl)$$

Where :
- l is the average absorption length,
- σ is the transition cross-section (3)

Then the general formula is:

$$\tau(N) = \frac{\tau(rad)(1+\sigma Nl)}{1+(9/2\pi)(N/N_0)^2} \tag{4}$$

This model fits well experimental data as can be seen in Figure 6 for Yb^{3+}: CaF_2 grown by LHPG with:
τ(rad) = 2.05 ms, $\sigma l = 1.7 \times 10^{-22}$ cm^3 (0.041 cm^3/%) and $N_0 = 7.47\times10^{21}$ cm^{-3} (32 %).
For Yb^{3+}: CaF_2 grown by simple melting we found:
τ(rad) = 2.05 ms, $\sigma l = 2.9\times10^{-22}$ cm^3 (0.071 cm^3/%), $N_0 = 7.47\times10^{21}$ cm^{-3} (32 %).

In the case of the self-quenching by fast diffusion process [16] the model does not fit experimental data.

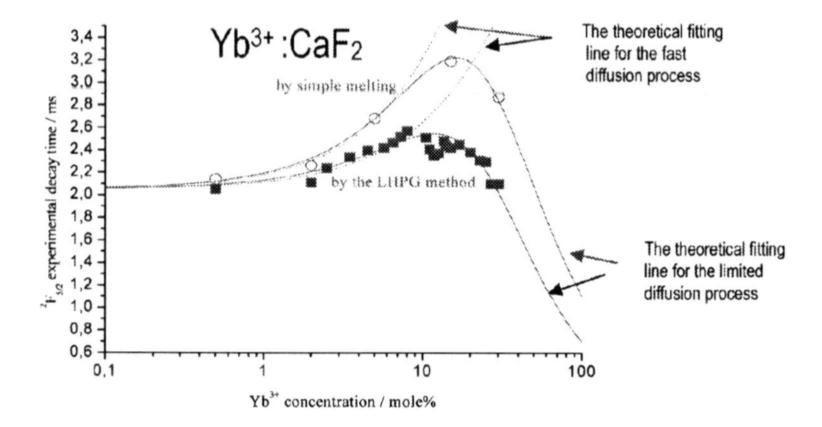

Figure 6. The best fits of theoretical curves for limited diffusion process (continuous lines), and for fast diffusion process (broken lines)

6. Proposition of the Laser Crystal Optimization

6.1. NEW MODELING

As an application, a simple quantitative method for optimizing the gain material concentration for amplifiers and lasers is proposed and performed. Since we now have continuous reliable mathematical curves for self-quenching, it is possible to determine in a simple and unambiguous way the material optimum concentration for its active optical use. From steady state rate equation, the material gain is simply given by:

$$G = \exp[\sigma_g \sigma_a N\tau\,(N)\,l] \tag{4}$$

where σ_g is the gain cross-section taking care of the quasi three-level situation for laser between first excited and ground state, σ_a is the pump absorption cross-section for the pumping wavelength, N is the chemical concentration of active ions, $\tau\,(N)$ is the ions excited state lifetime at the considered concentration N; l is the amplification length. From Equation (4), the product $\tau\,(N)N$, can be optimized easily.

Since this maximum value is unique, we propose to consider such maximum value in cm^{-3} as an absolute scale for self-quenching characterizing any given host-doping couple.

Interestingly, it is verified that the optimum concentration for gain, N_m, is found equal to $0.83N_0$. Then the critical concentration itself, which from Eq. (4) can be simply defined as the concentration reducing τ_w to $0.41\tau_w$ is a good indication of the self-quenching magnitude and can also easily provide the optimum concentration. This simply comes from the fact that the derivative:

$$\frac{d(N\tau/N_0)}{dN} = \frac{[1 - 9(N/N_0)^2/2\pi}{[1 + (9/2\pi)(N/N^0)^2]^2}$$

is zero for :

$$N/N_0 \quad 2\pi/9 = 0.83.$$

It is plotted, as an example, in Figure 5 for Yb^{3+} in CaF_2. The optimized concentration for gain is found relatively high: $6.39\ .10^{21}$ cm^{-3} (26.5 mole %). We had found 5.47 × 1021 cm^{-3} (35 mole %) for KY3F10 host and $1.6 \times 10^{21} cm^{-3}$ (6 mole %) in Y_2O_3 crystalline fibres.

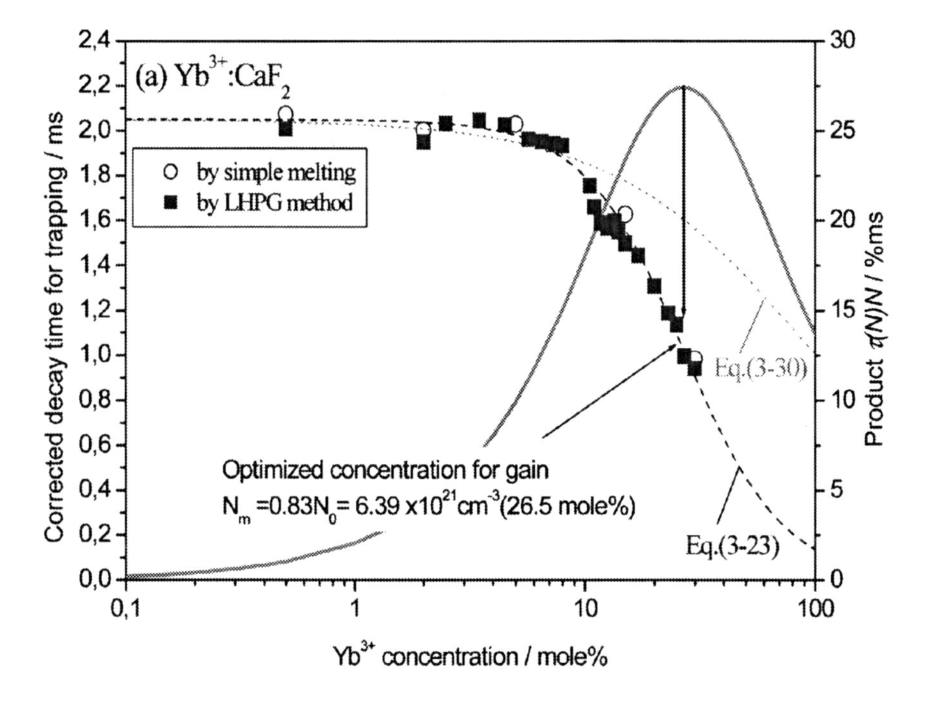

Figure 7. Theoretical approach of the laser optimization in Yb^{3+}: CaF2. Eq. (3-23) and Eq.(3-30) are connected with reference [29].

6.2. GAIN CROSS-SECTION

Usually we can evaluate gain and absorption in doped-crystals depending on the β parameter which is between 0 and 1.

$$\sigma_{gain}(\lambda) = \beta\, \sigma_{emission}(\lambda) - (1-\beta)\, \sigma_{absorption}(\lambda)$$

At a glance this is clear that cubic Yb^{3+}: CaF_2 is much more broader than cubic Yb^{3+}: KY_3F_{10} as a result of the disordered character of the Ca^{2+} sites in CaF_2 crystal with respect of KY_3F10 in which Yb^{3+} is substituted in trivalent Y^{3+} site.

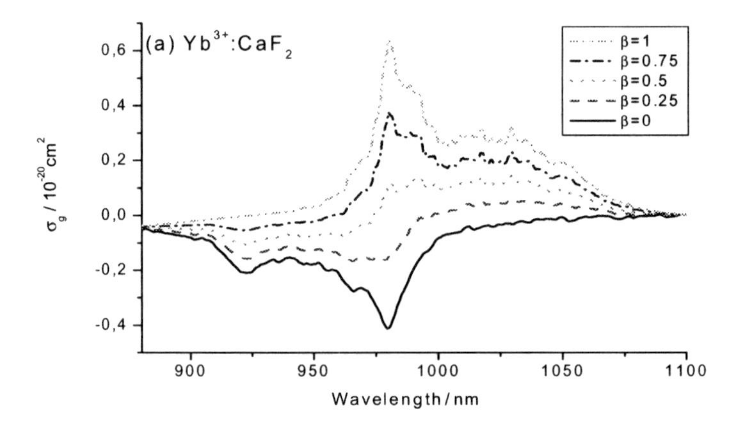

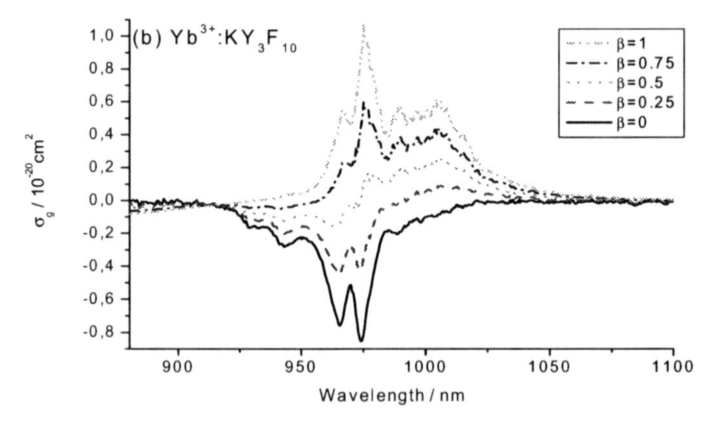

Figure 8. Comparison between $\sigma_{gain}(\lambda)$ and $\sigma_{emission}(\lambda)$ for different values of the parameter β between 0 and 1 in the two cubic crystals CaF_2 and KY_3F_{10}.

In Yb^{3+}: CaF_2 cross-sections are:
$\sigma_{absorption}$:2.0 x 10-21 cm2 at 920 nm and $\sigma_{emission}$:3.0 x 10-21 cm2 at 1030 nm

6.3. LASER TESTS UNDER DIODE PUMPING

Previous results on Yb^{3+}:CaF_2 laser emission and the tuneability have been recently shown under Ti-doped sapphire laser source [30]. Figure 9 shows our first output laser emission under high power diode pumping on Czochralski 5%Yb^{3+}:CaF_2 sample grown by the Czochralski technique at Tohoku university. The main characteristics of the pump beam are: Laser diode:975 nm; maximum output power:15 W; waist:220 μm. The observed threshold is 0.75 W and the efficiency is 22.8% without AR coating. Such results should be improved with selective coatings of samples.

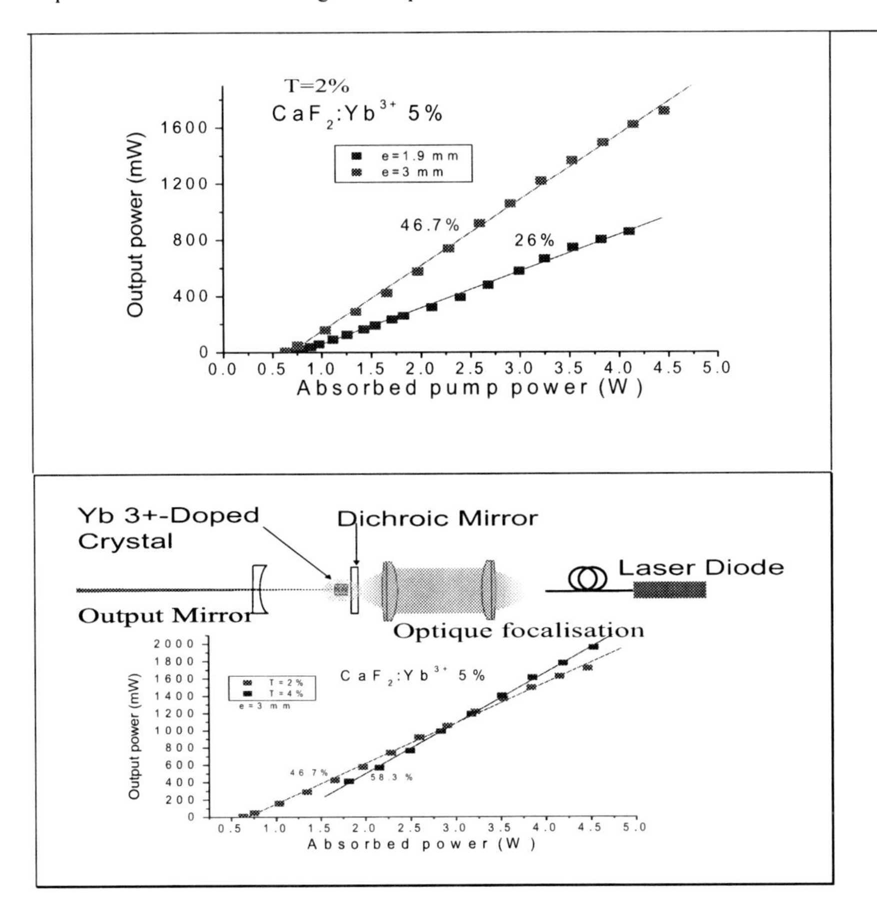

Figure 9. Output power (in mW) with respect to absorbed pump power (in W) as observed in CZ samples with different thicknesses "e" and different transmission factors "T".

7. Thermal Conductivity Measurements

7.1. GENERAL METHOD TO MEASURE THERMAL CONDUCTIVITY

Thermal properties are very important for the choice of new laser crystals. Usually, we characterize thermal conductivity Ḱ (in J cm-1 K-1) or (in W m-1 K-1) by the product of:

Density (g cm-3) X Heat capacity (J g-1 K-1) X Thermal Diffusivity (cm2 s-1)

The new technique used in this work [31] is based on optical parameter variations induced by temperature changes. In the set-up, thermal waves are produced by absorption of a modulated argon laser beam focalized on the sample surface. However, oxide materials do not absorb photons at the excitation wavelength (λ_p = 514.5 nm) then a gold coating (about 150 nm thick) is used and a chromium layer (about 50 nm thick) is inserted between the gold layer and the substrate in order to have good adhesion. In this work, all the compounds were coated together to have exactly the same layers thickness. With a laser diode probe beam (λ_{ld} = 670 nm) one can scan about 30 μm around the pump beam. The amplitude and phase shift of the reflected beam were measured and temperature profile along thermal wave allows to determine the thermal diffusivity of the substrate. The experimental set-up is shown in Fig. 10.

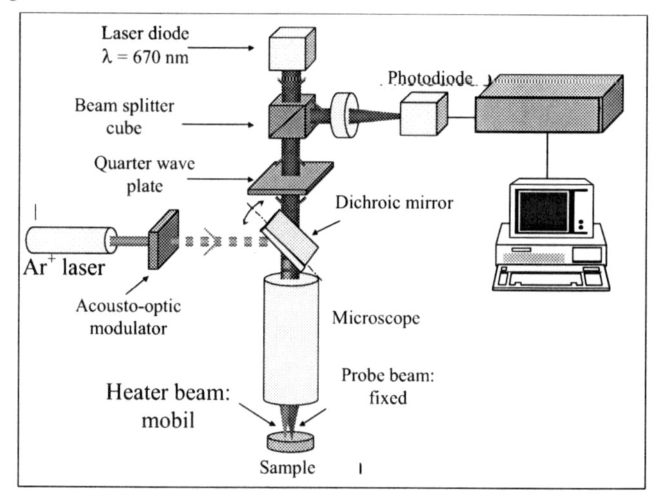

Figure 10. Photoreflectance microscope scheme used to determine thermal diffusivity of samples.

From phase shift and amplitude of reflected beam, the thermal diffusion length is deduced knowing the modulation frequency ω. The temperature profile is:

$$T(r) = \frac{T_0}{r}\, e^{-r/\mu} \cos(\omega t - r/\mu)$$

with T, the temperature at a distance r from the absorption point of the pump beam and μ the thermal diffusion length:

$$\mu = \sqrt{\frac{2D}{\omega}}$$

D is the thermal diffusivity.
This indicates that when the frequency increases, the thermal diffusion length decreases. So, for too high frequency, the scanned zone is too reduced for good measurements because of weak signal on noise ratio. Figure 11 presents simulated amplitude and phase curves for different thermal conductivity values at 100 kHz.

Following, the equation of thermal diffusion corresponds to the second Fourier law:

$$\frac{\partial T}{\partial t} - D\nabla^2 T = \frac{H}{\rho C}$$

where ρ is the density; C is the thermal capacity; D is the thermal diffusivity tensor and H is the modulated excitation of the laser source on the sample surface (in $W.cm^{-3}$). This equation has simple solutions in the Fourier space when the substrate is heated with periodic waves of pulsation ω:

$$T(z, \omega) = T_1 e^{-\sigma z} + T_2 e^{+\sigma z}$$

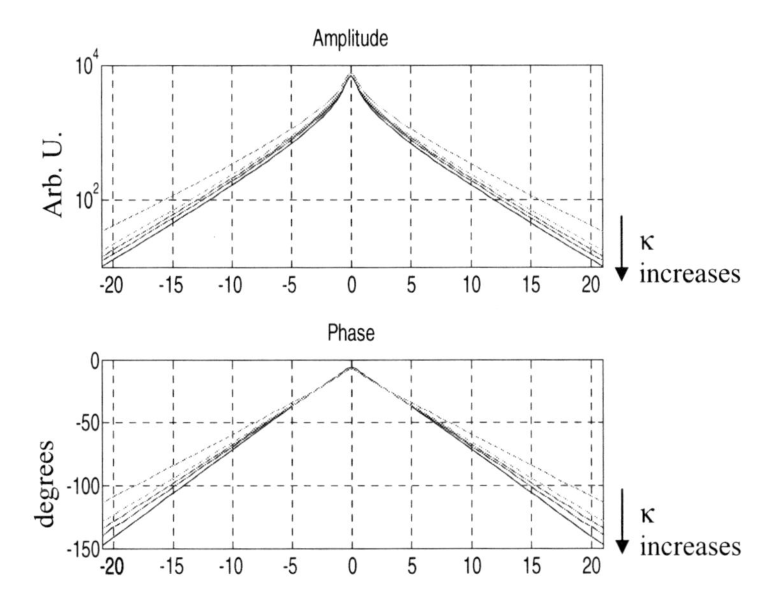

Figure 11. Simulated amplitude and phase curves for different thermal conductivity values (3 ; 6 ; 7.5 ; 9 ; 12 $W.m^{-1}.K^{-1}$) with a constant ω /D = 3 x 106 (USI) ratio. The frequency of the modulated beam is 100 kHz.

with T_1 and T_2 depending on boundary conditions and σ is the complex constant of propagation called thermal wave number:

$$\sigma = \sqrt{\frac{\omega}{2D}(1+i)}$$

As undoped YAG is a well known compound, we have used it as a standard. A piece of undoped YAG crystal was coated with the other crystals in order to have the same gold and chromium layers. Many κ values were given in the literature from 10.5 [33] to 13 [34] $W.m^{-1}.K^{-1}$. A value of 10.7 $W.m^{-1}.K^{-1}$ is chosen in this work and corresponds to the best simulation.

7.2. THERMAL EXPANSION COEFFICIENT MEASUREMENTS

Thermal expansion coefficients α were calculated from X ray diffraction data at different temperatures between room temperature and 1100°C using the equation of linear thermal expansion coefficient :

$$\alpha_l = \frac{1}{l}\frac{\partial l}{\partial T}$$

For CaF_2, the linear thermal expansion coefficients α is $18.7 x 10^{-6} K^{-1}$. For comparison, the values are respectively 6.7 and $8 x 10^{-6} K^{-1}$ for YAG and GGG hosts respectively.

7.3. HEAT CAPACITY MEASUREMENTS

Heat capacities were determined by Differential Scanning Calorimetry using a METTLER TOLEDO STARe equipment (DSC822e) with α-Al_2O_3 as standard. The temperature range varies between 0°C to 50°C. So measured values correspond to room temperature heat capacities.

7.4. THERMAL DIFFUSIVITY, THERMAL CONDUCTIVITY AND HEAT CAPACITY OF CaF2

Figure 12 presents the thermal conductivity measurements for Yb^{3+}-doped CaF_2 CaF_2 is well known for its good thermal properties. The thermal conductivity of undoped compound is around 10 $W.m^{-1}.K^{-1}$. This is explained by the simplicity of its formula [35]. For laser experiments, the optimal Yb^{3+} concentration should be around 5%. Then in the following Table are reported the values for three doping rates: 0.5%, 5% and 8.9% as compared with other Yb^{3+} -doped laser hosts.

Figure 12 presents the thermal conductivity of fluorite compounds versus absorption coefficient. This is a more accurate presentation than in %Yb as the absorption and the dopant in % vary according to laser hosts. It appears that κ decreases as quickly as in YAG because of weak calcium mass (M_{Ca} = 40.1 g.mol^{-1}) and high ytterbium mass. Moreover, substituting Ca^{2+} by Yb^{3+} involves charge compensations. Then vacancies and interstitial fluorides can interact with phonon diffusion and consequently affect thermal conductivity. For the 5%doped-CaF_2, κis equal to 5.2 W.m^{-1}.K^{-1} which is a value quite similar to 5%Yb^{3+}doped-YAG (5.7 W.m^{-1}.K^{-1})

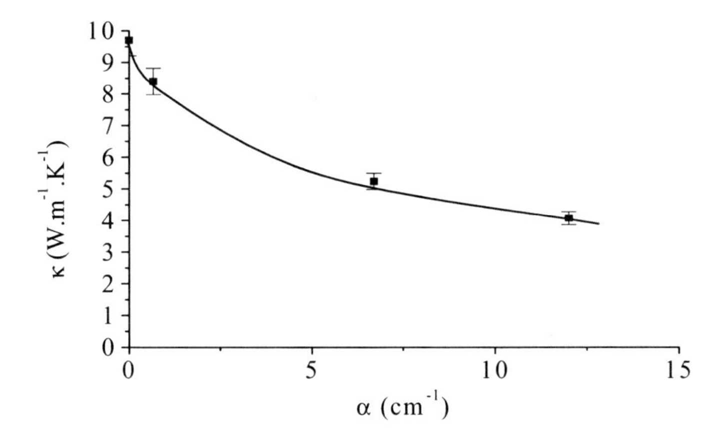

Figure 12. Influence of Yb^{3+}-doping rate converted here in absorption coefficient on thermal conductivity of CaF_2.

Finally, in CaF_2, as the thermal expansion value is important' thus, the κ/αratio which represents the thermal shock parameter is quite small, about 3 times smaller than the Yb:YAG value.

Table 2

Compound	Doping rate	D (10^{-6} m^2.s^{-1}) +/–5%	K (W.m^{-1}.K^{-1}) +/–5%	Cp (J.kg^{-1}.K^{-1})	ρ (kg.m^{-3})
CaF_2	8.9% Yb^{3+}	1.5	4.1	859	3180
	5%Yb^{3+}	1.9	5.2		
	0.5%Yb^{3+}	3.0	8.4		

8. Conclusion

We have chosen to comment detailed analysis of Yb^{3+}-doped cubic CaF_2 fluoride as typical example of Yb^{3+}-doped host crystals which is considered among important high average power solid-state lasers under high power laser diode pumping due to unusual combination of favourable properties. A combinatorial chemistry approach has been applied on concentration gradient crystal fibres which were grown by the LHPG method. Gradient concentration fibres are really unique tool to make the correlation between Yb^{3+} concentration and lifetime measurements which have been measured *in situ* of the same sample in relation to the distance from the top of the crystallized rod. This fast and simple combinatorial method allows us to measure the Yb^{3+} intrinsic radiative lifetime which is one of the basic parameters to know for laser emission. Then, this is possible to measure stimulated emission cross-section and to analyze both, the influence of the radiation trapping by the presence of resonant transition between Stark levels of Yb^{3+} $^2F_{5/2} \leftrightarrow {}^2F_{7/2}$ near 980 nm and the non-radiative energy transfer to uncontrolled rare earth impurities in doped hosts giving a notable contribution to the quenching processes since hydroxyl groups are absent in the fluoride host. Finally, a limited diffusion process model within the doping ion subsystem towards impurities analogous to the doping ions themselves has been given to interpret such mechanisms and a new approach of the laser crystal optimization has been proposed. Important parameters found for laser application are the Yb^{3+} radiative lifetime close to 2.05 ms and the critical concentration $N_0=7.47\times10^{21}$ cm^{-3} (32 mole %), optimum concentration for gain $N_m= 6.39\times10^{21}$ cm^{-3} (26.5 mole %), respectively. Large N_0 and N_m values, several times bigger than oxides, is agreeable property for laser application as tuneable solid-state laser and luminescent applications in general. Such preliminary optical and also thermal analysis allows us to ask for the growth of optimized Czochralski bulky laser samples.

Acknowledgements

This chapter is the result of the fruitful international co-operation between different researchers belonging to the 3 following laboratories:

1. M. Ito, A. Bensalah, Y. Guyot, C. Goutaudier , K. Lebbou, A. Brenier, G. Boulon Physical Chemistry of Luminescent Materials, Claude Bernard /Lyon1 University, UMR 5620 CNRS, 69622 Villeurbanne, France
2. J. Petit, P. Goldner, B. Viana, D. Vivien, D. Fournier LCAES-ENSCP, UMR 7574 CNRS, Paris 6 University, 11 rue Pierre & Marie Curie 75231 Paris, France
3. A. Jouini, H. Sato, A. Yoshikawa, T. Fukuda IMRAM, Tohoku University, 2-1-1 Katahira, Aoba-ku, Sendai 980-8577, Japan

References

[1] A. Brenier, J. of Luminescence 92 (2001) 199
[2] A. Brenier, G. Boulon, Europhysics Letters 55 (2001) 647
[3] A. Brenier, G. Boulon, J. Alloys and Compounds 210 (2001) 323-324
[4] G. L. Bourdet, Optics Communications 198 (2001) 411
[5] G. Boulon, Combinatorial chemistry to grow single crystals and analysis of concentration quenching processes: application to Yb^{3+}-doped laser crystals. "Frontiers of Optical Spectroscopy". NATO Science series. Ed.B.Di Bartolo, Kluwer Academic Publishers-Dordrecht, The Netherlands (2005) 689-714
[6] B. P. Sobolev, P. P. Fedorov, J. Less-Common Met. 60 (1978) 33
[7] J. A. Campbell, J.-P. Laval, M.-T. Fernandez-Diaz, M. Foster, J. Alloys and Compounds 323-324 (2001) 111
[8] N. B. Grigor'eva, L. P. Otroshchenko, B.A. Maksimov, I. A. Verin, B. P. Sobolev, V. I. Simonov, Crystallography Reports 41 (1996) 607
[9] J. P. Lavel, A. Abaouz, B. Frit, A. Lebail, J. Solid State Chim. 85 (1990) 133
[10] J. P. Lavel, A. Abaouz, B. Frit, J. Solid State Chim. 81 (1989) 271
[11] Sharon E. Ness, D. J. M. Bevan, H. J. Rossell, Eur. J. Solid State Inorg. Chem. 25 (1988) 509
[12] D. J. M. Bevan, Maxine J. McCall, Sharon E. Ness, Max. R. Taylor, Eur. J. Solid State Inorg. Chem. 25 (1988) 517
[13] M. Bouffard, J. P. Jouart, M.-F. Joubert, Opt. Mater. 14 (2000) 73
[14] J. Kirton, S. D. McLaughlan, Physical Review 155 (1967) 279
[15] J. Kirton, A. M. White, Physical Review 178 (1969) 543
[16] M. Ito, C. Goutaudier, Y. Guyot, K. Lebbou, T. Fukuda, G. Boulon, J.Phys.: Condens. Matter 16 (2004) 1501-1521
[17] R.S. Feigelson, J. Crystal Growth, 79 (1986) 669-680
[18] M.T. Cohen-Adad, L. Laversenne, M. Gharbi, C. Goutaudier, G. Boulon, R. Cohen-Adad, J. Phase Equil., 22 (2001) 379-385
[19] L. Laversenne, C. Goutaudier, Y. Guyot, M. Th. Cohen-Adad, G. Boulon, J. Alloys Compounds, 341 (2002) 214
[20] G. Boulon, A. Brenier, L. Laversenne, Y. Guyot, C. Goutaudier, M.T. Cohen-Adad, G. Métrat, N. Muhlstein, J. Alloys compounds, 341 (2002) 2-7
[21] G. Boulon, L. Laversenne, C. Goutaudier, Y. Guyot, M.T. Cohen-Adad, J. of Lum., 102- 103(2003) 417-425
[22] A. Yoshikawa, G. Boulon, L. Laversenne, K. Lebbou, A. Collombet, Y. Guyot, T. Fukuda, J. Appl. Phys. : Cond. Mater. 94 (2003) 5479- 5488
[23] A. Bensalah, M. Ito, Y. Guyot, C. Goutaudier, A. Jouini, A. Brenier, H. Sato, T. Fukuda, G. Boulon, J. of Luminescence (accepted on September 2005)
[24] D. S. Sumita and T. Y. Fan, Optics Letters 19 (1994) 1343-1345
[25] N. Uehara, K. Ueda, and Y. Kubota, Jpn. J. Appl. Phys. 35 (1996) L499
[26] M. P. Hehlen, in OSA TOPS on Advanced Solid-State Lasers, S. A. Payne and C. Pollock, Eds. 1 (1996) 530
[27] H. P. Christensen, D. R. Gabbe, and H. P. Jenssen, Phys. Rev. B 25 (1982) 1467
[28] S. M. Kaczmarek, T. Tsuboi, M. Ito, G. Boulon, G. Leniec J.Phys.: Condens. Matter 17 (2005) 3771-3786

[29] F. Auzel, F. Bonfigli, S. Gagliari, G. Baldacchini, J.of Lumin., 94-95, (2001) 293-297
[30] F. Auzel, G. Baldacchini, L. Laversenne, G. Boulon, Optical Materials 24 (2003) 103-109
[31] A. Lucca, M. Jacquemet, F. Druon, F. Balembois, P. Georges, P. camy, J.L. Doualan, R. Moncorgé, Optics Letters 29, (2004)1879-1881
[32] J. Petit, Ph. Goldner, B. Viana, J.P Roger and D. Fournier, submitted to IEEE Journal of Quantum Electronics on October 2005
[33] A. Tünermann, H. Zellmer, W. Schöne, A.Giesen, K. Contag, Topics Appl. Phys. **78**, 369-408 (2001)
[34] L. Fornasiero, E. Mix, V. Peters, K. Petermann, G. Huber, Cryst. Res. Technol., 34 255- 260 (1999)
[35] R. Gaume, Annales de Chimie Science des Matériaux, 28 (2003) 88

TERAHERTZ SENSING AND MEASURING SYSTEMS

JOHN W. BOWEN
Cybernetics
School of Systems Engineering
The University of Reading
Whiteknights, Reading
RG6 4EU, United Kingdom
cybjb@cyber.reading.ac.uk,

1. Introduction

The terahertz (THz) region of the electromagnetic spectrum extends from 100 GHz to 10 THz, a wavelength range of 3 mm to 30 μm. While it is the last part of the electromagnetic spectrum to be fully explored, systems operating in this frequency range have a multitude of applications, in areas ranging from astronomy and atmospheric sciences to medical imaging and DNA sequencing.

As the terahertz range lies between the microwave and infrared parts of the spectrum, techniques from both of these neighbouring regions may be extended, specially adapted and hybridised for the generation, detection and analysis of terahertz radiation. However, operation at these frequencies has traditionally been difficult and expensive because of the low power available from sources and the precision machining required in their fabrication. Recently, new techniques based on the generation and detection of terahertz radiation using ultra-fast pulsed lasers have been developed and these have led to exciting advances in terahertz imaging and spectroscopy. The output frequencies of quantum cascade lasers have also been steadily moving down into the terahertz range and may lead to compact solid-state terahertz sources in the near future.

This paper will cover techniques for the generation, detection and analysis of terahertz radiation. After a discussion of quasi-optical techniques for the design of terahertz systems, the operation of three exemplar terahertz systems will be explained. Following a summary of terahertz sources and detectors, an overview of the operation of terahertz systems for a wide range of sensing and measuring applications will be given.

2. Waveguides *versus* Quasi-optics

At microwave frequencies, systems are often based around hollow metal pipe waveguides, typically of rectangular or circular cross-section, which provide a means of controlled propagation of the radiation. This approach may be carried over into the terahertz region, although,

B. Di Bartolo and O. Forte (eds.), Advances in Spectroscopy for Lasers and Sensing, 103–118.

as the frequency increases, the skin depth in metals decreases and obtaining a surface finish on the inside walls of the waveguide sufficient to keep propagation loss at an acceptable level becomes increasingly difficult. Exacerbating this is the fact that the cross-sectional dimensions of a single-moded (i.e. dispersionless) waveguide must be reduced as the frequency is increased. This makes hollow metal waveguide difficult and expensive to manufacture for the terahertz region using conventional machining techniques. Terahertz hollow metal waveguide structures fabricated using thick resist photolithographic micro-machining [1] have been demonstrated to frequencies as high as 1.6 THz, but the fabrication technique is still in its infancy. The use of other types of transmission line have been explored, for example dielectric waveguide and planar transmission lines such as microstrip, but losses confine their use to the lower frequency end of the terahertz range, most having an upper usable frequency of about 150 GHz.

As an alternative, many systems operating at terahertz frequencies make use of optical components, such as lenses and reflectors, to control and manipulate beams travelling through free-space. The advantages of this approach in comparison to waveguide-based systems include: lower loss, wider (multi-octave) bandwidth, easier fabrication and, therefore, lower cost. However, unlike optical systems in the visible part of the spectrum, where optical components typically have lateral dimensions which are tens of thousands times the wavelength, practical size constraints limit the size of terahertz optical components to only a few tens of wavelengths. Therefore, diffraction becomes a significant aspect of propagation and must be taken into account in the design of optical systems, particularly as optical components often have to be used in the transition region between the near- and far-fields. Optical systems operating in this regime, where geometrical optics ceases to hold, are termed quasi-optical.

Some of the sources and detectors used at terahertz frequencies are small compared to the wavelength and so, in order to efficiently couple power between them and a well directed free-space beam, it is necessary to use an antenna. Often, the best performance is achieved when the source or detector is mounted inside a hollow metal waveguide and a horn antenna is used to launch a beam through a quasi-optical system. Therefore, while quasi-optical propagation is to be preferred over long runs of waveguide, it is quite common for systems to include some short lengths of waveguide as well as quasi-optics.

3. Quasi-optics

This section outlines the most important points in the design and analysis of terahertz quasi-optical systems. More detailed treatments can be found in references [2,3,4].

3.1. BEAM-MODES

One convenient way to describe a diffractively spreading paraxial beam is as a superposition of a set of beam-modes, each of which maintains a characteristic form as it propagates. The field in any cross-section through the beam will then be given by the superposition of the beam-modes in that plane, taking account of their relative amplitudes and phases. The beam-modes are based around Hermite-Gaussian or Laguerre-Gaussian functions and are known as Hermite-Gaussian and Laguerre-Gaussian beam-modes, respectively. In both cases, the lowest order or fundamental beam-mode has the form

$$\psi_{00}(r,z)=\left(\frac{2}{\pi}\right)^{\frac{1}{2}}\frac{1}{w}\cdot\exp\left(\frac{-r^2}{w^2}\right)\cdot\exp\left(\frac{-ikr^2}{2R}\right)\cdot\exp i\Theta\cdot\exp -ikz \tag{1}$$

In the above, the beam is propagating in the z direction and has a wavevector magnitude $k = \omega/c = 2\pi/\lambda$. The fundamental beam-mode is axially symmetric, with $r = (x^2 + y^2)^{1/2}$. The quantity $\psi(r,z) \equiv \psi(x,y,z)$ represents any one Cartesian component of the electromagnetic fields $\mathbf{E}(x,y,z)$ or $\mathbf{H}(x,y,z)$, e.g. $E_x(x,y,z)$. The subscripts on ψ are mode numbers: 00 in this case to indicate the fundamental beam-mode. The fields all vary with time as exp $i\omega t$, but this factor is omitted here.

The first exponential term in equation (1) indicates that the beam-mode has a Gaussian transverse amplitude distribution, with a half-width of w at the $1/e$ amplitude level (Figure 1).

The second exponential term indicates that the beam-mode has a spherical phase-front with a radius of curvature R.

The third exponential term indicates an on-axis phase slippage of Θ relative to an on-axis plane-wave exp-ikz.

The variation of the parameters w, R and Θ with z are given by the following equations:

$$\hat{w}^2 = \hat{w}_0^2\left(1+\hat{z}^2\right) \tag{2}$$

$$\hat{R} = \hat{z} + \hat{z}^{-1} \tag{3}$$

$$\Theta = \tan^{-1}\hat{z} + \Theta_0 = \sin^{-1}\left(1+\hat{z}^{-2}\right)^{-1/2} + \Theta_0 \tag{4}$$

where ˆ indicates division by $kw_0^2/2$ and $\hat{z} = 2(z - z_0)/kw_0^2$. In the above, w_0, z_0 and Θ_0 are constants for a given set of beam-modes.

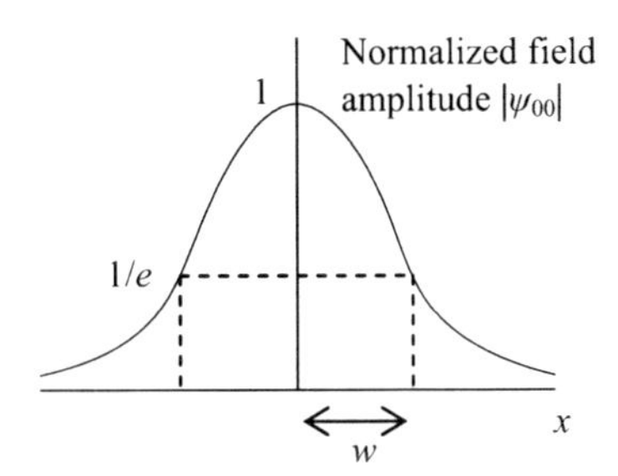

Figure 1. Transverse amplitude distribution of fundamental beam-mode.

The changing form of the beam-mode along the length of the beam is illustrated in Figure 2. At all z distances, it maintains its Gaussian transverse amplitude distribution, but this takes a minimum width w_0 at the beam-waist, which is located at z_0. Away from the beam-waist, the beam spreads hyperbolically, asymptotically reaching the far-field spread angle $\theta_0 = \tan^{-1}(2/kw_0)$. The normalization factor pre-multiplying the exponential terms in equation (1) ensures that the power remains constant, at unity, at all z. The radius of curvature of the phase-front changes with z, being infinite at the beam-waist and taking its minimum value at $\pm\hat{z} = 1$. In the far-field, the centre of curvature moves in towards the beam-waist.

As mentioned above, a paraxial beam of arbitrary form can be described by a superposition of beam-modes of differing order. The higher order beam-modes are obtained by pre-multiplying the fundamental beam-mode by a factor involving both a mode-dependent phase-slippage relative to the fundamental and a modulation of the Gaussian amplitude distribution by Hermite or Laguerre polynomials. For example, the general expression for a Hermite-Gaussian beam-mode with mode numbers m and n is

$$\psi_{mn}(x, y, z) = \left(2^{m+n-1}\pi m!\, n!\right)^{-1/2} \frac{1}{w} \cdot \left\{ H_m\left(\sqrt{2}\frac{x}{w}\right) \cdot H_n\left(\sqrt{2}\frac{y}{w}\right) \cdot \exp - \frac{x^2 + y^2}{w^2} \right\} \cdot \left\{ \exp - \frac{ik\left(x^2 + y^2\right)}{2R} \right\} \left\{ \exp i(m + n + 1)\Theta \right\} \cdot \exp ikz \tag{5}$$

where $H_N(X)$ is the Hermite polynomial of X of mode or order number N. An arbitrary beam can then be described by the superposition of beam-modes

$$\psi(x, y, z) = \sum_{mn} C_{mn} \psi_{mn}(x, y, z) \tag{6}$$

where the complex weighting coefficients C_{mn} can be found by carrying out the following integral over the field in any cross-sectional plane $z = z_S$ through the beam:

$$C_{mn} = \iint_{z_S} \psi^*_{mn}(x, y; z_S)\, \psi(x, y; z_S)\, dxdy \tag{7}$$

In carrying out this integral, w, R and Θ can be chosen for maximum simplicity and computational economy; each choice giving a different set of C_{mn}. Thus, if the form of the field in the aperture plane of an antenna, say, is known, the beam-mode composition of the beam launched by the antenna can be determined. For example, about 98% of the power launched by a corrugated feed horn [5] is carried in the fundamental beam-mode, and so this type of antenna has found widespread use in coupling active devices to free-space beams in terahertz systems.

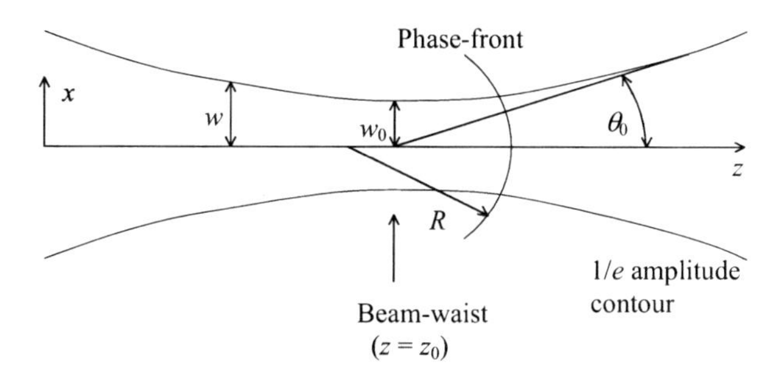

Figure 2. Changing form of the fundamental beam-mode along the beam axis.

3.2 BEAM CONTROL

As indicated in Figure 2, beams in terahertz optical systems spread under diffraction and so it is necessary to introduce beam control components, such as lenses or reflectors, to periodically refocus the beam to a new beam-waist so that it can be contained within a maximum width as it propagates through a terahertz system. The focussing action of an ideal thin lens or reflector is described by the thin lens equations:

$$d_e - f = \left[\frac{f^2}{(d_i - f)^2 + \left(kw_{0i}^2/2\right)^2}\right](d_i - f) \tag{8}$$

$$w_{0e}^2 = \left[\frac{f^2}{(d_i - f)^2 + \left(kw_{0i}^2/2\right)^2}\right]w_{0i}^2 \tag{9}$$

$$\tan(\Theta_e - \Theta_i) = \frac{-kw_{0i}^2/2}{d_i - f} = \frac{-kw_{0e}^2/2}{d_e - f} \tag{10}$$

where the subscripts i and e relate to the incident and emergent beams respectively, f is the focal length of the optical element and d is the distance between the beam-waist and the plane of the optical element.

There are certain special choices of the input and output planes for an optical system for which the beam transformation is a Fourier transformation [2]. This can be used to good effect in systems design, as it can circumvent the need to carry out a beam-mode decomposition.

There are some special cases of particular importance. Notably, if the beam-waist of the incident beam lies in the front focal plane of a lens (so that $d_i = f$), the output beam-waist will lie in its back focal plane and will have a size $w_{0e} = 2f/(kw_{0i})$. If two lenses are separated by the sums of their focal lengths and the input beam-waist is in the front focal plane of the first lens, the output beam-waist will be formed in the back focal plane of the second lens and will have a size $w_{0e} = (f_2/f_1)w_{0i}$. Note that this latter case is frequency independent and so is important in the design of wide-band systems.

There is a maximum distance from a lens at which a beam-waist may be formed, regardless of the focal length of the lens. If the width of the beam at the lens is w_L, this distance is $d_{max} = kw_L^2/4$ and the beam-waist formed at this distance will have a width $w_{0m} = w_L/\sqrt{2}$. This places constraints on the maximum distance between optical elements if the width of the beam is to be maintained within a given size and on the degree of collimation achievable.

In order to avoid loss of power and the generation of higher order beam-modes, the lens and reflector apertures in quasi-optical systems have to be large enough to avoid significant truncation of the beam. It can be shown that as long as the apertures in the system are at least $3w$ in diameter, about 99% of the power in an incident fundamental beam-mode will be transferred into an on-going fundamental beam-mode.

The lenses used in terahertz systems are usually made from high density polyethylene (HDP), which has a refractive index of 1.52 and a relatively low loss through the terahertz range. As a real lens is of finite thickness, the phase-front curvature and phase slippage will change over its thickness and this must be taken into account in its design. The lens profiles on each side of the lens will be aspheric and are designed to equalise the phase elapsed along all equal amplitude contours in the beam between the beam-waist and the central plane of the lens.

In order to overcome absorption and frequency-dependent reflection losses from lenses, many systems use off-axis ellipsoidal or parabolic metallic reflectors. For proper operation, it is necessary to ensure that the distances from the centre of the reflector to the geometrical foci of its parent ellipsoid are equal to the radii of curvature of the phase-fronts of the incident and emergent beams [6]. A single reflector can introduce aberrations through amplitude distortion; however it is possible to use a second mirror to cancel the aberrations introduced by the first, as long as the two reflectors have the correct relative orientation [7].

3.3 WIRE GRID POLARISING BEAM-SPLITTERS

The wire grid polarising beam-splitter is a key component in many terahertz optical systems, operating with close to 100% efficiency over a multi-octave range. Its operation is illustrated schematically in Figure 3. It consists of an array of free-standing wires, typically 10 μm diameter tungsten wires wound onto a frame at a pitch of 25 μm. The grid resolves an input beam into two orthogonal linearly polarised components: that with its E-vector parallel to the wires is reflected, while that with its E-vector perpendicular to the wires is transmitted. The grid may also be used to combine two orthogonal linearly polarised beams to produce an emergent elliptically polarised beam.

4. Exemplar System 1: Null-balance Transmissometer

Figure 4 illustrates the layout of a null-balance transmissometer [8] designed for high precision measurement of the complex transmission coefficient of material samples at discrete frequencies around 100 GHz. The null-balance technique is immune to source fluctuations and standing wave effects resulting from reflections between the sample and components in the measurement system. However, as it makes use of a coherent source, to measure the transmission coefficient over a range of frequencies, it is necessary to retune the source to each frequency. The spectral resolution is determined by the linewidth of the source.

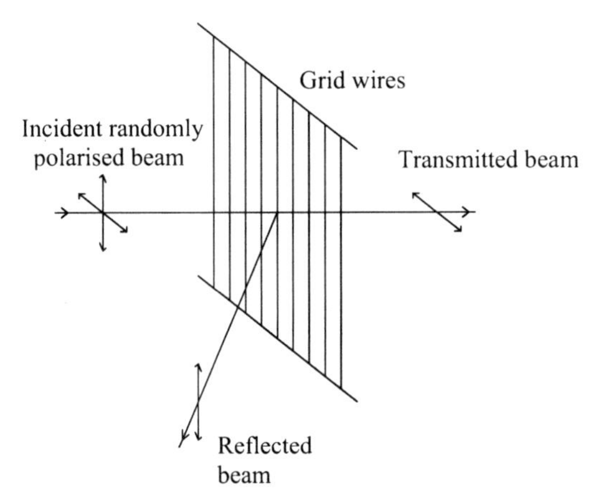

Figure 3. Wire grid polarising beam splitter.

The source is a continuous wave Gunn oscillator, consisting of a Gunn diode in a waveguide mount, delivering around 10 mW. The detector is a waveguide mounted Schottky diode. Both source and detector are coupled to the quasi-optical system via corrugated feed horns. Beam control is provided by 42 mm diameter HDP lenses. Lock-in detection is employed.

The operation of the transmissometer is as follows. Grid G1, with its wires oriented vertically to the plane of the instrument, defines the polarisation of the input beam. Grid G2, with wires at 45°, splits the beam into two orthogonal linearly polarised components. One of these components is transmitted, via two of the plane mirrors M, through the sample and on to grid G3. The second of the component beams travels to grid G3 via a reference grid RG, which can be rotated about the beam axis, and a roof mirror RM that can be translated to change the path difference between the two beams. Thus the amplitude and phase of the beam in this reference arm can be changed by adjusting RG and RM respectively. The two component beams are recombined at grid G3 and travel on to grid G4, which acts as a polarisation analyser. If the amplitude and phase of the beam in the reference arm exactly matches that in the sample arm, the beam incident on grid G4 will be linearly polarised such that it is completely reflected from grid G4 and a null signal will be recorded at the detector. If, on the other hand, there is an imbalance in the amplitude and/or phase of the two beams, there will be a polarisation component that is transmitted through grid G4 and detected by the detector. To determine the transmission coefficient of the sample, RG and RM are adjusted to achieve a null at the detector. The amplitude of the transmission coefficient is determined from the angle of the reference grid between nulls and its phase is determined from the displacement of the roof mirror.

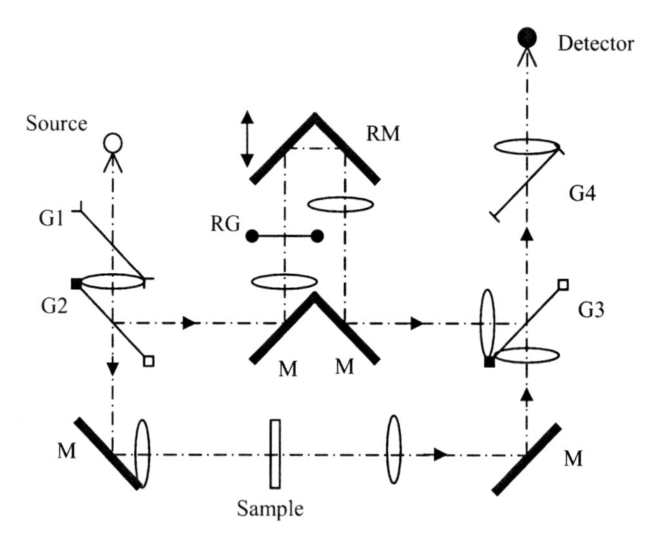

Figure 4. Null-balance transmissometer.

5. Exemplar System 2: Dispersive Fourier Transform Spectrometer

Figure 5 illustrates the layout of a dispersive Fourier transform spectrometer (DTFS) system. This instrument is capable of measuring the complex transmission spectrum, and thus the complex optical constants, of samples between 100 GHz and 1 THz. The major difference in function between this and the transmissometer system discussed in Section 4 is that this system measures at all frequencies within this bandwidth simultaneously, but with a poorer spectral resolution. In this instrument, the spectral resolution is inversely proportional to the maximum path difference that can be achieved between its two arms.

The source is a high pressure mercury arc lamp, which acts as a noise source, i.e. a temporally incoherent source producing power over a broad instantaneous bandwidth. However, the power from the source is low and so the detector is an InSb hot electron bolometer, housed in a cryostat and cooled to 4.2 K with liquid helium.

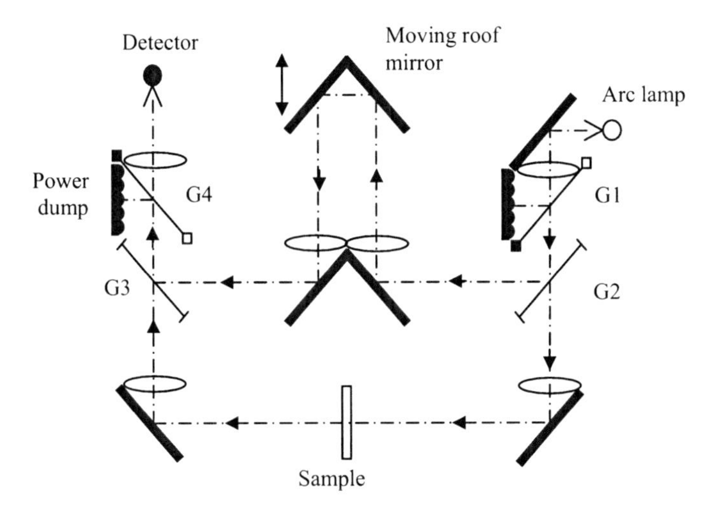

Figure 5. Dispersive Fourier transform spectrometer.

Again, the beam from the source is split into two polarisation components, one of which passes through the sample. The second beam propagates via a roof mirror which is continually moving back and forth to modulate the path difference between the two arms. After recombination at grid G3, the beam has an elliptical polarisation, the ellipticity of which changes as the mirror moves. The grid G4 separates one linear component of this polarisation which falls on the detector.

Consider first the situation when the sample is removed. In this case, when the path difference is zero all frequency components will be in phase, the polarisation falling on grid G4 will be purely linear, and a large signal will be seen at the detector. As the roof mirror is moved away from this position the frequency components in the two arms will go in and out of phase with each other and the signal at the detector will fall away rapidly in an oscillatory manner. In fact, the signal, or interferogram, recorded at the detector as a function of path difference is the squared auto-correlation function of the source. Fourier transformation of the interferogram yields the power spectrum of the source, usually referred to as the background spectrum.

If the sample is inserted, the resulting interferogram will be modulated by the sample's amplitude and phase transmission spectrum. If the sample is dispersive, the interferogram will no longer be symmetrical. Fourier transformation of this interferogram and ratioing it to the background spectrum yields the frequency dependent complex insertion loss, which is the factor by which the amplitude and phase of the beam is changed when the sample replaces an equivalent volume of free space. From this, the amplitude and phase transmission coefficients, complex dielectric constant and complex refractive index of the sample may be determined [9].

6. Exemplar System 3: Terahertz Pulsed Imaging System

Femtosecond lasers provide an alternative means to generate and detect broad bandwidth terahertz signals. Before discussing how a pulsed imaging system can be implemented using such techniques, we need to consider the basic generation and detection mechanisms.

6.1 PHOTOCONDUCTIVE GENERATION – THE AUSTON SWITCH

A pulse shorter than 100 fs from a near-infrared (typically 780 nm) femtosecond laser is incident on a piece of photoconductive material, such as GaAs or radiation damaged silicon on sapphire. The radiation is directed towards a gap between two halves of an antenna, such as a dipole, that has been printed as a metallized pattern on the surface of the semiconductor. A bias voltage of about 1 kV cm^{-1} is applied across the gap. When the near-infrared pulse is incident on the semiconductor, carriers are created which are accelerated by the electric field to produce an emitted pulse of terahertz radiation launched by the antenna. As a result of various effects internal to the semiconductor, the emitted terahertz pulse will have a duration somewhat longer than the incident visible pulse, typically about 400 fs, but is still short enough to have a frequency spectrum extending to around 3 THz. The lowest frequencies emitted are governed by antenna and optics considerations, but are usually around 100 GHz.

6.2 OPTOELECTRONIC GENERATION VIA RECTIFICATION

In an alternative approach to the Auston switch, the femtosecond laser pulse is incident on a non-linear crystal, typically ZnTe or GaAs. The component frequencies of the incident pulse mix with each other in the crystal, producing outputs at their difference frequencies, which lie in the terahertz range. Phase matching considerations limit the efficiency and structure the spectrum of the output pulse, but bandwidths extending as high as 35 THz have been achieved using this technique. Unlike the Auston switch, it is generally necessary to use an amplified femtosecond laser pump.

6.3 PHOTOCONDUCTIVE DETECTION

A structure similar to the Auston switch described in Section 6.1 can be used for detection, although in this instance no bias is applied. Again, a femtosecond laser is incident on the photoconductive material, although in this case it acts as a gating or probe signal, making the antenna receptive to incident terahertz radiation just for the duration that photo-excited carriers exist. An incident terahertz pulse overlapping in time with the near-infrared gating pulse will induce a measurable current in the antenna.

6.4 ELECTRO-OPTIC DETECTION VIA THE POCKELS EFFECT

In an alternative gated detection scheme, co-propagating terahertz signal and near-infrared gating pulses are incident on a crystal, such as ZnTe or $LiTa_3$, that exhibits the linear Pockels effect. An incident terahertz pulse electrically polarises the crystal, inducing birefringence at the near-infrared probe wavelength. If there is no incident terahertz radiation, the near-infrared beam leaves the crystal linearly polarised. On passing through a quarter wave plate it is converted to circular polarisation, which is analysed into two components via a Wollaston prism, each being detected by a photodiode. In this case the difference in the photodiode currents will be zero, indicating that there is no incident terahertz signal. If a terahertz signal is present, the probe beam will exit the crystal with an elliptical polarisation. The difference in the resulting photodiode currents is proportional to the amplitude of the terahertz electric field.

6.5 IMAGING SYSTEM

Figure 6 illustrates the layout of a typical terahertz pulsed transmission imaging system. A terahertz pulse is generated, as described above, by illuminating either an Auston switch or an optoelectronic crystal with the pulse from a near-infrared femtosecond laser. The terahertz pulse is transmitted through the sample, via focussing optics, to arrive at the detector. Either of the detection schemes mentioned above may be used. Some of the energy from the near-infrared pulse is split off and sent to the detector as a gating pulse. Indium tin oxide (ITO) glass, which reflects terahertz radiation but transmits near-infrared, can be used as a dichroic to overlay the terahertz and gating pulses. A variable path difference allows the delay between the terahertz and gating pulses to be changed. The transmitted terahertz pulse can be mapped out as a function of time by scanning the optical delay line, each point on the recorded pulse resulting from the average signal generated in the detector by a train of pulses at a given setting of the delay line. Insertion of a sample modifies the pulse amplitude and shape and, by Fourier transforming the pulses recorded with and without a sample *in situ* and ratioing the results, the complex transmission coefficient and complex optical constants of the sample can be measured, as in DFTS. The size of the beam-waist at the sample can be of the order of a millimetre, or a fraction of a millimetre, and so, by raster scanning the sample through the beam, an image can be built up. Each pixel in the image contains spectral information and so a variety of parameters can be used to construct the image, some giving better contrast in certain circumstances than others. For example, possible imaging parameters include: pulse attenuation, pulse delay, pulse width, amplitude or phase delay over a given bandwidth, refractive index, absorption coefficient.

Similar systems can be constructed to detect the terahertz signal reflected from an object. Given penetration of the terahertz radiation inside the object, signals from internal boundaries will be reflected with a time delay which depends on the depth of the boundary, and so this information can be used to reconstruct a three-dimensional image of the object and its internal structure.

The femtosecond time resolution that is available means that it is possible to use the technique to explore rapidly changing phenomena. For example, a further portion of the femtosecond laser pulse may be split off to generate an optical pump beam to allow optical pump – terahertz probe spectroscopy for the exploration of chemical reaction dynamics.

In order to increase the spatial resolution of terahertz pulsed imaging, terahertz near-field microscopy techniques are under development. These depend on the detection of evanescent terahertz fields in the vicinity of sub-wavelength apertures or scanning probe tips. Spatial resolution better than 10 μm has been demonstrated [10].

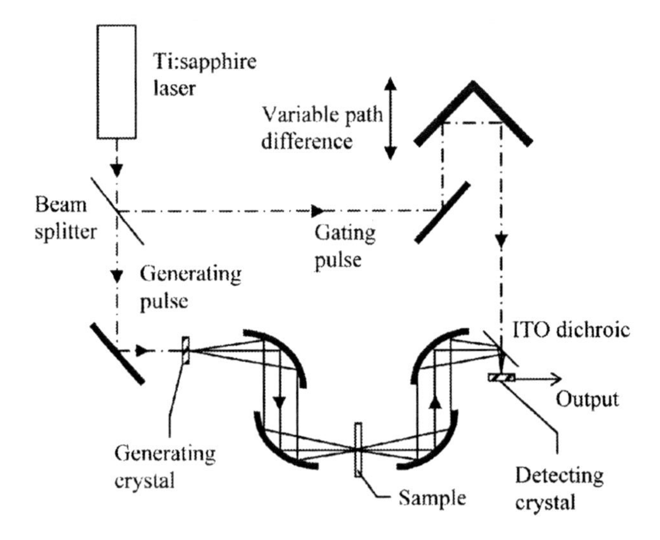

Figure 6 Terahertz pulsed imaging system.

7. Terahertz Sources

Some sources, namely the Gunn oscillator, mercury arc lamp and generation via femtosecond pulses, have been mentioned above; this section will briefly summarise other sources that are commonly available.

Solid-state coherent sources, such as the Gunn oscillator and IMPATT oscillator are attractive because they can be powered from simple laboratory power supplies, deliver useful power levels and can be readily tuned over a range of frequencies. However, their power falls off with frequency and their useful operating range is confined to below 300 GHz. For operation to frequencies beyond 1 THz, they can be used to pump solid-state frequency multipliers, which use varactor diodes as non-linear elements. However, the efficiency of these multipliers is low.

Several types of vacuum tube oscillators can be used. These include reflex klystrons, which can be used up to about 220 GHz, and backward wave oscillators which are available up to about 1 THz. These oscillators are tuneable and deliver higher output powers than solid-state devices but are more expensive and require high voltage power supplies.

Optically pumped far infrared (FIR) lasers can be tuned to deliver over 2000 different frequencies in the 300 GHz – 6 THz range, dependent on the gas used in the FIR laser cavity (which can readily be changed). However, they are relatively large and expensive devices, are more complicated to operate and are less stable in open loop operation than the other oscillators mentioned here.

Terahertz parametric oscillators are recently developed pulsed sources. An example has been demonstrated that is tuneable from 0.7-2.6 THz and has a 1.3 mJ, 4 ns pulse with greater than 200 mW peak power [11].

The operation of quantum cascade lasers is also now being extended from higher frequencies down into the terahertz range, with significant power at frequencies as low as 2.8 THz being reported. However, they currently have to be cooled to 50 K for continuous wave operation (75 K for pulsed operation).

8. Detectors

Apart from the femtosecond pulse gated detection already mentioned, there are two types of detection employed at terahertz frequencies: direct detection and heterodyne detection. These are described briefly below.

8.1 DIRECT DETECTION

In direct or video detection the response is proportional to the power in the incident signal. Therefore the detection is incoherent and all phase and frequency information is lost. However, response extends over a very large range of input wavelengths.

In rectifying detectors, such as the Schottky diode, the output signal is a rectified version of the input signal which is smoothed to give an output voltage proportional to the input power.

In thermal detectors an absorbing element is heated by the incident radiation and the resulting temperature change measured by some form of temperature transducer. Examples are various types of bolometer, pyroelectric detectors and photo-acoustic detectors, such as the Golay cell. At frequencies down to about 1.36 THz it is also possible to use photoconductive detectors.

Bolometers and photoconductive detectors give the lowest noise equivalent powers but require cooling to 4.2 K (1.5 K for some) to operate. The other types can be operated at room temperature.

8.2 HETERODYNE DETECTION

In heterodyne detection the incident signal is downconverted to a lower frequency, retaining all the phase and frequency information that was present in the original signal. Therefore it is a form of coherent detection. It works by diplexing (i.e. superimposing) the incident signal with the signal from a fixed, but slightly different frequency, and fixed amplitude source (the local oscillator or LO), and applying this combined signal to a non-linear device called a mixer. The output from the mixer contains a component at the difference in frequency between the incident signal and the local oscillator (typically in the MHz range and called the intermediate frequency or IF) and has an amplitude directly proportional to the amplitude of the incident signal. If the incident signal contains a range of frequencies, the output signal will contain the same range of frequencies, each with preserved amplitude and phase, they will just all have been downconverted to much lower frequencies. Thus a heterodyne spectrometer can be constructed by feeding the output from the mixer into a narrow-band radio frequency filter and rectifying diode detector, so that the output signal corresponds to effectively a single frequency in the input signal, and then sweeping the local oscillator to sequentially scan over a range of input frequencies. In practice, the output of the mixer at a given frequency contains contributions from signals at both the LO frequency plus the IF, and the LO frequency minus the IF, and so, for spectroscopy, it is necessary to use a quasi-optical terahertz filter at the input, called a sideband filter, to reject frequencies either higher of lower than that of the LO. Non-linear devices used as mixers include Schottky diodes, superconductor-insulator-superconductor (SIS) tunnel junctions and superconducting hot electron bolometers. The spectral resolution of this technique can be very high and surpasses that of the DFTS and pulsed imaging techniques, although it is of more limited bandwidth and, because it is coherent, is less well suited to the measurement of radiation from extended thermal sources than techniques involving direct detection.

9. Applications

9.1 ASTRONOMY AND ATMOSPHERIC SCIENCES

Thermal emission from cool bodies, such as dust, at a few Kelvin to tens of Kelvin peaks in the terahertz range. Furthermore molecular emission lines from cool gas occur throughout the terahertz range. Therefore terahertz observations are important for probing cool regions of the universe, particularly regions where star and planet formation are occurring. Spectra obtained using heterodyne spectrometers on telescopes give chemical specificity and velocity information. Observations of comets and planetary atmospheres can also be made. Very distant, high red-shift objects, such as young galaxies, also emit strongly in the terahertz range. The cosmic background radiation peaks in this range and so observations of its anisotropy are important for probing conditions in the early universe.

Major telescopes, either in operation or under construction, include the 15 m James Clerk Maxwell Telescope (JCMT) in Hawaii, the Herschel Space Observatory (448 GHz – 5 THz, due in 2006), and the Atacama Large Millimetre Array (ALMA) in Chile (30 – 857 GHz, due in 2010). Because of atmospheric water-vapour absorption, the ground-based telescopes are situated at the highest, driest sites in the world.

Satellite, balloon and aircraft-borne terahertz spectroscopy is used in the detection of ozone destroying pollutants, and greenhouse gases, and in medium range weather forecasting, through the measurement of atmospheric water vapour profiles.

9.2 BIOLOGICAL SCIENCES

Terahertz pulsed imaging is being explored for applications in dentistry [12], studies of skin hydration and burn characterisation, and the diagnosis of skin cancers, such as basal cell carcinoma [13]. Penetration depths of 5-6 mm in moist tissue, 9-11 mm in dry tissue and 20-30 mm in fatty material are possible.

A terahertz label-free DNA sequencing technique, based on the differences in refractive index of hybridised and denatured DNA is currently under development [14]. The technique offers femtomolar sensitivity levels and single base mutation detection capability.

Other work is directed towards using terahertz transmissometry for non-contact, non-destructive measurement of the water content and nutrient uptake in plants [15].

9.3 SECURITY AND NON-DESTRUCTIVE TESTING

The ability of terahertz radiation to penetrate materials that are opaque at visible and infrared wavelengths, together with its ability to identify substances through spectroscopic signatures, has led to a number of applications in the security and non-destructive testing areas.

It is being explored for use in the detection of illicit drugs [16], biological threats such as anthrax, and explosives concealed in packages and envelopes [17], and in the detection of weapons hidden under clothing [18, 19]. Imaging systems for detecting concealed weapons are often passive systems, which rely on the sky as an extended thermal source to illuminate the subject, with differences in emissivity of different materials providing the image contrast [20]. Because of the greater penetration of frequencies at the lower end of the terahertz range through fog, smoke and dust than visible and infrared radiation, similar systems can be used for all-weather navigation and for mapping volcanoes using radar.

The use of terahertz systems in non-destructive testing is being explored in the detection of voids, defects and delaminations in materials (e.g. space shuttle insulation [18, 21], microleaks in the seals of flexible plastic packages [22], plastic tubes on the production line [23]); production line monitoring of protrusions, scratches, voids and other defects on the surface of rolled steel and other metals [24]; the non-contact mapping of carrier density and mobility in semiconductor wafers via the terahertz Hall effect [25]; defects in integrated circuits via laser terahertz emission microscopy [26]; and polymorphism [27] and tablet coating thickness in pharmaceuticals [28]. Other important application areas include radar scale modelling and plasma diagnostics for thermonuclear fusion research.

10. Conclusion

Terahertz sensing and measuring systems are evolving rapidly and show great promise for a very wide range of applications. A great deal of work remains to be done, not only in the development of hardware, but in understanding image contrast mechanisms and in developing new signal processing techniques to efficiently exploit the large amount of information contained within a terahertz image.

11. References

1. Digby, J.W., McIntosh, C.E., Parkhurst, G.M., Towlson, B.M., Hadjiloucas, S., Bowen, J.W., Chamberlain, J.M., Pollard, R.D., Miles, R.E., Steenson, D.P., Karatzas, L.S., Cronin, N.J. and Davies, S.R. (2000) Fabrication and characterization of micromachined rectangular waveguide components for use at millimetre-wave and terahertz frequencies, *IEEE Transactions on Microwave Theory and Techniques* **48**, 1293-1302.
2. Martin, D.H. and Bowen, J.W. (1993) Long-wave optics, *IEEE Transactions on Microwave Theory and Techniques*, **41**, 1676-1690.

3. Lesurf, J.C.G. (1990) *Millimetre-wave Optics, Devices and Systems*, Adam Hilger, Bristol and New York.
4. Goldsmith, P.F. (1998) *Quasioptical systems: Gaussian beam quasioptical propagation and applications*, IEEE Press, New York.
5. Wylde, R.J. (1984) Millimetre-wave Gaussian beam-mode optics and corrugated feed horns, *Proceedings of the IEE*, Part H, **131**, 258-262.
6. Martin, R.J. and Martin, D.H. (1996) Quasi-optical antennas for radiometric remote-sensing, *Electronics and Communications Engineering Journal*, **8**, 37-48.
7. Murphy, J.A. and Withington, S. (1996) Perturbation analysis of Gaussian-beam-mode scattering at off-axis ellipsoidal mirrors, *Infrared Phys. Tech.*, **37**, 205-219.
8. Hadjiloucas, S. and Bowen, J.W. (1999) The precision of null-balanced bridge techniques for transmission and reflection coefficient measurements, *Review of Scientific Instruments*, **70**, 213-219.
9. Birch, J.R. and Parker, T.J. (1979) Dispersive Fourier transform spectrometry, in K.J. Button (ed.), *Infrared and Millimeter Waves, vol. 2: Instrumentation*, Academic Press, Orlando and London, 137-271.
10. van der Valk, N. C. J. and Planken, P. C. M. (2004) Towards terahertz near-field microscopy, *Philosophical Transactions of the Royal Society of London A*, **362**, 315-321.
11. Kawase, K., Shikata, J., and Ito, H. (2001) Terahertz wave parametric source, *Journal of Physics D*, **34**, R1-R14.
12. Crawley, D., Longbottom, C., Wallace, V.P., Cole, B., Arnone, D.D. and Pepper, M. (2003) Three-dimensional Terahertz Pulse Imaging of dental tissue, *Journal of Biomedical Optics*, **8**, 303-307.
13. Woodward, R.M., Wallace, V.P., Pye, R.J., Cole, B.E., Arnone, D.D., Linfield, E.H. and Pepper, M. (2003) Terahertz pulsed imaging of ex vivo basal cell carcinoma, *Journal of Investigative Dermatology*, **120**, 72-78.
14. Haring Bolıvar, P., Nagell, M., Richter, F., Brucherseifer, M., Kurz, H., Bosserhoff, A. and Büttner, R. (2004) Label-free THz sensing of genetic sequences: towards 'THz biochips', *Phil. Trans. R. Soc. Lond.* A (2004) **362**, 323-335
15. Hadjiloucas, S., Karatzas, L.S. and Bowen, J.W. (1999) Measurement of leaf water content using terahertz radiation, *IEEE Transactions on Microwave Theory and Techniques*, **47**, 142-149.
16. Kawase, K., Ogawa, Y. and Watanabe, Y. (2003) Non-destructive terahertz imaging of illicit drugs using spectral fingerprints, *Optics Express*, **11**, 2549-2554.
17. Choi, M.K., Bettermann, A. and van der Weide D.W. (2004) Potential for detection of explosive and biological hazards with electronic terahertz systems, *Philosophical Transactions of the Royal Society of London A*, **362**, 337-349.
18. Karpowicz, N., Zhong, H., Zhang, C., Lin, K-I., Hwang, J-S., Xu, J. and Zhang, X-C. (2005) Compact continuous-wave subterahertz ststem for inspection applications, *Applied Physics Letters*, **86** , 054105-1-3.
19. Bjarnason, J.E., Chan, T.L.J., Lee, A.W.M., Celis, M.A. and Brown, E.R. (2004) Millimeter-wave, terahertz, and mid-infrared transmission through common clothing, *Applied Physics Letters*, **85**, 519-521.
20. Appleby, R. (2004) Passive millimetre-wave imaging and how it differs from terahertz imaging, *Philosophical Transactions of the Royal Society of London A*, **362**, 379-394.
21. Zhong, H., Xu, J., Xie, X., Yuan, T., Reightler, R., Madaras, E. and Zhang, X-C. (2005) Nondestructive Defect Identification With Terahertz Time of Flight Tomography, *IEEE Sensors Journal*, **5**, 203-208.
22. Morita, Y., Dobroiu, A., Kawase, K. and Otani, C. (2005) Terahertz technique for detection of microleaks in the seal of flexible plastic packages, *Optical Engineering*, **44**, 019001-1-6.
23. Dobroiu, A.Y., M., Ohshima, Y.N., Morita, Y., Otani, C. and Kawase, K. (2004) Terahertz imaging system based on a backward-wave oscillator, *Applied Optics*, **43**, 5637-5646.
24. Loffler, T., Siebert, K.J., Quast, H., Hasegawa, N., Loata,G., Wipf, R., Hahn, T., Thomson, M., Leonhardt, R. and Roskos, G. (2004) All-optoelectronic continuous-wave terahertz systems, *Philosophical Transaction of the Royal Society of London A*, **362**, 263-281.
25. Mittleman, D.M., Cunningham, J., Nuss, M.C. and Geva, M. (1997) Noncontact semiconductor wafer characterization with the terahertz Hall effect, *Applied Physics Letters*, **71**, 16-19.
26. Yamashita, M., Kawase, K., Otani, C., Kiwa, T. and Tonouchi, M. (2004) Imaging of large-scale integrated circuits using laser-terahertz emission microscopy, *Optics Express*, **13**, 115-120.
27. Taday, P. (2004) Applications of terahertz spectroscopy to pharmaceutical sciences, *Philosophical Transactions of the Royal Society of London A*, **362**, 351-364.
28. Fitzgerald, A.J., Cole, B.E. and Taday, P.F. (2005) Nondestructive analysis of tablet coating thickness using terahertz pulsed imaging, *Journal of pharmaceutical sciences*, **94**,177-183.

ULTRAFAST SPECTROSCOPY OF BIOLOGICAL SYSTEMS

SELMA SCHENKL, GORAN ZGRABLIC, FRANK VAN MOURIK, STEFAN HAACKE AND MAJED CHERGUI

Ecole Polytechnique Fédérale de Lausanne
Laboratoire de Spectroscopie Ultrarapide
ISIC-BSP, CH-1015 Lausanne-Dorigny
Switzerland
Majed.Chergui@epfl.ch,

Abstract: We present recent developments on the ultrafast dynamics of biological systems, which address the dynamics of intra-molecular energy relaxation, of medium rearrangements due to excitation of a chromophore and reveal in "real-time" light-induced electric field changes within a protein (bacteriorhodopsin).

1. Introduction

Observing atoms in the process of chemical reactions, biological functions or motion in solids has been the dream of physicists, chemists and biologists for decades until the 1980s. The implementation of femtosecond laser spectroscopies brought entirely new possibilities for the study of light-induced phenomena in condensed matter physics, in chemistry and in biology. In particular, the ultrafast studies of chemical reactions have given birth to the field of Femtochemistry [1], which was recognized by the Nobel Prize to Ahmed Zewail in 1999.

For the first time, it was possible to conduct observations on time scales that are shorter than single nuclear oscillation periods in solids, molecules and proteins. It therefore became possible, in principle, to monitor molecules at various stages of vibrational distortion, recording "stop-action" spectroscopic events corresponding to well-defined molecular geometries far from equilibrium, including stretched and/or bent unstable transient structures. Molecular structures corresponding to such unstable intermediates between reactant and product could be followed in real-time[1,2]. Distorted crystal lattices and other specific out-of-equilibrium structures have been characterized in real-time[3-5].

To get a feeling why femtosecond time resolution is needed for these "real-time" observations, it is useful to note that the order of magnitude of the speed of atoms in matter is given by the speed of sound: 300 m/s-1000m/s, which translate to 0.3-1.0 Å in 100 fs. Such length scales are precisely the ones one deals with in dynamical processes of molecules, biological systems, and condensed matter.

The key to observing moving structures in real-time is based on using suitable light sources for excitation and detection. The link between the optical spectroscopy and the molecular or condensed phase structures is made via the knowledge of potential energy surfaces, which is a prerequisite if one wants to get the structural *dynamics*. However, ultrafast *kinetics*, which is based on the evolution of populations, is also a fundamental tool to not only observe evolving structures, but also to understand and pinpoint the pathways of energy flow, which play such a fundamental role in biology, e.g. vision, photosynthesis, phototaxis, etc.

In a previous review[6] in this series, we dealt with the basic physics of ultrafast phenomena, and showed some applications, on quantum systems of increasing complexity, ranging from

B. Di Bartolo and O. Forte (eds.), Advances in Spectroscopy for Lasers and Sensing, 119–127.

small molecules to proteins. Here we will mainly deal with recent results that have been obtained on biological systems, mainly retinal protein systems, which concern not only structural changes, but also the measurement of electric fields and their changes within biological systems.

Retinal proteins, of which rhodopsin (the protein responsible for vision) and bacteriorhodopsin form the most important representatives, are membrane proteins. The main chromophore (retinal is embedded in the protein pocket) and is surrounded by a number of polar and non-polar amino-acid residues, which play a crucial role in the photobiological reaction (Fig. 1).

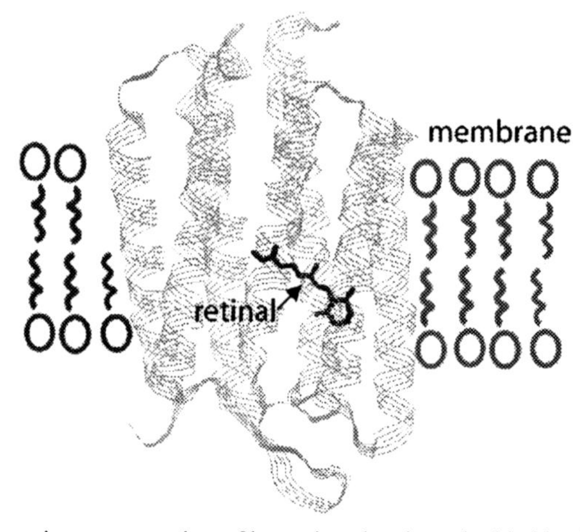

Figure 1: Schematic representation of bacteriorodopsin embedded in the membrane.

In this review, we will mainly deal with bacteriorhodopsin (bR, fig. 1). Franck-Condon excitation of *all-trans* retinal results in an immediate change of permanent dipole moment by more than 12 D [7,8], causing a sudden polarization of retinal, and isomerization of the all-trans form to the 13-cis-form of retinal. This isomerization has been considered the primary event of the photobiological cycle. Compared to retinal in organic solvents, where the isomerization yield is < 20% [9,10] and is not selective around a specific C-C bond, the specific environment of the protein enhances the yield of isomerized retinal to 65%, and ensures stereo-selectivity of the photo-product since the isomerization occurs selectively around the C13-C14 bond [11]. The debate is still going on, as to whether the origin of the enhanced isomerization efficiency in the protein is due to dielectric effects in the polar environment of the protein [12,13] and/or steric effects. We have addressed these issues in two ways, using ultrafast spectroscopic techniques: a) The first consists in tuning the properties of the environment of retinal by using a series of solvents (both protic and aprotic) differing by as much as an order of magnitude in viscosity, dielectric constant, and by a factor of three in density [14]. b) The second consists in probing the ultrafast response of the tryptophan amino-acid residues that sit in the direct vicinity of retinal in the protein, upon excitation of retinal [15].

2. Solvent Effects on the Isomerization Efficiency of Retinal

A crucial issue in chemistry and biology is the time scale of: a) the rearrangement of the environment species, after excitation of a chromophore and; b) the redistribution of the energy

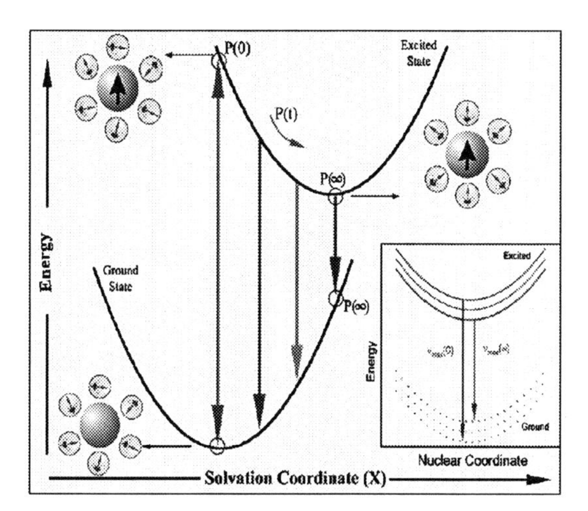

Figure 2: Schematic representation of solvation dynamics in polar environments. The pump pulse induces a redistribution of charge in the solute (red circle) at time t = 0. The solvent species respond to the new field of forces and minimise the free energy, up to a new equilibrium configuration P(∞) from which fluorescence occurs. The dynamical Stokes shift is the evolution of the emission spectrum in the course of solvent reorganisation (i.e. from P(0) to P(∞)).

in the system. This process, termed electronic *solvation dynamics*, can best be investigated by means of femtosecond fluorescence up-conversion techniques, which allow following the emission spectrum in the course of relaxation. This delivers the so-called dynamical Stokes shift, as depicted in Fig. 2.

In polyatomic systems, the dynamics Stokes shift is not only due to the solvent but also to intramolecular processes, and the distinction between the intra- and intramolecular contributions is difficult, especially in the very short time domain, if the solute contains high frequency modes.

In order to address the issue of the solvent effect on the isomerization dynamics of retinal, we carried out a systematic fluorescence up-conversion study [14] of the retinal chromophore in its protonated Schiff base form (PSBR) in various solvents. Our up-conversion set-up in polychromatic mode allows us to record entire spectra per time delay. A sketch of the set-up is show in figure 3.

Briefly, part of the 810 nm fundamental light of an amplified high repetition rate (250 kHz) TiSa femtosecond system is frequency doubled to provide the pump pulse, the rest is used as gate pulse. Fluorescence is collected by large angle optics and focussed onto the BBO non-linear crystal where it is mixed with the gate pulse, at variable pump-gate time delay to generate the up-converted light. As the BBO crystal is rotated, a broadband sum frequency signal is generated that translates the entire fluorescence spectrum (450 – 900 nm). The near-UV signal is dispersed in a monochromator and imaged onto a CCD camera, so that nearly the entire range of the emission is detected per laser shot. This allows the simultaneous recording of the kinetics and time evolution of the fluorescence, which are obtained in 3-D plots, as shown in figure 4 for the case of PSBR in methanol.

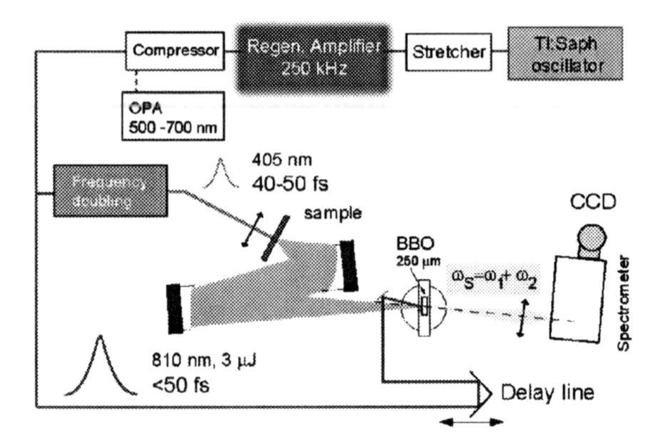

Figure 3: Experimental set-up for broadband fluorescence up-conversion spectroscopy.

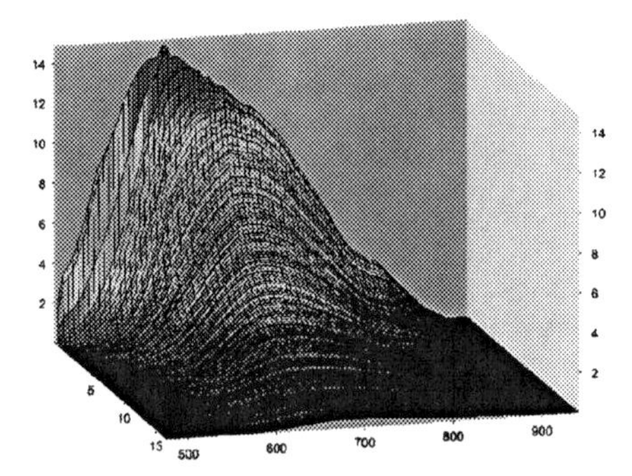

Figure 4: Typical 3-D plots (intensity-time delay-wavelength) for PSBR in methanol.

The analysis of such data allows us to extract two main components, shown as kinetic traces in figure 5. The high energy component (II) is very short lived (~100 fs) and is due to the S_2 state which is also excited at our pump wavelength of 400 nm. The lower energy component (I) rises on the time scale of the component II decay and exhibits a 0.5-0.65 ps decay in nearly all solvents (Table I). This component reflects the direct repopulation of the ground state, in line with the small isomerization yield of PSBR in solvents. Two longer decay components are observed, which are attributed to torsional motion leading to photo-isomerization about different C-C bonds.

Surprisingly, it can be seen from table I that the various decay channels show little or no dependence with respect to the various solvents. In particular, the lack of dependence with the dielectric constant points to the fact that our experiment mainly probes intramolecular dynamics in line with the almost identical steady-state absorption-emission Stokes shifts found in the various solvents [14]. This suggests that dielectric effects and the protic vs aprotic character of the environment do not play a crucial role in the protein dynamics. The lack of an effect of a viscosity-dependence can be explained by the fact that the solvent cages, for all the solvents

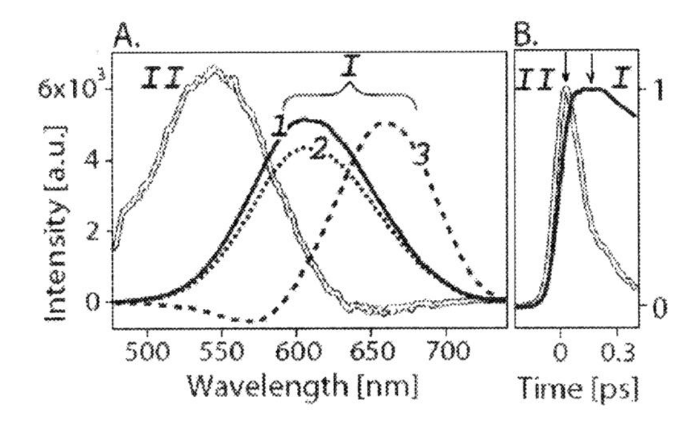

Figure 5: A. Decomposition of the time-dependent fluorescence data: Band I is composed of three sub-bands. Spectrum II shows the maximum intensity of the band II (t = 0 fs). **B.** Time traces I and II taken at the central wavelengths of the band I (640 nm) and band II (540 nm), respectively. A bi-exponential fit on the trace II gives $\tau_{II,1} = (20 \pm 10)$ fs (76%) and $\tau_{II,2} = (170 \pm 20)$ fs (24%).

studied, are significantly larger than the protein pocket and that the chromophore in solution does not "sense" the environment by contact. These two combined observations suggest that in the protein, the bond selectivity of isomerization and its efficiency are mainly governed by steric effects.

Table I: Decay constants τ_i, and relative amplitudes a_i (in parentheses) of bands I and II of PSBR in different solvents.

Band / Solvent		MeOH	ProOH	OctOH	DCM	cHex
I	$\tau_{I,1}$/ps	0.64 (0.35)	0.64 (0.42)	0.50 (0.45)	0.63 (0.46)	0.56 (0.48)
	$\tau_{I,2}$/ps	1.5 (0.30)	2.2 (0.37)	2.5 (0.36)	2.0 (0.32)	2.1 (0.40)
	$\tau_{I,3}$/ps	4.6 (0.35)	5.5 (0.21)	6.3 (0.19)	4.7 (0.22)	3.7 (0.12)
II	$\tau_{II,1}$ /fs	< 30 (0.76)	< 30 (0.87)	< 30 (0.89)	< 30 (0.85)	< 30 (0.88)
	$\tau_{II,2}$ /fs	170 (0.24)	140 (0.13)	150 (0.11)	140 (0.15)	150 (0.12)

In order to further gain insight in the protein dynamics itself, we set out to measure the protein response to excitation of retinal by probing nearby amino-acid residues, as explained in the next section.

3. Excitonic Coupling and Transient Electric Fields within Proteins

We have investigated the response of Tryptophan (Trp) amino-acids located in the vicinity of retinal chromophore in bacteriorhodopsin, after excitation of the latter (bR, fig. 6). Four tryptophans are in the vicinity of retinal in the binding pocket of bR: Trp86 and Trp182 sandwich retinal, while Trp138 and Trp189 are located in the vicinity of the β-ionone ring (fig. 6A).

The absorption spectrum of bR (fig. 6B) exhibits a band in the visible, due to retinal, and a band near 280 nm, predominantly due to the eight Trps present in the protein. The Trp absorption is due to transition from the ground state to the lowest two close-lying excited states (labelled L_a and L_b) of the indole moiety. The transition into the L_a state implies a large difference dipole moment with respect to the ground state[16]. Trp residues are therefore

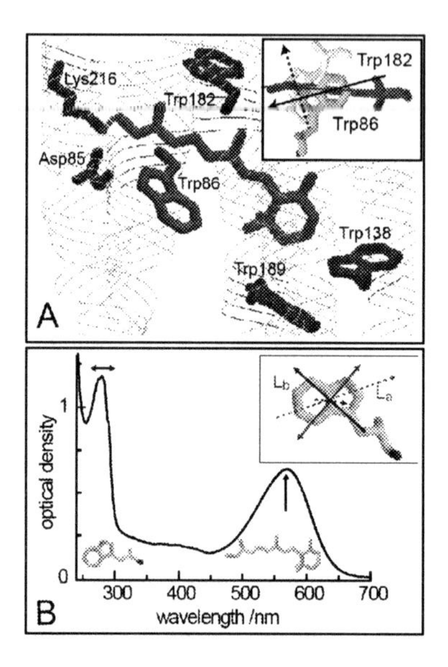

Figure 6: Structure and optical properties of Bacteriorhodopsin. A) Retinal binding pocket of bacteriorhodopsin. Retinal (purple) is covalently bound to Lys216 through a Schiff base linkage. The four nearest Tryptophan (Trp) residues are shown. *Inset:* Relative orientations of the difference dipole moments for retinal (red arrow), of Trp 86 (full black arrow) and Trp182 (dashed black arrow). B) Absorption spectrum of bR in the purple membrane. The band with maximum at 568 nm is due to the S_0-S_1 transition of the protonated Schiff base form of retinal, while the near-UV band at 280 nm is dominated by the S_0-L_a and S_0-L_b transitions of eight Trp's. The vertical arrow indicate the excitation wavelength and the horizontal one the range of probe wavelengths. *Inset:* Orientations of the transition (full arrows) and difference (dashed arrows) dipole moments of tryptophan for $S_0 - L_a$ (red) and $S_0 - L_b$ (blue) [16]. Arrows point to the positive end of the difference dipole moment.

particularly well-suited molecular-level sensors of electric field changes within the protein, as the latter are expected to cause large changes in their spectral features. In particular, Trp86 is closest to retinal, and its difference dipole moment is almost parallel with respect to the retinal backbone (inset, fig. 6A), so that it is particularly sensitive to the Coulomb force fields close to retinal. To probe the changes in the force fields associated with the excitation, we excited retinal at 560 nm, and interrogated the Trp absorption changes with 80-90 fs time resolution, using tuneable near-UV pulses [15].

Figure 7A shows the transient absorption changes, detected between 265 nm and 280 nm. The traces are normalized, highlighting their similar shape over the whole range of probe wavelengths. The presence of a rise time is directly deduced from the temporal derivative of the bleach transient (fig. 7B). Its full width at half maximum (FWHM) of 150 fs is significantly larger than the 80-90 fs pump-probe cross-correlation. For longer times, we observe a biexponential recovery of the absorption with time scales: $\tau_1 = 420$ fs, $\tau_2 = 3.5$ ps, which have been reported in experiments that probe retinal directly [17,18].

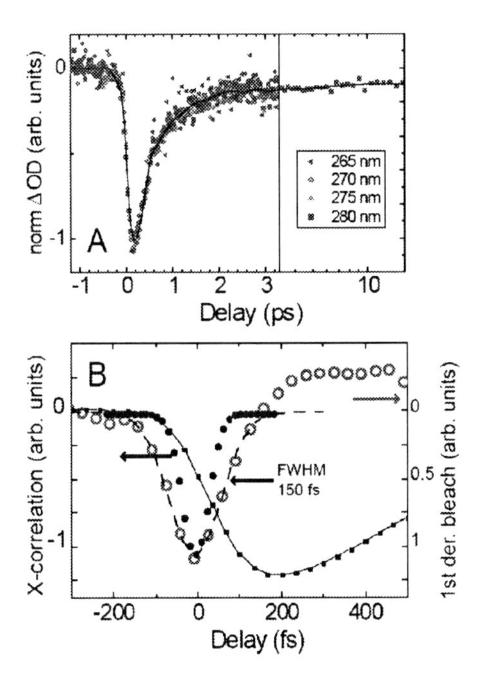

Figure 7: Ultrafast response of bacteriorhodopsin in the near-UV. A) Transient bleach signals at 265-280 nm after excitation with 25 fs pulses at 560 nm at 1 kHz repetition rate. UV probe pulses (80-90 fs) were obtained by second harmonic generation of the output of a tuneable non-collinear OPA. The transients are normalized in amplitude, demonstrating that they are independent of the probe wavelength. Each data point represents the average of 10.000 laser shots, allowing for a ΔOD sensitivity of 10^{-4}. The time axis of the left panel is linear, while the right panel has a logarithmic one. B) The early time portion of the bleach transient in fig. 7A (black squares), and its temporal derivative (red circles) showing a full width at half maximum (FWHM) of 150 fs. The latter is clearly larger than the experimental time resolution of 85 fs (black dots) determined by the pump-probe cross-correlation.

The most notable features of fig. 7 are that: a) the signal appears as a bleach in the region of Trp absorption, while no Trp was excited. b) The bleach does not rise with the cross-correlation but on a longer time scale. As discussed in ref. [15], the bleach transients (fig. 7A) are mainly due to the response of the Trp86 residue.

The reason the signal appears as a bleach results from effects of excitonic coupling, due to the resonance interaction of retinal and the Trps at ~280 nm, implying that intensity borrowing can occur between retinal and the Trp chromophores. We developed a simple model (see details in ref.[15]), considering a system of three coupled chromophores, having three electronic levels each: S_0, S_1, S_n for retinal and, S_0, L_a and L_b for Trp86 and Trp182. They form an excitonic complex, whose proper electronic states are linear combinations of the singly-excited near UV states of $|L_{a,b},0,0\rangle$, $|0,L_{a,b},0\rangle$ and $|0,0,S_n\rangle$ (product basis notation |Trp86, Trp182, retinal>), of which three are important (X_1 trough X_3, see fig. 8), as a result of favourable relative orientations of the transition dipole moments. The doubly excited states (XX_1 and XX_{-2}) with retinal in the S_1 state, arise from linear combinations of $|L_{a,b},0,S_1\rangle$ and $|0,L_{a,b},S_1\rangle$.

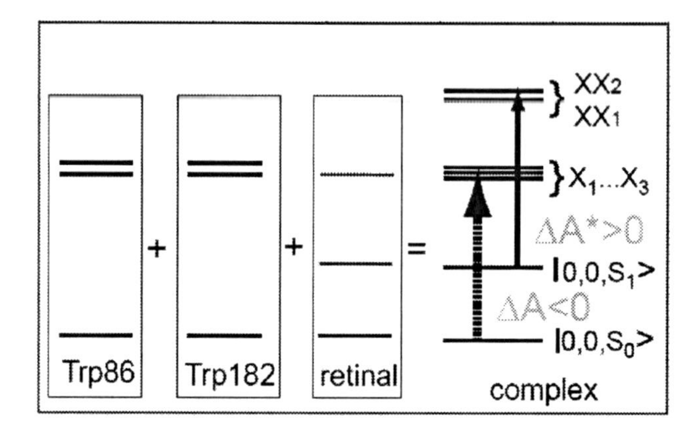

Figure 8: Dipole-dipole coupling of Trp86, Trp182 and retinal leads to the formation of an excitonic complex with proper electronic states [15]. The X_i's represent singly excited states of the excitonic complex, with one excitation per chromophore, while the XX_i's represent doubly excited states with retinal in the S_1 state. Photo-excitation of retinal attenuates the transition to the $X_1 \ldots X_3$ states ($\Delta A < 0$), due to the bleach of the $|0,0,S_0\rangle$ (product basis notation |Trp86, Trp182, retinal>) ground state of the excitonic complex. The photo-induced absorption from $|0,0,S_1\rangle$ reaches the XX_1 and XX_2 states (in which both Trp and retinal are excited), giving rise to an absorption ($\Delta A^* > 0$). As the difference dipole moment $\Delta\mu$ increases, ΔA^* red-shifts due to a lowering of the XX_1 and XX_2 levels (attractive interaction between retinal and Trp86). This leads to an increasing bleach around 280 nm, which is a measure of the retinal difference dipole moment $\Delta\mu$

The simulated absorption (not shown here) changes as a function of the retinal difference dipole moment $\Delta\mu$. The photo-induced signal in the region of Trp absorption indeed appears as a bleach of the excitonic transitions into the X_i levels (the bleach character arises from the depletion of ground state retinal due to excitation with the pump pulse), while transitions from $|0,0,S_1\rangle$ to XX_i levels show up as a positive signal on the red side.

As the retinal difference dipole moment $\Delta\mu$ acts only when the system is in the $|0,0,S_1\rangle$ state, the bleach ΔA is $\Delta\mu$-independent, while the position of the induced absorption ΔA^* red-shifts with increasing difference dipole moment. The red shifting is due to an attractive interaction between the excited state dipole moments of Trp86 and retinal (inset fig. 6A), with the latter's varying in time. The signal is the sum of ΔA and ΔA^*, and it results in an increasing bleach signal with increasing $\Delta\mu$, in accordance with our results. Therefore, we explain that the bleach is due to formation of an excitonic complex between the Trps and retinal, and that the time evolution of the bleach rise, is due to the changes in retinal difference dipole moment. Trp86, therefore appears as a reporter of the light-induced electric field changes within the protein. We can follow them over the whole course of the dynamics from the Franck-Condon region to the vibrationally relaxed photoproduct. The time evolution of the difference dipole moment is due to the long hypothesized, but never observed, translocation of charge along the retinal chain. In addition, our observations establish a connection between the latter and the skeletal changes of the conjugate chain [15].

While the studies in solution pointed to a lack of influence of the dielectric environment on the dynamics of retinal isomerization (see above), this work shows that in return, the huge dipole moment changes in the chromophore do affect its environment in the protein. Whether this has a consequence on the bond selectivity and efficiency of the isomerization process is unclear at this stage, but it is very likely, in the sense that modifying the environment, which occurs within the first 200 fs, sets the stage for the isomerization, which occurs in 400-500 fs. Recent work probing with UV pulses the retinal and its transition state *en route* to isomerization seem to point in this direction. In this sense this work stresses the role of the dynamic force fields, which drive structural dynamics and govern enzymatic reaction.

Acknowlegements

We are deeply indebted to our colleagues: G. van der Zwan (VU Amsterdam), N. Friedman and M. Sheves (Weizmann Institute), R. Schlesinger and G. Büldt (KFZ Jülich), without whom these results would not have been made possible. We also thank the Swiss NSF for its support via grants 2155830.98, 2153-65135.01, 2000-67912.02 and the NCCR: "Quantum Photonics"

References

(1) Zewail, A. H. *J Phys Chem A* **2000**, *104*, 5660-5694.
(2) Chergui, M. *Femtochemistry ultrafast chemical and physical processes in molecular systems Lausanne, Switzerland, September 4-8, 1995*; World Scientific: Singapore etc., 1996.
(3) Dhar, L.; Rogers, J. A.; Nelson, K. A. *Chem Rev* **1994**, *94*, 157-193.
(4) Chergui, M. *Cr Acad Sci Iv-Phys* **2001**, *2*, 1453-1467.
(5) Bonacina, L.; Larregaray, P.; van Mourik, F.; Chergui, M. *Phys. Rev. Lett.* **2005**, *95*, -.
(6) Chergui, M. In *Frontiers in Optical Spectroscopy*; Di Bartolo, B., Forte, O., Eds.; Springer: **Berlin**, 2005, p 497.
(7) Mathies, R. A.; Stryer, L. *Proc. Nat. Acad. Sci.* **1976**, *73*, 2169-2178.
(8) Huang, J.; Chen, Z.; Lewis, A. *J. Phys. Chem.* **1989**, *93*, 3314-3320.
(9) Becker, R. S.; Freedman, K. *J. Am. Chem. Soc.* **1985**, *107*, 1477-1485.
(10) Koyama, Y.; Kubo, K.; Komori, M.; Yasuda, H.; Mukai, Y. *Photochem. Photobiol.* **1991**, *54*, 433-443.
(11) Wald, G. *Science* **1968**, *162*, 230-&.
(12) Xu, D.; Martin, C.; Schulten, K. *Biophys. J.* **1996**, *70*, 453-460.
(13) Kennis, J. T. M.; Larsen, D. S.; Ohta, K.; T, F. M.; Glaeser, R. M.; Fleming, G. R. *J. Phys. Chem B* **2002**, *106*, 6067-6080.
(14) Zgrablic, G.; Voitchovsky, K.; Kindermann, M.; Haacke, S.; Chergui, M. *Biophys J* **2005**, *88*, 2779-2788.
(15) Schenkl, S.; van Mourik, F.; van der Zwan, G.; Haacke, S.; Chergui, M. *Science* **2005**, *309*, 917-920.
(16) Callis, P. R. *Meth. Enzym.* **1997**, *278*, 113-150.
(17) Mathies, R. A.; Brito Cruz, C. H.; Pollard, W. T.; Shank, C. V. *Science* **1988**, *240*, 777-779.
(18) Herbst, J.; Heyne, K.; Diller, R. *Science* **2002**, *297*, 822-825.

LUMINESCENCE SPECTROSCOPY OF SOLIDS: LOCALIZED SYSTEMS

JOHN COLLINS
Department of Physics and Astronomy
Wheaton College
Norton, MA 02766, USA
jcollins@wheatonma.edu,

Abstract

In the introduction we consider the nature of luminescence research and emphasize its material-related character. We also recount briefly the history of the research in this field. In the subsequent section the theoretical bases for the treatment of the interaction of radiation with material systems is established. The next two sections discuss the treatment of localized luminescent systems in which the radiative processes are associated with optically active centers.

1. Introduction

1.1 BASIC CONCEPTS

A material, after absorbing some energy, can become a source of light by two processes:

1. The absorbed energy is converted into *low* quantum-energy heat that diffuses through the material, which then emits *thermal radiation.*
2. The absorbed energy is in part *localized* as high-quantum energy of atoms, which then emit radiation called *luminescence radiation.*

We note the following:

1. The quantity and quality of thermal radiation depend primarily on the temperature rather than on the nature of the body.
2. The quality and quantity of luminescence are strongly dependent on the *nature* of the emitting material.

Therefore the results of investigations in the field of luminescence are relevant to the study of material bodies.

The process of luminescence is started by exciting the material with UV radiation, x-rays, electrons, alpha particles, electric fields or energy that is liberated in chemical reactions. Accordingly luminescence is qualified as photoluminescence, Roentgen luminescence, electro-luminescence, etc. Regardless of the type of excitation or the nature of the system, the common character of the phenomenon of luminescence resides in the fact that, once excited, the system is

B. Di Bartolo and O. Forte (eds.), Advances in Spectroscopy for Lasers and Sensing, 129–155.

left alone, without any external influence and, in this condition, it emits radiation that we may appropriately call *spontaneous*.

We shall limit ourselves to examining the process of photoluminescence in solids, i.e. the luminescence emission that follows the absorption of UV or optical radiation. In particular, we consider luminescent systems that can be termed *localized*. That is, the absorption and emission processes are associated with quantum states of optically active centers that are spatially localized to particular sites in the solid. Delocalized luminescent systems are considered in the following article.

In this section we discuss different classes of luminescent systems that we call "localized luminescent centers." Such systems include any luminescent system in which radiative transitions occur between states that are associated with a particular center. The most important classes of localized luminescent centers are *Transition Metal Ions* (TMI) and *Rare-Earth Ions* (REI) that have been intentionally doped into ionic insulating host materials. The luminescence properties of these systems depend on both the dopant ion and the host. Another class of localized centers is color centers in solids. One such center is an electron trapped at a vacant lattice site.

One theme that emerges continually throughout is that of luminescence in solids as a materials-related phenomenon. That is, the observed luminescence has much to say regarding not only the emitting center, but about the material as a whole.

The host materials for the optically active centers considered here are large band gap (usually > 6 eV) ionic solids. The large gap renders the host transparent to visible radiation, and also precludes any electrons from bridging the gap thermally, so that an un-doped host material is optically and electrically inert. Such host materials include alkaline halides, oxides, chlorides, fluorides, chlorides, tungstates, molybdates, phosphates, and garnets, to name but a few.

1.2 HISTORY

The first reported observation of luminescence light from glowworms and fireflies is in the Chinese book Shih-Ching or "Book of Poems" (1200-1100 B.C.). Aristotle (384-322 B.C.) reported the observation of light from decaying fish. The first inquiry into luminescence dates c.1603 and was made by Vincenzo Cascariolo. Cascariolo's interests were other than scientific: he wanted to find the so-called "philosopher's stone" that would convert any metal into gold. He found some silvery white stones (barite) on Mount Paderno, near Bologna, which, when pulverized, heated with coal and cooled, showed a purple-blue glow at night. This process amounted to reducing barium sulfate to give a weakly luminescent barium sulfide:

$$BaSO_4 + 2C \rightarrow BaS + 2CO_2$$

News of this material, the "Bologna stone," and some samples of it reached Galileo, who passed them to Giulio Cesare Lagalla. Lagalla called it "lapis solaris" and wrote about it in his book "De phenomenis in orbe lunae" (1612).

It is rare the case when one can insert in a scientific article a literary citation. This happens to be the case here. In "The Sorrows of Young Werther," of Goethe, the protagonist is unable to see the woman he loves because of an engagement he cannot refuse, and sends a servant to her, "only so that I might have someone near me who had been in her presence... " This is then his reaction when the servant comes back: [1.1]

> It is said that the *Bologna stone*, when placed in the sun, absorbs the sun's rays and is luminous for a while in the dark. I felt the same with the boy. The consciousness that her eyes had rested on his face, his cheeks, the buttons of his jacket and the collar of his overcoat, made all these sacred and precious to me. At that moment I would not have parted with him for a thousand taler. I felt so happy in his presence.

The second important investigation on luminescence is due to Stokes and dates the year 1852. Stokes observed that the mineral fluorspar (or fluorite) when illuminated by blue light gave out yellow light. Fluorite (CaF_2) is colorless in its purest form, but it absorbs and emits light when it contains such impurities as *Mn, Ce, Er,* etc. The term "fluorescence " was coined by Stokes and has continued to be used to indicate short-lived luminescence. A *Stokes' law* has been formulated according to which the wavelengths of the emitted light are always longer than the wavelength of the absorbed light.

2. Spectra and Energy Levels of Localized Systems

2.1 THE HAMILTONIAN OF AN ION IN A SOLID

For an optically active ion in a solid, the Hamiltonian can be written as:

$$H = H_{FI} + H_L + H_{CF} \tag{2.1}$$

where H_{FI}, H_L and H_{CF} are the Hamiltonians for the free ion, lattice, and crystal field interaction, respectively.

The free ion Hamiltonian, H_{FI}, includes all electric and magnetic interactions in the ion, which for many-electron atoms is quite cumbersome. Extensive treatments on the various interactions contained in H_{FI} are given in references [2.1-2.3]. Our goal here is to consider the most important of these interactions so as to gain insight into the nature of the energy states of the lowest unfilled (4f) configuration.

The free ion Hamiltonian is given by:

$$H_{FI} = \sum_i \frac{p_i^2}{2m} - \frac{Ze^2}{r_i} + \sum_{i>j} \frac{e^2}{r_{ij}} \tag{2.2}$$

where the first sum is over all electrons, and includes the kinetic energy of the electrons and the electrostatic interaction between the nucleus and the electrons. The second sum represents the interaction among the electrons.

As a first approximation, we replace H_{FI} with a Hamiltonian, H_0, that has a spherically symmetric potential term, U(r):

$$H_0 = \sum_i \frac{p_i^2}{2m} + U(r_i) \tag{2.3}$$

Because $U(\vec{r}_i)$ is spherically symmetric, the angular part of its eigenfunctions of H_0 are given by spherical harmonics, and the radial part depends on $U(\vec{r}_i)$. The quantum numbers describing the eigenstates of H_0 are n, l, m_l, and m_s. The eigenvalues are degenerate with respect to m_l and m_s, i.e. the quantum numbers n and l define the configuration and the levels within each configuration are degenerate.

The difference between H_{FI} and H_0 gives the perturbation term:

$$H_{FI} - H_0 = \sum_i \left(-\frac{Ze^2}{r_i} - U(r_i) \right) + \sum_{i>j} \frac{e^2}{r_{ij}} \tag{2.4}$$

The first sum in (2.4) shifts the energy of each level within any given configuration by the same amount. As we are interested in the energy differences within a configuration, we drop the first sum and consider only the second.

$$H_C = \sum_{i>j} \frac{e^2}{r_{ij}} \tag{2.5}$$

We refer to this term as the configuration interaction, H_C. The sum is over all electrons in the atom. It is common to assume, at this point, that the configurations are well separated in energy from one another, allowing one to consider only the interaction of the electrons within each configuration. Since we are interested in the optical properties of the solid, the sum in (2.5) can be restricted to those electrons in the partially-filled configuration (e.g. the 4f configuration for lanthanide ions). This interaction destroys the spherical symmetry, and splits the ground configuration into different spectral terms. Such terms are identified using the ^{2S+1}L notation. L^2, S^2, L_z and S_z all commute with H_C and so the corresponding quantum numbers (L, S, M_L, and M_S, respectively) are valid. The energies of the states are independent of M_L and M_S, and so have a degeneracy of (2L + 1)(2S + 1).

For rare earth ions, the spin-orbit term (H_{SO}) is the next most important interaction, followed by the crystal field interaction, H_{CF}. This is the so-called weak crystal field scheme. For transition metal ions H_{CF} is larger that H_{SO}, so that either the medium or strong field scheme is appropriate. We first consider the rare earth ions.

2.2 RARE EARTH IONS IN SOLIDS

2.2.1 *Energy Levels of Rare Earth Ions in Solids*

The rare earth elements include the lanthanides and the actinides. By far the most important of these as luminescent centers are the lanthanides, particularly when incorporated into a solid in the trivalent state. The configuration of the trivalent lanthanides is

$$1s^2 2s^2 2p^6 3s^2 3p^6 3d^{10} 4s^2 4p^6 4d^{10} \underline{4f^n} 5s^2 5p^6$$

with the number of 4f electrons running from 1 (Ce^{3+}) to 13 (Yb^{3+}). The 4f electrons are responsible for the optical activity of the center. The 5s and 5p shells are located farther from the nucleus than the 4f shell, and partially shield the 4f electrons from the nearby ligands. This shielding renders the crystal field interaction (H_{CF}) much smaller than the spin orbit interaction, (H_{SO}).

For REI, then, the next most important term in the Hamiltonian is H_{SO}, which is given by:

$$H_{SO} = \sum_i \xi_i \vec{l}_i \bullet \vec{s}_i \tag{2.6}$$

This splits each spectral term into different levels, called J-multiplets. The operators $J^2 = (\mathbf{L} + \mathbf{S})^2$ and $J_z = (L_z + S_z)$ commute with H_{SO}, so that J and M_J are good quantum numbers. Each J-multiplet has a degeneracy of 2J + 1. For rare earth ions, these splittings are large enough to cause some mixing of states having different L and S and of equal J-values. Nevertheless, the J-multiplets are commonly labeled using S, L, and J quantum numbers according to the usual spectral designation $^{2S+1}L_J$, where J runs from L + S to L-S.

The remaining interaction is that of the crystal field with the rare earth ion. H_{CF} consists of a static term ($H_{CF\text{-}static}$) and a dynamic term ($H_{CF\text{-}dymanic}$). We consider only the static term here since it makes a more important contribution to the energy.

Due to the shielding of the 4f electrons, there is little or no overlap of their wavefunctions with those of the ligands. Thus, we consider the ion to be under the influence of an external field - this is the crystalline field approximation. In this approximation, $H_{CF\text{-}static}$ is the interaction of the electrons with the potential due the ligands, $V(\vec{r}_i)$.

$$H_{CF-static} = \sum_i eV(\vec{r}_i) \tag{2.7}$$

The sum is over all 4f electrons, and $V(\vec{r}_i)$ reflects the symmetry at the site of the REI. This interaction splits each J-multiplet into no more than (2J + 1) levels for ions with an even number of f-electrons and no more than (J + 1/2) levels for ion with an odd number of f-electrons. The number of levels into which each multiplet splits depends on the symmetry of the crystal field - for higher ion site symmetries there are fewer levels. The splittings due to H_{CF} typically range from ten to a few hundred cm^{-1}.

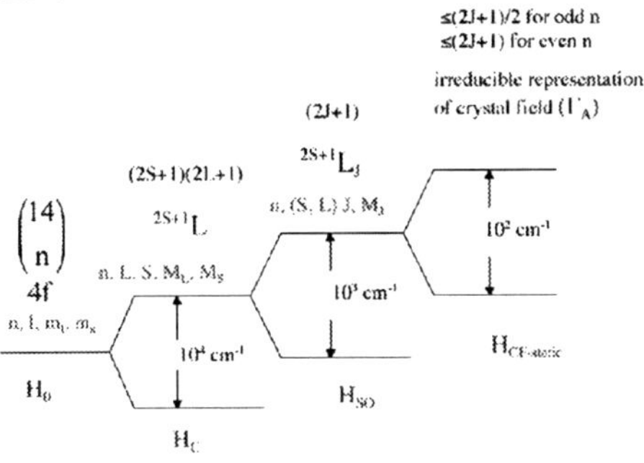

Figure 2.1 A summary of the interactions and their effect on the energies of trivalent lanthanide ions in solids. Associated with each term in the Hamiltonian are the corresponding energy splittings, the valid quantum numbers, the spectral designation of the states, and the degeneracy of the levels.

A summary of these interactions is shown in Figure 2.1. The observed energy levels of the rare earth ions in $LaCl_3$ are shown in Figure 2.2. Because of the small crystal field interaction, the energy levels are similar in other ionic solids [2.2, 2.3]. Most of the energy levels of the REI are known up to about 40,000 cm^{-1}.

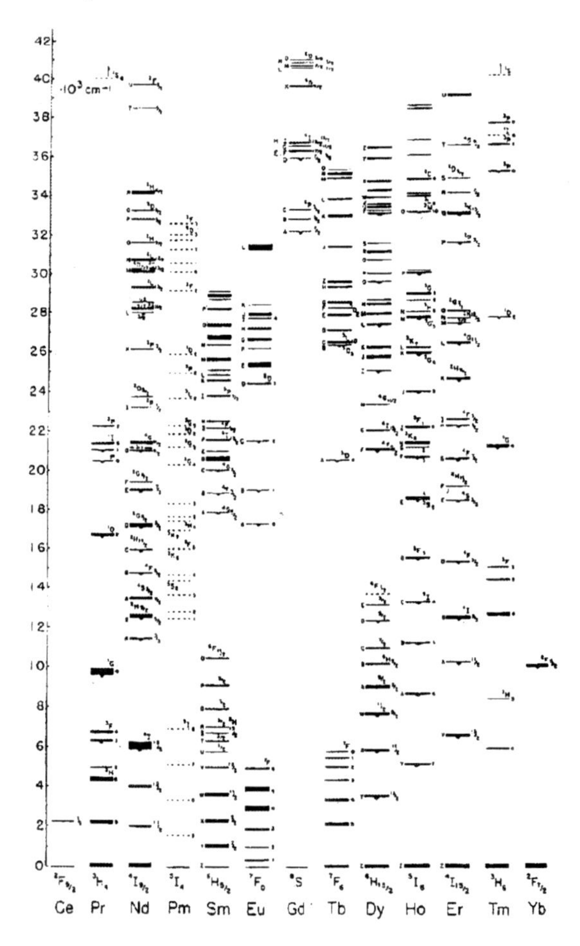

Figure 2.2 Energy levels of trivalent rare earth ions. The width of the levels indicates the total splitting due to the crystalline field in anhydrous $LaCl_3$. [2.1].

2.2.2 *Spectral Features of Rare Earth Ions in Solids*

The small crystal field interaction implies that the energy levels of REI are not very sensitive to the motion of the lattice. Thus, f→f transitions are characterized by sharp lines, with line widths on the order of 1 cm^{-1} at low temperatures. A sample REI emission spectrum is shown in Figure 2.3. Typical radiative lifetimes range from a few μsec to a few msec.

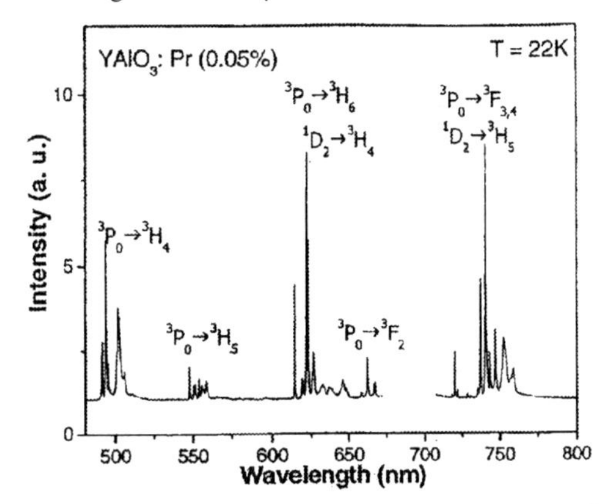

Figure 2.3 Luminescence spectrum of YALO : Pr (1%) at 22K. Excitation was into the $^3P_0$ level at 460 nm.

Although radiative transitions do occur between f-levels, luminescence spectra show certain lines very weak or missing entirely. This is due to one or more of the following factors.

1. Certain transitions may be only weakly allowed due to the symmetry of the crystal field.
2. An excited REI can decay via the emission of phonons instead of a photon. Such transitions are called multiphonon transitions, and are discussed in section 6.3.
3. The REIs are in thermal equilibrium; each ion reaches thermal equilibrium with the lattice, and so they are in equilibrium with one another. Thus, the levels of the ground multiplet (or of an excited metastable multiplet) are populated according to the Boltzmann distribution.

REI spectra also exhibit sidebands to the zero-phonon line. These are due to vibronic transitions (section 3.4), which involve the emission of a photon and the simultaneous emission or absorption of one or more phonons. For REI, such sidebands usually involve a single phonon, and are much weaker than the zero-phonon line. The structure of the sideband reflects somewhat the phonon spectrum of the host lattice.

In many systems, the REI's are situated at two or more types of lattice sites. Due to variations in the crystal field, the same transition at the different sites will generally appear at different energies, complicating the spectra considerably. In glasses, the random positioning of the ions lead to a continuum of crystal field interactions and to broad bands on the order of 100 cm^{-1} wide.

We conclude this section with the observation that the spectral features of REI can be traced back to their peculiar charge distribution, namely, that the optically active 4f electrons are shielded from the crystalline field by the outer 5s and 5p shells. The next group of ions to be considered, the Transition Metal Ions, has valence electrons that are exposed to the crystalline field, and exhibit a vastly different behavior.

2.3 TRANSITION METAL IONS IN SOLIDS

2.3.1 *Introduction*

The class of transition metals consists of those elements with unfilled d-shells. In pure form they are metals, with electrical and thermal properties usually associated with metals. When doped into insulators, however, such similarities are less evident. Their spectroscopic properties vary significantly ion to ion and host to host. We restrict our discussion to those elements most commonly used as dopants in optical materials – the 3d elements.

When doped into a solid, these elements form positive ions, the valency of which is host-dependent. The outer 4s electrons are stripped from the ion and reside closer to the anions of the solid. Common valencies and shell configurations of each ion are shown below.

Transition Metal Ion	Electronic Configuration
Ti^{3+}, V^{4+}	$1s^2\ 2s^2\ 2p^6\ 3s^2\ 3p^6\ 3d^1$
V^{3+}, Cr^{4+}	” $3d^2$
V^{2+}, Cr^{3+}, Mn^{4+}	” $3d^3$
Cr^{2+}, Mn^{3+}	” $3d^4$
Mn^{2+}, Fe^{3+}	” $3d^5$
Fe^{2+}, Co^{3+}	” $3d^6$
Co^{2+}	” $3d^7$
Ni^{2+}	” $3d^8$
Cu^{2+}	” $3d^9$

2.3.2 *Energy Levels of Transition Metal Ions in Solids*

Optical transitions in these centers involve the 3d electrons, which are in the outermost shell and so are exposed to the crystal field. Consequently, the crystal field interaction is larger than the spin-orbit interaction.

To begin the discussion of the TMI, we choose the case where there is only one 3d electron, so that H_C is zero. Following the interaction with the central field, the next strongest term in the Hamiltonian is the crystal field interaction. Though several crystal field symmetries are possible, we examine the case of a TMI situated at a site with octahedral (or near octahedral) symmetry (Figure 2.4).

At an octahedral site the angular parts of the wavefunctions are linear combinations of spherical harmonics [2.6], the plots of which are shown in the Figures 2.5 (a)-(e). The following observations can be made regarding these wavefunctions.

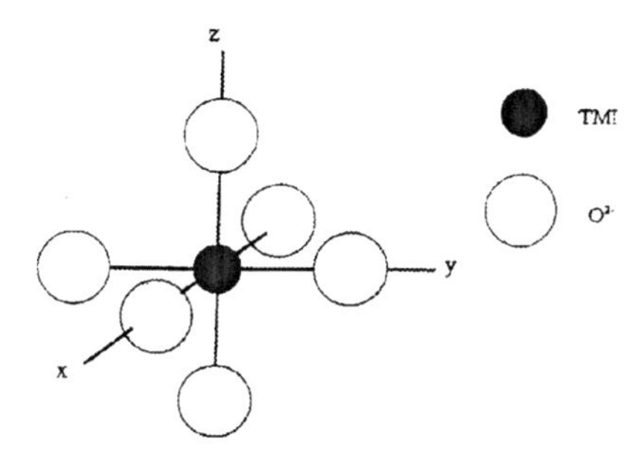

Figure 2.4 Transition metal ion in an oxide at a site of octahedral symmetry.

1. The wavefunctions are even. In fact, each wavefunction is invariant under all symmetry operations of the octahedral field.
2. The wavefunctions d_{xy}, d_{xz}, and d_{yz} are identical both in shape and in their relative orientation with respect to the surrounding ions, and so have the same energy. These states are labeled as t_{2g}.
3. The wavefunctions $d_{r^2-z^2}$ and $d_{x^2-y^2}$ are directed toward the O^{2-} ions and so have higher interaction energies than the t_{2g} states. The energy these two states are identical. These states are labeled e_{2g}.

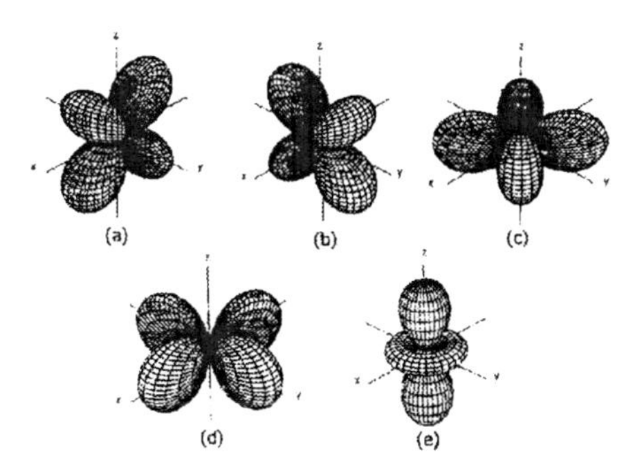

Figure 2.5 The wavefunctions of a $3d^1$ system in an octahedral environment. The ground state wavefunctions are labeled (a) d_{xz} (b) d_{yz} and (c) d_{xy}, and the excited states are (d) $d_{r^2-z^2}$ and (e) $d_{x^2-y^2}$.

Given observations 2 and 3 above, this system exhibits two energy levels as shown in Figure 2.6. The difference in energy between the two levels is commonly designated as 10Dq, where Dq is a measure of the strength of the crystal field. Note that a similar behavior is exhibited by TMI when the site symmetry is cubic. For example, in Ti^{3+}-doped Al_2O_3, the crystal field is mainly cubic, with a slight trigonal distortion. The splitting between the ground state ($^2T_2$) and the excited state (^2E) is around 19000 cm^{-1} [2.7]. The absorption band occurs in the blue-green region, accounting for the crystal's pink color.

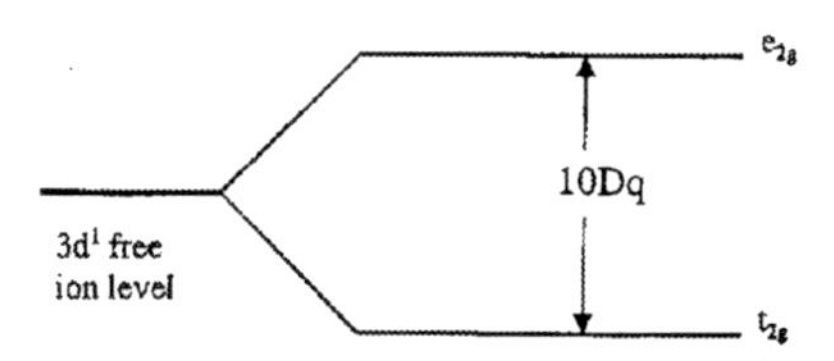

Figure 2.6 Splitting of the ground configuration due to an octahedral field for a 3d^1 system.

Now we consider the system having two 3d electrons. Taking into account the interaction with the crystal field, but before including H_C, each 3d electron will be in either the t_{2g} or e_{2g} state. The state of the system, then, is a product of one-electron states. For the case of a $3d^2$ system (Cr^{4+} or V^{3+}) the states are $(t_{2g})^2$, $t_{2g}e_{2g}$, and $(e_{2g})^2$.

When H_C is included, the one-electron product states are split. For a $3d^2$ system, the new states are assigned the group theoretical labels: A_1, A_2, E, T_1, and T_2. Since the spins of this system are parallel or anti-parallel, the total spin quantum number, S, is either 1 or 0, and 2S+1 is 3 and 1, respectively. Including spin, the following labels describe the states: $^1A_1$, $^3A_2$, ^1E, $^3T_1$, $^1T_1$, $^3T_2$, and $^1T_2$.

The energy of each state depends on the strength of the crystal field. This is usually depicted in a Tanabe-Sugano diagram [2.8] of the type shown in Figure 2.7, which shows the $3d^2$ splittings in an octahedral field. At zero crystal field, the levels are designated according to the free ion spectral terms, shown to the left of the diagram. In the strong field limit, the slopes of the lines approach one of three possible values, which are associated with the $(t_{2g})^2$, $t_{2g}e_{2g}$, and $(e_{2g})^2$ orbitals.

2.3.3 *Spectral Features of Transition Metal Ions in Solids*

Absorption and emission spectra of TMI in solids generally show broad bands, though sharp lines are occasionally observed. The absorption and emission spectra of Ti^{3+}:Al_2O_3 are shown in Figure 2.8. The broad bands result from the 3d electrons' exposure to the neighboring ions, making the wavefunctions sensitive to the position of the nearby ions.

Note the large Stokes shift between the peaks of the absorption and emission bands. This can be understood in the following way. Absorption of a photon results in a redistribution of the charge of the 3d electrons. The neighboring ions then "see" the new potential and relax to a new equilibrium position. During relaxation, the center gives up energy to the lattice, usually by emitting phonons. A similar relaxation process occurs following the emission of a photon. Since part of the absorbed energy is given up to the lattice, less is available for the emitted photon.

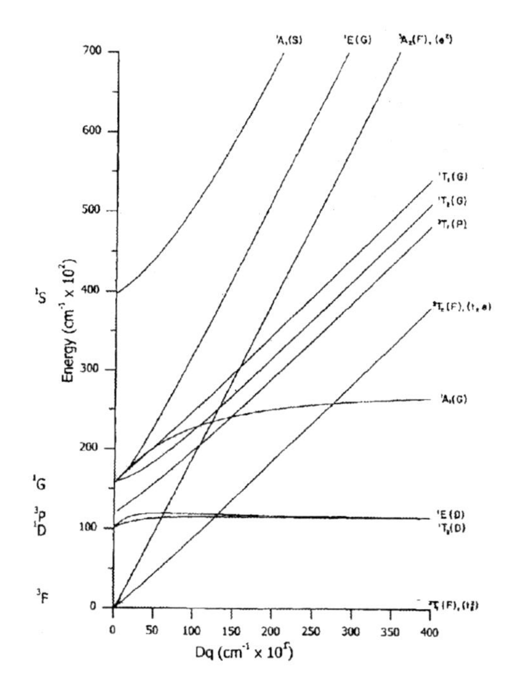

Figure 2.7 Tanabe Sugano diagram for a $3d^2$ ion in an octahedral field.

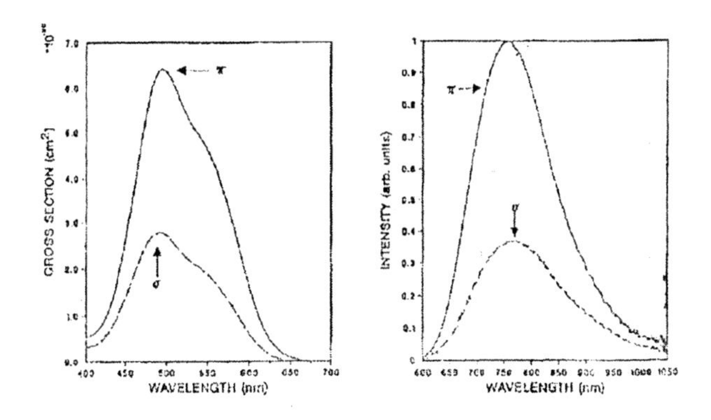

Figure 2.8 Polarized (a) absorption spectra and (b) emission spectra of Ti^{3+} in Al_2O_3.

2.3.4 *The Configurational Coordinate Model*

The broad bands and large Stokes shifts can be understood using a *Configurational Coordinate* (CC) model [2.9, 2.10]. In this model, each electronic energy level is represented by a parabola on an energy vs. position diagram (Figure 2.9). The position coordinate can be considered one of the normal coordinates of the crystal, though is often associated with the "breathing mode " of the ligands about the central TMI. The equilibrium position depends on the electronic state of the system. Horizontal lines within each parabola represent vibrational levels.

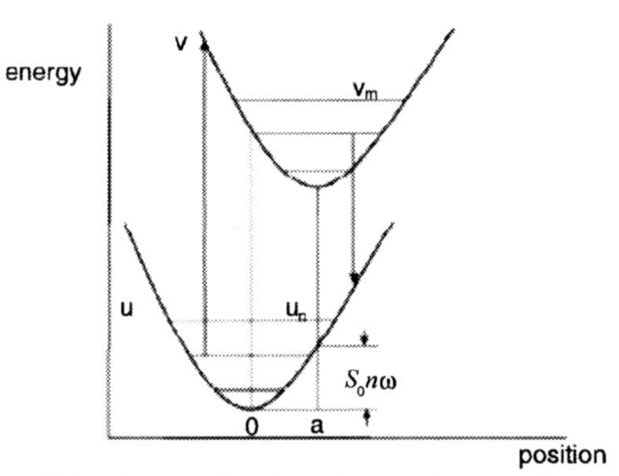

Figure 2.9 Configuration Coordinate diagram of a two-level system.

The assumptions that are included in the model are the following.

1. *The Adiabatic Approximation*: The electrons and nuclei are separated into fast and slow subsystems, respectively. The electronic wavefunctions depend parametrically on the position of the nearby nuclei. The motion of the nuclei is determined by a potential defined by an average position of the electrons. Physically, this is justified by the fact the mass of the electron is much less than the nuclear mass.
2. *The Harmonic Approximation*: The electronic states that result from the adiabatic approximation have potential curves that are parabolic. Physically, this assumes that the amplitude of the TMI vibrations is small.
3. *The Franck-Condon Principle*: This asserts that during a radiative transition the electronic charge is redistributed before the nuclei can react. Thus, transitions are represented by vertical arrows in the CC diagram. This is justified on the same basis as the adiabatic approximation.

Further discussions of these assumptions can be found in references 2.9 and 2.11.

In Figure 2.9 the ground and excited electronic levels are labeled, u and v respectively, and their parabolas are offset from one another by a distance a. We assume identical force constants for each parabola. The vibrational levels of u and v are labeled as n and m, respectively, and the associated vibrational wavefunctions are ϕ_n and ϕ_m. These are transitions in absorption (upward) or emission (downward). In absorption, the electron moves from the state $|u_n\rangle$ to $|v_m\rangle$, creating (m-n) phonons in the process. After a transition (absorption or emission), the ion relaxes to a new equilibrium position by releasing phonons to the surrounding lattice.

The Stokes shift between absorption and emission and the shape of the spectral bands are determined by both the force constant of the vibration, k, and the offset, a. The most important parameter in the model is the Huang-Rhys parameter, S_0, which is defined as

$$S_0\hbar\omega=\frac{1}{2}ka^2 , \tag{2.8}$$

where ω is the angular vibrational frequency of the system. S_0 is also the average number of phonons created or destroyed in the transition [2.12]. The physical meaning of S_0 can be understood in the following manner. The lattice motion induces a local strain, which depends on the electronic state of the ion. If the strain is similarly affected for two electronic states, the value of S_0 between those two states is small. If two states affect the strain very differently, then S_0 will be large. The constant S_0, then, describes the *relative* coupling of the electronic wavefunctions to the vibrationally-induced strains in the lattice.

Enforcement of the Franck-Condon principle leads to the result that the strength of a transition from $|v_m\rangle$ to $|u_n\rangle$ is proportional to the square of the overlap between the vibrational wavefunctions of the two states:

$$W_{m,n} \propto \left|\langle\phi_m|\phi_n\rangle\right|^2 \tag{2.9}$$

These overlap integrals are called the Franck-Condon factors. Since the vibrational levels of the initial electronic state are populated according to the Boltzmann distribution, $W_{m,n}$ has an inherent temperature dependence.

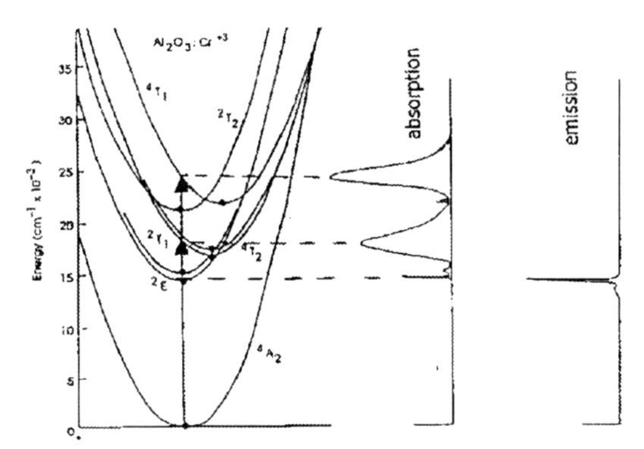

Figure 2.10 The configurational coordinate diagram and absorption and emission spectra of Cr^{+3} : Al_2O_3. [2.15]

A CC model for ruby (Cr^{3+}:Al_2O_3) is shown in Figure 2.10, along with the absorption and emission spectra. Note that the transitions from the ground $^4A_2$ state to the upper $^4T_1$ and $^4T_2$ states, are spin-allowed. The $^4T_1$ and $^4T_2$ states are offset from the $^4A_2$ state, and so exhibit broad absorption bands. In emission, the situation is different. Following absorption into the $^4T_2$ state, the ion relaxes to the base of the $^4T_2$ parabola emitting phonons in the process. From that point, there is a feeding into the ^2E state, which is accompanied by the further emission of phonons. Because there is small offset between the ^4A and ^2E states, the $^2E \rightarrow ^4A$ emission is a sharp line. (This is the 694 nm ruby laser emission line.) This transition is between states of different spin, and so it is weak, and the lifetime of the ^2E state is correspondingly long – 3 ms.

2.4 COLOR CENTERS

2.4.1 *Introduction*

Even the purest crystals are found to have defects in the crystalline structure. The defect centers of concern in this section are color centers, some of which are given in the table below. These are most widely found in alkaline halides, though they also occur in oxides and other miscellaneous hosts. Unless otherwise stated, our discussion assumes alkaline halide host materials.

Color Center	Description in Alkaline Halides
F	one electron trapped at an anion site
F^-	two electrons trapped at an anion site
F^+	vacant anion site
F_2 (M)	two neighboring F-centers along the 110 axis
F_2^+	two neighboring vacant anion sites (along 110 axis) containing a single electron
F^3 (R)	three neighboring F-centers forming an equilateral triangle
F_A	F-center adjacent to a smaller alkali impurity ion
V_K	singly ionized halogen-ion molecule formed from two adjacent halides

2.4.2 *Energy Levels of F Centers*

A single crystal often contains more than one type of color center. Consequently, the spectra often show several broad bands in absorption. The simplest of these centers is the F center, which consists of a single electron trapped at an anion vacancy (Figure 2.11). F centers can be formed either chemically or by bombardment with particles (e^-, p^+, neutrons) or high-energy radiation (ultraviolet, x-ray, γ-ray). A more complete discussion of F centers can be found in Fowler [2.14].

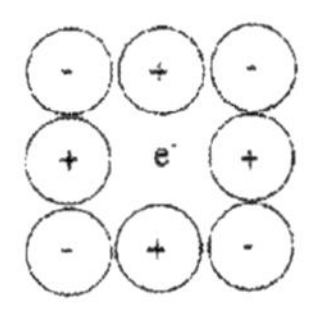

Figure 2.11 The F center, a trapped electron at an anion vacancy.

In cubic lattices, such as alkaline halides, the obvious physical model for the F center is that of a particle in a three dimensional box of length $2l$, where l is the distance form the center of the vacancy to the nearest alkaline cation. The energy difference between the two lowest levels of such a particle is

$$\Delta E = \frac{3\pi^2\hbar^2}{8ml^2}, \tag{2.10}$$

where we note the l^{-2} dependence. Data on the low energy absorption band of F centers in several alkaline halides is found to obey the following (Mollwo-Ivey) law: $\Delta E = 17.7(l)^{-1.84}$. The nearly l^{-2} dependence lends some credibility to the particle-in-a-box model. Generally, absorption spectra show only one excited level associated with an isolated F center before the onset of the conduction band.

The states of the color center are labeled according to the irreducible representations of the site symmetry. For alkaline halides the ground and first excited states are Γ_1^+ and Γ_4^-, respectively. It is found that the lowest electronic states exhibit character consistent with 1s and 2p states the of hydrogen atom. As can be seen in Figure 2.12, even the simple particle-in-a-box model yields ground and first excited state wavefunctions that show similar charge distribution patterns to the s and p hydrogenic wavefunctions, respectively.

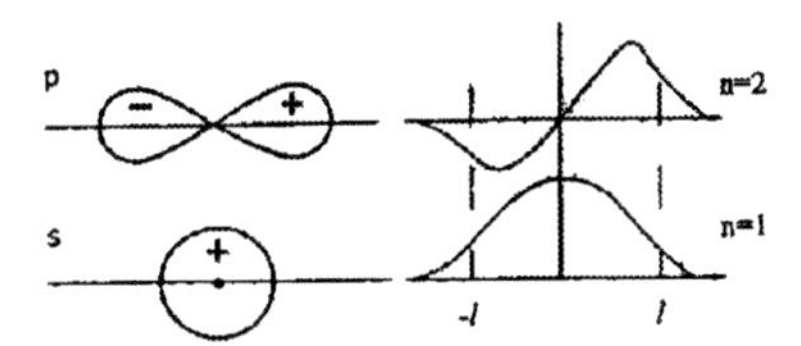

Figure 2.12 Diagram comparing the s- and p-type hydrogenic wavefunctions to the n = 1 and n = 2 wavefunctions of a particle in a finite well.

2.4.3 *Spectral Features of F Centers*

The potential in which the electron is situated is determined completely by the charge distribution of the nearby ions. As a result, the coupling between the F center and the lattice is strong, and absorption and emission spectra exhibit broad bands and large Stokes shifts, on the order of 1 eV. Figure 2.13 shows the absorption and emission of an F center in KBr. No zero phonon line is observed, and absorption and emission bands show no structure, even at low temperature. A configurational coordinate diagram of an F center has a large offset between the ground and excited state parabolas, and a Huang Rhys parameter, S_0, on the order of 30.

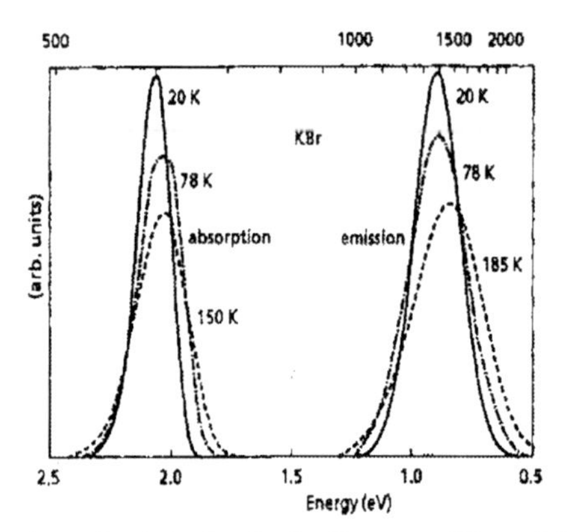

Figure 2.13 Absorption and emission bands of an F center in KBr at different temperatures. [2.15]

Absorption from the 1s state to the 2p state is an allowed transition with an f-number on the order of unity. Because of the relaxation of the nearby ions following excitation, the matrix element driving the emission process can be very different from that responsible for absorption. If the transition is from a p state to an s state, the f-number will again be close to 1.

The quantum efficiency of emission from the 2p state is close to unity at low temperatures. As the temperature is raised, the quantum efficiency decreases due to the presence of one or more quenching processes. One such process is non-radiative decay from the 2p to the 1s state. Another quenching mechanism is thermal activation from the occupied 2p state into the conduction band. Following absorption, the 2p state moves to within a few tenths of an eV of the conduction band. Photoconductivity experiments show thermal activation into the conduction band at temperatures around 100K, depending on the host. Other quenching processes, including the interaction with other centers, have also been proposed [2.16]. At room temperature, the luminescence from F centers in alkaline halides is often severely quenched.

3. Processes in Localized Systems

3.1 INTRODUCTION

The processes that affect the spectral characteristics of optically active centers include both radiative and non-radiative types. In this section, we discuss the origin of these processes and their effects on the luminescence properties of the system.

Of the non-radiative processes, many of them include the emission and/or absorption of phonons. In fact, phonon-related processes are important to the luminescence properties of virtually all optically active centers, even those where the ion is only weakly coupled to the lattice.

To understand why this is so, recall that the density of phonons for a monatomic solid in the frequency range between ω and $\omega + d\omega$ is:

$$n(\omega)d\omega = \left[\frac{1}{\exp(\hbar\omega/kT)-1}\right]\frac{3\omega^2}{2\pi^2 c_s^3}d\omega \qquad (3.1)$$

With the velocity of sound in a solid (c_s) on the order of 5×10^5 cm/sec, assuming T = 300K, and a Debye temperature of 1000K, the total number of phonons is on the order of $10^{23}/cm^3$. This high phonon density makes phonon-related processes ubiquitous, affecting all the spectral characteristics (e.g. intensity, lineshape, linewidth, and lifetime) of optical centers in solids.

3.2 RADIATIVE DECAY

3.2.1 *Radiative Transitions between f-Levels of REI in Solids*

The probability of absorption or emission of radiation is determined by matrix elements of the form $\langle\psi_a|\mathbf{D}|\psi_b\rangle$, where **D** is the operator responsible for the transition. For rare earth ions in gaseous form, the 4f-states are of odd parity. If **D** is the electric dipole operator, which is odd, then the matrix element vanishes. Thus, electric dipole transitions between f-states are forbidden. In a crystal, however, such transitions are frequently observed. In this section, we show that this is because the crystal field interaction mixes some of the character of the upper nearby configurations of opposite parity (e.g. 5d and 6s) into the 4f states.

Because $H_{CF\text{-}static}$ is small, we treat it as a perturbation. We define $|u\rangle$ and $|u'\rangle$ to be the free ion states of the 4f and 5d configurations, respectively. To the first order in the perturbation, the 4f state of the ion in a solid, $|\psi\rangle$, is

$$|\psi_k\rangle = |u_k\rangle + \sum_j \frac{\langle u'_j|V_{odd}|u_k\rangle}{E_k - E'_j}|u'_j\rangle \,. \qquad (3.2)$$

The sum is over all free ion states belonging to the 5d configuration, V_{odd} represents the odd components of $H_{CF\text{-}static}$, and E_k and E'_j are the energies of $|u_k\rangle$ and $|u'_j\rangle$, respectively. When there is inversion symmetry at the rare earth ion site, $V_{odd} = 0$ and no mixing occurs. If the ion site lacks inversion symmetry, mixing is always present.

The matrix element of the electric dipole operator between two states of the type given by (3.2) is

$$\langle\psi_l|e\mathbf{r}|\psi_k\rangle = \sum_m \frac{\langle u_l|V_{odd}|u'_m\rangle}{E_l - E'_m}\langle u'_m|e\mathbf{r}|u_k\rangle + \sum_j \frac{\langle u'_j|V_{odd}|u_k\rangle}{E_k - E'_j}\langle u_l|e\mathbf{r}|u'_j\rangle \,. \qquad (3.3)$$

Transitions are driven by the square of (3.3) Since the terms of the form $|\langle u_f|e\mathbf{r}|u'_j\rangle|^2$ are nonzero, electric dipole transitions between 4f states can occur whenever $V_{odd} \neq 0$. Transitions of this type have f-numbers on the order of 10^{-6}.

We note that magnetic dipole and electric quadrupole transitions are allowed even among pure f-states, and obey the selection rules, ΔL, $\Delta J \leq 1$ and ΔL, $\Delta J \leq 2$, respectively. Transitions are frequently observed, however, with ΔL and ΔJ as large as six. Based on both theoretical and experimental evidence, it is generally agreed that the vast majority of f $\rightarrow$ f transitions are electric dipole in nature. A method of calculating these transitions is presented by Dr. Brian Walsh later in this volume.

3.2.2 *Radiative Transitions between d-levels of TMI in Solids*

In section 2.4.2 it was noted that for a TMI at a location of inversion symmetry (e.g. octahedral), the 3d orbitals are even. Thus electric dipole matrix elements between states belonging to the 3d configuration are zero. For electric dipole transitions to occur, there must be a distortion to the inversion symmetry. This distortion introduces odd components into H_{CF}, and mixes the character of crystal field states of higher configuration (e.g. $3d^{n-1}4p$) into those of the $3d^n$ configuration. This can occur through vibrational motion of the nearby ligands, but we assume that the crystal contains a static distortion, and that it can be treated using perturbation techniques.

Considering the $^3T_2$ state of V^{3+} in an octahedral field, for example, the corrected wavefunction, $|\psi(^3T_2)\rangle$ to the first order, is

$$\left|\psi\left(^3T_2\right)\right\rangle = \left|u\left(^3T_2\right)\right\rangle + \sum_j \frac{\left\langle u'_j\right|V_{odd}\left|u\left(^3T_2\right)\right\rangle}{E\left(^3T_2\right)-E'_j}\left|u'_j\right\rangle \tag{3.4}$$

where $|u(^3T_2)\rangle$ is the state of the system in the octahedral field, $|u'_j\rangle$ represents a state of an upper (odd) configuration in the same field, and $E(^3T_2)$ and E'_j are their respective energies. The odd components of $H_{CF\text{-}static}$ are labeled V_{odd}. A similar expression can be obtained for $|\psi(^3T_1)\rangle$. The matrix element of the electric dipole operator between the corrected $^3T_2$ and $^3T_1$ states is:

$$\left\langle\psi\left(^3T_1\right)\right|e\mathbf{r}\left|\psi\left(^3T_2\right)\right\rangle = \sum_m \frac{\left\langle u'_m\right|V_{odd}\left|u\left(^3T_2\right)\right\rangle}{E\left(^3T_2\right)-E'_m}\left\langle u\left(^3T_1\right)\right|e\mathbf{r}\left|u'_j\right\rangle + \sum_j \frac{\left\langle u\left(^3T_1\right)\right|V_{odd}\left|u'_j\right\rangle}{E\left(^3T_1\right)-E'_j}\left\langle u'_m\right|e\mathbf{r}\left|u\left(^3T_2\right)\right\rangle \tag{3.5}$$

Typically, the energy denominators in (3.5) are large (several eV) and V_{odd} is small, so that the transitions are weak: f-numbers are on the order of 10^{-4}.

We note that since the spin-orbit interaction is small in TMI, to a large degree spin remains a good quantum number. Thus, transitions between states of different spin values are very weak.

Magnetic dipole transitions are allowed between 3d states, even in a perfect octahedral field. In spite of this, they are not as important as electric dipole transitions, unless the TMI is at a site having perfect octahedral symmetry, such as in the host MgO, where electric dipole transitions are strictly forbidden. Magnetic dipole transitions in TMI have f-numbers on the order of 10^{-6}.

3.3 MULTIPHONON DECAY

The process considered here is the decay of an excited ion by the emission of phonons, commonly called multiphonon decay (Figure 3.1). In this section, we find an expression for the multiphonon decay rate of ions in solids, considering first the case of rare earth ions.

The first step is to express the crystal field as a Taylor expansion about the equilibrium position of the ion:

$$H_{CF} = H_{CF\text{-}static} + \sum_i \frac{\partial V}{\partial Q_i} Q_i + \text{ higher order terms} \tag{3.6}$$

In (3.6) V is the potential seen by the electrons and the Q_i are the normal modes of vibration of the lattice. At a lattice site with no inversion symmetry,

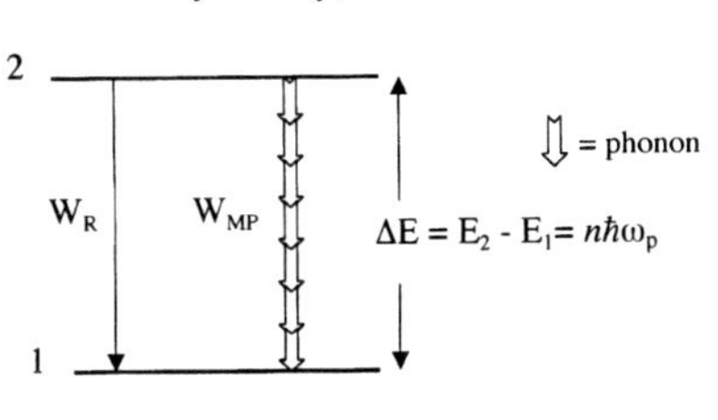

Figure 3.1. Two level system showing the radiative (W_R) and multiphonon (W_{MP}) decay processes.

the zero[th] order term, $H_{CF\text{-}static}$, is responsible for breaking the f$\rightarrow$f parity selection rule, allowing radiative electric dipole transitions within the 4f configuration. The first order and higher order terms describe modulations of the crystal field due to vibrations of the ligands, and is what we have termed $H_{CF\text{-}dynamic}$. It is also called the *ion-lattice* or *electron-phonon interaction*. This couples the motion of the electron to that of the lattice, allowing for phonon transitions between electronic states.

To obtain an expression for the multiphonon decay rate, one may take the first order term in (3.6) to the nth order of the perturbation. Alternatively, one can use the higher ordered term in (3.6) to a lower order of perturbation. No matter the approach, however, the result takes the following form:

$$W_{MP}(n) = W_{MP}(n-1)C \ , \tag{3.7}$$

where $W_{MP}(n)$ and $W_{MP}(n\text{-}1)$ are the multiphonon decay rates of an n and n-1 phonon process, respectively, and C is a constant much less than unity. This leads to the so-called energy gap law [3.1].

$$W_{MP}(n) = W_0 e^{-\alpha\Delta E} \tag{3.8}$$

W_0 is a constant, and ΔE is the energy gap to the next lowest level. If $\hbar\omega$ is the energy of the phonons involved, then $\Delta E = n\hbar\omega$. The high-energy phonons play the dominant role since fewer phonons are required, thereby lowering the order of the process. The constant α in (3.8) describes the ion-lattice coupling strength, and is determined experimentally. The energy gap law has been found to be valid for REI in solids [3.2].

An alternative starting point for describing the multiphonon decay rate is to express a non-adiabatic interaction Hamiltonian in the adiabatic approximation. This leads to a similar energy gap law for the case of REI [3.3].

The temperature dependence of the multiphonon decay process is related to the number of available phonons in the host, particularly the high energy phonons. The average occupation number of phonons in the ith mode is $N(\omega_i) = [\exp(\hbar\omega_i/kT) - 1]^{-1}$, and the emission of a phonon is proportional to $(N(\omega_i)+1)$. The nth-order multiphonon decay rate goes as $(N(\omega_i) + 1)^n$. This temperature dependence has been verified [3.4].

Observations have lead to the following rule of thumb for REI: If the energy difference to the next lowest level requires seven phonons or more, the dominant decay mechanism is radiative, otherwise it prefers to emit non-radiatively.

For TMI, a perturbation approach to multiphonon decay is not feasible. The multiphonon decay process and its temperature dependence are well described, however, using the Configurational Coordinate model. In this model, the ion first undergoes a non-radiative transition from state $|v_m\rangle$ to state $|u_n\rangle$, as shown in Figure 3.2. This transition is driven by the ion-lattice interaction, and the rate at which it occurs can be accurately calculated by summing by the Frank-Condon factors (2.9), weighted of course by the population of the initial state, $|v_m\rangle$. The transition is favored at the point where two parabolas cross, since that is where the Franck-Condon factor is greatest. Once in the electronic state $|u\rangle$, the system decays rapidly by giving up phonons to the lattice, resulting in the thermalization of $|u\rangle$. The work of Struck and Fonger [3.5] provides an in-depth discussion of this method.

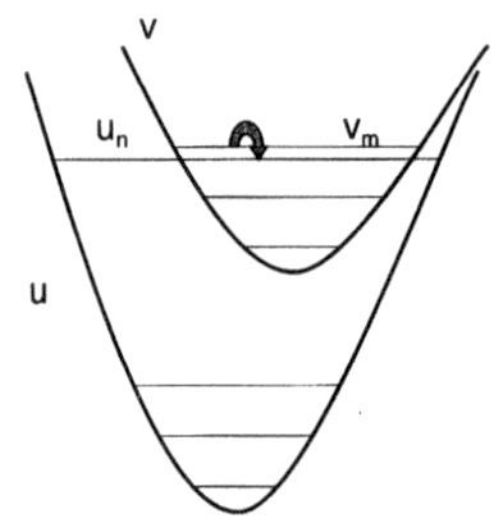

Figure 3.2. Configurational Coordinate diagram showing a non-radiative transition from $|v_m>$ *to* $|u_n>$.

3.4 VIBRONIC TRANSITIONS

A vibronic transition in emission involves the creation of a photon and a simultaneous creation or absorption of one or more phonons. Schematic diagrams of the vibronic transitions are shown in Figure 3.3. At low temperatures, the vibronic sidebands appear on the high (low) energy side of the zero-phonon line in absorption (emission). At high temperatures, vibronic lines appear on both the high- and low-energy sides of the zero-phonon line. The vibronic spectrum of Pr-doped YAG is shown in Figure 3.4.

As with multiphonon decay, vibronic transitions are driven by the ion-lattice interaction. These transitions occur via two different types of processes, commonly labeled the M and the Δ processes [3.6]. Whereas both processes apply to the rare earth ions, the Δ process is most relevant to transition metal ions and color centers.

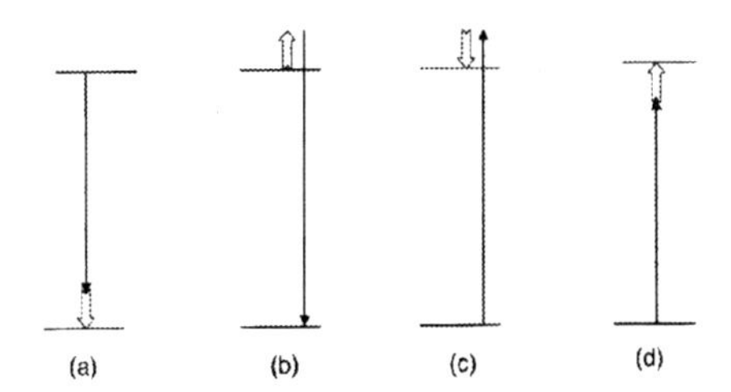

Figure 3.3. Vibronic transitions in (a,b) emission and (c,d) absorption. (b) and (d) are anti-Stokes processes and occur only at high temperatures. (a) and (c) can occur at all temperatures.

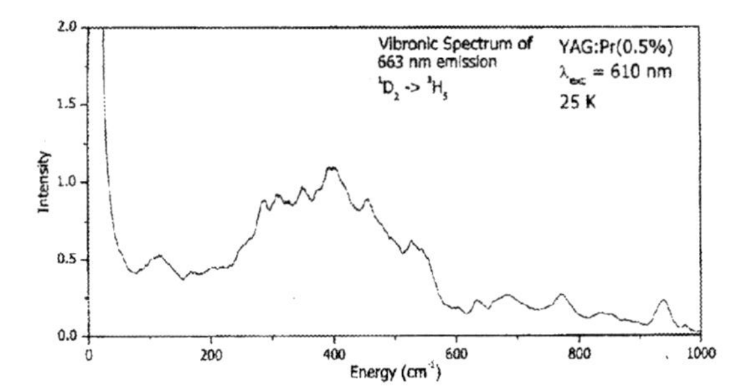

Figure 3.4. Vibronic emission spectrum of the $^1D_2$ level of Pr : YAG at 25K.

In eq. (3.2) we saw that the odd components of $H_{CF\text{-}static}$ cause the mixing of opposite parity states into the 4f-states of REI, allowing radiative transitions to occur. The M process occurs when the vibrations of the nearby ligands introduce odd components into $H_{CF\text{-}dynamic}$, thus leading to radiative transitions. Because they are vibrationally induced, these transitions are accompanied by a phonon. The M process is especially important in systems where the ion is at a site of inversion symmetry.

The Δ process occurs in systems where the equilibrium position of the optical center with respect to the ligands changes following an electronic transition. In the Configurational Coordinate model, this means that S_0 is greater than zero. For Δ processes, the intensity, I, of the vibronic transitions is given by:

$$I = I_0 \frac{e^{-S_0} S_0^N}{N!} \tag{3.9}$$

where I_0 is the intensity of the zero-phonon line and N is the number of phonons involved in the process. This function is called a Pekarian distribution, and is plotted in Figure 3.5 for different values of S_0. For REI, $S_0 \approx 0.1$, so that that most of the energy is in the zero phonon line.

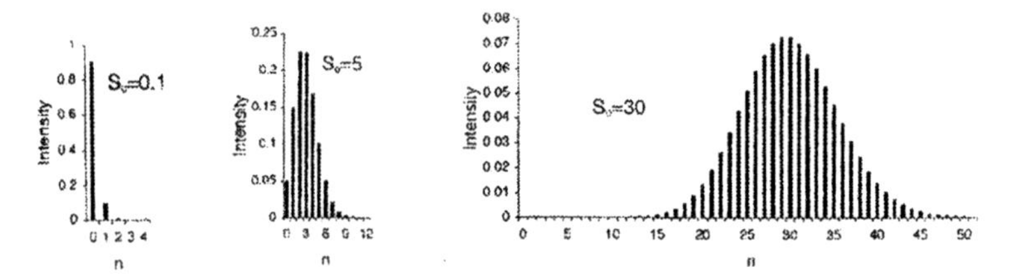

Figure 3.5. Plots of vibronic emission intensity versus number of phonons for various values of S_0 according to the Pekarian distribution in equation 3.9.

Vibronic transitions of REI can exhibit a strong host dependence, with the ratio of the area under the zero-phonon line to the area under the vibronic sidebands changing by as much as two orders of magnitude depending on the host [3.7].

Equation (3.9) is also applicable to transition metal ions and color centers. For TMI S_0 is between 2 and 10, so they often exhibit a zero phonon line and strong vibronic sidebands. In color centers S_0 is often on the order of 30, and the transitions are almost always purely vibronic.

3.5 ENERGY TRANSFER

When more than one optically active center is present, an excited center may impart all or part of its energy to a nearby center, a process called energy transfer. This can happen radiatively or non-radiatively.

In the radiative process (Fig. 3.6 (a)), the excited ion emits a photon and another ion absorbs that photon before it leaves the crystal. This process is shows a $1/R^2$ dependence and has little effect on the lifetime of the level from which the radiation originates.

Some non-radiative energy transfer processes are shown in Figure 3.6. The ion from which the energy is being transferred is called the sensitizer, S. The ion to which the energy is transferred is called the activator, A. We first consider the case of resonant energy transfer, shown is Fig. 3.6 (b).

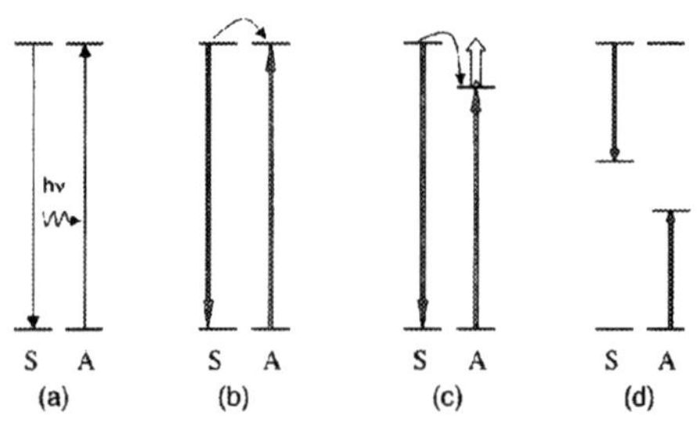

Figure 3.6. Energy transfer processes: (a) radiative, (b) resonant, non-radiative, (c) phonon-assisted, non-radiative and (d) non-radiative cross-relaxation.

The initial state of the system is an excited sensitizer and the activator in the ground state, and the final state of the system is the sensitizer in ground state and an activator in excited state. The interaction Hamiltonian is the coulombic interaction between the electrons of the sensitizer and those of the activator. Forster [3.8] and Dexter [3.9] developed the following equation for the resonant energy transfer rate, W_{SA}, between S and A:

$$W_{SA} = K \frac{Q_A}{R^n \tau_S} \int \frac{g_S(E) g_A(E)}{E^4} dE \; . \tag{3.10}$$

In (3.10), K is a constant, Q_A is the area under the absorption band of the activator, τ_S is the lifetime of the sensitizer, and R is the distance between the S and the A. g_S and g_A are the normalized shape functions of the emission bands of the sensitizer and the absorption bands of the activator, respectively. The integral includes the overlap of these two functions, and is essentially a conservation of energy statement.

The R^n term is determined by the multipolar interaction between S and A. In most cases, the dipole-dipole (n = 6) mechanism is dominant. In cases where the dipole-dipole term vanishes due to symmetry, then the dipole-quadrupole (n = 8) or quadrupole-quadrupole (n = 10) terms may dominate. The exchange interaction may also drive the transfer, but is important only when S and A are in close proximity to one another [3.10].

Experimentally, one can gain insight into the nature of the interaction by examining the response of the system to pulsed excitation. Based on the work of Förster [3.11], assuming an even distribution of sensitizers and activators, the evolution in time of the population density (n_S) of excited sensitizers goes as

$$n_S(t) = n(0) \exp\left(-\frac{t}{\tau_S} - \Gamma\left(1 - \frac{3}{q}\right) \frac{C_A}{C_0} \left(\frac{t}{\tau_S}\right)^{3/q} \right) . \tag{3.11}$$

In (3.11), C_A is the acceptor concentration, C_0 is the sensitizer concentration at which energy transfer is equally likely as spontaneous deactivation, τ_S is the lifetime of the sensitizer in that absence of the acceptor, and q = 6, 8, or 10 depending on whether the multipolar interaction is dipole-dipole, dipole-quadrupole, or quadrupole-quadrupole, respectively. A best fit of the decay curve to this equation can be used to determine the type of interaction.

In phonon-assisted energy transfer (Figure 3.7 (c)) the overlap integral in (3.10) is close to zero, and energy difference between the emission of the sensitizer and absorption of the activator must be made up for by the absorption or creation of one or more phonons. The following observations can be made regarding phonon-assisted transfer.

1. The transfer rate obeys an exponential energy gap law if the gap is much larger than the phonon energy [3.12], that is, if more than one phonon is required. This situation is frequently encountered in REI.
2. If W_{SA} is much faster than the decay rate of the upper levels of both S and A, the two levels become thermalized according to the Boltzmann distribution.

3. For TMI, the phonon-assisted energy transfer rate generally increases with temperature. For REI, the rate can increase or decrease with temperature depending on the energy levels involved [3.13].

3.6 UPCONVERSION

The term "upconversion " applies to processes whereby a system excited with photons of energy $\hbar\omega_1$ emits photons at a frequency $\hbar\omega_2$ where $\hbar\omega_2 > \hbar\omega_1$. The importance of this phenomenon is due, in part, to the fact that it can be useful for the generation of short wavelength emission without having to rely on UV excitation, which can have deleterious and secondary effects (such as creation of color centers and excitation of deep traps).

The scheme in Figure 3.7(a) is called excited state absorption (ESA) and involves a single ion. In this process two photons are absorbed sequentially, and the intermediate state is real. Non-radiative relaxation to a lower level may or may not occur following the absorption of the first photon.

Figure 3.7(b) shows a process called two-photon absorption (TPA). In this process, the two photons are absorbed simultaneously, and the intermediate state is a virtual one.

Another type of upconversion is a two-ion energy transfer process in which the activator is initially in an excited state (Figure 3.7(c)). This is referred to as upconversion by energy transfer (ETU).

Three-ion upconversion, an example of which is shown in Figure 3.7(d), has been observed, but such processes are very weak.

All upconversion processes are non-linear, since they require the absorption of at least two photons. ESA and ETU are the most common mechanisms. Experimentally, the two can be distinguished from one another by examining the system's response to pulsed (laser) excitation. Both systems show a rise followed by a decay. For ESA-type upconversion, the rise time is within the lifetime of the pump (laser) pulse. For ETU, the luminescence peak occurs after the laser pulse is over, with the rise time depending on the decay time of the luminescent state and the ETU rate. The reader is referred to a recent review of upconversion processes by Auzel [3.14].

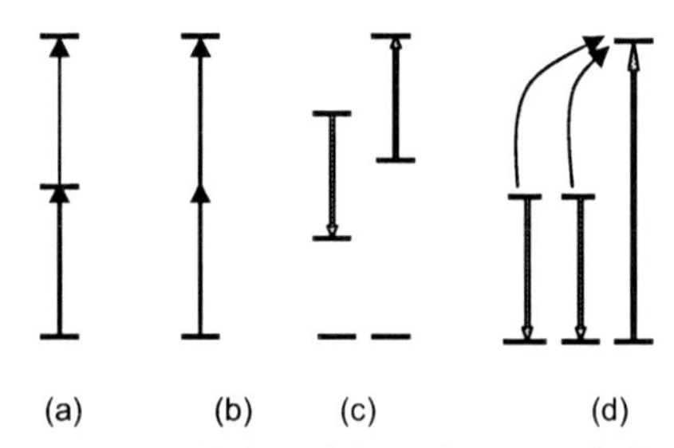

Figure 3.7. Upconversion processes (a) Excited State Absorption, (b) Two-Photon Absorption, (c) Upconversion by Energy Transfer, and (d) three-ion cooperative energy transfer.

3.7 LINE BROADENING AND SHIFTING WITH TEMPERATURE

In this section, we discuss the affect of temperature on the width and position of the zero-phonon lines, particularly in REI-doped solids.

3.7.1 *Spectral Line Broadening with Temperature*

The width of a spectral line is affected by several mechanisms, including inhomogeneous broadening, direct one-phonon processes, multiphonon decay, radiative decay, and Raman scattering. Inhomogeneous broadening is caused by variations in the crystal field local to the REI. This effect is not temperature dependent, and so will not be discussed here. Direct one-phonon processes, multiphonon decay, and radiative decay all affect width of the level by reducing its lifetime. With lifetimes of luminescent REI levels generally longer than a 10^{-7} sec, the sum of the contributions of these processes, as given by the uncertainty principle, is less than 0.01 cm^{-1}. Raman scattering involves the simultaneous absorption and emission of phonons, usually of two different frequencies. The electronic state of the ion is unchanged in the process, so that the lifetime is not affected. Experience has shown that the main contribution to the linewidth is from Raman processes, and is on the order of a few cm^{-1}.

Raman scattering is a second order process. Using a perturbation approach, the probability of the ion absorbing an ion at frequency ω_i and emitting a phonon of ω_j is [3.15]

$$P_{ij} = A\omega_i\omega_j N(\omega_i)(N(\omega_j)+1). \tag{3.12}$$

$N(\omega_i)$ and $(N(\omega_j) + 1)$ refer to the absorption and emission of a phonon, respectively, where $N(\omega_i) = [\exp(\hbar\omega_i /kT)-1]^{-1}$ is the number of phonons of the ith mode. A is a constant describing the ion-lattice interaction strength.

Assuming a phonon density of states given by the Debye distribution and integrating over all allowed phonon frequencies, the contribution to the linewidth becomes

$$\Delta E = A'\left(\frac{T}{T_D}\right)^7 \int_0^{T_D/T} \frac{x^6 e^x}{(e^x-1)^2}\,dx. \tag{3.13}$$

T_D is the Debye temperature of the solid. Experimental data of linewidth versus temperature are usually fit to (3.13) treating A' and T_D as adjustable parameters [3.16].

3.7.2 *The Shift of Spectral Lines with Temperature*

In section 4 we discussed the effect of $H_{CF\text{-}static}$ to the energy of the system. The contribution of $H_{CF\text{-}dynamic}$ to the energy of an ion in a solid is revealed in the shift of the spectral lines with temperature. As in the case of line broadening, Raman scattering is found to be the dominant process.

The contribution of Raman processes to the energy of the ith level is

$$\delta E_i = \sum_{j,k} \frac{\langle i,k | H_{CF-dyn} | j,k\pm 1 \rangle \langle j,k\pm 1 | H_{CF-dyn} | i,k \rangle}{E_i - E_j + \hbar\omega_k} + \langle i,k | H_{CF-dyn} | i,k \rangle \tag{3.14}$$

The sum over j includes all the electronic states of the system except the ith state. The sum over k includes all phonon modes.

Assuming a Debye phonon distribution, and also that the energy difference between electronic levels is greater than the phonon energies, (3.14) reduces to

$$\delta E_i = A''\left(\frac{\mathrm{T}}{\mathrm{T_D}}\right)^4 \int_0^{\mathrm{T_D}/T} \frac{x^3}{e^x - 1}\,dx\,, \tag{3.15}$$

where A" is a constant. A fit of this equation to lineshift data allows one to determine the Debye temperature of the solid.

We note the following:

1. Raman scattering involves the participation of real intermediate states. For REI, these states include all unoccupied 4f levels as well as those of the upper configurations.
2. The experimentally observed shift is the difference in the energy shifts of the two levels involved in the transition.

Experimental data show that lines shift generally toward the red, though blue shifts are sometimes observed. Typically, the observed shift is on the order of 10 cm^{-1} as the temperature varies from 4K to 500K [3.16].

Acknowledgements

The author would like to thank Professor B. Di Bartolo for his many helpful discussions in the preparation of this work and for the support of NASA under grant number NNL04AA30G. Much of the contents of this work can also be found in the *Handbook of Solid State Spectroscopy*.

References

1.1. Goethe J. W., *The Sorrows of Young Werther*, trans Mayer E and Brogan L, The Modern Library, New York

2.1. Dieke G. H. (1968) *Spectra and Energy Levels of Rare Earth Ions in Crystals*, eds. Crosswhite H M and Crosswhite H, Interscience, New York

2.2. Carnall, W. T., Goodman, G. L., Rajnak, K., and Rana, R. S. (1988) *A Systematic Analysis of the Spectra of the Lanthanides Doped into Single Crystal LaF_3* (Argonne National Laboratory Report AN-88-8)

2.3. Morrison, C. A., and Leavitt, R. P. (1982) in *Handbook on the Physics and Chemistry of Rare Earths*, Gschneidner, K. A. and Eyring, L. eds., North Holland, Amsterdam

2.4. Wegh, R. T., Meijerink, A., Lamminmaki, R. J., Holsa, J. (2000) *J. Luminescence* **87-89** 1002-4
2.5. Wegh, R. T. Van Loef , E. V. D., Burdick, G. W. and Meijerink, A., (2003) *Molecular Physics* **101** (7) 1047-56
2.6 Imbusch. G. F. (1978) in *Luminescence Spectroscopy*, Lumb, M. D. ed. Academic Press, London, New York p. 31
2.7 Moulton, P. F. (1986) *J. Opt. Soc. Amer. B* **3** 125-32
2.8 Tanabe, Y. and Sugano, S. (1954) *J. Phys. Soc. Japan* **9**, 753 and 766
2.9 Rebane, K. (1970) *Impurity Spectra of Solids* (New York: Plenum) Chapter 1
2.10 Struck, C. W. and Fonger, W. H. (1991) *Understanding Luminescence Spectra and Efficiency Using W_p and Related Functions*, Springer-Verlag, Berlin ,Heidelberg
2.11 Di Bartolo, B. (1978) in *Luminescence of Inorganic Solids*, Di Bartolo, B. ed: Plenum, New York
2.12 Toyazawa, Y. and Kamimura, H. (1967) *Dynamical Processes in Solid State Lasers*, Kubo, R. ed. W A Benjamin, New York p. 90
2.13 Struck, C.W. and Fonger, W.H. (1991) *Understanding Luminescence Spectra and Efficiency Using W_p and Related Functions*, Springer-Verlag, Berlin Heidelberg p. 165
2.14 Fowler, W.B. (1968) in *Physics of Color Centers*, Fowler, W.B. ed., Academic, New York p. 55
2.15 Gebhardt, W. and Kuhnert, H. (1964) *Phys. Letters* **11** 15
2.16 Baldacchini, G. (2001) *Advances in Energy Transfer Processes* Di Bartolo, B. ed (New Jersey London Singapore Hong Kong: World Scientific)
3.1 Imbusch, G. F. (1978) *Luminescence Spectroscopy* Lumb M D ed (London, New York: Academic)
3.2 Weber, M. J. (1973) *Phys. Rev. B* **8** 54-64
3.3 Auzel, F. (1978) *Luminescence of Inorganic Solids*, Di Bartolo, B. ed., Plenum, New York
3.4 Moos, H. W. (1970) *J. Luminescence* 1,2 106
3.5 Struck, C. W. and Fonger, W. H. (1991) *Understanding Luminescence Spectra and Efficiency Using W_p and Related Functions*, Springer-Verlag, Berlin, Heidelberg
3.6 Auzel, F. 1980 *Radiationless Processes* Di Bartolo, B. ed., Plenum, New York
3.7 de Mello Donega, C. Meijerink, A. and Blasse, G. (1992) *J. Phys. Condens. Matter* **4** 8889-902
3.8 Förster, T. (1948) *Ann. Phys.* **21** 55
3.9 Dexter, D. L. (1953) *J. Chem. Phys.* **21** 836-850
3.10 Inokuti, M. and Hirayama, F. (1965) *J. Chem. Phys.* **43** 1978
3.11 Förster, T. (1949) *Z. Naturforch.* **4a** 321
3.12 Yamada, N., Shionoya, S., and Kushida, T. (1972) *J. Phys. Soc. Japan* **32** 1577
3.13 Collins, J. M. and Di Bartolo, B. (1996) *J. Luminescence* **69** 335-341
3.14 Auzel, F. (2004) *Chem. Rev.* **104** 139-173
3.15 Di Bartolo, B. (1968) *Optical Interactions in Solids*, Wiley, New York, Chapter 15
3.16 Chen, X., Di Bartolo, B., Barnes, N. P. and Walsh, B. M. (2004) *Physica Status Solidi (b)* **241** (8) 1957-76

LASER SPECTROMETERS FOR ATMOSPHERIC ANALYSIS

F. D'AMATO
CNR – Istituto Nazionale di Ottica Applicata
Largo E. Fermi 6, 50125 Firenze - Italy
damato@interfree.it,

1. Introduction

This chapter will deal with practical applications of spectroscopic techniques. We can speak of a practical application when a scientific discovery has been found suitable to carry out a specific task, either better, or at a lower cost (or both) than existing devices.

Since many years [1-4] absorption spectroscopy has proven to be a powerful tool for the measurement of the concentrations of gaseous species. Yet, nowadays only very few people have seen commercial gas analyzers at work.

This means that either the principle of the technique is not good, or that the technological development is insufficient, or the benefit/cost ratio is lower than that of already existing devices. We'll try to understand which one(s) of the above conditions has prevented a wide-spread diffusion of laser spectrometers.

The available space is not sufficient to enter all details of the described techniques and devices. To overcome this, an extensive bibliography will be provided.

2. Available Analytical Techniques

We are dealing with instruments that can measure gas concentrations in different kinds of samples (air, breath, industrial emission, etc.). Of course, many measurements are already carried out in different ways. In order to understand which are the competitors to laser spectrometers, let's examine the main classes of devices.

Four methods will be described. For each of them the advantages and drawbacks will be highlighted, in order to understand for which kind of measurement they are most suitable. The four methods are Mass Spectrometry, Gas Chromatography, Chemical Methods and Solid State Sensors.

2.1. MASS SPECTROMETRY

This technique is about one century old. It's a sensitive and very precise method. The sample under investigation is ionized, then accelerated by an electric field. The ions are then deviated by a magnetic field. The deviation is inversely proportional to their mass. An array of sensors detects the ions. The signals from the detectors form the "spectrum" of the sample. Of course, each substance has its typical spectrum, so that it is possible to separate the contributions due to

B. Di Bartolo and O. Forte (eds.), Advances in Spectroscopy for Lasers and Sensing, 157–186.

the different constituents. In order the device to work, the sample must be kept at very low pressure (~10^{-4} Torr), and the acceleration requires electric fields of the order of some kV. A high vacuum pump and a high voltage supply are necessary. There can be some uncertainties, like for $^{12}C^{16}O_2$ and $^{14}N_2{}^{16}O$, both having mass 44. If so, a pretreatment of the sample is necessary, in order to remove the unwanted component (by means of physical or chemical methods) [5], and this slows down the measurement. Nevertheless, the mass spectrometry is the most used technique to measure the isotopic ratio $^{13}CO_2/^{12}CO_2$ in the human breath, for medical purposes.

2.2. GAS CHROMATOGRAPHY

This is another one century old technique. It is based on the different transit times of molecules in a capillary column, due to different mobility, under the effect of an inert carrier gas. A detector at the end of the column yields a signal proportional to the number of molecules per unit time. Very small samples are necessary (down to nanoliters), and devices of this type have proved to be suitable for airborne applications [6].

Yet, the measurement is intrinsically slow, due to the transit time. Moreover, due to the spread of the molecular velocities, the signal produced by each species has a finite, non zero time duration. It can occur that the signals from two different species overlap. By adjusting the initial temperature and pressure of the sample, and the length and diameter of the column, and by applying suitable data analysis routines it's possible to minimize (but not to suppress) this effect. The net result is a limited selectivity.

2.3. CHEMICAL METHODS

A very wide variety of chemical analyses can be performed to determine concentrations. They can be very precise and selective, suitable for a very large number of different elements or compounds, but have two drawback. They are not "on-line" methods, which means that samples must be taken out of the experiment to be analyzed. This can be a serious problem for highly reacting species. Moreover, they require a physical technique to yield a result: reaction products must be weighted, or the opacity of a solution must me measured by optical technique, etc., so they are not direct methods.

2.4. SOLID STATE SENSORS

This class of sensors is based on adsorbing materials which change one of their physical properties in presence of a particular molecule. In the simplest case, the sticking of the molecule increases the weight of the adsorbing film, in other cases the resistivity of the film varies.

In order to reveal very small changes in weight, the film is put onto an oscillating quartz. The change in the resonance frequency of the quartz is an indicator of how many molecules were adsorbed. In the second case, the film is inserted in one arm of a Wheatstone bridge.

These sensors are very simple, cheap and robust. They are being used for electronic noses [7] or for methane sensors.

Yet, they have some drawbacks. They are slow, sometimes they can be poisoned by the adsorbing molecules, which means that the molecules are not released any more from the film. But the worst problem is that their selectivity is very poor. In the case of a methane leak sensor, this implies false warnings, which are not dangerous indeed, but are surely tedious and time wasting.

3. Available Optical Techniques

There are several kinds of optical techniques. We have two criteria to divide them. The first division is among those techniques which require well defined frequencies, named "dispersive", which use lasers as light sources, and those which require broadband sources, like lamps or blackbodies, which are called "non dispersive".

The second division is among those methods which rely on the measurements of the effects of the light on the sample, against those which determine the effects of the sample on the light.

Some examples are reported.

3.1. OPTOACOUSTIC DETECTION

This is a technique in which the power left by the light beam in the sample is detected. It's a dispersive technique. When a laser beam at the right frequency crosses the sample (Fig. 1), some power is absorbed. While the absorption takes place, the energy acquired by the absorbers is converted into kinetic energy, i.e. the molecules increase their speed and exert a greater pressure on the walls of the container. If the beam is blocked, the sample thermalizes, and the pressure comes back to its original value.

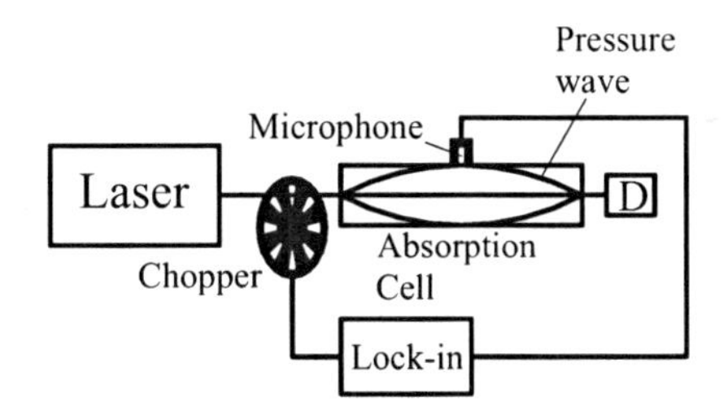

Figure 1. General setup for opto-acoustic detection. The detector D is only for power reference.

If this cycle is repeated, by periodically blocking the laser with a chopper, the pressure wave becomes an acoustic wave. A microphone inserted in the sample container yields a periodical signal, which can be easily detected by a lock-in amplifier.

This technique is very simple, doesn't require expensive detectors or electronics, the volume of the sample is very small. Of course, it's not suitable for open path measurements. In principle the signal is easy to be amplified, as the background is zero.

In order to insure the latter feature, it is necessary that no power is lost into the sample, apart that directly related to the measurement. Light scattered by dust, or reflected by the windows, causes a non-zero background baseline, so limiting the amplification of the signal.

When a very high sensitivity is required, the absorption cell must be carefully designed, and the microphone(s) must be located where the acoustic standing wave has a maximum. As the wavelength of sound in gases depends on pressure and temperature, these two parameters must be kept constant, in order to insure the match between cell dimensions and wavelength.

In order to suppress external noise, a careful profile of the gas line must be adopted. Very useful for this purpose are big volumes located at the entrance and at the exit of the cell. The problem is that they dramatically increase the gas exchange time. Finally, as the signal is proportional to the laser power, high power lasers (like CO_2 laser) are preferred.

Further details and useful references can be found in [8].

3.2. CAVITY RING DOWN/ENHANCED SPECTROSCOPY

When a light beam is confined between two high reflectivity mirrors (~99.999% or higher), like in Fig. 2,

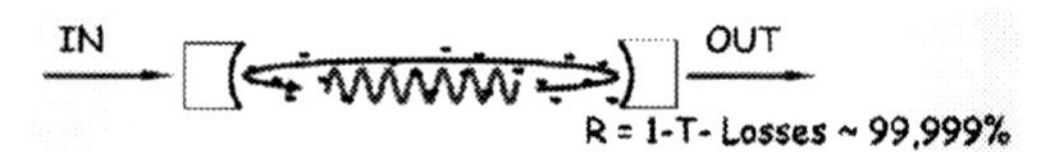

Figure 2. Principle of CRDS (Courtesy of D. Romanini).

photons can travel back and forth up to several km before being absorbed or transmitted. A detector outside the cavity yields a decreasing signal, whose decay time depends only on reflectivity. If inside the cavity there is an absorber, the decay time reduces. By measuring the decrease of the decay time versus laser frequency the absorbers concentration can be retrieved.

This relatively simple technique, due to the high interaction length, is very sensitive. When the laser frequency sweeps, only those frequencies which match the cavity length are transmitted. This means that the signal from the detector is available at regularly spaced frequencies. This is very useful for the identification of the absorptions.

Yet, if the reflectivity of the mirrors drops, the minimum detectable level increases. The same occurs if dust enters the cavity, as any loss of photons inside the cavity causes a lower decay time.

There are problems when working at low pressure. The cavity modes are spaced by c/2L, where L is the length of the cavity. For a cavity length of 1 m, this means 150 MHz. The absorption width of CO_2 lines at 2 μm, in the lower stratosphere, is about 300 MHz, so only two measurement points are available for each line. Higher frequency resolution would request too a high cavity length. The available power onto the detector is very low, so high quality, very low noise detector are necessary.

A technique to increase the power in the cavity, based on optical feedback into the laser, is described in [9]. In this case, the cavity transmission is measured, rather than the decay time. The latter is only used to determine the effective cavity length.

3.3. LASER INDUCED FLUORESCENCE

It's a one century old technique [10]. A light beam is absorbed by a sample, which then emits light at frequency lower than the previous one (Fig. 3). Each molecule has its own emission/absorption spectrum. By looking at the emitted light it is possible to determine the concentration of the absorbers. Without absorption, the signal from the detector is zero. The detector can be a photomultiplier, so a very low emission level can be detected. For these reasons a very high sensitivity can be achieved. An example is the detection limit of $1.8 \cdot 10^6$ molecules/cm^3 of OH- in the upper troposphere obtained in [11]. Another positive feature is that by means of a suitable optics, different points of the sample can be investigated.

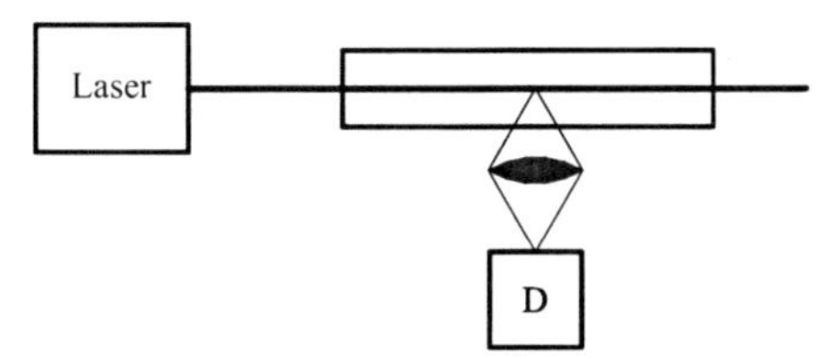

Figure 3. Principle of LIF.

Like for opto-acoustic detection and CRDS, this technique is only suitable for point measurements. It is limited to those species which have absorption/emission bands in the UV/Visible. A laser is not necessary, as lamps (like mercury lamp, a very cheap source) can fullfil the measurement requirements.

3.4. DIRECT ABSORPTION

Opto-acoustic detection and LIF are based on the measurement of the effects caused by light to the sample. CRDS belongs to the other group, like direct absorption. Direct absorption is in principle the most straightforward technique.

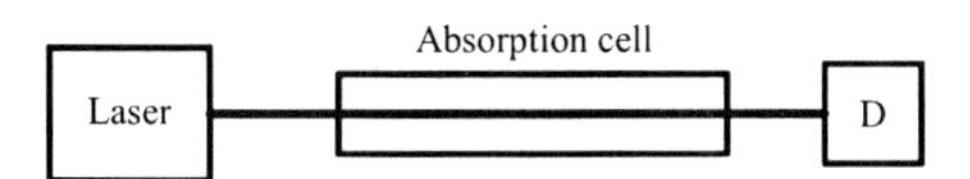

Figure 4. Principle of direct absorption.

The laser frequency is swept across a molecular absorption. Beyond the sample (Fig. 4), a detector converts the received power into an electrical signal. By measuring which fraction of the input beam escaped the sample it is possible to retrieve the concentration of the absorbers. This technique is also suitable for broadband sources. In this case a great care must be put in limiting the emission bandwidth of the blackbody, in order to avoid as much as possible regions where more than one gas have absorptions. Moreover, direct absorption can be used for both point and open path measurements.

The drawbacks are that a precise alignment is required and that, usually, a very small variation must be appreciated over a large signal.

This is the technique which all the instruments, described in the following chapters, are based on. One important remark must be made at this point: since now on, whatever the method or the instrument, the target molecules will exhibit discrete, not continuous absorptions.

4. Scientific Background

The starting point of the scientific survey is the Beer-Lambert law, which links the decrease in light power when crossing a sample, to the physical characteristics and the concentration of the absorber(s), and to the interaction length between light and sample:

$$I_{OUT}(\nu)=I_{IN}(\nu)\cdot e^{(-Sg(\nu)NL)} \tag{1}$$

where $I_{OUT/IN}$ is the power at the exit/entrance of the sample, S is the line-strength, $g(\nu)$ is the frequency profile of the absorption, N the density of the absorbers and L the interaction length.

Let's discuss the effect of pressure and temperature on the absorptions.

Pressure affects first of all the number of absorbers, which rises linearly with increasing pressure. The frequency shape of the absorption, with increasing pressure, changes from a Doppler profile at low pressures:

$$g_D(\nu-\nu_0)=\frac{\sqrt{\ln\left(\frac{2}{\pi}\right)}}{\gamma_D}\exp\left[\frac{-\ln(2)(\nu-\nu_0)^2}{\gamma_D^2}\right], \qquad \gamma_D=\frac{\nu_0}{c}\sqrt{\frac{2RT\ln(2)}{M}} \quad \text{(HWHM)} \tag{2}$$

to a Lorentz profile at high pressures:

$$g_P(\nu-\nu_0)=\frac{\frac{\gamma_P}{\pi}}{(\nu-\nu_0)^2+\gamma_P^2} \tag{3}$$

through a Voigt profile at intermediate pressures:

$$g_V(\nu-\nu_0)=\int_{-\infty}^{+\infty} g_D(\nu'-\nu)g_L(\nu_0-\nu')d\nu' \tag{4}$$

A minor effect is a shift of the absorption frequencies. Temperature affects the absorptions under many points of view. The density of the sample is inversely proportional to temperature.

The shape of the absorptions depends on temperature both because of Doppler width (Eq. 2), and because the pressure broadening depends on temperature [12]. The strengths of the lines depend on temperature because the populations of the levels, from which the transitions take place, follow the Boltzmann distribution. In most cases this is a problem. An example is given by the measurement of isotopic ratios, see Table I. A uncertainty of 0.1 K in the knowledge of the temperature is prohibitive for this kind of measurement.

TABLE I. Temperature dependence of two CO_2 lines suitable for the measurement of isotopic ratio

Iso.	Frequency (cm^{-1})	Strength (@ 296 K)	Dependence
$^{13}CO_2$	2295.84559	3.576E-20	−2.8‰/K
$^{12}CO_2$	2296.05649	4.562E-20	+15.9‰/K

Sometimes, on the contrary, this effect can be used to realize thermometers.

5. Lasers in Optical Techniques

The best sources for dispersive optical techniques are lasers. The reason lies in the fact that very often, in order to retrieve concentrations, a fit of the absorption profile versus theoretical profiles is necessary. If the source frequency width is small, compared to the absorption width, then the shape of the latter is almost unperturbed. Otherwise the instrumental effect must be kept into account during data reduction.

Among the lasers sources, one of the best kind of lasers for this purpose are diode lasers. They emit in the range visible – middle InfraRed (IR), where the molecular ro-vibrational absorption bands lie, they are tunable and compact. Their emission is normally multimode, except at most few regions, but they can be forced to emit one mode only by means of external tools, or by inserting a dispersive element inside their cavity. Moreover, their tunability with bias current allows the use of frequency modulation techniques by just modulating the injection current, so reducing the measurement noise.

Of course, they have some limits. Not all frequencies are available. If a special emission frequency is needed, the laser is realized ad hoc, at a prohibitive cost. Most times emission power and spectral purity are alternative: as an example, lasers emitting 250 mW for DVD writers have a multimode emission. In the near infrared lasers operate at room temperature. In the middle infrared, apart from recent Quantum Cascade Lasers (QCL), which operate at room temperature in pulsed regime, it is necessary to cool down the lasers by using either liquid nitrogen or closed-cycle helium refrigerators. The divergence of diode lasers is high (~tens of degrees), and for all diodes (but Vertical Cavity Surface Emitting Lasers, VCSEL's) it is different in two orthogonal directions. This makes it difficult to focus the laser on very small targets. In the following the key features of the different kind of sources will be described.

5.1. FABRY PEROT LASER (FPL)

These lasers have no special system to select the emission frequency. Their cavity is encompassed between the two facets of the active medium, one of which has usually a high reflectivity coating, in order to emit through the opposite one. They are available from visible to middle IR, where power is at most few hundreds μW. They are the cheapest sources in each emission region.

5.2. EXTERNAL CAVITY LASER (ECL)

They are FPL with an anti-reflection coating on one facet. The cavity is closed by one (or more) mirror, or by a dispersive element, like a grating. Their tunability is some tens of nanometers, but current and cavity parameters must be changed simultaneously. The laser is relatively cheap, but it must be included in a frame containing several parts, among which the driver of the actuator, for the movement of the mirror/grating.

5.3. DISTRIBUTED FEEDBACK LASER (DFB)

A diffraction grating is deposited on the laser substrate, with its groves perpendicular to the direction of propagation of the laser radiation. The emission wavelength is forced to match the grove spacing, so only one emission mode is allowed. In the middle IR only QCL are available in this configuration. They are quite expensive, and their power is in the range of 1-30 mW.

5.4. DISTRIBUTED BRAGG REFLECTOR (DBR)

VCSEL's (see Section 5) belong to this class lasers. The active medium is embedded between two structures of dielectric layers which provide a selective reflectivity. The emission occurs across one of these structures, whose reflectivity has been lowered by means of a chemical etching. As the "hole" drilled by the etching is circular, the shape of the emitted beam is circular too, and the divergence is the same in all directions. These lasers are suitable for mass production, are the cheapest ones, among those exhibiting single mode emission, require a low bias current ≤10 mA, but their emission power is always below 1 mW, and the maximum wavelength around two microns.

6. Other Components

As a general rule, the higher the wavelength, the higher the laser price. The same holds for the other components, like detectors and optical materials. From visible to middle IR, useful detectors are based on silicon, germanium, InGaAs-PIN, InSb, PbS, HgCdTe. In the middle IR they need to be cooled down to liquid nitrogen. With increasing wavelength, suitable optical materials are (but not only!) glass, BK7, quartz, CaF_2, ZnSe.

7. Detection Techniques

Several features would suggest to work in the near IR. Low prices for all components, availability of high quality, high power, room temperature lasers, easy alignment, due to simple tools to visualize near infrared beams. Unfortunately the most intense molecular absorptions, belonging to fundamental ro-vibrational bands, lie in the middle IR, while in the near IR only transitions, with variations of vibrational quantum numbers greater than one, are available.

An example of the difference is given in Fig. 5. Despite the factor of 0.3 in the path-length and the factor 0.1 in the pressure, the absorbance is six times higher at lower frequency.

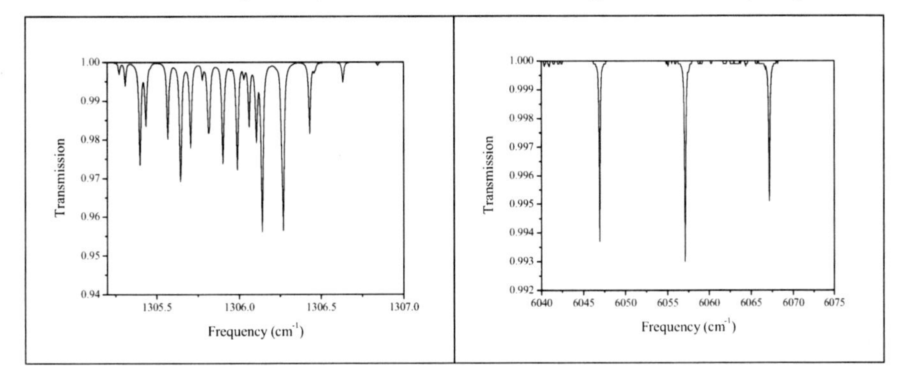

Figure 5. Methane absorptions at different frequencies at atmospheric concentration (1.8 ppm). Left: L = 30 m, P = 0.1 atm. Right: L = 100 m, P = 1 atm. Source HITRAN [13].

While it is relatively easy to detect and precisely quantify absorptions of a few percents, when the ratio I_{OUT}/I_{IN} is above 99%, it is often preferable to adopt noise reduction techniques.

7.1. DERIVATIVE TECHNIQUES

The simplest techniques to increase the signal to noise ratio are the so called derivative techniques (Fig. 6).

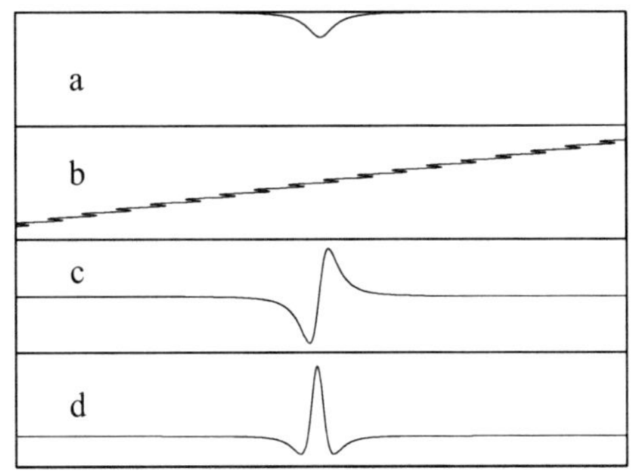

Figure 6. a) absorption profile; b) behaviour of the laser emission frequency; c) first derivative profile; d) second derivative profile.

By scanning the laser frequency around an absorption line, a typical profile like in Fig. 6a is recorded. Let's assume that the laser frequency is modulated like in Fig. 6b. A sinusoidal current at frequency ν is added to the bias current, with ν >> ramp frequency. Let's call modulation depth the maximum elongation of the laser frequency from the value due to the bias current only. The detector yields a constant signal at frequencies far from the absorption, and a periodically varying signal when close to the absorption frequency. The modulation signal is sent as a reference to a lock-in amplifier. If the demodulation occurs at ν, the shape of the signal is like in Fig. 6c, and we speak of "first derivative". When demodulating at 2ν, the lock-in output is like in Fig. 6d, and we obtain a "second derivative" of the absorption signal. As a matter of fact, the shapes of the two derivative signals are exactly the mathematical derivatives of the absorption shape, in the limit of modulation depth equal to zero.

7.2. FREQUENCY MODULATION TECHNIQUES

When a higher S/N is necessary, because either the absorbance is very low ($\approx 10^{-4}$ or less), or a high resolution is required, it's possible to use the "frequency modulation techniques". In this case, a modulation is added to the laser frequency, at a frequency ν such that ν ≈ width of the absorption line and $\nu > f$, where f is the characteristic frequency of the laser excess noise power spectrum [14]. The resulting electric field is given by:

$$E_1(t) = E_0(t) \cdot \sum_{n=-\infty}^{\infty} r_n \cdot \exp(i2\pi\, n\nu t) \qquad r_n(\beta, M, \psi) = \sum_{k=-1}^{1} a_k J_{n-k}(\beta) \tag{5}$$

We can visualize the laser emission as a carrier at the unperturbed laser frequency, plus two sidebands (one on each side of the carrier), opposite in phase, at a frequency ν from the carrier. Without absorptions, the beat notes between the sidebands and the carrier cancel, otherwise they give a net signal. The problem is that the demodulation must be done at a frequency ν. At atmospheric pressure, the constraint about the absorption width yields ν ≈ 1 GHZ, which requires very small detectors (diameter ≈50 μm, very difficult to be maintained aligned in field measurements!) and very fast electronics. To overcome this problem, the so called Two Tone Frequency Modulation (TTFM, [15-17]) can be adopted. Two modulations are superimposed to the laser, at frequencies ν_1 and ν_2 such that:

$$\nu_1, \nu_2 > \frac{1}{f}, \omega_L \qquad \frac{1}{f} < |\nu_1 - \nu_2| << \omega_L \tag{6}$$

where ω_L is the absorption linewidth. The resulting electric field will be:

$$E_2(t) = E_0(t) \cdot \sum_{n,m} r_n r_m \cdot \exp\left[i2\pi(n\nu_1 + m\nu_2)\, t\right] \tag{7}$$

and the resulting detector photocurrent:

$$I_\Omega(t)=\frac{c}{8\pi}|E_0(t)|^2\cdot 2\cos(2\pi\Omega t)\sum_{n,m} r_n r_m r_{n-1}^* r_{m+1}^*\cdot\exp[-2\alpha(\nu_0+(n+m)\nu_m)] \tag{8}$$

$$=\frac{c}{8\pi}|E_0(t)|^2\cdot 2\cos(2\pi\Omega t)[Q(\alpha)+M^2] \tag{9}$$

Without any absorption:

$$I_\Omega(t)=\frac{c}{8\pi}|E_0(t)|^2\cdot 2M^2\cos(2\pi\Omega t) \tag{10}$$

This term is called "Residual Amplitude Modulation" (RAM), and sets a limit to the maximum amplification of the signal. In case of absorptions:

$$I_\Omega(t)=\frac{c}{8\pi}|E_0(t)|^2\cdot\cos(2\pi\Omega t)\cdot[2M^2+2\alpha_0(\beta^2-M^2)-\alpha_+(\beta^2+2M\beta\sin\psi+M^2)-\alpha_-(\beta^2-2M\beta\sin\psi+M^2)] \tag{11}$$

The laser emission and the corresponding absorption profile are shown in Fig. 7.

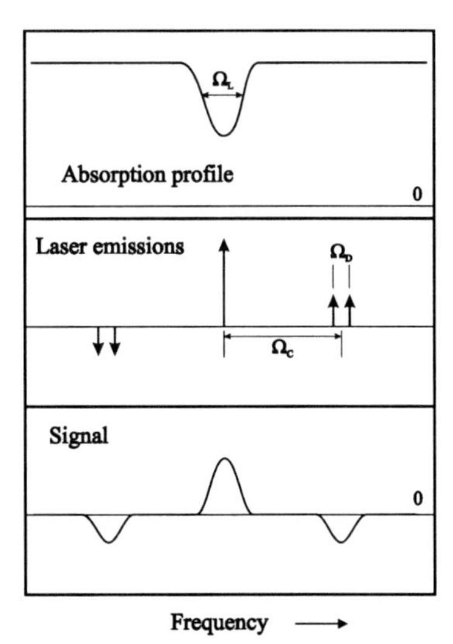

Figure 7. Typical features of TTFM.

7.3. FURTHER TRICKS

Sometimes even the most sensitive technique is not sufficient, because either the signal is still too small, or it is hidden under noise which the technique cannot suppress. Without changing the detection technique, some improvements can be done, in order to reduce the noise.

7.3.1. *Reference Arm*

One possible improvement is to introduce a reference arm, like in Fig. 8 [18]. Let's suppose that the detection technique is direct absorption. In the absence of absorbers, the optoelectronics is adjusted in such a way, that the difference between the signals coming from D1 and D2 is zero. When an absorption signal occurs, it raises above a zero background. If any modulation technique has been adopted, the signals from D1 and D2 must be equal in amplitude and opposite in phase, without absorptions. In both cases, the noises originating

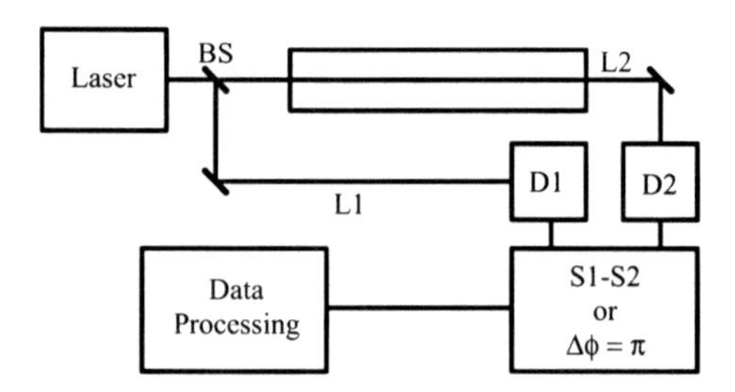

Figure 8. Experimental setup with a reference arm.

before the Beam Splitter BS, like laser excess noise, optical fringes before the beam splitter, RAM (if the case) will cancel. There is a constraint about the path-lengths difference. For direct absorption, the difference in the light transit time, between L1 and L2, must be smaller than the conversion time of the ADC which samples the signal. This is not a serious problem, unless a very fast ADC (≈1 MHz) is in use, and the measurement path is an open path, some hundreds meters in length. In case of modulation techniques, we should have:

$$|L1-L2|=\left(n+\frac{1}{2}\right)\frac{c}{\nu} \tag{12}$$

in order to obtain two perfectly out-of-phase signals. This solution is very elegant, but requires very stable optics and electronics. Nothing can be done in this way against noise originating after the beam splitter. The setup is complicated by the presence of another detector and various opto-mechanical components.

7.3.2. *Stark Effect*

When the noise is originated after the beam splitter, it is very difficult to subtract it. In principle, a measurement profile should be obtained without absorbers, to be compared with the profile acquired in operation. This works fine if the noise is just, for instance, a permanent perturbation of the measurement baseline. If there are fringes, whose positions change with time, this method cannot be applied. It would be necessary to periodically remove the absorption signal, leaving everything else unchanged. In some cases, this can be obtained by using Stark effect [19]. A periodic electric field is applied to the sample. The period is much smaller than the period of the current ramp. The modulation frequency is sent as a reference to a lock-in amplifier, together with the signal coming from the experiment, whatever the detection technique (Fig. 9).

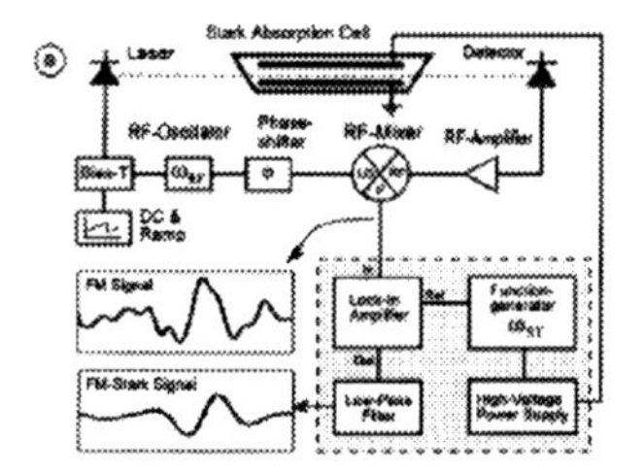

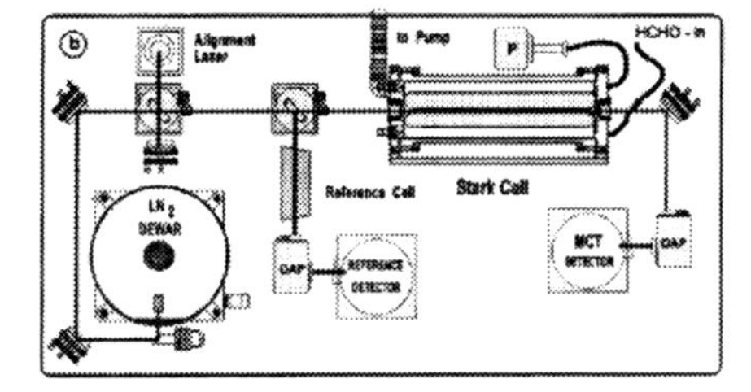

Figure 9. Electronic setup (a) and opto-mechanical setup (b) for noise suppression by Stark effect (courtesy of P.W. Werle).

The effect is shown in Fig. 10. A dramatic increase in the stability of the signal is obtained.

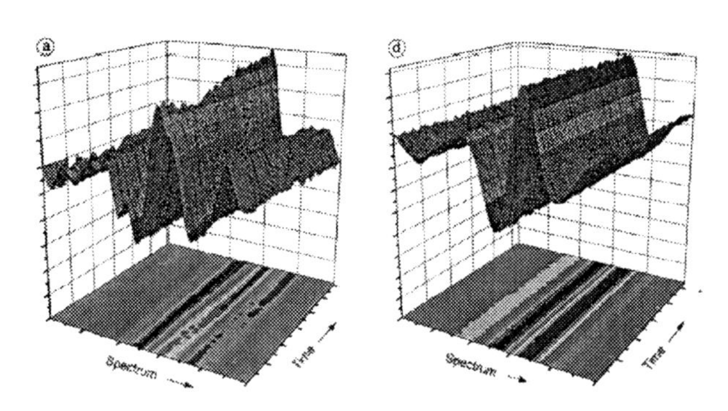

H_2CO 4.6 ppm
25 mBar
35 cm
1757.94 cm^{-1}

Figure 10. Time behaviour of the experimental signal without (a) and with (b) the application of the modulated electric field.

This technique can be used only for those molecules which exhibit a large Stark shift of their absorptions, and is possible only in those experimental conditions in which high voltages (≈kV) are allowed.

8. Lengthening the Optical Path

As we have seen in the Beer-Lambert law, the absorbance it linearly related to the length of the interaction between light and sample. If it is required, and according to the kind of measurement to be carried out, there are two ways of increasing the optical path. When dealing with open path measurements, a corner cube reflector can be used to fold the optical line (Fig. 11a). In the case of point measurements, a multi-pass cell can give up to some tens of meters in few liters (Fig. 11b).

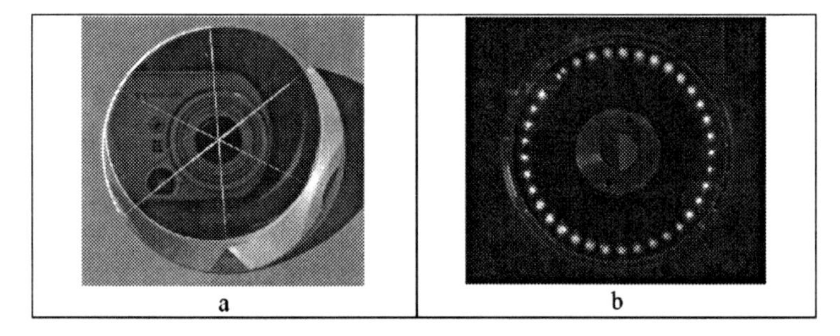

Figure 11. a) photo of a corner cube. Of course, the camera looks upside-down, with its objective in the center of the cube; b) photo of the laser spots onto the in/out mirror of a spherical Herriott cell [20.]

9. Electronics

Electronics must always include a circuitry for the temperature stabilization of the laser, and a current supply. Then, according to the detection technique and to the way of generating a frequency sweep, it's necessary to add some ADC's, some DAC's, waveforms synthesizer(s), mixers, etc. The most general case is described by the electronic setup for TTFM (Fig. 12).

Let's suppose to have as a target gas a molecule whose absorptions, in the experimental conditions, are so weak that it's necessary to adopt a reference arm for the active frequency locking of the laser. In order to produce the two frequencies for TTFM, we can start from Ω_D, the demodulation frequency. This signal is sent, through two phase shifters, to the mixers for the final signal analysis. A divider leads the signal to another mixer, in which $\Omega_D/2$ is added to the emission Ω_C of a VCO, so producing $\Omega_C \pm \Omega_D/2$. This modulation signal is mixed to the laser bias signal. All the detector chain operates at Ω_D. In this scheme, if we remove Ω_D, and we send Ω_C to all the three mixers, we have the frequency modulation technique. If we remove all the radio frequency circuitry, and the signals from the preamplifiers are sent directly to the

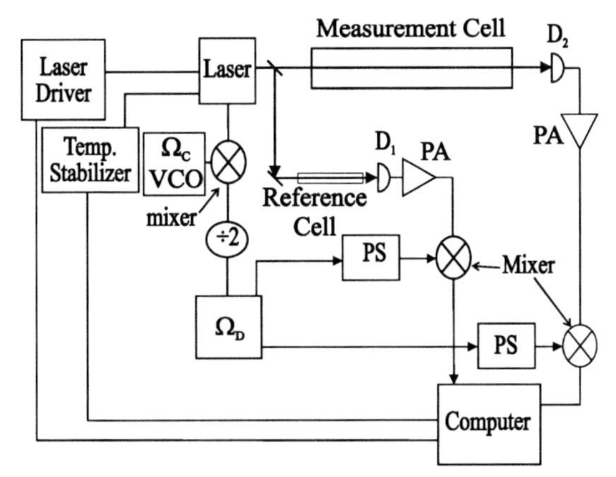

Figure 12. Electronic setup for TTFM. PA Pre Amplifier, PS Phase Shifter.

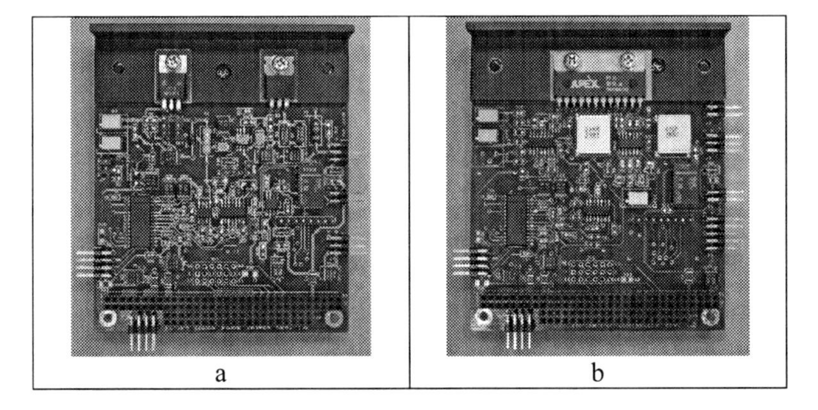

Figure 13. Example of circuitry for laser spectrometers: a) laser current supply; b) temperature stabilizer. Courtesy of SIT S.r.l.

computer, we have direct absorption. The tendency is anyway to reduce the dimensions and power consumption. An example of this is shown in Fig. 13, where a power supply, a temperature stabilizer and a lock-in boards are shown, designed in the industrial standard PC-104.

10. Calibration

This is a serious issue for all practical instruments, in particular those built for medical purposes, or those for environmental measurement, whose results are the basis for political decisions (in Italy, stop of traffic due to air pollution).

In principle, devices based on direct absorption don't need any external calibration, provided that the spectroscopic data of the absorption, temperature, pressure and length of the sample are known. For any other technique, zero and full scale reading must be periodically checked, when possible in the same experimental conditions.

Very often this cannot be realized easily. When measuring across the chimney of an industrial plant, zero could be obtained by switching off the plant, which is unbearable. Full scale check would require a known concentration of the target gas to be in the chimney. This hypothesis is absurd as well.

The problem has solutions. As for the zero, there are different possibilities. If the target gas is normally absent in the atmosphere, an alternative path can be realized outside the chimney. Otherwise, the temperature of the laser can be changed, in order to move its emission far from absorptions. It's important to remember that RAM can vary when varying the work point of the laser.

As for the full scale, the simplest trick is to short circuit the optical line directly from emitter to detector, inserting between them a sealed absorption cell with a reference mixture, chosen in such a way to give the right absorption signal.

11. Environmental Constraints

It's worth discussing at this point which kind of environment the instruments must face outside the laboratory. This will help us to assess the suitability of lasers spectrometers in practical applications. In one's laboratory everything is optimized for measurements: power supply, temperature, vibrations, cleanliness, and so on. Unfortunately the world outside is quite harsher and hostile. Let's see how.

11.1. POWER SUPPLY

While those devices operating in fixed locations can enjoy a power supply of about the same quality as in the labs, the instruments operating on board of vehicles or aircrafts have very different possible line voltages: a vehicle usually offers 12 VDC, on board of an aircraft 28 VDC and 110 VAC @ 400 Hz are available. Sometimes portable generators are used.

The problems don't arise from the available voltage, but from its quality. In these conditions there are often voltage drops or spikes, and the available power is limited.

The power supplies must face these kinds of troubles. In principle linear power supplies should be better, but in practice they are more sensitive to the quality of the line. On the contrary, switching supplies work in a wide range of input voltage, so offering an intrinsic safety. Moreover, capacitors can be used to clean the output of suppliers.

11.2. ELECTROMAGNETIC INTERFERENCES

It's not so infrequent to find Faraday cages in the labs, in order to shield the most delicate instruments from electromagnetic (EM) disturbances (i.e. those coming from high energy, pulsed lasers like TEA CO_2). In the field the problem is twofold. It's necessary to protect the instruments, and at the same time the instruments must avoid electromagnetic pollution. The latter feature is important mainly on board of aircrafts, where many devices can suffer problems from EM disturbances (one example for all, ejectable seats on board of military planes).

The starting point to cope with this kind of problem is the EM safety rule typical for the specific application. The rule sets the limits to the EM power spectrum coming out from the device. Once the instrument satisfies the rule, it is necessary to identify, if any, the external sources of noise which can disturb the measurement, and to suppress them.

The usual tools are first of all a good electronic design, then filtering, shielding, and grounding. A good technique, when signals must be carried for long distances in EM polluted environments, is to either carry them as currents, instead of voltages, or to use electro-optic couplers, in order to use optic fibers.

11.3. AMBIENT TEMPERATURE

Remote, stand-alone instruments can undergo very wide temperature ranges. For example, those located on chimneys, or in unpressurized bays in high altitude aircrafts, or in front of vehicles, and so on.

If we take into account all these possibilities, we have a temperature excursion from – 60°C to + 80 °C. Several kinds of problems arise. The standard operating range of electronics is 0 ÷ 70°C. CPU's stop working outside this range, unless extended temperature ranges (–40 ÷ 85°C) are explicitly requested. There is also a problem of thermal drift of many components. Temperature also affects mechanics and optics. There are obvious changes of size of all components, which are very dangerous for multi-pass cells. Moreover curved mirrors increase their curvature radii with increasing temperature. Finally, the strength of absorption lines depends on temperature.

Countermeasures are both active and passive. Temperature stabilization belongs to the first group. Otherwise components and materials with intrinsic thermal stability must be adopted. For mechanics and optics, materials like INVAR, ZERODUR, quartz or carbon fibers are the best solution, though often expensive.

As for stabilization, heating is very simple, providing that a source of power is available. Cooling looks a bit more difficult. Yet, some very simple and cheap cooling devices exist, based on compressed air, which are widely adopted in industrial plants, where compressed air is always available.

Should temperature still affect the instrumental reading, a calibration versus temperature would be the final solution.

11.4. AMBIENT PRESSURE

When instruments are located in unpressurized bays of high altitude aircrafts, the pressure excursion is a serious problem too. Hard disks require air to prevent crashing of the head onto the disk, large electrolytic capacitors or batteries can explode in absence of air, cooling of the electronics must be arranged by using heat conduction, instead of convection.

There are some physical problems too. Molecular absorption strengths and widths depend on pressure, so the dynamic range of the instruments must be suitable, and the data analysis procedures must be flexible enough.

Finally, when the lasers are cooled down to liquid nitrogen temperature, the Dewars must be pressurized. This because the boiling temperature is related to pressure. Without pressurization, the cold finger temperature would fluctuate in such a way to create serious troubles.

In the latter case pressurization is absolutely necessary, unless liquid nitrogen is replaced by a Stirling closed-cycle cooler, more simple but definitively much more expensive.

As for the electronics, solid state memories replace common hard disks, and the risky components must be checked before use.

11.5. VIBRATIONS

While in the labs optical tables with tunable vibrations absorbers are adopted, any instrument located on a moving platform undergoes mechanical stresses. In the case of a vehicle, we all know which are the effects of movements. Even on chimneys there are two kinds of vibrations: those coming from the apparatus inside the plant itself, and those coming from the effect of the wind. The worst conditions occur on airborne platforms. According to the specifications of the "Radio Technical Commission for Aeronautics 1997", a device to be located in the wings of a jet aircraft must pass this vibration test: 11.9 g rms in the frequency range 10 Hz ÷ 1 kHz along each axis, three hours per axis. It's not required for the instrument to be working during the test, it must "survive" it.

In this case, a very smart mechanical design is absolutely necessary. Light materials must be used, and computer simulations must be performed, in order to check the validity of the design. Finally, a prototype of the instrument must anyway undergo the vibration test.

11.6. NO CLEAN ROOMS

Sometimes the measurements must be carried out in special environment like volcanic emissions, chimneys, chemical plants, hazardous areas. The instruments must not only respect the standard safety rules (low voltages, no sparks, etc.), but also be able to maintain their optical lines efficient. Whenever possible, the instrument must be insulated from the atmosphere. Otherwise, inert materials must be used (e.g. AISI 316 and Teflon), the optics must adopt protective coatings, or be easily replaceable (by using kinematic mountings), or heated, according to the different environments. If necessary, a detection technique intrinsically less sensitive to dirty or dust (like direct absorption) must be adopted.

11.7. SIZE AND WEIGHT

The reasons for which size and weight must be as small as possible lie in quite a wide range. In the following (12.7) we'll see the limits on board of an aircraft. On the ground, let's think about a portable device: it's something that a man should be able to carry on his shoulders for one whole work day. This means that the frame, the optical breadboard, including all components, the electronics, the battery must weight in total few kg. The case of space platforms is even worse. Apart from the available space and payload, sending a device to the Space Station has a cost of 30 ÷ 40 k€/kg.

In order to solve the problem a careful design is necessary, which avoids oversized components and adopts light materials. Lead batteries must be replaced by NiCd ones, and so on.

11.8. UNATTENDED OPERATION

In several cases of those listed above the instruments must work unattended: chimneys, stratospheric aircrafts or space platforms, volcanic emissions. In some cases the personnel attending the instruments is not qualified to service the device, but only to operate it, like for leak finders. This means that the ON/OFF switch must be the only interface with the operators.

To do this, a careful design of the software must be implemented, taking into account all the possible sources of problems and adopting the right countermeasures. The use of actuators is very helpful in many cases.

12. Examples: From the Laboratory to the Space

After describing all the above problems, we should be quite surprised to know that some devices worked properly outside the laboratories. Yet, in the following we will see how several instruments have carried out their tasks from laboratories to the most distant locations which the human brain can imagine.

12.1. MEDICAL DIAGNOSTICS WITH CH_4

Our tour starts from the laboratories of ENEA (Italian Committee for New technologies, Energy and Environment) in Frascati (Rome), in 1994. A methane analyzer has been developed and used for medical diagnostics. This devices was based on a lead salts laser emitting at 1306 cm^{-1}, adopting the second derivative as detection technique. The laser was multimode, so a monochromator was necessary. The laser frequency was locked onto an absorption by using a reference arm and the first derivative detection. Another second derivative measurement on the reference arm gave a power reference for the laser. The optical and electronic set-up are shown in Fig. 14 a and b.

This instrument was very large and expensive, its power consumption was about 3 kW, it required liquid nitrogen for the two detectors, so it was very far from our concept of practical application. Yet, it fulfilled its requirements [21] and was a very useful trainer.

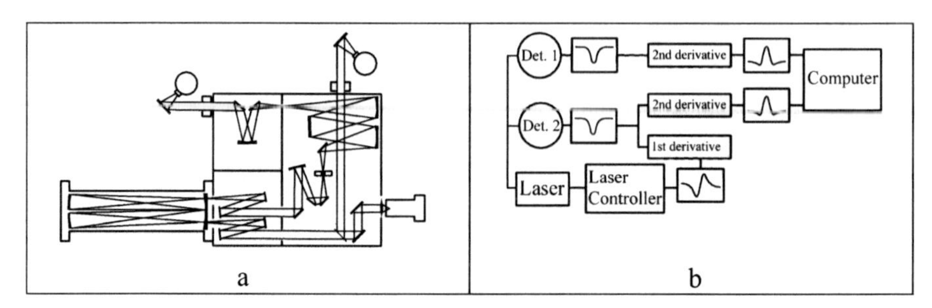

Figure 14. a) optical and b) electronic setup of the methane analyzer.

12.2. BEES' BREATH ANALYSIS WITH CO_2

Another example of laboratory equipment is a CO_2 sensor for the non-invasive measurement of the metabolism of a variety of bee, "osmia cornuta", whose way of living is useful for the plants impollination in greenhouses.

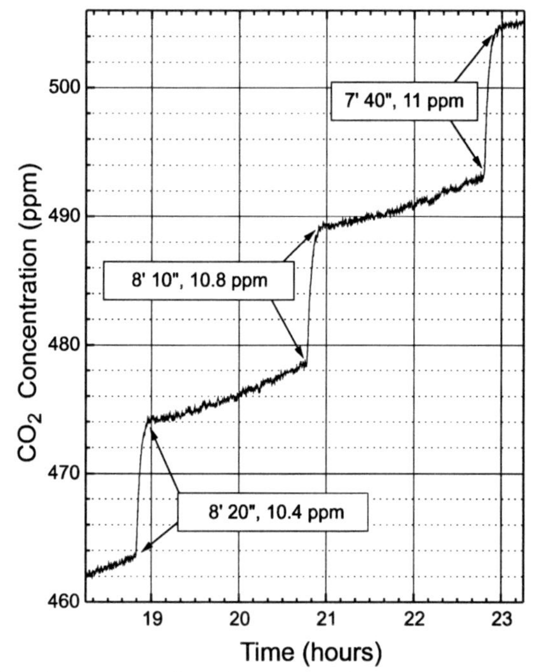

Figure 15. Measurement of the CO_2 emission by a cocoon of osmia cornuta.

In this case the optical scheme was very simple. The optical chain was formed by: laser (room temperature, DFB, λ = 2.004 μm), collimator, single pass absorption cell, focusing mirror, detector. The detection technique was TTFM. The cocoon of a bee was put into the cell. The metabolism was deduced by the increase of the CO_2 concentration inside the cell. A result of this measurements is shown in Fig. 15 [22]. The steps in the concentration of CO_2 correspond to the breaths of the insect inside the cocoon. The noise level of about 0.5 ppm corresponds to an absorbance of about $0.8 \cdot 10^{-5}$.

12.3. HIGH RESOLUTION MEASUREMENT OF AMBIENT CO_2

In the frame of the EC Contract "High Resolution Diode Laser Carbon Dioxide Environmental Monitor", a high resolution carbon dioxide analyzer has been realized [23]. This device was based on a diode laser of the same kind of that in chapter 12.2.

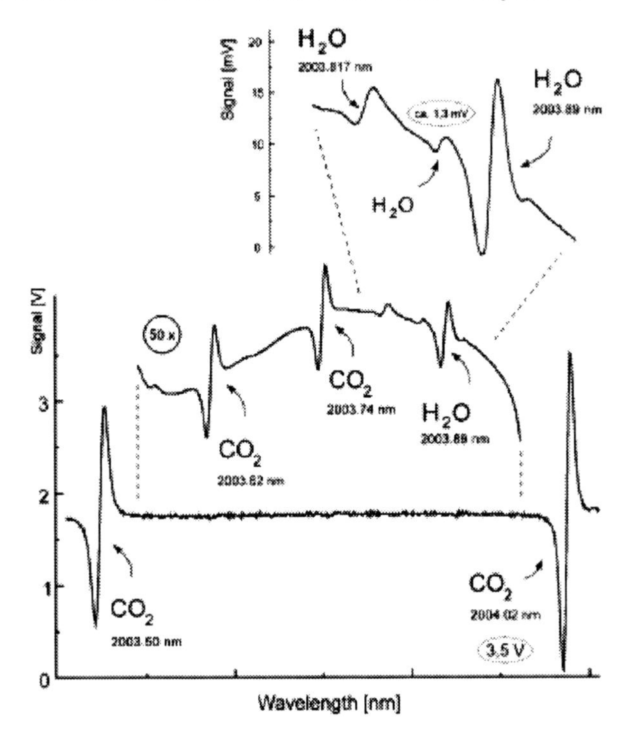

Figure 16. Measurement of the CO_2 air content: the central part of the absorption spectrum has been expanded, resolving weaker and weaker absorptions. The line of water at about 2003.85 nm proves a S/N for the most intense CO_2 absorptions better than 10^{-4}.

The optical and electronic setups were quite different. The absorption path was 100 m, obtained by using a commercial astigmatic Herriott cell with 182 passes. There was a reference cell, at the same temperature and pressure of the main cell, used to check the laser frequency and to provide a reference absorption signal. The detection technique was either single- or two-tone frequency modulation. The capabilities of the instruments are shown in Fig. 16. The instrument, though not so small and simple (0.5 m^2 for the optical breadboard, with a high level, but complicate electronics), was able to perform the task for which it had been built.

12.4. OUTDOOR STATIC MEASUREMENT OF CO FOR ENVIRONMENTAL DIAGNOSTICS

It is usual to hear that instruments based on lead salts lasers are complicate and large. I'll try to demonstrate that this is not so obvious. The CO analyzer described in the following was designed around a lead salts, pulsed diode laser. This laser, emitting around 2120 cm^{-1}, was not steadily single mode during the current pulse, but was single mode time by time. So, by adjusting the current pulse temporal shape it was possible to select the right emission mode, without any monochromator. The application of the instrument is the environmental monitoring [24]. The measurements are integrated along an optical path limited by a retro reflector. The detection technique was direct absorption, and the path length was ~100 m. The setup of the instrument is shown in Fig. 17.

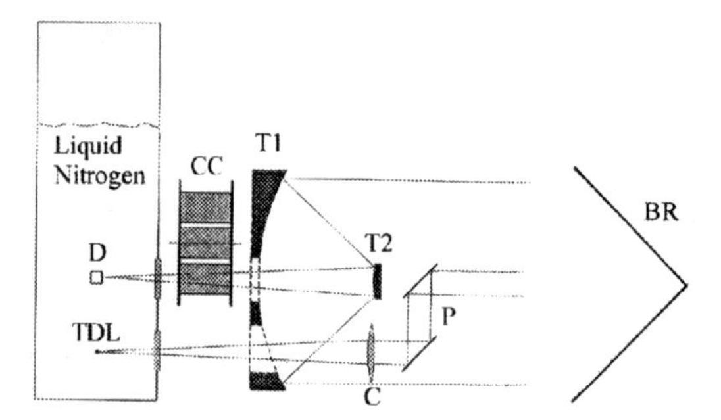

Figure 17. measurement of the CO concentration. D Detector; TDL Tunable Diode Laser; CC Calibration Cells; T1,2 Telescope mirrors; Periscope; C Collimator; BR Back Reflector.

12.5. CH_4 PIPELINES INSPECTION ON BOARD OF VEHICLES

In this chapter a commercial device is described. Its aim is that of detecting leaks of methane from citizen pipelines under the road surfaces [25]. The requirements are very tight:
resolution/precision 1 ppm in the concentration range 1 ÷ 100 ppm, 10% in the range 100 ÷ 10000 ppm, time resolution ≤1 sec., suitable for operation on board of vehicles for at least one year, operated by personnel without any specific knowledge of optoelectronics, selectivity at least 10000.

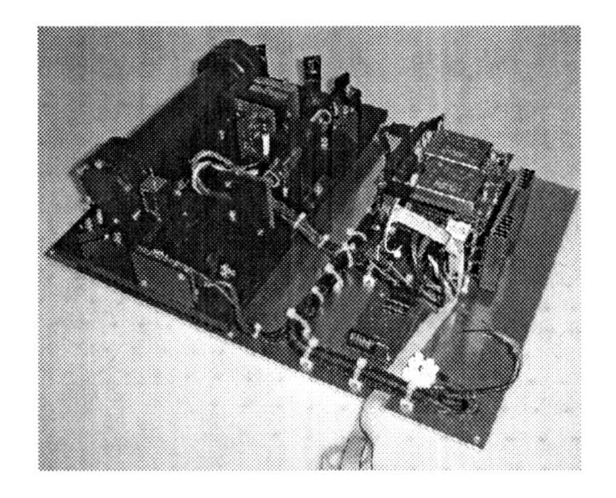

Figure 18. photo of the leak finder. The base fits a 19" rack module, with depth 35 cm. Courtesy of SIT S.r.l.

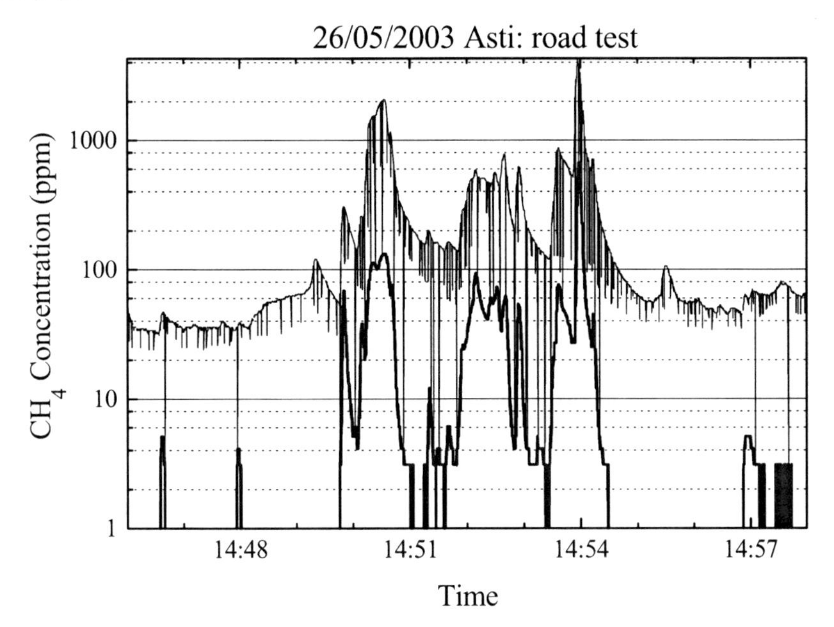

Figure 19. Inter comparison between the laser spectrometer (thick trace) and an electronic nose (thin trace).

The latter parameter means that if the concentration of any other molecule is 10000 ppm, the instrument must read at most 1 ppm. The instrument is shown in Fig. 18. It is based on a DFB, room temperature laser emitting around 1.65 μm. The detection technique is the second derivative. The interaction path length is 6 m, obtained with a multi-pass cell, whose volume is 94 cm^3. The difference in the readings between a laser spectrometer and an electronic nose is shown in Fig. 19. Two instruments of the two different kinds were operated together in Asti (I). The electronic nose never reads zero, because of poor selectivity. The exhaust gases in the traffic are able to yield a reading of several tens of ppm.

12.6. STRATOSPHERIC AIRCRAFTS 1: CO

Let's take a plane. It's not a common one, it's a stratospheric aircraft produced in Russia, named Geophysica, used since about ten years to carry several different instruments at the border between troposphere and stratosphere, for scientific campaigns all over the world. In order to be installed on board, any instrument must satisfy mechanical and electronic constraints, and pass the related safety tests.

Figure 20. Optical setup of the CO analyzer for stratospheric measurements.

The laser spectrometer we are dealing with measures the concentration of CO [26]. It is based on a lead salts laser like in 9.4, but in this case the laser is specially selected to emit single mode in the region of interest. The device is required to provide a time resolution of 2 sec., with a concentration resolution of ±10 ppb.

Due to what mentioned in Par. 11.4, the need of liquid nitrogen forced to pressurize the optical breadboard.

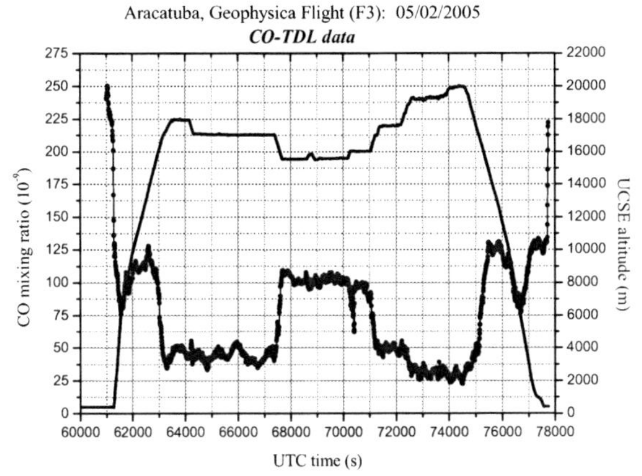

Figure 21. Measurement of CO concentration vs time during the third flight of the TROCCINOX 2 campaign (Brazil, Jan.-Feb 2005).

The optical layout is shown in Fig. 20. An example of practical measurements is given in Fig. 21, where the CO concentration is plotted versus UTC time, together with the flight altitude.

12.7. STRATOSPHERIC AIRCRAFTS 2: CH_4

On board of the same aircraft of Par. 12.6 we have installed a methane analyzer, named ALTO (Airborne Laser Tunable Observer) [27], based on a DFB, room temperature laser @ 1.651 μm, like in 9.5. The detection technique, in this case, is the TTFM, as the required resolution/ precision is 30 ppb (the ambient concentration of CH_4 is about 1.8 ppm).

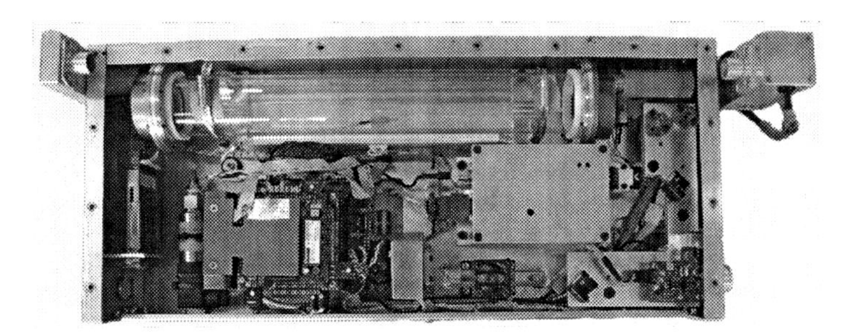

Figure 22. Methane analyzer ALTO.

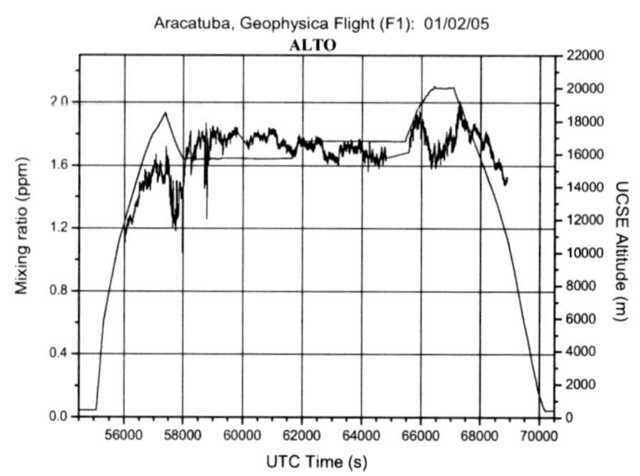

Figure 21. Measurement of methane concentration vs time during the first flight of the TROCCINOX 2 campaign (Brazil, Jan.-Feb 2005).

The time resolution is 1 sec. The layout of the instrument is shown in Fig. 22. In the upper right corner the valve is visible, which controls the air flow inside the multi-pass cell. This valve is kept closed below 5000 m, to prevent too humid air from entering the inside the cell. This instrument is located in front of the left wheel, were constraints due to the available space had to be satisfied. A measurement taken during last campaign in Brazil is shown in Fig. 23.

12.8. SPACE

The adventure of space missions for laser spectrometer was proposed many years ago [28]. A gas analyzer based on lead salts laser was proposed to perform measurements during the final approach of Titan. The final choice fell towards a combined instrument, a Gas Chromatograph Mass Spectrometer. The first real attempt was made with Mars Surveyor '98. A double laser spectrometer was on board [29], with two measurement channels, one for water (1.37 μm, DFB laser), and one for CO_2 (2.004 μm, DFB laser). This instrument was the first of his kind to be qualified for space. Unfortunately, the mission was not successful. The spacecraft was lost during landing, probably because of a crash onto the surface of Mars. The next opportunity will occur in 2007, again towards Mars, for a hygrometer at 1.877 μm [30]. This instrument summarizes all the above discussions about size, weight, power consumption, simplicity, reliability, performances. In addition to this, another feature must be considered. In the space all the electronics must be radiation hardened. This reason, even more important than power consumption, prevents the use of sophisticated CPU, like Pentium, on such devices.

13. Costs and markets

We have seen that a great care must be taken in building instruments which must operate in very different conditions. Yet, the listed examples show that a correct design and a smart engineering allow to realize useful and reliable devices. So, again, the question arises about the reason for which laser spectrometers are not so widespread: as the scientific bases are well established, as the technology is ready, this reason must then lie in the market competition.

All people who own a computer have a diode laser: it is located in the CD/DVD reader/writer. It's a FP laser, with the only real constraint to be focusable. Its cost is less than one Euro, its lifetime up to millions of hours. This result is due to the very large investments in this field and to the simplicity of the source. No special thermal or current stabilization is required, no tool for single mode emission.

Unfortunately, this is not the case of lasers for spectroscopy. They need to be single mode, they need a very effective temperature stabilizer, a suitable current supply, very often an active frequency stabilization circuitry and/or software. All this makes these sources very expensive. The cheapest lasers for spectroscopic applications, the VCSEL, has a cost of about 1100 Euro/piece, which can drop to few hundreds Euro for large quantities (>1000). It's absolutely unpredictable, nowaday, a cost less than 100 Euro/piece.

The problem of electronics is minor, in comparison to the laser. Mobile phones have now a very low cost (few tens of Euro), though they include a high capacity battery and a display. This means that the microprocessor driving the phone, and all the RF electronics, have a negligible cost.

As an exercise, we can verify if a methane leak detector for kitchens (in most Italian houses foods are heated by using methane, instead of electricity), usually an electronic nose, can be replaced by a laser spectrometer. The cost of a standard commercial device is some more than 100 Euro + VAT. It is just an ON/OFF device, it doesn't yield any value for the methane concentration, its aim is that of giving a warning when methane is approaching the explosion threshold of 5% mixing ratio. As a compromise between the poor sensitivity and the need for a really early warning, the alarm level is usually set at a concentration of 5‰.

For a laser spectrometer this means a great deal of simplification. The absorbance of methane at 1.651 μm, at 5‰ mixing ratio, over a path length of 8 cm, is 1.7%. A very simple optical line is necessary in this case, consisting only of a focusing mirror, imaging the laser onto the detector (Fig. 24).

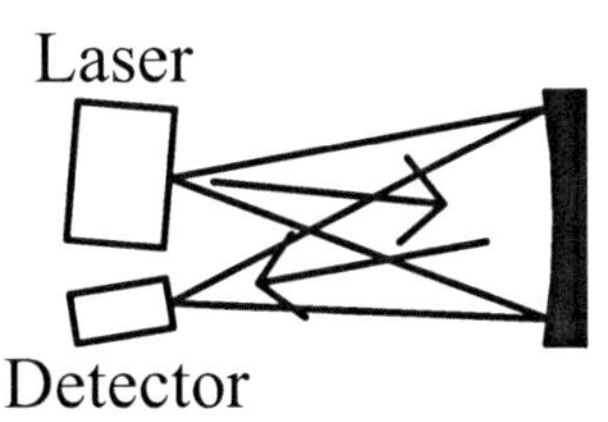

Figure 24. Scheme of the simplest optical configuration for the methane leaks warner.

A good thermal stabilization is necessary anyway. There is no possibility of introducing a reference arm for active frequency stabilization, so the simplest way to match the absorption is to scan periodically the laser frequency over the widest possible range. Second derivative is the best choice for the detection technique. All this can be realized with a very simple electronics.

Again, the problem is in the laser. Its cost is higher than that of the whole alternative device. Moreover, some manpower is needed to align it and to find its precise operating point (temperature and bias current).

14. Conclusions

Before starting the discussion, it's worth reminding that the target molecule for laser spectrometer is a light molecule, exhibiting discrete absorptions in the near- or middle-IR. Benzene-type molecules are not good targets, unless they include some radicals, having the above type of behaviour. Most hydrocarbons, even having linear structures, have quasi-continuum absorptions, so are not visible with the described devices.

Coming back to the initial question, hopefully the reader should have got convinced that the scientific bases of laser spectrometers are firmly established, and that whatever the experimental condition, the technology is able to provide the right solution. The "best" technique doesn't exist, but for each kind of application it is possible to select that one, which is most suitable.

This leads to one conclusion: the quality/price ratio of laser spectrometers is still too low, compared to other kinds of devices. The reason lies most times in the high cost or in the requirements of the lasers (e.g. liquid nitrogen). So, even if the device is simple, even if the prices of electronics and optics can be made almost negligible, a laser spectrometer is often very far from being practically useful for a particular purpose.

The competition is open when the price of alternative devices is high, or when they lack some useful features, like selectivity, or when the safety regulations require such high sensitivities or resolutions, or such low detection limits, that they cannot be achieved by standard devices.

This is the case of the leak finder in chapter 12.5. Its higher cost, compared to that of standard devices, can be acceptable because it allows a much higher speed of inspection. Moreover, the lack of false positive avoids waste of manpower.

A significant impulse towards the diffusion of laser spectrometers could arrive from the governments. New regulations about pollution could require more sensitive analyzers, beyond the capabilities of standard techniques.

Of course, the producers of lasers should develop both cheaper production methods, and new kinds of sources, like QCL.

15. References

1. Webster C.R., Menzies R.T., Hinkley E.D. (1988) Infrared Laser Absorption: Theory and Applications, in: R.M. Measures (ed.), *Laser Remote Chemical Analysis* - Chemical Analysis Series Vol. 94, John Wiley & Sons, New York pp. 163-532.
2. Schiff H.I., Mackay G.I., Bechara J. (1994) The Use of Tunable Diode Laser Absorption Spectroscopy for Atmospheric Measurements In: M.W. Sigrist (ed.), *Air Monitoring by Spectroscopic Techniques*, Wiley, New York pp. 239-333.
3. Brassington D.J. (1995) Tunable Diode Laser Absorption Spectroscopy for the Measurement of Atmospheric Trace Species, in: R.E. Hester (ed.) *Advances in Spectroscopy*, Vol. 24, Wiley, New York pp. 85-148.
4. Sigrist M.W. (1998) Air monitoring, optical spectroscopic methods. In: Meyers RA (ed.) *Encyclopedia of environmental analysis and remediation*. Wiley, New York 84-117.

5. Mion F., Ecochard R., Guitton J., Ponchon T. (2001) $^{13}CO_2$ breath tests: comparison of isotope ratio mass spectrometry and non-dispersive infrared spectrometry results, *Gastroenterol Clin Biol.* 25, 375-379.
6. Elkins J.W., Fahey D.W., Gilligan J.M., Dutton G.S., Baring T.J., Volk C.M., Dunn R.E., Myers R.C., Montzka S.A., Wamsley P.R., Hayden A.H., Butler J.H., Thompson T.M., Swanson T.H., Dlugokencky E.J., Novelli P.C., Hurst D.F., Lobert J.M., Ciciora S.J., McLaughlin R.J., Thompson T.L., Winkler R.H., Fraser P.J., Steele L.P., Lucarelli M.P. (1996) Airborne gas chromatograph for in situ measurements of long-lived species in the upper troposphere and lower stratosphere, *Geophys. Res. Lett.* 23, 347-350.
7. Zampolli S., Elmi I., Ahmed F., Passini M., Cardinali G.C., Nicoletti S., Dori L. (2004) An electronic nose based on solid state sensor arrays for low-cost indoor air quality monitoring applications, *Sensors and Actuators* B101, 39-46.
8. Sigrist M.W. (2003) Trace Gas Monitoring by Laser Photoacoustic Spectroscopy and Related techniques, *Rev. Sci. Instrum.* 74, 486-490.
9. Morville J., Kassi S., Chenevier M. and Romanini D. (2005) Fast, low-noise, mode-by-mode, cavity-enhanced absorption spectroscopy by diode laser self-locking, *Appl. Phys.* B80, 1027-1038.
10. Wood R.W., Phil. Mag. (6th ser.) 10, p. 513 (1905).
11. Creasey D.J., Halford-Maw P.A., Heard D.E., Pilling M.J. and Whitaker B.J. (1997) Implementation and initial deployement of a Field instrument for measurement of OH and HO_2 in the troposphere by laser-induced fluorescence, *J. Chem. Soc., Faraday Trans.* 93, 2907-2914.
12. Baldacchini G., D'Amato F., De Rosa M., Buffa G., Tarrini O. (1995) Temperature Dependence of Self Broadening of Ammonia Transitions in the ν_2 Band, *J. Quant. Spectrosc. Radiat. Transfer* 53, 671-680.
13. HITRAN PC 92, Copyright University of South Florida, 1992.
14. Werle P.W. (1995) Laser excess noise and interferometric effects in frequency-modulated diode-laser spectrometers, *Appl. Phys.* B60, 499-506.
15. Janik G.R., Carlisle C.B., Gallagher T.F. (1986) Two-tone frequency-modulation spectroscopy, *J. Opt. Soc. Am.* B3, 1070-1074.
16. Cooper D.E., Warren R.E. (1987) Frequency modulation spectroscopy with lead-salt diode lasers: a comparison of single-tone and two-tone techniques, *Appl. Opt.* 26, 3726-3732.
17. D'Amato F., De Rosa M. (2002) Tunable diode lasers and two-tone frequency modulation spectroscopy applied to atmospheric gas analysis, *Opt. & Lasers in Eng.* 37, 533-551.
18. Durry G., Pouchet I., Amarouche N., Danguy T., Megie G. (2000) Shot-noise-limited dual-beam detector for atmospheric trace-gas monitoring with near-infrared diode lasers, *Appl. Opt.* 39, 5609-5619.
19. Werle P.W., Lechner S. (1999) Stark-modulation-enhanced FM-spectroscopy, *Spectrochim. Acta* A55, 1941-1955.
20. Herriott D.R., Schulte H.J. (1965) Folded optical delay lines, *Appl. Opt.* 4, 883-889.
21. Bianchi M., Papi C., D'Amato F., Koch M., Capurso L. (1996) Concentration of Methane in Breath of Colonic Polyps Patients Measured with a New Technique, 5th United European Gastroenterology Week, Paris (F), 2-6/11/1996, *Gut* 39 (suppl. 3), p. A206.
22. Felicioli A., D'Amato F., De Rosa M. & Pinzauti M. : "Preliminary investigation on individual respiration of adults and pupæ of Osmia Cornuta during diapause by means of a CO_2 analyzer based on an infrared tunable diode laser", book of Abstracts of the XXI International Conference of Entomology, Foz do Iguassu (Brazil), 20-26/08/2000, p. II-612.
23. Werle P.W., Muecke R., D'Amato F., Lancia T. (1998) Near-infrared trace-gas sensors based on room-temperature diode lasers, *Appl. Phys.* B67, 307-315.
24. Baldacchini G., D'Amato F., De Rosa M., Nadezhdinskii A.I., Lemekhov N., Sobolev N. (1996) Measurement of Atmospheric CO Concentration with Tunable Diode Lasers, *Infrared Phys. & Technol.* 37, 1-5.
25. Chiarugi A., D'Amato F., Fogale D., Finardi G. (2003) A mobile methane pipeline inspection system, Book of Abstracts of TDLS 2003, Zermatt (CH), 14-18/07/2003, available at "tdls.conncoll.edu".
26. D'Amato F., Mazzinghi P., Viciani S., Werle P.W., Castagnoli F., De Pas M., Giuntini M. : "Stratospheric carbon monoxide in tropical convections: in-situ measurements with a mid-IR airborne spectrometer", Book of Abstracts of TDLS 2005, Firenze (I), 11-15/07/2005.
27. D'Amato F., Mazzinghi P., Viciani S., Werle P.W., Castagnoli F., De Pas M., Giuntini M. : "Airborne measurement of CH_4 profiles during the TROCCINOX-2 campaign with the near infrared TDL instrument "ALTO", Book of Abstracts of TDLS 2005, Firenze (I), 11-15/07/2005.
28. Webster C.R., Sander S.P., Beer R., May R.D., Knollenberg R.G., Hunten D.M., Ballard J. (1990) Tunable diode laser IR spectrometer for in situ measurements of the gas phase composition and particle size distribution of Titan's atmosphere, *Appl. Opt.* 29, 907-917.

29. May R.D., Forouhar S.F., Crisp D., Woodward W.S., Paige D.A., Pathare A., and Boynton W.V. (2001) The MVACS tunable diode laser spectrometers, *J. Geophys. Res.* 106, 17673-17682.
30. Webster C.R., Flesch G.J., Mansour K., Haberle R., and Bauman J. (2004) Mars laser hygrometer, *Appl. Opt.* 43, 4436-4445.

SENSITIVE DETECTION TECHNIQUES IN LASER SPECTROSCOPY

W. DEMTROEDER, M. KEIL, TH. PLATZ AND H. WENZ
University of Kaiserslautern
D67663 Kaiserslautern, Germany
demtroed@physik.uni-kl.de,demtroed@rhrk.uni-kl.de

1.Introduction

For many experiments in spectroscopy the sensitivity of detection techniques is of great importance. In particular for applications in atomic and molecular spectroscopy, analytical chemistry, atmospheric research or in environmental studies techniques with the minimum still detectable number of molecules will win the competition among different methods.

In this paper several of these sensitive detection techniques developed in laser spectroscopy are presented which are illustrated by some examples. They are either based on modified absorption spectroscopy using frequency-or phase modulation, on opto-acoustic and opto-thermal spectroscopy, on laser-induced fluorescence or on resonant two-photon (RTPI) or multi-photon ionization (REMPI). The advantages and limitations of some of these techniques are discussed and possible further improvements are proposed.

2. Absorption Spectroscopy

If a light beam with intensity I_0 passes through a sample of absorbing molecules, the transmitted intensity is for small absorptions ($\alpha \cdot L << 1$)

$$I_t = I_0 e^{-\alpha L} \approx I_0(1-\alpha L) \tag{1}$$

Where $\alpha = \sigma_{ik} \cdot N_i$ is the absorption coefficient, which depends on the absorption cross section per molecule σ_{ik}, on the absorbing transition $|i\rangle \rightarrow |k\rangle$ and the number density N_i of absorbing molecules in level $|i\rangle$. The total absorption after a path length L is then

$$(I_0 - I_t) = I_0 \cdot \alpha \cdot L = \sigma \cdot N_i \cdot L I_0 \tag{2}$$

B. Di Bartolo and O. Forte (eds.), Advances in Spectroscopy for Lasers and Sensing, 187–206.

The detected signal is $S = a\cdot(I_0-I_t)$ where the factor a depends on the sensitivity of the detector. The minimum detectable number of molecules N_i^{min} depends on the achievable signal-to-noise ratio $R = S/N$ which should be larger than one, i.e. the minimum still detectable difference I_0-I_t must be at least as large as the noise. This gives

$$N_i^{min} \geq 1/(a\cdot R\cdot\sigma_{ik}\cdot L\cdot I_o) \quad (3)$$

and illustrates, that N_i^{min} depends on the absorption cross section σ_{ik}, on the realized absorption pathlength L and on the incident intensity I_0. The different techniques for sensitive absorption spectroscopy are mainly based on minimizing $R = S/N$ and maximizing L. The detector should have optimum sensitivity, i.e. the maximum value of a. If possible, an absorbing transition with a large cross section σ is chosen

The main noise source comes from intensity fluctuations of the laser intensity and pointing instabilities of the laser beam. Here different stabilization techniques have been developed, which stabilize the laser intensity and also the laser wavelength on the center of the absorption line.

Most absorption methods for detecting molecules use the fundamental vibrational transitions in the mid infrared region. Since the frequencies of these transitions are characteristic for any molecular species, this frequency range is called the *fingerprint region* for molecular detection. The absorption cross sections are fairly large. However, in this spectral region reliable and easy to handle tunable laser sources for high resolution spectroscopy had been lacking until recently before cw optical parametric oscillators or reliable difference frequency laser spectrometers were developed.

Overtone transitions, on the other hand, have much smaller absorption coefficients but since they range from the near infrared into the visible region, widely tunable cw semiconductor lasers are here available, which can be modified into tunable single mode lasers by using external cavity devices. Also the detectors are more sensitive in this range. Therefore the disadvantage of smaller absorption cross sections might be outweighed by the better experimental conditions.

For illustration Fig. 1 shows a typical experimental setup for sensitive molecular overtone spectroscopy. The widely tunable cw single mode Al/Ga/As diode laser has an anti-reflection coating on one end face of the diode and is operated in an external cavity of the Littman type, consisting of a grazing incidence optical grating and a highly reflecting flat mirror [1]. The mirror can be tilted by a piezo-device around an axis, located at the intersection of the grating and the mirror surface planes. In this case the two_conditions for the grating equation

$$d(\sin\alpha+\sin\beta) = \lambda \quad (4)$$

(where d is the groove spacing on the grating, α is the angle of incidence and ß the diffraction angle) and the cavity length

$$L_c = m\cdot\lambda/2 \quad (5)$$

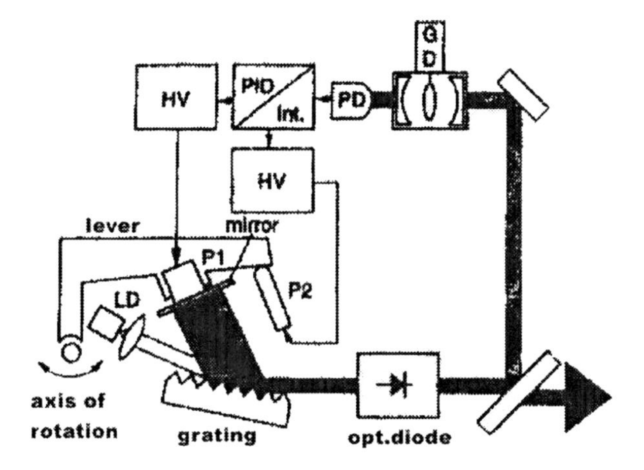

Fig. 1. Single-mode diode laser with external Littman cavity.

(where m is a large integer) can be simultaneously fulfilled over a wide tuning range of the laser wavelength λ. In our setup a frequency tuning range $\Delta\nu$ (with $\nu = c/\lambda$) of about 400 GHz can be realized without mode hops [2].

The zero-order diffraction from the grating is used as output beam, which is then sent through an electro-optic crystal where an applied radio frequency voltage produces a change of the refractive index and thus a phase modulation of the transmitted laser wave. This modulation generates two sidebands with equal amplitudes but opposite phases and with frequencies about ±1GHz apart from the laser frequency. The frequency separation just matches the Doppler width of the absorption lines. If there is no absorption, the detected signal will be zero, independent of possible intensity fluctuations of I_0 (Fig. 2a). If one of the sidebands is weakened by absorption, the lock-in detects the difference in the transmitted intensity of the two sidebands (Fig. 2b).

The improvement in sensitivity by about two orders of magnitude is illustrated by recording a rotational line in the overtone band of the H_2O molecule with and without modulation (Fig. 3).

The lock-in cannot handle the high frequency of 1 GHz. Therefore the frequency is down-converted in the following way: The amplitude of the rf voltage driving the phase modulator is modulated at a lower frequency in the kilohertz range and the lock-in detector amplifies the detected signal at this frequency.

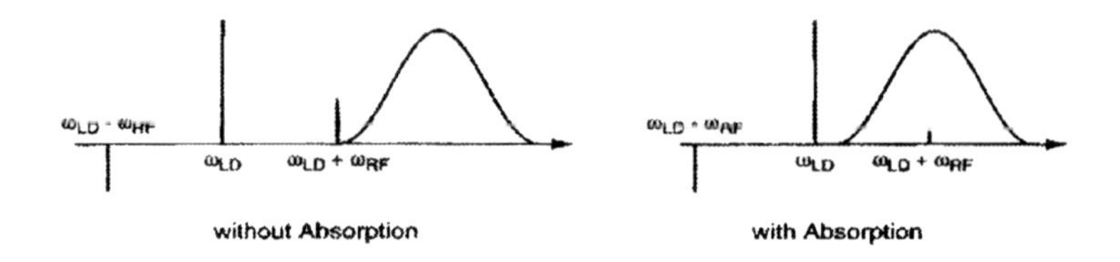

Fig. 2. Principle of phase-modulated absorption spectroscopy.

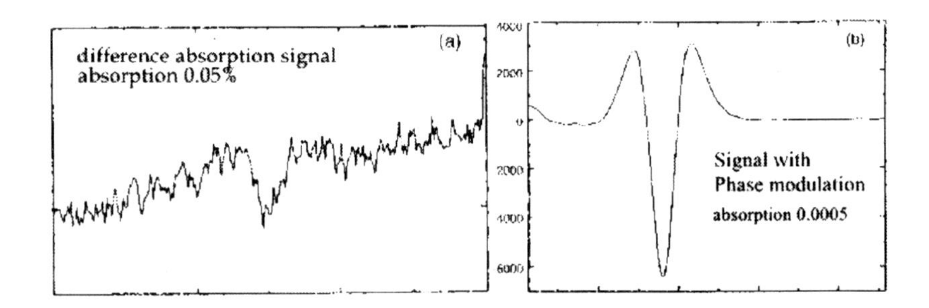

Fig.3. Rotational absorption line in the overtone-band (1,2,1)←(0,0,0) of the H_2O molecule measured with unmodulated laser beam (a) and with phase modulation (b).

Instead using phase modulation the laser wavelength can be modulated by periodic changes $\Delta L(f)$of the external cavity length. The modulation depth $\Delta\lambda$, which equals about two times the Doppler width, is determined by the magnitude of ΔL, and the modulation frequency f is limited by the mass of the cavity mirror mounted on the piezo.

The absorbing sample is placed within a nearly confocal multipass cell consisting of two spherical highly reflecting mirrors. The number of passes through the cell can be adjusted by slightly changing the mirror separation. The main problem is the partial overlap of laser beam profiles from successive reflections. They result in interference effects which appear as unwanted structures in the measured absorption spectrum when the laser wavelength is tuned. Here the wavelength modulation is helpful because the modulation $\Delta\lambda$ is large compared to the free spectral range of the multipass cell and therefore it smears out these interferences. The transmitted intensity I_t is compared with the intensity I_r of a reference beam, which is split off by a beam splitter before the multi-pass cell (Fig. 4). The choice of the beam splitter is crucial, because any tiny change of the splitting ratio while the wavelength is tuned, produces a changing background which can hide weak lines and can shift the line positions. We found, that a Wollaston prism was the best beam splitter. A matched pair of two detectors on the same chip for monitoring the reference and the signal beam provides the ratio

$$S = (I_r - I_t)/I_r = 1 - (I_0//I_r)(1-\alpha\cdot L) \quad (6)$$

and allows the determination of the absorption coefficient α.
Part of the laser beam is sent through a reference cell filled with acetylene which provides reference lines for wavelength calibration. An etalon with a free spectral range of 2.5 GHz gives frequency markers to correct for non-uniform scans of the laser wavelength.

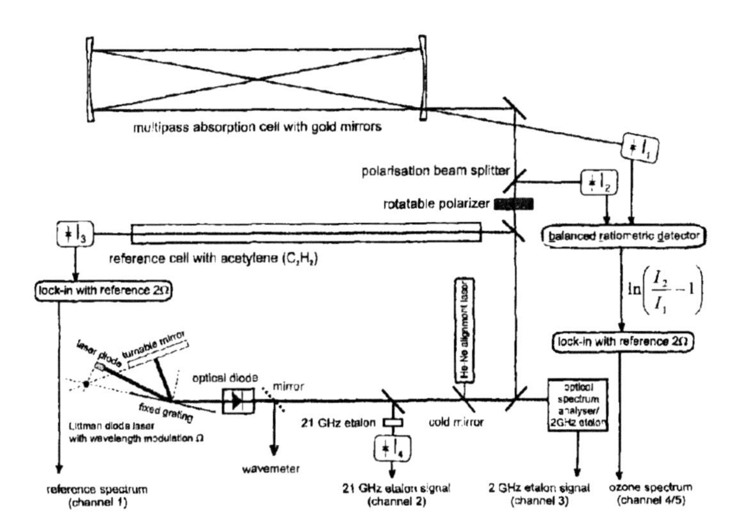

Fig. 4. Complete experimental setup for modulated absorption spectroscopy.

With this system the overtone spectrum of ozone O_3 was measured within the wave-number range between 6000-6500cm^{-1}, where higher_vibrational_levels ($v_1 < 6, v_2 < 8, v_3 < 3$) are reached which are closely below the dissociation limit of the electronic ground state [3]. In this energy range many vibronic and Coriolis couplings between nearby vibrational levels occur and the spectrum is very dense (Fig. 5). Since the absorption cross section for these high overtone transitions are very weak, they could not have been detected with conventional absorption spectroscopy.

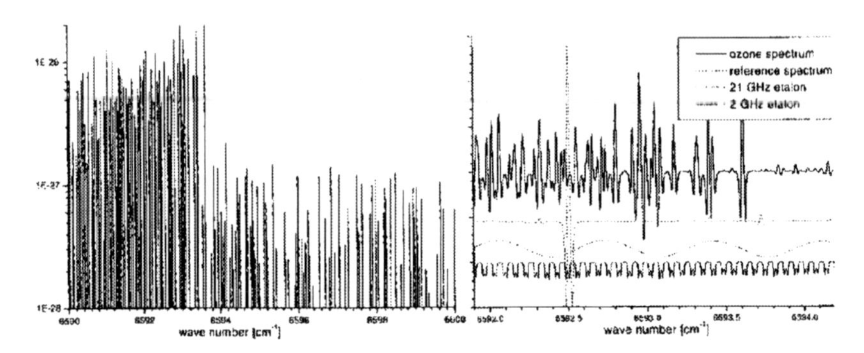

Fig. 5. Section of the overtone spectrum of ozone around $\lambda = 1.5\mu m$.

The left side shows the compressed rotational lines of one overtone band. The right side gives an expanded view of part of the band. Superimposed as dashed spectrum is a line of the overtone band of acetylene and the frequency marks of a Fabry-Perot interferometer

There are other sensitive absorption techniques, such as cavity ring down spectroscopy [4] or noise immune cavity enhanced optical heterodyne spectroscopy (NICE OHMS) [5]. In particular the latter technique, where the sample is placed inside an enhancement cavity with highly reflecting mirrors and the laser intensity is modulated at a frequency that equals the free spectral range of the enhancement cavity, has proven to be extremely sensitive with a detection limit for the absorption coefficient $\alpha \approx 10^{-14} cm^{-1}$.

3. Opto-Acoustic Spectroscopy

Instead of detecting weak absorption as the small difference between incident and transmitted intensities, the absorbed power can be directly detected by its effect on the absorbing molecules. If molecules in a closed cell absorb photons their excitation energy can be transferred by collisions into translational energy of the collision partners (Fig. 6a). This raises the temperature and because of the constant density increases the pressure [6]. The incident laser beam is chopped at a frequency that matches an acoustic resonance of the absorption cell. Then standing acoustic waves are generated which are detected by a sensitive microphone, located at a place where the pressure change is maximum (Fig. 6b).

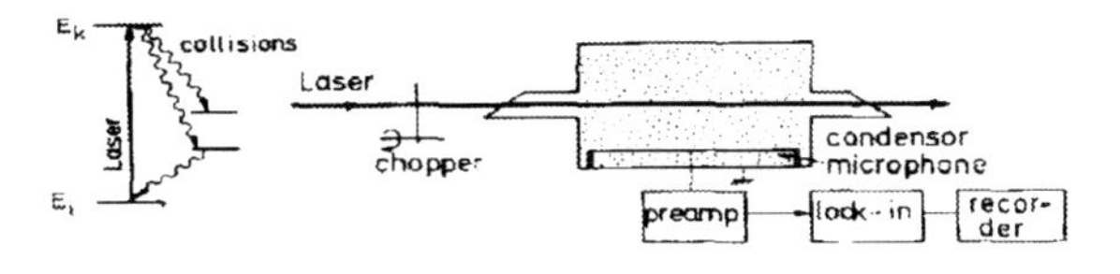

Fig. 6. a) Level diagram for opto-acoustic spectroscopy, b) experimental arrangement with resonant acoustic cell and condenser microphone.

The sensitivity of this technique reaches values, where a relative absorption $\alpha L = 10^{-5}$ to 10^{-6} can still be detected. Placing the acoustic resonance cell inside an optical multi-pass cell (Fig.7) absorption path lengths of 100 m can be readily realized, yielding detectable absorption coefficients of $\alpha = 10^{-9}$ to 10^{-10}cm^{-1}. The sensitivity is demonstrated in Fig. 8 by a section of the overtone spectrum of acetylene, where the signal to noise ratio is very good in spite of the small absorption probability. [7].

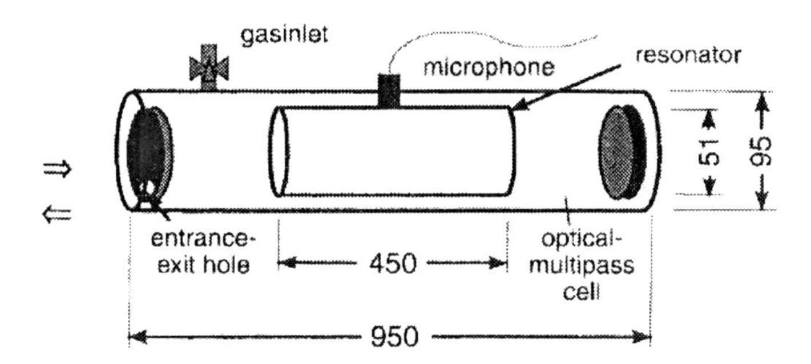

Fig. 7. Opto-acoustic resonant cell inside an optical multi-pass cell.

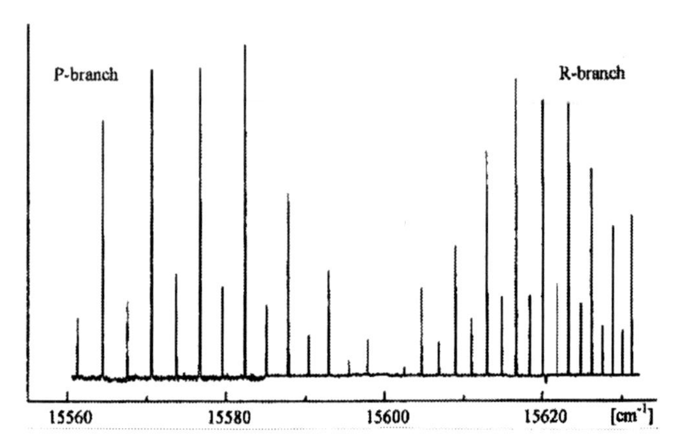

Fig. 8. Rotational lines in the overtone band (5,1,0,0,0,0) ← (0,0,0,0,0,0) of acetylene.

4. Opto-Thermal Spectroscopy

The techniques discussed above were applied to molecules in cells, with absorption line widths limited by Doppler broadening. With opto-thermal spectroscopy, first reported by Scoles and coworkers [8], molecules in cold collimated beams are excited, thus reducing the Doppler width by more than two orders of magnitude. Its basic principle is depicted in Fig. 9. A molecular beam is crossed under 90° by a laser beam that excites the molecules into higher vibrational levels in the electronic ground state. After a flight path of about 30 cm the molecules reach the detector, which consists of a cooled bolometer at about 1.5K. If the lifetime of the excited states is longer than the flight time to the detector, the excited molecules will transfer their excitation energy to the bolometer, thus increasing its temperature slightly. This decreases the electrical resistance of the bolometer consisting of a small doped silicon disk. The resistance change can be detected as a voltage change, when a small electric current is sent through the silicon disk.

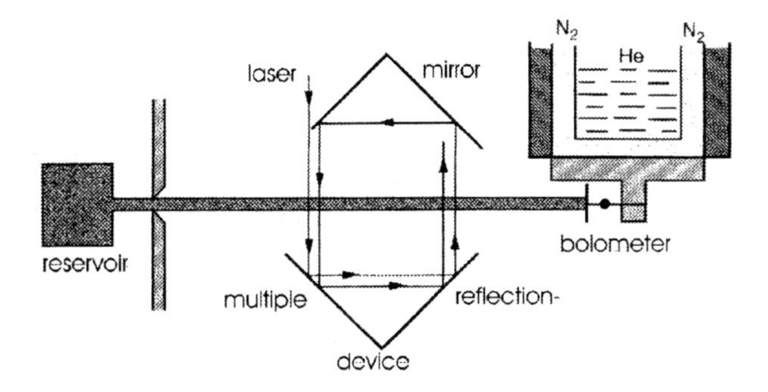

Fig. 9. Experimental arrangement for opto-thermal spectroscopy.

With a sensitive device even such small powers as 10^{-14}W, transferred by the molecules to the bolometer, can be detected.

The sensitivity combined with the high spectral resolution makes this technique very attractive, as can be seen from Fig. 10, which shows a comparison of the same section in the overtone spectrum of ethylene, measured by Fourier transform spectroscopy, opto-acoustic spectroscopy and opto-thermal spectroscopy. The Doppler-free resolution of the latter illustrates that many lines, hidden within the Doppler profile can be clearly resolved and identified.

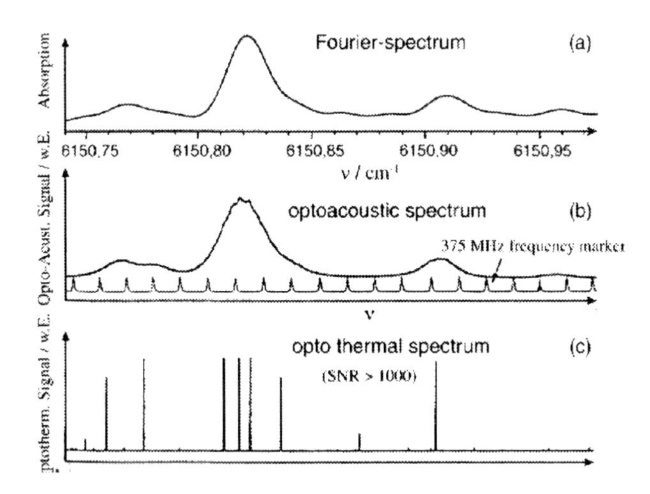

Fig. 10. Comparison of the same section of the overtone spectrum of ethylene, recorded with a) modulated absorption, b) Fourier-spectroscopy and c) opto-thermal spectroscopy.

Furthermore the accuracy of measured line positions is much higher than for Doppler-broadened line profiles. This allows the determination of shifts and briadenings of spectral lines due to perturbations of high lying vibronic levels in polyatomic molecules. Overtone spectra of even larger molecules, such as $CHCl_3$ can be resolved and analyzed [9].

Since the absorption cross section of overtone transitions decreases with increasing vibrational excitation, for the detection of weak lines the intensity of the laser should be as high as possible. When the crossing point of molecular and laser beam is placed inside an optical enhancement cavity with highly reflecting mirrors (Fig. 11), the sensitivity can be increased by at least two orders of magnitude. In our lab an intra-cavity power of 100W was achieved for an incident laser power of 500mW. This gives an enhancement factor of 200!

5. Cavity Ringdown Spectroscopy

This technique uses the temporal decay of a laser pulses passing through an absorbing sample inside a high finesse enhancement cavity [4;10]. Its principle is illustrated in Fig. 11.The output pulses of a laser are mode-matched into a high finesse optical cavity (mirror reflectivity $R > 0.9998$, finesse $F > 15.000$, mirror transmission $T = 10^{-4}$, mirror separation L, absorption losses of the empty cavity $A = 10^{-4}$). Each pulse travels back and forth inside the cavity. Each time it hits the end mirror, the small fraction $T \cdot P$ of the intra-cavity power P is transmitted to the detector. If the roundtrip time of the pulses is $t_r = 2L/c$, the detector receives a sequence of pulses with decreasing amplitude and a decay constant for the empty cavity

$$\tau_1 = (L/c)/(T+A) \sim (L/c)/(1-R) \tag{7}$$

When the absorbing sample with absorption coefficient α is inserted into the cavity, the losses are increased and the decay time of the pulses

$$\tau_2 = (L/c)/(T+A+\alpha \cdot L) \sim (L/c)/1-R+\alpha L) \tag{8}$$

becomes shorter. The difference of the two reciprocal decay times

$$1/\tau_2 - 1/\tau_1 = c \cdot \alpha \tag{9}$$

becomes independent of the cavity length L and the mirror reflectivity R and yields directly the absorption coefficient α.

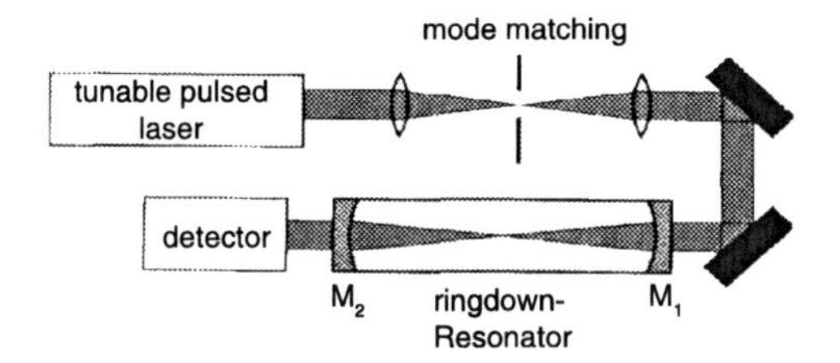

Fig. 11. Schematic Setup for Cavity Ringdown Spectroscopy.

The accuracy of this method depends on the accuracy of the time measurements. Their relative error decreases with increasing values of τ_i. Since the decay times are inversely proportional to the finesse of the cavity, high finesse cavities give higher sensitivity.

The laser wavelength λ is now tuned over the spectral range of absorbing transitions and the decay time $\tau_2(\lambda)$ is measured. From this measurement the computer calculates the absorption coefficient $\alpha(\lambda)$ according to equ. (9).The achievable sensitivity reaches limits of $\alpha < 10^{-10}\text{cm}^{-1}$ [11].

6. Resonant Two-Photon Ionization

The most sensitive detection technique for measuring molecular absorption spectra is the resonant two-photon or multi-photon ionization (REMPI). Here a tunable laser L_1 excites the molecules into higher electronic states and photons from a second laser L_2, generally with fixed wavelength, ionize the excited molecules. The ions are extracted from the interaction region by a weak electric field, accelerated by a stronger field onto an ion multiplier and the ion rate is counted as a function of the wavelength of laser L_1. The measured signal $S(\lambda_1)$ represents the absorption spectrum of L_1 (except some possible intensity variations due to the dependence of the ionization rate on the intermediate excited level) but has such a high sensitivity, that single absorbed photons of L_1 (this means also single absorbing atoms or molecules!) can still be detected.

Because of their higher intensity, generally pulsed lasers are used. However this has some disadvantages:

a) The spectral resolution is limited by the laser bandwidth

b) The duty cycle (ratio of pulse duration and time interval between successive pulses) is very low. For example for a pulse width of 10ns and a repetition frequency of 100Hz the duty cycle is only 10^{-6}!). Thus most molecules escape detection between two successive laser pulses.

In many cases therefore single mode cw lasers are more advantageous. Combined with excitation in a collimated molecular beam the spectral width of absorption lines is only limited by time of flight broadening and possible saturation broadening. However, in order to compete with the radiative decay of the intermediate level the second laser has to ionize the excited molecules *before* they decay. This implies, that the spatial overlap between the beams of L_1 and L_2 has to be optimized and the laser beams have to be collimated by a cylindrical lens, to produce a beam cross section in the focal plane which matches in the cross section of the molecular beam and restricts the time of flight of molecules through the laser beam to a time less than the lifetime of the excited state.

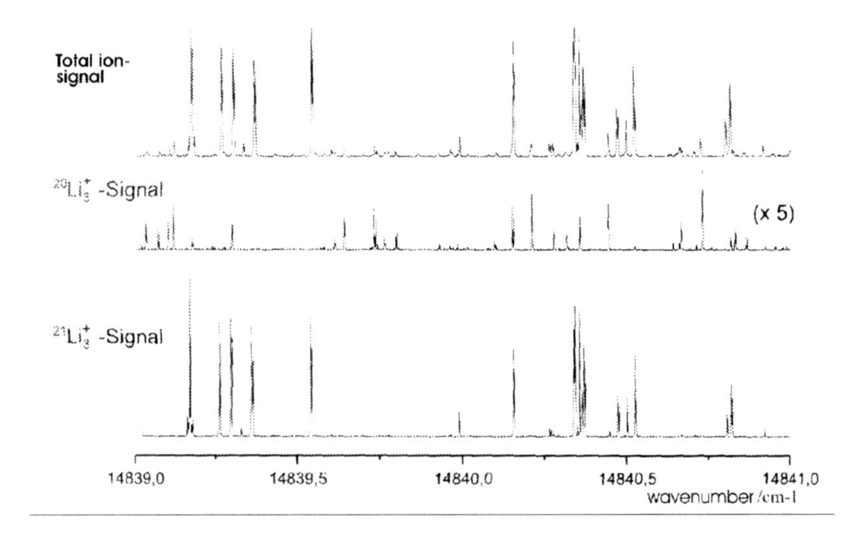

Fig. 12. RTPI spectrum of Li_3: a) without mass selection b) Spectrum of the isotopomer $^7Li^7Li^6Li$ c) of $^7Li^7Li^7Li$.

The REMPI-technique can be combined with mass spectrometry, if the excitation volume is placed inside the ion source of a mass spectrometer. With pulsed lasers time-of flight mass spectrometers are useful, while for cw lasers quadrupole mass spectrometers are more adequate. This allows laser spectroscopy of selected isotopomers in molecular beams or of specific clusters in a beam containing many different cluster sizes. As an example Fig. 12 shows the same spectral section of the RTPI spectrum for the two isotopomers $^{21}Li_3$ ($^7Li^7Li^7Li$) and $^{20}Li_3$ ($^6Li^7LI^7Li$). The two spectra look quite different. This is not only due to the isotope shift of the vibrational and rotational levels but mainly due to the nuclear spin statistics. The isotope ^7Li with a nuclear spin of 3/2 and an electron spin of ½ is a Boson, while the isotope ^6Li is a Fermion. Therefore different symmetry selection rules apply to the two cases.

With higher resolution it appears that every line consists of a substructure, caused by the hyperfine interaction, which is different for the different isotopomers (Fig. 13).

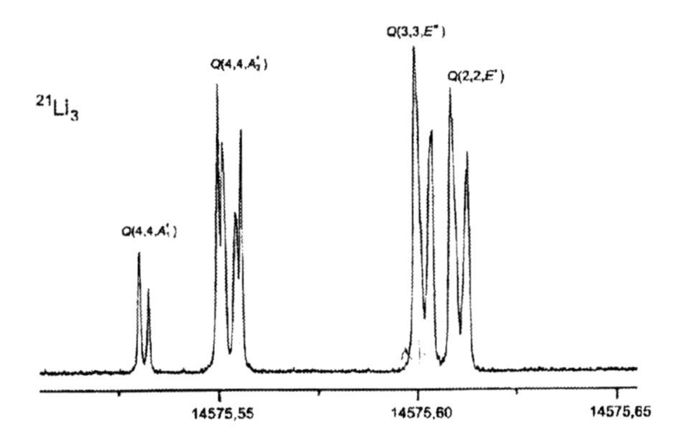

Fig. 13. Hyperfine structure of the lithium trimer $^7Li^7Li^7Li$ [12].

7. Comparison of the Different Techniques

Each of the different techniques discussed above, has its merits and limitations. While the modulated absorption spectroscopy can be applied to molecules in cells as well as in molecular beams, opto-acoustic spectroscopy needs a buffer gas in order to transfer the laser excitation energy sufficiently fast into translational energy of the gas molecules. It is therefore restricted to cells and its spectral resolution is limited by the Doppler-width of the absorption lines. From the experimental side the opto-acoustic cell is easy to install and the equipment is cheap. The modulated absorption spectroscopy needs an opto-acoustic or electro-optic modulator with driver.

Opto-thermal spectroscopy requires a molecular beam apparatus and a good bolometer at low temperatures. It is therefore more expensive, but very sensitive and allows Doppler-free spectral resolution.

All three techniques gain in sensitivity when an enhancement cavity or a multi-pass cell is used. Besides REMPI the most sensitive absorption technique is NICE OHMS. However, its technical realization demands excellent laser stabilization techniques and a more sophisticated experimental setup.

Resonant two-photon ionization needs a second laser for ionization which has sufficient power, because the bound-free transition from the excited state into the ionization continuum has a low transition probability. Regarding the minimum still detectable number of absorbing molecules this is the optimum techniques. It allows, in combination with mass spectrometers mass selective spectroscopy for example of selected isotopomers or of specific molecules in a mixture of different species. It can be used in molecular beams, allowing Doppler-free spectroscopy, or in cells where the Doppler-width is generally the limiting factor. However, with optical double-resonance techniques, such as RTPI, even in cells Doppler-free spectral resolution can be achieved. The reason is the following: If the laser L_1, passing through the absorption cell into z-direction, is stabilized onto the center of a molecular line, only molecules are excited with velocity components $v_z < \gamma$ are excited, where γ is the homogenous line width of the transition. The second laser L_2 then ionizes only these molecules within a narrow velocity interval around $v_z = 0$, yielding Doppler-free signals.

8. Spectroscopy in Cold Molecular Beams

When a gas expands from a high pressure reservoir at temperature T_0 through a narrow nozzle into vacuum, the internal energy $E = E_{kin} + E_{pot} = N \cdot f \cdot kT/2 + p \cdot V$ (N = number of molecules in the reservoir, f = number of degrees of freedom of each gas molecule) is partly transferred into directional flow energy $E_{fl} = N \cdot (m/2)u^2$, since after the expansion the molecules move with a mean flow velocity u through the vacuum, where the spread of their velocities has greatly decreased which means that their relative velocities are very small. In a reference frame moving with the flow velocity u the kinetic energy of the molecules is very low (typical a few Kelvin). By collisions during the expansion the internal energy (rotational or vibration energy is partly transferred to this cold "heat-bath" and the population of the molecular levels is compressed into the lowest rotational-vibrational levels.This simplifies the absorption spectrum drastically compared to a spectrum taken in a cell at room temperature. Cold molecular beams are therefore well suited to record the absorption spectra of even large molecules with rotational resolution [13].

If the molecular beam is collimated the transverse velocity components of the molecules are reduced and Doppler-free spectral resolution can be achieved. Because of the low kinetic energy, loosely bound molecular complexes, such as van-der Waals molecules or clusters can be stable in these cold molecular beams.Thus a whole new class of interesting objects can be studied by high resolution laser spectroscopy [14,15] .

One example is the investigation of anisole-water complexes in cold beams [16]. The rotationally resolved spectra give information about the geometric structure of the complexes and can answer the question, whether the water molecules dock within the plane of the mother molecule M or perpendicular to it. Looking for the change of the geometry of the complex $M(H_2O)_n$ with increasing number n of water molecules, gives information on the hydrization of molecules in liquids.

9. Applications of Laser Spectroscopy

There are meanwhile numerous applications of laser spectroscopy in basic sciences such as chemistry and biology, but also in medicine and astronomy. Furthermore many technical problems can be solved more readily with modern optical and spectroscopic methods. Here only a few examples can be given:

For the separation of specific molecules out of a mixture of many species selective excitation of the wanted molecules by a narrow band laser with subsequent ionization by a second laser produces ions of the selected molecule, which can be separated by a mass spectrometer. One example is the sensitive detection of rare plutonium isotopes in the surrounding of recycling facilities for nuclear fuel under conditions where the background of other atoms with the same mass is larger by many orders of magnitude. Another example is the isotope separation of rare isotopes used for medical applications. Often the excited molecules can undergo chemical reactions with other admixed atoms or molecules, producing new reaction products which can be separated by chemical means. This combination of laser excitation with specific chemical reactions generally enhances the efficiency of the separation by several orders of magnitude.

A rapidly growing field of laser applications in Chemistry is coherent control of chemical reactions with shaped femto-second laser pulses (Fig. 13). Its basic idea is, that by a proper choice of the time-and frequency profile of the exciting laser pulse the excited molecule preferably decays into a wanted reaction channel, while unwanted decay channels arte suppressed [17]. The reason for this effect is the dependence of phase and spatial extension of the electronic wave function in the excited molecular state on the shape of the exciting laser pulse. The frequency distribution and time profile of the laser pulse can be controlled by a

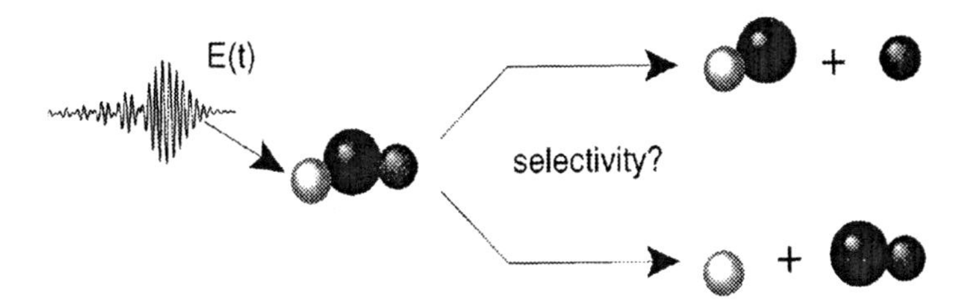

Fig. 14. Control of reaction channels by excitation with shaped femto-second pulses [17].

liquid crystal mask (Fig. 15). The frequency components of the femto-second pulse are spatially separated by an optical diffraction grating. The different frequency components pass through different pixels of the liquid crystal mask and can be independently controlled by applying specific voltages to these pixels. A second grating recombines the spatially separated components and the shape of the superposition depends on the different phase shifts. A feedback control with a learning algorithm changes voltages to the different pixels and thus the shape and phase distribution of the laser pulse, until the wanted reaction has become optimal.

This technique has been meanwhile used in many laboratories [17-20] and in favorable cases the efficiency of the wanted reaction channel could be increased up to 90%. The method has been applied even to large biological molecules, such as the light harvesting antenna molecules in photosynthesis. Although here the achieved reaction efficiency of the photo-synthesis reaction could not surpass that developed by nature in the course of evolution, it was found that it does depend on the shape and spectral distribution of the exciting pulse and the different reaction channels could be separated [19].

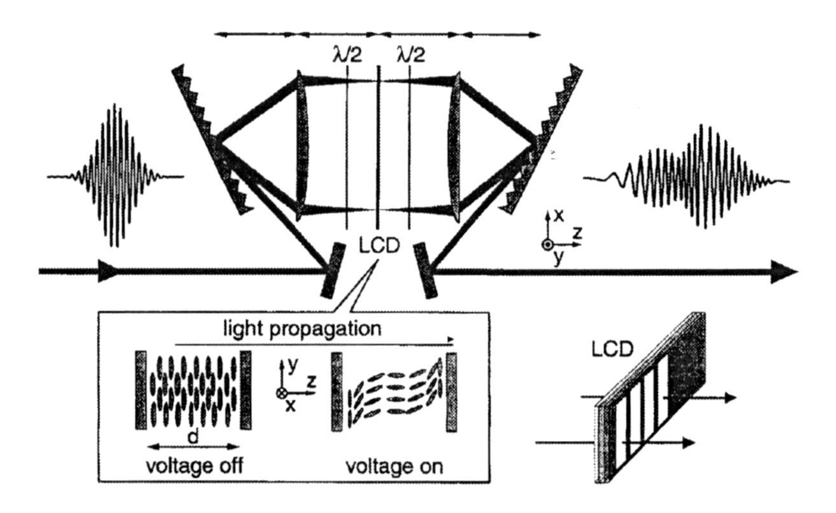

Fig. 15. Schematic arrangement for shaping of femto-second pulses [17].

One example for biological applications of modern optical technology is the confocal microscopy, which can overcome the limit for spatial resolution of conventional microscopes. Its basic principle is explained by Fig. 16. A laser beam is focused into the biological sample and the laser-induced fluorescence is viewed perpendicular to the laser beam direction. If the excitation volume is imaged onto a narrow aperture in front of the fluorescence detector, the size of the aperture determines the size of the observed volume. If the two-photon fluorescence in the UV is monitored, which is proportional to the square of the exciting laser intensity, most of the detected signal comes from a spatial zone around the maximum of the intensity profile of the exciting laser and the spatial resolution can be enhanced by nearly a factor of two.

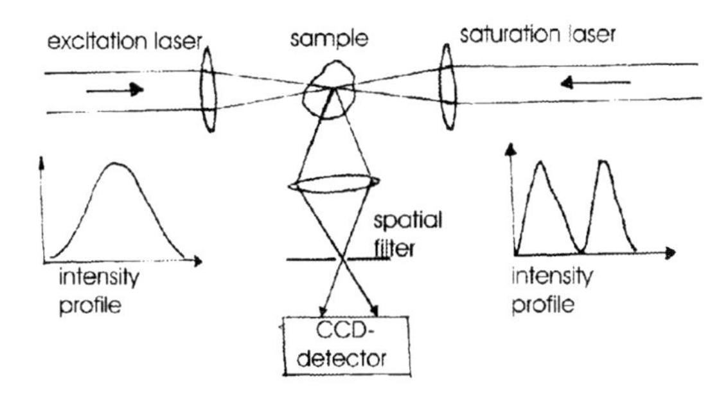

Fig. 16. Schematic arrangement for confocal microscopy.

With a special trick the spatial resolution can be further enhanced. If a second laser with another non-resonant wavelength and a donut mode TEM_{11} for its intensity profile is superimposed onto the exciting laser operating in the Gaussian TEM_{00} mode, this second laser can deplete the ground state population of the investigated bio-molecules in an annular ring around the circular excitation zone. This second laser induces Raman scattering into the direction of the inducing laser, thus restricting the fluorescence induced by the first laser to a small circular zone with a diameter, which depends on the saturation of the second laser and can be much smaller than the wavelength of the exciting laser is chosen. With this technique biological structures in the submicron range can be resolved [21]. The confocal microscopy has also found many applications for the solution of technical problems. One example is the contact-free inspection of the flatness of surfaces of grinding wheels (Fig. 17). They consist of pressed diamond powder and are burned in a furnace to form a very hard material. After burning they cannot be formed any longer. If they have a non-uniform surface which would result in an unbalanced rotation, they are useless and have to be thrown away. The inspection therefore has to be performed before the burning process. An optical inspection offers the best solution, because it is contact-free and does not change the sandy surface. The laser beam hits the rotating sandy disc from above and is focused in the plane of the surface by a cylindrical lens. The focal line averages over the grain structure. The back-scattered light is imaged onto a narrow slit in front of the detector. If the average surface plane deviates from the ideal focal plane, the back-scattered intensity passing through the slit will change and the detector receives less intensity. With this technique deviations of less than 0.1mm can be detected.

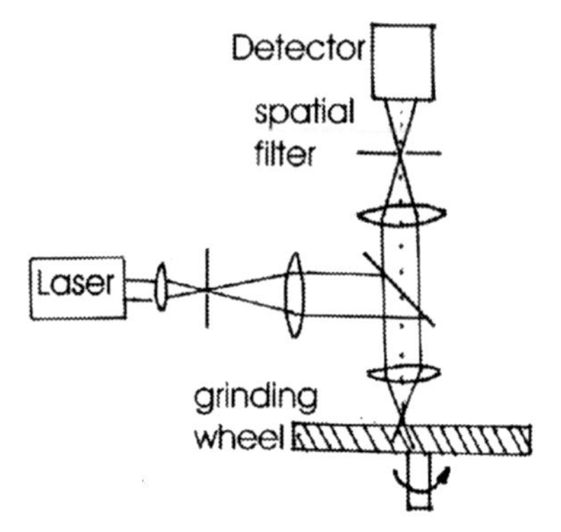

Fig. 17. Inspection of the flatness of grinding wheels with confocal microscopy.

An interesting application of laser spectroscopy in astronomy is the realization of large telescopes with adaptive optics. The angular resolution of any telescope is in principle limited by diffraction. A mirror with diameter D receiving radiation with wavelength λ from distant stars cannot resolve angular separations $\Delta\alpha < \lambda/D$. For earth-bound telescopes, however, this limit is never reached because density fluctuations in the earth`s atmosphere cause distortions of the optical wave-fronts, which results in a broadening and blurring of the star images in the detector plane and results in limits of about 0.5-1 seconds of arc for the angular resolution. The method of adaptive optics can greatly reduce this influence of the density fluctuations. Its principle is explained in Fig. 17. The radiation received from the stars by the telescope is imaged onto the detector. Before it reaches the detector it is reflected by a thin plane mirror with piezo-elements on its backside and passes through a beam splitter. A wave-front sensor monitors the deviations from plane wave fronts. A feedback loop gives correction signals to the piezo-elements, which change the mirror surface until the reflected wave forms an ideal plane wave.

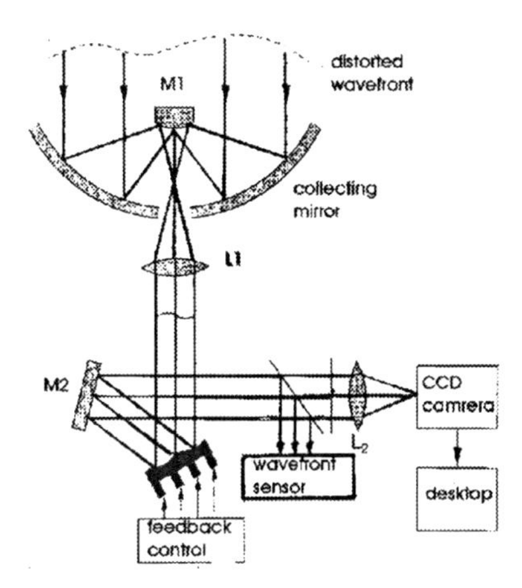

Fig. 18. Adaptive Optics for astronomical telescopes.

Since the feedback loop has to operate in a sufficiently short time (typically about 10-100ms) in order to compensate the density fluctuations in the atmosphere, the distant stars do not give enough intensity. Therefore an artificial star is created in the following way: A dye laser beam is send through the telescope into the direction of star observation. In an altitude of about 40-70km in our atmosphere a layer with a small concentration of sodium atoms is found. If the laser wavelength is tuned to the resonant transition of sodium at 590nm a brilliant yellow fluorescence is observed which looks like an artificial star. This radiation which passes through the same atmospheric layers as the star light, is received by the telescope and is used for the wave front correction [22].

10. Conclusion and Acknowledgements

These lectures, which are mainly restricted to the spectroscopy of free atoms and molecules, could only cover a small part of the many sensitive detection techniques in laser spectroscopy, where most of the examples were taken from experiments in our labs. There are many more interesting applications, such as the detailed investigation of collision processes with laser-spectroscopic techniques or applications in solid state physics, which were not included here. For more detailed information the reader is referred to the cited literature.

I would like to thank my former PhD students for their contributions, Prof. DiBartolo for the invitation to Erice and Dr. Ottavio Forte for his the editing efforts.

References

1) K. Liu, M.G. Littman: Opt. Lett. 6, 117 (1981)
2) H. Wenz, R. Grosskloß, W. Demtröder: Laser und Optoelektronik 28, 58 (1996)
3) H. Wenz, W. Demtröder and J.M. Flaud: J. Molec. Spectrosc. 209, 267 (2001)
4) K.W. Busch, M.A. Busch: Cavity Ringdown Spectroscopy (Oxford University Press, Oxford 1999)
5) Jun Ye, Long-Sheng Ma, J.L. Hall: Using FM Methods in a High Finesse Cavity: A demonstrated Path to $<10^{-12}$ Absorption Sensitivity. In ref. [4], page 233
6) J. Xiu, R.Stroud: Acousto-optic Devices: Principle, Design and Applications (Wiley, New York 1994)
7) Ch. Hornberger, W. Demtröder: Chem. Phys. Lett. 190, 1712 (1994)
8) M. Zen: Cryogenic Bolometers, in: G. Scoles, (ed): Atomic and Molecular Beam Methods, Vol.1 (Oxford University Press, London 1988)
9) Th. Platz, W. Demtröder: Chem. Phys. Lett. 294, 397 (1998)
10) B.A. Baldus, R.N. Zare et.al: Cavity –locked Ringdown Spectroscopy; J. Appl. Phys. 83, 3991 (1998)
11) G. Berden, R. Peters: Int. Rev. Phys. Chem. 19, 565-607 (2000)
12) W. Meyer, M. Keil,A. Kudell, M.A. Baig, W. Demtröder: J. Chem. Phys 115, 2590 (2001).
13) M. Ito, T. Ebata, N. Mikami: Ann. Rev. Phys. Chemistry 39, 123 (1988)
14) M. Havenith: Infrared Spectroscopy of Molecular Clusters. Springer Tracts in Modern Physics 176 (2002)
15) D.J. Wales (ed): Intermolecular Forces and Clusters.(Springer, Berlin, Heidelberg 2005)
16) M. Becucci et.al: J. Chem. Phys. 120, 5601 (2004)
17) A. Assion,T. Baumert, M. Bergt, T. Brixner, B. Keifer, V. Seyfried, M. Strehle, G. Gerbrer: Science 282, 919 (1998)
18) D. Zeidler, S. Frey, K.L. Kompa, M. Motzkus: Phys. Rev. A64, O23420 (2001)
19) W. Wohlleben, T. Buckup, J.L. Herek, R.J. Cogdell, M. Motzkus: Biophysical Journal 85, 442 (2003)
20) E. Gershgoren et al: Opt. Lett. 28, 361 (2003)
21) C. Bastiaens, S.W. Hell: Light Microscopy on the Move- An Introduction. Structural Bioölogy 117, 147 (2004)
22) D.T. Gavel: Adaptive Optics, Proc. SPIE Vol. 4494, p. 215 (2002)

LUMINESCENCE SPECTROSCOPY OF SOLIDS: DELOCALIZED SYSTEMS

BALDASSARE DI BARTOLO
Department of Physics
Boston College
Chestnut Hill, MA 02467
dibartob@bc.edu, rinodiba@comcast.net,

Abstract

The purpose of this article is to describe the luminescence properties of solid state systems that we call delocalized. This designation derives from the fact that radiative processes that take place in these systems are associated with quantum states of the entire solid.

The first section consists of a classification and characterization of these systems divided into "intrinsic" and "doped" semiconductors. Models are presented for each type of them. The doped semiconductors are separated into p-type and n-type.

The second section deals with the radiative processes that take place in delocalized systems, starting with excitations yielding excitons and including absorption and emission for both pure and doped systems. Nonradiative processes are then examined. Finally, p-n junctions are considered and their properties examined; particular attention is given to radiative processes in such junctions.

1. Delocalized Systems

1.1 DENSITY OF ONE-ELECTRON STATES AND FERMI PROBABILITY DISTRIBUTION

Given a volume $V = L^3$ the number of one-particle states in the range $dp_x dp_y dp_z$ is

$$\frac{V}{8\pi^3} dk_x dk_y dk_z = \frac{V}{h^3} dp_x dp_y dp_z \tag{1.1}$$

The number of one-particle states in the range $(p, p + dp)$ is

$$g(p)dp = \frac{V}{h^3} \int_0^{\pi} \int_0^{2\pi} p^2 \sin\theta d\theta d\varphi dp = \frac{4\pi V}{h^3} p^2 dp \tag{1.2}$$

and, if the particles are electrons, taking the spin into account,

B. Di Bartolo and O. Forte (eds.), Advances in Spectroscopy for Lasers and Sensing, 207–227.

$$\begin{aligned}2g(E)dE &= 2g(p)\frac{dp}{dE}dE \\ &= 2\frac{4\pi V}{h^3}p^2\frac{1}{2}\sqrt{\frac{2m}{E}}dE = \frac{4\pi V}{h^3}2mE\frac{1}{2}\sqrt{\frac{2m}{E}}dE \\ &= \frac{4\pi V}{h^3}(2m)^{3/2}E^{1/2}dE\end{aligned} \tag{1.3}$$

Given a system of Fermions at temperature T, the probability distribution that specifies the occupancy probability is

$$F(E) = \frac{1}{e^{(E-E_F)/kT}+1} \tag{1.4}$$

In metals E_F, the *Fermi energy*, is the energy of the most energetic quantum state occupied at $T = 0$. At $T \neq 0$, E_F is the energy of a quantum state that has the probability 0.5 of being occupied.

The number of available states in $(E, E + dE)$ for a system of electrons is

$$2g(E)dE = \frac{4\pi V}{h^3}(2m)^{3/2}E^{1/2}dE = \frac{8\sqrt{2}\pi m^{3/2}}{h^3}VE^{1/2}dE \tag{1.5}$$

The Fermi energy at T = 0 is determined by

$$\begin{aligned}N = \int_0^\infty 2g(E)dE &= \frac{8\sqrt{2}\pi m^{3/2}}{h^3}V\int_0^{E_F}E^{1/2}dE \\ &= \frac{16\sqrt{2}\pi m^{3/2}}{3h^3}VE_F^{3/2}\end{aligned} \tag{1.6}$$

Then

$$E_F = \frac{\hbar^2}{2m}\left(3\pi^2\frac{N}{V}\right)^{2/3} = \frac{0.121h^2}{m}n^{2/3} \tag{1.7}$$

where n = N/V.

1.2 CLASSIFICATION OF CRYSTALLINE SOLIDS

Crystalline solids are arranged in a repetitive 3-d structure called a *lattice*. The basic repetitive unit is the *unit cell*. Prototypes of crystalline solids are

Copper-metal
Diamond-insulator
Silicon-semiconductor

We can classify the solid according to three basic properties:

1. Resistivity ρ at room temperature

$$\rho = \frac{E}{J} \qquad (\Omega * m)$$

where E = electric field, J = current density

2. Temperature coefficient of resistivity

$$\alpha = \frac{1}{\rho}\frac{d\rho}{dt} \qquad \left(K^{-1}\right)$$

3. Number density of charge carriers, n (m^{-3})

The resistivity of diamond is greater than 10^{24} times the resistivity of copper. Some typical parameters for metals and undoped semiconductors are reported in the following table.

	Unit	**Copper (Metal)**	**Silicon (Semiconductor)**
n	m^{-3}	9×10^{28}	1×10^{16}
ρ	Ωm	2×10^{-8}	3×10^{3}
α	K^{-1}	4×10^{-3}	-70×10^{-3}

If we assemble N atoms each level of an isolated atom splits into N levels in the solid. Individual energy levels of the solid form *bands*, adjacent bands being separated by *gaps*. A typical band is only a few eV wide. Since the number of levels in one band may be on the order $\sim 10^{24}$, the energy levels within a band are very close.

1.2.1 Insulators

The electrons in the filled upper band have no place to go: the vacant levels of the band can be reached only by giving an electron enough energy to bridge the gap.

For diamond the gap is 5.5 eV. For a state at the bottom of the conduction band, the energy difference $E - E_F$ is 0.5 E_g, since, as we shall see later, the Fermi energy for undoped semiconductors is approximately at the middle of the gap. Therefore we have $(E - E_F) \gg kT$, and we can write for the probability that one electron occupies a quantum level at the bottom of the *conduction band* as

$$P(E) = \frac{1}{e^{(E-E_F)/kT} + 1} \approx e^{(E-E_F)/kT} = e^{-\frac{E_g}{2kT}} \qquad (1.8)$$

At room temperature such a probability has the value

$$P(E) = e^{-5.5/2(0.026)} \approx 1.2 \times 10^{-46}$$

and is negligible.

1.2.2 Metals

The feature that defines a metal is that the highest occupied energy level falls near the middle of an energy band. Electrons have empty levels they can go to!

A classical free electron model can be used to deal with the physical properties of metals. This model predicts the functional form of Ohm's law and the connection between the electrical and thermal conductivity of metals, but does not give correct values for the electrical and thermal conductivities. This deficiency can be remedied by taking into account the wave nature of the electron.

1.2.3 Semiconductors

In this section we shall treat semiconductors that do not contain any impurity, generally called *intrinsic semiconductors*. We shall see later how the presence of impurities affects greatly the properties of semiconductors.

The band structure of a semiconductor is similar to that of an insulator: the main difference is that a semiconductor has a much smaller energy gap E_g between the top of the highest filled band (*valence band*) and the bottom of the lowest empty band (*conduction band*) above it. For diamond $E_g = 5.5eV$ whereas for *Si*, $E_g = 1.1eV$.

The charge carriers in *Si* arise only because, at thermal equilibrium, thermal agitation causes a certain (small) number of valence band electrons to jump over the gap into the conduction band. They leave an equal number of vacant energy states called *holes*. Both electrons in the conduction band and holes in the valence band serve as charge carriers and contribute to the conduction.

The resistivity of a material is given by

$$\rho = \frac{m}{e^2 n \tau} \tag{1.9}$$

where m is the mass of the charge carrier, n is the number of charge carriers/V, and τ is the mean time between collisions of charge carriers. Now, $\rho_C = 2 \times 10^{-8}\ \Omega$, $\rho_{Si} = 3 \times 10^3\ \Omega m$, and $n_C = 9 \times 10^{28}\ m^{-3}$, $n_S = 1 \times 10^{16}\ m^{-3}$, so that

$$\frac{\rho_{Si}}{\rho_{Cu}} = 10^{11} \quad \text{and} \quad \frac{n_{Cu}}{n_{Si}} = 10^{13}$$

The vast difference in the density of charge carriers is the main reason for the great difference in ρ. We note than the temperature coefficient of resistivity is positive for *Cu* and negative for *Si*.

The atom *Si* has the following electronic configuration

$$Si : \underbrace{1s^2 2s^2 2p^6}_{core} 3s^2 3p^2$$

Each *Si* atom has a core containing 10 electrons and contributes its $3s^2 3p^2$ electrons to form a rigid two-electron covalent bond with its neighbors. The electrons that form the *Si*-*Si* bonds constitute the valence band of the *Si* sample. If an electron is torn from one of the four bonds so that it becomes free to wander through the lattice, we say that the electron has been raised from the valence to the conduction band.

1.3 INTRINSIC SEMICONDUCTORS

We shall now present a model for an intrinsic semiconductor. In general the number of electrons per unit volume in the conduction band is given by

$$n_c = \int_{E_c}^{top} N(E)F(E)dE \tag{1.10}$$

where $N(E)$ = density of states and E_c = energy at the bottom of the conduction band.

We expect that E_F to lie roughly halfway between E_v and E_c: the Fermi function decreases strongly as E moves up in the conduction band. To evaluate the integral in (1.10) it is sufficient to know $N(E)$ near the bottom of the conduction band and integrate from $E = E_c$ to $E = \infty$.

Near the bottom of the conduction band, according to (1.3), the density of states is given by.

$$N(E) = \frac{4\pi}{h^3}\left(2m_e^*\right)^{3/2}\left(E - E_c\right)^{1/2} \tag{1.11}$$

where m_e^* = effective mass of the electron near E_c. Then

$$n_c = \frac{4\pi}{h^3}\left(2m_e^*\right)^{3/2} \int_{E_c}^{\infty} \frac{\left(E - E_c\right)^{1/2}}{e^{(E-E_F)/kT} + 1} dE \rightarrow$$
$$\xrightarrow[(E_c - E_F)\rangle\rangle kT]{} \frac{4\pi}{h^3}\left(2m_e^*\right)^{3/2} \int_{E_c}^{\infty} \frac{\left(E - E_c\right)^{1/2}}{e^{(E-E_F)/kT}} dE \tag{1.12}$$

The integral may then be reduced to one of type

$$\int_0^{\infty} x^{1/2} e^{-x} dx = \frac{\pi^{1/2}}{2} \tag{1.13}$$

and we obtain the number of electrons per unit volume in the conduction band:

$$n_c = 2\left(\frac{2\pi m_e^* kT}{h^3}\right)^{3/2} e^{(E_F - E_c)/kT} \tag{1.14}$$

Let us now consider the number of holes per unit volume in the valence band:

$$n_h = \int_{bottom}^{E_V} N(E)\left[1 - F(E)\right]dE \tag{1.15}$$

where E_v = energy at the top of the valence band. $1 - F(E)$ decreases rapidly as we go down below the top of the valence band (i.e. holes reside near the top of the valence band). Therefore, in order to evaluate n_h, we are interested in $N(E)$ near E_v:

$$N(E) = \frac{4\pi}{h^3}\left(2m_h^*\right)^{3/2}\left(E_v - E\right)^{1/2} \tag{1.16}$$

where m_h^* = effective mass of a hole near the top of the valence band.

For $E_F - E_v >> kT$

$$1-F(E)=1-\frac{1}{e^{(E-E_F)/kT}+1}\approx e(E-E_F)/kT \tag{1.17}$$

Substituting (1.16) and (1.17) into (1.15) we obtain

$$\begin{aligned} n_h &= \int_{\text{bottom}}^{E_V} N(E)[1-F(E)]dE \\ &= \frac{4\pi}{h^3}(2m_h^*)^{3/2}\int_{-\infty}^{E_v}(E_v-E)^{1/2}e^{(E-E_F)/kT}dE \\ &= 2\left(\frac{2\pi m_h^* kT}{h^2}\right)^{3/2} e^{(E_v-E_F)/kT} \end{aligned} \tag{1.18}$$

We now use the fact that

$$n_c = n_h \tag{1.19}$$

and equate the two expressions for n_c and n_h given by (1.14) and (1.18), respectively. We find

$$E_F = \frac{E_c+E_v}{2}+\frac{3}{4}kT\ln\frac{m_h^*}{m_e^*} \tag{1.20}$$

If $m_e^* = m_h^*$, E_F lies exactly halfway between E_c and E_v. Replacing (1.20) in the expression for n_c = n_h we find

$$n_c = n_h = 2\left(\frac{2\pi kT}{h^2}\right)^{3/2}(m_e^* m_h^*)^{3/4} e^{-\frac{E_g}{2kT}} \tag{1.21}$$

At room temperature

$$2\left(\frac{2\pi kT}{h^2}\right)^{3/2} m^{3/2} \approx 10^{19} cm^{-3}$$

where m = mass of the electron.

1.4 DOPED SEMICONDUCTORS

1.4.1 n-type Semiconductors

Consider the phosphorus atom's electronic configuration:

$$P: 1s^2 2s^2 2p^6 3s^2 3p^3 \qquad (Z=15)$$

If a P atom replaces a Si atom it becomes a *donor*. The fifth extra electron is only loosely bound to the P ion core: it occupies a localized level with energy $E_d \ll E_g$ below the conduction band.

By adding donor atoms, it is possible to increase greatly the number of electrons in the conduction band. Electrons in the conduction band are *majority carriers*. Holes in the valence band are *minority carriers*.

Example: In a sample of pure *Si* the number of conduction electrons is $\approx 10^{16} m^{-3}$. We want to increase this number by a factor 10^6. We shall dope the system with *P* atoms creating an *n*-type semiconductor. At room temperature the thermal agitation is so effective that practically every *P* atom donates its extra electron to the conduction band.

The number of *P* atoms that we want to introduce in the system is given by

$$10^6 n_0 = n_0 + n_P$$

or

$$n_P = 10^6 n_0 - n_0 \approx 10^6 n_0 = 10^6 \times 10^{16} = 10^{22} m^{-3}$$

The number density of *Si* atoms in a pure *Si* lattice is

$$n_{Si} = \frac{N_a \rho}{A} = 5 \times 10^{28} m^{-3}$$

because N_a = Avogadro number, ρ = density of *Si* = 2,330 kg/m^3 and A = molar mass = 28.1*gm/mole* = 0.028*kg/mole*. The fraction of *P* atoms we seek is approximately

$$\frac{n_P}{n_{Si}} = \frac{10^{22}}{5 \times 10^{28}} = \frac{1}{5 \times 10^6}$$

Therefore if we replace only one *Si* atom in five million with a phosphorous atom, the number of electrons in the conduction band will be increased by a factor of 10^6.

1.4.2 p-type Semiconductors

Consider the electronic configuration of an aluminum atom:

$$Al:\ 1s^2 2s^2 2p^6 3s^2 3p \qquad (Z = 13)$$

If an *Al* atom replaces a *Si* atom it becomes an *acceptor*. The *Al* atom can bond covalently with only three *Si* atoms; there is now a missing electron (a hole) in one *Al-Si* bond. With a little energy an electron can be torn from a neighboring *Si-Si* bond to fill this hole, thereby creating a hole in that bond. Similarly, an electron from some other bond can be moved to fill the second hole: in this way the hole can migrate through the lattice. It has to be understood that this simple picture should not be taken as indicative of a hopping process, since a hole represents a state of the whole system.

Holes in the valence band are now *majority carriers*. Electrons in the conduction band are *minority carriers*. We compare the properties of an *n*-type semiconductor and of a *p*-type semiconductor in the following table.

Property	**Type of Semiconductor**	
	n	*p*
Matrix material	Silicon	Silicon
Matrix nuclear charge	14*e*	14*e*
Matrix energy gap	1.2*eV*	1.2*eV*

Dopant	Phosphorous	Aluminum
Type of dopant	Donor	Acceptor
Majority carriers	Electrons	Holes
Minority carriers	Holes	Electrons
Dopant energy gap	$0.045eV$	$0.067eV$
Dopant valence	5	3
Dopant nuclear charge	$+15e$	$+13e$

1.5 MODEL FOR A DOPED SEMICONDUCTOR

Most semiconductors owe their conductivity to impurities, i.e. either to foreign atoms put in the lattice or to a stoichiometric excess of one of its constituents.

1.5.1 n-type Semiconductors (see Figure 1a)

At $T = 0$ all the donor levels are filled. At low temperatures only a few donors are ionized: the Fermi level is halfway between donor levels and the bottom of the conduction band.

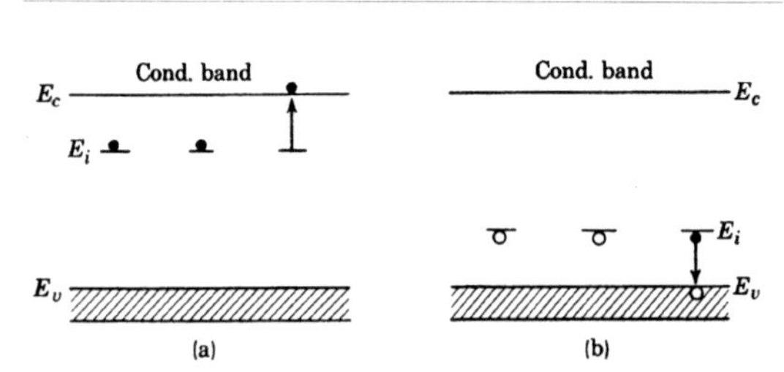

Figure 1. Energy Level Scheme for a) an n-type semiconductor, and b) a p-type semiconductor. E_i is the energy of the donor level (a) or the acceptor level (b).

If we assume that E_F is below the bottom of the conduction band by more than a few kT, then we can use in this case a formula similar to (1.14) and find

$$n_c = 2\left(\frac{2\pi m_e^* kT}{h^2}\right)^{3/2} e^{(E_F - E_c)/kT} \tag{1.22}$$

If E_F lies more than a few kT above the donor level at E_d, the density of empty donors is equal to

$$n_d\left[1 - F(E_d)\right] \approx n_d e^{(E_d - E_F)/kT} \tag{1.23}$$

Equating (1.22) and (1.23) we obtain

$$E_F = \frac{1}{2}(E_d + E_c) + \frac{kT}{2}\ln\left[\frac{n_d}{2}\left(\frac{2\pi m_e^* kT}{h^2}\right)^{-3/2}\right] \tag{1.24}$$

At $T = 0$, E_F lies halfway between the donor level and the bottom of the conduction band. As T increases E_F drops (see Figure 2). Putting the expression (6.24) of E_F in n_c given by (6.22) we find:

$$n_c = (2n_d)^{1/2} \left(\frac{2\pi m_e^* kT}{h^2} \right)^{3/4} e^{-\frac{E_c - E_d}{2kT}} \tag{1.25}$$

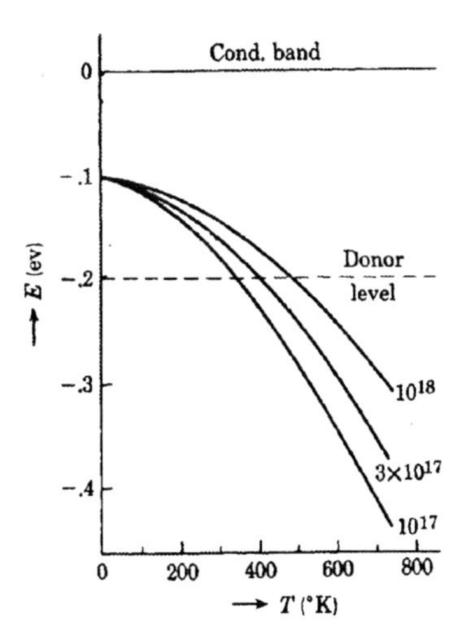

Figure 2. The variation of the position of the Fermi level with temperature with a donor level 0.2eV below the bottom of the conduction band for three different values of n_d *[1].*

1.5.2 p-type Semiconductors (see Figure 1b)

The case of *p*-type semiconductors can be treated in a similar way. n_h has an expression similar to that for n_c. The Fermi level lies halfway between the acceptor level and the top of the valence band at $T = 0$. As T increases E_F rises.

Figure 3 represents schematically the behavior of the Fermi level for an *n*-type and for a *p*-type semiconductor. The figure illustrates the fact that as the temperature increases the Fermi level for an *n*-type semiconductor does not drop indefinitely as indicated by (1.24). As the temperature increases the intrinsic excitations of the semiconductor become more important and the Fermi level tends to set in the middle of the gap. Similar effects take place for the *p*-type semiconductor. For additional considerations, the reader is referred to the book by Dekker (see Bibliography at the end of this article).

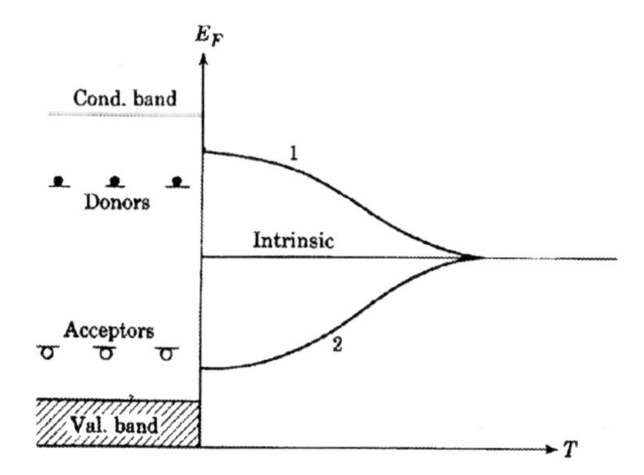

Figure 3. The variation of the position of the Fermi level with temperature. Curve 1 relates to insulators with donors and curve 2 relates to insulators with acceptors.

2. Processes in Delocalized Systems

2.1 DIRECT GAP AND INDIRECT GAP SEMICONDUCTORS

The energy of the band gap of a semiconductor determines the spectral region in which the electronic transitions, both in absorption and emissions, take place. For visible or near infrared transitions we need materials with gaps ~*1-1.7eV*. A list of such materials is provided below.

Material	Type	Band Gap (eV)
Si	Indirect	1.16
InP	Direct	1.42
GaAs	Direct	1.52
GaP	Indirect	2.3
AlP	Indirect	2.5
SiC	Indirect	3.0

Direct gap transitions take place when the maximum energy of the valence band and the minimum energy of the conduction band both occur in correspondence to a value of the linear momentum equal to zero or at the same $\vec{k} \neq 0$. Such semiconductors are called *direct gap semiconductors*.

In other materials, the maximum of the valence band and the minimum of the conduction band occur at different values of $\vec{k}$. Such materials are called *indirect gap semiconductors*.

It is interesting to consider the case of the semiconductor *GaAs*. By changing the chemical composition of this material according to the formula $GaAs_{1-x}P_x$ it is possible to change the band gap from *1.52eV* with $x = 0$ to *2.3eV* with $x = 1$. In addition, for $x > 0.4$ the material changes its character from direct gap to indirect gap semiconductor.

Mixtures of *InP* and *AlP* can also yield gaps from *1.42* to *2.5eV*.

2.2 EXCITATION IN INSULATORS AND LARGE BAND GAP SEMICONDUCTORS

If a beam of light, with photons exceeding in energy the energy gap, goes through an insulator or a semiconductor it raises an electron from the valence band into the conduction band for each photon absorbed, leaving behind a hole. The electron and the hole may move away from each other contributing to the *photoconductivity* of the material. On the other hand, they may combine producing an *exciton*, a hydrogen-like or a positron-electron pair-like structure.

Excitons are free to move through the material. Since the electron and the hole have opposite charge, excitons are neutral and, as such, are difficult to detect.

When an electron and a hole recombine, the exciton disappears and its energy may be converted into light or it may be transferred to an electron in a close-by atom, removing an electron from this atom and producing a new exciton.

Excitons are generally more important in insulators and in semiconductors with large gaps, even if some excitonic effects in small gap materials have been observed.

Excitons do not obey the Fermi-Dirac statistics and therefore it is not possible to obtain a filled band of excitons.

Excitons may also be created in doped semiconductors. In these, however, the free charges provided by the impurities tend to screen the attraction between electrons and holes and excitonic levels are difficult to detect.

Two models are generally used to deal with excitons in solids. They are more than two different ways of looking at the same problem, but, rather, they reflect two extreme physical situations:

a) A model in which the electron, after its excitation, continues to be bound to its parent atom, and
b) a model where the electron loses the memory of its parent atom and binds together with a hole.

The first case corresponds to the so-called *Frenkel exciton* and the second case to the *Wannier exciton.*

Experimentally the Frenkel exciton is in principle recognizable because the optical transitions responsible for the production of the exciton occur in the same spectral region of the atomic transitions.

Experimentally the transitions responsible for the production of a Wannier exciton fit a hydrogen-like type of behavior.

2.3 RADIATIVE TRANSITIONS IN PURE SEMICONDUCTORS

2.3.1 Absorption

The absorption optical spectra of pure semiconductors generally present the following features (see Figure 4):

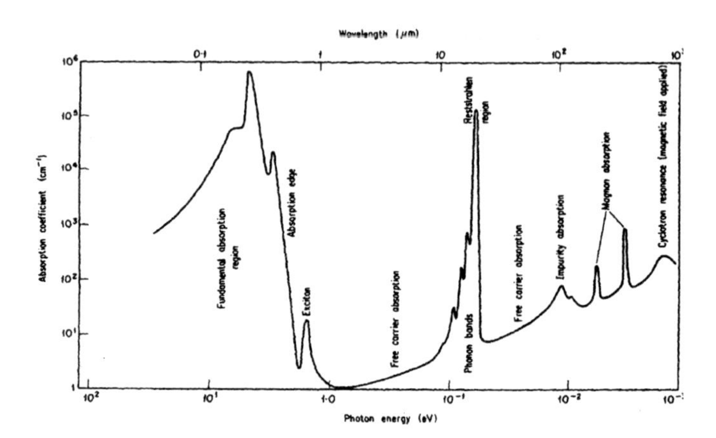

Figure 4. Absorption spectrum of a hypothetical semiconductor [2].

a) A region of strong absorption is present in the ultraviolet with a possible extension to the visible and infrared, due to electronic transitions from the valence to the conduction band. These *interband transitions* produce mobile electrons and holes that contribute to the photoconductivity. The value of the absorption coefficient is typically 10^5–10^6cm^{-1}. On the high energy side, the absorption band *(~20eV)* decreases in value smoothly in a range of several *eV*. On the low energy side, the absorption decreases abruptly and may decrease by several orders over a range of a few tenths of an *eV*. In semiconductors, this region of the absorption spectrum is referred to as the *absorption edge.*

b) The low energy limit of the absorption edge corresponds to the photon energy necessary to move an electron across the minimum energy gap E_g. The exciton structure appears in the absorption edge region. It is more evident in insulators such as ionic crystals than in semiconductors.

c) At longer wavelengths the absorption rises again due to *free-carrier absorption*, i.e. electronic transitions within the conduction or valence bands. This absorption extends to the infrared and microwave regions of the spectrum.

d) A set of peaks appear at energies *0.02 to 0.05eV* (λ *= 50-20* μm*),* due to the interaction between the photons and the vibrational modes of the lattice. In ionic crystals, the absorption coefficient may reach $10^5 cm^{-1}$; in homopolar crystals the absorption coefficient is generally much lower, such as $10\text{-}10^2 cm^{-1}$.

e) Impurities, if present in the semiconductor, may be responsible for absorption in the region of, say, $10^{-2} eV$ or so. This absorption is observable for kT lower than the ionization energy.

f) If the semiconductor contains paramagnetic impurities the absorption spectrum will present absorption lines in the presence of a magnetic field that splits the Zeeman levels.

g) An absorption peak in the long wavelength region may be present in the presence of a magnetic field, due to *cyclotron resonance* of the mobile carriers.

We want to make some additional considerations regarding the features a) and b) of the absorption spectrum.

Interband transitions can take place subject to the two conditions of energy and conservation of wave vector:

$$\begin{cases} E_f - E_i = \hbar\omega(\vec{k}') \\ \vec{k}_f - \vec{k}_i = \vec{k}' \end{cases} \qquad (2.1)$$

where the subscripts f and i refer to the final and initial one electron states and $\vec{k}'$ is the wave vector of the absorbed photon of energy $\hbar\omega(\vec{k}')$. Since the wavelength of the radiation is much longer than the lattice constant, $\vec{k}'$ is much smaller than the size of the reciprocal lattice and $\vec{k}'$ can then be neglected in the second equation in (2.1). This means that in a $(E,\vec{k})$ diagram we should rely on *vertical* transitions.

The interband absorption is restricted by the conditions (2.1) and shows a structure that depends on the density of final states. The peaks can be presumably associated with values of $\vec{k}$ about which the empty and the filled bands run parallel:

$$\vec{\nabla}_k E_v(\vec{k}) = \vec{\nabla}_k E_c(\vec{k}) \qquad (2.2)$$

In such a case there is a large density of initial and final states available for the transitions in a small range of energies.

In section 2.1 we have made a distinction between direct gap and indirect gap semiconductors. For the former, the maximum of the valence band energy and the minimum of the conduction band energy occur frequently at $\vec{k} = 0$, (but not always, e.g. not for Ta-doped halides of lead salts), whereas for the latter they occur at different values of $\vec{k}$. In indirect gap semiconductors, the absorption transitions at the band edge are phonon-assisted and have probability smaller than that of the direct gap absorption transitions. The absorption edge of the indirect gap absorption may show features related to the available phonon energies.

We now turn our attention to the excitonic structure of the absorption band. We have two models at our disposal: the Frenkel model and the Wannier model.

In the Frenkel model, an excited electron describes an orbit of atomic size around an atom with a vacant valence state; this model is more appropriate for ionic insulators.

The Wannier model represents an exciton as an electron and a hole bound by the Coulomb attraction, but separated by several lattice sites. This model is more appropriate for semiconductors.

An example of the Frenkel exciton is given by the crystal MnF_2 [3] in which the excited state may be considered to consist of an electron and a hole residing on the same ion. The excitation can travel throughout the system via the energy transfer mechanism.

A good example of the Wannier exciton is given by Cu_2O which presents absorption lines up to $n = 11$. [4]

2.3.2 Emission

Following the absorption process, an excited electron can decay radiatively by emitting a photon (possibly accompanied by a phonon) or non-radiatively by transforming its excitation energy entirely into heat (phonons). The following reasons make the emission data relevant:

1) Emission is not simply the reversal of absorption. In fact, the two phenomena are thermodynamically irreversible and therefore, emission spectroscopy furnishes data not available in absorption.
2) Emission is easier to measure than absorption, since its intensity depends on the intensity of excitation.
3) The applications of emission from solids, such as those of fluorescent lights and television, far outnumber the application of absorption.

In the section 2.5, we discuss the photon emission processes in solids.

2.4 DOPED SEMICONDUCTORS

Two types of impurities are particularly important when considering the optical behavior of semiconductors.

Donors. As we have seen in section 6.4, when a material made of group IV atoms, like *Si* or *Ge*, is doped with a small amount of group V atoms like As, the extra electrons of these atoms continue to reside in the parent atoms, loosely bound to them. The binding energy, called E_D, is typically around *0.01eV*. It is in fact *0.014eV* for *As*, *0.0098eV* for *Sb* and *0.0128eV* for *P*. E_D is also called the *ionization energy of the donor atom.* The electrons that, because of thermal excitation, a donor puts in the conduction band cannot produce luminescence, because this process needs, besides an excited electron, a hole where the electron can go and the valence band, being filled with electrons, has no holes.

Acceptors. If a material made of group IV atoms, such as *Si* or *Ge*, is doped with group III atoms, such as *Ga* or *Al,* a hole for each of these atoms forms, and remains loosely bound to the parent atom. The amount of energy necessary to move an electron from the top of the valence band to one of these holes is labeled E_A and is typically around 0.03eV. E_A may also be called the *ionization energy of the acceptor.*

Both types of impurities can be doped into the same crystal, deliberately, or they may be due to the fact that it is practically impossible to fabricate semiconductor crystals of perfect purity.

2.5 RADIATIVE TRANSITIONS ACROSS THE BAND GAP

We shall now examine, following Elliott and Gibson [5], the radiative processes that can take place across the band gap of a semiconductor (see Figure 5).

<u>Processes *A* and *B*</u>

An electron excited to a level in the conduction band will thermalize quickly with the lattice and reside in a region ~*kT* wide at the bottom of the conduction band. Thermalization is generally achieved by phonon emission, but also, less frequently, by phonon-assisted radiative transitions.

If such photons have energy exceeding E_g, they can be reabsorbed and promote another electron to the conduction band.

<u>Process *C*</u>

The recombination of electrons and holes with photon emission, the reverse process to absorption, is possible, but not very likely, because of competing processes. It may be present only in high purity single crystals. The widths of the related emission bands are expected to be ~kT because the thermalized electrons and holes reside at the band edges in this range of energy.

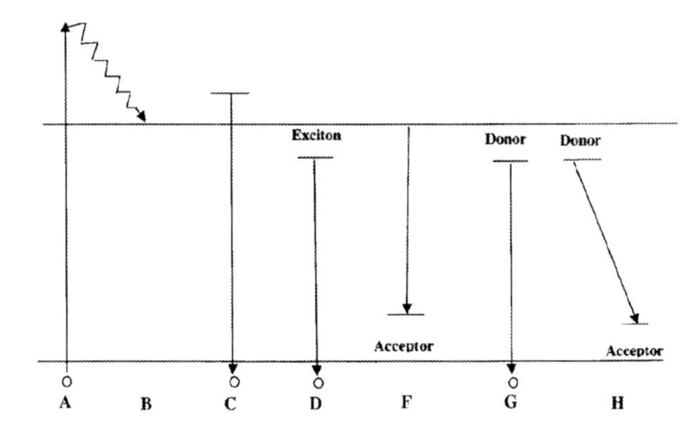

Figure 5. Transitions producing emission of photons in solids.

<u>Process *D*</u>

The radiative decay of the exciton can be observed at low temperature in very pure crystals. There are two types of decay:

1) the decay of the free exciton, and
2) the decay of an exciton bound to an impurity.

Transitions of the first type are observed at low temperatures. Since the exciton levels are well defined, a sharply structured emission can be expected.

As for the transitions of the second type, they may be observed in materials of high purity into which impurities are purposely doped. An exciton may bind itself to one such impurity; the energy of the *bound exciton* is lower than the $n=1$ energy by the binding energy of the exciton to the impurity. It may be noted that emission from bound excitons in indirect gap materials can take place without the assistance of phonons because the localization of a bound exciton negates the requirement of wave vector conservation. Bound electron emission is observed at low temperature and is generally much sharper than free electron emission.

<u>Processes *F* and *G*</u>

The transitions related to these processes are between the band edges and donors and acceptors and are commonly observed in solids. In particular, we may have conduction band to neutral acceptor (*F*) and neutral donor to valence band (*G*) transitions. They may be phonon assisted.

Process *H*

In the transitions related to these processes, an electron leaves a neutral donor and moves to a neutral acceptor. After such a transition both donor and acceptor are ionized and have a binding energy equal to

$$E_b = -\frac{e^2}{4\pi\varepsilon_o kr} \tag{2.3}$$

where r is the donor-acceptor distance. The energy of the transition is then

$$\hbar\omega(r) = E_g - E_A - E_D + \frac{e^2}{4\pi\varepsilon_o kr} \tag{2.4}$$

An example of such a transition is given by *GaP* containing sulfur donors and silicon acceptors, both set in phosphor sites.

2.6 NON-RADIATIVE PROCESSES

In the great majority of cases, recombination of electrons and holes takes place by emission of phonons. Since the probability of such processes decreases with the number of phonons emitted these processes are favored by the presence of intermediate levels between the valence and the conduction bands produced by impurities or defects.

An additional mechanism, known as the *Auger process* could be responsible for the non-radiative recombination of electrons and holes. In an Auger process, an electron undergoes an interband transition and gives the corresponding energy to another conduction band electron, which is then brought to a higher level in the same band. The latter electron decays then to the bottom of the band with the phonon emission facilitated by the near continuum of states. In most cases, however, the mechanism of non-radiative decay has not been identified with certainty.

2.7 p-n JUNCTIONS

2.7.1 Basic Properties

A *p-n* junction consists of a semiconductor crystal doped in one region with donors and in an adjacent region with acceptors.

Assuming, for simplicity's sake, that the junction has been formed mechanically by pushing towards each other a bar of *n*-type semiconductor and a bar of *p*-type semiconductor, so that a *junction plane* divides the two regions. (See Figure 6.)

Let us now examine the motion of the electrons (majority carriers of the *n*-type bar) and of holes (majority carriers of the *p*-type bar).

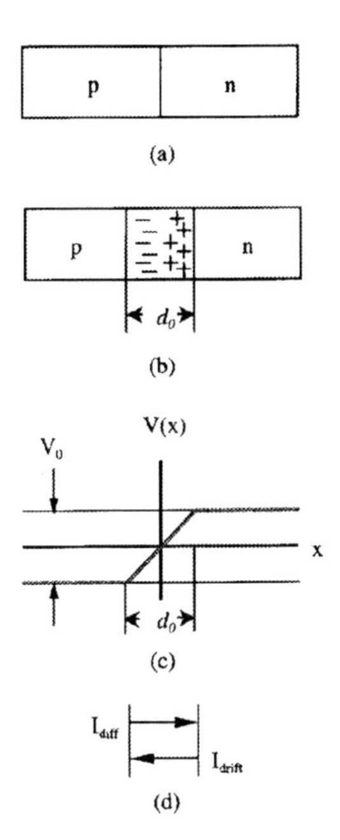

Figure 6. a) An n-type material and a p-type material joined to form a p-n junction, b) Space charge associated with uncompensated donor ions at the right of the junction plane and acceptor ions at the left of the plane, c) Contact potential difference associated with the space charge, d) Diffusion current I_{diff} made up by majority carriers, both electrons and holes, compensated in an isolated p-n junction by a current I_{drift} made up by minority carriers.

Electrons on the *n*-side of the junction plane tend to diffuse (from right to left in the figure) across this plane and go to the *p*-side where there are only very few electrons. On the other hand, holes on the *p*-side tend to diffuse (from left to right in the figure) and go to the *n*-side where there are only very few holes.

The *n*-side region is full with positively charged donor ions. If this region is isolated, the positive charge of each donor ion is compensated by the negative charge of an electron in the conduction band. But, when an *n*-side electron diffuses towards the *p*-side, a donor ion, having lost its compensating electron, remains positively charged, thus introducing a fixed positive charge near the junction plane. An electron arriving to the *p*-side quickly combines with an acceptor ion and introduces a fixed negative charge near the junction plane on the *p*-side. Holes also diffuse, moving from the *p*-side to the *n*-side, and have the same effect as the electrons.

Both electrons and holes, with their motion, contribute to a *diffusion current* I_{diff}, that is conventionally directed from the *p*-side to the *n*-side.

An effect of the motion of electrons and holes across the junction plane is the formation of two space charge regions, one negative and one positive. These two regions together form a *depletion zone*, of width d_o in Figure 6, so called because it is relatively free of mobile charge carriers. The space charge has associated with it a *contact potential difference*, V_o, across the depletion zone, which limits the further diffusion of electrons and holes.

Let us now examine the motion of the minority carriers: electrons on the *p*-side and holes in the *n*-side. The potential V_o set by the space charges represents a barrier for the majority carriers, but favors the diffusion of minority carriers across the junction plane. Together both types of minority carriers produce with their motion a drift current I_{drift} across the junction plane in the sense contrary to that of I_{diff}.

An isolated *p-n* junction in equilibrium presents a contact potential difference V_o between its two ends. The average diffusion current I_{diff} that moves from the *p*-side to the *n*-side is balanced by the average drift current I_{drift} that moves in the opposite direction.

Note the following:

a) The net current due to holes, both majority and minority carriers, is zero.
b) The net current due to electrons, both majority and minority carriers, is zero.
c) The net current due to both holes and electrons, both majority and minority carriers included, is zero.

2.7.2 The Junction Rectifier

When a potential difference is applied across a *p-n* junction, with such a polarity that the higher potential is on the *p*-side and the lower potential on the *n*-side, an arrangement called *forward-bias connection* (Figure 7a), a current flows through the junction.

The reason for this phenomenon is that the *p*-side becomes more positive than it was before and the *n*-side more negative, with the result that the potential barrier V_o decreases, making it easier for the majority carriers to move through the junction plane and increasing considerably the diffusion current I_{diff}. The minority carriers sense no barrier and are not affected, and the current I_{drift} does not change.

Another effect that accompanies the setting of a forward bias connection is the narrowing of the depletion zone, due to the fact that the lowering of the potential barrier must be associated with a smaller space charge. The space charge is due to ions fixed in their lattice sites, and a reduction of their number produces a reduction of the width of the depletion zone.

If the polarity is reversed in a *backward-bias* connection (Figure 7b) with the lower potential on the *p*-side and the higher potential on the n-side of the *p-n* junction, the applied voltage increases the contact potential difference and, consequently, I_{diff} decreases while I_{drift} remains unchanged. The result is a very small back current I_B.

2.7.3 Radiative Processes in p-n Junctions and Applications

In a simple semiconductor, one electron-hole pair may combine with the effect of releasing an energy E_g corresponding to the band gap. This energy in silicon, germanium and other simple

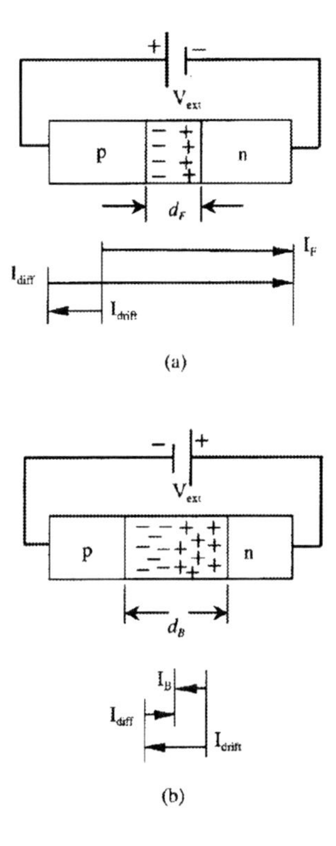

Figure 7. a) Forward-bias connection of a p-n junction, showing the narrowed depletion zone and the large forward current I_F, b) Backward-bias connection of a p-n junction, showing the widened depletion zone and the small back-current.

semiconductors, is transformed into thermal energy, i.e. vibrational energy of the lattice. In certain semiconductors, such as *GaAs*, the energy of a recombined electron-hole pair can be released as a photon of energy E_g. However, due to the limited number of electron-hole possible recombinations at room temperature, pure semiconductors are not apt to be good emitters.

Doped semiconductors also do not provide an adequate large number of electron-hole pairs, with the *n*-type not having enough holes and the *p*-type not enough electrons.

A semiconductor system with a large number of electrons in the conduction band and a large number of holes in the valence band can be provided by a heavily doped *p-n* junction. In such systems, a current can be used in a forward-bias connection to inject electrons in the *n*-type part of the junction and holes in the *p*-part. With large dopings and intense currents the depletion zone becomes very narrow, perhaps a few microns wide, and a great number of electrons are in the *n*-type material and a large number of holes in the *p*-type material.

The radiative recombination of electrons and holes produces a light emission called *electroluminescence*, or, more aptly, *injection electroluminescence*.

The materials used for *Light Emitting Diodes* (LED's) comprise such alloys as $GaAs_{1-x}P_x$, in which the band gap can be varied by changing the concentration x of the P atoms. For $x \simeq 0.4$ the material is a direct-gap semiconductor and emits red light. Almost pure *GaP* produces green light, but, since it is an indirect-gap semiconductor, it has a low transition probability.

The passage of current through a properly arranged *p-n* junction can generate light. The reverse process is also possible, where a beam of light impinging on a suitable *p-n* junction can generate a current. This principle is at the basis of the *photo-diode*.

A remote TV control consists of an LED that sends a coded sequence of infrared light pulses. These pulses are detected by a photo-diode that produces the electrical signals that perform such various tasks as change of volume or channel.

In a forward biased *p-n* junction, a situation may be created in which there are more electrons in the conduction band of the *n*-type material than holes in the valence band of the *p*-type material. Such a situation of *population inversion* is essential for the production of *laser action*. Of course, in addition to this condition the appropriate geometry for the *p-n* junction is necessary in order to allow the light to be reflected back and forth and produce the chain reaction of stimulated emission. In this way, a *p-n* junction can act as a *p-n junction laser*, with a coherent and monochromatic light emission.

Acknowledgement

The author would like to acknowledge the critical reading of the manuscript of this paper by Prof. C. Klingshirn and the benefit of discussions with Prof. J. Collins and Dr. N Barnes.

He also acknowledges the NASA Sponsorship through Grant NNL04AA30G.

References

1. Dekker, A J 1957 Solid State Physics (Englewood Cliffs, NJ: Prentice Hall) p 312
2. Elliott, R J and Gibson, A F 1974 An Introduction to Solid State Physics and It's Applications (London and Basingstoke: MacMillan) p 208
3. Flaherty, J M and Di Bartolo, B 1973 Phys. Rev. B 8 5232
4. Baumesteir, P W 1961 Phys. Rev. 121 359
5. Elliott, R J and Gibson, A F 1974 An Introduction to Solid State Physics and It's Applications, (London and Basingstoke: MacMillan) p 229

Bibliography

1. Dekker, A J 1957 Solid State Physics (Englewood Cliffs, NJ: Prentice Hall)
2. Di Bartolo, B 1968 Optical Interactions in Solids (John Wiley & Sons, New York, London, Sydney, Totonto)

3. Elliott, R J and Gibbon, A F 1974 An Introduction to Solid State Physics and Its Applications, (Macmillan, London and Basingstoke)
4. Lumb, M D, ed 1978 Luminescence Spectroscopy (New York: Academic)
5. Klingshirn, C F 1997 Semiconductor Optics (Springer, Berlin, Heidelberg, New York)
6. Yu, P Y and Cardona, M 2001 Fundamentals of Semiconductors (Springer, Berlin, Heidelberg, New York)

LASER INDUCED BREAKDOWN SPECTROSCOPY (LIBS)

The process, applications to artwork and environment

R. FANTONI, L. CANEVE, F. COLAO, L. FORNARINI, V. LAZIC, V. SPIZZICHINO*
ENEA – Applied Physical Technologies Dept.
Laser Applications Section
V. E. Fermi 45, I-00044 Frascati (Rome) Italy
** ENEA Fellow*
fantoni@frascati.enea.it,

Abstract

In this chapter the laser ablation process which is involved in the laser induced breakdown occurrence, with plasma formation and relaxation is shortly described. The fundamentals of laser induced breakdown technique are revised, presenting current laboratory instrumentation, and methods for qualitative, semi-quantitative, and quantitative analyses in air and low pressure gasses and underwater recognition and analysis of different materials.

Examples of applications of LIBS technique to artwork characterization and environmental analyses are reviewed and discussed, with special emphasis to fragile substrates, ceramics and marbles in the first case, to rocks relevant to planetary exploration and to underwater measurements of environmental interest in the second.

1. Introduction

Laser Induced Breakdown Spectroscopy (LIBS) is not really a new technique. First experiments aimed to detect atomic line emission after the laser induced plasma formation at metal surfaces in air date back to the early eighties [1]. However before considering the technique for quantitative analytical applications, substantial instrumental and data analyses methods development was required. Due to the intense background plasma emission, dominating the process shortly after the laser pulse, fundamental steps in this directions were achieved in the nineties, with the development of intensified optical multichannel analyzers as detectors, offering the combined possibilities of fine time gating and high spectral resolution, the latter with the coupling to a high resolution monochromator. Since then, the process of laser ablation at high laser power has been studied in detail in order to establish the fundamentals of the technique, especially in relation to the need of retaining the sample stoichiometry during surface evaporation and the occurrence of local thermal equilibrium conditions in plasma for selected time windows.

B. Di Bartolo and O. Forte (eds.), Advances in Spectroscopy for Lasers and Sensing, 229–254.

1.1. THE PROCESS

The laser interaction with matter always involves an energy exchange. Once a laser beam is focused onto a dense matter, nearby a surface, the laser energy can be partially reflected or scattered at the same wavelength, re-emitted at different wavelengths, absorbed or transmitted in crossing the surface. The interaction of a pulsed laser beam with the surface may lead to it a partial evaporation originating a process known as laser induced ablation, whose main characteristics are:

- Localized heating at the surface, both in space and in time;
- Fast process evolution, with a complete cycle shorter than 10 ms (for pulse widths <100 ns) [2].

For high irradiances (e.g. >1MW/cm^2, depending on the target composition) the evaporation composition) the evaporation occurs with a plasma formation, involving several successive processes such as: surface melting with hydrodynamic flow, vaporization, vapor ionization, surrounding and evaporating gas breakdown, free electron acceleration within the evaporated cloud, plasma heating.

The plasma growth and decay implies: expansion, shock waves formation and propagation, deceleration of free electrons with inverse Bremsstrahlung emission, collisions in the gas cloud with excitation and relaxation of atoms/ions, chemical recombination, radiative recombination.

Occasionally conditions associated to a Local Thermal Equilibrium (LTE) can occur, whenever electrons, atoms and ions in the plasma are associated to the same temperature. The LTE is a dynamic status reached for plasma parameters slowly varying in a suitable temporal range, during the plasma relaxation. Initially, the plasma temperature is very high (typically T >30000K), and auto-ionization accompanied by a continuum emission spectrum is dominant [3]. Fast collisional processes bring the plasma to conditions of LTE, where the electrons have a Maxwell energy distribution, all the species have the same temperature, the population of energy levels follow a Boltzmann distribution, and Saha equation describes the concentration ratio among the same species at different degrees of ionization. Upon LTE conditions, characteristic line emission spectra are detected mostly from the atomic and first ionic excited species produced. Note that a high electron density is needed to reach LTE: electron collisional rates must exceed radiative rates by at least one order of magnitude [4]. Further plasma cooling and recombination cause plasma departure from LTE. In its final stages the process is characterized by cluster and nanoparticles ejection in the dark slow expanding tail of the plume [5].

During the laser ablation at high energy densities, accompanied by the plasma formation which gives rise to line emissions recorded in LIBS, the bright plume is observed in combination with the formation of a crater at the surface. In the evaporation atoms are ejected perpendicularly to the sample surface, the enhancement along this direction is due to crater shape which may convey material towards laser direction, thereby increasing excitation. In summary, LIBS emission intensity is maximum for normal incidence, and the central plume region has the most intense emission, due to the higher density of the emitting species.

1.2. DOUBLE PULSE EXCITATION

LIBS can be used in characterizing targets at any phase (solid, liquid, gas, aerosol, nanostructure) since the same process, above described at a surface, can occur in different media (from vacuum to high density gases or liquids), even in the absence of the surface.

Thermal, mechanical and chemical properties of the medium however strongly affect the time evolution of the process and may limit its analytical applications. In fact the LTE time window useful for data acquisition is dramatically reduced at increasing density of the surrounding medium. In order to overcome this limitation, which initially prevented quantitative analyses in liquid phase, the possibility to use trains of laser pulses, suitably delayed, in order to initiate and sustain the process, has been successfully exploited [6]. In particular the double pulse technique, consisting of focusing a pair of pulses, delayed by 1 - 100 μs, onto the target, was demonstrated for analytical applications to water solutions and submersed samples [7]. The first pulse creates a bubble in the dense medium, the second pulse, coming during its expansion, creates a secondary plasma in the rarefied medium formed. The main characteristics of the double pulse LIBS for application in dense media are the summarized in the following:

- The LIBS signal is detected after the second pulse, during the expansion of the secondary plasma in the low density medium generated by the first pulse, the latter causing only a weak plasma.
- The expansion of the secondary plasma contained in the low pressure bubble occurs on a time scale longer than what expected in the high density medium.
- Narrow emission lines are detected due to the relatively low pressure inside the bubble.

The overall process for a solid target submersed in a liquid is schematized in Figure 1, note that the laser pulses are usually collinear, especially when emitted from the same laser source.

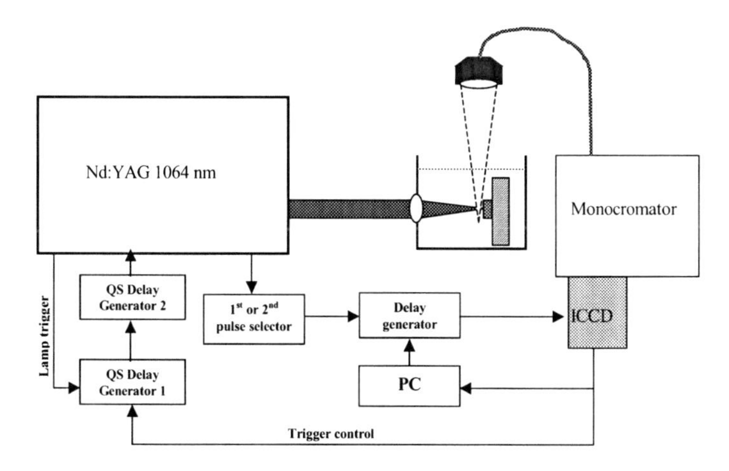

Figure 1. Schematic of the set-up for double pulse LIBS on submersed solid samples.

1.3. GENERATION OF LIBS SIGNALS

Once a high irradiance pulsed laser beam is focused in a dense medium or onto a surface, the laser induced plasma is formed and LIBS signal can be detected during the plasma expansion and decay. In case of double (multiple) pulse excitation, LIBS is signal is always relevant to the second laser pulse, which sets the zero delay time for data acquisition.

In order to allow for reliable analytical applications the proper LTE conditions must be selected for data acquisition, which consists of recording spectrally resolved signal averaged in the pre-selected time window (with a time delay and width fulfilling as much as possible the LTE constraints).

2. LIBS Instrumentation

Before describing with some detail how a standard LIBS set-up for analytical applications is built, it is worthwhile to remind that one of the main advantages of the technique is its characteristic micro-sampling. In fact due to the *in-situ* laser evaporation and ionization, requirements for sample preparation and handling are minimized. It is sufficient to ablate less than 1 μg of material at the sample surface and the technique can be considered as only micro-destructive.

Main constituents of a LIBS set-up, as sketched in figure 1, are a pulsed laser source and a spectrometer (dispersive element plus multichannel detector) with suitably designed optical systems (laser focusing and collecting plasma emission) and electronics (trigger and delay generators) components. The experiments is usually fully computer controlled.

2.1. LASER CHOICE

A major issue in building a LIBS set-up is the laser choice, i.e. the choice of the main parameters (wavelength and pulse duration) for the high power (>10 MW, at any repetition rate compatible with the use of the spectrometer) source.

2.1.1 Wavelength selection

In the wavelength choice the following points must be kept in mind:

- Q-switched Nd:YAG laser sources (with fundamental emission at 1064 nm in the near IR) are usually utilized for plasma generation; their operation at second (532 nm), third (355 nm) or fourth (266 nm) harmonics allows to reach shortest wavelength regions, up to the UV, but reduces the available power.
- The UV generated plasma is characterized by more intense emissions at equivalent laser photon density, shorter plasma durations, narrower emissions and less intense continuum, in comparison with the situation encountered for IR generation.
- Since in a plasmas generated by an UV laser the temperature and the electron density reached are usually lower [8] than in the case of IR generation, a less intense ion emission is observed with narrower emission lines (figure 2) [9].

- The UV laser generation is best for substrates either characterized by a high evaporation threshold (ceramics, stones) or by high IR reflectivity (metals).
- Lower thermal effect are induced at the sample surface by the UV radiation, with a reduced probability of secondary chemical reactions (oxidation in air), which would alter the surface composition.
- Generally, better plasma stoichiometry is achieved with UV excitation due to the reduced matrix effect.

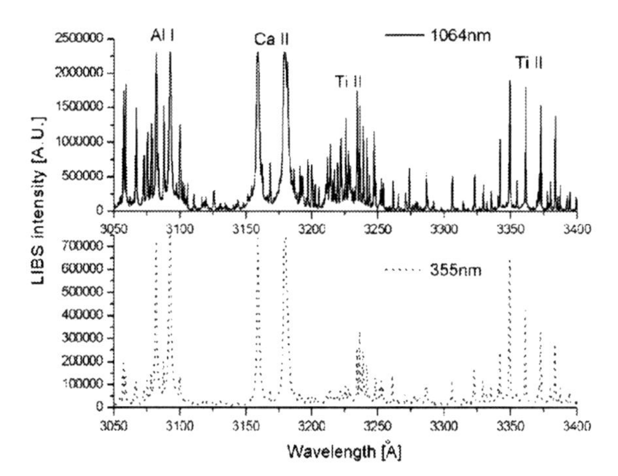

Figure 2. An UV portion of LIBS spectra taken on an andesite (rock) sample upon excitation at different laser wavelengths at approximately the same photon density.

For in water analyses the following factors must be taken in to account:

- The plasma ignition is easier to achieve by higher incident photon energy, i.e. UV excitation, which is required to break the water bond.
- The strong absorption of water in the UV and red regions suggests use of visible (400-500 nm) excitation and attenuate the emission spectra with analytical limitations.
- The energy per pulse available from commercial lasers suggests to work only in the near IR with the Nd:YAG fundamental.

A current compromise in water is the use of near IR (1064 nm) or visible (532 nm) sources.

2.1.2 Pulse duration

A further possibility of choice in the laser characteristics is the pulse duration, from nanoseconds to femtoseconds. To this respect the following points are now well established:

- Stoichiometric evaporation of the sample is preserved at best when using ultrashort laser pulses; less material is ablated, a cleaner crater is burnt on the substrate (figure 3).
- Atomic lines should be favored at the same irradiance (note that higher irradiance are easily reached with fs for the very effective pulse compression); however Stark effect distorts atomic line profiles at high irradiances (figure 4).

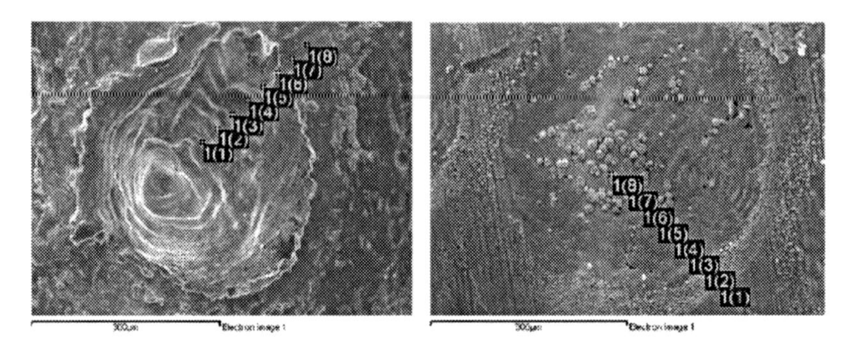

Figure 3. Craters observed at SEM on a bronze surfaces after 200 shots of ns (left) and fs (right) laser pulses. Laser parameters: 90 mJ, 8 ns at 532 nm (l); 2.4 mJ, 250 fs at 527 nm (r). Operation at 10Hz with the same focusing.

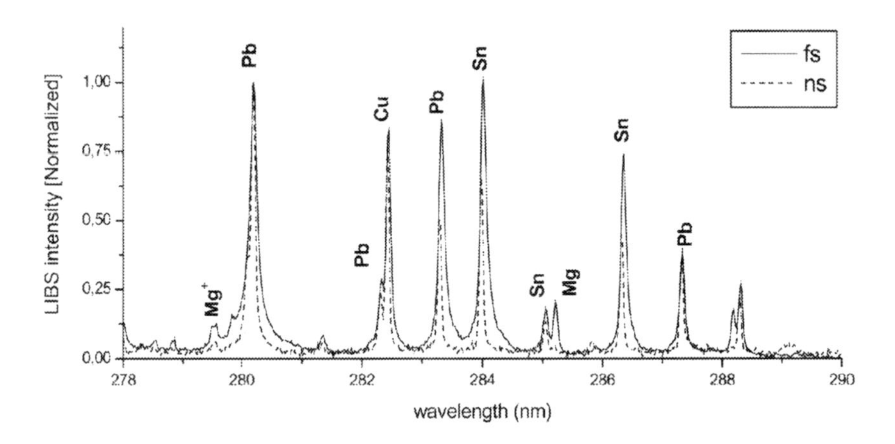

Figure 4. Comparison ns and fs LIPS spectra in the region near 284 nm on the same bronze sample. Laser parameters as in figure 2. Main Cu, Sn and Pb line assignments are indicated.

2.2. SPECTROMETER CHOICE

Two options are generally available for laboratory systems, either a high resolution monochromator coupled to an Intensified CCD detector, or an Echelle spectrometer coupled to a large ICCD. The latter, which is the most expensive combination, presents the advantage that different spectral regions are dispersed at different high orders on selected areas of the detector, thus allowing for a single shot collection of the entire UV-near IR spectrum at medium-high resolution, for example 200-1000 nm. However, the signal output is lower than in the case of the monochromator, which might be insufficient for minor/trace element detection, particularly when considering low intensity and short-duration plasmas as in the case of underwater LIBS analyses.

For field systems have often mandatory requirements to stand harsh environments, and/or to employ of very compact systems. To this respect, whenever possible, series of very small linear arrays detectors are preferred, although this choice implies a reduced sensitivity in the absence of the intensifier.

2.3. EXPERIMENTS UNDER CONTROLLED ATMOSPHERE AND IN WATER

2.3.1 LIBS in controlled atmosphere

Whenever LIBS experiments were carried on in controlled atmosphere at lower than ambient pressure we obtained a less confined plume. The plasma decay is faster in the rarefied atmosphere due to the faster expansion (figure 5). The emission intensity is lower due to the less efficient inverse Bremsstrahlung heating during the laser pulse and consequent lower plasma temperature (figure 6).

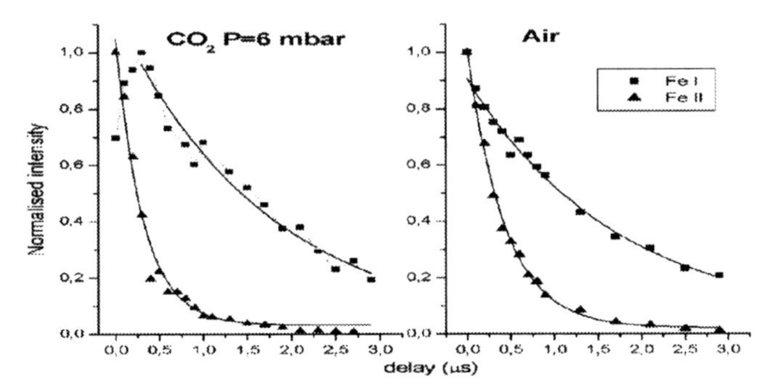

Figure 5. Time evolution of the plasma integral line intensities for Fe I and Fe II lines in air and CO_2 atmosphere, as measured on a Mars analogue rock sample (Andesite). Laser parameters: wavelength 355 nm, pulse energy 21 mJ, pulse duration 8 ns; acquisition parameters: gate width 100 ns.

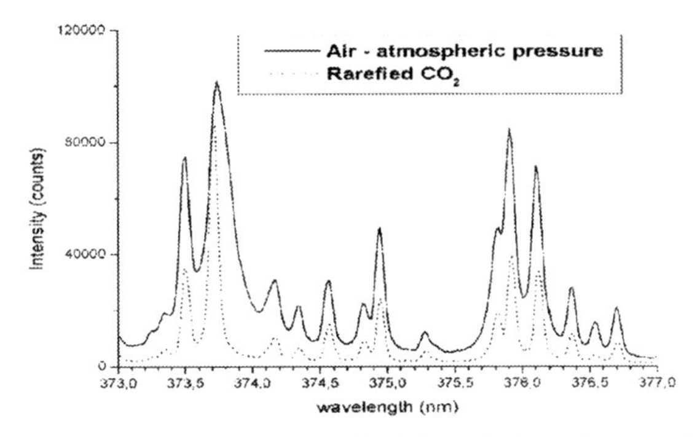

Figure 6. Part of LIBS spectra detected in CO_2 and air on the same Mars analogue sample. Laser parameters as in figure 4, acquisition parameter: delay 1 µs, gate 0.1 µs.

No enhancement in the intensity of weak lines could be observed by multiple pulse techniques applied on samples examined in rarefied atmosphere, since the bubble confinement mechanism is effective at high pressures [10].

2.3.2 LIBS in water

The LIBS laboratory set-up for surface analysis of submersed samples has been already shown in figure 1. Its peculiar optical characteristics can be summarized as follows:

- the double pulse emitted from the laser source is horizontally entering the cuvette through the lens sealing one of its sidewalls;
- the signal collection is perpendicular to the laser beam across the solution;
- the signal is guided to the entrance of the monochromator by a fiber glass and detected at its exit by the ICCD.

The use of the double pulse LIBS technique in water is mandatory whenever the detection of minor/trace elements is required, as it allows to generate the plasma in a reduced medium density (see sect. 1.2). The occurrence of an intense plasma after the second laser pulse can be visualized in collecting LIBS images at a selected wavelength.

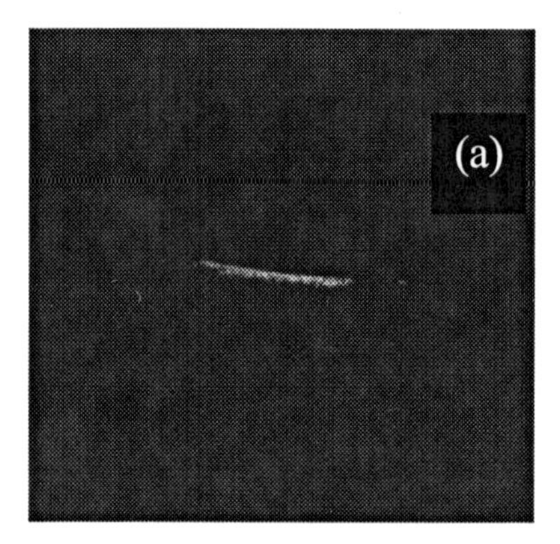

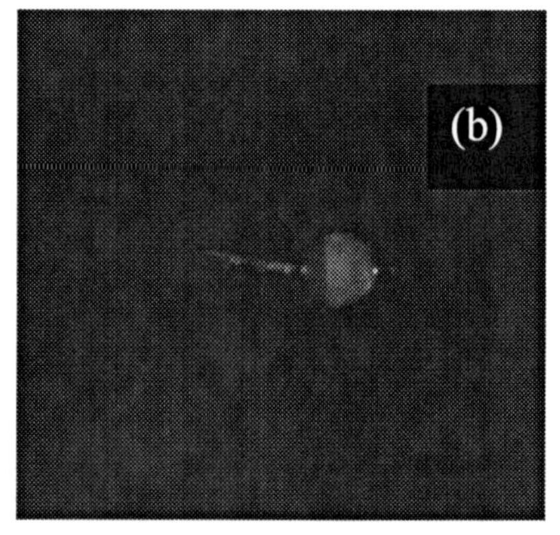

Figure 7. Images of laser induced luminescence in water. Images were acquired with a bandpass filter at 450 nm +/–20 nm: (a) single pulse; (b) double pulse followed by the intense emission (laser pulse energies $E_1 = E_2 = 60$ mJ, pulse duration 8 ns, wavelength excitation 1064 nm, interpulse delay 45 μs).

As an example LIBS images collected 45 μs after respectively the first pulse (a) in a single pulse mode, or the second pulse (b) in the double pulse mode are shown in Figure 7.

3. LIBS Data Analysis Procedure

A complete LIBS spectrum, recorded in all UV – visible range (typically from 200 nm to 850 nm) consists of hundreds to thousands spectral features, even for rather simple alloys with a few (2 to 4) main constituents, such as commercial aluminum or bronzes.

Single line shape can be largely affected by the plasma formation and expansion (i.e. by experimental parameters), as well as by instrumental factors. In particular for strong atomic or ionic transitions, arising from highly populated energy levels, the line-width may show saturation and self-reversal effects which usually point towards a high space in-homogeneity (mostly temperature and species density in-homogeneity) in the plasma observed during the selected time window. Extreme examples of line distortions observed on the strong yellow Na doublet in soil samples are shown in Figure 8, where at increasing Na concentration the space in-homogeneity is responsible first for the broadening associated to the self absorption and successively for the deep formation, known as self reversal. Both effects are related to an optical thick condition of the plasma with respect to this element, which suffers from emission/absorption of atoms at different temperature in different region of the plume.

In general LIBS spectral lines suffer of both homogenous and in-homogenous broadenings, thus the most appropriate line fitting is obtained using a Voight profile, although Lorentzian line fits are suitable to most practical purposes in high density plasma, anyway heavily absorbed lines are usually discarded in the quantitative data analysis (see sect. 3.2).

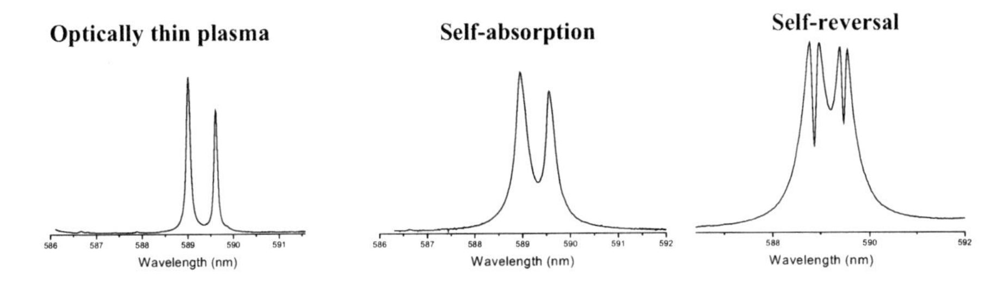

Figure 8. Line distortion effects on resonant sodium lines at 589.0 nm and 589.6 nm on different soil samples.

For sake of completeness let's summarize the main broadening mechanisms in LIBS: collisional due to the surrounding gas pressure, Stark due to the electrical field, Doppler due to the temperature, plasma in-homogeneity due to the mentioned optical thickness, and finally instrumental factors related to the overall spectrometer resolution.

With a zero line intensity at the sample surface and a negligible induced emission [11], the line intensity at frequency ν which escapes from a homogeneous plasma of thickness L, is:

$$I_{\alpha}^{ki}(\nu) = \frac{2h\nu^{3}}{c^{2}} \frac{1}{1-e^{-h\nu/kT}} (1-e^{-k_{\nu}L}) \tag{1}$$

LIBS plasma are usually handled at first approximation by assuming a simplified model which implies to observe an optically thin plasma ($k_{\nu}L \ll 1$) during a time window when the LTE assumption holds and Saha equation rules the atom to ions distribution for the different species.

The LTE assumption implies that the electron temperature can be used to describe the population distribution, following Boltzmann statistics, for all species in the plasma during the observation window. Under this assumption the integrated line intensity for a selected α species derived from eq. (1) gives:

$$I_\alpha = KC_\alpha^{(n)} \frac{g_k A_{ki} e^{-E_{ki}/kT}}{U(T)} \tag{2}$$

with:

K	system constant (optical efficiency, etc.)
$C^{(n)}{}_\alpha$	element concentration in the (n-th) ionization state
A_{ki}	transition rate coupling k and i levels
g_k	degeneracy of the k-th initial level
$U(T)$	partition function at the equilibrium temperature T

T is determined by means of a Boltzmann plot of the line intensities versus the initial state energy for a given species.

In order account for atom/ion distribution for each α species, the sum over all the n-th ionic states of the species must be performed, which in turn depends on the electron density relevant to the plasma in the observed time window. The electron density N_e can be retrieved with an accuracy of about 20-30% from the measured Stark broadening Δw_{Stark}:

$$\Delta w_{Stark} = 2W \frac{N_e}{10^{16}} + 3.5\ W \left(\frac{N_e}{10^{16}}\right)^{5/4} \left(1 - 0.75 N_D^{1/3}\right) A \tag{3}$$

Stark parameters W and A weakly depend on temperature [12], N_D is the Debye shielding parameter:

$$N_D = 1.72\ 10^9 \frac{T^{3/2}}{N_e^{1/2}} \tag{4}$$

Atom (a) to ion (i) ratio's for the same s species are obtained from the electron density assuming the validity of Saha equation [13]:

$$\frac{n_e n_{i,s}}{n_{a,s}^1} = 2 \frac{U_{i,s}}{U_{a,s}} \frac{(2\pi\ m_e k_B T_e)^{3/2}}{h^3} \exp[-E_s^\infty / k_B T_e] \tag{5}$$

In which T_e is the electron temperature, which coincides with the atomic temperature T in eq. (2) upon validity of LTE.

Eqs. (1) to (5) allow to handle LIBS spectra and to support LIBS applications as analytical technique.

Should non equilibrium phenomena occur, the concentration values retrieved from ionic and atomic lines coupled in the plasma decay maybe affected by large errors [14]. Data reported in figure 9 for atomic iron lines in a stainless steel sample demonstrate that the LTE time window might be quite narrow and non negligible deviations can be observed both at short and long time delays after plasma ignition.

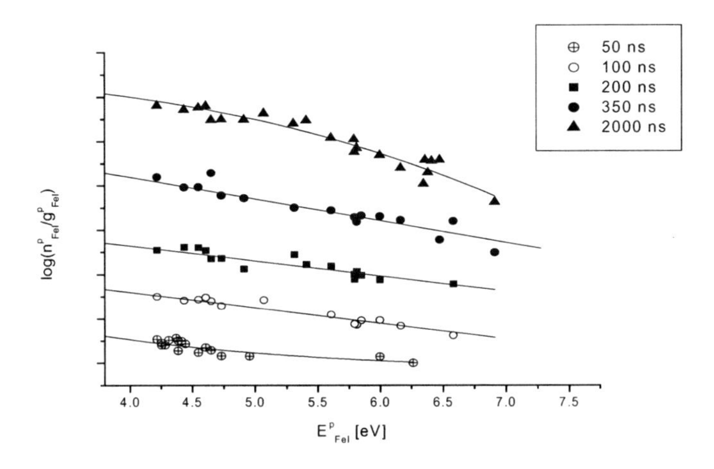

Figure 9. Atom state distribution function for iron at different delay from plasma ignition. Different data series are shifted on the ordinate axis avoiding data superposition. data collected on a stainless steel sample excited at 355 nm.

3.1. ELEMENT RECOGNITION: QUALITATIVE ANALYSIS

Qualitative analysis is based on the recognition of most intense characteristic lines of each element with the help of current line emission data bases [15]. The procedure can be conducted in a semi-automatic way, by means of custom designed software (base on LabView programs in our case), after identification of a major element (usually a metal, e.g. Iron) with a sufficient number of well identified lines which can be used in a Boltzmann plot to estimate the plasma temperature. Different portions of the experimental spectrum are compared with a simulation obtained from the data base line positions, where the intensities are recalculated at the estimated temperature and the instrumental width is assumed.

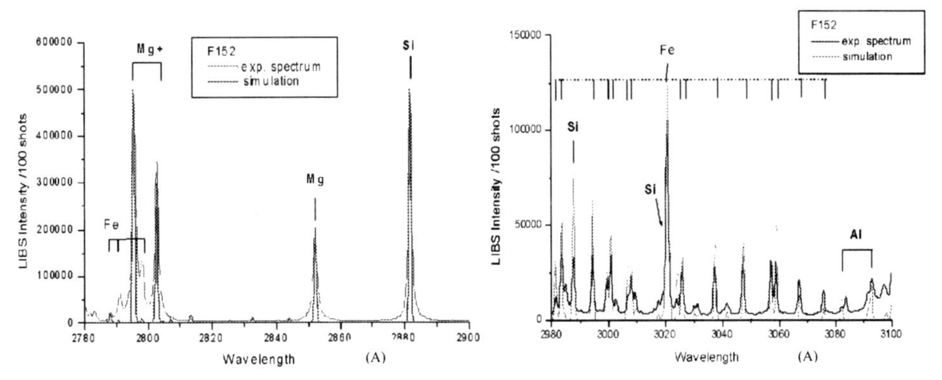

Figure 10. Identification of different elements Fe, Mg, Si, Al found in suspended organic matter collected from Antarctic seawater on a polycarbonate filter, other regions showed the presence of Na, Cr and Cu, too.

An example of qualitative analysis performed on material suspended in seawater, as collected on a polycarbonate film after filtration, is given in figure 10, where labels relevant to the elements identified in different spectral regions free from substrate C^+, C and C_2 emissions are given. Spectra are assigned on the basis of a simulation at 8400 K, consistent with Fe line intensities.

3.2. ELEMENT DETERMINATION: SEMI-QUANTITATIVE AND QUANTITATIVE ANALYSIS

The use of RU lines (*Rayon Ultimes* – well known intense emission lines used for each element quantification in emission spectroscopy) for analytical determinations is strongly recommended. In any case significant lines for quantitative analysis must be selected on LIBS spectra according to the following criteria:

- Well resolved lines – e.g. isolated lines.
- Free from self-absorption and self-reversal lines – e.g. lines with a medium-low intensity.
- Related to transition not involving the ground state – mostly to avoid the onset of non linearity due to spatial in-homogeneities phenomena.The lines peaking at the selected wavelengths for quantitative analysis must be integrated numerically or by fitting the experimental data with an appropriate line shape function and the background must be subtracted to the integrated intensity. To this aim the background can eventually be evaluated by means of a numerical integration performed on the same number of nearby channels, where no peak emerges. *Calibration curves*

In order to achieve quantitative results, the most followed approach implies the construction of calibration curves by means of measurements performed on suitable reference materials with known composition (standards). Unwanted matrix effects are limited, if not fully avoided, in selecting standards with a similar matrix, both in major constituents chemical composition and gross morphology, the latter mainly affecting the sample physical properties. A few points must be kept in mind in constructing calibration curves:

- Calibration curves are built in order to normalize the line intensities measured with respect to the same ones relevant to reference samples.
- Calibration curves may include saturation effects [16] (Figure 11).
- In order to reduce the matrix effect in substrates for which standards are not available, calibration curves may be corrected to account for significant differences in temperature and electron density [17]. In combination with calibration curves, internal standardization to one major element may help in the case of fractionation, when the plume has not the same stoichiometric composition as the target [18]. An example of normalization to internal standards is shown in figure 12 for a quaternary bronze alloy containing different amounts of Cu, Pb, Sn and Zn. In this case the standardization to Zinc is more effective in reducing fractionation problems than the standardization on Copper.

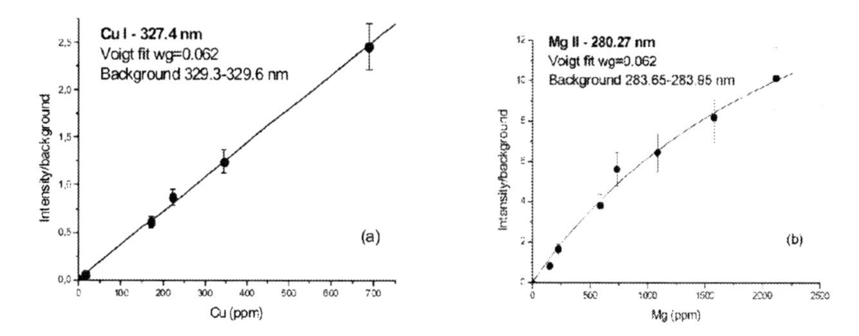

Figure 11. Examples of calibration curves on $CaCO_3$ doped references samples, to be used for marble quantitative analyses. (a) Calibration for Cu I line at 327.4 nm; calculated ratio Cu I/Cu II is 8.20; (b) Calibration for Mg II line at 280.3 nm; calculated ratio Mg I/Mg II is 2.24.

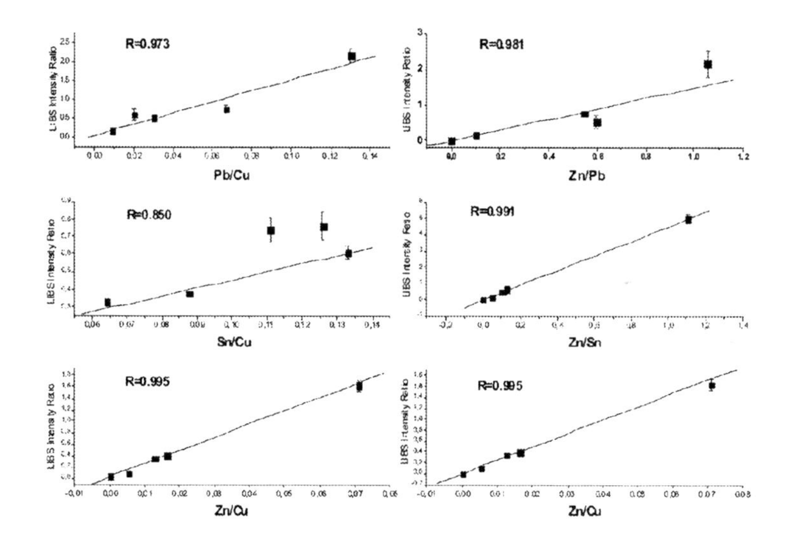

Figure 12. Standardization to Copper (left) and to Zinc (right) of LIBS intensities from quaternary bronze ref. samples. Laser parameters: wavelength 1064 nm, fluence 258 J/cm^2, pulse width 8 ns, rep. rate 10 Hz.

3.2.2 The calibration free approach

This approach implies the contemporary determination of all species present in a sample, and usually work well in the case of well known matrices (e.g. alloys). As proposed by the group at IPCF-CNR [19], the unknown experimental constant K in eq. (2) can be estimated from the closure relation:

$$q_\alpha = \ln\left(\frac{C_\alpha K}{U_\alpha(T)}\right) \tag{6}$$

$$\sum_\alpha C_\alpha = \frac{1}{K}\sum_\alpha U_\alpha(T)e^{q_\alpha} = 1 \tag{7}$$

where C_α is the concentration of each atom and q_α is determined by a Boltzmann plot for each element, after assuming the temperature corresponding to the LTE determined on one of the main constituents, as the intercept on the y axis once the slope is fixed.

3.3. ELEMENT DISTRIBUTION: SPACE RESOLVED PROFILES AND STRATIGRAPHY

One of major advantages of LIBS technique is its capability to perform space resolved measurements at a target surface with a resolution limited only by the laser focusing capability, typically in the micrometer range both in the x-y plane and in depth. Furthermore depth profiles can be obtained in an almost non destructive way (micro-destructive indeed), once the former LIBS ablating pulses are used to reach an inner layer at the target surface. This application is especially of interest in the field of cultural heritage preservation and examples will be given in section 4.

Stratigraphy is often utilized to assist surface cleaning processes, e.g. the mechanical or laser removal of a deteriorated crust. Whenever stratigraphy is performed by recording LIBS spectra on the crusts during drilling a hole by laser, the disappearance of surface impurities is monitored. The depth profile analysis of sample crusts can be aimed to ascertain the laser cleaning effectiveness and to determine the right point when the process must be interrupted. Quantitative profiles can be obtained for most elements detected in encrustations, either as major components or traces [17].

In order to estimate the ablation rate during stratigraphy measurements, a thin layer of bulk sample can be cut and the number of laser shots necessary to drill a hole in it can be measured. The average ablation rates are useful when comparing data collected during different experiments on the same kind of substrates (stones, metals, etc.).

3.4. UNDERWATER ELEMENT RECOGNITION AND ANALYSIS

The use of LIBS for analysis of submersed samples or of liquids, mostly water solutions, still suffers of severe limitations both in sensitivity and reproducibility, due to the nature of the process itself. In fact the plasma emission produced in bulk water is generally lower than at the water-air interface, with a consequent degradation of the analytical detection limit. This is due to several factors, e.g. water absorption of the laser and plasma emission and their scattering on suspended particles and micro-bubbles [20], radiation shielding by the high density plasma [21] and fast quenching in the dense medium, even for the secondary plasma from dual pulse excitation. The complex processes involving laser induced breakdown in water are extensively described in [22]. Experiments aimed to collect quantitative results in water must be designed with a proper data handling suitable to overcome these limitations.

3.4.1 The filtering algorithm

The process of generation of LIBS signal in water has been carefully examined in order to improve the achievable S/N ratio [23]. Strong plasma intensity shot-to-shot oscillations occur both for water solutions and for immersed samples. For different laser pulses the LIBS emission is sometimes missing, even at high laser energy. The plasma generation in water is in fact a statistic process, with a predefined probability of breakdown generation [24]. Signal to noise ratio can be improved after identification and elimination of the pulses not leading to the breakdown.

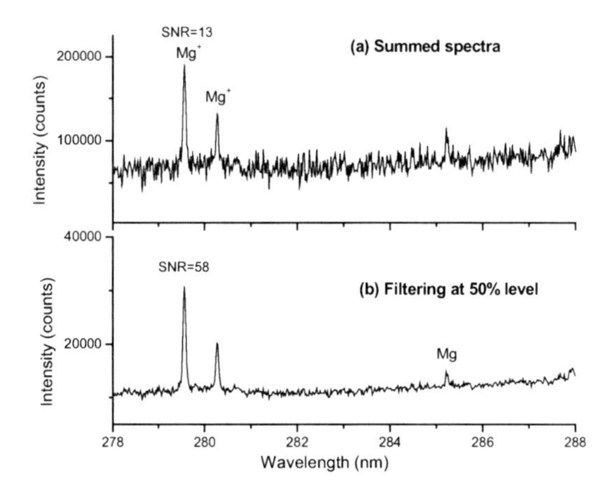

Figure 13. Comparison of DP-LIBS spectra obtained after: (a) summing over 1000 shots and (b) filtering at 50% of maximum peak; Mg concentration (from $MgSO_4$) is 5 mg/l, laser pulse energies are $E_1 = 37$ and $E_2 = 156$ mJ.

In order to take into account for process statistics a filtering algorithm has been developed which retains for averaging only sufficiently intense spectra, rejecting both those not showing breakdown at all and the weakest ones. In the spectra acquired at a fixed monochromator position, the discrimination factor for data filtering was the peak intensity P of a selected element line. The reference intensity over a single measurement is the spectrum with the highest detected peak intensity (P_{max}). Filtering at 40% level, corresponds to the discarding of all spectra with the peak intensity $P < 0.4\ P_{max}$. An example of the effectiveness of the filter algorithm used in analyzing a water reference solution is reported in figure 13.

3.5. THE DOUBLE PULSE TECHNIQUE

After acknowledging the effectiveness of double pulse (DP) plasma generation with LIBS detection of the emission relevant to the last generated plasma in dense media, several efforts have been made to understand the phenomena involved in the process. However a satisfactory theoretical description is not yet available in air, and semi-empirical models appear to work fine in water, where the bubble confinement effect dominates. Experiments are conducted both in air and in water with the main purpose to increase LIBS sensitivity, especially when trace detection is attempted.

3.5.1 DP LIBS in air

Peculiar features are observed in DP excitation in air with respect to single pulse excitation at the same laser fluence [25]:

- Most intense overall spectra.
- Most intense ionic lines.
- Most intense and more persistent weak lines.
- Longer time decay of the line emission spectra.
- Lager LTE window for data acquisition.
- Narrower lines.
- Lowering of the ablation threshold.
- Higher ablation rates.

Some of the observed features, e.g. the narrower lines, are consistent with the bubble mechanism proposed for explaining the process in water. Experiments have been conducted mostly on metallic targets so far, and IR excitation has been used whenever the double pulse was emitted by the same laser source.

3.5.2 DP LIBS in water

The DP technique has been developed for the detection of lines both from immersed targets and from cations in water solutions, quantitative analysis can be achieved in both cases. The bubble confinement improves the accuracy, and ensures a large enough LTE time window with respect to single pulse excitation. Although non equilibrium effects appear less noxious in DP LIBS in water, as a consequence of the effective plasma confinement in the bubble, space in-homogeneities, due to the same factor, may give rise to peculiar perturbation on the spectra related to the initial plasma formation at the target surface and to its successive expansion in the water.

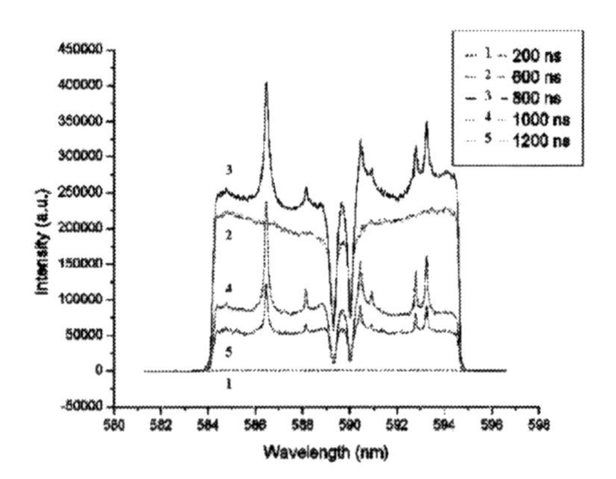

Figure 14. A section of spectrum of Ti-laser induced plasma in sea-water taken in the spectral region of Na by DP LIBS at different delays from the second pulse (gate width 100 ns).

As an example, a section of the DP LIBS spectrum of a Titanium target immersed in artificial sea water is shown in figure 14. The kinetic analysis shows the rich Ti emission structure emerging from plasma breakdown at the surface combined with the Na absorption doublet from colder atoms at the edge of the bubble.

4. Applications to Artwork

Due to its minimally invasive and only micro-destructive character, coupled to the possibility of combination with other laser surface diagnostics, such as Laser Induced Fluorescence (LIF) and Raman Spectroscopy, to the availability of portable instruments and to the possibility stratigraphy with on-line monitoring cleaning processes, LIBS has found in the Cultural Heritage preservation a fertile field of application. As several examples in the following show, LIBS technique resulted to be suitable for qualitative and quantitative multi-elemental analyses on a large variety of samples surfaces in most different stages of preservation.

4.1. QUALITATIVE ANALYSIS: FRAGILE SUBSTRATES

Qualitative analysis on ceramics and fragile substrates of completely unknown composition was attempted at Histria (RO) museum during the CULTURE 2000 "Advanced On-Site Laboratory for European Antique Heritage Restoration" (April 24-26, 2004), a few results obtained on sample analyzed in air are reported in the following.

Figure 15 shows a fragment of a Roman glass examined, together with a significant part of the LIBS spectrum recorded (accumulation of 50 laser pulses, laser parameters: wavelength 355 nm, energy 15 mJ, pulse width 6 ns, repetition rate 20 Hz). Identified elements in all the UV- visible range were Mg, Ca, Si, Al and Mn. The abundance of the latter element points towards a local manufacture of the glass starting from Mn rich sands characteristic of the Black Sea.

Figure 16 shows a fragment of gilded cloth together with a significant part of the LIBS spectrum recorded upon the same experimental conditions as for the Roman glass. Identified elements in all the UV- visible range were Ag, Cu, Ca, and Sr, thus indicating that the gilding color was due to a Copper/Silver alloy without use of any real gold.

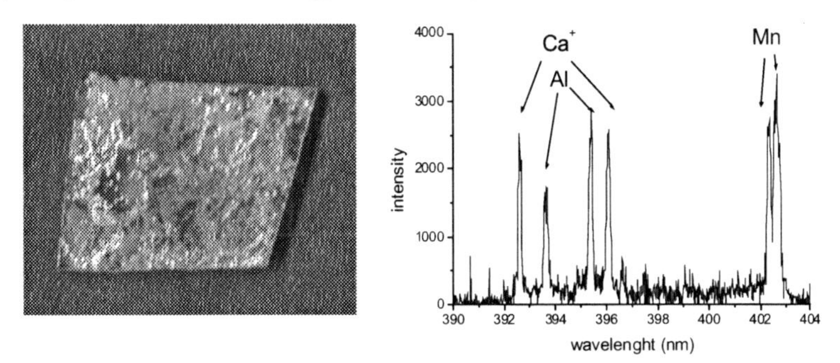

Figure 15. Fragment of a Roman glass examined by LIBS (left), a section of its LIBS spectrum (right).

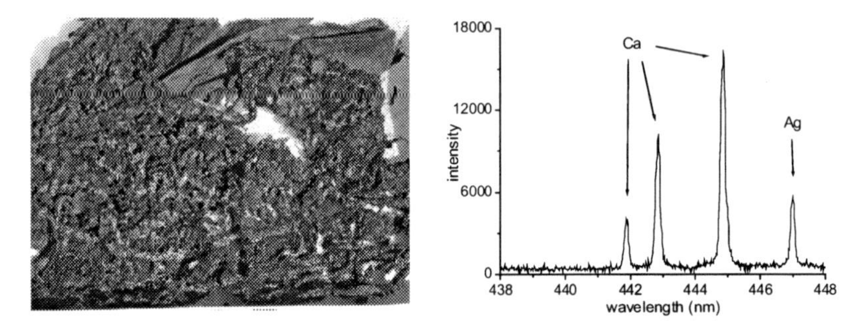

Figure 16 .Fragment of a gilded cloth examined by LIBS (left), a section of its LIBS spectrum (right).

Underwater recognition of precious alloys and marbles could be demonstrated as well [26].

4.2. SEMI-QUANTITATIVE ANALYSIS: DECORATED CERAMICS

Samples of Renaissance Umbrian ceramics (XV-XVI century), decorated by glaze, red and gold lustre, and different tonalities of blue pigments where examined by LIBS [27]. Since decorations were overlapping the bisque in successive layers, whose contribution could no be disentangled by stratigraphy, the composition of each layer was determined by subtraction, which led to a semi-quantitative result, related also to sample in-homogeneities and missing surface planarity. Single shot measurements were performed both on glaze and pigments. The glaze composition was obtained after subtracting the bisque contribution, measured on the back face of the ceramics. A significant amount of Sn and Pb were always observed, thus indicating the use of glazing mixture based of silica sands, where soda and lead oxides were added to decrease the melting point, and cassiterite (SnO_2) was used as an opacifier. Different elements also found in the glaze, such as Cr, Ti, Mn and Zn, could be considered as impurities.

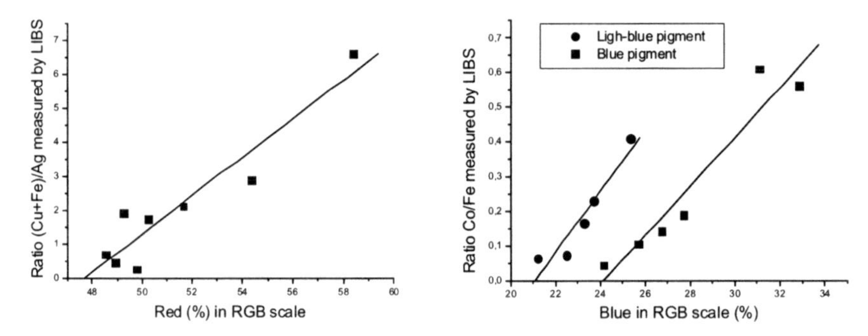

Figure 17. Correlation between LIBS concentrations and results of RGB colorimetric analysis: Ratio of Copper plus Iron concentration and Silver in gold luster versus the percentage of Red colour (left); ratio Cobalt/Iron versus Blue colour (right).

Lusters and blue pigments composition was obtained after subtracting the glaze spectra. It resulted that each luster contained Silver and Copper, which were found in glaze only in traces, while blue pigments contained Cobalt and Nickel, in agreement with the ancient receipt for gold luster and for a blue smalt. As figure 17 shows, semi-quantitative LIBS data correlated well with results of a colorimetric analysis on lusters and blue pigments.

4.3. QUANTITATIVE ANALYSIS AND STRATIGRAPHY: MARBLE

Quantitative analysis on white marble crust and bulk samples, coming from ancient quarries, were based on calibration curves similar to those already shown in figure 11. Due to difference in temperature and electron density between the marble samples and the standards ($CaCO_3$ pellets pressed from ultra pure powders doped with the addition of known quantities of reference soils), calibration curves were corrected according to the procedure described in sect. 3.2.1 [17].

After quantitative analysis of crust and bulk, LIBS stratigraphy has been performed on the same sample in order to monitor the disappearance of surface impurities at increased depth during the laser ablation process. Some significant results collected on surface contaminants for two different Greek marble samples are reported in Figure 18, elemental concentrations (% or ppm in weight) are plotted as a function of the number of laser shots which is in turn proportional to the reached depth.

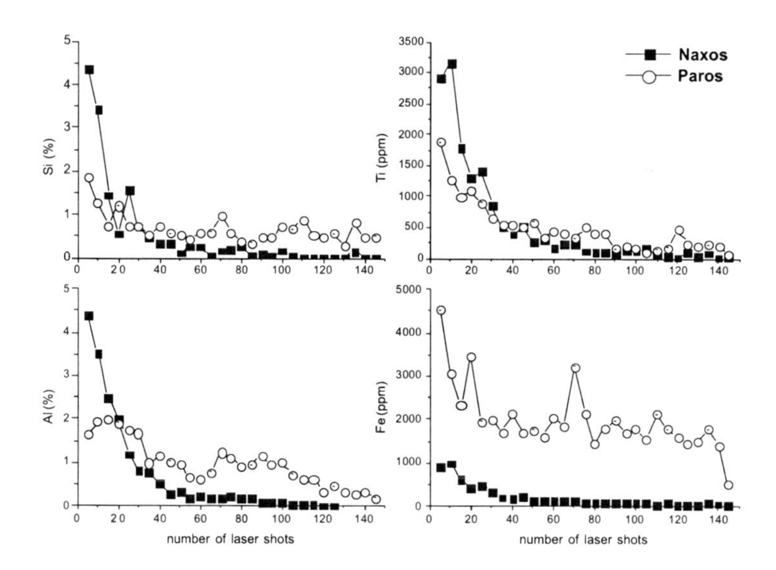

Figure 18. LIBS stratigraphy on marble samples from ancient Greek quarries. Laser ablation and LIBS detection have been performed upon UV excitation at 355 nm (ablation rate about 5 μm/shot).

The removal of surface impurities can be indirectly related to the sulphated layer ablation, since it is well known that transition metals catalyze the transformation of marble into gypsum upon the action of atmospheric pollutants (e.g. SO_2). However to an effective monitoring of sulphated crust, direct stratigraphy on Sulphur is advisable.

Difficulties in Sulphur detection, related to the existence of a few useful weak lines from S and S^+ in the usual LIBS range, have been recently overcome by means of DP LIBS in air. The results obtained for a heavily encrusted (up to 10% S on the crust) white marble sample coming from Veneto are shown in figure 19, where data for Sulphur and Iron are reported as ratio to Calcium, the latter utilized as internal standard. Although Iron emission reaches stable value after about same number of applied shots as Sulphur, the correlation between Sulphur and Iron emission is very weak, namely 0.58 for the reported example. This means that iron might play role in the surface sulphation, but its distribution in the crust is not an appropriate indicator for Sulphur distribution.

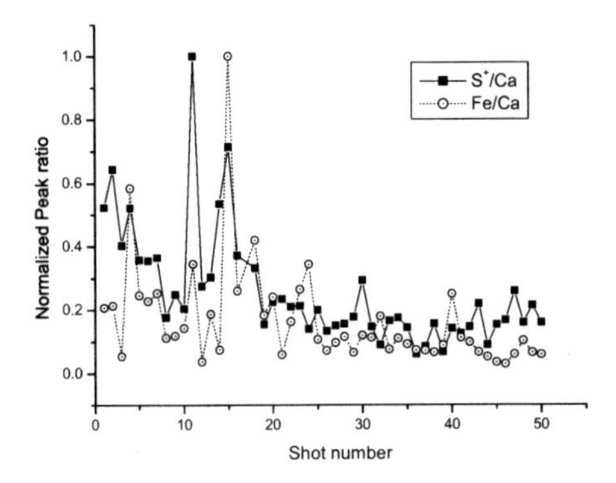

Figure 19. LIBS stratigraphy on an encrusted marble sample. Laser excitation at 1064 nm, the number of shot refers to each pair of double pulses (ablation rate about 50 μm/shot).

LIBS diagnostics was also successfully applied during laser cleaning of marble [28] and bronzes is air [29], automatic algorithms suitable to switch off the cleaning action as soon as the original surface is reached are under development in several laboratories.

5. Applications to Environment

LIBS was successfully applied in laboratory analyses of soils and sediments, in order to trace the presence of heavy metals as pollutants [30] or to investigate sedimentation processes in natural ecosystems (Ross Sea, Antarctica) [31]. To this respect main advantages of the technique were the minimal sample preparation required (dried powders pressed in pellets were usually examined), and the portability of the set-up, which could be installed on board of the oceanographic vessel.

In laboratory it was also recently demonstrated to the possibility of asbestos recognition by LIBS [32]. To this aim an algorithm based on a couple of LIBS intensity ratio's (Fe/Si and Mg/Si) as asbestos identifier has been developed.

Environmental applications however push towards developments of the technique for remote in situ analyses. The laser beam directionality offers the possibility of *in situ* and semi-remote LIBS applications for environment and planetary exploration. A few results of laboratory experiments preliminary to the design and realization of field systems are reported in the following.

5.1. PLANETARY EXPLORATION (ROCKS, DUST)

As a result of LIBS development in environmental analysis, a LIBS/Raman spectrometer is planned on board of the Rover which will land on Mars soil in the next unmanned mission. ESA and NASA are running projects to realize the instrument. Preliminary laboratory work has been performed to ascertain LIBS capabilities in analyzing Mars analogue rock samples upon the rarefied Mars atmosphere.

To this aim in our laboratory we optimized the LIBS signal generation/acquisition upon Martian conditions [33]. The equipment shown in Figure 20 was used to conduct experiments in simulated Mars atmosphere, consisting of CO_2 (95.3%), N_2 (2.7%), Ar (1.6%) e O_2 (0.13%), at low pressure 6.3 mbar, with controlled substrate temperature.

Figure 20. LIBS laboratory set-up for experiments on Mars analogue rocks. From left to right: delay generator, ICCD, laser source, monochromator, low pressure chamber containing the sample holder.

After optimization of the experimental conditions (laser parameters, data acquisition window) LIBS spectra were analyzed automatically by utilizing the calibration free (CF) method (described in sect. 3.2.2). LIBS results obtained on different "andesite" samples are compared in Table I with the atomic composition derived from SEM-EDX analysis performed on the same samples. Results obtained by both techniques were mostly in agreement, although the CF-LIBS accuracy resulted to be element and concentration dependent, ranging from a few percent up to 20-30% for major constituents. Larger discrepancies were obtained mostly for Ca, whose concentration was systematically underestimated by LIBS due to saturation (not yet included in the CF model used). Matrix effects and sample space in-homogeneity could be responsible for other discrepancies. In spite of limitations in the accuracy achieved, CF-LIBS seems to be a suitable method of analysis for a system operating remotely at the pre-selected experimental parameters and without any reference, and producing real time data, as it should be the case during a space mission. Alternative an automatic Calibration Curve LIBS analysis can be carried on even in space, providing standards on board and suitably built data set library.

TABLE I. Major element concentration in andesite (silicate minerals). SEM-EDX and CF-LIBS concentration comparison. All data are % in weight

Sample	A1		A2		A3	
	EDX	LIBS	EDX	LIBS	EDX	LIBS
Fe	5.4 ± 1	3.6 ± 1.3	5.2 ± 1	4.7 ± 1.7	2.5 ± 1	1.5 ± 0.5
Si	22 ± 2	31 ± 6	27 ± 2	28 ± 5	31 ± 3	31 ± 6
Mg	<0.5	0.5 ± 0.14	1.9 ± 0	1.2 ± 0.4	0.6 ± 0	0.3 ± 0.1
Ca	14 ± 1	0.5 ± 0.2	3.1 ± 0	0.5 ± 0.2	1.4 ± 0	0.1 ± 0.04
Ti	<0.5	0.2 ± 0.04	0.5 ± 0	0.4 ± 0.1	<0.5	0.1 ± 0.02
Mn	<0.5	0 ± 0.01	<0.5	0.1 ± 0.012	<0.5	0 ± 0.002
Al	3.5 ± 0	4.1 ± 1.1	9 ± 1	6.4 ± 1.7	8 ± 1	6.7 ± 1.8
Na	<0.5	–	3.3 ± 1	0.9 ± 0.3	3.1 ± 0	2 ± 0.5
K	1.5 ± 0	1.4 ± 0.2	0.9 ± 0	–	3.1 ± 0	0.7 ± 0.12
Cu	0.5 ± 0	–	–	–	–	–
Ba	<0.5	–	–	0.04 ± 0.014	< 0.5	0 ± 0.011

5.2. SUB-GLACIAL LAKE BOTTOM (ROCKS, SEDIMENTS, WATER)

After developing the DP-LIBS another challenging task is to perform analysis of environmental interest directly in water, either on the water itself or solid materials at the basin bottom, by using a compact set-up installed on a submersible carrier (e.g. a ROV remotely operated vehicle). With the aim to design an apparatus to be used for sub-glacial lakes exploration in Antarctica, preliminary laboratory experiments were carried on, analyzing rocks and sediments immersed in water, the latter after pressing in pellets to avoid the immediate dispersion of the sample.

Underwater rock analysis was straightforward, as already performed for marble recognition [26]. Due to the tendency of the sediments to dissolve in water upon the laser action, single shot spectra were collected. Some examples generated in DP-LIBS experiments aimed to the qualitative (semi-quantitative) analysis of Antarctic marine sediments are shown in Figure 21. On the investigated sediments Ca, C, Mg, Mn, Si, Al, Fe, Ba, Ti, Na and K were detected in the investigated spectral range (240-750 nm). Results are consistent with former analysis conducted in air on the dried sample [31].

Water analyses were attempted as well on standard solutions, calibration curves were built for several elements (Mg, Mn, Cr). By applying the filter algorithm described in sect. 3.4.1, it was possible to reach a sensitivity at the ppb level for the investigated metals dissolved in water.

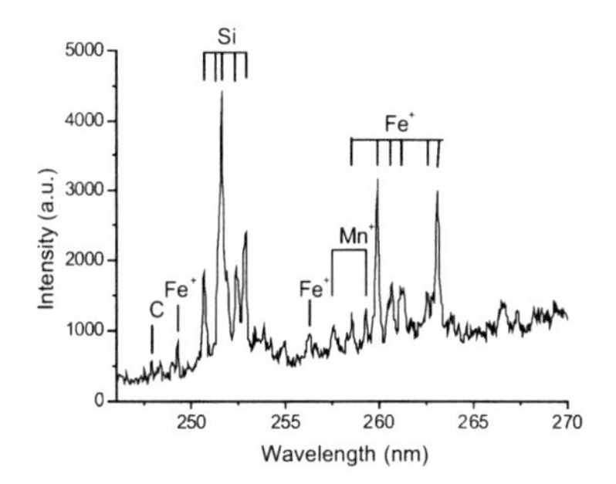

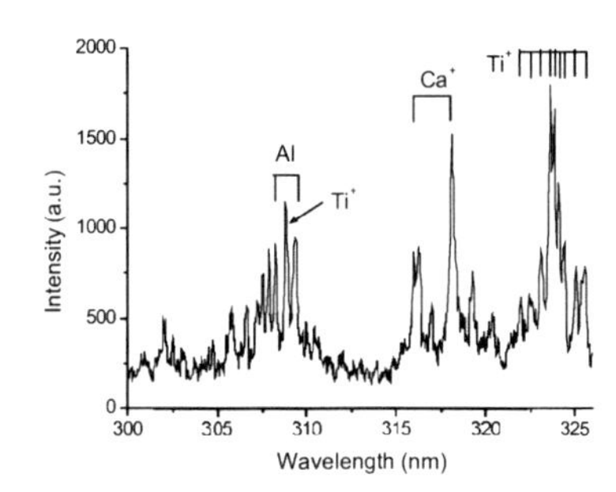

Figure 21. Selected spectral region of single shot DP-LIBS spectra of an Antarctic sediment collected during the XVI campaign and analyzed in water.

5.3. TRACE ELEMENTS IN INDUSTRIAL PROCESSING

Applications very close to environmental analysis concern material contamination in industrial processing. The capability of LIBS in monitoring main material constituents and impurities have been already demonstrated, mostly for alloys [34], during industrial processing. The possibility of a remote analysis of metals (up to 100 m distance), during or immediately after solidification was also ascertained for steel [35]. The possibility to cover even longer distances by remote LIBS has already been demonstrated by use of femtosecond laser pulses [36], due to the extreme self-focussing effects which may occur in air.

LIBS determination of material contamination from different elements in industrial processing might be of interest in view of in-situ, or even semi-remote applications. To this respect we investigated the possibility to detect Boron and Phosphorus contamination in a Zircona bricks used in industrial ovens. As in case of Sulphur, LIBS detection of non metallic species is rather difficult.

Two refractory samples labeled as A and O, containing different quantities of contaminants, were investigated. The standard qualitative analysis revealed the presence of: Na, Al, Si, Ca, Fe, Zr, Ti, Hf, Sc, Sr, Ba, Mg and B traces (the latter found only on sample A), while the identification of K, B and P required special instrumental setting (wavelength range extension, high resolution, proper time window choice). Mass content for P was 7000 ppm and 300 ppm for samples A and O, respectively, while Boron concentration was less than 100 ppm for sample A and even lower for O.

For the quantitative analysis aimed to trace detection the large number of Zr and Zr^+ emission lines were utilized to derive the plasma parameters (temperature, electron density) as described in sect. 3. Plasma parameters had to be optimized to observe at least a suitable emission line from each non metallic species. Results reported in Fig. 22 show the successful detection of B and P traces, achieved with specific acquisition time windows: delay 1.0 μs, width 1.0 μs for Boron and delay 2.5 μs, width 10.0 μs for phosphor, respectively. Note that the LTE condition could not be fulfilled in the latter case.

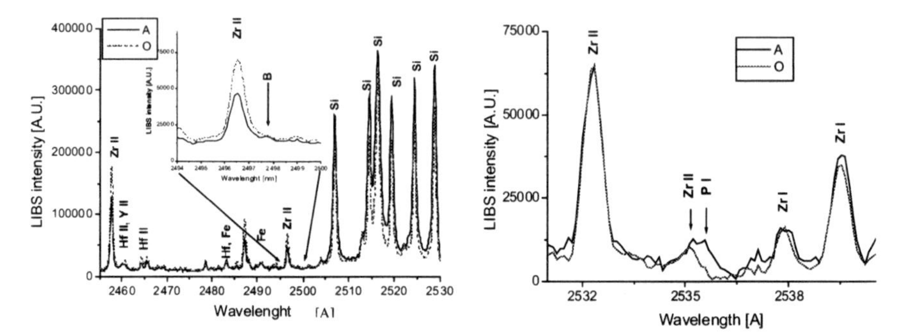

Figure 22. Sections of high resolution LIBS spectra were boron and phosphorous traces could be detected. Laser parameters: wavelength 355 nm, energy 20 mJ, pulse duration: 8 ns, repetition rate 10 Hz.

6. Conclusions

LIBS technique is nowadays well developed as far as the understanding of the laser matter interaction processes involved in the plasma generation and evolution are concerned, at least whenever a single pulse is applied. This allowed to support its analytical applications in a wide range of problems where the advantages of a minimal and easy sampling are of importance and a fast multi-elemental analysis is required, even if its sensitivity and accuracy cannot compete with other assessed laboratory tools.

The possibilities of in situ and remote application of the LIBS technique are further supporting the development of automatic field systems suitable to operate unmanned also in harsh environments.

7. References

1. L.J. Radziemski, T.R. Loree (1981), Laser-induced breakdown spectroscopy: time-resolved applications, *J. Plasma Chem. Plasma Proc.*, **1**, 281-293.
2. S. Amoruso, R. Bruzzese, N. Spinelli, R. Velotta (1999), Charaterization of laser-ablation plasmas, *J. Phys. B: At. Mol. Opt. Phys.*, **32**, 131-172.
3. R.E. Russo, X. Mao, H. Liu, J. Gonzalez, S.S. Mao (2002), Laser ablation in analytical chemistry—a review, *Talanta* **57**, 425-451.
4. H.R. Griem (1964), Plasma Spectroscopy, McGraw-Hill.
5. S.I. Dolgaev, A.V. Simakin, V.V. Voronov, G.A. Shafeev, F. Bozon-Verduraz (2002), Nanoparticles produced by laser ablation of solids in liquid environment, *Appl. Suf. Sci.* **186**, 546-551.

6. R. Sattmann, V. Sturm, R. Noll (1995), Laser induced breakdown spectroscopy of steel samples using multiple Q-switch Nd:Yag laser pulses, J. Phys D, **28**, 2181-2187.
7. A. De Giacomo, M. Dell'Aglio, F. Colao, R. Fantoni (2004), Double pulse laser produced plasma on metallic target in seawater: basic aspects and analytical approach, *Spectrochimica Acta B*, **59**, 1431-1438.
8. F. Dahmani (1992), Experimental study of laser and wavelength dependences of laser–plasma coupling, transport, and ablation processes in planar gold targets, Physics of Fluids B: Plasma Physics, **4**, 1943-1952.
9. Y-I. Lee, K. Song, H-K. Cha, J-M. Lee, M-C. Park, G-H. Lee, J. Sneddon (1997), Influence of Atmosphere and Irradiation Wavelength on Copper Plasma Emission Induced by Excimer and Q-switched Nd:YAG Laser Ablation, *Appl. Spectrosc.* **51**, 959-964.
10. G. Cristoforetti, S. Legnaioli, V. Palleschi, A. Solvetti, E. Tognoni (2004), Influence of ambient gas pressare on laser induced breakdown spectroscopy technique in the parallel double-pulse configuration, Spectrochim Acta B, **59**, 1907-1917.
11. H. Conrads, Diagnostics of high temperature high density plasma by radiation analysis, in Applied Spectroscopy Reviews Vol. 6, 135-188, E.G. Brame Ed., Marcel Dekker Inc, New York, (1973)
12. H.R. Griem (1974), Spectral Line Broadening by Plasmas, London, Academic Press.
13. E.H. Pipemeier (1986), Laser ablation for atomic spectroscopy, in Analytical Applications of Lasers, E.H. Pipemeier Ed. (John Wiley and Sons, New York) 627-669.
14. Colao, F., Fantoni, R., Lazic, V. (2002), LIPS as an analytical technique in non-equilibrium plasma, in V. Ochkin Ed. SPIE Special Issue on *Spectroscopy of non equilibrium plasma at elevated pressures*, SPIE, Bellinghan, **4460**, 339-348.
15. NIST electronic database, available at http:\\physlab.nist.gov\physrefdata\contents-atomic.html; Kurucz database, available at http:\\cfa-www.harvard.edu\amdata\ampdata\kurucz23\sekur.html.
16. V. Lazic, R. Barbini, F. Colao, R. Fantoni, A. Palucci (2001), Self-absorption model in quantitative laser induced breakdown spectroscopy measurements on soils and sediments, *Spectrochim. Acta B*, **56**, 807-820.
17. V. Lazic, R. Fantoni, F. Colao, A. Santagata, A. Morone, V. Spizzichino (2004), Quantitative laser induced breakdown spectroscopy analysis of ancient marbles and corrections for the variability of plasma parameters and of ablation rate, *J. Anal. At. Spectrom.*, **19**, 429-436.
18. Fornarini, L., Colao, F., Fantoni, R., Lazic, V., Spizzichino, V. (2005), Calibration Analysis of bronze samples by nanosecond Laser Induced breakdown spectroscopy: A theoretical and experimental approach, *Spectrochim. Acta B* **60**, 1186 – 1201.
19. Corsi, M., Cristoforetti, G., Palleschi, V., Salvetti, A., Tognoni, E. (2001), A fast and accurate method for the determination of precious alloys caratage by LIPS, *Eur. Phys. J. D*, **13**, 373-377.
20. M. Kompitsas, F. Roubani-Kalantzopoulou, I. Bassiotis, A. Diamantopoulou, A. Giannoudakos, (2000), LIPS as an efficient method for elemental analysis of environmental samples, *EARSeL Proceedings of 4th Workshop on lidar remote sensing of Land and Sea.* Dresden June 16-17, 2000. EARSeL e-Proceeedings http://las.physik.uni-oldenburg.de/, vol. 1, pp 130-138.
21. N.F. Bunkin, A.V. Lobeyev (1997), Influence of dissolved gas on optical breakdown and small-angle scattering of light in liquids, Phys. Letters A **229**, 327-333.
22. D.X. Hammer, E.D. Jansen, M. Frenz, G.D. Noojin, R.J. Thomas, J. Noack, A. Vogel, B.A. Rockwell, A.J. Welch (1997), Shielding properties of laser induced breakdown in water for pulse durations from 5 ns to 125 fs, Appl. Optics **36**, 5630-5640.
23. V. Lazic, F. Colao, R. Fantoni, V. Spizzichino (2005), Laser Induced Breakdown Spectroscopy in water: improvement of the detection threshold by signal processing, *Spectrochim. Acta B* **60**, 1002-1013.
24. P.K. Kennedy, D.X. Hammer, B.A. Rockwell (1997), Laser-induced breakdown in aqueous media, Prog. Quant. Electr. **21**, 155-248.
25. F. Colao, V. Lazic, R. Fantoni, S. Pershin (2002), A comparison of single and double pulse laser-induced breakdown spectroscopy of aluminum samples, *Spectrochimica Acta B* **57**, 1167-1179.
26. V. Lazic, F. Colao, R. Fantoni, V. Spizzichino, "Recognition of archeological materials under water by Laser Induced Breakdown Spectroscopy", Spectrochim. Acta Part B **60**, 1014-1024 (2005).
27. V. Lazic, F. Colao, R. Fantoni, A. Palucci, V. Spizzichino, I. Borgia, B. Brunetti, A. Sgamellotti (2003) Characterization of lustre and pigment composition in ancient pottery by laser induced fluorescence and breakdown spectroscopy, *J. Cultural Heritage*, **4**, 303s-308s.
28. P. Maravelaki-Kalaitzaki, D. Anglos, V. Kilikoglou, V. Zafiropulos (2001) Composition characterization of encrustation on marble with laser induced breakdown spectroscopy, Spectrochim Acta B **56**, 887-903.
29. F. Colao, R. Fantoni, V. Lazic, L. Caneve, A. Giardini, V. Spizzichino (2004), LIBS as a diagnostic tool during the laser cleaning of copper based alloys: experimental results, *J. Anal. At. Spectrom.*, **19**, 502-504.
30. F. Capitelli, F. Colao, M.R. Provenzano, R. Fantoni, G. Brunetti, N. Senesi (2002), Determination of heavy metals in soils by laser induced breakdown spectroscopy, *Goederma*, **106**, 46-62.

31. R. Barbini, F. Colao, V. Lazic, R. Fantoni, A. Palucci, M. Angelone (2002), On board LIBS analysis of marine sediments collected during the XVI Italian campaign in Antarctica, *Spectrochimica Acta Part B: Atomic Spectroscopy,* **57**, 1203-1218.
32. L. Caneve, F. Colao, F. Fabbri, R. Fantoni, V. Spizzichino, J. Striber (2005), LIBS Analysis of asbestos, Spectrochimica Acta B **60**, 1115-1120.
33. F. Colao, R. Fantoni, V. Lazic, A. Paolini (2004), LIBS application for analyses of Martian crust analogues: search for the optimal experimental parameters in air and CO_2 atmosphere, *Appl. Phys. A* **79**, 143-152.
34. C. Lopez-Moreno, K. Amponsah-Manager, B.W. Smith, I.B. Gornushkin, N. Omenetto, S. Palanco, J.J. Laserna and J.D. Winefordner (2005), Quantitative analysis of low-alloy steel by microchip laser induced breakdown spectroscopy, *Journal of Analytical Atomic Spectrometry*, **20**, 552-556.
35. S. Palanco, S. Conesa and J.J. Laserna (2004), Analytical control of liquid steel in an induction melting furnace using a remote laser induced plasma spectrometer, *J. Anal. At. Spectrom.*, **19**, 462-467.
36. K. Stelmaszczyk, P. Rohwetter, G. Méjean, Jin Yu, E. Salmon, J.E. Kasparian, R. Ackermann, J.-P. Wolf, L. Wöste (2004), Long-distance remote laser-induced breakdown spectroscopy using filamentation in air, **85**, 3977-3979.

OPTICAL SPECTROSCOPY OF SEMICONDUCTOR QUANTUM STRUCTURES

H. KALT
Institut für AngewandtePhysik, Universität Karlsruhe (TH) and DFG-Center for Functional Nanostructures (CFN)
D-76128 Karlsruhe (Germany)
Heinz.Kalt@phys.uni-karlsruhe.de,

This contribution intends to give an introduction into the optical spectroscopy of semiconductor nanostructures. Rather than comprehensively covering this large area we will give some instructive examples for various spectroscopic methods applied to test the properties of excitons in the model system ZnSe quantum wells. The described methods are modifications of photoluminescence spectroscopy which are able to reveal the relaxation and transport dynamics of excitons.

1. Introduction to Optical Properties of Semiconductor Quantum Structures

1.1. OPTICAL TRANSITIONS AT THE FUNDAMENTAL BANDGAP

We want to start with some basic principles of semiconductor optics.[1,2] The latter is (at least in the visible and adjacent spectral regions) inseparably linked to the electronic states of the semiconductor since the electromagnetic wave couples to the dipole moment of transitions between these states. In the case of crystalline materials the electronic states form a band structure due to the overlap of equivalent electronic orbitals of the constituent atoms. Typical features of the band structure – parabolic maxima and minima of bands at high symmetry points in reciprocal space as well as forbidden areas (band gaps) – are a direct consequence of the lattice potential seen by the electrons.

Calculations of such band structures consider a single electron moving in the periodic potential of the ion rumps, feeling the Coulomb interaction with all other electrons present and obeying the Pauli principle. An example (the semiconductor ZnSe crystallizing like GaAs in the zincblende structure) is shown in Figure 1. The energy of a single electron is here plotted for various directions in quasi-momentum (or k-) space. The symbols given at points of high symmetry (Γ, X, L etc.) within the first Brillouin zone are representations (groups of matrices) which characterize the symmetry of the electronic wave functions and define optical selection rules. The electronic band

B. Di Bartolo and O. Forte (eds.), Advances in Spectroscopy for Lasers and Sensing, 255–276.

structure of an intrinsic semiconductor like ZnSe features fully occupied valence bands and empty conduction bands separated by an energy gap – the fundamental band gap.

The most important semiconductors have their uppermost valence-band maximum at the Γ-point of the Brillouin zone, i.e. at zero quasi-momentum. In direct-gap semiconductors (like GaAs, InAs, ZnSe, GaN, ZnO etc.) one finds the lowest minimum of the conduction band at the same point in k-space, while it is situated at large wave-vector k (e.g. at the X- or L-point) for indirect-gap semiconductors (Si, Ge etc.). Electrons and holes (missing electron in a fully occupied band), which are thermally generated from dopant states or across the band gap, optically excited or injected via electrical contacts, accumulate at the fundamental gap. Here also optical transitions in emission and absorption set in. So we will focus in the following on the band structure properties close to the fundamental gap.

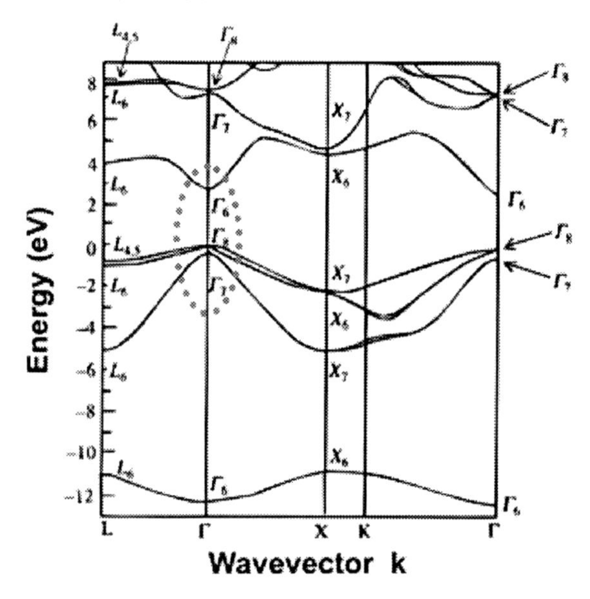

Figure 1. Band structure of the direct-gap semiconductor ZnSe according to [3].

Zooming in (see Fig. 2) into the band structure at the Γ-point of a zincblende semiconductor like GaAs or ZnSe directly visualizes an important property of the electrons and holes with small k-vectors: they have parabolic dispersion relations (dispersion relation: E(k)), i.e. the kinetic energy with respect to the band extremum depends quadratically on k-vector (or quasi-momentum):

$$E(k) = \frac{(\hbar k)^2}{2m_{eff}} \tag{1}$$

This is the same relation as for a free particle, except that the mass is modified. The carriers have an effective mass m_{eff} which is determined by the interactions of the carrier with the ions and all other carriers. Since the inverse of the effective mass is directly given by the curvature of the dispersion relations (here: parabolic) one finds a quasi-free movement of the electrons and holes with a constant m_{eff}.

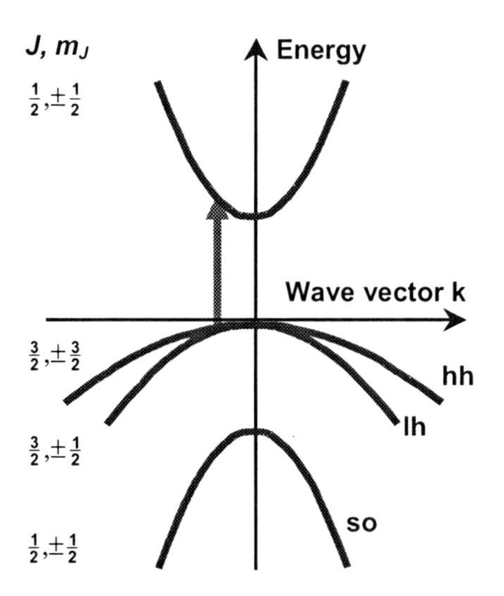

Figure 2. Band structure of a zincblende semiconductor close to the Γ-point. The point of zero-energy is arbitrarily fixed to the maximum of the valence band. The conduction band as well as the valence bands (hh: heavy-hole, lh: light-hole, so: split-off valence band) have parabolic dispersion relations. Also listed are the total angular momentum J and its projection onto a selected axis m_j. The arrow indicates an optical transition (absorption).

Optical transitions close to the fundamental gap are subject to selection rules which derive from symmetry arguments and have to be calculated within group theory. However, for the zincblende semiconductors considered here one can use more intuitive rules involving the total angular momentum J of the related band wave functions. Let us use the II-VI compound ZnSe as an example. When the Zn and Se atoms bind into the ionic ZnSe lattice, the two electrons of the $4s^2$ shell of Zn are transferred to the $4p^4$ shell of Se. The conduction band in Fig. 2 can than be assumed to derive from a linear combination of Zn 4s orbitals (unoccupied) while the highest valence-bands derive from the (filled) Se 4p orbitals. The respective bands can now be characterized by the quantum numbers J (total angular momentum) and their projection m_j (see Fig. 2). The total angular momentum results from the coupling of orbital momentum (0 for s-orbitals and 1 for p-orbitals) and spin. Since a photon which is absorbed or emitted in an optical transition carries a spin of +1 (σ^+ circular polarization) or –1 (σ^- circular polarization), the quantum number m_j has to change accordingly. E.g., an electron can be lifted from the hh valence band (3/2, –3/2) to the conduction band (1/2, –1/2) by absorption of a σ^+ photon. In the language of group theory (and treating the electromagnetic field in the so-called dipole-approximation) one would say: the interband transitions at the fundamental gap of ZnSe (and also GaAs) are dipole-allowed.

The momentum ∇k of the photon taking part in the optical transition is negligible with respect to the momentum of the involved electrons and holes. This means that the optical transitions are vertical in k-space (as indicated by the vertical arrow in Fig. 2). Such transitions are called direct transitions and are the basic mechanism in active optical devices like light emitting diodes (LEDs) or laser diodes. The much less efficient indirect transitions involve additionally the participation of phonons (or disorder) and are dominant in indirect-gap semiconductors like Si or Ge.

1.2. EXCITONS

The model of interband transitions as described above yields a sufficient description of optical absorption only for narrow-gap semiconductors at room temperature. Inspecting the absorption spectra of ZnSe or GaAs in particular at low temperature reveals a much more complicated situation. The reason is, that the Coulomb interaction between the electron (negative charge) transferred to the conduction band and the hole (acts as a positive charge) left behind in the valence band has to be accounted for. This Coulomb interaction results in so-called correlation effects like enhanced scattering and the formation of bound electron-hole pair states called excitons. [1, 2, 4]

The adequate Schrödinger equation for this two particle situation has to consider the kinetic energy operators of electrons and holes. Then one has to add the Coulomb potential (depending on the relative coordinate r_{eh} of electron and hole) to the single-particle potential (periodic lattice potential and Coulomb interaction among same carrier type) of both carriers. When introducing relative ($\vec{r}$) and center-of-mass ($\vec{R}$) coordinates, this Schrödinger equation becomes separable leading to a quasi-free motion of the center of mass (COM) and an Hydrogen- (or better positronium-) atom like motion of the electron with respect to the hole. The resulting two-particle wavefunction Ψ of the exciton reads:

$$\Psi_{n_B,\vec{K}} \propto \Phi_{n_B}(\vec{r}_{eh})\, e^{i\vec{K}\vec{R}}\, w_{CB}(\vec{r}_e) w_{VB}(\vec{r}_h) \tag{2}$$

where K is the COM wave-vector and w_{CB} and w_{VB} are Wannier functions of the conduction or valence band, respectively. The latter are linear combinations of Bloch functions and describe spatially localized electrons/holes (in contrast to Bloch functions which extend through the whole crystal). The complex exponential function describes the plane-wave behavior of the COM. The envelope function Φ is a modified H-atom wavefunction of main (Bohr's) quantum number n_B.

The eigenenergy of the exciton depends on n_B and COM wave-vector K:

$$E(n_B,\vec{K}) = E_g - Ry^* \frac{1}{n_B^2} + \frac{\hbar^2 K^2}{2M} \tag{3}$$

The resulting dispersion relation is shown in Figure 3. Starting from the band-gap E_g marking the onset of the continuum of unbound states one finds a Rydberg series of excitonic states which have a parabolic dispersion related to their kinetic energy (M: exciton mass = sum of electron and hole masses). The Rydberg energy Ry^* itself is modified due to the fact that the hole mass is comparable to the electron mass (in contrast to the proton mass) and due to the screened Coulomb interaction in the solid:

$$Ry^* = \frac{e^2\mu}{2(4\pi\varepsilon_0)^2\varepsilon_r^2\hbar^2} = \frac{\mu}{\varepsilon_r^2} Ry \tag{4}$$

with μ and ε_r being the exciton reduced mass and the relative dielectric constant, respectively; Ry ist the atomic Rydberg of 13.6eV. Typical excitonic binding energies are in the order of few meV to several tens of meV (GaAs: 4.2 meV, ZnSe: 20 meV). The exitonic Bohr radius amounts to 14 nm for GaAs and 3.5 nm for ZnSe, i.e. the excitons extend over several hundreds or even thousands of unit cells. Such excitons, for which the effective mass concept as described above is still a valid, are called Wannier excitons.

The influence of these excitonic states on the optical properties can be illustrated for the case of the absorption spectrum typical for bulk GaAs or ZnSe at low temperatures. We already discussed that the band-to-band transitions are dipole allowed at the fundamental gap of zincblende semiconductors. A transition to an excitonic state via photon absorption is limited to the cases of s-type excitonic envelope functions since the photon spin is already used up for changing the electron's total angular momentum. Further, the excitonic states can only couple to photons for nearly zero momentum, i.e., at points where the linear photon dispersion crosses the excitonic parabolas (see Figure 3). [5] The result is a series of absorption lines below the energy of the band gap. These lines are inhomogeneously broadened by fluctuations of sample parameters (strain, disorder, etc.) as well as homogenously by interactions e.g. with phonons. The pronounced excitonic features (as schematically shown in Fig. 3 for low temperatures) are thus washed out with increasing temperature. Still, an enhanced absorption remains with respect to pure band-to-band transitions (dotted line in Fig. 3).

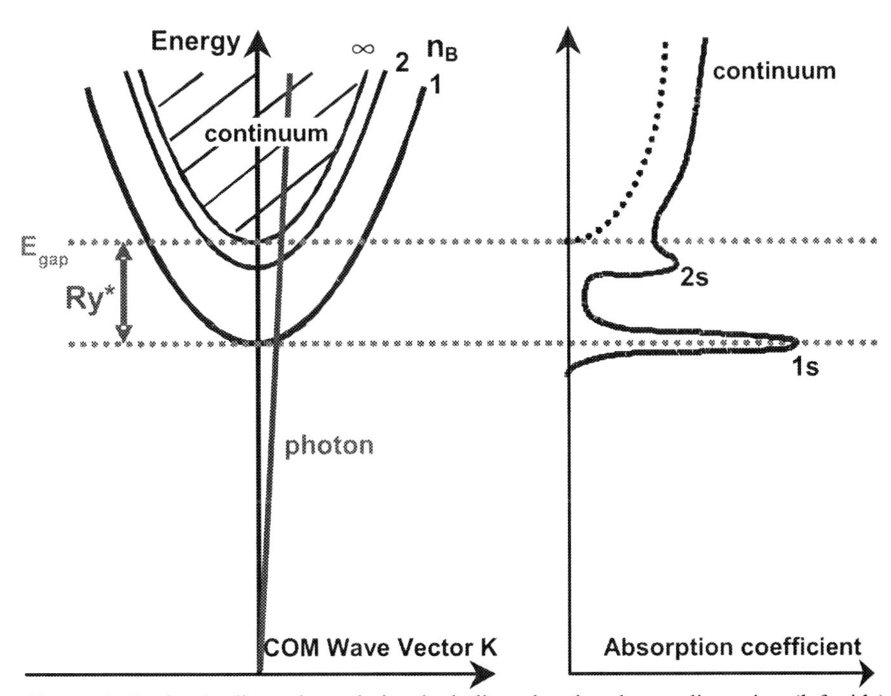

Figure 3. Excitonic dispersion relation including also the photon dispersion (left side) and schematic of an excitonic absorption spectrum (solid line) in comparison to pure band-to-band absorption (black doted line). (right side). according to [1]

1.3. QUANTUM STRUCTURES

Modern optoelectronic devices often incorporate low-dimensional semiconductor structures. Here one takes advantage of the influence of energy quantization e.g. on the optical transition strength or on the density of states (DOS) of the carriers. The latter is shown for the two-, one- and zero-dimensional case in Fig. 4. The electronic DOS changes from a square root-like dependence on energy in bulk semiconductors with parabolic bands to a step-like function in two-dimensional structures. An inverse square root DOS is found in one-dimensional and discrete states in zero-dimensional systems. The implications of these modifications are numerous. One example is the threshold reduction in laser diodes due to the fact that population inversion is more easily achieved when the DOS does not tend to zero at the band extrema. [6] Another one is the atom-like DOS in zero-dimensional structures which can be used for long-lived storage of electronic spin. [7]

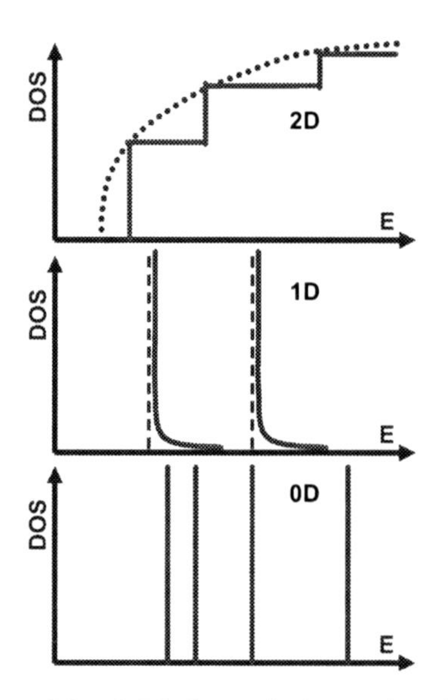

Figure 4. Density of electronic states (DOS) for various dimensions. The energy dependence of the DOS is $E^{1/2}$ for parabolic bands in bulk (dotted line), step-like for 2D, $E^{-1/2}$ for 1D and discrete in 0D systems.

The DOS in realistic semiconductor structures like the quasi-twodimensional quantum wells (QWs), the quasi-onedimensional quantum wires or the quasi-zero-dimensional quantum dots is closely related to the idealized DOS shown in Fig. 4. But the features are broadened (in)homogeneously and the electronic states are modified since the walls of the confining potential are not infinitely high.

We will concentrate in the following on the case of quantum-well structures based on the II-VI semiconductor compound ZnSe. The Figure 5 shows the band-gap energy of ZnSe and related materials plotted versus the lattice constant. From this plot one can explore the possibilities to form quantum wells by epitaxial growth of a thin ZnSe layer sandwiched between two barrier layers with higher band-gap energies. The natural choice of a substrate for epitaxial growth is GaAs, which is available with high surface quality and is nearly lattice matched to ZnSe. That means that the lattice constant is of the two materials is nearly equal so that the strain introduced in the ZnSe layer when grown on GaAs is not too large. Large strain would lead to defect formation leaving the layer optically dead. Similar arguments have to be used for the choice of barrier layers.

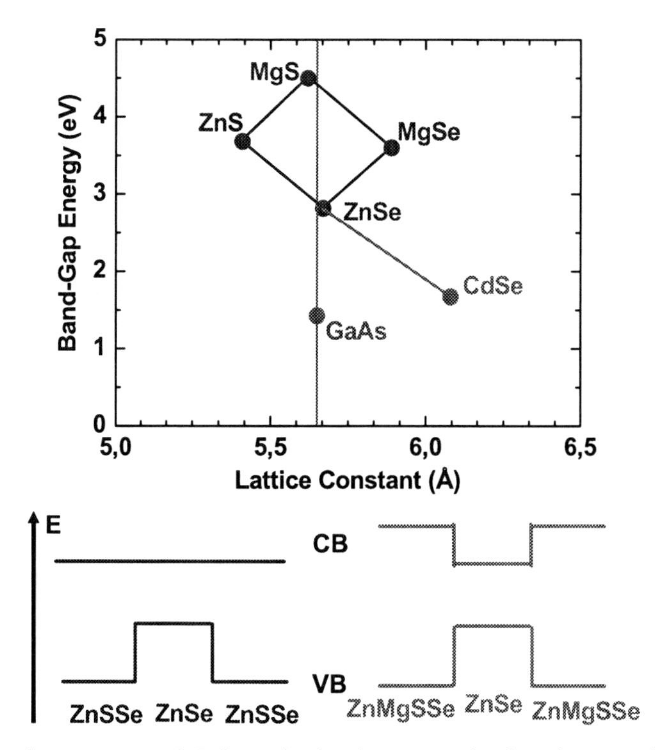

Figure 5. Band-gap energy of ZnSe and related compounds plotted versus the lattice constant of the respective material. The vertical (red) line marks the lattice constant of GaAs. The other lines indicate in a simplified way the change of lattice constant and gap energy for alloys made from the compounds at the respective line ends. The lower part of the figure shows schematically the line-up of conduction (CB) and valence bands (VB) for ZnSe/ZnSSe and ZnSe/ZnMgSSe QWs. For more details on ZnSe-based quantum wells. see [8]

On first sight MgS with its large band gap and a lattice constant close to ZnSe and GaAs seems to be a good choice for the barrier layers. But unfortunately, MgS does not form smooth interfaces with ZnSe. If one uses an ZnSSe alloy where only few percent of the Se atoms are substituted by Sulphur ones, quantum wells are formed which have nearly no confinement of the electron in the conduction band. The reason is that the substitution of Se by S mainly leads to a modification of the valence band states which derive from anion states ('common anion rule'). If one substitutes additionally a few percent of the Zn atoms by Mg (i.e. the barrier is ZnMgSSe) one achieves confinement for electrons and holes (see Fig. 5). Further, by suitable choice of the alloy composition one finds the lattice constant of ZnMgSSe to lie on the vertical line which indicates the GaAs lattice constant. This assures growth without significant strain and high crystalline quality.

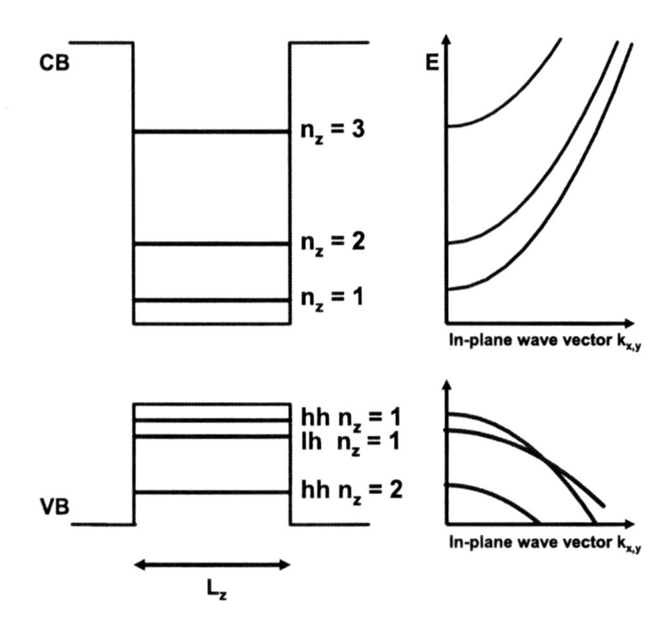

Figure 6. Confinement of electron and hole states within a quantum well (left drawing) and in-plane dispersion for the related subbands (right drawing).

The confinement of electrons and holes in a quantum well leads to the formation of so-called subbands (see Fig. 6) since the energies of the electrons (holes) are quantized perpendicular to the layer plane. The quantization energy (for the idealistic case of infinitely high barriers) is given by:

$$E_{n_z,q} = \frac{\hbar^2 n_z^2 \pi^2}{2 m_q L_z^2} \tag{5}$$

where the new quantum number $n_{z,q}$ stands for e or h, and L_z is the QW width. This confinement leads also to a lifting of the heavy-hole (hh) and light-hole (lh) band degeneracy. The motion within the layer plane (x-y plane) is still unhindered resulting in a parabolic in-plane dispersion relation of the electron and hole subbands.

1.4. EXCITONS IN QUANTUM WELLS

Optical transitions in quantum wells are dipole allowed between valence and conduction subbands with same quantum number n_z. Again, Coulomb interaction has to be accounted for which is even stronger in low-dimensional structures due to reduced screening and increased overlap of electron and hole wavefunctions. The exciton energy is given for an idealized 2D system by:

$$E_{n_B}^{2D} = E_g + E_{n_z,e} + E_{n_z,h} - Ry^* \frac{1}{(n_B - 1/2)^2} + \frac{\hbar^2 (K_x^2 + K_y^2)}{2M}. \quad (6)$$

Starting from the energy of the inter-subband transition one finds the exciton groundstate ($n_B = 1$) lowered by 4Ry*. Note that the excitonic parameters in realistic quasi-2D quantum wells have to be calculated numerically using effective dimensions [1]. Typical exciton binding energies in QWs are thus 2 -3Ry*. The last term in Eq. (6) is the in-plane kinetic energy of the exciton.

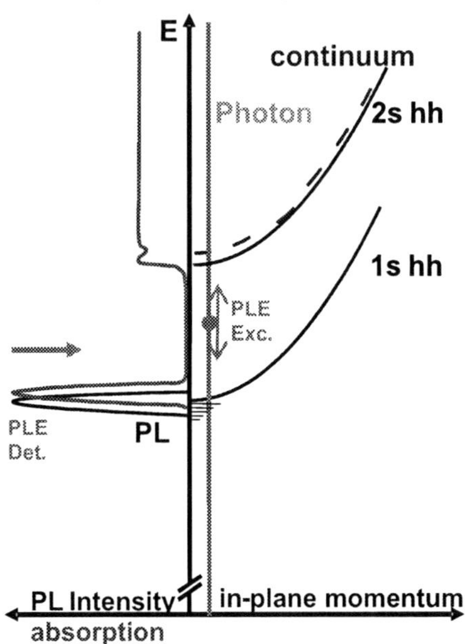

Figure 7. Dispersion for the lowest QW exciton states typical for zincblende semiconductors like GaAs and ZnSe (right hand side). The dashed area below the 1s hh parabola indicates the presence of localized states due to disorder. The exciton continuum states are found in the area enclosed by the dashed parabola. Also plotted for comparison is the dispersion of a photon in the weak coupling approximation. The left hand side shows schematically the photo-luminescence and absorption spectra related to these exciton states.

The resulting dispersion relation and the related optical spectra are illustrated in Figure 7. The parabolic dispersions of the 1s and 2s exciton states of the heavy-hole exciton situated below the lowest inter-subband gap are plotted. The onset of the unbound electron-hole states (excitonic continuum, dashed parabola) coincides at zero momentum with the energy of the inter-subband gap. Due to fluctuations of the quantum-well thickness (interfacial disorder), which are in GaAs/AlGaAs or ZnSe/ZnMgSSe quantum wells on the order of 1-2 monolayers, one finds localized exciton states at the low-energy edge of the 1s parabola.

The left-hand side of Fig. 7 shows schematically the excitonic photoluminescence and absorption spectra which can be measured in QWs of reasonable quality at He-temperature. The excitonic states can couple to the photons only in the area of the light cone (between the linear photon dispersion and the vertical axis). The result is a distinct absorption line related to the 1s exciton which is inhomogeneously broadened by the disorder. There is no significant absorption between the 1s and 2s exciton states. The later is situated very close to the onset of the inter-subband transitions and clearly visible only in high-quality samples. The absorption into the excitonic continuum (inter-subband transitions enhanced by the Coulomb correlation) reflects the flat density of states in a quasi-2D system. This absorption spectrum is smeared out when disorder is large (inhomogeneous broadening) or when temperature is increased (homogeneous broadening due to exciton-phonon interaction).

Since photo-excited excitons relax to the bottom of the parabola via emission of phonons one observes low-temperature excitonic photoluminescence only from the dispersion minimum and the related localized tail states. Due to the relaxation the PL line has a 'Stokes shift' with respect to the absorption line on the order of the inhomogeneous line width. The presence of defects in the semiconductor quantum well leads to 'extrinsic' recombination like PL from bound-excitons situated below the 1s exciton PL. For a review on such extrinsic PL. see e.g. [9]

The exciton/carrier relaxation to the radiative states opens up the possibility for a convenient investigation of excitonic (including excited states) and band-to-band transitions in semiconductor nanostructures. Such information can in principle also be gained from the absorption spectrum (see Fig. 7). In practice however, this typically requires to remove the substrate on which the nanostructure has been epitaxially deposited. Further, the absorption of a nanostructure having a thickness of a few nanometer is quite small and difficult to measure. These problems are overcome by the method of photoluminescence excitation spectroscopy (PLE). With this method one records the PL intensity as a function of the exciting light wavelength. The detection wavelength is chosen to be at the low-energy side of the excitonic PL within the tail-state emission. Thus the detection point is situated below the maximum of the close-by excitonic absorption resonance (see arrow in Fig. 7). The wavelength of the excitation source (typically a narrow-linewidth cw laser or a narrow line filtered by a monochromator from a white light source; in both cases the light intensity is kept strictly at a fixed level) is now tuned along the photon dispersion in Fig. 7 across the excitonic resonances and the continuum. Around the intersections of the photon line and the exciton dispersion the excitation light is absorbed. The photo-generated carriers/ excitons relax to the radiative states within the PL line. Provided this relaxation is equally efficient for all excess energies one finds that the PL intensity, measured at the detection point, is directly proportional to the absorbance. Thus the PLE spectrum resembles directly the absorption spectrum.

2. Exciton Formation and Relaxation in ZnSe Quantum Wells

In the former chapter we have demonstrated that excitonic states in quantum wells show up as resonances in absorption spectra. Real excitons can form after photo-excitation from free electron-hole pairs. In the case that the electron-hole pair is excited with some excess energy, the exciton can have significant kinetic energy and momentum. This 'hot exciton' then travels in the semiconductor according to its momentum, and relaxes to its band minimum by giving its excess energy to the lattice through phonon-scattering processes. Relaxed excitons, also called 'cold excitons', have an average excess energy comparable to the thermal energy of the lattice. At this stage, a quasi-equilibrium state of exciton and phonon system is reached. Eventually, the relaxed exciton recombines radiatively, resulting in photoluminescence, or non-radiatively, converting the energy into heat. At high temperatures, thermal dissociation processes can break the excitons into electron-hole pairs. When the temperature is low enough that the phonon population is not sufficient for these processes, the exciton is very stable and dominates many optical processes of semiconductors.

This picture of the fate of excitons in semiconductors is a well accepted concept since many decades and has been nicely reviewed e.g. in [1, 4]. But, recently this concept has been challenged and the traditional interpretation of PL experiments has been questioned. (see e.g. [10]) In particular, theory and experiments indicate that a PL signal at the exciton resonance in III-V semiconductor structures might not necessarily be a signature for a real exciton population. [11, 12] The situation turns out to be quite different in semiconductors with high excitonic binding energy and strong polar coupling allowing for an efficient exciton formation after non-resonant excitation. We will demonstrate in this chapter that hot-exciton formation and relaxation can nicely be observed in ZnSe-based quantum wells.

2.1 PHOTOLUMINESCENCE EXCITATION SPECTROSCOPY: FORMATION OF EXCITONS

The exciton-formation processes in polar II-VI semiconductors can be quite different from the case of GaAs. The excitonic binding energy of bulk ZnSe is five times larger than in GaAs (20 versus 4.2 meV), and the Fröhlich coupling between electrons and LO phonons is seven times stronger (0.43 versus 0.06). These differences suggest that in II-VI systems, the LO-phonon assisted exciton-formation processes can be very efficient, thus dominating over the individual relaxation of electrons and holes. The latter mechanism is proposed for GaAs. And indeed a variety of experiments reflects an efficient hot-exciton formation in II-VI systems. [13-15] The relaxation of excitons via LO-phonon emission is evidenced impressively by the LO-phonon cascade found in the PLE spectra of numerous II-VI systems. [16]

An example for such a cascaded relaxation found in ZnSe/ZnSSe quantum wells is given in the figure 8. The PLE spectrum shows a series of sharp peaks spaced by the energy of the LO phonon in ZnSe. This spectrum is in strong contrast to PLE spectra found in GaAs QWs which resemble the spectrum shown in Fig. 7. This difference has it origin in the completely different relaxation process in both material systems. For a fit to the experimental PLE spectrum in Fig. 8 including emission of LO phonons and (the much less efficient emission) of acoustic phonons. see [16]

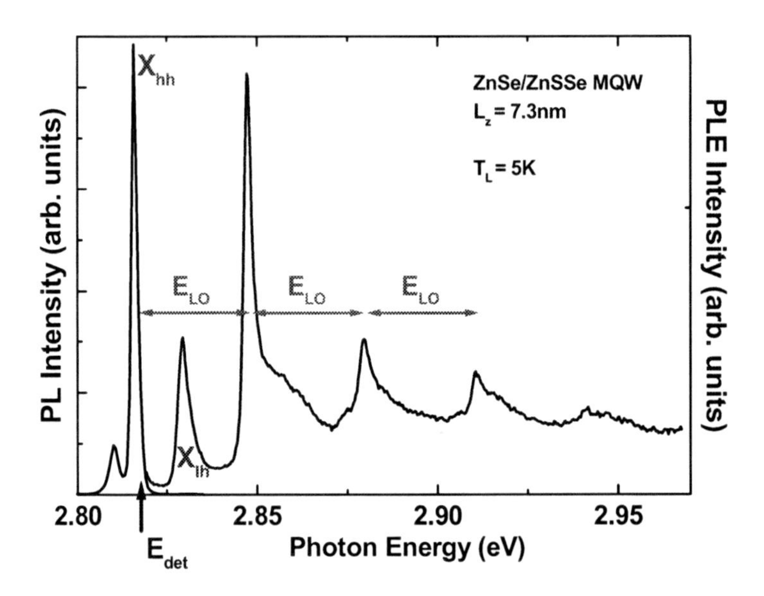

Figure 8. Photoluminescence and photoluminescence excitation spectra from a ZnSe/ZnSSe quantum well sample. Efficient formation of heavy-hole excitons via cascades-like LO-phonon emission is evidenced by the sharp peaks in the PLE spectrum. [16]

2.2 PHONON-SIDEBAND SPECTROSCOPY: ENERGY DISTRIBUTION OF HOT EXCITONS

Hot excitons cannot couple to photons directly due to their much larger momentum. But this limitation can be overcome if other quasi-particles like phonons are involved in the recombination and take away the excess momentum of hot excitons. Indeed, phonon-assisted recombination and absorption of excitons have been observed in many materials. The related peaks in PL or absorption spectra are called phonon sidebands (PSB) or phonon replica. Due to the multi-particle feature of these processes, the PSB is typically much weaker than the emission band without a phonon involvement (zero-phonon line: ZPL). But still, due to the strong coupling of excitons to LO phonons in polar semiconductors, the PSB has been frequently exploited for the study of hot excitons. For a summary of early results see e.g. [17], very recent results on II-VI compounds and Group III-Nitrides are found e.g. in [13-15] and [18].

The figure 9 shows schematically the hot-exciton dynamics in ZnSe quantum wells and the resulting PL spectrum composed of ZPL and PSB. The excitons are formed with high center-of-mass kinetic energy and momentum after the emission of one (or more) LO phonons. In the zero-phonon recombination process, they need to relax to the band minimum by emitting acoustic phonons before they contribute to the ZPL. However, in the phonon-assisted recombination process, the hot exciton can recombine by emitting one photon together with one (or more) LO phonons, to fulfill the conservation of energy and momentum. The lineshape of the PSB is thus given by the product of the exciton distribution function and the probability for the phonon-assisted recombination. For a detailed discussion of the latter. see [19, 20] The analysis of the PSB emission gives therefore a direct access to the distribution of the excitons within the parabolic band.

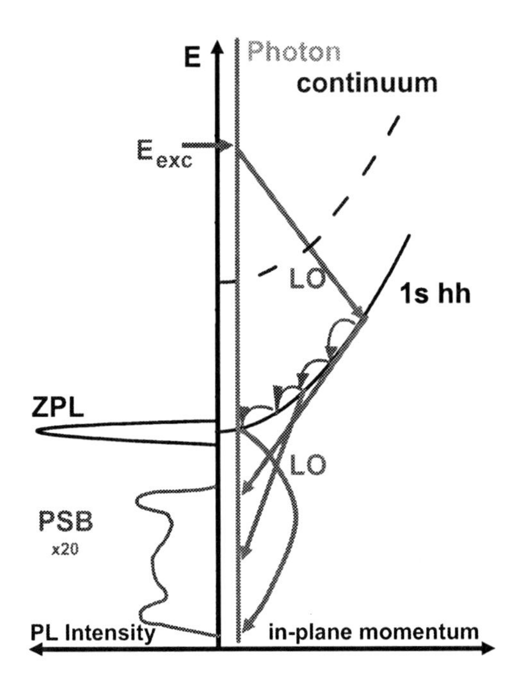

Figure 9. Schematic drawing of the excitonic dynamics (right panel) and the resulting PL spectrum (left panel) in ZnSe-based quantum-well samples at low lattice temperature. The shown excitonic dynamics include the laser excitation, the hot-exciton formation achived by LO-phonon emission, the energy relaxation of hot excitons by acoustic-phonon (AC) emission, and the LO-phonon-assisted recombina-tion of hot excitons. The PL spectrum is composed of a ZPL as well as the first order PSB.

In the PSB spectrum, we observe a sharp peak at its upper energy limit. The peak locates at twice the LO-phonon energy below the excitation photon energy. When we tune the laser photon energy, the peak shifts to keep this energy difference unchanged. We can attribute this peak to LO-phonon assisted recombination of hot excitons right after their formation and before the first step of the energy relaxation, i.e., before the first acoustic-phonon scattering event. [20] Hot excitons that have already emitted or absorbed one or more acoustic phonons cannot contribute to this peak due to the involved change in energy. In this sense, this peak monitors specific excitonic states with a kinetic energy well defined by the excitation photon energy.

During its lifetime an exciton experiences numerous elastic and inelastic scattering events. The former ones only change the direction of the exciton momentum leaving the energy of the exciton unchanged. During an elastic scattering event, the exciton wavefunction retains its coherence. An inelastic scattering event, however, changes the exciton energy and thus the phase of its wavefunction. It destroys the coherence of the individual exciton. One can view at excitonic relaxation in a sense that the wavefunction of an individual exciton is in a coherent state between each successive inelastic scattering events. Since the described sharp peak monitors the excitons before the very first inelastic scattering event, i.e. acoustic-phonon scattering, it reflects the presence of individually coherent excitons in that energy state. Thus this peak has been exploited to deduce the spatial coherence of excitons during non-classical transport.[21]

2.3 TIME-RESOLVED PHONON-SIDEBAND SPECTROSCOPY: ENERGY RELAXATION OF HOT EXCITONS

In order to understand the temporal and spatial dynamics of excitons one needs to monitor their momentary energy distribution given by the statistical distribution function and the density of states. In quasi-equilibrium the steady-state energy distribution of a low-density exciton gas obeys Boltzmann statistics.[1] When the exciton gas is in quasi-equilibrium with the lattice, the temperature of the exciton distribution equals the lattice temperature. The density of states in quantum wells is given by a step function in energy so that within each subband the density of states is constant.

In the case of monochromatic laser excitation followed by a fast hot-exciton formation process as we have discussed in the previous section for ZnSe QWs, the exciton distribution formed right after the excitation is very far from the quasi-equilibrium distribution. Instead, the distribution is a delta function (ignoring homogeneous broadening) at one LO-phonon energy below the excitation energy. Such an initial energy distribution evolves towards quasi-equilibrium via interactions among the excitons and with the lattice.

Here, we want to distinguish two processes during this evolution, thermalization and relaxation. By thermalization, we describe the process in which the excitons exchange energy within the exciton subsystem through exciton-exciton interaction. In this process, the average energy of the exciton subsystem keeps constant, but the energy distribution evolves from the delta function towards the Boltzmann function. The time scale of the thermalization is determined by the rate of exciton-exciton scattering and thus by the density of the excitons. Only for an exciton population with high density, the excitons exchange their energy rapidly, and the thermalized distribution can be established within picoseconds. However, the exciton temperature of this distribution is still different from the lattice temperature. When the sample is kept at liquid-He temperature, the exciton temperature will typically be higher. The evolution of the exciton temperature towards lattice temperature is achieved by the second process: relaxation.

In the energy-relaxation process, the excitons lose their excess energy to the lattice. This is done by phonon scattering processes, including optical-phonon and acoustic-phonon emissions. Thus, the time scale of the relaxation is determined by the scattering rates of these processes. In polar materials, the rate of optical-phonon emission is usually orders of magnitudes higher than that of acoustic-phonon emission. For example

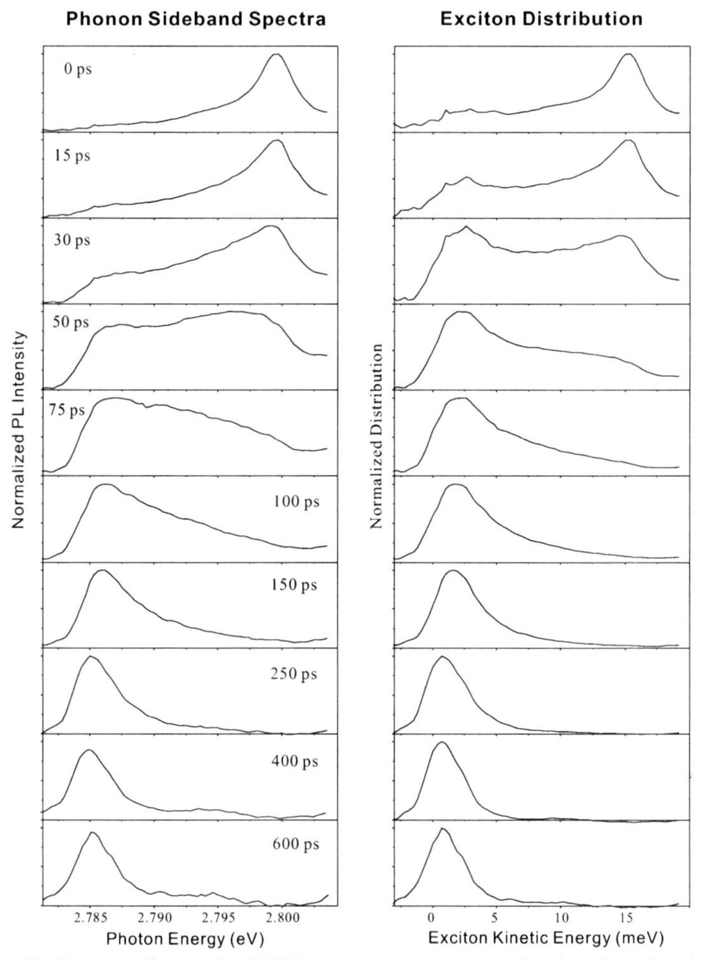

Figure 10. Temporally resolved PSB spectroscopy on excitonic relaxation in a 7.3 nm ZnSe/ZnSSe quantum well. The left panel shows the measured PSB spectra at various time delays after pulsed laser excitation. The corresponding energy distribution functions deduced from these spectra are shown in the right panel.[20, 22]

in ZnSe quantum wells, calculations proved that the LO-phonon emission time is about 100 fs, while the acoustic-phonon emission time is in the range of few 10 ps.[20, 22] Therefore, if the hot exciton has enough energy to emit optical phonons, it follows this efficient path of relaxation rapidly before any acoustic-phonon emission event can take place. The number of optical phonons that a hot exciton emits depends on its excess energy. This sequence of relaxation is confirmed e.g. by the modeling of details of the LO-phonon cascades in the photoluminescence excitation spectra of ZnSe quantum wells discussed above.[16] The rest of the relaxation is then achieved by much slower acoustic-phonon emission. Considering that the average energy of the acoustic-phonon emitted in each scattering event is only a few meV, one anticipates a relaxation time of several hundred picoseconds.

Experimental investigation of these thermalization and relaxation processes in conventional ZPL spectroscopy is difficult because the hot excitons reside mainly in the dark high-momentum states. One typically measures the rise time and decay time of the ZPL after a pulsed excitation.[13, 14, 23-30] The interpretation of the data is not straightforward since the contributions of exciton formation, thermalization, relaxation, and recombination processes are hard to separate. In contrast to this, the time-resolved PSB spectroscopy provides an ideal way to study the energy relaxation process since the energy distribution of excitons can be directly monitored in this technique.

In our experiments the samples of ZnSe multiple quantum wells are excited by picosecond laser pulses from a mode-locked Ti:Sapphire laser. The pulse energy is reduced to 4 pJ, corresponding to a carrier density of 10^9 cm^{-2}, to exclude the influence of exciton-exciton scattering. The sample is installed in a cryostat and cooled down to 4 K by liquid helium. The PL from the sample is spectrally dispersed by a spectrometer and then temporally resolved by a streak camera.

On the left panel of Fig. 10 we show a series of the PSB spectra measured at different time delays after pulsed excitation, as indicated beside each curve.[20, 22] We deduce from these PSB spectra the exciton distribution function at each time delay, as shown in the right panel of Fig. 10. In this experiment, the excitation laser photon energy is 2.8313 eV defining the initial kinetic energy of excitons to be 15 meV. In Fig. 10 we see that the initial exciton distribution is narrow and centered at 15 meV. The width of the initial distribution can be attributed to the spectral resolution of the system. The peak at 15 meV gradually disappears, showing that the excitons are driven out of the initial state on a time scale of few 10 ps. Thermalization occurs on a timescale of 100 ps. After 400 ps, the distribution function keeps unchanged, indicating the end of the relaxation process.

These experiments nicely reveal the relaxation dynamics of hot excitons. The observed timescales of the relaxation processes are well reproduced by detailed theoretical analysis of the exciton-phonon interaction.[20, 22]

3. Non-classical Transport of Excitons in ZnSe Quantum Wells

3.1 TRANSPORT REGIMES OF EXCITONS

In a semiconductor quantum well, the vertical motion of excitons along the growth direction is confined by the potential barriers. Within the quantum-well plane, the exciton can move freely unless it is localized by potential fluctuations.[31] This lateral transport of quantum-well excitons is an important part of the overall excitonic dynamics since the hot excitons have rather high kinetic energies. But only recently this coupling of energy relaxation and transport has been studied on the relevant scales namely the timescale of exciton-phonon interaction and the length scale of quasi-ballistic transport.[20, 21, 32-35]

Generally, the excitonic transport process is determined by the interplay between the free propagation of the exciton wavepacket according to its group velocity and various scattering processes. Since the former can be easily modeled, the knowledge of the influence of scattering processes on the transport is essential in understanding the transport behavior. We have already mentioned the scattering processes that a quantum-well exciton encounters during its lifetime. In particular, elastic scattering processes conserve the energy of the exciton and thus do not destroy the coherence of the exciton wavefunction.[36] For excitons in a realistic semiconductor quantum well, the elastic scattering mechanisms are the scattering by interface roughness, lattice defects and impurities. The scattering by acoustic phonons, optical phonons and the inter-exciton scattering are the main inelastic scattering mechanisms.

Excitonic transport displays quite different features when one observes it on different length and time scales. The average distance the exciton travels between two successive scattering events is called mean free path. Transport processes observed on a length scale shorter than the mean free path are called ballistic transport since there is no disturbance by scattering. On a length scale larger than this ballistic transport length, the exciton undergoes scattering. However, if the scattering events are all of elastic nature, the exciton energy is conserved. This transport regime is called quasi-ballistic. More importantly, since the elastic scattering doesn't change the phase of its wave-function, the exciton stays in the regime of coherent transport. Enlarging the observation length scale further, inelastic scattering occurs, and the exciton loses the coherence during the transport. The details of the transport on such a length scale depend on the energy of the exciton. In the case that the exciton is still hot with a kinetic energy larger than the thermal energy, it relaxes the energy and decreases the velocity during the transport. This phase is called hot-exciton transport. After the completion of the energy relaxation, the exciton system reaches quasi-equilibrium in terms of energy. But the spatial inhomogeneity of the exciton distribution leads to further transport in terms of a classical diffusion that can be described by the diffusion equation with a single constant diffusivity.

To summarize, the ballistic transport occurs between two successive scattering events, no matter whether they are inelastic or elastic. Between two successive inelastic scattering events, the exciton shows quasi-ballistic transport. Both of these processes are coherent transport processes. Transport of hot excitons over a large scale is coupled to energy relaxation, while cold excitons diffuse in real space. The length and time scales of these processes depend on the rate of the involved scattering mechanisms.

A well suited method to study excitonic transport on the length scale of the light wavelength and on the time scale of the exciton-phonon interaction is based on confocal photoluminescence microscopy (typically called μ-PL).[37-42] Figure 11 shows a typical configuration of this kind of setup.[43] The laser beam is focused to a tiny spot on the sample through a microscope objective. The PL is collected by the same objective. A shiftable pinhole is installed in the image plane of the microscope to choose the detection spot, which is independent of the excitation spot. The signal is then dispersed by a spectrometer and detected by a CCD camera. By scanning the pinhole one can obtain the PL spectrum at different locations, even far outside the excitation spot. Alternatively, the pinhole and the spectrometer can be removed so that the PL image is directly imaged on a CCD camera. For time-resolved experiments, a pulsed laser excitation is used and a streak camera is attached to the spectrometer to provide temporal resolution of about 5 ps.

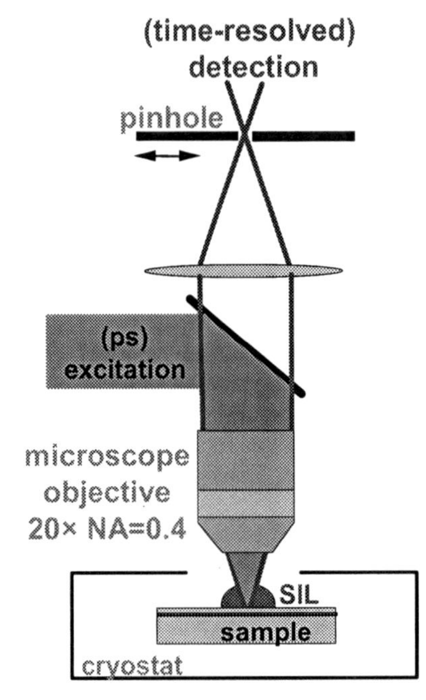

Figure 11. Schematic drawing of a confocal μ-PL setup enhanced by a solid immersion lens (nano-PL). One can extend this scheme by pulsed laser excitation and time-resolved detection (streak camera) to time-resolved nano-photoluminescence (nano-TRPL).

With this kind of system, the spatial distribution of the photoluminescence signal PL after a spatially narrow excitation can be measured directly. However, the spatial resolution is restricted to the diffraction limit and is on the order of 1 μm. This prevents the observation of ballistic and coherent transport processes occurring on a sub-μm length scale. To enhance the spatial resolution, we put a hemispherical solid-immersion lens (SIL) on the sample surface. It modifies the numerical aperture of the microscope and thus enhances the resolution to about 200 nm in terms of the half width at half maxima (HWHM) of the laser spot.[43] Additionally, the SIL also improves the collection efficiency by about five times.[43] This feature is very important for the case of weak signals like in PSB spectroscopy. In the following, we will call this experimental arrangement nano-photoluminescence (nano-PL) or in the time-resolved modification nano-TRPL.

3.2 SPATIALLY-RESOLVED EXCITATION SPECTROSCOPY: NON-DIFFUSIVE TRANSPORT OF EXCITONS

Using the SIL-enhanced nano-PL we study the transport properties of excitons in ZnSe quantum wells at a temperature of 7 K. We start with experimental results obtained from zero-phonon-line (ZPL) spectroscopy under continuous-wave (cw) laser excitation.

In the left panel of Fig. 12 we show the profile of the PL spot (zero phonon line) measured by scanning the pinhole in the image plane of the microscope. The profile of the cw laser spot is also measured by the same method. The PL spot is significantly larger than the laser spot, reflecting the occurrence of excitonic in-plane transport.

To characterize the excitonic transport, we measure the average transport length, which can be deconvoluted from the half width at half maximum (HWHM) of the PL spot, as function of the excitation excess energy. The right panel of Fig. 12 shows a general trend of increasing transport length with excess energy. But more importantly, we find a pronounced oscillatory feature with a period equal to the LO-phonon energy. Recalling the fast relaxation assisted by LO-phonon emission and the slow relaxation assisted by acoustic-phonon emission, we can understand this feature as follows: Increasing the excitation excess energy, the velocity of hot excitons formed by the first LO-phonon emission (see Fig. 9) becomes larger. This results in the increase of the transport length. However, when the excess energy is further increased such that the hot exciton has enough energy to emit one additional LO-phonon, it does so rapidly to relax the kinetic energy. Consequently, the initial velocity of excitons for the transport process is quenched.

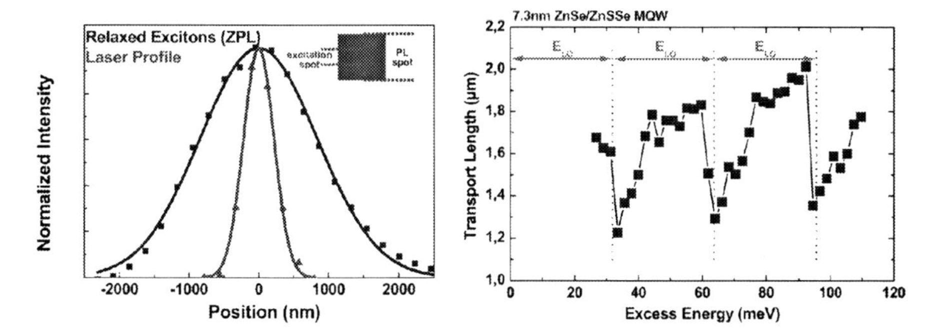

Figure 12. Time-integrated ZPL spectroscopy of excitonic transport in ZnSe quantum wells. Left panel: Spatial profiles of the ZPL intensity (squares) and the laser spot (triangles) measured by using the pinhole-scanning technique. The solid lines are Gaussian fits to the profiles. Right panel: The transport length deconvoluted from the PL-spot size as function of the excitation excess energy. [20, 32]

Our observations show the importance of the initial velocity on the transport process. This implies that the transport even in this cw experiment is dominated by non-diffusive processes involving hot excitons. [20, 32]

3.3 TEMPORALLY RESOLVED NANO-PHOTOLUMINESCENCE: SPATIO-TEMPORAL DYNAMICS AND COHERENT TRANSPORT OF EXCITONS

In order to reveal the dynamics of excitons simultaneously in space and time, we perform time-resolved nano-PL (nano-TRPL) experiments (see 3.1). We have chosen the laser photon energy such that the hot excitons are formed with a small kinetic energy within the inhomogeneous PL linewidth. The evolution of the PL signal from various locations on the sample with different distances from the laser spot are measured by scanning the pinhole. From these time-resolved PL data we obtain the spot diameter of the PL intensity as a function of delay time as shown as shown in Fig. 13.

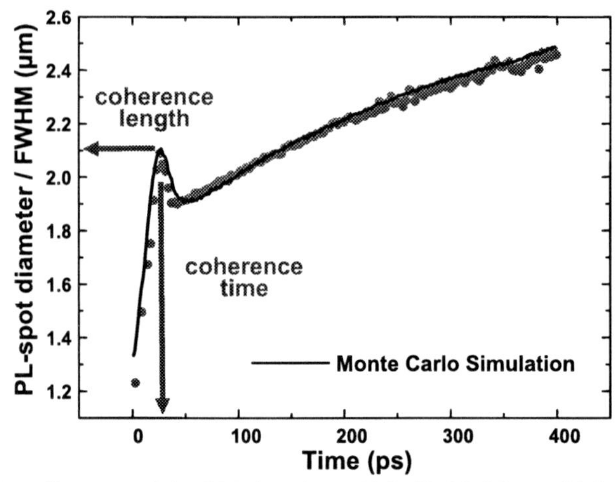

Figure 13. Spot diameter of the ZPL in a 8 nm ZnSe/ZnMgSSe multiple quantum well as a function of time. The dots are deduced from the experiment for excitation of the excitons within the inhomogeneous linewidth. The solid line is a one-parameter fit using a Mont-Carlo simulation of the exciton transport.[35]

If one assumes that the transport was a classical diffusion process, the diffusion equations predict that the square of the FWHM should increase linearly with time, with the slope related to the diffusivity.[44] In contrast, the experimentally deduced behavior shows an oscillatory feature around 30 ps followed by a nonlinear expansion of the spot size.

To reveal the details of the spatio-temporal dynamics of the hot excitons, we reproduce this dynamics including exciton formation, relaxation, real-space motion and recombination by Monte Carlo simulations.[33, 35] Microscopic processes including acoustic-phonon absorption, acoustic-phonon emission, and interface-roughness scattering are calculated explicitly. By adjusting the sole fit parameter (the correlation length of the interface roughness), the experimental data are nicely reproduced (solid curve in Fig. 13).

The most striking feature in Fig. 13 is actually the peak in the exciton density distribution at about 30 ps. It indicates a breathing-like behavior of the hot-exciton transport at early times. After a rapid expansion of the exciton density profile due to the high initial kinetic energy of the excitons, the spot size reduces to finally increase again. The Monte Carlo simulations reveal the origin of this spectacular behavior. While interface roughness scattering, which is much faster than the inelastic scattering, is dominantly directed towards small scattering angles, the emission of acoustic phonons has the largest matrix element for a backward scattering of the excitons. Starting their quasi-ballistic movement out of the excitation spot after their LO-phonon assisted generation, a significant number of excitons reverse their direction of propagation when the first acoustic phonon is emitted after an average time of about 30 ps.[33, 35]

Since the first inelastic scattering event marks the end of the quasi-ballistic or coherent transport phase, we can determine from the maximum in Fig. 13 simultaneously the coherence length and time of the excitons. The deduced values of 800 nm (deconvoluted from the half width at half maximum of the PL spot) and 29 ps are consistent with independent measurements using cw nano-PL [21] and four-wave mixing [45], respectively. These quantities are of crucial importance for the assessment of excitons in semiconductors for possible applications like coherent information processing exploiting the spatial part of excitonic wavefunction.

4. Acknowledgements

This review contains results obtained in a very fruitful collaboration with the following colleagues and students: Hui Zhao, B. Dal Don, D. Lüerßen, S. Wachter, M. Umlauff, J. Hoffmann, W. Langbein, R. Bleher, S. Moehl, Th. Unkelbach, G. Schwartz, C. Klingshirn, K. Ohkawa, B. Jobst, D. Hommel, J. Söllner, M. Scholl and M. Heuken. The research was supported by the DFG-Center for Functional Nanostructures (CFN) and the DFG-Graduate School 'Collective Excitations in Solids'.

5. References

1. For a comprehensive introduction see: C. Klingshirn: *Semiconductor Optics*, 2nd edn. (Springer, Berlin Heidelberg New York 2005)
2. For a review on latest developments see H. Kalt and M. Hetterich (eds.): *Optics of Semiconductors and Their Nanostructures,* (Springer Berlin Heidelberg 2004)
3. J.R. Chelikowsky, M.L. Cohen: Phys. Rev. B **14**, 556 (1976)
4. C. Klingshirn and H. Haug: Phys. Rep. **70**, 315 (1981)
5. This is the picture of weak coupling which treats the light-matter interaction in the framework of time-resolved perturbation theory. It can be applied for linear absorption and emission phenomena in typical semiconductors sample. The picture of strong coupling, which results in polariton modes, applies to propagation phenomena in samples with small (in)homogeneous broadening of the excitonic resonances. This scenario is described in the contribution to this book by C.Klingshirn and in [1]
6. M. Grundmann (ed.): *Nano-Optoelectronics: Concepts, Physics and Devices* (Springer, Berlin Heidelberg New York 2002)
7. D.D. Awschalom, D. Loss, N. Samarth (eds): *Semiconductor Spintronics and Quantum Computation* (Springer, Berlin Heidelberg New York 2002)
8. Heinz Kalt: *Low-Dimensional Structures of II-VI Compounds: General Properties, Quantum-Well Stuctures, Superlattices and Coupled Quantum-Well Structures, Quantum-Wire Structures,* Landolt-Börnstein Vol. **III,34** *Semiconductor Quantum Structures*, Subvolume C2: *Optical Properties* (Part 2), edited by C.Klingshirn, pp. 1-219 (Springer, Berlin Heidelberg New York 2004)
9. R. Sauer and K. Thonke in [2]
10. S.W. Koch and M. Kira in [2]
11. M. Kira, W. Hoyer, T. Stroucken, and S.W. Koch: Phys. Rev. Lett. **87**, 176401 (2001)
12. R.A. Kaindl, M.A. Carnaban, D. Hägele, R. Lövenich, and D.S. Chemla: Nature **423**, 734 (2003)
13. R.P. Stanley, J. Hegarty, R. Fischer, J. Feldmann, E.O. Göbel, R.D. Feldman, and R.F. Austin: Phys. Rev. Lett. **67**, 128 (1991)

14. N. Pelekanos, J. Ding, Q. Fu, A.V. Nurmikko, S.M. Durbin, M. Kobayashi, and R.L. Gunshor: Phys. Rev. B **43**, 9354 (1991)
15. J.H. Collet, H. Kalt, L.S. Dang, J. Cibert, K. Saminadayar, and S. Tatarenko: Phys. Rev. B **43**, 6843 (1991)
16. H. Kalt, M. Umlauff, J. Hoffmann, W. Langbein, J.M. Hvam, M. Scholl, J. Söllner, M. Heuken, B. Jobst, and D. Hommel: J. Cryst. Growth. **184/185**, 795 (1998)
17. S. Permogorov: *Optical emission due to exciton scattering by LO phonons in semiconductors*. In: *Excitons*, ed by E.I. Rashba, M.D. Sturge (North-Holland Publishing Company, Amsterdam New York Oxford 1982) pp. 177-204
18. X.B. Zhang, T. Taliercio, S. Kolliakos, and P. Lefebvre: J. Phys.: Condens. Matter **13**, 7053 (2001)
19. S. Permogorov: Phys. Status Solidi B **48**, 9 (1975)
20. Hui Zhao and Heinz Kalt in [2]
21. H. Zhao, S. Moehl, and H. Kalt: Phys. Rev. Lett. **87**, 097401 (2002)
22. M. Umlauff, J. Hoffmann, H. Kalt, W. Langbein, J.M. Hvam, M. Scholl, J. Söllner, M. Heuken, B. Jobst, and D. Hommel: Phys. Rev. B **57**, 1390 (1998)
23. T.C. Damen, J. Shah, D.Y. Oberli, D.S. Chemla, J.E. Cunningham and J.M. Kuo: Phys. Rev. B **42**, 7434 (1990)
24. H. Kalt, J. Collet, S.D. Baranovskii, R. Saleh, P. Thomas, L.S. Dang, and J. Cibert: Phys. Rev. B **45**, 4253 (1992)
25. F. Yang, G.R. Hayes, R.T. Phillips, and K.P. O'Donnell: Phys. Rev. B **53**, 1697 (1996)
26. Z.L. Yuan, Z.Y. Xu, W. Ge, J.Z. Xu, and B.Z. Zheng: J. Appl.Phys. **79**, 424 (1996)
27. J. Kusano, Y. Segawa, Y. Aoyagi, S. Namba, and H. Okamoto: Phys. Rev. B **40**, 1685 (1989)
28. Y. Masumoto, S. Shionoya, and H. Kawaguchi: Phys. Rev. B **29**, 2324 (1984)
29. A. Schülzgen, F. Kreller, F. Henneberger, M. Lowisch, and J. Puls: J. Cryst. Growth **138**, 575 (1994)
30. U. Neukirch, D. Weckendrup, J. Gutowski, D. Hommel, and G. Landwehr: J. Cryst. Growth **138**, 861 (1994)
31. see e.g. W. Langbein, G. Kocherscheidt, and R. Zimmermann in [2]
32. H. Zhao, S. Moehl, S. Wachter, and H. Kalt: Appl. Phys. Lett. 80, 1391 (2002)
33. H. Zhao, B. Dal Don, S. Moehl, and H. Kalt: Phys. Rev. B **67**, 035306 (2003)
34. H. Zhao, S. Moehl, and H. Kalt: Appl. Phys. Lett. **81**, 2794 (2002)
35. H. Zhao, B. Dal Don, G. Schwartz, and H. Kalt: Phys. Rev. Lett. **94**, 137402 (2005)
36. V.V. Mitin, V.A. Kochelap, and M.A. Stroscio: *Quantum heterostructures*, (Cambridge University Press, UK 1999) pp. 18-20
37. A.F.G. Monte, S.W. da Silva, J.M.R. Cruz, P.C. Morais, and H.M. Cox: J. Appl. Phys. **85**, 2866 (1999)
38. A.F.G. Monte, S.W. da Silva, J.M.R. Cruz, P.C. Morais, and A.S. Chaves: Phys. Rev. B **62**, 6924 (2000)
39. L.-L. Chao, G.S. Gargill-III, E. Snoeks, T. Marshall, J. Petruzzello, and M. Pashle: Appl. Phys. Lett. **74**, 741 (1999)
40. F.P. Logue, D.T. Fewer, S.J. Hewlett et al: J. Appl.Phys. **81**, 536 (1997)
41. A. Vertikov, I. Ozden, A.V. Nurmikko: J. Appl. Phys. **86**, 4697 (1999)
42. V. Malyarchuk, J.W. Tomm, V. Talalaev, C. Lienau, F. Rinner, and M. Baeumler: Appl. Phys. Lett. **81**, 346 (2002)
43. S. Moehl, H. Zhao, B. Dal Don, S. Wachter, and H. Kalt: J. Appl. Phys. **93**, 6265 (2003)
44. L.M. Smith, D.R. Wake, J.P. Wolfe, D. Levi, M.V. Klein, J. Klem, T. Henderson, and H. Morkoc: Phys. Rev. B **38**, 5788 (1988)
45. S. Wachter, M. Maute, H. Kalt, and I. Galbraith: Phys. Rev. B **65**, 205314 (2002)

ZnO REDISCOVERED – ONCE AGAIN!?

C. KLINGSHIRN, R. HAUSCHILD, H. PRILLER, J. ZELLER,
M. DECKER, AND H. KALT
Institut für Angewandte Physik and Center for Functional Nanostructures (CFN) der Universität Karlsruhe, Wolfgang-Gaede-Str. 1, 76131 Karlsruhe
claus.klingshirn@phys.uni-karlsruhe.de,

In an introductory chapter we shortly review the history of ZnO research and the motivations for the present renaissance of the worldwide interest in this II-VI compound. Then we concentrate after a few comments on growth, doping, transport, and deep centers on the following topics: band structure, excitons and polaritons, luminescence dynamics, high excitation effects like biexcitons or the transition to an electron-hole plasma, and finally lasing. We finish with a short conclusion and outlook.

1. Introduction and Historic Remarks

ZnO is a II-VI semiconductor with a direct, dipole allowed band gap around 3.4 eV i.e. in the near UV and a relatively large exciton binding energy of about 60 meV.

ZnO has seen in the past periods of intensive research, which started gradually in the fifties of the last century [1, 2] and peaked around the end of the seventies and the beginning of the eighties [3, 4, 5, 6]. Then the interest faded away, partly because it was not possible to dope ZnO both n- and p-type, partly because the interest moved to structures of reduced dimensionality, which were at that time almost exclusively based on the III-V system $GaAs/Al_{1-y}Ga_yAs$. The emphasis of research at that time was essentially on bulk samples covering topics like growth, doping, transport, deep centers, band-structure, excitons, bulk- and surface-polaritons, luminescence, high excitation or many particle effects and lasing. Results of the first research period are reviewed e.g. in [1, 2, 3, 4, 5, 6] and entered in data collections [7] or in textbooks on semiconductor optics [8].

The present renaissance on ZnO research started in the mid nineties and is documented by numerous conferences, workshops, and symposia and by more than 1000 ZnO-related papers per year, compared to slightly beyond 100 in 1970.

The present renaissance is based on the possibility to grow epitaxial layers, quantum wells, nano rods and related objects or quantum dots and on the hope to obtain:

- a material for blue/UV lasers and optoelectronics in addition to (or instead of) the GaN based structures
- a radiation hard material for electronic devices in a corresponding environment
- a diluted or ferromagnetic material, when doped with Co, Mn, Fe, V etc. for spintronics
- a transparent, highly conducting oxide (TCO), when doped with Ga, Al, In etc. as a cheaper alternative to ITO.

B. Di Bartolo and O. Forte (eds.), Advances in Spectroscopy for Lasers and Sensing, 277–293.

For several of the above mentioned applications a stable, high, and reproducible p-doping is obligatory, which is however still a major problem.

The emphasis of the present active period of ZnO research is essentially on the same topics as before, but including nanostructures. For first reviews of this new ZnO research period, see e.g. [9]. We shall present or cite in the following deliberately both old and new results covering six decades of ZnO research.

2. Growth, Doping, Transport, and Deep Centers

Bulk samples in the form of hexagonal needles or platelets can be grown by gas transport and/or oxidation of zinc. Examples are found e.g. in [10]. Hydrothermal growth under high pressure in LiOH + KOH solutions [11] may result in samples of several ten cm^3. For growth from melt or flux, see [12].

Textured or epitaxial layers and quantum wells (either ZnO or $Cd_{1-y}Zn_yO$ wells between $Zn_{1-y}Mg_yO$ or ZnO barriers, respectively) have been grown by a variety of techniques. For early examples of oxidation of Zn or for vapour phase transport, see [13, 14]. For modern techniques we mention only a few examples out of the tremendous number of publications and refer the reader otherwise to [9]. For pulsed laser deposition (PLD) see [15], for molecular beam epitaxy (MBE) [16] or for metalorganic chemical vapour deposition (MOCVD, MOVPE) [17].

A hot topic is presently the growth of nano-rods, -combs, -brushes, -nails, -tubes, -rings, -belts, -wool, -walls, -tetrapods or -flowers by various techniques, see [9, 18]. It became already possible to produce $Zn_{1-y}Mg_yO/ZnO$ superlattices in the nanorods [19]. Quantum dots have been prepared e.g. by sol-gel techniques or by spray combustion [20].

While p-doping is still a major problem, n-type doping with Ga,In or Al is possible beyond $n = 10^{20}$ cm^{-3}. Such highly doped samples look bluish, because the tail of the free carrier absorption extends into the red and the plasmon frequency approaches values close to 0.5 eV [8, 21]. In Fig. 1 we show from top to bottom the reflection spectra of a weakly and a highly doped ZnO:Ga sample, respectively, (note the different scales on the x-axis!) and the resulting energies of the transverse $\hbar\omega_-$ and $\hbar\omega_+$ branches of the plasmon phonon mixed states as a function of $n^{1/2}$, where n is the carrier density.

The diffusion constants of various dopants, including Mn and Co, have been summarized in [22]. The value of the Hall mobility of electrons at room temperature is long known and amounts about 200 cm^2/Vs [23]. It is limited by the intrinsic process of LO-phonon scattering, resulting in a relaxation time τ of 2×10^{-14} s.

Dislocations in ZnO can be decorated by various etching techniques [14, 24]. Electron transmission microscopy showed a lot of dislocations which are higher in density for epitaxial layers than for bulk samples [25]. For elastic and piezoelectric properties see [26], and for phonons [7].

As most wide gap semiconductors and insulators, ZnO shows a large variety of absorption and emission bands, situated energetically deep in the gap. Two different luminescence bands in the green are known, one is related to Cu, the other to oxygen vacancies [8, 27]. Emission bands in the yellow range are due to deep Li or Na acceptors [28]. Further emission- and/or absorption bands exist in the red or IR partly connected with iron or OH^- ions. Recently a band in the orange spectral range has been observed, which expands after the steplike onset of band-to-band excitation gradually over distances of several mm in times of several hundreds of ms [29]. More details on

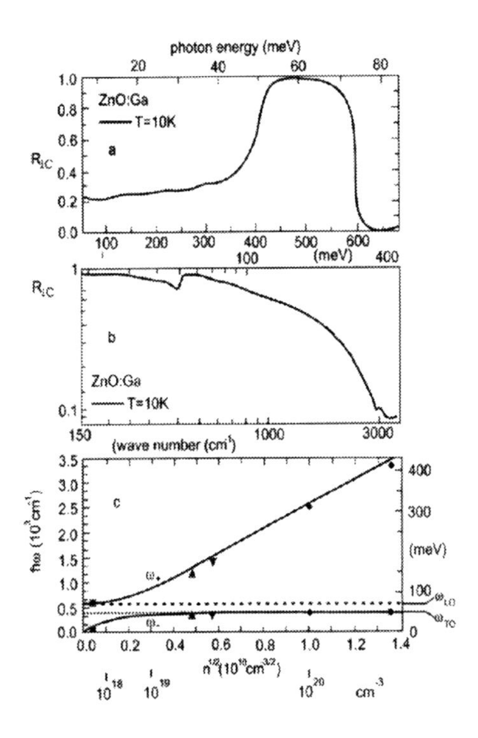

Figure 1. Reflection spectra of a weakly ($N_{Ga} \leq 10^{17}$ cm) and a highly $N_{Ga} \leq 3 \cdot 10^{20}$ cm^{-3} doped ZnO:Ga sample and the resulting branches of the plasmon-phonon mixed states as a function of $n^{1/2}$. From [21].

deep centers including the para- or ferromagnetic properties induced by dopants like Mn, Co, V, Fe are reviewed in [8, 9].

3. Band Structure, Excitons, Polaritons, and Dynamics

ZnO has rather strong ionic binding. Consequently, the conduction band (CB) arises essentially from the Zn^{++} 4s orbitals (symmetry Γ_7) and the upper valence bands (VB) from the O^{--} 2p states with an admixture of Zn^{++} 3d levels. The VB splits due to the hexagonal crystal field Δ_{cr} and the spin orbit coupling Δ_{so} into three subbands always labelled A, B, C from higher to lower energies. The gaps are at low temperature approximately $E^A_g = 3.437$eV, $E^B_g = 3.442$eV, $E^C_g = 3.481$eV [4, 7]. In contrast to other wurtzite II-VI compounds $\Delta_{cr} >> \Delta_{so}$ due to the small nuclear charge of oxygen. This fact in turn results in selection rules such that the transitions from the A and B VB to the CB have considerable oscillator strength only for $\boldsymbol{E} \perp \boldsymbol{c}$ (Γ_5) and from the C VB for $\boldsymbol{E} \parallel \boldsymbol{c}$ (Γ_1). The other, dipole allowed transitions involve a spin flip with a drastic reduction of their oscillator strength. Group theory allows a k-linear term for Γ_7 bands for $\boldsymbol{k} \perp \boldsymbol{c}$, but not for $\boldsymbol{k} \parallel \boldsymbol{c}$ and not for Γ_9 bands.

Exciton series exist for all three combinations of VB and CB. The exciton binding energies E^b_x are very similar with $E^b_x = (60 \pm 1)$ meV [5, 6, 7, 8]. The lower values in [30] for B and C excitons arise from the rather unusual procedure to count from the reflection minima, i.e. from the longitudinal eigenenergies.

The value of Δ_{so} in ZnO is not only small, but the interaction with the close lying Zn^{++} 3d states shifts one Γ_7 VB above the Γ_9 resulting in a VB-ordering A Γ_7, B Γ_9, C Γ_7 in contrast to the usual ordering A Γ_9, B Γ_7, C Γ_7 for II-VI compounds with higher nuclear anion charge like CdS(e). This concept has been introduced for ZnO in [31] and later also for the cubic CuCl [32]. Theoretical justifications have been given in [33]. The unusual ordering has been questioned in [34]. Since then many experiments have confirmed the inverted VB structure e.g. in absorption [35], luminescence, ***k***-space spectroscopy, under ***B*** or strain fields [36, 37, 38]. Recently the inverted assignment has been questioned again e.g. by [30, 39, 40]. Some arguments against this approach have been given already in [41]. Others will be given below.

In Fig. 2 we show various optical spectra in the exciton region. Fig. 2a shows the room temperature absorption spectrum of a textured ZnO layer on a glass substrate produced by oxiding an evaporated Zn layer. The experience is that such layers grow with ***c*** normal to the surface. The peak at 3.3 eV corresponds to the A and B exciton absorption and is possibly the first experimental observation of an excitonic feature in a semiconductor, though the author of [13] was most probably not aware of this fact.

The low temperature absorption spectra of ZnO have been investigated by many authors, both in the allowed and forbidden orientations [35, 36, 38]. The temperature dependent absorption following the Urbach-Martienssen rule [8] was measured up to 800K and allowed to deduce the temperature dependence of the band gap. See Fig. 3. These data are in good agreement with [43].

In Fig. 2b we show reflection spectra for the $n_B = 1$ A and B Γ_5 and the C Γ_1 excitons in their respective polarisations together with fit curves. A decent fit involves the problems of additional boundary conditions (abc), of spatial dispersion and possibly of an exciton free surface layer. For more information see e.g. [8, 35, 36, 37, 44] and references therein. Such spectra, including also states with higher main quantum numbers $n_B > 1$ of the envelope function have been reported by many authors [4, 7, 30, 31, 37, 40, 44].

The longitudinal transverse splittings are approximately $\Delta^A_{LT} = 2$ meV, $\Delta^B_{LT} = 10$ meV, $\Delta^C_{LT} = 12$ meV fulfilling the claim [31] that the sum of the oscillator strength of A and B excitons should equal that of the C exciton. In [40], the different values of Δ^A_{LT} and Δ^B_{LT} have been used as an argument to doubt the band structure.

However, in the case of close lying resonances the value of Δ_{LT} of the lower one is reduced, that of the upper one increased due to their interaction as has been detailed for exciton resonances already in [44] and in a more didactical manner in [8].

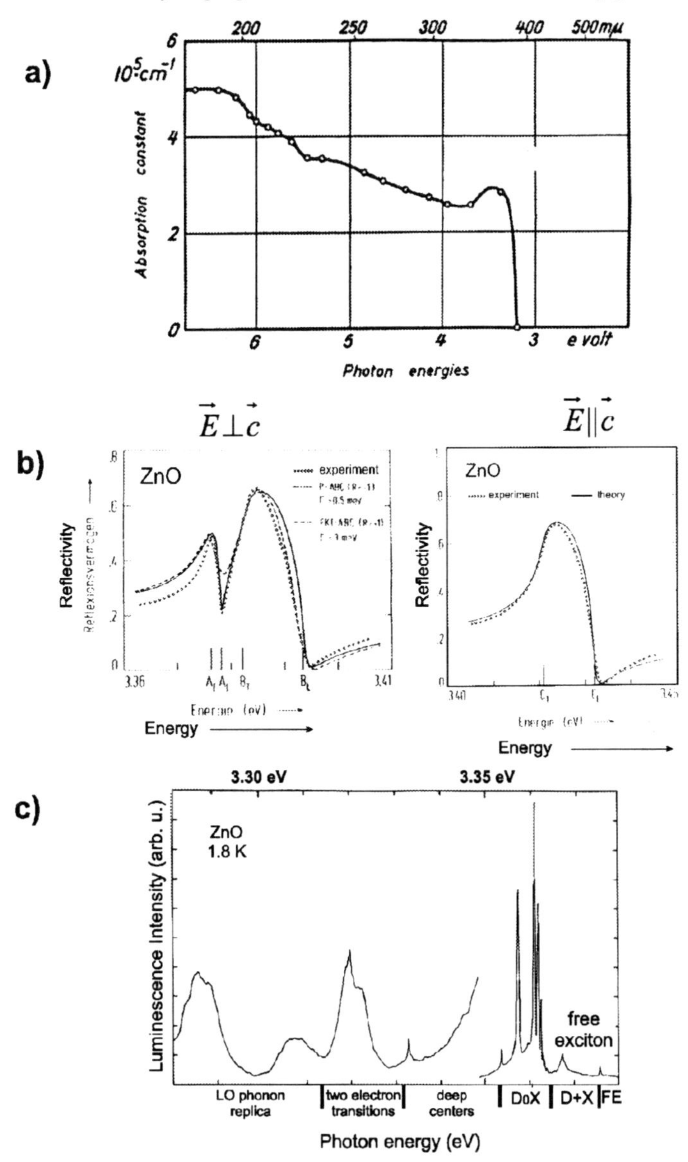

Figure 2. Room-temperature absorption spectrum of a thin, textured ZnO layer [13, 2] (a), low temperature reflection from K. Hümmer et al. [4, 37] (b) and luminescence spectra by R. Helbig et al. [4, 27] (c).

A small reflection feature seen for $\boldsymbol{E} \parallel \boldsymbol{c}$ in [30] at the position of B Γ^L_5 is attributed to a B Γ_1 state. Actually this feature is a mixed mode polariton [4, 8, 45] resulting from the slight mismatch to $\boldsymbol{k} \perp \boldsymbol{c}$ (<15°). The n_B=1 A or B $\Gamma_1 \oplus \Gamma_2$ or Γ_6 exciton states are in ZnO spin triplets, independent of the band structure. The triplet states are always situated energetically slightly below the corresponding transverse eigenstates.

For the investigation of the exciton resonances in reflection and absorption in a ***B***-field see e.g. [4, 8, 36, 38, 41], and for the Landau levels and the resulting effective masses [46]. Though the effective masses may be anisotropic in uniaxial materials [47] like ZnO (and in fact they are e.g. for CdS or CdSe [7]), the observation of the Landau levels indicate rather isotropic values for the conduction band but also for the valence bands [46].

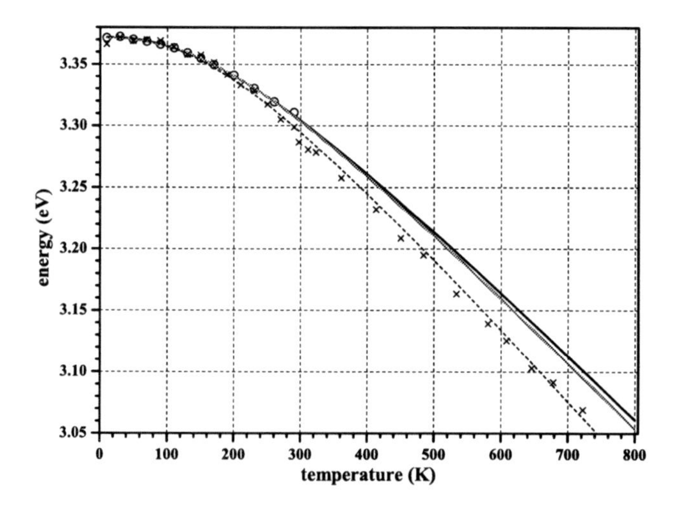

Figure 3. The temperature shift the A exciton deduced from the Urbach tail (x) and from luminescence (o) compared to a theoretical model including the thermal expansion and the electron phonon coupling. From [42].

The luminescence spectra of high quality ZnO samples are dominated at low temperature and excitation density by spectrally very narrow (<0.1 meV) emission lines of bound exciton complexes (BEC), especially neutral donor bound excitons (D^0X). See e.g. [4, 8, 27, 41]. We give in Fig. 2 an example. For photoluminescence excitation spectra (PLE) of various BEC emission lines see [51]. This technique allows to monitor excited states of BEC, e.g. ones which contain a hole from the BVB instead of the AVB.

The zero-phonon luminescence of the free excitons is at low temperatures very weak but increases relative to the BEC when the latter become thermally ionized from their defect centers in the temperature range around (80 ± 20) K. The zero phonon line of the free A exciton luminescence has been investigated thoroughly in [37] and later e.g. in [30, 39, 42]. See [48]. The influence of the $\boldsymbol{k}$-linear term is nicely observed for the A Γ_5 excitons in $\boldsymbol{E} \perp \boldsymbol{c}$, for $\boldsymbol{k} \perp \boldsymbol{c}$ but not for $\boldsymbol{k} \parallel \boldsymbol{c}$ as expected and the luminescence of the A Γ_1 is seen in $\boldsymbol{E} \parallel \boldsymbol{c}$ [37]. This emission is also reported in [30], but is used only to claim "…that the crystal is of good quality.", but ignoring that this observation involves the presence of Γ_1 symmetry.

In Fig. 5 we show time resolved unpolarized emission spectra of the $n_B = 1$ A exciton polaritons. The features due to emission from the lower A Γ_5 polariton branch (LPB) around 3.3755 eV (superimposed by the A Γ_1 exciton), from the intermediate A Γ_5 polariton branch caused by the $\boldsymbol{k}$-linear term around 3.377 eV and the upper A Γ_5 polariton branch (UPB) around 3.378 eV are clearly visible. The luminescence dynamics shows the relaxation from the UPB and the feeding into the LPB and the A Γ_1 state. Further examples for the luminescence dynamics of excitons in ZnO can be found in [42, 52, 53] and references given therein.

It has been found that the luminescence of the LO-phonon replica (see also below) of the free exciton polariton decays at all temperatures more slowly than the zero phonon band [42]. The explanation is the following: To the zero phonon luminescence contribute only exciton polaritons, which hit during their lifetime the surface with $\boldsymbol{k}_{\parallel} \leq \omega/c$. In contrast the LO phonon replica luminescence originates from excitons scattered on the photonlike polariton branch. This scattering includes all exciton like polaritons and the escape depth of the resulting photon-like polaritons is large. Therefore the LO-phonon replica monitor to a better extend the behaviour of all excitons, more precisely of exciton polaritons. For details see [8]. The temperature dependence of the near edge luminescence spectra have been observed by many groups e.g. [27, 41, 42, 52] and a consistent picture arose. At low temperature one observes depending on the sample quality more or less resolved BEC luminescence and its LO-phonon replica. Around (80 ± 20) K BEC are thermally ionized from their centers and the free exciton luminescence, including the LO-phonon replica takes over. In Fig. 4 we show a fit of the zero phonon band and of its first two LO-phonon replica for different temperatures. The fit includes the thermal distribution of the exciton-like polaritons on their parabolic dispersion curve, $\boldsymbol{k}_{\parallel}$ conservation in the above mentioned sense and a temperature dependent homogenous broadening $\gamma_h(T)$ as well as the temperature dependence of the gap. See Fig. 3.

At room temperature these bands merge to an unstructured emission feature with a width of approximately 100 meV and γ_h (300K) ≈ 40 meV. This value exceeds the RT values of excitons in many other semiconductors, including GaAs and GaAs based quantum wells and micro cavities. Furthermore it exceeds Δ_{LT} of all exciton resonances and the resulting dephasing time falls below the round trip time of most micro cavities. This high damping value is again a consequence of the strong coupling of excitons to LO-phonons. It will make room temperature lasing on Bose-condensed exciton polaritons in a micro cavity predicted in [49] highly unlikely, though micro cavities with ZnO will show at low temperatures interesting features due to the complex structure of the exciton polariton dispersion and their large oscillator strengths.

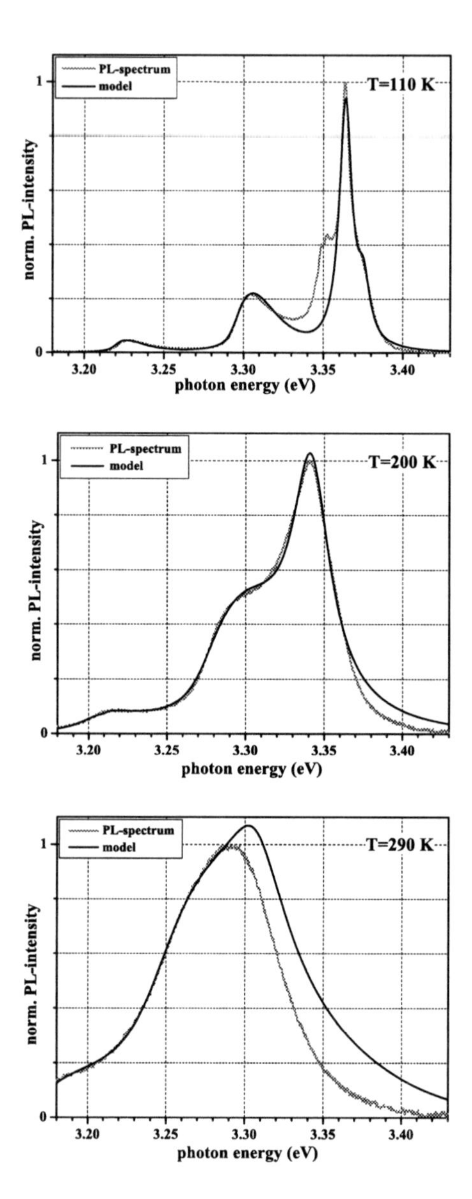

Figure 4. Luminescence spectra of ZnO for different temperatures with fit curves [42].

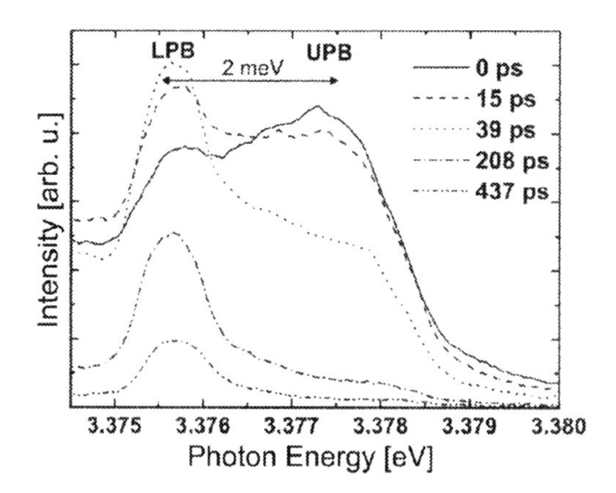

Figure 5. Fine structure and luminescence dynamics in the exciton polariton. From [42, 52].

To conclude this section, a few words on the luminescence of ZnO structures of reduced dimension. In Fig. 6 we show luminescence spectra of a ZnO QW between $Zn_{1-y}Mg_yO$ barriers. One sees for the lowest excitation emission from the ZnO buffer layer at 3.36 eV due to BEC recombination, from the barrier in the region from 3.44 to 3.48 eV and from the QW features peaking at 3.417 eV and at 3.40 eV. The higher one corresponds as usual in QW [8] to localized and defect bound excitons, the lower one is attributed in [50] to a mesoscopic crater-like defect. The new emission band appearing for higher, pulsed excitation around 3.39 eV will be discussed below.

The luminescence of nanorods looks similar to that of bulk samples apart form some (in)homogenous broadening.

It has been found in bulk materials, that surfaces || **c** are aging and deteriorating, if kept at ambient atmosphere. To obtain e.g. good reflection spectra, it was necessary to use freshly cleaved or etched surfaces. In contrast, naturally grown surfaces perpendicular to **c** showed even after many years in ambient atmosphere perfect A and B reflection features [42]. Nanorods, which have a significant smaller ratio of volume to surface || **c** tend do show a significant deterioration of the luminescence over a few months.

4. High Excitation Phenomena and Lasing

In this section we treat first so-called many particle- or high excitation phenomena in the intermediate density regime where excitons are still good quasi particles, then the transition to an electron-hole plasma at the highest density. The transition between the two regimes is not sharp but for didactic reasons this separation is useful. Finally

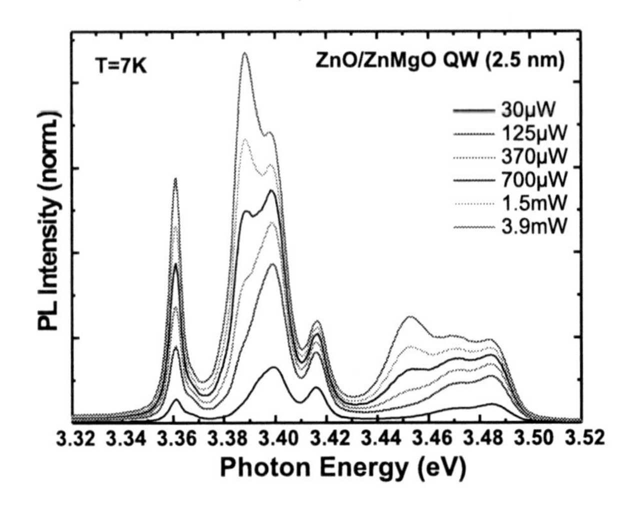

Figure 6. Luminescence spectra of a single 2.5 nm wide ZnO QW grow between Zn_{1-y} Mg_yO barriers on a ZnO buffer layer, deposited in turn on a GaN substrate [42].

we consider some aspects of lasing in ZnO. For some reviews of these topics in general and more specifically for ZnO, see [5, 6, 7, 8, 55].

Typical processes in the first regime are inelastic scattering processes and biexciton formation and decay. While elastic scattering between excitons results essentially in an excitation induced increase of homogeneous broadening, the inelastic processes give frequently rise to new emission bands. In the inelastic exciton-exciton (X-X) or more precisely polariton-polariton scattering, two excitonlike polaritons interact via their dipoles. One of them is scattered under energy and momentum conservation onto the photon-like part of the dispersion curve while the other reaches a state with higher quantum number $n_B = 2,3,\ldots\infty$, resulting in new emission bands labelled P_2, P_3, …, P_∞ [5, 6, 7, 55].

Other inelastic scattering processes are known between free excitons and free carriers (X-e), LO-phonons (X-nLO) or between bound exciton complexes and free carriers or acoustic phonons [5, 6, 7, 55]. In Fig. 5 we show examples for the luminescence from nanorods and from an epilayer. The emission is dominated at low excitation by BEC recombination. With increasing ns excitation intensity I_{exc}, this line broadens on its low energy side due to the appearance of the so-called M-band, which may be due to biexciton decay, but also due to the process involving BEC mentioned above [5]. With further increasing excitation the epilayer shows at 3.32 eV the appearance of the P_∞ band. Surprisingly this band is absent in thin (≈50 nm) but not in thick (≈200 nm) nanorods (see below). These rods are bulk like for excitons because both their length

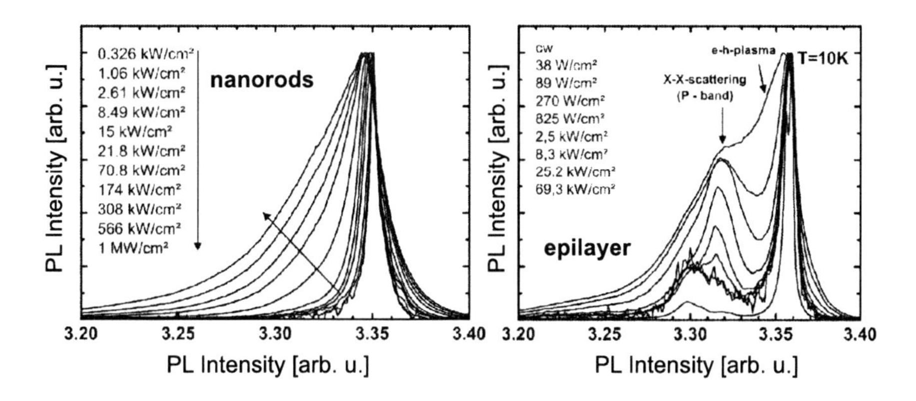

Figure 7. The luminescence of ZnO nanorods (a) and of an epilayer (b) for increasing excitation. From [42].

(>1μm) and their diameters are large compared to the excitonic Bohr radius $a_B \approx 1.8$ nm. But for photon-like polaritons the thin rods act as one dimensional wave-guides.

This reduction of the phase-space in one of the outgoing channels in the X-X process may quench this process. Further work is necessary to verify or falsify this hypothesis. With increasing temperature, the X-X process goes over to the X-e process, if a sufficient number of excitons are thermally ionized to favour this dipole-monopole interaction [5, 8, 55].

In Fig. 8 we show the emission of thick nanorods after excitation with frequency tripled 100 fs pulses from a TiSa laser. The rather broad emission band around 3.32 eV could be due to either the P-bands mentioned above or to the beginning formation of an EHP discussed below. The narrow spikes appearing at 3.34 eV indicate the onset of stimulated emission. Their spectral position is not fully compatible with X-X scattering and needs further investigation.

For the biexciton various contradicting values of its binding energy relative to two free A Γ_1 or A Γ^T_5 excitons between 10 and 20 meV have been deduced from luminescence or ns four wave mixing (FWM) experiments. The solution came, when two groups observed in samples from different sources independently three biexciton levels in a modified two photon absorption experiment namely luminescence assisted two photon spectroscopy (LATS) [57]. In a joint publication [58] the binding energy of the ground-state biexciton has been found to be 15 meV and two higher states were observed, which are due to biexcitons, in which one or two holes from the A VB are replaced by B-holes. See also [59].

In table 1 we give a listing of the respective biexciton energies and of the binding relative to two A excitons or to the involved A and / or B excitons from [58, 60]. The binding energies of the AA Γ_1 and AB $\Gamma_{5,6}$ biexcitons agree within experimental error [61]. Since the data of both excitons are very similar, the deviation for the BB Γ_1 exciton claimed in [60] is difficult to understand. As mentioned in [58], FWM data may

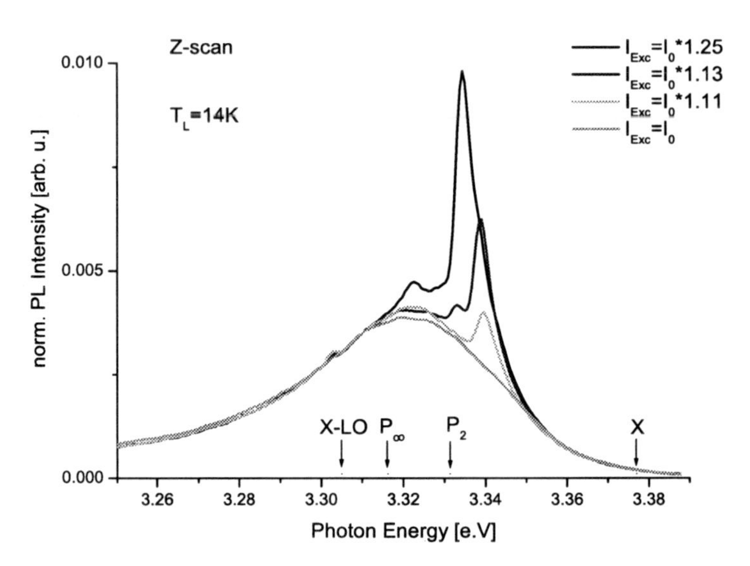

Figure 8. luminescence of thick nanorods under band-to-band excitation with 100 fs pulses for increasing fluence [56].

result in erroneous results. Similar arguments may hold for the claimed biexciton binding energy of a CC Γ_1 biexciton of only 1.4 meV [60].

In $ZnO/Zn_{1-y}Mg_yO$ quantum wells a new emission band appears with increasing excitation at 3.385 eV (see Fig. 6). If it is due to a biexciton decay, a binding energy around 20 meV results since the biexcitonic M-band luminescence has ideally the shape of an inverted Baltzmann distribution [8], so that the binding energy can be deduced from the distance between the exciton energy and the high energy slope of the M-band. This value would make some sense bearing in mind table 1 and the enhancement of exciton and biexciton binding energies in quantized systems [8]. Further examples of ZnO QW, nanorod and nanodot luminescence under high excitation are found e.g. in [62].

Some selected results on two photon absorption (TPA), TPA-spectroscopy or second and third harmonic generation are found in [38, 63], the dynamics of the luminescence under high excitation has been investigated e.g. in [42, 52, 53, 64]. For the investigation of effects in the regime of extreme nonlinear optics, see [65].

The dephasing time T_2 and/or the homogeneous broadening have been investigated at low temperatures by an analysis of the reflection spectra [4, 36, 37, 44, 54] yielding values of the order of 1 meV and also in time resolved FWM experiments [60] with comparable results.

TABLE 1. Biexciton energies and binding energies with respect to two A excitons and with respect to two A/B excitons. The last column gives corresponding data for A/B excitons from [60], see [61].

Biexciton energy and symmetry [58]	Binding energy with respect to two A excit [58]	Binding energy with respect to A/B excit. [58]	[60]
$E_{AA\Gamma_1} = 6.7355$ eV	14.7 ± 1 meV	14.7 ± 1 meV	15.6 meV
$E_{AB\Gamma_{5,6}} = 6.7407$ eV	9.5 "	14.7 "	16.6 meV
$E_{BB\Gamma_1} = 6.7469$ eV	3.3 "	13.7 "	4.7 meV

If the generation rate is increased to sufficiently high values, excitons cease to exist as individual quasi-particles and a new collective phase is formed, the electron-hole plasma (EHP). For details of this transition, see e.g. [5, 8]. In ZnO the EHP has been observed in gain spectroscopy, e.g. in [52, 66]. Simultaneously with the appearance of EHP gain, the excitonic reflection features start to disappear. Densities exceeding 10^{19} cm^{-3} have been deduced. For more information on the EHP, including its dynamics and structures of reduced dimensionality, see [42, 52, 67] and references therein.

The investigation of stimulated emission and lasing of ZnO started in the mid sixties. For a few early examples, see e.g. [68] and references therein and for tunable laser emission [69]. The understanding of the processes developed in parallel (see [5, 55]). It turned out, that all of the inelastic scattering and recombination processes mentioned above, including the two-electron transitions [41], may result in stimulated emission. Most of them can be mapped on a four level laser system, with correspondingly low thresholds for inversion. The X-LO emission process is inverted when one exciton is created and no LO-phonon present i.e. at $\hbar\omega_{LO} >> k_BT$. Similarly the X-X processes, e.g. the P_2 band is inverted if two excitons are present and none of them in the state $n_B=2$ [5, 55]. Nowadays such processes come also with names like thresholdless lasing or lasing without inversion.

The band - to - band recombination in an EHP requires $\mu_{eh}(n_p,T_p) > E_g(n_p,T_p)$ for population inversion [5, 8, 66, 67] and is therefore more like a three level system. The quantities $\mu_{eh}(n_p, T_p)$ and $E_g(n_p, T_p)$ stand for the chemical potential of the electron-hole pair system and the reduced or normalized band-gap depending both on the plasma density and temperature n_P and T_p, respectively. However, inelastic scattering processes between carriers or the emission of phonons or plasmons may lead to thresholds below that given above. Generally it has to be stated that the transition from excitons to the EHP is continuous in direct gap semiconductors [70]. Recent theories including this continuous transition are found e.g. in [71] and references given therein.

In the present phase of ZnO activity, stimulated emission and gain are generally attributed to the processes known already in bulk, but now for epitaxial layers [42, 52, 62], in quantum wells obtaining partly gain up to 550 K [62, 72] and most recently in nanorods and related structures [73]and Fig. 8.

A new topic is lasing in random media i.e. here in ZnO powders, relying on weak localization of light due to enhanced backscattering [74].

5. Conclusion and Outlook

Evidently ZnO sees a very vital renaissance of basic and applied research. As is usual in such situations, beautiful new effects are found, old mistakes are straightened out, new ones are made and also long known phenomena are being reinvented. We tried to give a good mixture of all these aspects. The question how long this renaissance will last, depends to a large extend on the question if a stable, high, and reproducible p-doping can be realized. Since prophecy is a notoriously difficult task, especially if it concerns the future, the authors do not want to give an answer here.

Note Added in Proof
A group in Sendai reported in May 2005 on a first blue/UV ZnO based LED.

Acknowledgements

The authors are grateful to the Deutsche Forschungsgemeinschaft and the Landeskompetenznetz Baden-Württemberg for financial support. The high quality bulk samples have been grown by R. Helbig (Erlangen), the epilayers, quantum wells and nanorods by T. Yao (Sendai), A. Waag (Braunschweig), R. Sauer, and K. Thonke (Ulm), M. Grundmann and Lorenz (Leipzig), H.J. Fan, and M. Zacharias (Halle)

References

[1] M. E. Brown (Ed.): ZnO - Rediscovered (The New Jersey Zinc Company, New York (1957))
[2] G. Heiland et al.: Solid State Physics **8**, 191 (1959)
[3] W. Hirschwald et al.: Current Topics in Materials Science **7**, 143 (1981)
[4] R. Helbig: Freie und gebundene Excitonen in ZnO, Habilitation Thesis, Erlangen (1976); K. Hümmer: Excitonische Polaritonen in einachsigen Kristallen, Habilitation Thesis, Erlangen (1978)
[5] C. Klingshirn, H. Haug: Physics Reports **70**, 315 (1981)
[6] B. Hönerlage et al.: Physics Reports **124**, 161 (1985)
[7] Landolt - Börnstein: *New Series Group III, Vols. 41 B* and *C2*
[8] C. Klingshirn: *Semiconductor Optics*, 2nd ed. (Springer, Heidelberg, Berlin, 2005)
[9] Ü. Özgür et al.: J. Appl. Phys. Rev. (2005) (in press); C. Klingshirn et al.: Physik Journal (2005) (in press) and Advances in Solid State Physics (in press)
[10] E. Scharowski: Z. Physik **135**, 318 (1953); E. M. Dodson, J. A. Savage: J. Mat. Sci. **3**, 19 (1968); R. Helbig: J. Crystal Growth **15**, 25 (1972)
[11] R. A. Laudise, A. A. Ballmann: J. Phys. Chem. **64**, 688 (1960); E. Ohshima et al.: J. Crystal Growth **260**, 166 (2004)

[12] J. W. Nielsen, E. F. Dearborn: J. Phys. Chem. **64**, 1762 (1960); K. Oka et al., J. Chrystal Growth **237-239** 5 (2002); N. Ohashi et al., J. Appl. Phys. **91**, 3658 (2002); D. C. Reynolds et al.: J. Appl. Phys. **95**, 4802 (2004)
[13] E. Mollwo: Reichsber. Physik **1**, 1 (1943)
[14] H. Schneck, R. Helbig: Thin Solid Films **27**, 101 (1975)
[15] M. Lorenz et al.: Annalen der Physik **13**, 59 (2004)
[16] T. Yao et al.: J. Vac. Sci. Technol. B **18**, 2313 (2000) and J. Crystal Growth **209**, 816 (2000)
[17] W. I. Park et al.: Appl. Phys. Lett. **79**, 2022 (2001); Th. Gruber et al.: Phys. Stat. Sol. a **192**, 166 (2002) and Appl. Phys. Lett **84**, 5359 (2004); T. Riemann et al.: Phys. Stat. Sol. b (to be published)
[18] Z. W. Pan et al.: Science **291**, 1947 (2001); H. Yan et al.: Adv. Mater. **15**, 402 (2003); D. Banerjee et al.: Appl. Phys. Lett. **83**, 2061 (2003); X. Y. Kong et al.: Science **303**, 1348 (2004); R. Kling et al.: Nanotechno. **15**, 1043 (2004); G. Prinz et al.: to be published, T. Nobis et al., Nano Letters **4**, 797 (2004); H. J. Fan et al., submitted Appl. Phys. Lett.
[19] W. I. Park et al.: Adv. Mater. **15**, 526 (2003)
[20] L. Spanhel et al.: J. Am. Chem. Soc. **113**, 2826 (1991) and J. Sol-Gel Sci. and Technol. **26**, 499 (2003); L. Mädler et al.: J. Appl. Phys. **92**, 6537 (2002)
[21] M. Göppert et al.: J. Luminesc. **72-74**, 430 (1997)
[22] F. W. Kleinlein, R. Helbig, Z. Physik **266**, 201 (1974)
[23] A. R. Hutson: Phys. Rev. **108**, 222 (1957); H. Rupprecht: J. Phys. Chem. Sol. **6**, 144 (1958); M. A. Seitz, D. H. Whitmore: J. Phys. Chem. Sol. **29**, 1033 (1968); J. A. Savage, E. M. Dodson: J. Mat. Sci. **4**, 809 (1968); P. Wagner, R. Helbig: J. Phys. Chem. Sol. **35**, 327 (1974); E.M. Kaidashev et al., Appl. Phys. Lett. **82**, 3901 (2003)
[24] G. Heiland et al. Z. Physik **176**, 485 (1963) and A. Klein ibid. **188**, 352 (1965)
[25] D. Gerthsen et al. Appl. Phys. Lett. **81**, 3972 (2002)
[26] D.F. Crisler et al. Proc. IEEE **56**, 225 (1968), W. Glück, Sol. State Commun. **8**, 1831 (1970)
[27] Chr. Solbrig: Z. Physik **211**, 429 (1968); R. Dingle: Phys. Rev. Lett. **23**, 579 (1969); D. C. Reynolds et al.: Phys. Rev. **140**, A1726 (1965) and Phys. Rev. **185**, 1099 (1969); E. Tomzig, R. Helbig: J. Luminesc. **14**, 403 (1976); R. Kuhnert, R. Helbig: J. Luminesc. **26**, 203 (1981); K. A. Vanheusden et al.: Appl. Phys. Lett. **68**, 403 (1996)
[28] D. Zwingel, F. Gärtner: Sol. State Commun. **14**, 45 (1974)
[29] H. Priller et al., Appl. Phys. Lett. **86**, 111909 (2005)
[30] S. F. Chichibu et al.: J. Appl. Phys. **93**, 756 (2003)
[31] D. G. Thomas: J. Phys. Chem. Sol. **15**, 86 (1960); J. J. Hopfield: J. Phys. Chem. Sol. **15**, 97 (1960)
[32] M. Cardona: Phys. Rev. **129**, 69 (1963) and J. Phys. Chem. Sol. **24**, 1543 (1963)
[33] K. Shindo et al.: J. Phys. Soc. Japan **20**, 2054 (1965); B. Segall: Phys. Rev. **163**, 769 (1967); U. Rössler: Phys. Rev. **184**, 733 (1969)
[34] D. C. Reynolds, C. W. Litton, T. C. Collins: Phys. Rev. **140**, A1726 (1965)
[35] J. J. Hopfield, D. G. Thomas: Phys. Rev. Lett. **15**, 22 (1965)
[36] G. Blattner et al.: Phys. Rev. B **25**, 7413 (1982)

[37] K. Hümmer, R. Helbig, M. Baumgärtner: Phys. Stat. Sol. b **86**, 527 (1978); R. Kuhnert, R. Helbig, K. Hümmer: Phys. Stat. Sol. b **107**, 83 (1981)
[38] M. Fiebig et al.: Phys. Stat. Sol. b **177**, 187 (1993); J. Wrzesinski, D. Fröhlich: Phys. Rev. B **56**, 13087 (1997) and Sol. State Commun. **105**, 301 (1998)
[39] D. C. Reynolds et al.: Phys. Rev. B **60**, 2340 (1999)
[40] B. Gil: Phys. Rev. B **64**, 201310 R (2001)
[41] W. R. L. Lambrecht et al.: Phys. Rev. B **65**, 075207 (2002); B. K. Meyer et al.: Phys. Stat. Sol. b **241**, 231 (2004); A. V. Rodina et al.: Phys. Rev. B **69**, 125206 (2004); R. Sauer, K. Thonke in *Optics of Semiconductors and Their Nanostructures*, H. Kalt and M. Hetterich (Eds.), Springer Series in Solid State Sciences **146**, 73 (2004)
[42] H. Priller, PhD Thesis, Karlsruhe (2005); M. Decker, Diplom Thesis, Karlsruhe (2005)
[43] F. J. Manjon et al.: Sol. State Commun. **128**, 35 (2003)
[44] J. Lagois, K. Hümmer: Phys. Stat. Sol. b **72**, 393 (1975); J. Lagois: Phys. Rev. B **16**, 1699 (1977)
[45] R. I. Weiher, W. C. Tait: Phys. Rev. **185**, 1114 (1969) and Phys. Rev. **B5**, 623 (1972); K. Hümmer, P. Gebhardt: Phys. Stat. Sol. b **85**, 271 (1978)
[46] K. Hümmer: Phys. Stat. Sol. b **56**, 249 (1973)
[47] J. R. Chelikowsky, Sol. State Commun. **22**, 351 (1977), M.L. Cohen and J.R. Chelikowsky, Springer Series in Solid State Sciences **75**, (1988)
[48] The references [37] show, that the claim in [39] to observe "...free exciton emission... for the first time" is clearly unjustified.
[49] A. Kakovin, M. Zamfirescu et al.: Phys. Stat. Sol. a **192**, 212 (2002) and Phys. Rev. B **65**, 161205 R (2002)
[50] J. Christen and A. Waag, private communication
[51] G. Blattner et al., phys. stat. sol. b **107**, 105 (1981), J. Gutowski et al., Phys. Rev. B **38**, 9746 (1988)
[52] H. Priller et al., phys. stat. sol. b **241**, 587 (2004)and Proc. EXCON, J. Luminesc. **112**, 173 (2005)
[53] T. Makino et al.: Appl. Phys. Lett. **77**, 975 (2000); H. D. Sun et al.: Appl. Phys. Lett. **78**, 2464 (2001); S. Giemsch et al.: To be published
[54] J. Lagois, B. Fischer: Phys. Rev. Lett. **36**, 680 (1976); J. Lagois: Phys. Rev. B **23**, 5511 (1981); I. Hirabayashi et al.: J. Phys. Soc. Japan **51**, 2934 (1982); M. Fukui et al.: J. Phys. Soc. Japan **53**, 1185 (1984); F. DeMartini et al.: Phys. Rev. Lett. **38**, 1223 (1977); M. Fukui et al.: Phys. Rev. B **22**, 1010 (1980)
[55] J. M. Hvam: Sol. State Commun. **12**, 95 (1973) and Phys. Stat. Sol. b **63**, 511 (1974); C. Klingshirn: Phys. Stat. Sol. b **71**, 547 (1975); S. W. Koch et al.: Phys. Stat. Sol. b **89**, 431 (1978)
[56] R. Hauschild et al.: to be published
[57] H. Schrey et al.: Sol. State Commun. **32**, 897 (1979)
[58] J. M. Hvam et al.: Phys. Stat. Sol. b **118**, 179 (1983)
[59] H. J. Ko, Y. F. Chen, T. Yao: Appl. Phys. Lett. **77**, 537 (2000)
[60] K. Hazu et al.: Phys. Rev. B **68**, 332051 (2003); J. Appl. Phys. **95**, 5498 (2004) and J. Appl. Phys. **96**, 1270 (2004)

[61] The absolute values of the exciton and biexciton energies differ e.g. in [4-8, 35-38, 58, 60] and other papers by one or two meV. This is no serious discrepancy. If spectrometers are not frequently calibrated with low pressure spectral lamps, they may easily show deviations of the order of 1 meV. Furthermore, the wave length of spectral lines are sometimes given in vacuum, sometimes in standard atmosphere. The inclusion of the refractive index of air gives in the near UV a shift of photon energy of almost 1 meV.
[62] H. D. Sun et al.: Appl. Phys. Lett. **77**, 4250 (2000) and Phys. Stat. Sol. b **229**, 867 (2002); K. Miyajima et al.: Intern. J. Mod. Physics B **15**, 28 (2001); N.T.T. Lieu et al.: Acta Phys. Polonica A **103**, 67 (2003); Ch. Chia et al.: Appl. Phys. Lett. **82**, 1848 (2003); L. Bergman et al.: MRS Proc. **789**, 251 (2004); B. P. Zhang et al.: Nanotechnol. **15**, 382 (2004)
[63] E. Mollwo, G. Pensl: Z. Physik **228**, 193 (1969); R. Dinges et al.: Phys. Rev. Lett. **25**, 922 (1970); C. Klingshirn: Z. Physik **248**, 433 (1971); W. Kaule: Sol. State Commun. **9**, 17 (1971); G. Pensl: Sol. State Commun. **11**, 1277 (1972); G. Koren: Phys. Rev. B **11**, 802 (1975); G. Wang et al.: Appl. Optics **40**, 5436 (2001); C. Y. Liu et al.: Optics Com. **237**, 65 (2004) and Appl. Phys. B **79**, 83 (2004); U. Neumann et al.: Appl. Phys. Lett. **84**, 170 (2004)
[64] J. Collet, T. Amand: Phys. Rev. B **33**, 4129 (1986); J. Gutowski, A. Hoffmann: Adv. Mat. Opt. and Electr. **3**, 15 (1994)
[65] O. D. Mücke, T. Tritschler, M. Wegener: Opt. Lett. **27**, 2127 (2002)
[66] J. Heidmann and T. Skettrup, Sol. State. Commun. **23**, 27 (1977)
K. Bohnert, G. Schmieder, C. Klingshirn: Phys. Stat. Sol. b **98**, 175 (1980);
[67] Y. Toshine et al.: Phys. Stat. Sol. C **4** 839 (2004); Ü. Özgür et al.: Appl. Phys. Lett. **84**, 3223 (2004); K. Takagi et al., Molecular Crystals and Liquid Crystals Sci. and Technol. **29**, 427 (2002) and J. Takeda ibid. p. 521;
[68] F. H. Nicoll: Appl. Phys. Lett. **9**, 13 (1966); J. R. Packard et al.: J. Appl. Phys. **38**, 5255 (1967); S. I. Wai, S. Namba: Appl. Phys. Lett. **16**, 354 (1970); J. M. Hvam: Phys. Rev. B **4**, 4459 (1971); W. D. Johnston jr.: J. Appl. Phys. **42**, 2731 (1971); C. Klingshirn: Sol. State Commun. **13**, 297 (1973)
[69] W. Wünstel, C. Klingshirn: Optics Commun. **32**, 269 (1980)
[70] K. Bohnert et al.: Z. Physik B **42**, 1 (1981)
[71] M. F. Pereira, K Henneberger: Phys. Stat. Sol. b **202**, 751 (1997) and Phys. Stat. Sol. b **206**, 477 (1998); T. J. Inagaki, M. Aihara: Phys. Rev. B **65**, 205204 (2002) and references therein
[72] G. Tobin et al.: Physica B **340-342**, 245 (2003); D. M. Bagnall et al.: Appl. Phys. Lett. **73**, 1038 (1998); P. Yu: J. Crystal Growth **184/185**, 601 (1998) and references therein
[73] M. H. Huang et al.: Science **292**, 1897 (2001); J. C. Johnson et al.: J. Phys. Chem. B **105**, 11387 (2001) and Nano Letters **4**, 197 (2004); Y. G. Wang et al.: Chem. Phys. Lett. **377**, 329 (2003); Th. Nobis et al.: Phys. Rev. Lett. **93**, 103903 (2004)
[74] H. Cao: Waves Random Media **13**, R 1 (2003); G. Hackenbroich: Physik Journal **3** (7), 25 (2004)

COHERENT SPECTROSCOPY OF SEMICONDUCTOR NANOSTRUCTURES

V. G. LYSSENKO[1], J. ERLAND ØSTERGAARD[2]
[1]*Institute of Microelectronics Technology, RAS, Chernogolovka, Moscow district, 142432 Russia*
[2]*University of Southern Denmark, Campusvej 55, 5230 Odense M, Denmark*
lyss@ipmt-hpm.ac.ru,lyss@pprs1.phy.tu-dreseden.de,

1. Introduction

Coherent optics [1] of semiconductor structures has been developed tremendously over the recent decades, partly due to the very encouraging developments in the ultrashort laser. With such lasers, it has been possible to study ultrafast optical processes in semiconductor low-dimensional or periodic structures. Today, semiconductor structures [2] as e.g. quantum well and superlattices are produced routinely using Molecular Beam Epitaxy, which mark a breakthrough in material science. In such materials, the size of the material in one or more dimensions can be of the order or less than the extension of wavefunctions associated with optical excitations, typically around 100 Å. This quantum confinement effect changes the physics, which must be reconsidered in relation to the optical properties. As in bulk semiconductors, it is the formation of coulomb bound electron-hole pairs; i.e. excitons; which determines the optical properties near the band-gap. The growth techniques are so developed, that semiconductor structures can be engineered with specific optical properties [3]: 1) The (excitonic) band-gap is changed due to the quantum confinement. 2) The linewidth of the exciton resonance, which is usually one of the measures of the quality of the structure, related to the degree of localization of the excitons on rough interfaces or other defects. 3) The oscillator strength of the exciton transmission is increased, leading to e.g. faster radiative recombination, due to an increased overlap of electron-hole wavefunctions.

From an applied point of view, it is important to mention, that strong excitonic features are observed even at room temperatures in quantum confined structures [4]. This observation has given more attention to exciton physics, since excitonic features are only observable at low temperatures in bulk semiconductors due to

B. Di Bartolo and O. Forte (eds.), Advances in Spectroscopy for Lasers and Sensing, 295–331.

thermal ionisation by LO-phonons. In lower-dimensional structures, the binding energy of an exciton is increased due to the quantum confinement, which however does not change the exciton-phonon coupling.

Interactions among optical excitations in semiconductors are stronger, and for that reason also order of magnitude faster, as compared to atomic system. It has therefore only lately been possible to extend the field of semiconductor optics to time scale on which these interactions take place. Another important issue is the large nonlinear optical coefficients, which can be found in such strongly interacting media. In semiconductors, a number of additional sources of optical nonlinearity have been identified due to Coulomb and exchange interactions, which do not play a role in atomic systems. In fact, much research has been devoted over the recent decades to understand the differences in optical properties of semiconductors and atoms. The optical properties of atoms have been studied by nanosecond or microsecond pulsed lasers during the recent decades and are well understood. Therefore, a lot of basic understanding and terminology could be carried over to semiconductor optics from optics of atoms. The advances in semiconductor optics have led to several interesting applications such as quantum well lasers [5] and fast optical gates [6].

1.1. NONLINEAR OPTICS

The invention of the laser in the early sixties [7] started the field of Nonlinear Optics, which relates to effects where light and matter does not interact linearly. The induced polarization P, by an applied optical field E, is traditionally expanded in a power series of E

$$P = \chi_{ij}^{(1)} E_j + \chi_{ijk}^{(2)} E_j E_k + \chi_{ijkl}^{(3)} E_j E_k E_l + \dots \tag{1.1}$$

where the susceptibilities $\chi^{(n)}$ describe the microscopic displacement of charges [8]. The linear susceptibility $\chi^{(1)}$ was calculated by Lorentz in 1909 treating the electron as a harmonic oscillator. In this linear theory, the well-known Lorentzian absorption line shape is found, which is the line shape observed experimentally for a homogeneously broadened transition at excitation densities where the response is linear. A homogeneously broadened transition is characterized by only one transition frequency. At higher excitation densities, corresponding to a laser power of a few kW/cm^2 typically, new optical phenomena are possible due to the nonlinear terms in the expansion in Eq. (1). In the spirit of the Lorentzian model, an anharmonic oscillator model can be used to calculate the higher order susceptibilities, and this forms the basis of classical nonlinear optics [8,9]. Optical nonlinearities are normally classified due to the order of the susceptibility. A variety of phenomena like *n*-harmonics generation, optical rectification, sum and difference generation have been identified through nonlinear mixing of electromagnetic waves [9].

Three requirements must be meet in a nonlinear wave mixing process. 1) The nonlinear susceptibility corresponding to the order of the wave mixing process must be at least non-zero. This can be predicted in condensed matter from the crystal symmetries, which are widely tabulated. A general example is prediction about the second order susceptibility: the $\chi^{(2)}$ is only non-zero in media which lack inversion symmetry. Often, nonlinear optical experiments are performed resonantly or near resonantly due to the resonance-enhancement of the optical nonlinearity. 2) Energy must be conserved in the wave-mixing process. The generated wave must therefore have a frequency $\omega_{mix} = \sum \pm \omega_j$, where the summation is over the applied optical fields taking part in the wave mixing process. Here the plus (minus) sign refers to an absorption (emission) process. 3) Momentum must also be conserved and the generated wave k_{mix} has a wave vector $k_{mix} = \sum \pm k_j$ equal to the sum of wavevectors of the electric fields. One speaks about phase-matched directions for the nonlinear signal in which the nonlinear polarizations and the interacting optical field co-propagate with the same speed over a certain distance. This ensures enough energy transferred between waves to produce a nonlinear signal. The simultaneous requirements 2) and 3) are normally quite severe due to the dispersion relation, $\omega = \omega(k)$.

Nonlinear susceptibility can in general be derived in a quantum-mechanical calculations based on the time-evolution of the density matrix. This is convenient since the density-matrix elements relate directly to physical observables. For the two-level system [10] characterized by a single transition frequency and a transition dipole moment, the time evolution of the density matrix reduces to two dynamical equations for the transition amplitude and the excited-state population, which are coupled through the optical field. The induced polarization is proportional to the transition amplitude and the transition dipole moment, and it oscillates with the transition frequency after an excitation with a short pulse. In these two equations, resembling the Optical Bloch Equations (OBE), phenol-menological decay constants T_1 and T_2 are introduced to describe relaxation of the population and the optically induced dipole, respectively. T_2 relates to the linewidth 2γ in this model by $T_2 = 1/\gamma$. OBE have been useful to describe a number of transient coherent optical phenomena, including the free polarization decay, Rabi flopping, quantum beats and photon echo. Quantum beats [11] refer to the modulation of the polarization from a coherent superposition of more excited levels. The modulation frequency is given by the energy splitting of the excited levels. The photon echo [12] can be observed in the media which are inhomogeneously broadened using two excitation pulses. Inhomogeneous broadening describes a distribution of transition frequencies, which is found when e.g. Doppler broadening or crystal fields are important. In this case, the macroscopic polarization created by pulse # 1, decays in real time inverse proportional to the width of the distribution due to destructive interference. On a time scale shorter than T_2, each individual polarization component has not yet dephased, however, and a second pulse can reverse the time evolution of the polarization. If this pulse #2 is delayed with respect to the pulse #1, the time reversal of the polarization results in a reconstruction of the macroscopic polarization at a time after pulse #2, corresponding to this delay. This reconstructed macroscopic polarization results in the emission of a photon echo.

Four-wave mixing is a widely used technique for studying optical non-linearities in solids. Referring to Eq. (1), two incident laser beams with wavevector $\boldsymbol{k}_1$ and $\boldsymbol{k}_2$ are mixed via the third-order susceptibility. This produce nonlinear signals in the "background free" directions $2\boldsymbol{k}_1 - \boldsymbol{k}_2$ and $2\boldsymbol{k}_2 - \boldsymbol{k}_1$, due to near-phase matching, with amplitudes related to the field strength of both laser beams. If pulsed laser beams are used, information about dynamical processes, governing the linewidth of the transmission, can be obtained as a function of the time delay between pulses in the two beams. In this case one refers to transient four-wave mixing (TFWM). A theoretical model for TFWM based on the OBE is presented in [13]. The nonlinearity in this model is due to saturation of the two-level transition amplitude. The response is asymmetric in time ordering of the two incident pulses #1 and #2 with wavevectors $\boldsymbol{k}_1$ and $\boldsymbol{k}_2$, respectively. If pulse #1 precedes pulse #2 (positive delay τ), a nonlinear signal is only observed in the direction $2\boldsymbol{k}_2 - \boldsymbol{k}_1$ and no nonlinear signal is observed for negative delay $\tau < 0$. The TFWM process can be understood as an interaction of the induced polarization of pulse #1 with the optical field of pulse #2 creating a polarization grating, on which pulse #2 scatter in phase matched direction $2\boldsymbol{k}_2 - \boldsymbol{k}_1$. In real time t, the nonlinear signal appears in the form of a free-polarization decay. If scattered signal is time-integrated by slow detector, the TFWM signal I_{TFWM} decays exponentially as a function of time delay τ between the pulses $I_{TFWM}(\tau) \propto \exp(-a\tau)$ with $a = 2/T_2$ for a homogeneously broadened transition. If the transition is inhomogeneously broadened, the nonlinear signal appears as a photon echo. In the time-integrated version, the nonlinear signal decays exponentially with $a = 4/T_2$ for delays much larger than the inverse inhomogeneous linewidth [13].

Nonlinear TFWM signal has been observed in semiconductors at negative delay τ [14] and attributed to exciton-exciton interactions [15]. These excited state interactions include phase-space filling (PSF), Coulomb and exchange interactions, which are otherwise hidden in the static linewidth. These interactions have been included in the OBE resulting in Semiconductor Bloch Equations [16].

1.2. SEMICONDUCTOR OPTICS

In atoms, sharp transitions and narrow linewidths are typical due to weak interactions and the two-level approximation is therefore usually good. This picture must be modified in semiconductors where the electronic properties are governed by the corresponding electron-hole bands and electronic excitation is possible at photon energies above the bandgap. Coulomb interaction can not be neglected, and the hydrogen-like exciton formation modifies the optical properties near the band edge considerably. Moreover, excitons are mobile crystal excitations [17]. The effective mass approximation is applied in the region around zero wave vector, therefore exciton bands are parabolic. The exciton formation results in a large enhancement of the oscillator strength due to the increase of the electron-hole wavefunctions overlap [18]. This large oscillator strength makes the concept of the exciton-polaritons, as a new quasi-particle with both photon character and exciton character, important in bulk crystals [19]. The possible transformation of the exciton-polaritons to the external photons happens at the sample surface. The radiative lifetime of exciton-polaritons is therefore determined by the reflectivity at the surface, the group velocity of the polaritons and the thickness of the sample [20].

Exciton linewidth is determined by processes in which the exciton is scattered, including exciton-exciton scattering, therefore the exciton lifetime is density-dependent [21]. Wang *et al.* [22] measured the very initial radiative dynamics of resonantly excited excitons in a GaAs quantum well. From the rise time of the time-resolved photo-luminescence, it was concluded, that momentum relaxation plays a dominant role in emission processes of resonantly excited excitons.

Exciton linewidth broadening by exciton-exciton and exciton-carrier collisions has been also measured by TFWM [23,24]. It was found that exciton-carrier scattering is an order magnitude more efficient than exciton-exciton scattering. At higher excitations Coulomb screening becomes dominant mechanism of exciton-exciton interaction and the optical nonlinearity [25], resulting in strong bleaching as observed experimentally in [26].

In TFWM experiments in quantum wells, it becomes evident, that interactions beyond a two-level model are significant. Signal for negative delay with rise-time half of the corresponding decay-time [14] was attributed to local field corrections (LFC). It was treated on an equal footing as the applied optical field in a many-body calculation [15].

A clear interpretation of TFWM experiments in the GaAs quantum wells is often complicated, since the influence from the interface roughness can dominate the response. The inhomogeneous broadening of exciton transitions in the GaAs quantum wells was found [27] by resonant Rayleigh scattering and hole burning experiments. In quantum wells at low excitation densities, it is possible to observe distinct free polarization decay and a stimulated photon echo in time-resolved TFWM experiments [28]. The observation of the photon-echo signal is attributed to localized excitons at the well interface.

The properties of the interface, and the relation to the growth process, have been an active research area since Weisbuch *et al.* [29] proposed, that island-like structures at the interface are responsible for the exciton broadening with decreasing well-width. The interface roughness can be characterized by a bimodal roughness spectrum [30], what has been supported by spectrally resolved TFWM experiments [31]. Dependence of transition energy on fluctuation of the well thickness result in polarization interference leading to a time modulation of the TFWM signal [32]. The modulation period was found to be inversely proportional to the energy splitting of the interfering polarizations. Polarization interference should be distinguished from quantum beats, e.g. between the heavy-hole and light-hole excitons split by the confinement in the quantum wells [33-35].

Two methods have been developed to distinguish between polarization interference from non-interacting two-level systems, and quantum beats from a three-level system. First [36] uses real-time resolution of the TFWM signal, observed as a function of a real-time t and time delay τ, has maxima due to the beat and can be analysed in a fan-chart plotting these maxima as a function of t and τ. Constant trajectories for the signal maxima at (t-2τ) are evidence for polarization interference. Quantum beats give a more complicated behaviour between t and τ, which makes it distinguishable from polarization interference. Influence of inhomogeneous broadening on quantum beats has been discussed in [37]. Another method [38,39] to discriminate between polarization interference and quantum beats is to spectrally resolve the TFWM signal. For polarization interference, a phase shift and a modulation-amplitude minimum is found at the line center of one of the interfering resonances. For quantum beats, no phase-shift and a modulation-amplitude maximum are found at the line-center of one of the beating resonances. The influence of inhomogeneous broadening on spectrally resolved TFWM, including the cases of polarization interference and quantum beat, was discussed in [40].

Excitation induced dephasing (EID) can act as a significant source of optical nonlinearity as measured in TFWM experiments, and was found to be independent of the relative polarization angle between the driving linearly polarized optical fields. This demonstrated the spin-independent nature of the EID nonlinearity. The EID and local field affects are the origin of the phase shifts of the quantum beats in a three-level system [41] at high excitation level.

In spectrally resolved TFWM, strong nonlinear quantum beats due to exciton-biexciton transition [42] allows to determine biexciton binding energy [43]. The biexciton binding energy increases in quantum confined structures, and several groups have confirmed the existence and the increased binding energy of quantum well biexcitons in different experiments including quantum beats [42, 43,44], TFWM [45-47] and pump-probe [48] investigations [49].

2. Spectrally Resolved Transient Four-Wave Mixing

Linear optical spectroscopy has contributed immensely to the development of quantum mechanics in the beginning of this century. One observation was that the emission from free atoms occurs at discrete wavelengths. From spectroscopic data like the Balmer series from hydrogen spectrum, Bohr made his famous postulates. With light intensities available at those times, it was a good assumption that the optical susceptibility was independent of the electric field strength. However, with the invention of the laser in the early sixties, a variety of new spectroscopic techniques were made possible, since the optical parameters become dependent on the electric field strength. With such techniques, it is possible to extract specific information, which would be hidden in a broad inhomogeneous line with linear spectroscopy. Inhomogeneous broadening can be due to Doppler broadening in an atomic vapour or random electric fields in a crystal.

Today, narrow band cw lasers are used to do high-resolution spectroscopy with resolution down to only a few kHz. This indicates a frequency stability of a factor of 10^{-12} compared to the optical frequency. In the time-domain, ultrafast lasers with pulse durations of a few femtoseconds, corresponding to a few optical periods, are used to probe the fast dynamical processes in solids. One of the important nonlinear techniques is four-wave mixing, which is discussed below in Chapter 2 and 3.

On the theoretical side, much work has been done using the density matrix formalism. The advantage of the density matrix is that physical observables like occupation densities and transition amplitudes are directly proportional to the density matrix elements. This proportionality makes a close connection between experiment and theory. One other advantage is that dissipation can conveniently be incorporated in the density matrix description by adding relaxation terms to the equation of motion for the density matrix. The simplest system to consider is the two-level system. This is reviewed in the next section, where it is used to describe the four-wave mixing process perturbatively. This has two purposes: 1) it is a simple model from which it is possible to develop intuition since it contains the basic important physics and 2) it is a good basis for extension, which include more complicated interactions.

2.1. TWO-LEVEL MODEL

The simplest way to describe near-resonant interaction between light and matter is to consider a classical field $E(t) = E_0 e^{i\omega t} + c.c.$ having a frequency ω correspon-ding to the energy difference between two isolated quantum levels with energies $E_a = \hbar\omega_a$ and $E_b = \hbar\omega_b$, $\omega \approx \omega_{ba} = \omega_b - \omega_a$. The light-matter interaction is described in the dipole approximation, which neglects variation in the field over atomic distances, by interaction energy V

$$V = -\mu E(t), \tag{2.1}$$

where $\mu = -er$ is the dipole moment. In Eq. (2.1) only the external field, interaction with the material system is included, and atom-atom interaction is neglected. In order to find the dipole moment μ for the transition $a \rightarrow b$, the atomic wave functions for the two states are assumed to have definite parity resulting in $\mu_{aa} = \mu_{bb} = 0$ and

$$V_{ba} = V_{ab}^{*} = -\boldsymbol{\mu}_{ba} \cdot \boldsymbol{E}(t). \tag{2.2}$$

It turns out, that it is much more convenient to introduce a new operator

$$\rho = \sum_{\Psi} P_{\Psi} |\Psi><\Psi|, \tag{2.3}$$

yielding the density matrix formulation of the quantum mechanics. In Eq (2.3), P_Ψ represents the classical, rather than the quantum mechanical probability, that the system is in the state Ψ. The diagonal elements ρ_{nn} of density matrix $\boldsymbol{\rho}$ give the probability, that the system is in eigenstate n. The off-diagonal elements $\rho_{mn} = \rho^*_{mn}$ give the coherence between the states n and m. ρ_{mn} is nonzero, if the system is in coherent superposition of the states n and m. The time evolution of the density matrix ρ is given by:

$$\dot{\boldsymbol{\rho}} = -\frac{i}{\hbar}[H, \boldsymbol{\rho}], \tag{2.4}$$

where $H = H_0 + V(t)$ is the Hamiltonian for the system. H_0 is the Hamiltonian for the unperturbed system with $H_{0,nm} = E_n \delta_{nm}$. In Eq. (2.4,) P_Ψ was assumed to be time independent. In systems, where P_Ψ is time dependent and relaxation phenomena play a role, a decay rate γ_{nm} is phenomenologically introduced in Eq. (2.4)

$$\dot{\rho}_{nm} = -\frac{i}{\hbar}[H, \rho] - \gamma_{nm}\left(\boldsymbol{\rho_{nm}} - \boldsymbol{\rho_{nm}^{eq}}\right), \tag{2.5}$$

where ρ_{nn}^{eq} describes the (thermal) equilibrium value. An important result for the density matrix is the calculation of expectation values for observable quantities A

$$<A> = tr(\boldsymbol{\rho}A). \tag{2.6}$$

For the simple two-level system, the expectation value for the dipole operator is

$$<\mu> = \rho_{ab}\mu_{ba} + \rho_{ba}\mu_{ab}. \tag{2.7}$$

To obtain the general two level equations i.e. the equation of motion for the density matrix, Eqs. (2.2) and (2.3) are inserted in Eq. (2.4):

$$\begin{aligned}
\rho'_{ba} &= -i\omega_{ba}\rho_{ba} + \frac{i}{\hbar}V_{ba}\left(\boldsymbol{\rho_{bb}} - \boldsymbol{\rho_{aa}}\right), \\
\rho'_{bb} &= -\frac{i}{\hbar}\left(V_{ba}\boldsymbol{\rho_{ab}} - V_{ab}\boldsymbol{\rho_{ba}}\right), \\
\rho'_{aa} &= -\frac{i}{\hbar}\left(V_{ab}\boldsymbol{\rho_{ba}} - V_{ba}\boldsymbol{\rho_{ab}}\right).
\end{aligned} \tag{2.8}$$

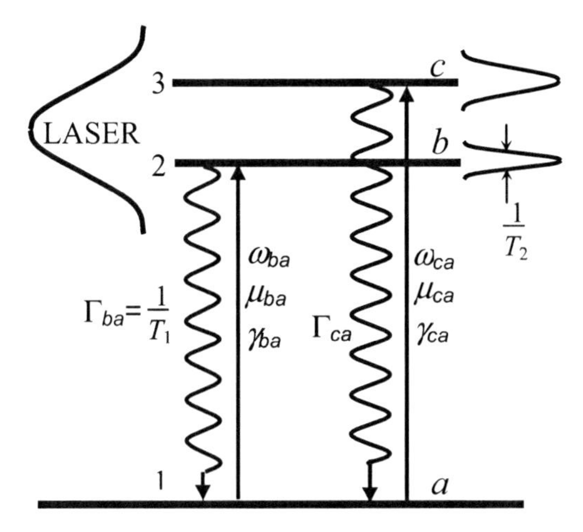

Figure 1. Transitions in two- or three-level systems.

It is evident from Eqs. (2.8), that the probability of occupation, for the closed two level system, is conserved in $\rho_{aa} + \rho_{bb} = 1$.

To include relaxation processes in the system, the situation in Fig. 1 is considered. It is assumed, that an excitation of level b decays to level a at a rate $1/T_1$ typically due to spontaneous emission. It is also assumed, that the atomic dipole moment is dephased in the coherence time T_2, leading to a transition linewidth $2\gamma_{ba} = 2/T_2$, for example, due to atom-atom collisions. As in Eq. (2.5), these decay terms are introduced phenomenologically in the two-level Eqs. (2.8)

$$
\begin{aligned}
\dot{\rho}_{ba} &= -\left(i\omega_{ba} + \frac{1}{T_2}\right)\rho_{ba} + \frac{i}{\hbar}V_{ba}\left(\rho_{bb} - \rho_{aa}\right), \\
\dot{\rho}_{bb} &= -\frac{\rho_{bb}}{T_1} - \frac{i}{\hbar}\left(V_{ba}\rho_{ab} - V_{ab}\rho_{ba}\right), \\
\dot{\rho}_{aa} &= \frac{\rho_{bb}}{T_1} - \frac{i}{\hbar}\left(V_{ab}\rho_{ba} - V_{ba}\rho_{ab}\right).
\end{aligned}
\tag{2.9}
$$

Due to the conservation of the occupation probability, the equation for ρ_{aa} and ρ_{bb} can be expressed as

$$
\dot{\rho}_D = -\frac{\rho_D - \rho_D^{eq}}{T_1} - \frac{2i}{\hbar}\left(V_{ba}\rho_{ab} - V_{ab}\rho_{ba}\right), \tag{2.10}
$$

where $\rho_D = \rho_{bb} - \rho_{aa}$ is the population difference with an equilibrium value $\rho_D^{eq} = (\rho_{bb} - \rho_{aa})^{eq}$. The equation for ρ_{ba} in Eqs. (2.9) and the equation for the population difference ρ_D in Eq. (2.10) constitute the two coupled equations for a closed two level system, known as the Optical Bloch Equations (OBE).

To investigate the relaxation terms introduced in Eqs. (2.9), the two level equations are solved when the applied field is switched off at $t = 0$ giving $V_{ba} = 0$. The solution to Eq. (2.10) is:

$$\rho_D(t) = \rho_D^{eq} + [\rho_D(0) - \rho_D^{eq}]e^{-t/T_1}. \qquad (2.11)$$

It is evident, that the population difference ρ_D decays to equilibrium in a characteristic time of T_1, which is therefore referred as the population relaxation time. The solution for the off-diagonal elements ρ_{ba} under the same conditions, $V_{ba} = 0$, is:

$$\rho_{ba}(t) = \rho_{ba}(0)e^{-(i\omega_{ba}+1/T_2)t}. \qquad (2.12)$$

With Eq. (2.12) the expression for expectation value of the dipole moment Eq. (2.7) becomes:

$$\langle \mu(t) \rangle = \{\mu_{ab}\rho_{ba}(0)e^{-i\omega_{ba}t} + c.c\}e^{-t/T_2}. \qquad (2.13)$$

This shows, that the induced dipole moment for an undriven atom oscillates at a frequency of ω_{ba} and decays in a characteristic time T_2 known as the dipole dephasing time. The polarization $\boldsymbol{P}$ of a media is found by adding contributions, described by Eq. (2.13), from all atoms. To relate T_1 and T_2 it is noticed, that T_1 is the decay time related to the intensity, whereas T_2 is the decay time of the macroscopic polarization i.e. related to the field itself. Thus, a factor of 2 is introduced between T_1 and T_2

$$\frac{1}{T_2} = \frac{1}{2T_1} + \gamma_c \equiv \gamma, \qquad (2.14)$$

where a dephasing rate γ_c due to atom-atom collisions is included.

In the following, the transient behaviour of four-wave mixing is addressed in a two-beam configuration. The two incident fields $E_1(t)\exp[i(\boldsymbol{k}_1 \cdot \boldsymbol{r} - \omega t)] + c.c.$ and $E_2(t)\exp[i(\boldsymbol{k}_2 \cdot \boldsymbol{r} - \omega t)] + c.c.$ with degenerate frequency ω, have wave vectors $\boldsymbol{k}_1$ and $\boldsymbol{k}_2$, where E_1 and E_2 describe the amplitude of the pulses in real-time with duration τ_p. The pulses can be shifted with a delay τ between the two. Utilising the rotating wave approximation, the third order off-diagonal density matrix element with wavevector $2\boldsymbol{k}_2 - \boldsymbol{k}_1$ from the system in Eqs. (2.9) can be calculated in a straightforward perturbative approach as done in the pioneering work by Yajima and Taira [13]. The initial population difference, $\rho_D^{(0)} = \rho_D^{eq}$, is inserted in the first equation of Eqs. (2.9). This can give a first order term $\rho_{ba}^{(1)}$ with wavevector $\boldsymbol{k}_1$

$$\hat{\rho}_{ba}^{(1)} = \rho_{ba}^{(1)}e^{it} = \frac{i\mu_{ba}}{\hbar}e^{-[i(\omega_{ba}-\omega)+1/T_2]t}e^{i\boldsymbol{k}_1\cdot\boldsymbol{r}}\int_{-\infty}^{t} dt'''E_1(t''')e^{[i(\omega_{ba}-\omega)t+1/T_2]t'''}. \qquad (2.15)$$

The complex conjugate of Eq. (2.15) can be used to calculate the population difference $\rho_D^{(2)}$ with wavevector $\boldsymbol{k}_2$-$\boldsymbol{k}_1$ with the field E_2. This gives

$$\rho_D^{(2)}(t) = -\frac{2i\mu_{ba}}{\eta} e^{-t/T_1} e^{i\boldsymbol{k}_2\cdot\boldsymbol{r}} \int_{-\infty}^{t} dt'' E_2(t'')\hat{\rho}_{ba}^{(1)*}(t'') e^{t''/T_1} + c.c.. \quad (2.16)$$

This second order term, $\rho_D^{(2)}$, is a grating term due to wavevector $\boldsymbol{k}_2 - \boldsymbol{k}_1$, which modulates the population in real space. The grating constant is $\Lambda = 2\pi/|\boldsymbol{k}_2-\boldsymbol{k}_1| = \lambda/2\sin(\theta/2)$ with an angle θ between the incident beams. This population grating can scatter the field E_2 in the direction $\boldsymbol{k}_3 = 2\boldsymbol{k}_2 - \boldsymbol{k}_1$ via the third order polarization. This is calculated with Eq. (2.16) and the first equation of Eqs. (2.9) with the general result

$$\hat{\rho}_{ba}^{(3)}(t) = -2i\rho_D^{(0)}\left(\frac{\mu_{ba}}{\text{h}}\right)^3 e^{i[\boldsymbol{k}_3\cdot\boldsymbol{r}-(\omega_{ba}-\omega)t]-t/T_2} \int_{-\infty}^{t}\int_{-\infty}^{t'}\int_{-\infty}^{t''} dt'''dt''dt'$$

$$= e^{i(\omega_{ba}-\omega)t'''} e^{(1/T_2-1/T_1)(t'-t'')+t'''/T_2} \begin{Bmatrix} E_2(t')E_2(t'')E_1(t''')e^{i(\omega_{ba}-\omega)(t'-t'')} + \\ E_2(t')E_1(t'')E_2(t''')e^{i(\omega_{ba}-\omega)(t'''-t')} \end{Bmatrix}. \quad (2.17)$$

The last term in Eq. (2.17) is the term from pulse-overlap in the sample. This is calculated in a similar way. The first-order polarization in Eq. (2.15) is induced by E_2, which together with the field E_1, create a polarization grating as in Eq. (2.16). The same field E_2 can then scatter on this grating via the third-order polarization during the pulse overlap. Both terms in (2.17) are nonzero, when the pulses overlap, whereas only the first term is nonzero, when the delay between the pulses is larger then pulse-duration τ_p. Since the pulses used in the

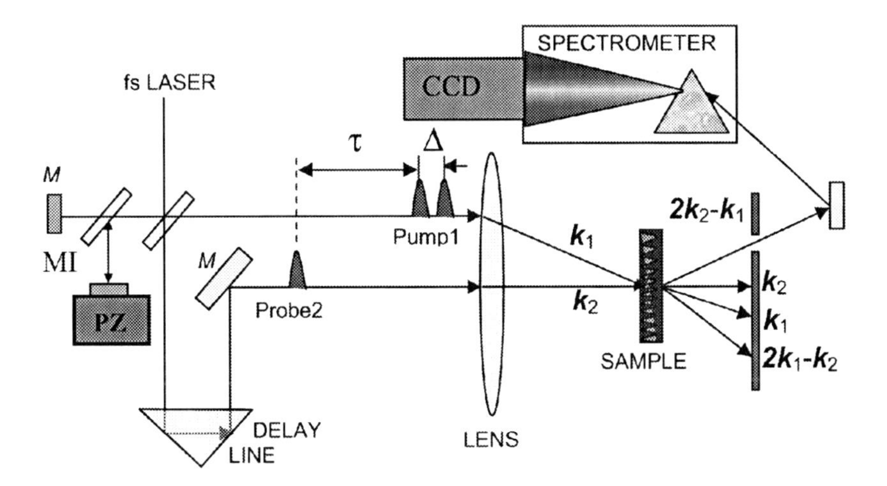

Figure 2. Sketch of spectrally resolved TFWM. Spectrally broad pulses of fs LASER split by 50% beam-splitter BS into pulse #1, propagating with wavevector k_1 and exciting SAMPLE in time $t = 0$. Second pulse, propagating in k_2 direction, delayed by DELAY LINE, excite sample at time τ, creating in sample density grating with wavevector k_2-k_1 and self-defracts into direction $2\boldsymbol{k}_2-\boldsymbol{k}_1$. Spectrum of the nonlinear signal I_{TFWM} Eq. (2.21) is measured by SPECTROMETER as a function of the delay τ.

experiments usually are much shorter than the dephasing time ($\tau_p << T_2$), contributions to the TFWM signal resulting from the pulse-overlap term are neglected in the present calculations. The macroscopic polarization giving rise to the outgoing signal is given by

$$P^{(3)}(\boldsymbol{k}_3,t) = \hat{P}^{(3)}e^{-i\omega t} + c.c., \qquad \hat{P}^{(3)} = N\int_0^\infty \mu_{ba}\hat{\rho}_{ba}^{(3)}(\boldsymbol{k}_3,t,\omega_{ba})g(\omega_{ba})d\omega_{ba}, \quad (2.18)$$

where the distribution of resonance frequencies is described by g. In this simple two-level model, a nonlinear signal is only observed when pulse #1 precedes pulse #2, which we refer to as positive delay.

The spectrally resolved TFWM signal (detected with a spectrometer) is $I_{TFWM}(\tau,\omega) \propto |P^{(3)}(\tau,\omega)|^2$, where $P^{(3)}(\tau,\omega)$ is the Fourier transform with respect to t of $P^{(3)}(t,\tau)$. This form the basis of spectrally resolved TFWM and is an important point, since the Fourier-transform, i.e. the nonlinear spectrum contains information about real-time behaviour of the nonlinear signal. This is illustrated in Fig. 2. The nonlinear polarization emits an electric field, which must satisfy Maxwell's wave equation. If the polarization amplitude varies negligibly in an optical period, then the emitted electric field is proportional to the polarization, as indicated in Fig. 2. As will be evident in the following, this form of spectral resolution makes it possible to use TFWM for detailed spectroscopic studies. It is important to notice that in this particular way, the timing in the experiment is performed before the sample, whereas the spectrum is obtained after the sample. This has important implications in relation to nonlinear beat spectroscopy, which is the subject of Chapter 3.

We make use of the complex frequencies $\Omega_{ij} = \omega_{ij} - i\gamma_{ij}$, where γ_{ij} is damping and $i, j = 1,2,3\ldots$ in general for multi-level systems. In order to develop some intuition, we first consider the response from one homogeneously broadened resonance. With $t''' = 0$ and $t'' = t' = \tau$ in Eq. (2.17), the result for the third-order polarization, in the limit of short δ-pulses, is

$$P_{\text{hom}}^{(3)}(t,\tau) \propto N\mu_{21}^4 e^{-i\Omega_{21}(t-\tau)}e^{i\Omega_{21}^*\tau}\theta(t-\tau)\theta(\tau), \quad (2.19)$$

where t is the time after the arrival of the first pulse (at $t = 0$) and the second pulse arrives at the delay $t = \tau$. The spectral overlap of the resonance with the incident fields as well as various integrations involving the electric fields of the incident pulses have been suppressed. N denotes the number of the two-level "atoms" and $\theta(t)$ is Heaviside step function. Hence, the real-time evolution is a simple exponential decay with rate γ.

Fourier transformation of (2.19) gives

$$P_{\text{hom}}^{(3)}(\omega,\tau) \propto N\frac{2\mu_{21}^4 e^{i\Omega_{21}^*\tau}}{\Omega_{21}-\omega}\theta(\tau). \quad (2.20)$$

It can immediately be seen that the correlation trace decays as

$$I_{TFWM}(\tau) \propto \Theta(\tau)\exp(-2\tau/T_2)/[(\omega-\omega_{21})^2+\gamma_{21}^2] \quad (2.21)$$

and that the spectrum of I_{TFWM} is a Lorentzian. It is interesting to notice that the decay of the signal in real time and the decay of the correlation trace in the time-delay domain are similar. This correspondence is present only in this simple model, and e.g. interference effects or exciton-exciton interactions changes the behavior of the signal in real-time versus time-delay.

2.2. INFLUENCE OF INHOMOGENEOUS BROADENING

The FWM technique in the time-delay domain has been applied in various bulk semiconductors, such as CdSe [50,51], GaAs [52,53] or quantum wells GaAs/AlGaAs [54] and InGaAs [55]. These studies have been demonstrated that inhomogeneous broadening is present in many solid state materials. The inhomogeneous broadening can be due to environmental resonance shifts caused by crystal fields or strain as well as interface-disorder [56] in the quantum well structures. Compared to the homogeneous line width, reflecting the phase distorting (phonon or carrier-carrier) scattering processes, the inhomogeneous line width is often much broader. Within the inhomogeneous line, the homogeneous line width can be determined by spectral hole burning experiments [9], or photon echo experiments [57,58] in the time-domain. In case the inhomogeneous line width is comparable

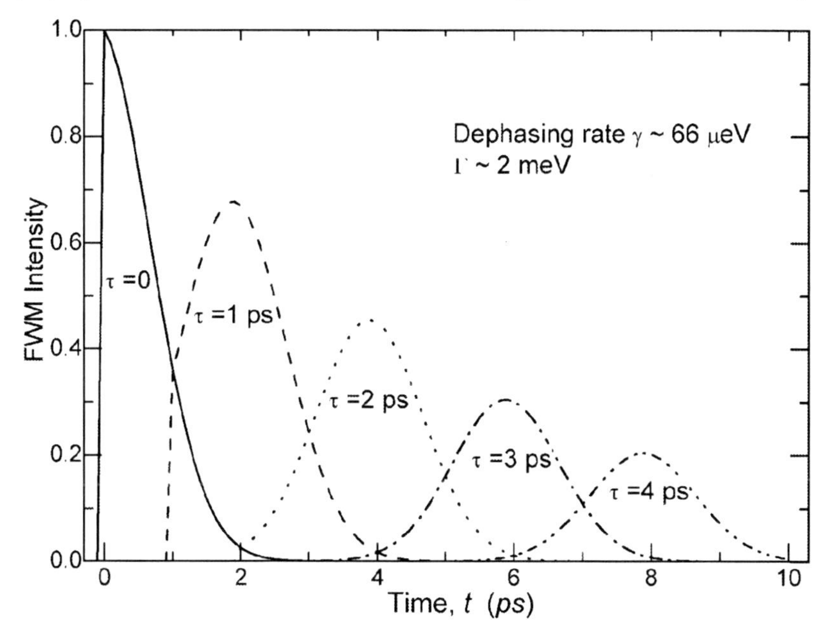

Figure 3. The normalized absolute square of Eq. (22) vs. real time t fir different delays $\tau = 0$ (solid); 1ps (dashed); 2ps (dotted); 3ps (dash-dotted) and 4pd (dash-dotted-dotted). For this calculation $\gamma \approx 66$ μeV, $\Gamma \approx 2$ meV.

or smaller than the homogeneous line width, it is normally difficult to get accurate information on the two broadening mechanisms.

If a resonance is homogeneously broadened, the result in Eq. (2.19) must be integrated over e.g. a Gaussian distribution $g_N(\omega_{21})$ of transition frequencies ω_{21} with central frequency ω_{21}^c and FWHM $=\Gamma\sqrt{8\ln 2}$:

$$g_N(\omega_{21}) = \frac{1}{\Gamma\sqrt{\pi}}\exp\left[-\frac{(\omega_{21}-\omega_{21}^c)}{2\Gamma^2}\right], \tag{2.21}$$

The polarization evolution in real time t is now found from

$$P_{inh}^{(3)}(\tau,t) \propto \int g_N(\omega_{21})P_{\mathrm{hom}}^{(3)}(\tau,t,\omega_{21})d\omega_{21} = P_{\mathrm{hom}}^{(3)}(\tau,t,\omega_{21}^c)\cdot\exp\left[-\frac{\Gamma^2(t-2\tau)}{2}\right]. \tag{2.22}$$

The result obtained in Eq. (2.22) shows, that the photon echo emitted at $t = 2\tau$ is the real time Fourier transform of the inhomogeneous line. The θ-function clips the Gaussian in real time for delays smaller than the inverse line width $\tau < 1/\Gamma$.

In Fig. 3 is shown the normalized absolute value of Eq. (2.22) as a function of real time t for different delays. It illustrates, that the photon echo comes out without being clipped only after a certain delay as explained above. For delays shorter than Γ^{-1} the response will have a broader frequency content due to the θ- function, which, in the experiments, has a rise time given by the duration of the laser pulse. Fourier transforming Eq. (2.22) gives

$$P_{inh}^{(3)}(\tau,\omega) \propto$$

$$\frac{iN\mu_{21}^4\sqrt{\pi}}{\sqrt{2}\Gamma}e^{2(i\omega-\gamma_{21})\tau}\exp\left[-\frac{(\omega_{21}^c-i\gamma_{21}-\omega)^2}{2\Gamma^2}\right]erfc\left[i\frac{\omega_{21}^c-i\gamma_{21}-\omega}{\sqrt{2}\Gamma}-\frac{\Gamma\tau}{\sqrt{2}}\right]. \tag{2.23}$$

In this case $I_{TFWM}(\tau) \propto \exp(-4\tau/T_2)$ for about $\tau > 5/\Gamma$ [59]. The resonance enhancement is now a Gaussian, which also reflect the spectrum of I_{TFWM} at sufficiently large delay $\tau \gg 1/\Gamma$, when the *erfc* becomes independent of the delay. This illustrated in Fig. 4, where the absolute square of Eq. (2.23) is shown versus delay and for different detuning $\delta = (\omega - \omega_{21}^c)/\Gamma$.

In Fig. 4(c), for a small inhomogeneous broadening $\Gamma = 3\gamma$, there is a small dependence on detuning, reflecting that the spectrum becomes narrower for increasing delay. This is due to the discontinuity in the real time behaviour for delays shorter than the inverse inhomogeneous broadening, as shown in Fig. 3. This is expressed in Eq. (2.23) in the *erfc*-function. For a larger inhomogeneous broadening, as in Fig. 4(d) with $\Gamma = 10\gamma$, the dependence on detuning is much stronger and, for a small detuning, the correlation traces even show a distinct maximum. This behaviour reflect how inhomogeneous broadening and resulting photon echo is revealed in spectrally resolved four-wave mixing, allowing a simultaneously determination of the homogeneous linewidth and the inhomogeneous

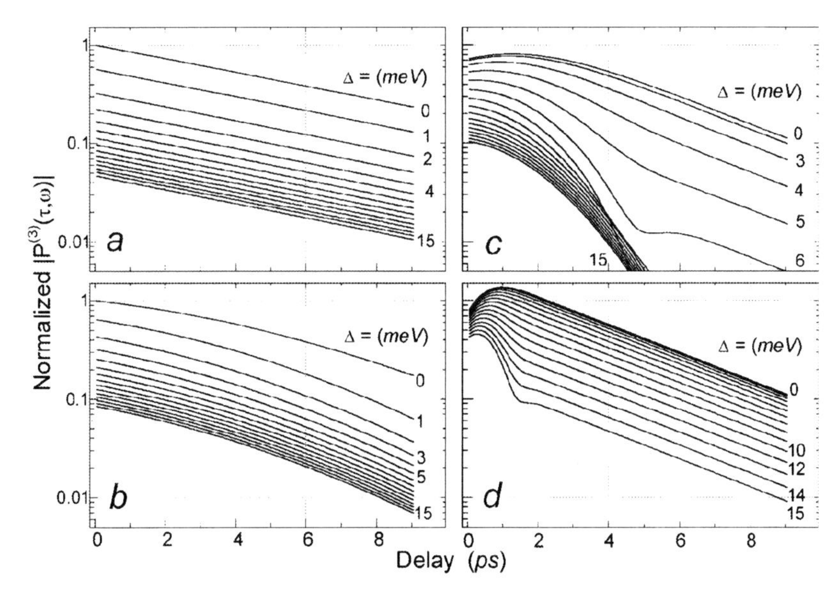

Figure 4. The normalized absolute value of Eq. (2.23) showing the calculated decay of the TFWM signal versus delay τ for $\gamma = 0.67$ meV, different detunings $\Delta = |\omega-\omega_{res}|$ from 0 to 15 meV and different inhomogeneous broadenings Γ: (a) < 0.1 meV, (b) 0.67meV; (c) 2 meV; (d) 6.7 meV.

linewidth. The expression in Eq. (2.23) also contains a phase factor, where the magnitude of the phase shift through the resonance is determined by ratio $\beta = \Gamma/\gamma$. This is the equivalent of the phase-shift introduced by the resonance denominator in the case of homogeneous broadening in Eq. (2.20).

3. Three-Level Systems

3.1. QUANTUM BEATS VERSUS POLARIZATION INTERFERENCE

Nonlinear beat spectroscopy can be used to infer level diagrams, which involve several excited transitions. The beat can, however, be of different nature depending on the microscopic structure of the levels. Quantum interferences are seen if the beating levels are coupled as e.g. a three level system. As such, it is a beat term in the wavefunction, which is detected. Polarization interference, observed as the coherent superposition of the responses from independent transitions, i.e. non-interacting two-level systems, is also a possibility. Spectrally resolved TFWM can be used to distinguish between the two cases.

Consider a thin sample illuminated by two pulsed beams propagating nearly normal to the sample. The atomic system in the sample is characterized by two distinct dipole-allowed optical transitions with nearly the same transition frequency. This can be realised in the following model systems:

Two independent two-level systems with nearly equal transition frequencies. This shall be called a II-system as the double I symbolises the structure of the transition.

A V-type three-level system with two allowed transitions between a common ground state and two closely lying excited states.

A Λ-type three-level system with two allowed transitions between a higher state and two closely lying lower states.

A cascade system with almost equal frequencies of lower and upper transition. Since the letter E symbolises the level diagram, this three-level system is called E-type. The Optical Bloch Equations for a general three-level system are analogous to those in Eqs. (2.8). The 1-2 an2 2-3 transitions are dipole allowed with transition dipole moment μ_{21} and μ_{32}. The calculation of the third-order polarization is done fore E-type three-level system, and the result will then be modified for V-type and Λ-type three-level systems. With the complex frequencies $\Omega_{ij} = \omega_{ij} - i\gamma_{ij}$ and neglecting relaxation of the occupation densities towards equilibrium ($T_1 \rightarrow \infty$), the equation of motion can be written [8]

$$\dot{\rho}_{21} + i\Omega_{21}\rho_{21} = \frac{i}{\hbar}\left[\mu_{21}\left(\rho_{11} - \rho_{22}\right) - \mu_{32}\rho_{31}\right] E \tag{3.1}$$

$$\dot{\rho}_{32} + \Omega_{32}\rho 32 = \frac{i}{\hbar}\left[\mu_{32}\left(\rho_{22} - \rho_{33}\right) - \mu_{21}\rho_{31}\right] \tag{3.2}$$

$$\dot{\rho}_{31} + i\Omega_{31}\rho_{31} = \frac{i}{\hbar}\left[\mu_{32}\rho_{21} - \mu_{21}\rho_{32}\right] \cdot E, \tag{3.3}$$

$$\dot{\rho}_{11} = \frac{i}{\hbar}\mu_{21}\left[\rho_{21} - \rho_{21}^{*}\right] \cdot E, \tag{3.4}$$

$$\dot{\rho}_{33} = \frac{i}{\hbar}\mu_{32}\left[\rho_{32} - \rho_{32}^{*}\right] \cdot E, \tag{3.5}$$

with $N = \rho_{11} + \rho_{22} + \rho_{33}$ as the common density of three-level atoms giving the last equation. Above equations are solved using Green-function techniques similar to those employed by Yajima and Taira [13]. If level 2 is initially occupied $\rho_{22}(t = 0) = N$, then $\rho_{11}(t = 0) = \rho_{33}(t = 0) = 0$, and we assume no initial coherence in the system:

$$\rho_{21}(t = 0) = \rho_{32}(t = 0) = \rho_{31}(t = 0) = 0.$$

The first – order transition amplitude $\rho_{21}^{(1)}$ and $\rho_{32}^{(1)}$ are calculated using Eqs. (3.1) and (3.2) with density matrix at their initial values. Next the electric field and the first-order matrices produce second-order quantities $\rho_{11}^{(2)}, \rho_{22}^{(2)}, \rho_{33}^{(2)}$ and $\rho_{31}^{(2)}$ with

Eqs. (3.3-3.5). Finally these second-order quantities and the electric field create the third-order transition amplitudes $\rho_{21}^{(3)}$ and $\rho_{32}^{(3)}$ via Eqs. (3.1) and (3.2). The final expression for third-order polarization can be written as

$$P^{(3)}(t) = \mu_{21}\rho_{21}^{(3)} + \mu_{32}\rho_{32}^{(3)} = \int_{-\infty}^{t} dt' \int_{-\infty}^{t'} dt'' \int_{-\infty}^{t''} dt''' E(t')E(t'') \cdot \{A(t,t',t'',t''')E^*(t''') + B(t,t',t'',t''')E(t''')\}, \tag{3.6}$$

where A and B contain appropriate Greens functions. When the electric field is considered as a sum $E(t) = E_1(t) + E_2(t)$, then the above expression multiplies 16-fold. However, we keep only surviving terms due to wave vector conservation, we neglect terms due to pulse overlap and retain only resonant terms. Then we get the following selection: In V- and Λ–type systems the only relevant term is $E_2(t')E_2(t'')E_1^*(t''')$ whereas in E-type systems an additional term is relevant via the response from $\rho_{31}^{(2)}$: $E_1^*(t')E_2(t'')E_2(t''')$. Therefore result for A is

$$\begin{aligned} A &= iN\mu_{21}^2 e^{-i\Omega_{21}(t-t')}\left[2\mu_{21}^2 e^{i\Omega_{21}^*(t''-t''')} + \mu_{32}^2 e^{i\Omega_{32}^*(t''-t''')}\right] \\ &\quad - iN\mu_{32}^2 e^{-i\Omega_{32}(t-t')}\left[2\mu_{32}^2 e^{i\Omega_{32}^*(t''-t''')} + \mu_{21}^2 e^{i\Omega_{21}^*(t''-t''')}\right] \end{aligned} \tag{3.7}$$

and for B it is

$$B = iN\mu_{21}^2\mu_{32}^2 e^{-i\Omega_{31}(t'-t'')} \cdot \left[e^{-i\Omega_{21}(t-t')} + e^{-i\Omega_{32}(t-t')}\right] \times \left[e^{-i\Omega_{21}^*(t''-t''')} + e^{-i\Omega_{32}^*(t''-t''')}\right]. \tag{3.8}$$

In the δ-pulse limit, we get

$$P^{(3)}(t,\tau) \propto A(t,\tau,\tau,0)\Theta(t-\tau)\Theta(\tau) + B(t,0,\tau,\tau)\Theta(t)\Theta(-\tau). \tag{3.9}$$

Function A (B) describes the response for positive (negative) delay. Slight modifications are necessary for treating V- and Λ-type systems (also with level 2 initially occupied): There is no signal for negative delay ($B = 0$), the sign before γ_{21} and γ_{32} should be reversed, and minus in Eq. (3.7) should be changed to a plus sign.

For V-type systems (in which B is zero) we get

$$\begin{aligned} P_{QB}^{(3)}(t,\tau) &\propto \Theta(\tau)\Theta(t-\tau) \\ &\times\left\{[2\mu_{21}^4 e^{-i\Omega_{21}(t-2\tau)} + 2\mu_{32}^4 e^{-\Omega_{32}(t-2\tau)}] + \mu_{21}^2\mu_{32}^2[e^{-i\Omega_{21}(t-\tau)}e^{\Omega_{32}\tau} + e^{-i\Omega_{32}(t-\tau)}e^{i\Omega_{21}\tau}]\right\} \end{aligned} \tag{3.10}$$

If the TFWM-signal is spectrally resolved, it is relevant to calculate the Fourier transform of $P^{(3)}(t,\tau)$ with the above expressions for A and B. For positive delay ($\tau > 0$), we get

$$P^{(3)}(\omega,\tau) \propto \frac{2\mu_{21}^4 e^{i\Omega_{21}^*\tau} + \mu_{21}^2\mu_{32}^2 e^{i\Omega_{32}^*\tau}}{\Omega_{21}-\omega} - \frac{2\mu_{32}^4 e^{i\Omega_{32}^*\tau} + \mu_{21}^2\mu_{32}^2 e^{i\Omega_{21}^*\tau}}{\Omega_{32}-\omega}. \tag{3.11}$$

Assuming equal dipole matrix elements of the transitions and distinctly different damping of the two levels, e.g. $\mu_{21} \approx \mu_{32}$, $\gamma_{31} >> \gamma_{21}$, a simple expression for TFWM signal near one resonance is

$$I_{TFWM} \propto \frac{e^{-2\gamma_{21}\tau}\{1+e^{-\gamma_{31}\tau}\cos([\omega_{21}-\omega_{31}]\tau)\}}{(\omega_{21}-\omega)^2+\gamma_{12}^2}, \tag{3.12}$$

showing that the average signal and the beat modulation decay with the slowest and fastest damping rates.

The signal for negative delay $\tau < 0$ is given by

$$P^{(3)}(\omega,\tau) \propto 2\mu_{21}^2\mu_{32}^2\Theta(-\tau)\left[\frac{1}{\Omega_{21}-\omega}+\frac{1}{\Omega_{32}-\omega}\right]e^{-i\Omega_{31}\tau}. \tag{3.13}$$

Results for II system can easily be derived from the above calculations. Considering Eq. (14) at $\mu_{32}=0$ we get the contribution from a single two-level system. Two such transitions give the third-order polarization for II-type systems:

$$P_{PI}^{(3)}(\omega,\tau) \propto \Theta(\tau)\left[\frac{2\mu_{21}^4 e^{i\Omega_{21}^*\tau}}{\Omega_{21}-\omega}+\frac{2\mu_{43}^4 e^{i\Omega_{43}^*\tau}}{\Omega_{43}-\omega}\right]. \tag{3.14}$$

It is appropriate to express any of the above results Eqs. (14), (16) or (17) as

$$\boldsymbol{I}^{(3)} = \boldsymbol{I}_{ave}[1+\boldsymbol{I}_m\cos(\Delta\omega\cdot\tau+\phi)], \tag{3.15}$$

where $\Delta\omega$ is the difference between the two transition frequencies. $\boldsymbol{I}_{ave}$, $\boldsymbol{I}_m$ and ϕ can be expressed as analytical functions of ω and/or τ. Overview of the above-considered characteristics is:

Quantum beats are characterized by a small change of the phase ϕ and the modulation amplitude $\boldsymbol{I}_m$ when tuning through one of the resonances. The reason for this is that the beating between the two terms with different frequency denominator is unimportant. Between or outside the resonances ϕ changes from $-\pi/2$ to $+\pi/2$.

Polarization interference must include two terms with different frequency denominators to give a beat term. ϕ changes from $-\pi/2$ to $+\pi/2$ and $\boldsymbol{I}_m$ goes through a minimum when the detector frequency is tuned through one of the resonances.

Another method to distinguish between polarization interference and quantum beats based on real-time resolution of the nonlinear signal has been demonstrated by Koch *et al.* [36]. Results for third order polarization a s a function of real time t (after the arrival of pulse #1) and time delay τ is for II-system

$$P_{PI}^{(3)}(t,\tau) \propto \Theta(t)\Theta(t-\tau)i[N_{21}\mu_{21}^4 e^{-i\Omega_{21}(t-2\tau)}+N_{43}\mu_{43}^4 e^{-i\Omega_{43}(t-2\tau)}], \tag{3.16}$$

where the TFWM signal intensity is proportional to the absolute square of the polarization as shown in Fig. 5a. The temporal interference results from the cross term, which has a phase that varies as (t -2τ) in agreement with the experiments in ref. [36]. The results for a V-type three-level system is

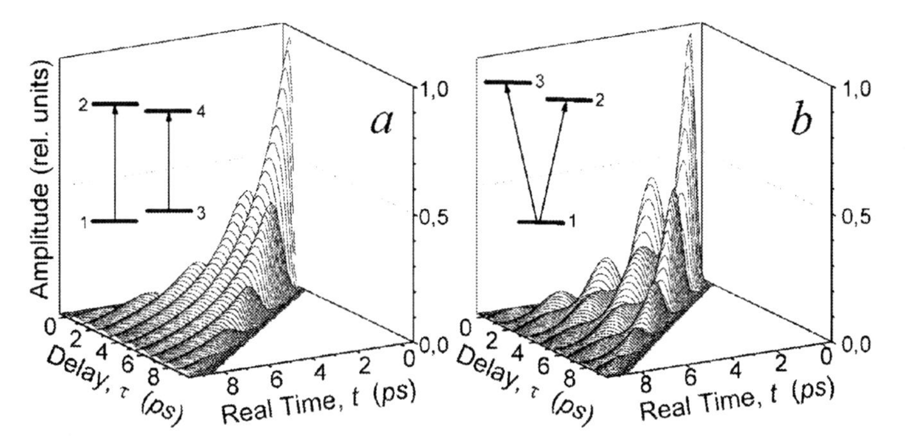

Figure 5. Calculated intensity of the TFWM signal in real time t versus delay τ for (a) for two independent two-level systems from Eq. (3.16) and (b) a V-type three -level system from Eq. (3.17). Both calculations are performed with parameters $\hbar\omega_1$ = 1.5 eV γ_1 = 2 meV, $\hbar\omega_2$ = 1.502 eV, γ_2 = 2 meV, $\mu_1 = \mu_2$.

$$P_{QB}^{(3)}(t,\tau) \propto iN\theta(t-\tau)\theta(\tau)$$
$$\times \mu_{21}^2 \left\{2\mu_{21}^2 \exp[-i\Omega_{21}(t-\tau) + i\Omega_{21}^*\tau] + \mu_{31}^2 \exp[-i\Omega_{21}(t-\tau) + i\Omega_{31}^*\tau]\right\}$$
$$+ \mu_{31}^2 \left\{2\mu_{31}^2 \exp[-i\Omega_{31}(t-\tau) + i\Omega_{31}^*\tau] + \mu_{21}^2 \exp[-i\Omega_{31}(t-\tau) + i\Omega_{21}^*\tau]\right\}. \quad (3.17)$$

In this case, the intensity of the TFWM signal, as shown in Fig. 5b behave in a more complicated fashion, than previously predicted in [36] with quantum beats varying as τ, t, $t-\tau$ and $t-2\tau$. There are indications in the previously reported experiments of such a complex behavior, although the term varying as $t-\tau$ was emphasised.

3.2. INFLUENCE OF INHOMOGENEOUS BROADENING

Inhomogeneous broadening has the strong influence on the modulated TFWM signal, when more than one resonance are coherently excited. With spectral resolution it is possible to deduce the homogeneous linewidth γ and the inhomogeneous line width Γ simultaneously. When the interfering resonances share a common level, it is necessary, in addition to the inhomogeneous broadening, to describe how the inhomogeneous broadenings of the different resonances are correlated [60]. In an atomic gas system, which normally is Doppler broadened, one expects a priori full correlation, whereas this condition could be relaxed in a condensed matter system.

If several independent resonances are coherently excited by the same laser pulse, the total polarization is calculated as a sum of terms like Eq. (3.23), for positive delay $\tau > 0$,

$$P_{PI}^{(3)}(\tau,\omega) \propto \sum_j P_{inh,j}^{(3)}(\tau,\omega,\omega_{21},j). \tag{3.18}$$

The subscript *PI* indicates, that Eq. (3.18) result in polarization interference, when the total emitted field from $P_{PI}^{(3)}(\tau,\omega)$ is squared in the external detector. Note, that in case of inhomogeneous broadening, it is the phase of the *erfc*, which determines the beat period, see Eq. (2.23). The signature of polarization interference is a phase change of the beats around a resonance, because the beating is between a resonant term and a non-resonant term. The *erfc* depends strongly on the ratio $\beta = \Gamma/\gamma$, which sets the limits for the possibility of observing polarization interference. The emitting polarizations do not interfere for a delay exceeding the inverse inhomogeneous broadening. This behavior can be seen in the calculation in Fig. 6, where the modulation persist longer for small inhomogeneous broadening (Fig. 6a) than for a larger inhomogeneous broadening (Fig. 6b). A small phase shift of the modulation as a function of detuning can be observed, and this again depends on β, as seen in Eq. (3.18). For small inhomogeneous broadening it is clear, that another signature of polarization interference is a minimum modulation amplitude at the resonance frequency, $\delta = 0$.

We introduce a correlation distribution of ω_{21} and ω_{31} to describe the

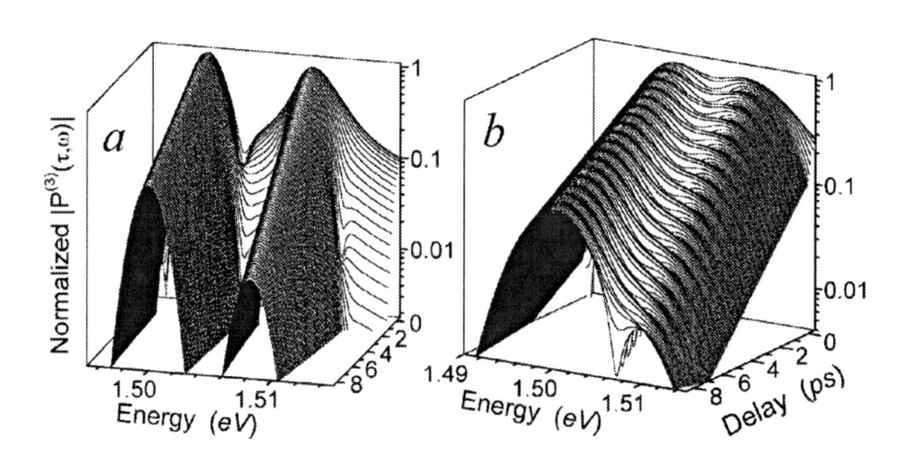

Figure 6. Polarization interference calculated from Eq. (3.18) for $\hbar\omega_{21} = 1.5$ eV, $\hbar\omega_{31} = 1.5083$ eV corresponding to a beat period of 0.5 ps. $\hbar\gamma_2 = 0.5$ meV, $\hbar\gamma_{21} = 1$ meV, $\Gamma_{21} = \Gamma_{21} = 3\ \gamma_{21}$ (a) and 10 γ_{21}(b).

inhomogeneous broadening in a three-level system. If there is no correlation, $\delta\omega_2$ is independent of $\delta\omega_2$. On the other hand, full correlation describes the situation, where $\delta\omega_2$ is a function f of $\delta\omega_1$, i.e. $\delta\omega_2 = f(\delta\omega_2)$. The two-dimensional Gaussian is [60] $g_N(\omega_{21},\omega_{21},\lambda) =$

$$\frac{\sqrt{1-\lambda^2}}{2\tilde{\Gamma}_{21}\tilde{\Gamma}_{31}}\exp\left\{-\frac{(\omega_{21}-\omega_{21}^c)^2}{2\tilde{\Gamma}_{21}}-\lambda\frac{(\omega_{21}-\omega_{21}^c)(\omega_{31}-\omega_{31}^c)}{\tilde{\Gamma}_{21}\tilde{\Gamma}_{31}}-\frac{(\omega_{31}-\omega_{31}^c)^2}{2\tilde{\Gamma}_{31}}\right\}, \quad (3.19)$$

where λ is the correlation parameter and ω_{j1}^c is the center frequency of the inhomogeneously broadened line. $\lambda = 0$ ($\lambda = 1$) means no (full) correlation. In Eq. (3.19), $\tilde{\Gamma}_{j1} = \Gamma_{j1}\sqrt{1-\lambda^2}$. It is seen, that the correlation makes the effective linewidth smaller, when λ increases from 0 towards 1.

Calculating the polarization from an inhomogeneously broadened three-level system we notice, that the first term in Eq. (3.17) is similar to the result for a single resonance. Therefore, we focus on the second term from Eq. (3.17), which we denote the beating term

$$P^{(3)}_{beat,\hom}(\tau,t) = iN\mu_{21}^2\mu_{31}^2 \exp[-i\Omega_{21}(\tau-t)]\exp(i\Omega_{31}^*\tau) \quad (3.20)$$

and we consider only contribution from one resonance, that is $\omega \approx \omega_{21}$. From Eq. (3.19) and Eq. (3.20) we get

$$\begin{aligned} P^{(3)}_{beat,\hom}(\tau,t) &= \iint d\omega_{21}d\omega_{31}g_N(\omega_{21},\omega_{31},\lambda)P^{(3)}_{beat,\hom}(\tau,t,\omega_{21},\omega_{31}) = \\ &i\theta(t-\tau)N\mu_{21}^2\mu_{31}^2 \exp[-i\omega_{21}^c(t-\tau)]\exp[i\omega_{31}^c\tau-\gamma_{21}(t-\tau)-\gamma_{31}\tau] \\ &\times\exp\left[-\frac{\Gamma_{21}^2(t-\tau)^2}{2}+\lambda\Gamma_{21}\Gamma_{31}(t-\tau)\tau-\frac{\Gamma_{31}^2\tau^2}{2}\right] \end{aligned} \quad (3.21)$$

i.e. the beating term gives rise to a signal in real time emitted at $t = \tau$. Notice, that it is the real inhomogeneous broadening, Γ_{j1}, which enters in Eq. (3.21). This signal from Eq. (3.21) will, in addition to T_2-dephasing, be damped by a Gaussian with a fixed FWHM related to the inhomogeneous broadening of one un-correlated resonances ($\lambda = 0$). In case of correlation between resonances ($\lambda \neq 0$) additional FWHM is time- (t) and delay (τ)-dependent and relates to both inhomogeneous broadenings. In Eq. (3.21) the phase of both resonances appears.

Fourier transforming Eq. (3.21). We obtain the result for the beating term:

$$P^{(3)}_{beat,inh}(\tau,\omega) =$$

$$iN\frac{\mu_{21}^2\mu_{31}^2}{\Gamma_{21}}\sqrt{\frac{\pi}{2}}\exp[i(\omega_{31}^c+\omega)\tau]erfc\left\{\frac{i(\omega_{21}^c-\omega)+\gamma_{21}}{\sqrt{2}\Gamma_{21}}-\lambda\frac{\Gamma_{31}\tau}{\sqrt{2}}\right\}\times$$

$$\exp\left\{-\gamma_{31}\tau-\frac{(1-\lambda^2)\Gamma_{31}^2\tau^2}{2}-\frac{(\omega_{21}^c-i\gamma_{21}-\omega)^2}{2\Gamma_{21}^2}-i\lambda\frac{(\omega_{21}^c-i\gamma_{21}-\omega)\Gamma_{31}\tau}{\Gamma_{21}}\right\}$$

(3.22)

In Eq. (3.22) the effect of the correlation becomes evident. The beating term will decay both from a T_2 term and Gaussian term in case of uncorrelated broadenings. For large correlation the Gaussian decays very slowly and therefore in a quantum beat experiment, the modulation persists as long as the signal goes. The two first resonant terms ($\omega \approx \omega_{21}$) from Eq. (3.14) then give the polarization:

$$P_{QB,inh}^{(3)}(\tau,\omega)=4iN\frac{\mu_{21}^2}{\Gamma_{21}}\sqrt{2\pi}e^{i\omega\tau}\exp\left[-\frac{(\omega_{21}^c-i\gamma_{21}-\omega)^2}{2\Gamma_{21}^2}\right]\times \tag{3.23}$$

$$\left\{\begin{array}{l} 2\mu_{21}^2e^{(i\omega-2\gamma_{21})\tau}erfc\left[\frac{i(\omega_{21}^c-\omega)+\gamma_{21}}{\sqrt{2}\Gamma_{21}}-\frac{\Gamma_{21}\tau}{\sqrt{2}}\right]+erfc\left[\frac{i(\omega_{21}^c-\omega)+\gamma_{21}}{\sqrt{2}\Gamma_{21}}-\lambda\frac{\Gamma_{31}\tau}{\sqrt{2}}\right] \\ \times\exp\left[\left(i\omega_{31}^c-\gamma_{31}-\lambda\frac{\gamma_{21}\Gamma_{31}}{\Gamma_{21}}\right)\tau-i\lambda\frac{(\omega_{21}^c-\omega)\Gamma_{31}\tau}{\Gamma_{21}}-(1-\lambda^2)\frac{\Gamma_{31}^2\tau^2}{2}\right] \end{array}\right\}$$

If the two beating resonances are separated much more then their respective inhomogeneous broadenings, the main contribution to the detected signal near one resonance is given by Eq. (3.23). If not, one needs to add a contribution that can be obtained from Eq. (3.23) with exchange of indexes (2⇔3).

4. Exciton-Exciton Interaction

In the treatment so far, it has been assumed, that the field interacting with the two-level transitions is the externally applied field. This assumption is valid only in diluted media, where no local field effects exist. In a dense medium like a semiconductor, however, the effective field, interacting with an exciton, must ainclude then correction from the local field due to the coherent exciton population. In the Lorentz model, the local field correction is given by a geometric factortimes the polarization of the medium. Several authors have analysed the extension of the two-level equations with such local field correction [61]. It is show theoretically [62] that the Optical Bloch Equations including local field effects predict intrinsic optical bistability since the resonance ifrequency becomes inversion-dependent. A similar result is shown in calculation for ultrashort light pulses, that local-field effects give a strong inversion-dependent transmission leading to fast optical bistability. Mike *et al.* [63] showed experimentally in a series of linear and nonlinear measurements, that local effect are important in a dense atomic vapour.

In the spirit of this work, an approximation to the Semiconductor Bloch Equations can be written

$$\begin{aligned}\frac{\partial P}{\partial t} &= -\frac{i}{\text{h}}\omega_{21} - \gamma_{21}P - \frac{i}{\text{h}}(2n-1)(\mu E + VP) \\ \frac{\partial n}{\partial t} &= -\gamma_1 n - \frac{2V}{\text{h}}|P|^2 - \frac{2i\mu}{\text{h}}(EP^* - E^*P)\end{aligned}, \tag{4.1}$$

where P is the polarization, E is the electric field and n is the exciton population. V represents the local field in the Hartree-Fock approximation. The Eqs. (4.1) mediate new nonlinear mechanisms in the modified two-level equations through the coupling of the polarization and the population. Notice, that the Optical Bloch Equations are obtained if $V = 0$. Following Wegener *et al.* [15], the third order polarization can be derived perturbatively as in Sections 2.1 and 3.1. The first and second order equations are similar to those without the local field correction. The equation for the third order term can then be written

$$\left(\frac{\partial}{\partial t} + \gamma_2 + \frac{i}{\text{h}}\omega_{21}\right)P^{(3)} = -\frac{i}{\text{h}}2n^{(2)}[\mu E + VP^{(1)}], \tag{4.2}$$

where (#) indicates the order of the term. The local field renormalizes the transition energy slightly, and this renormalization has been included in the trann-sition frequency ω_{21}. In Eq. (4.2), the first order polarization enters via the Coulomb field as a local field correction. In a short pulse limit and for a homogeneously broadened resonance, the third order polarization with the exciton-exciton interaction can be found from Eq. (4.2)

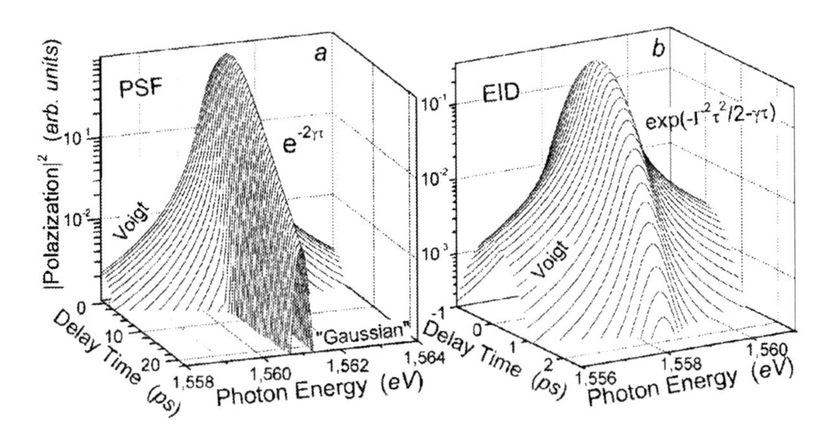

Figure 7. Calculated with. Eq. (4.3) spectrally-resolved TFWM for (a) Phase-State-Filing and (b) Excitation-Induced-Dephasing mechanisms. In both cases TFWM signal reach maximum at delay $\tau_{max} \cong \gamma/\Gamma^2$, then decay exponentially with homogeneous dephasing rate γ for PSF or Gaussian-like with inhomogeneously decay rate Γ. At τ_{max} spectral shape of TFWM signal is Voigt-function in both cases, changing to Gaussian at large delays $\tau >> \tau_{max}$.

$$\left\{\theta(\tau)\theta(t-\tau)+i\frac{2V}{\gamma_1}\left[\theta(\tau)\theta(t-\tau)\left(1-e^{-\gamma_1(t-\tau)}\right)+\theta(-\tau)\theta(t)e^{2\gamma_2\tau}(1-e^{-\gamma_1 t})\right]\right\}. \quad (4.3)$$

The first term of (4.3) is equal to Eq. (2.19), the second term describes the extra contributions to the nonlinear polarization from the local-field correction (LFC). The first LFC acts for positive delay with amplitude determined by the exciton population decay rate γ_1. If $\gamma_1 << \gamma_2$, the LFC term and the usual two-level term for positive delay are inseparable, since both terms decay with γ_2 in that limit. Since the terms are $\pi/2$ out of phase, they do not interfere.

In the real-time resolved TFWM-experiments, a detailed line shape analysis by Weiss *et al.* [64] showed the relative importance of these terms, because the LFC term has a different time-behavior, compared to the pure two-level term, for finite γ_1. The other LFC term acting for negative delay is new. The rise time of this term is $0.5/\gamma_2$, i.e. the signal rises twice faster for negative delay than it decays for positive delay. This is in accordance with the observation of Leo *et al.* [14] and is due to the fact, that for negative delay, the linear polarization from pulse 2 creates a grating with pulse #1 on which the linear polarization from pulse #2 is diffracted. The linear polarization from pulse #2 is therefore is used twice at a latter time-delay giving the factor of 2. The analysis presented above has been extended from Eq. (4.2) to introduce stochastic modulation of the exciton frequency due to coupling with a reservoir by Schmitt-Rink *et al.* [65]. An extensive analysis of stochastic processes in coherent optical transients has been given by Hanamura [66].

As discussed earlier, another source for the third-order polarization is excitation induced dephasing (EID). It has been demonstrated that EID is a significant nonlinear process in GaAs [67].

5. Optical Coherent Control in Semiconductors

The developments with coherent control techniques in optical spectroscopy have recently demonstrated population control and coherence manipulations when the induced optical phase is explored with phase-locked laser pulses. Fascinating demonstrations were obtained with atoms and molecules such as localization of electronic wave packets in atomic sodium [68] and control of molecular chemical reactions [69,70]. More recently, the full characterisation of the amplitude and phase of an electronic Rydberg wave packet in Cs [71] was obtained using quantum state holography with tailored laser pulses [72]. These developments have been guiding the new research field of quantum control and also the recent applications of coherent control techniques to semiconductors and nanostructures.

Population and orientation control of excitons in GaAs quantum wells and quantum dots, as well as control of electron-phonon scattering is GaAs, have provided advancements in ultrafast spectroscopy of semiconductors [73–76]. In experiments recording the secondary emission, after resonant excitation with phase-locked pulses, with a spectral filtering technique [77] or time-resolved [73,78] using up-conversion, the coherent emission due to resonant Rayleigh scattering with optical fringes could be distinguished from incoherent photoluminescence without fringes. In the interpretation of these experiments in Gurioli *et al.* [77] and Marie *et al.* [73], and other experiments based on differential reflection measurements by Heberle *et al.* [73], the exponential decay of the fringe contrast has been assigned to the homogeneous polarization decay with decay time T_2 in an analogy with the concepts developed in atomic and molecular spectroscopy. This, however, neglects the inhomogeneous broadening in semiconductor samples due to stochastic alloy and size fluctuations. On the other hand, Wörner and Shah [78] and Garro *et al.* [79] state that, in general, it is not possible to interpret the decay of the coherent control in terms of T_2. Wörner and Shah have experimentally observed an almost exponential decay of the fringe contrast versus delay between two pulses and ascribed this to the Fourier transform of the ''real'' exciton absorption line. In virtue of this, they come to the conclusion, that the measured dependence does not contain information about whether the investigated transition is homogeneously or inhomogeneously broadened. Moreover, in a number of recent papers by Garro *et al.*, [79] the experimentally measured coherent control (CC) signals have been compared with three-level optical Bloch calculations, taking into consideration the inhomogeneous broadening of excited levels while no dephasing was included. On the basis of such a comparison, these authors come to the conclusion that the CC decay is inversely proportional to the inhomogeneous broadening. These obvious contradictions in the interpretation of similar CC results can be resolved by detecting phase-dependent spectrally resolved four-wave mixing that demonstrates how inhomogeneous broadening contributes to the phase dynamics.

Even small inhomogeneous broadenings, comparable to the homogeneous broadening, reveal in spectrally resolved or spectrally integrated recordings of the coherent emission in coherent control experiments. The model for the optical fringe contrast, developed on the basis of these experiments, describes in a simple way the spectrally resolved detection of the coherent emission, as well as the detection of time-integrated photoluminescence used in most experiments. This spectroscopy technique in semiconductor optics, using the phase information in coherently excited ensembles, is very important for extracting information on electronic dynamics in nanostructures and for coherent manipulations of populations, such as those demonstrated previously in single quantum dot spectroscopy [75].

5.1. EXPERIMENTAL TECHNIQUE

The coherent control (CC) experiments are performed in a configuration where the photoluminescence (PL) and/or the coherent emission in four-wave mixing (FWM) are recorded after resonant excitation with phase-locked laser pulses with wave vector $\boldsymbol{k}_1$. For the latter, a second beam with wave vector $\boldsymbol{k}_2$ is incident on the sample resulting in the coherent scattering of the emission in the direction $2\boldsymbol{k}_2 - \boldsymbol{k}_1$. The experimental setup is illustrated in Fig. 2. Pulses from a 150 fs 80 MHz Ti:Sapphire laser are split by a 50% mirror into the „pump" and "probe" beams, the latter travelling through a delay line acquiring a time delay τ in the TFWM, and focused on the sample surface by a lens. The pump and probe beams have comparable intensities in the range of 0.1–1 mW with a spot diameter ~100 μm giving exciton densities of $\sim 10^9$ cm^{-2} for each pulse. All measurements were performed at two temperatures of 12 and 120 K. The collinearly polarized phase-locked laser pump pulses, with an intensity ratio a^2, are generated in a Michelson interferometer (MI) with time delay Δ. Time-integrated incoherent (PL) or coherent (TFWM, Rayleigh scattering) emission can be detected either spectrally integrated, or spectrally resolved with an optical multichannel analyser (OMA) after a spectrometer (with spectral resolution 0.1 meV), as a function of Δ and τ, revealing the spectral information necessary for the interpretation of the CC in the inhomogeneous ensembles. Experimental measurements have been performed on a well characterized narrow 28 Å GaAs single quantum well (SQW), surrounded by 250 Å $Al_{0.3}$ $Ga_{0.7}As$ barriers, grown by molecular-beam epitaxy with growth interruptions on the barrier-well interfaces [80]. The PL spectrum of this sample at $T = 12$ K consists of two peaks with an inhomogeneous linewidth ~2.5 meV and separation ~13 meV. The resonant excitation allows for the simultaneous detection of the coherent and incoherent signals in the CC experiments. The results reported here do not depend critically on the particular SQW and similar results are obtained from other multiple QW and superlattice samples with different ratios between the homogeneous and inhomogeneous broadenings.

5.2. THEORY

The theoretical basis for spectrally resolved FWM [81,82] can be easily extended on two levels (ground and excited states) with optical transitions, characterized by the frequency ω_{12} and the damping $\gamma_{12} = 1/T_2$, and excited by two phase-locked δ-pulses:

$$E(t) = \delta(t)e^{-i\omega_L t} + a\delta(t-\Delta)e^{-i\omega_L(t-\Delta)} \tag{5.1}$$

with the laser frequency ω_L. The a accounts for differences in the field strength, and the delta functions $\delta(t)$ are used for convenience in the short-pulse limit compared to T_2.

5.2.1 *Homogeneous broadening*

The first-order polarization $P^{(1)}_{\text{hom}}(t,\Delta)$ induced by the phase-locked pulses for homogeneously broadened two-level systems is

$$P^{(1)}_{\text{hom}}(t,\Delta) \propto iN\mu 2_{10}[e^{-i\Omega_{10}t}\theta(t) + ae^{-i\Omega_{10}(t-\Delta)}\theta(t-\Delta)]\,, \tag{5.2}$$

where the complex frequency $\Omega_{10} = \omega_{10} - i\gamma_{10}$, the density of levels N, the optical matrix element μ_{10} , and the Heaviside function $\theta(t)$. From Eq. (5.1) follows the fringe contrast (FC) for coherent control experiments detecting the time-integrated (i.e., spectrally integrated (SI)) reradiated emission from the polarization, [84] with $I_{\max,\min} = \max,\min[\int \left|P^{(1)}_{k_1}(t,\Delta)\right|^2 dt]$ as

$$F^{(1)}_{SI}(\Delta) = \frac{I_{\max} - I_{\min}}{I_{\max} + I_{\min}} = \frac{2ae^{-\gamma_{10}|\Delta|}}{1+a^2}\,. \tag{5.3}$$

This shows that $F^{(1)}_{SI}(\Delta)$ decays with T_2 for increasing time separation between the CC pulses. When the observed PL is proportional to the excited carrier density we can conclude that detecting the incoherent PL or the coherently reradiated emission in CC experiments reveals the same result.

Following [82] the third-order polarization $P^{(1)}_{\text{hom}}(t,\tau,\Delta)$ emitting the FWM signal in the direction $2\boldsymbol{k}_2 - \boldsymbol{k}_1$, using Eq. (1) as the starting point, is

$$P^{(1)}_{\text{hom}}(t,\tau,\Delta) \propto N\mu^4_{10}[\theta(\tau) + ae^{-i\Omega_{10}\Delta}\theta(\tau-\Delta)]e^{-i\Omega_{10}(t-\tau)}e^{i\Omega^*_{10}\tau}\theta(t-\tau), \tag{5.4}$$

where the "*" denotes complex conjugation and the Heaviside functions $\theta(t)$ account for the emission of the TFWM signal after the probe pulse. The dependence of the CC term in the square bracket in Eq. (5.4) on Δ is similar to the first order polarization in Eq. (5.1), proving the statement that the TFWM signal as function of the phase-delay Δ is proportional to the first-order polarization. The spectrally resolved TFWM is in this case:

$$P^{(1)}_{\text{hom}}(\omega,\tau,\Delta) \propto `N\mu^4_{10}[\theta(\tau) + ae^{-i\Omega^*_{10}\Delta}\theta(\tau-\Delta)]\frac{e^{i\Omega^*_{10}\tau}}{\Omega_{10}-\omega}, \tag{5.5}$$

showing that all the time-integrated spectral components have the same dependence on the phase-delay Δ. We can therefore define the fringe contrast FC in TFWM coherent control experiments from Eq. (5.4) in the following way:

$$F^{(3)}_{\text{hom},SR}(\Delta) = F^{(3)}_{\text{hom},SI}(\Delta) = \frac{|P^{(3)}|^2_{\max} - |P^{(3)}|^{(2)}_{\min}}{|P^{(3)}|^2_{\max} + |P^{(3)}|^{(2)}_{\min}} = \frac{2ae^{-\gamma_{10}|\Delta|}}{2ae^{-\gamma_{10}|\Delta|} + a^2}, \tag{5.6}$$

with the decay of the $F^{(3)}_{\text{hom}}(\Delta)$ for increasing delay Δ according to the definition given above.

From Eqs. (5.2) and (5.6) it can be concluded that the coherent control signals for homogeneously broadened systems will decay with T_2 as a function of $|\Delta|$ under typical conditions with $a \sim 1$ and $|\Delta| > T_2$. Notice that a delayed maximum of the $F^{(3)}_{\text{hom}}(\Delta)$ (4.3) at $|\Delta_{\max}| = \ln(a)/\gamma_{10}$ can be found when $a \neq 1$. The intuitive explanation of this is that the $F^{(3)}_{\text{hom}}(\Delta)$ is maximum when the remaining polarization amplitude due to the first CC pulse is equal to the polarization amplitude induced by the second CC pulse. This occurs for nonzero Δ when the intensity of the first pulse is largest, i.e., $a < 1$. Notice also that $\max[F^{(3)}_{\text{hom}}(\Delta)] = 1$. The decay of the FC with T_2 is the interpretation of CC experiments first developed in atomic and molecular spectroscopy.

5.2.2 *Inhomogeneous broadening*

With inhomogeneous broadening, described by a Gaussian distribution $G(\omega_{10})$ with central frequency ω^c_{10} and with a broadening parameter $\Gamma = \Gamma_{FWHM}\sqrt{8\ln 2}$ scaled to the full width at half maximum Γ_{FWHM}, the first order polarization is

$$P^{(1)}_{inh}(t,\Delta) = \int P^{(1)}_{\text{hom}}(\omega_{10})G(\omega_{10})d\omega_{10}$$

$$\propto N\mu_{10}^2[e^{-i\Omega^C_{10}t}e^{-\Gamma^2 t^2/2}\theta(t) + ae^{-i\Omega^C_{10}(t-\Delta)}e^{-\Gamma^2(t-\Delta)^2/2}\theta(t-\Delta)], \quad (5.7)$$

i.e., the first-order polarization decays faster because of the destructive interference within the inhomogeneously broadened resonance. With Eq. (5.7) the spectrally integrated fringe contrast $F^{(1)}_{inhSI}$ is

$$F^{(1)}_{SI,inh}(\Delta) = \frac{2a\exp(-\Gamma^2\Delta^2/4)erfc(\gamma/\Gamma + \Gamma\Delta/2)}{(1+a^2)erfc(\gamma/\Gamma)} \approx \frac{2ae^{-\Gamma^2\Delta^2/4}}{1+a^2}, \quad (5.8)$$

where the *erfc* denotes the complementary error function. This shows that the decay of the fringe contrast in inhomogeneously broadened systems, detecting the reradiated emission, is determined solely by the inhomogeneous broadening Γ for $\Gamma >> \gamma_{10}$. Note furthermore that when expanding the *erfc* for $\Gamma \to 0$, the expression for the homogeneously broadened case in Eq. (5.5) is recovered. Thus, in semiconductor systems, with inhomogeneous broadening, the decay of the fringe contrast does not decay with the T_2 time [78,79]. Note also that $F^{(1)}_{SI,inh}(\Delta)$ has a delayed maximum, depending on the ratio γ_{10}/Γ.

A similar behavior can be found for TFWM-detected CC starting from the expression for the third-order polarization with inhomogeneous broadening,

$$P^{(3)}_{inh}(t,\tau,\Delta)$$

$$\propto N\mu^4 \left[\Theta(\tau) + ae^{-i[\Omega^* + \Gamma^2(t-2\tau)]\Delta}e^{\frac{\Gamma^2\Delta^2}{2}}\Theta(\tau-\Delta)\right]e^{-i\Omega^c(t-\tau)}e^{i\Omega^{c*}\tau}e^{\frac{\Gamma^2(t-2\tau)^2}{2}}, \quad (5.9)$$

where the last term is the third-order polarization with inhomogeneous broadening due to one $\boldsymbol{k}_1$ pulse at $t = 0$ followed by the $\boldsymbol{k}_2$ pulse resulting in a photon echo. The CC term in the square bracket is the extra term caused by the phase-locked $\boldsymbol{k}_1$ pulse pair. The CC term is now time dependent in contrast to the homogeneously broadened case Eq. (5.3), i.e., the different spectral components will behave differently in the CC FWM experiment. This will be transparent in the following expressions for the fringe contrast. For the spectrally integrated case, calculated like in the homogeneously broadened case, we find:

$$F^{(3)}_{SI,inh}(\tau,\Delta) = \frac{2ae^{-2\gamma\Delta}e^{-\frac{\Gamma^2\Delta^2}{4}}erfc\left[\frac{\gamma}{\Gamma}-\frac{\Gamma(\tau-\Delta)}{2}\right]}{e^{-4\gamma\Delta}erfc\left[\frac{\gamma}{\Gamma}-\Gamma\tau\right]+a^2 erfc\left[\frac{\gamma}{\Gamma}-\Gamma(\tau-\Delta)\right]} \xrightarrow[\text{for }\Gamma\to 0]{} F^{(3)}_{SI,\text{hom}}(\tau,\Delta), \quad (5.10)$$

which is complicated by the three *erfc* expressions. However, the correct limit is found for $\Gamma \to 0$, whereas in the other limit $\gamma_{10}/\Gamma \to 0$ the decay of the $F^{(3)}_{SI,inh}(\Delta)$ is determined by Γ due to the $F^{(3)}_{SI,inh} \to e^{-\Gamma^2\Delta^2/4}$ expression such as that in the first order case Eq. (5.8). Equation (5.10) shows that the time-integrated CC FWM in general is a complicated interplay between the homogeneous broadening and the inhomogeneous broadening. Note that for finite $\Delta \neq 0$, the $F^{(3)}_{SI,inh}(\Delta)$ is always less than 1 for any a and the so-called overshooting appearing at $F^{(3)}_{SI,inh}(\Delta) = 1$ is not possible in inhomogeneous systems [73]. If the TFWM signal is spectrally resolved, a different behavior, compared to the spectrally integrated case, is found in contrast to the homogeneously broadened case. First, the third-order polarization is found by a Fourier transformation of Eq. (5.9),

$$P^{(3)}_{inh}(\omega,\tau,\Delta) \propto N\mu^4 \frac{\sqrt{\pi}}{\Gamma} e^{i2\omega\tau} e^{-2\gamma\tau} e^{-(\Omega-\omega)^2/2\Gamma^2} \times \quad (5.11)$$

$$\left\{\Theta(\tau) erfc\left[\frac{i(\Omega-\omega)}{\sqrt{2}\Gamma}-\frac{\Gamma\tau}{\sqrt{2}}\right] + a\Theta(\tau-\Delta)e^{-i\omega\Delta-2\gamma\Delta} erfc\left[\frac{i(\Omega-\omega)}{\sqrt{2}\Gamma}-\frac{\Gamma(\tau-\Delta)}{\sqrt{2}}\right]\right\},$$

where again only the CC term in the curly bracket determines the fringe contrast. Note, that the period in the CC experiment is now determined by the detected frequency ω and different spectral components have different phases due the Δ term of the second erfc expression. From Eq. (5.11), the final general expression for the fringe contrast for spectrally resolved TFWM CC experiments is

$$F^{(3)}_{SR,inh}(\omega,\tau,\Delta) = \frac{2ae^{2\gamma\Delta} erfc\left[\frac{i(\Omega-\omega)}{\sqrt{2}\Gamma}-\frac{\Gamma(\tau-\Delta)}{\sqrt{2}}\right] erfc\left[\frac{i(\Omega-\omega)}{\sqrt{2}\Gamma}-\frac{\Gamma\tau}{\sqrt{2}}\right]}{\left|erfc\left[\frac{i(\Omega-\omega)}{\sqrt{2}\Gamma}-\frac{\Gamma\tau}{\sqrt{2}}\right]\right|^2 + a^2e^{4\gamma\Delta}\left|erfc\left[\frac{i(\Omega-\omega)}{\sqrt{2}\Gamma}-\frac{\Gamma(\tau-\Delta)}{\sqrt{2}}\right]\right|^2} \approx \frac{2ae^{-2\gamma\Delta}}{e^{-4\gamma\Delta}+a^2}, \quad (5.12)$$

i.e., each spectrally resolved component of the FWM signal will have a decay determined by T_2 in contrast to the spectrally integrated case Eq. (5.10).

In Fig. 8 the calculated FWM spectra from Eq. (5.11) with $\hbar\Gamma$ = 5, 2.5, 1.5, and 0.75 meV and $\hbar\gamma_{10}$ = 1 meV (T_2 = 0.66 ps) at a delay of τ = 0.5 ps are presented.

The curves are calculated for different phase delays between the two pump pulses around $\Delta = -2.5$ ps resulting in first-order polarizations with a phase differing by 0, $\pi/4$, $2\pi/4$, and $3\pi/4$. Notice that the chosen $a = 500$ compensates for the long Δ compared to T_2 to illustrate the interplay between the inhomogeneous and homogeneous broadenings. Notice also that this a is smaller than the optimal $a_{\text{opt}} = \exp(-2\gamma_{10}\Delta)$ from Eq. (5.11). At small Γ/γ_{10}, the FWM spectra are Lorentzian except for a narrow central Gaussian part corresponding to Γ. Changing the relative phase between the two pump pulses, the Lorentzian changes the amplitude without noticeable change in the shape of the FWM spectrum. At larger Γ/γ_{10}, the central Gaussian part becomes wider with more pronounced spectral modulations, while the remaining Lorentzian is without spectral modulations. From the plots of the largest Γ/γ_{10} it becomes evident that the Gaussian part is spectrally modulated by frequency $\Omega = 2\pi/\Delta$, while the modulation depth decreases again. This modulation depth is controlled by all the parameters in the FWM experiments i.e., a, T_2, Γ, τ, and Δ.

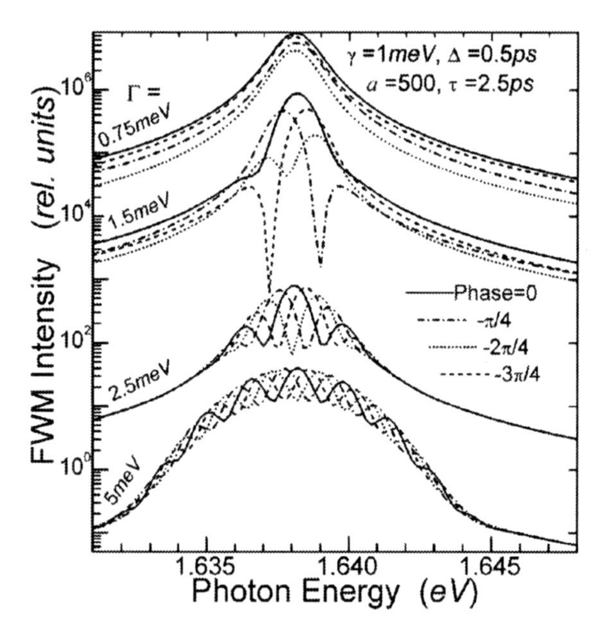

Figure 8. Calculated FWM spectra for different inhomogeneous broadenings $\hbar\Gamma$ = 0.75, 1.5, 2.5, and 5 meV using Eq. (5.11) with τ = 0.5 ps, $\hbar\gamma_{10}$=1 meV, $\Delta = -0.5$ ps, and a = 500. Solid, dashed, dotted, and dash-dotted curves are calculated for different relative phases between the phase-locked pump pulses.

5.2.3 *Spectrally Resolved Coherent Control*

To obtain a better understanding of how the interplay between the homogeneous and inhomogeneous broadening enters into the spectrally resolved FWM as described by Eq. (5.9), shown in Fig. 9 how the FWM spectra transform with decreasing inhomogeneous broadening. The TFWM spectra have been calculated with energies ($\hbar\omega_{10}^{c}$ = 1.638 eV) and broadenings ($\hbar\gamma_{10}$ = 0.75 meV) relevant for the experimental results to be presented in Fig. 10. For large inhomogeneous broadening, Fig. 9(a), the different spectral components have different phase $\omega\Delta$ within the chosen spectral window, where the dynamic range of the FWM signal is more than three orders of magnitude. In the other limit of small inhomogeneous broadening, as already discussed, the FWM CC fringes have no spectral variation as shown in Fig. 9(f) for Γ/γ_{10} = 1.3. However, in between these two limits the spectral wings of the FWM spectra have the same phase, whereas an increasing part of the center acquires the phase change. This occurs as a result of the mixing of the Lorentzian part due to the homogeneous broadening, with constant phase $\omega_{10}^{c}\Delta$, and the Gaussian part due to the inhomogeneous broadening in the complex *erfc* expression in Eq. (5.9). In Fig. 9, around a phase delay of $\Delta = -1$ ps the spectral range of the Gaussian part coincides with Γ. Therefore the transformation of the TFWM spectra with increasing inhomogeneous broadening

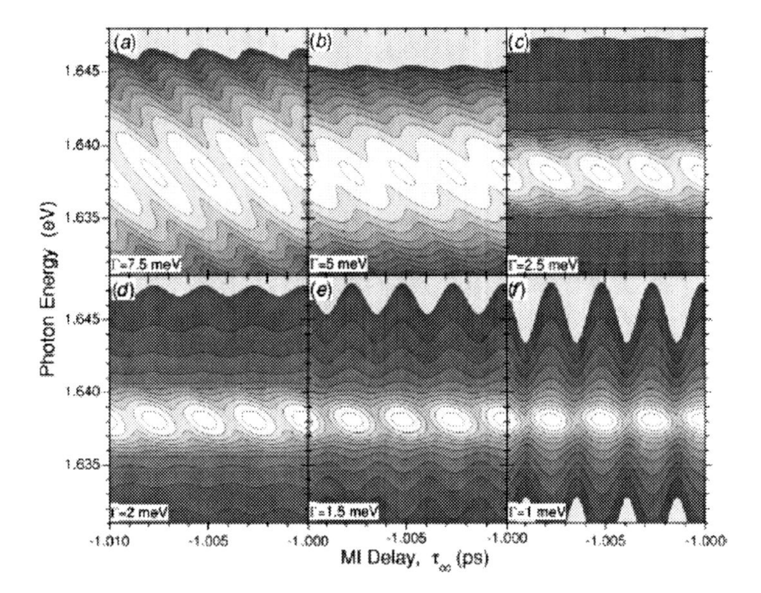

Figure 9. Calculated TFWM spectra with decreasing inhomogeneous broadening (*a*-*f*) with energy $\hbar\omega_{21}^{c}$ = 1.638 *eV* and γ_{21} = 0.75 *meV* versus delay Δ.

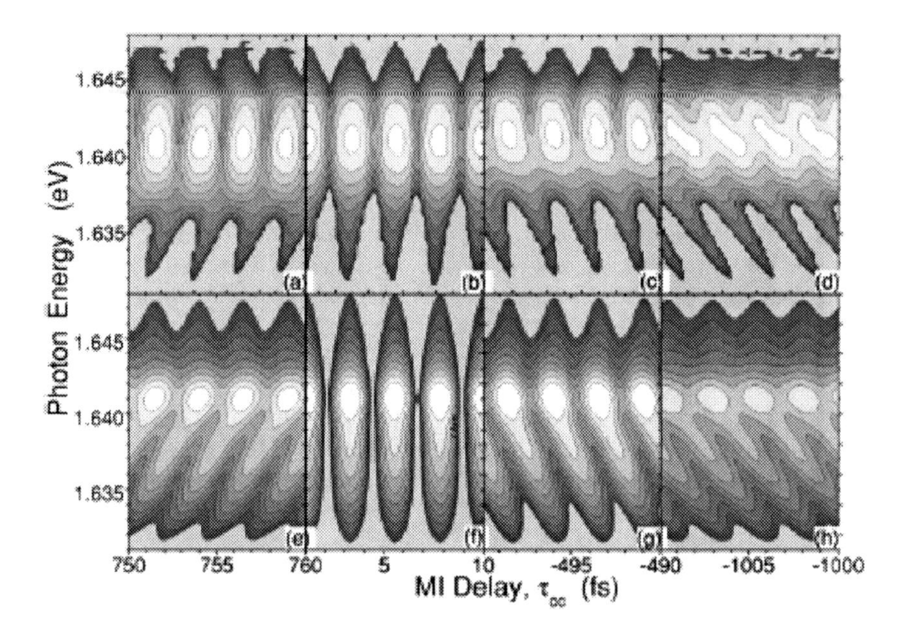

Figure 10. Experimentally measured at 120 K (a–d) and calculated (e–h) FWM spectra as a function of Δ around 0.75 ps (a,e), 0 ps (b,f), –0.5 ps (c,g), and -1 ps (d,h). The parameters for the calculation are given in the text.

is due to an increasing contribution of the Gaussian part with phase $\omega\Delta$ and not due to a change of the actual phases with inhomogeneous broadening. The spectral variation of the fringe contrast is thus a measure of the homogeneous (no variation) and inhomogeneous contributions to the electronic resonance.

This TFWM CC technique can be used in greater detail to investigate the nature of the SQW GaAs electronic resonance, discussed above, at a temperature of 120 K, where the homogeneous broadening is expected to have a significant contribution. Note that at high temperature the decay times in TFWM experiments are normally too fast to obtain much dynamic information on, e.g., scattering on interface roughness in the quantum wells. From previous investigations of similar samples, it has been concluded that the interface roughness is well described by the bimodal roughness model [85,86]. At low temperature, this results in a biexponential decay of the TFWM signal from a slightly wider GaAs SQW grown under similar conditions as the one studied here. The biexponential decay, however, is untraceable at temperatures above 50 K [87].

Here we show that the TFWM CC technique is very useful even at higher temperatures. The experimental results are presented in Figs. 10 (a–d) showing the variation of the TFWM spectra for different phase delays Δ. The spectral components on the high-energy side in this case has a constant phase, whereas on the low-energy side shows the spectral variation of CC fringes. From this, we can immediately conclude that the electronic resonance is composed of at least two parts with different inhomogeneous broadenings in agreement with previous studies [85,87]. From the calculated TFWM spectra [Figs. 11(e–h)], fitted to the

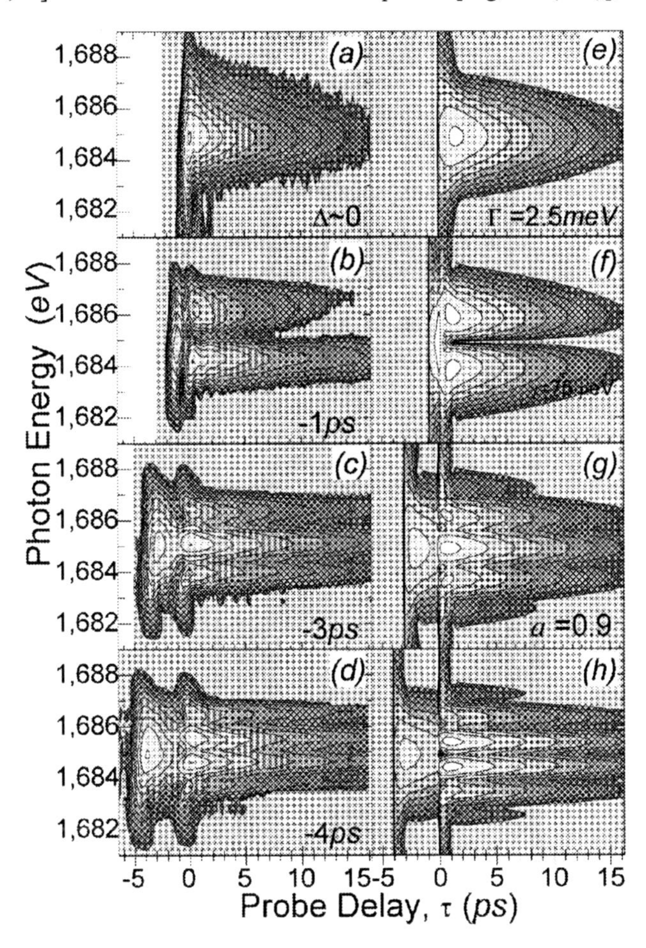

Figure 11. Comparison of experimentally measured at 12 K (a–d) and calculated (e–h) TFWM spectra as functions of probe delay τ for different delays Δ between the two phase-correlated pump pulses Δ = 0ps (*a,e*), –1 ps (*b,f*), –3 ps (*c,g*), and –4 ps (*d,h*). In the calculation $\hbar\Gamma = 2.5$ meV, $\hbar\gamma_{21} = 75$ μeV, and $a = 0.9$ were used.

experimental data, we obtain an overall spectral and phase agreement between the experiment and the modelling with two separate electronic resonances without mutual interference. From this we extracted the following parameters: The resonance at $\hbar\omega_{10}^{c}$=1.639 eV has $\hbar\Gamma$ =7 meV and $\hbar\gamma_{10}$ = 0.75 meV, while the one at $\hbar\omega_{10}^{c}$=1.641 eV has $\hbar\Gamma$ = 0 meV and $\hbar\gamma_{10}$ = 2.5 meV. This demonstrates the importance of this CC technique to study electronic dynamics in semiconductor nanostructures, also at conditions that can be inaccessible by other techniques.

Focusing on the phase-controlled TFWM spectroscopy, we present in Fig. 11 a comparison of the experimentally measured (a–d) and calculated (e–h) TFWM spectra as a function of probe delay τ for different delays between two phase correlated pump pulses Δ = 0 ps (a,e); –1 ps (b,f); –3 ps (c,g), and –4 ps (d,h). The experimental TFWM spectra have been recorded at T =12 K. The best agreement between the experimental and calculated spectra and their dynamics was obtained with the following parameters: $\hbar\omega_{10}^{c}$ = 1.6849 eV, $\hbar\gamma_{10}$ = 75 μeV (T_2 = 9 ps), $\hbar\Gamma$ = 2.5 meV, a = 0.9. One can see that without the second pump pulse or between the first and the second pump pulse, the TFWM spectra are unmodulated and decay in accordance with the homogeneous and inhomogeneous broadenings of the excited excitonic states. After the second pump pulse (in our experimental conditions at $\tau > 0$) the TFWM spectra are strongly sinusoidally modulated with the Gaussian spectral shape split into several bands decaying with different rates. By fine tuning the MI delay Δ, a maximum or a minimum of the modulated spectra can be shifted to any position in accordance with the interference term $\propto\cos(\omega\Delta)$ and $\Gamma/\gamma_{10}>>1$.

6. Conclusion

The two- or three-level model, described by the Optical Bloch equations, can be successfully applied to understand optical nonlinearities and coherent phenomena in atomic and semiconductor systems. In an excitonic system, a number of additional nonlinearities are significant. These include exciton-exciton interactions via a local field correction and excitation-induced dephasing.

A widely used technique to study optical nonlinearities is transient four-wave mixing, when nonlinear signal is detected "background-free" direction $2\boldsymbol{k}_2 - \boldsymbol{k}_1$ as a function of delay τ between "pump" and "probe" femtosecond pulses. Two phase-locked "pump" pulses allows get additional information in coherent control experiments.

The emitted electric field is a short transient on a time scale given by the coherence time of the polarization, typically a few picoseconds. The spectrometer produces Fourier transform of the emitted electric field, which in turn is proportional to the third-order nonlinear polarization. Detector records the absolute square of the Fourier transform as a spectrum of the nonlinear signal for each delay τ between the incident pulses. Quantum beats, polarization interferences, and basic properties of the spectrally resolved TFWM and coherent control experiment are described using Optical Bloch Equations taking into account inhomogeneous broadening of excitonic excited states

Acknowledgements

We thank J.M. Hvam and D. Birkedal for enlightening discussions, Rino Di Bartolo and his coworkers for a wonderful time in a friendly atmosphere. This work was supported by the Danish Natural Science Research Council, Otto Mønsteds Foundation, DFG and the Leibnitz prize.

References

1. Peyghambarian, N., Koch, S.W., and Mysyrowicz, A. (1993) *Introduction to Semiconductor Optics*, Prentice Hall, New Jersey.
2. Göbel, E.O., and Ploog, K. (1990) *Progr. Quant. Electr.* **14**, 289.
3. Bastard, G. (1988) *Wave Mechanics Applied to Semiconductor Structures*, Les Editions de Physique, Cedex.
4. Park, S.H., Morhange, J.F., Jeffery, A.D., Morgan, R. *et al.* (1988) *Appl. Phys. Lett.* **52**, 1201.
5. Zory, P.S. (1993) *Quantum Well Lasers*, Academic Press, London.
6. Hulin, D., Mysyrowicz, A., Antonnetti, A., Migus, A., *et al.* (1986) *Appl. Phys. Lett.* **49**, 749.
7. Javan, A., Bennet, W.B., Herriott, D.R. (1961) *Phys. Rev. Lett.* **6**, 106.
8. Meystre, P., and Sargent III, M., *Elements of Quantum Optics* (1990), Springer Verlag, Berlin.
9. Shen, Y.R. (1984) *Principles of Nonlinear Optics*, Wiley & Sons, New York.
10. Allen, L., and Eberly, J.H. (1975) *Optical Resonances and Two-Level Atoms*, Wiley, New York.
11. Haroche, S, Paisner, J.A., and Schawlow, A.L. (1973) *Phys. Rev. Lett.* **30**, 948.
12. Kurnit, N.A., Abella, I.D., and Hartmann, S.R. (1964) *Phys. Rev. Lett.* **13**, 567.
13. Yajima, T., and Taira, Y. (1979) *J. Phys. Soc. Japan* **47**, 1620.
14. Leo, K., Wegener, M., Shah, J., Chemla, D.S. *et al.* (1990) *Phys. Rev. Lett.* **65**, 1340.
15. Wegener, M., Chemla, D.S., Schmitt-Rink, S., and Schafer, W. (1990) *Phys. Rev. A.* **42**, 5675.
16. Haug, H., Koch, S.W. (1990) *Quantum Theory of the Optical and Electronic Properties of Semiconductors,* World Scientific, Singapore.
17. Thomas, D.G., and Hopfield, J.J. (1961) *Phys. Rev.* **124**, 657.
18. Elliott, R.J. (1957) *Phys. Rev.* **108**, 1384.
19. Hopfield, J.J. (1958) *Phys. Rev.* **112**, 1555.
20. Ulbrich, R.G., and Fehrenbach, G.W. (1979) *Phys. Rev. Lett.* **43**, 963.
21. Eccleston, R., Feuerbacher, B.F., Kuhl, J., Rühle, W.W. *et al.* (1992) *Phys. Rev. B* **45**, 11403.
22. Wang, H., Shah, J., Damen, T.C., and Pfeiffer, L.N. (1995) *Phys. Rev. Lett.* **74**, 3065.
23. Honold, A., Schultheis, L., Kuhl, J., and Tu, C.W. (1988) *Appl. Rev. Lett.* **52**, 2105.
24. Schultheis, L., Kuhl, J., Honold, A., and Tu, C.W. (1986) *Phys. Rev. Lett.*, **57**, 1635.
25. Schmitt-Rink, S., Chemla, D.S., and Miller, D.A.B. (1985) *Phys. Rev. B*, **32**, 6601.
26. Fehrenbach, G.W., Schäfer, W., Treusch, J., and Ulbrich, R.G. (1982) *Phys. Rev. Lett.* **49**, 1281.
27. Hegarty, J., Sturge, M.D., Weisbuch, C., Gossard, A.S., and Wiegmann (1982) *Phys. Rev. Lett.* **49**, 930.
28. Webb, M.D., Cundiff, S.T., and Steel, D.G. (1991) *Phys. Rev. Lett.* **66**, 934.

29. Weisbuch, C., Dingle, R., Gossard, A.C., and Wiegmann (1982) *Sol. State Commun.* **38**, 709.
30. Warwick, C.A., Jan, W.Y., Ourmazd, A., and Harris, T.D. (1990) *Appl. Phys. Lett.* **56**, 2666.
31. Birkedal, D., Lyssenko, V.G., Pantke, K.-H., Erland, J., and Hvam, J.M. (1995) *Phys. Rev. B* **51**, 7977.
32. Göbel, E.O., Leo, K., Damen, T.C., Shah, J. *et al.* (1990) *Phys. Rev. Lett.* **64**, 1801.
33. Leo, K., Damen, T.C., Shah, J,, Göbel, E.O., and Kohler, K. (1990) *Appl. Phys. Lett.* **57**, 19.
34. Feuerbacher, B.F., Kuhl, J., Eccleston, R., and Ploog. K. (1990) *Sol. State Commun.* **74**, 1279.
35. Leo, K., Göbel, E.O., Damen, Shah, J., Schmitt-Rink S. *et al.*, (1991) *Phys. Rev. B* **44**, 5726.
36. Koch, M., Feldmann, J., von Plessen, G., Göbel, E.O., *et al.*, (1992) *Phys. Rev. Lett.* **69**, 3631.
37. Cundiff, S.T. (1994) *Phys. Rev. A*, **49**, 3114.
38. Lyssenko, V.G., Erland, J., Balslev, I., Pantke, K.-H. *et al*, *Phys. Rev. B* **48**, 5720.
39. Erland, J., Balslev, I. (1993) *Phys. Rev. A*, **48**, 1765.
40. Erland, J., Pantke, K.-H., Mizeikis, V., Lyssenko, V.G., and Hvam, J.M. (1994) *Phys. Rev. B* **50**, 15047.
41. Paul, A.E., Sha, W., Patkar, S., and Smirl, A.L. (1995) *Phys. Rev. B* **51**, 4242.
42. Pantke, K.-H., Oberhauser, D., Lyssenko, V.G., Hvam, J.M..and Wiemann, G. (1993) *Phys. Rev. B* **47**, 2413.
43. Bar-Ad, S., and Bar-Joseph, I. (1992) *Phys. Rev. Lett.* **68**, 349.
44. Lovering, D.J., Phillips, R.T., Denton, G.J., and Smith, G.W. (1992) *Phys. Rev. Lett.* **68**, 1880.
45. Feuerbacher, B.F., Kuhl, J., and Ploog, K. (1991) *Phys. Rev. B* **43**, 2439.
46. Saiki, T., Kuwata-Gonokami, M., Matsusue, T., and Sakaki, H. (1994) *Phys. Rev. B* **49**, 7817.
47. Ferrio, K.G., and Steel, D.G. (1996) *Phys. Rev. B*, **54**, 5231.
48. Smith, G.O., Mayer, E.J., Kuhl, J., and Ploog, K. (1994) *Sol. State Commun.* **92**, 325.
49. Banyai, L., Galbraith, I., and Haug, H. (1988) *Phys. Rev. B*, **38**, 3931.
50. Pantke, K.-H, Lyssenko, V.G., Razbirin, B.S., Erland, J., and Hvam, J.M. (1993) *Proc. 21st Int. Conf. on the Physics of Semiconductors, Beijing*, 1992, World Scientific Publishing Co, Singapore.
51. Erland, J., Razbirin, B.S., Lyssenko, V.G., Pantke, K.-H., and Hvam J.M. (1994) *J. Crystal Growth*, **138**, 800-803.
52. Schultheis, L., Kuhl, J., Honold, A., and Tu, C.W. (1986) *Phys. Rev. Lett.* **57**, 1635-1638.
53. Lohner, A., Rick, K., Leisching, P., Leitenstofer, A., Elsaesser, T., Kuhn, T., Rossi, F., and Stolz, W. (1993) *Phys. Rev. Lett.* **71**, 77-80.
54. Schultheis, L., Honold, A., Kuhl, J., Köhler, K., and Tu, C.W. (1986) *Phys. Rev. B*, **34**, 9027.
55. Albrecht, T.F., Sandmann, J.H.H., Feldmann, J., Stolz, W., Göbel, E.O., Nebel, A., Fallnich, C., and Beigang, R. (1993) *Appl. Phys. Lett.* **63**, 1945.
56. Weisbuch, C., Dingle, R., Gossard, A.C., and Weigmann, W. (1981) *Sol. State Commun.* **38**, 709.
57. Noll, G., Siegner, U., Shevel, S.G., and Gobel, E.O. (1991) *Phys. Rev. Lett.* **64**, 792.
58. Schwab, H., Lyssenko, V.G., and Hvam, J.M. (1991) *Phys. Rev.* B **44**, 3999.
59. Abramowitz and Stedun (1972) *Handbook of Mathematical Functions*, chap. 7, Dover Publication Inc., New York
60. Cundiff, S.T. (1994) *Phys. Rev. A*, **49**, 3114.
61. Freidberg, R., Hartmann, S.R., and Manassah, J.T. (1989) *Phys. Rev. A*, **40**, 2446.
62. Ben-Aryeh, Bowden, C.M., and Englund (1986) *Phys. Rev. A*, **34**, 3917.
63. Maki, J.J., Malcuit, M.S., Sipe, J.E., and Boyd, R.W. (1991) *Phys. Rev. Lett.* **67**, 972.
64. Weiss, S., Mycek, M.-A., Bigot, J.-Y., Schmitt-Rink, S. et al. (1992) *Phys. Rev. Lett.* **69**, 2685.
65. Schmitt-Rink, S., Mukamel, S., Leo, K., Shah, J., *et al.* (1991) *Phys. Rev. A*, **44**, 2124.
66. Hanamura, E. (1983) *J.Phys. Soc. Jpn.* **52**, 3678.
67. Wang, H., Ferrio, K., Steel, D.G., Hu, Y.Z. *et al.* (1993) *Phys. Rev. Lett.* **71**, 1261.
68. Yeazell, J.A., and Stroud, Jr., C.R. (1988) *Phys. Rev. Lett.* **60**, 1494.
69. Warren, W.S. , Rabitz, H., and Dahleh, M. (1993) *Science* **259**, 1581.

70. Zare, R.N. (1998) *Science* **279**, 1875.
71. Weinact, T.C., Ahn, J., and Bucksbaum, P.H. (1998) *Phys. Rev. Lett.* **80**, 5508.
72. Leichtle, C., Schleich, W.P. , Averbukh, I. S., and Shapiro, M.(1998) *Phys. Rev. Lett.* **80**, 1418
73. Heberle, A.P., Baumberg, J.J., and Köhler, K. (1995) *Phys. Rev. Lett.* **75**, 2598.
74. Marie, X., Le Jeune, P., Amand, T., Brousseau, M. *et al.* (1997) *Phys. Rev. Lett.* **79**, 3222.
75. Bonadeo, N.H., Erland, J., Gammon, D., Park, D. *et al.*(1998) *Science* **282**, 1473.
76. Wehner, M.U., Ulm, M.H. Chemla, D.S. and Wegener, M. (1998) *Phys. Rev. Lett.* **80**, 1992.
77. Gurioli, M., Bogani, F., Ceccheroini, S., and Colocci, M. (1997) *Phys. Rev. Lett.* **78**, 3205.
78. Wörner, M., and Shah, J. (1998) *Phys. Rev. Lett.* **81**, 4208.
79. Garro, N., Snelling, M.J., Kennedy, S.P. Phillips, R.T. *et al.* (1999) *Phys. Rev. B*, **60**, 4497.
80. Gammon, D., Snow, E.S., Shanabrook, B.V., Katzer, D.S. *et al.* (1996) *Science* **273**, 87.
81. Erland, J., and Balslev, I. (1993) *Phys. Rev. A*, **48**, 1765.
82. Erland, J. Pantke, K.-H., Mizeikis, V., Lyssenko, V.G., and Hvam, J.M. (1994) *Phys. Rev. B* **50**, 15, 047.
83. Stolz, H. *Time-resolved Light Scattering from Excitons*, (1994) Springer Tracts in Modern Physics, **130** Spring-Verlag, New York.
84. Yee, D.S., Yee, K.J., Hohng, S.C., Kim, D.S. *et al.* (2000) *Phys. Rev. Lett.* **84**, 3474.
85. Gammon, D., Shanabrook, B.V., and Katzer, D.S. (1991) *Phys. Rev. Lett.* **67**, 1547.
86. Birkedal, D., Lyssenko, V.G., Pantke, K.-H., Erland, J., and Hvam, J.M. (1995) *Phys. Rev. B*, **51**, 7977.
87. Erland, J., Kim, J.C., Bonadeo, N.H., Steel, D.G. *et al.* (1999) *Phys. Rev. B,* **60**, 8497.

DYNAMICS OF UPCONVERSION IN LASER CRYSTALS

GONUL OZEN
Department of Physics
Istanbul Technical University
Maslak-Istanbul/TURKEY
gonul@bc.edu, gozenl@itu.edu.tr,

The Praseodymium ion in solids exhibits emission from a number of levels located in different regions of the spectrum and presents various spectral features that include anti-Stokes emission, otherwise called up-conversion. After a brief review of this phenomenon in general terms, we introduced in detail our experimental results for several laser crystals such as $LiLuF_4$ and $Y_3Al_5O_{12}$ doped with Praseodymium. Our experimental results included the effect of the selective excitation on the dynamics of the upconversion of the exciting energy by measuring decay patterns, emission spectra and excitation spectra as function of excitation wavelength and temperature in the range 24-300K. The experimental data indicate that the mechanism responsible for the upconversion can be controlled by selectively exciting the sample.

B. Di Bartolo and O. Forte (eds.), Advances in Spectroscopy for Lasers and Sensing, 333.

WAVEGUIDE FABRICATION METHODS IN DIELECTRIC SOLIDS

Optical channel waveguides in sapphire and Ti:sapphire

MARKUS POLLNAU
Ecole Polytechnique Fédérale de Lausanne
Institute of Imaging and Applied Optics
CH-1015 Lausanne
Switzerland
m.pollnau@ewi.utwente.nl,

1. Introduction

In recent years, the field of waveguide and integrated optics has attracted a significant amount of attention. While Silicon and III-V semiconductor technology in optics have matured to a certain extent and also dielectric waveguides in glasses have regained interest because of the development of simple waveguide fabrication methods such as ion-exchange/ion-in-diffusion techniques and femtosecond laser writing, crystalline dielectric materials have, perhaps with the exception of $LiNbO_3$, not lived up to the early expectations in the field. Nevertheless, this class of materials can offer distinct advantages over the aforementioned materials, most notably the simplicity of light generation, amplification, tunability, and modulation in a large wavelength range with high absorption and emission cross-sections. Such active waveguide devices profit from the tight confinement and excellent mode overlap of pump and signal beams.

In the following section, the investigation of different methods for the fabrication of surface and buried channel waveguides in an important crystalline material, sapphire – in the majority of the cases activated by Ti^{3+} ions – will be described. These methods include direct laser ablation, ion beam implantation followed by wet chemical etching, reactive ion etching or Ar ion milling, light ion beam implantation, and femtosecond laser irradiation. Potential applications of Ti:sapphire channel waveguides like low-threshold, widely tunable channel waveguide lasers and broadband luminescent emitters as white-light sources for interferometry will be discussed in section 3.

2. Fabrication Methods

If the question were asked why crystalline dielectric materials have been largely neglected for use as channel waveguides in integrated optics, the most obvious answers would probably be the lack of availability of suitable substrates and/or the difficult layer growth processes as well

B. Di Bartolo and O. Forte (eds.), Advances in Spectroscopy for Lasers and Sensing, 335–350.

as the painstaking efforts that have to be undertaken when structuring their surfaces. As we will see in the following subsection, it is indeed difficult to structure the surface of a hard crystalline material like sapphire with high accuracy and to find a suitable mask material that is not removed much faster than the hard crystalline layer. In the second subsection, other routes are pointed out that can avoid the initial layer growth process and work in the bulk of the material, thereby enabling three-dimensional (3D) structures. None of these methods has been developed to perfection, but researchers are learning to distinguish their advantages and disadvantages for waveguide fabrication in specific crystalline dielectric materials.

2.1. SURFACE CHANNEL WAVEGUIDES

Pulsed laser deposition (PLD) has proven to be a reliable technique for the growth of waveguiding films of various laser host materials, albeit with rather high propagation losses due to the inclusion of particulates ablated from the target into the deposited layer, and to be suitable for tailoring the concentration and valence state of dopants in the films. Control of the trivalent oxidation state of the incorporated titanium dopant in layers of sapphire is of particular importance in order to avoid adverse effects such as parasitic absorption and subsequent limitations in the waveguide laser performance due to the presence of tetravalent titanium [1]. Laser operation of PLD-grown Ti:sapphire planar waveguides [2] was demonstrated several years ago [3]. Such Ti:sapphire planar waveguides serve as samples for the subsequent fabrication of rib channel waveguides. Suitable methods for surface structuring of sapphire and characterization of Ti:sapphire rib channel waveguides will be described in this subsection. Also ion in-diffusion of Ti^{3+} ions into sapphire can be employed as a method to produce Ti:sapphire channel waveguides [4].

2.1.1. *Direct Laser Ablation*

Structuring by direct laser ablation was performed [5] with a Lambda Physik LPX 205 ArF* excimer laser (λ = 193 nm, pulse duration ~20 ns) in a clean-room environment. The laser beam passed through a rectangular mask and was focused on the substrate using a 10x demagnifying refractive projection objective. The image of the mask on the substrate was 500 x 100 μm^2. The substrate was mounted on a high-precision computer-controlled X-Y-stage and was scanned during irradiation perpendicular to the incident laser beam. We created channels of 100 μm in width, separated by different distances (15 to 40 μm) which determined the widths of the rib structures. Our experiments were performed either in static or in scanning irradiation modes. The static irradiations were performed with different numbers of pulses (1, 10, 20, 40, 100, 200, and 400) and laser fluences (between 0.6 and 2.45 J/cm^2) at 10 Hz pulse-repetition rate. The average ablation depth per pulse measured for irradiation with 400 pulses varies between 2 nm/pulse for 1 J/cm^2 fluence and 15 nm/pulse for 2.45 J/cm^2 fluence. The minimum number of pulses required to initiate ablation decreases as the laser fluence increases, e.g. 20 pulses at 1 J/cm^2 and 10 pulses at 1.35 J/cm^2. For fluences higher than 1.7 J/cm^2, irradiation with a single pulse leads to ablation. The lower fluence ablations are accompanied by an incubation effect during which sufficient damaging is accumulated before ablation starts. For the ablation of parallel channels that define the rib structures, we scanned the substrate under the laser beam with a speed of 250 μm/s, at 50 Hz pulse-repetition rate and 2 J/cm^2 laser fluence. The obtained channels are ~2.5 μm in depth and possess a relatively smooth bottom surface as observed by scanning electron microscopy (SEM). However, the channel sidewalls have a rough surface, because extensive melting and re-solidification occurred at the boundaries of the ablated areas. In a number of regions, the sidewalls are even cracked due mainly to the large number of pulses needed to ablate sufficiently deep structures. A similar phenomenon was also observed for the static ablation with a similar fluence and 200 pulses. The re-deposition of the ablated material in the vicinity of the ablated areas can also increase the optical losses of the fabricated ribs. This problem can be overcome by either employing laser ablation under an inert gas flow or performing it through a pre-deposited ultra-thin polymer layer on top of the sapphire substrate and which can be easily washed away after the ablation process.

2.1.2. *Ion Beam Implantation and Wet Chemical Etching*

3D micro-structures that are of interest for many applications can be fabricated in sapphire by ion beam implantation (IBI) followed by wet chemical etching [6]. In a first step, IBI is used to tailor an appropriate damage profile within the crystal. In a second step, wet chemical etching is performed to selectively remove the damaged material. IBI of sapphire produces damage clusters and, depending on ion type, dose and energy, even complete amorphization of implanted regions. We tailored damage profiles into the sapphire substrates by masking regions (for example in the form of stripes) that were not exposed to IBI. Since wire-type proximity masks impair the resolution and edge quality between the implanted and non-implanted areas, we used lithographically patterned contact-resist masks with a thickness of 1.8 μm. After IBI, the masks were removed and the samples were etched in a bath of 85% H_3PO_4 at 165°C for varying times. This method exhibits a high etching selectivity between the implanted and non-implanted regions [7, 6] and results in rib structures with lower surface roughness. A continuous damage profile of the implanted regions up to the sample surface is required to obtain well-defined structures. For this reason, He^+ implantation was undertaken using a set of five energies (290, 190, 110, 50, and 25 keV) with a total implantation fluence of 5×10^{17} ions/cm^2. Etching resulted in ribs (corresponding to the non-implanted regions) with heights of up to 1.1 μm. The rib surface possesses a relatively low rms roughness of ~21 nm.

2.1.3. *Reactive Ion Etching or Ar Ion Milling*

Compared to the above mentioned methods, reactive ion etching (RIE) [8] and Ar ion milling [9] have proven to be more suitable for the fabrication of high-quality, low-roughness rib structures in sapphire and Ti:sapphire. Both methods deliver ribs of up to 5 μm in height, with smooth side-walls (Fig. 1) – provided that the mask edges are of sufficient high quality.

In particular, RIE has the advantage of reducing the surface roughness to values of a few nanometers [10, 11, 7]. We spin-coated a polyimide layer of 12 μm thickness onto the surface of the Ti:sapphire layer. This coating was then structured by laser ablation with an ArF* excimer laser at λ = 193 nm. In this way, we defined a polyimide mask with channels of 100-μm width and separated by 15 μm. RIE was then performed in a 1:1-BCl_3:Cl_2 atmosphere at 3-mTorr pressure by use of an inductive plasma system (STS Multiplex ICP) at 800 W of inductive power. The flow rate of each of the gaseous mixture components was 50 sccm. During the process, the sample was maintained at a constant temperature of approximately 20°C. The etch rate of the sapphire was 45 nm/min and the etching selectivity between the sapphire and the mask was 1:3.5. The remaining polyimide was removed by acetone. The structures obtained by this procedure were investigated by profilometry and SEM. Additional results obtained by contact-mode atomic force microscopy (AFM) showed that the roughness of the sapphire substrates was lowered from an rms value of 15.9 nm to an rms value of 3.7 nm for the etched regions (Fig. 1). The experimental conditions for Ar ion milling can be found in Ref. [9].

We have applied the latter two methods and demonstrated the first Ti:sapphire channel waveguides [8, 9]. The mentioned features lead to low-loss signal propagation and transverse

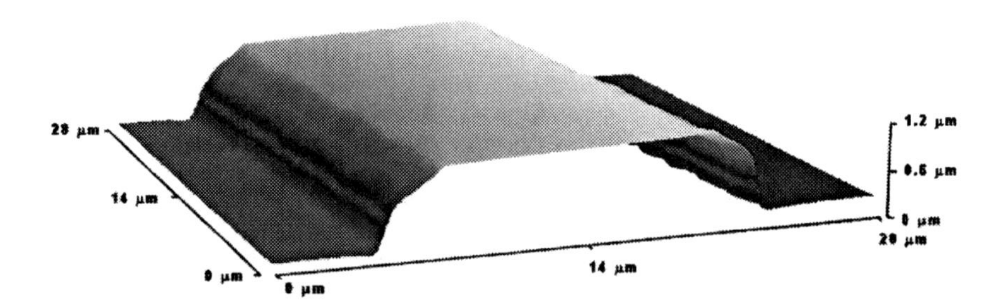

Figure 1. AFM contact-mode image of a reactive-ion-etched ~1.2 µm high by ~15 µm wide rib in a ~10 µm thick Ti:sapphire planar waveguide. Rib heights of up to 5 µm have been achieved. (Figure taken from [8]).

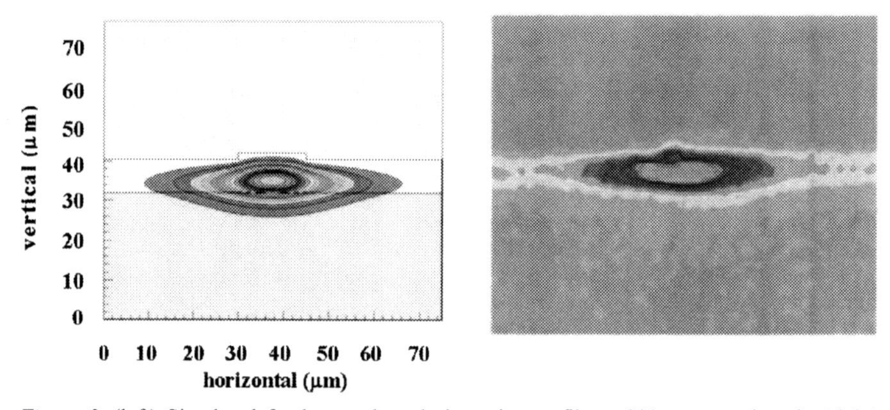

Figure 2. (left) Simulated fundamental-mode intensity profile at 800-nm wavelength. (right) Experimentally recorded channel-waveguide luminescence emission from the rib structure presented in Fig. 1. (Figure taken from [8]).

fundamental-mode luminescence output (Fig. 2) in a broad spectral range centered near 760 nm of several hundreds of μW. Further improvement in output power can be expected from exploiting gain to produce amplified spontaneous emission. However, spectral narrowing of the luminescent output with correspondingly reduced longitudinal resolution in the interferometric application may partly counteract the improvement in luminescence output power.

2.2. BURIED CHANNEL WAVEGUIDES

Generally, device performance of channel waveguides can be considerably improved by burying the guiding structure into the bulk of the sample. Advantages of buried waveguides derive not only from surface scattering losses being avoided, but also from a reduction in mode asymmetry compared to surface waveguides, thus providing higher efficiency for mode coupling to optical fibers.

Most of the methods reported for the fabrication of buried channel waveguides are appropriate for glasses. Ion exchange followed by a field-assisted burial step [12] was demonstrated in glass substrates containing high concentrations of alkali ions. UV writing techniques, also in combination with bonding [13], necessitate photosensitive material. Femtosecond laser irradiation [14] and focused IBI [15, 16] allow for direct writing of buried channel waveguides in amorphous materials, because a positive refractive index change is induced in the modified regions. In contrast, irradiation of crystalline matrices often produces negative index changes [17], although positive index changes induced by IBI have been observed in a few crystalline materials, such as $LiNbO_3$ [18], $KNbO_3$ [19], and Nd^{3+}:YAG [20].

Here, two methods are presented, which have produced the first buried channel waveguides in sapphire and Ti:sapphire. These methods are light IBI and femtosecond laser irradiation.

2.2.1. *Light Ion Beam Implantation*

We performed the fabrication of complex waveguiding structures in sapphire by high-energy proton implantation [21]. Compared to the significant damage created in sapphire by He^+ implantation (see subsection 2.1.2), which resulted in poor waveguiding quality, protons create less damage in the guiding region, thus assuring better waveguiding quality. Moreover, deeper damage profiles are obtained with protons, for the same incident energy, thus providing larger design flexibility. Good control of the implantation parameters enables writing of precisely localized optical barriers with well-defined decrease of refractive index in a hard crystalline material. By exploiting this fact, we demonstrated high-quality surface, buried, and stacked planar as well as buried single and parallel channel waveguides in sapphire.

Pure c-cut, optically polished sapphire substrates were irradiated by use of a Van de Graaf accelerator operating at beam currents in the range of 0.6-0.8 μA/cm^2. Incident ion energies of 0.4-1.5 MeV and doses of 10^{15}-10^{16} ions/cm^2 were applied. Implantations were performed under controlled sample temperature near 30°C, with the sample surface slightly tilted off axis to avoid channeling effects. The expected accumulated damage profiles were calculated by the "Stopping and Range of Ions in Matter" (SRIM) simulation code. The effective refractive indices of guided modes in surface planar waveguides were measured by m-line spectroscopy and the refractive index profiles were reconstructed by calculations based on the inverse

"Wentzel-Kramer-Brillouin" (WKB) method. Only negative refractive index changes were observed, with a decrease of typically 1% for doses in the order of 10^{16} ions/cm^2.

Double-energy implantation with similar energies widens the optical barrier and hence reduces waveguide losses due to light coupling to radiation modes. Successive implantations with significantly different energies produce an alternation of high and low refractive-index regions, resulting in a series of high-index layers, which according to the chosen implantation parameters, can permit or prohibit guiding. In the experimental example of Fig. 3(a), the dimensions and induced refractive-index changes were adjusted to permit fundamental mode propagation at 780 nm in a buried waveguide region, with a symmetric mode profile in the vertical direction optimized for coupling to a commercial 780-nm single-mode fiber, but prohibit light propagation in the superficial planar layer. A theoretical mode profile, Fig. 3(b), was calculated with a simplified step-index profile, Fig. 3(c). The experimental mode pattern, Fig. 3(a), is in accordance with our theoretical calculation, Fig. 3(b), both in terms of localization of the guided light in the buried planar waveguide and improvement of the mode symmetry in the vertical direction compared to surface waveguides. The mode shape and degree of asymmetry in the vertical direction can be varied by adjusting the implantation energies and doses. In m-line measurements, 780-nm light could not be coupled through the prism into the superficial layer at any incident angle, thus confirming that our design prohibits guiding in the superficial layer at this wavelength.

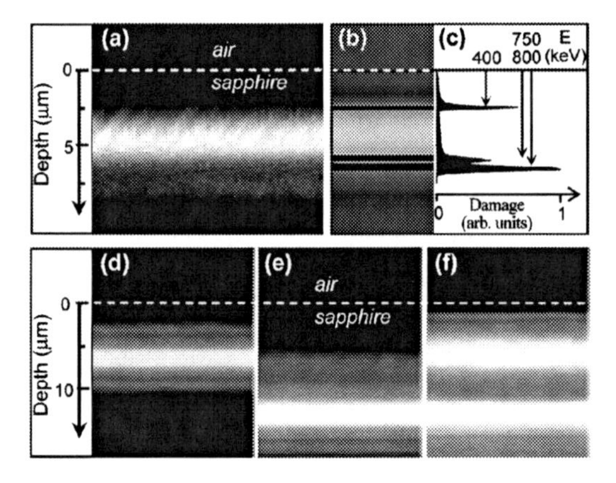

Figure 3. Mode patterns of 780-nm fundamental-mode laser light end-coupled to sapphire planar waveguides. (a) Experimental output profile recorded from a buried planar waveguide, (b) corresponding contour plot simulated with a simplified step-index profile estimated from (c) the accumulated damage profile calculated by SRIM simulations. Experimental output profiles recorded from two stacked planar waveguides. Individual excitation of (d) the upper and (e) the lower layer and (f) simultaneous excitation of both layers. The difference in the appearance of the upper guiding layer's width is due to different attenuation in front of the CCD camera. (Figure taken from [21]).

Following the same approach of multiple energy implantations, stacked planar waveguides were fabricated by two double-implanted barriers. This procedure creates a superficial and a buried planar waveguide. The output profiles recorded after end-coupling fundamental-mode laser light at 780 nm into the sample show that the upper and lower guiding layers can be excited individually, Figs. 3(d) and (e), or simultaneously, Fig. 3(f). Higher-order, stacked structures can be fabricated by implanting additional damaged layers with suitable incident ion energies and doses, however the procedure is ultimately limited by the maximum H^+ energy available from the ion accelerator.

Channel waveguides can be obtained by writing two horizontally confined, damaged areas into the guiding layer of a planar waveguide. We fabricated such sidewalls with lower refractive index by ion implantation through a slit [20]. Starting from the buried planar waveguide previously described, each of the two sidewalls was formed by, e.g., three vertically stacked, 20 μm wide, damaged areas produced inside the guiding layer by use of a 1-MeV proton beam with incident angles of 35°, 45°, and 55° and doses of 4×10^{15}, 1×10^{16}, and 4×10^{15} ions/cm^2, respectively. The approach of changing the angle of incidence rather than the ions' energy has the advantage of generating a wider optical barrier in the normal direction due to the radial dispersion of the ion beam. Figure 4 presents a comparison between (a) the experimental output profile of end-coupled, fundamental-mode laser light at 780 nm and (b) the corresponding calculated contour plot in a 6 μm wide buried channel waveguide defined by a buried planar waveguide with two horizontal sidewalls. Its fundamental-mode profile is elliptical as a result of the differences of channel dimensions and optical confinement in the horizontal and vertical directions. Cylindrical mode shapes or any desired degree of ellipticity between the two orthogonal directions can be obtained by adjusting the implantation energies, doses, and geometry. Finally, parallel buried channel waveguides were demonstrated by repeatedly writing horizontal optical barriers into a buried planar waveguide. Figure 4(c) presents the output profile of three parallel buried channel waveguides after excitation. The

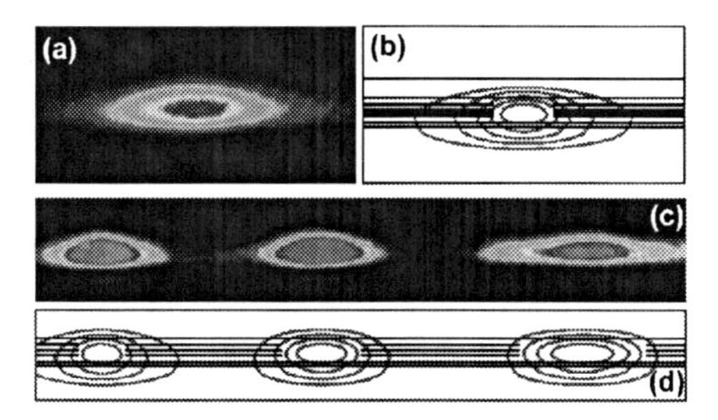

Figure 4. Mode patterns of 780-nm fundamental-mode laser light end-coupled to sapphire buried channel waveguides. (a) Experimental output profile recorded from a buried channel waveguide and (b) corresponding simulated contour plot. (c) Experimental output profile recorded from three parallel buried channel waveguides with different widths of 6, 10, and 15 μm and (d) corresponding simulated contour plot. (Figure taken from [21]).

difference in mode shapes results from the different widths (6, 10, and 15 μm) chosen for the three channel waveguides, Fig. 4(d).

The propagation losses in two 6 μm wide, 7.5-mm long channel waveguides at 633 nm were studied by a simple, not very accurate method. An upper limit of 2 dB/cm was determined for the propagation loss in a channel waveguide with three equidistant lateral barriers. In a channel waveguide fabricated with four equidistant lateral barriers, the propagation loss could be reduced to 0.7 dB/cm, indicating that the lateral light confinement is a crucial feature of ion-implanted channel waveguides.

2.2.2. *Femtosecond Laser Irradiation*

Nonlinear absorption of femtosecond laser pulses has been employed in order to induce structural changes by micro-explosions in numerous materials and also for fabricating waveguide structures. Femtosecond pulses can be used for precise micro-machining of materials, as the pulse energy is transferred to electrons and the optical excitation ends before the lattice is perturbed. The energy deposition is based on multiphoton absorption and avalanche ionization; the nonlinearity of the process can be exploited in order to write structures into the bulk of materials. Therefore, femtosecond writing provides the possibility of implementing 3D optical integrated circuits. Another advantage of this fabrication process is the capability of rapid prototyping of a device without the need for any photolithographic process.

We reported writing of waveguides by femtosecond laser pulses in a hard crystalline material, Ti:sapphire [22]. Waveguiding is observed around micro-damaged areas induced by femtosecond irradiation. We found that the threshold for creating micro-damage by femtosecond irradiation in sapphire is greatly decreased when the crystal is doped with Ti^{3+} ions. The process also shows promise for 3D integrated circuits in other hard crystalline materials when sensitized by an appropriate doping ion.

The writing system consisted of a Ti:sapphire laser at a repetition rate of 1 kHz, with a center pulse wavelength of 790 nm. The pulse duration was 150 fs and the pulse energy was varied from 0.5 to 6 μJ. For focusing the laser beam, a 0.3 NA microscope objective was used. The focal point was inside the bulk of the material (100 to 300 μm deep) with a spot-size of approximately 10 μm. The sample was held on a translation stage that moved in a direction perpendicular to the writing beam; the writing speed was varied from 9 to 17 μm/s. One sample of pure c-cut sapphire and one sample of Ti:sapphire with Ti^{3+} concentration of 0.21 at.% were irradiated. The undoped sapphire did not exhibit any damage or refractive-index modification, whereas channels approx. 10 mm in length were successfully written inside the bulk of the Ti:sapphire sample.

In Fig. 5(a), a microscope image of the end-face of two irradiated regions in the Ti^{3+}:sapphire sample is shown. Using the technique of digital holography [23, 24], the optical path differences in a 1.1 mm thick slice cut from the irradiated sample were monitored with the probing laser beam passing through the sample in the waveguide direction. Two line profiles were taken, line (I) slightly above and line (II) directly through the damaged regions, as indicated in Fig. 5(a). From the measured optical path differences, see Fig. 5(b), positive and negative refractive-index changes of approximately 1 and -2×10^{-4} obtain in the regions above the tips of the damaged regions and directly in the damaged regions, respectively.

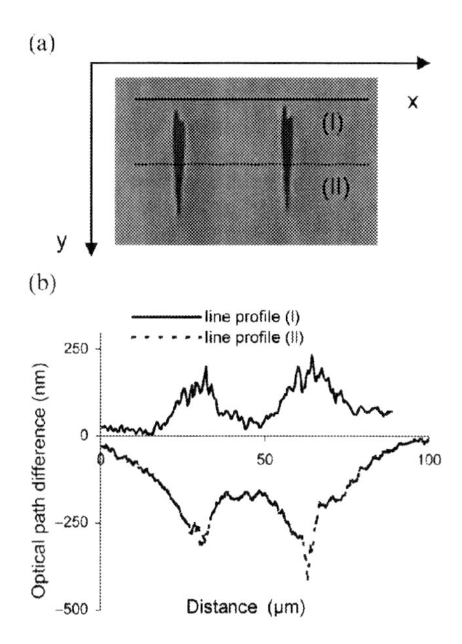

Figure 5. (a) Microscope image of the end-face of the irradiated Ti^{3+}:sapphire sample. (b) Optical path difference measurements using the technique of digital holography, performed in the waveguide direction while scanning along the lines (I) and (II) indicated in part (a). (Figure taken from [22]).

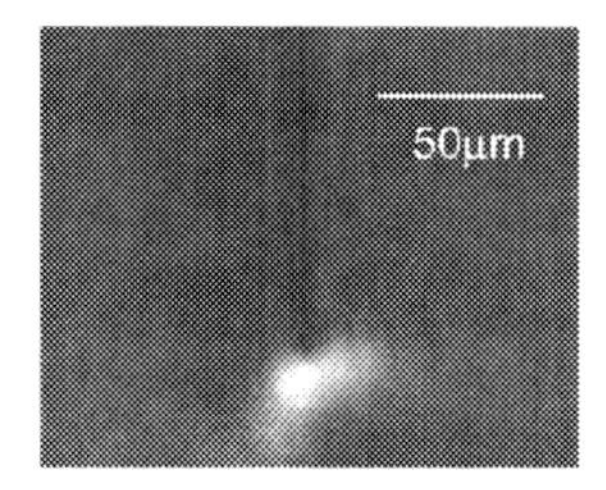

Figure 6. Mode profile (633 nm) of a waveguide written into the bulk of a Ti^{3+}:sapphire sample. (Visible diagonal fringes are due to a detection artifact). (Figure modified from [22]).

Consequently, the waveguides were always formed in those regions of the Ti:sapphire sample that are adjacent to the upper and lower tips of the regions damaged by the femtosecond irradiation. The light of a He-Ne laser at 633 nm was coupled into the polished end-face of the sample using a microscope objective. A mode image of guided light is shown in Fig. 6. Also an Ar-ion laser operating on all visible lines was launched into the waveguide in π-polarization in order to pump the active Ti^{3+} ions. The guided infrared fluorescence at the output of the waveguide was measured using an optical spectrum analyzer. The spectral shapes of fluorescence emitted from waveguide and bulk regions of the Ti:sapphire sample were identical, indicating that the Ti^{3+} fluorescence is not quenched by irradiation-induced strain.

The fact that femtosecond irradiation using the same parameters did not show any effect in undoped sapphire indicates that the threshold for creating micro-damage by femtosecond irradiation is greatly decreased when sapphire is doped with Ti^{3+} ions. A possible reason is that a double two-photon absorption, firstly into the dopant's absorption band in the blue-green spectral region and secondly, from there into the band gap of the host, has a higher probability than four-photon absorption directly into the band gap of the host. In addition, the lattice distortion initially introduced to the sapphire lattice by the Ti^{3+} dopant, which is larger than the substituted Al^{3+} ion, may support the mechanism that induces the observed micro-damage. This recipe of sensitizing the host material by an optically active ion may well be employed also for other hard crystalline materials.

3. Applications

In this section, two potential applications of Ti:sapphire channel waveguides are discussed, namely a low-threshold, broadly tunable channel waveguide laser and a broadband luminescent emitter as a white-light source for interferometry.

3.1. CHANNEL WAVEGUIDE LASER

Ti:sapphire is particularly attractive as a laser medium, because it offers a widely tunable output in the wavelength range 650-1100 nm (Fig. 7) and also exhibits a wide absorption band in the wavelength range 400-650 nm. However, an adverse consequence of this wide tunability is a low peak emission cross-section, which combined with a short fluorescence lifetime, necessitates high pump-power densities to achieve efficient CW lasing. The undesirable prerequisite of high pump-power levels can be overcome by adopting a planar or channel waveguide geometry. Channel waveguides, in particular, are characterized by lower laser thresholds than their planar counterparts due to the additional lateral confinement and excellent overlap between the laser and pump modes provided that the rib fabrication process does not introduce any additional losses over those present in the planar waveguide. Thresholds of the order of a few tens of mW would allow use of inexpensive pump sources such as frequency-doubled solid-state lasers and blue-green diodes. Moreover, a combination of Ti:sapphire channel waveguides with semiconductor saturable absorber mirrors and saturable Bragg reflector technologies would open up the possibility to develop a compact, portable, broadly tunable laser with the potential to produce high-repetition-rate mode-locked pulses.

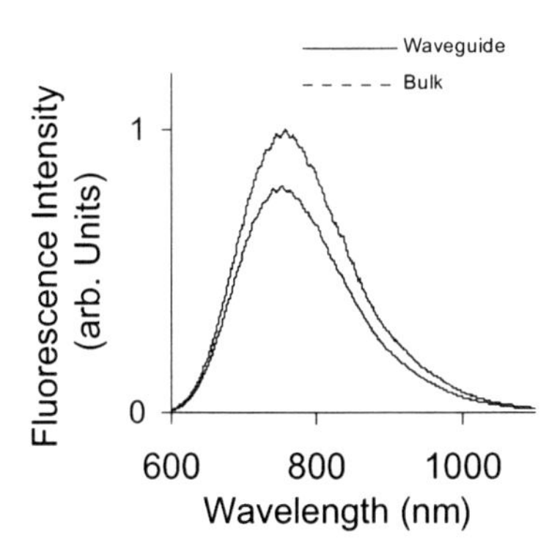

Figure 7. Comparison of spectral fluorescence output shapes from waveguide and bulk regions of a Ti^{3+}:sapphire sample. (Figure taken from [22]).

Ti:sapphire rib channel waveguides fabricated by Ar ion milling of pulsed-laser-deposited layers (section 2.1.2) have recently been operated as channel waveguide lasers [25]. Polarized laser emission was observed at 792.5 nm with an absorbed pump-power threshold of 265 mW, which is more than a factor of 2 lower in comparison to their planar counterparts [3]. Measured beam-propagation factors M_x^2 and M_y^2 of 1.3 and 1.2, respectively, indicated single-transverse-mode emission. A quasi-cw output power of 27 mW for an absorbed pump power of 1 W and a slope efficiency of 5.3% were obtained using an output coupler of 4.6% transmission with a pump duty cycle of 8%. Also in buried channel waveguides fabricated by light-ion implantation (section 2.2.1) lasing has been obtained [26].

3.2. BROADBAND LUMINESCENT EMITTER FOR INTERFEROMETRY

In recent years, broadband fiber interferometers have become very popular as basic instruments used in optical low-coherence reflectometers for diagnostics of fiber and integrated optics devices [27, 28] or in optical coherence tomography (OCT) for imaging applications in the biomedical field [29, 30]. Micrometer-resolution, tomographic images can be generated using a Michelson interferometer with a low-coherence light source. A major challenge in the further development and applicability of this method has been the improvement of both its spatial resolution and dynamic range.

The longitudinal resolution of such instruments is inversely proportional to the optical bandwidth of the light source. Standard, ~10 μm longitudinal-resolution OCT imaging can be performed using superluminescent diodes that have ~30 nm FWHM bandwidths centered near 800 nm. These sources are relatively inexpensive and have turn-key operation suitable for clinical use but provide limited resolutions due to their narrow bandwidths. Femtosecond Ti:sapphire lasers with a bandwidth of ~300 nm [31] have, therefore, been used as large-bandwidth, high-brightness light sources, and sub-cellular imaging with longitudinal resolution of ~1 μm has been demonstrated in this way [32]. Unfortunately, these systems are expensive and complex, limiting their widespread use.

Broadband luminescence from transition-metal-ion-doped crystalline materials such as, e.g., Ti:sapphire with luminescence gain bandwidths ranging from 670-1100 nm (Fig. 7) can also significantly improve the longitudinal resolution to < 2 μm [33], while avoiding the complexity of femtosecond light sources. However, the low brightness of luminescence from bulk materials has until recently been insufficient for achieving a useful dynamic range in OCT.

Output powers can be improved by an order of magnitude using a bulk crystal with high doping density and more efficient luminescent coupling. When pumping a Ti:sapphire crystal with 5 W of green laser power and retro-reflecting the pump and luminescent light, 40 μW of luminescence with a bandwidth of 138 nm can be coupled into a single-mode optical fiber [34]. Ultrahigh-resolution OCT imaging was performed with this light source using a dual-balanced fiber interferometer. A detection sensitivity of > 86 dB was achieved at an incident power of 7.3 μW on the sample. OCT imaging with < 2 μm longitudinal resolution in tissue was performed at low speed on an African frog (*Xenopus laevis*) tadpole *in vivo*. Individual cells and nuclei were easily resolved [34].

Although the obtained power levels do not permit high-speed imaging, this result shows that ultrahigh-resolution OCT imaging using a simple light source without the need for femtosecond Ti:sapphire lasers may be performed at low speed and is sufficient for many applications including imaging materials, *in vitro* specimens, or immobile *in vivo* specimens.

The brightness of such broadband light sources can be further improved by luminescence guiding. The suitability of Ti:sapphire planar waveguides as a high-brightness, low-coherence light source for interferometric applications has been demonstrated [35]. Spectrally broadband (~130 nm FWHM) luminescence with output powers of several hundreds of μW was obtained from a Ti(0.1%):sapphire planar waveguide with incident pump powers from an Ar-ion laser of up to 1 W. The detected output was fundamental-mode in the confined direction for a 10 μm-thick doped layer, whereas in the unconfined region it was multi-mode. While such power levels potentially lead to further significant improvement in interferometric applications, the fact that the output is multi-mode in one of the transverse directions leads to a low coupling efficiency to a single-mode fiber interferometer [36].

As we have seen in the previous section, guiding and fundamental-mode output in both transverse directions can be achieved by fabricating low-loss channel waveguides. Up to 300 μW of near-fundamental-mode luminescence output power has been obtained in this way [9, 37] (Fig. 8), which represents an order-of-magnitude improvement of power compared to a Ti:sapphire bulk sample. This transverse-fundamental-mode luminescence power can be coupled to a single-mode-fiber with high efficiency, if the numerical apertures of waveguide and fiber are matched.

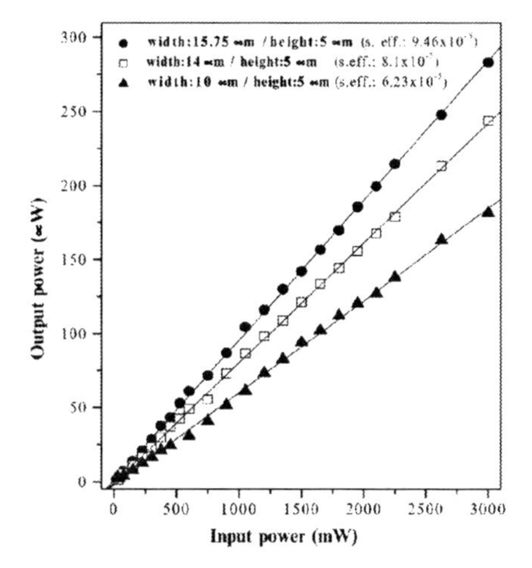

Figure 8. Luminescence output power versus pump input power at 488 nm obtained from a Ti^{3+}:sapphire channel waveguide. (Figure taken from [37]).

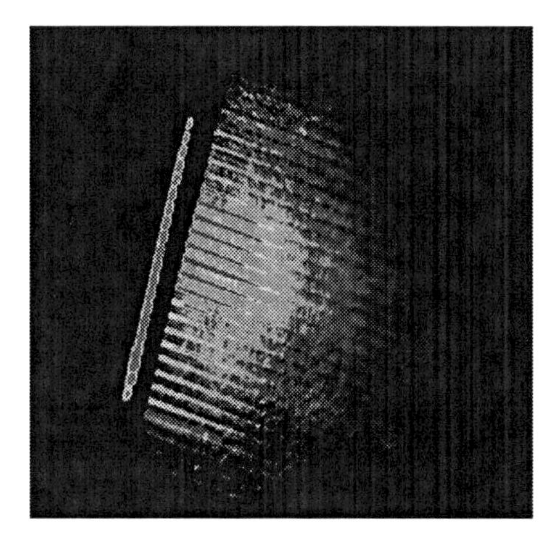

Figure 9. Top view of twenty 6 μm high microlenses pumping a parallel array of 10 μm Ti^{3+}:sapphire wide rib waveguides separated by 60 μm. (Figure taken from [37]).

Finally, parallel channel waveguides can be produced [9, 21] (Figs. 9 and 4), which can be excited via either a parallel micro-lens array [38] (Fig. 9), a coupled fiber bundle, or a channel-waveguide power splitter. In combination with a smart-pixel detector array [39], all-parallel OCT can potentially be achieved in one or possibly both lateral dimension, thereby avoiding lateral pixel-by-pixel scanning [40]. In such an experimental arrangement, a parallel light source can significantly improve the power per fundamental mode compared to a blackbody radiator, while avoiding the coherent cross-talk occurring when a femtosecond laser beam is used to illuminate the whole sample [41].

4. Conclusions

We have studied several different methods for micro-structuring Ti:sapphire planar waveguides to fabricate Ti:sapphire rib channel waveguides. Reactive ion etching and Ar^+ milling have so far proven to be the most suitable methods for the fabrication of high-quality, low-roughness rib structures. These methods have been successfully applied to the fabrication of Ti:sapphire channel waveguides. Buried channel waveguide structures have been obtained by light ion beam implantation or femtosecond laser irradiation, both allowing for 3D waveguide structures. Applications of Ti:sapphire channel waveguides include fundamental-mode luminescent emitters for white-light interferometry and low-threshold, widely tunable lasers.

5. Acknowledgments

The author thanks A. Crunteanu, L. Laversenne, V. Apostolopoulos, S. Rivier, G. Kulik, P. Hoffmann, T. Colomb, C. Depeursinge, and R.P. Salathé from the Ecole Polytechnique Fédérale de Lausanne, C. Grivas, T. Bhutta, T.C. May-Smith, D.P. Shepherd, and R.W. Eason from the University of Southampton, Ch. Buchal and A. Petraru from the Forschungszentrum Jülich, P. Moretti and J. Mugnier from the Université de Lyon, R. Osellame, G. Cerullo, and P. Laporta from the Politecnico di Milano, M. Flury, J. Vuille, and H.P. Herzig from the Université de Neuchâtel, and A.M. Kowalevicz, T. Ko, I. Hartl, and J.G. Fujimoto from the Massachusetts Institute of Technology for their contributions to the experiments.

6. References

1. Willmott, P.R., Manoravi, P., Huber, J.R., Greber, T., Murray, T.A., and Holliday, K. (1999) Production and characterization of Ti:sapphire thin films grown by reactive laser ablation with elemental precursors, *Opt. Lett.* **24**, 1581-1583.
2. Anderson, A.A., Eason, R.W., Jelinek, M., Grivas, C., Lane, D., Rogers, K., Hickey, L.M.B., and Fotakis, C. (1997) Growth of Ti:sapphire single crystal thin films by pulsed laser deposition, *Thin Solid Films* **300**, 68-71.
3. Anderson, A.A., Eason, R.W., Hickey, L.M.B., Jelinek, M., Grivas, C., Gill, D.S., and Vainos, N.A. (1997) Ti:sapphire planar waveguide laser grown by pulsed laser deposition, *Opt. Lett.* **22**, 1556-1558.
4. Apostolopoulos, V., Hickey, L.M.B., Sager, D.A., and Wilkinson, J.S. (2001), Gallium-diffused waveguides in sapphire, *Opt. Lett.* **26**, 1586-1588.
5. Crunteanu, A., Hoffmann, P., Pollnau, M., and Buchal, Ch. (2003) A comparative study on methods to structure sapphire, *Appl. Surf. Sci.* **208-209**, 322-326.

6. Crunteanu, A., Jänchen, G., Hoffmann, P., Pollnau, M., Buchal, Ch., Petraru, A., Eason, R.W., and Shepherd, D.P. (2003) Three-dimensional structuring of sapphire by sequential He^+ ion-beam implantation and wet chemical etching, *Appl. Phys. A* **76**, 1109-1112.
7. Xie, D., Zhu, D., Pan, H., Xu, H., and Ren, Z., (1998) Enhanced etching of sapphire damaged by ion implantation, *J. Phys. D: Appl. Phys.* **31**, 1647-1651.
8. Crunteanu, A., Pollnau, M., Jänchen, G., Hibert, C., Hoffmann, P., Salathé, R.P., Eason, R.W., Grivas, C., and Shepherd, D.P. (2002) Ti:sapphire rib channel waveguide fabricated by reactive ion etching of a planar waveguide, *Appl. Phys. B* **75**, 15-17.
9. Grivas, C., Shepherd, D.P., May-Smith, T.C., Eason, R.W., Pollnau, M., Crunteanu, A., and Jelinek, M. (2003) Performance of Ar^+-milled Ti:sapphire rib waveguides as single transverse mode broadband fluorescence sources, *IEEE J. Quantum Electron.* **39**, 501-507.
10. Sung, Y.J., Kim, H.S., Lee, Y.H., Lee, J.W., Chae, S.H., Park, Y.J., and Yeom, G.Y. (2001) High rate etching of sapphire wafer using $Cl_2/BCl_3/Ar$ inductively coupled plasmas, *Mat. Sci. Eng. B* **82**, 50-52.
11. Park, S.H., Jeon, H., Sung, Y.J., and Yeom, G.Y. (2001) Refractive sapphire microlenses fabricated by chlorine-based inductively coupled plasma etching, *Appl. Opt.* **40**, 3698-3702.
12. Jose, G., Sorbello, G., Taccheo, S., Cianci, E., Foglietti, V., and Laporta, P. (2003) Active waveguide devices by Ag-Na ion exchange on erbium-ytterbium doped phosphate glasses, *J. Non-Cryst. Solids* **322**, 256-261.
13. Gawith, C.B.E., Fu, A., Bhutta, T., Hua, P., Shepherd, D.P., Taylor, E.R., Smith, P.G.R., Milanese, D., and Ferrarris, M. (2002) Direct-UV-written buried channel waveguide lasers in direct-bonded intersubstrate ion-exchanged neodymium-doped germano-borosilicate glass, *Appl. Phys. Lett.* **81**, 3522-3524.
14. Davis, K.M., Miura, K., Sugimoto, N., and Hirao, K. (1996) Writing waveguides in glass with a femtosecond laser, *Opt. Lett.* **21**, 1729-1731.
15. Roberts A. and von Bibra, M.L.(1996) Fabrication of buried channel waveguides in fused silica using focused MeV proton beam irradiation, *J. Lightwave Technol.* **14**, 2554-2557.
16. Liu, K., Pun, E.Y.B., Sum, T.C., Bettiol, A.A., van Kan, J.A., and Watt, F. (2004) Erbium-doped waveguide amplifiers fabricated using focused proton beam writing, *Appl. Phys. Lett.* **84**, 684-686.
17. Townsend, P.D., Chandler, P.J., and Zhang, L. (1994) *Optical effects of ion implantation*, Knight, P.L. and Miller, A., eds., Cambridge University Press, Cambridge.
18. Rams, J., Olivares, J., Chandler, P.J., and Townsend, P.D. (2000) Mode gaps in the refractive index properties of low-dose ion-implanted LiNbO3 waveguides, *J. Appl. Phys.* **87**, 3199-3202.
19. Strohkendl, F.P., Fluck, D., Günter, P., Irmscher, R., and Buchal, Ch. (1991) Nonleaky optical wave-guides in $KNbO_3$ by ultralow dose MeV He ion-implantation, *Appl. Phys. Lett.* **59**, 3354-3356.
20. Moretti, P., Joubert, M.F., Tascu, S., Jacquier, B., Kaczkan, M., Malinowskii, M., and Samecki, J. (2003) Luminescence of Nd^{3+} in proton or helium-implanted channel waveguides in Nd:YAG crystals, *Opt. Mater.* **24**, 315-319.
21. Laversenne, L., Hoffmann, P., Pollnau, M., Moretti, P., and Mugnier, J. (2004) Designable buried waveguides in sapphire by proton implantation, *Appl. Phys. Lett.* **85**, 5167-5169.
22. Apostolopoulos, V., Laversenne, L., Colomb, T., Depeursinge, C., Salathé, R.P., Pollnau, M., Osellame, R., Cerullo, G., and Laporta, P. (2004) Femtosecond-irradiation-induced refractive-index changes and channel waveguiding in bulk Ti^{3+}:sapphire, *Appl. Phys. Lett.* **85**, 1122-1124.
23. Cuche, E., Bevilacqua, F., and Depeursinge, C. (1999) Digital holography for quantitative phase-contrast imaging, *Opt. Lett.* **24**, 291-293.
24. Cuche, E., Marquet, P., and Depeursinge, C. (1999) Simultaneous amplitude-contrast and quantitative phase-contrast microscopy by numerical reconstruction of Fresnel off-axis holograms, *Appl. Opt.* **38**, 6994-7001.
25. Grivas, C., Shepherd, D.P., May-Smith, T.C., Eason, R.W., and Pollnau, M. (2005) Single-transverse-mode Ti:sapphire rib waveguide laser, *Opt. Express* **13**, 210-215.
26. Grivas, C., Shepherd, D.P., Eason, R.W., Laversenne, L., Pollnau, M., and Moretti, P. (2006) to be published.
27. Youngquist, R.C., Carr, S., and Davies, D.E.N. (1987) Optical coherence domain reflectometry: A new optical evaluation technique, *Opt. Lett.* **12**, 158-160.
28. Gilgen, H.H., Novàk, R.P., and Salathé, R.P. (1989) Submillimeter optical reflectometry, *J. Lightwave Technol.* **7**, 1225-1233.
29. Huang, D., Swanson, E.A., Lin, C.P., Schuman, J.S., Stinson, W.G., Chang, W., Hee, M.R., Flotte, T., Gregory, K., Puliafito, C.A., and Fujimoto, J.G. (1991) Optical coherence tomography, *Science* **254**, 1178-1181.
30. Fercher, A.F. (1996) Optical coherence tomography, *J. Biomed. Opt.* **1**, 157-173.
31. Morgner, U., Kärtner, F.X., Cho, S.H., Chen, Y., Haus, H.A., Fujimoto, J.G., Ippen, E.P., Scheuer, V., Angelow, G., and Tschudi, T. (1999) Sub-two-cycle pulses from a Kerr-lens mode-locked Ti:sapphire laser, *Opt. Lett.* **24**, 411-413.

32. Drexler, W., Morgner, U., Kärtner, F.X., Pitris, C., Boppart, S.A., Li, X.D., Ippen, E.P., and Fujimoto, J.G. (1999) In vivo ultrahigh-resolution optical coherence tomography, Opt. Lett. **24**, 1221-1223.
33. Clivaz, X., Marquis-Weible, F., and Salathé, R.P. (1992) Optical low coherence reflectometry with 1.9 μm spatial resolution, *Electron. Lett.* **28**, 1553-1555.
34. Kowalevicz, A.M., Ko, T., Hartl, I., Fujimoto, J.G., Pollnau, M., and Salathé, R.P. (2002) Ultrahigh resolution optical coherence tomography using a superluminescent light source, *Opt. Express* **10**, 349-353.
35. Pollnau, M., Salathé, R.P., Bhutta, T., Shepherd, D.P., and Eason, R.W. (2001) Continuous-wave broadband emitter based on a transition-metal-ion-doped waveguide, *Opt. Lett.* **26**, 283-285.
36. Pollnau, M. (2003) Broadband luminescent materials in waveguide geometry, *J. Lumin.* **102-103**, 797-801.
37. Rivier, S. (2003) Characterization of active single and parallel Ti:sapphire rib channel waveguides for biomedical imaging, Ph.D. thesis, Ecole Polytechnique Fédérale de Lausanne, Switzerland.
38. Rivier, S., Laversenne, L., Pollnau, M., Flury, M., Vuille, J., Herzig, H.P., Grivas, C., Shepherd, D.P., and Eason, R.W., Ti:sapphire parallel channel waveguide broadband emitter, Conference of the Swiss Physical Society, Basel, Switzerland, 2003, in Bulletin SPG/SSP, Vol. **20**, 2003, paper 223, p. 55.
39. Bourquin, S., Seitz, P., and Salathé, R.P. (2001) Two-dimensional smart detector array for interferometric applications, *Electron. Lett.* **37**, 975-976.
40. Ducros, M., Laubscher, M., Karamata, B., Bourquin, S., Lasser, T., and Salathé, R.P. (2002) Parallel optical coherence tomography in scattering samples using a two-dimensional smart-pixel detector array, *Opt. Commun.* **202**, 29-35.
41. Karamata, B., Lambelet, P., Laubscher, M., Salathé, R.P., and Lasser, T. (2004) Spatially incoherent illumination as a mechanism for cross-talk suppression in wide-field optical coherence tomography, *Opt. Lett.* **29**, 736-738.

SPECTRAL PROPERTIES OF FILMS

ALEXANDER P. VOITOVICH
Institute of Molecular and Atomic Physics,
Academy of Sciences of Belarus,
Pr. Nezalejnosci 70, Minsk 220072, Belarus
Voitovich@imaph.bas-net.by,nssrd02@tut.by,

Films and film structures are widely used in optics, optoelectronics, microchip lasers and other fields and devices. Properties of films can differ significantly from the properties of bulk_solids with the same chemical compound. In films the physical phenomena which are absent or cannot be observed in bulk media take place. These phenomena cause the presence of specific features in the spectral properties of films. We shall consider the distinguishing features of films spectral properties and the reasons of their occurrence.

In the first section the interference phenomena in films are discussed. The quantum effects, caused by small thickness of films or small sizes of film structures, and the spectral properties corresponding to these effects are considered in Section 2. The influence of the superficial phenomena on films spectral properties is stated in Section 3. In the last part of the article all the considerations are summed up and the conclusions are presented.

1. Interference of Light in Films

Interference is a well known phenomenon described in textbooks on optics (see [1]). It has many applications.

In the given section the attention is payed only to the fact, that at the measurements of films spectral properties, for example, transmission or reflection spectra, occurrence and observation of interference picture are possible. It is necessary to distinguish the interference maxima (minima), for example, in the transmission spectra from the maxima (minima) caused by properties of a material, quantum-dimensional or other kind effects.

The reason of an easy observation of the interference in films is the circumstance, that thickness a of films is usually less than length l_{coh} of radiation coherence, i.e. $a < l_{coh}$. It means, that the phase of electromagnetic waves, emitted by the source in time τ during which these waves pass the film, changes a little or in general remains constant. In other words, the time τ is less than τ_{coh}, i.e. $\tau < \tau_{coh}$, where τ_{coh} — the coherence time of the oscillation, which is connected with the coherence length l_{coh} as follows:

$$l_{coh} = c\tau_{coh}, \tag{1}$$

where c- light speed.

For nonlaser sources of radiation $\tau_{coh} \approx 10^{-9} \div 10^{-10}$ s, i.e. $l_{coh} \approx 3 \div 30$ cm. For laser sources the quantities τ_{coh} and l_{coh} are much bigger, than for nonlaser sources. Thus, usually the thickness of films $a << l_{coh}$. Therefore, when a film is homogeneous enough and with a high quality surface, the interference is observed at the research of its spectra of absorption,

B. Di Bartolo and O. Forte (eds.), Advances in Spectroscopy for Lasers and Sensing, 351–363.

excitation or luminescence. It is necessary to undertake special measures in order to avoid the influence of interference on the correctness of measurements.

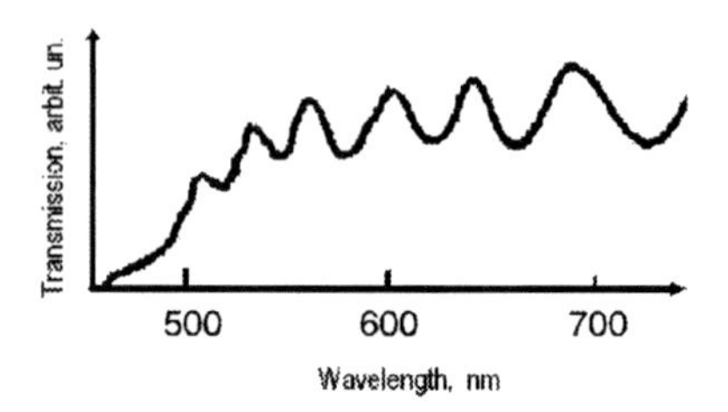

Fig. 1. The transmission spectrum of ZnS film.

As the example on Fig. 1 the transmission spectrum of ZnS film with the thickness of approximately 2μ, produced by vacuum evaporation, is presented. Alternating maxima and minima in the spectrum are caused by interference of light.

2. Quantum Effects in Films

2.1. QUANTUM DIMENSIONAL EFFECTS

These effects are caused by a change of properties of a solid state, when even one of its geometrical sizes becomes commensurable with the length of de Broglie wave of an electron, a hole or an exciton. Electrons represent the greatest interest as they determine the majority of practically interesting properties of solid states.

For a free electron the length λ_B of de Broglie wave in nonrelativistic approximation is equal

$$\lambda_B = \frac{h}{p} = \frac{h}{\sqrt{2m_o E}}, \tag{2}$$

where $\boldsymbol{h}$ - Planck's constant, $\boldsymbol{m_o}$ – rest electron mass, $\boldsymbol{E}$ – electron energy .

In solid states, crystals effective masses of an exciton, an electron and a hole are usually less than that of a free electron. For example, for ZnSe crystal they are equal accordingly:

$$m_{ex} = 0{,}13\, m_o,\ m_e = 0{,}17\, m_o,\ m_h = 0{,}6\, m_o \text{ and } m_{ex} = \frac{m_e m_h}{m_e + m_h}. \tag{3}$$

The lengths of de Broglie waves for an exciton, an electron and a hole in CdSe crystal are equal accordingly:

$$\lambda_{ex} = 5{,}3 \text{ nm},\ \lambda_e = 4{,}35 \text{ nm},\ \lambda_h = 1{,}25 \text{ nm}.$$

As we see, the lengths of de Broglie waves for particles in crystals lay in a nanometers range. Therefore, the objects, for which quantum dimensional effects are typical, should have sizes of the nanometers.order. In this connection about such objects they usually speak as about nanostructures or nanomaterials.

The objects, in which quantum dimensional effects take place, are classified on three types:

1) quantum dots — 3D-objects, which sizes in all three dimensions are less or about the length of de Broglie wave;

2) quantum wires — 2D-objects, which sizes in two dimensions are less or about the length of de Broglie wave;

3) quantum wells — 1D-objects, which size on one coordinate is less or about the length of de Broglie wave.

The films, which have thicknesses $a \sim \lambda_B$, are quantum wells independently of manufacturing kind: epitaxy, vacuum or chemical sedimentation. The film can also be a system of quantum points if it consists of grains with the sizes about length of de Broglie wave in all three dimensions.

Quantum dimensional effects are caused by a quantization of the movement of an electron in the direction in which a size of a crystal is commensurable with length λ_B. For example, movement of an electron in the plane of a film remains free, it is not localized. Its quasipulse projections on coordinate axes in the plane of a film can accept any values. However, in the direction, which is perpendicular to the plane of a thin film, movement of an electron is localized. It is limited to a potential barrier which arises on the border of a film with air or a substrate. In other words, the electron appears in a potential one-dimensional well. The electron wave function on the border of a film should equal zero. It means, that on a film thickness a the integer of half the lengths of de Broglie waves should be disposed, i.e.:

$$n\frac{\lambda_B}{2} = a, \tag{4}$$

where $n = 1,2,3, \ldots$.

From the equalities (2) and (4) the law of the quantization of an electron quasipulse projection p_z on axis $\mathbf{Z}$ perpendicular plane of a film follows:

$$p_z = \frac{\pi\hbar}{a}n, \tag{5}$$

where $\hbar = h/2\pi$. From the same equalities, values of the allowed energies for an electron with effective mass m_e follow:

$$E_n = \frac{\pi^2\hbar^2}{2m_e a^2}n^2 = E_o n^2. \tag{6}$$

Equation (6) is fair for rectangular indefinitely deep well.

As it follows from the equation (6) the electron energy in a perpendicular direction to a film is quantized. At the decrease of the thickness of a film the energy grows and the energy gap between dimensional zones increases. The quantization in semiconductors results in a shift of a bottom of a zone of conductivity and a ceiling of a valent zone. It changes the width of the forbidden zone of the semiconductor and shifts the border of an absorption spectrum in a short-wave spectrum range. The presence of subzones with the quantized values of energy $\boldsymbol{E_n}$ results in transitions from the filled $\boldsymbol{n}$-subzones in a valent zone into the empty $\boldsymbol{n}$-subzones in a zone of conductivity. It causes the occurrence of peaks in a spectrum of absorption.

Quantum dimensional effects can be observed experimentally in rather perfect films with a homogeneous thickness. This restriction arises from the requirement that level broadening due to dispersion on inhomogeneities and due to the fluctuations of thickness should not exceed the energy gap between the levels.

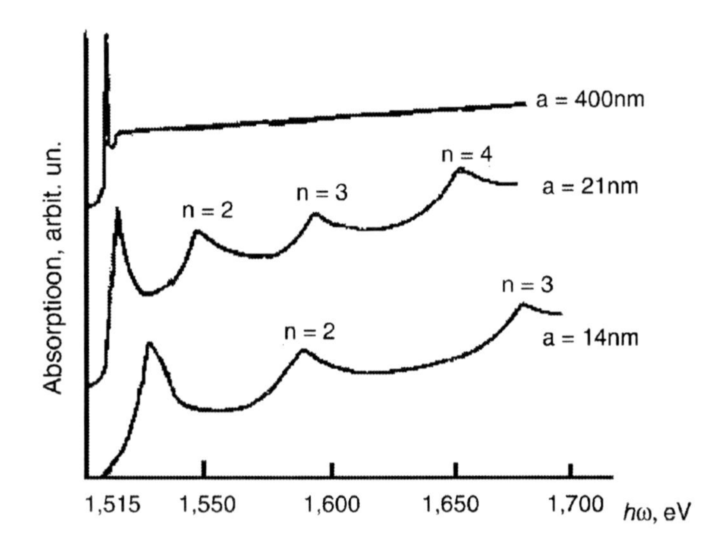

Fig. 2. Absorption spectra of $Al_xGa_{1-x}As$-GaAs-$Al_xGa_{1-x}As$ heterostructure, ***a*** *— thickness of GaAs layers.*

On Fig. 2 the absorption spectra taken from the literature are submitted for $Al_xGa_{1-x}As$-GaAs-$Al_xGa_{1-x}As$ heterostructure. Carriers of a charge move in GaAs layers, while $Al_xGa_{1-x}As$ films serve as potential barriers. From the data of Fig. 2 the high energy shift of an absorption edge follows. At a spectrum of absorption there are also the peaks caused by the quantization of energy.

Quantum dot electron energy is also quantized due to the dimensional effects and the expression for the energy is similar at the certain assumptions to the expression (6):

$$E_n = \frac{\pi^2 \hbar^2}{2m_e a^2} \chi_{ne}^2 = E_o \chi_{ne}^2, \qquad (7)$$

where χ_{ne} – Bessel function.

Quantum dimensional effects change not only energy spectra of electrons in semi-conductors, but also spectra and probabilities of optical transitions, probabilities of an electron scattering and other nanomaterials characteristics. They also change the dependence of the density of electronic states on energy. Such dependences are shown on Fig. 3 for the bulk semiconductor (shaped curve), a quantum well, a wire and a dot. In more details a reader can familiarize with these questions in the monography [2].

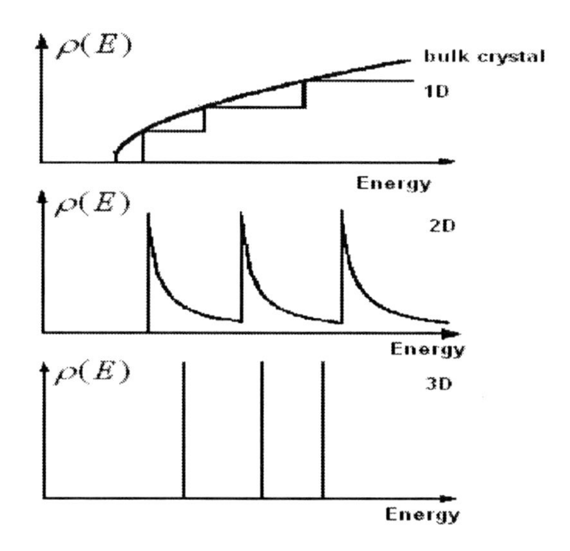

Fig. 3. Dependences of density of electronic states upon energy for the bulk semiconductor (shaped curve), a quantum well (1D), a wire (2D) and a dot (3D).

2.2. PHOTONIC CRYSTALS

In the previous subitem 2.1 the quantum dimensional effects are considered. They are connected to the spatial restriction of an electron movement (so-called "electron confinement") in the media, which sizes are commensurable to the length of de Broglie wave for an electron. These are sizes about 10 nanometers. In such media the density of electronic states cardinally changes in comparison with their density in bulk solids.

Other group of effects in periodic nanostructures, including film structures, is connected to the change in them the density of photon states. The similar periodic three-dimensional structures have received the name of "photonic crystals".

Such periodic nanostructures result in the spatial redistribution of an electromagnetic field (so-called "photon confinement"). For an optical range of electromagnetic waves the characteristic size of structures, in which the photon confinement is observed, is the length of a wave of light in the medium, i.e. the size is about 10^2 nanometers. This size approximately in 10^3 times more than the distance between atoms in crystals. Actually the photon confinement in nanostructures and Bragg's diffraction of X-rays in crystals are similar phenomena. However, the photon confinement, as against Bragg's diffraction, can result in the change of many processes. These changes can be controlled through the modification of the parameters of corresponding nanostructures. Since the time of the first pioneer works [3,4] in this direction the big progress was achieved. It was suggested to use one-, two- and three-dimensional photon crystals in many devices (see [5-7]).

The density of photon states $\rho(\omega)$ or $\rho(E)$, where ω and E — frequency and energy of a photon, determines a number of electromagnetic modes in the interval of values of frequencies ω, $\omega + d\omega$ or energies E, $E + dE$ in a single volume, and sometimes in a single space angle. For the electromagnetic waves in homogeneous isotropic three-dimensional space we have:

$$\rho(\omega) = \frac{\omega^2}{2\pi^2 u^3}, \tag{8}$$

where u — the speed of waves in the media. For non-uniform media the function $\rho(\omega)$ should be calculated in each concrete case, proceeding from dependence of medium dielectric permeability $\varepsilon(x, y, z)$ on coordinates. In nanostructures, in which the dependence $\varepsilon(x, y, z)$ is not periodic, i.e. in the structures, which are not being photonic crystals, it is necessary to enter only local density $\rho(\omega)$ for some area of space.

The probability of many optical processes is proportional to the density of photon states. Spontaneous emission, thermal radiation, Raman and Rayleigh scattering belong to such processes. All these processes can be operated, in principle, with the help of periodic nanostructures. For example, the contribution of change of photon states density $\rho(\omega)$ is necessary to take into account when considering huge Raman scattering. Opportunities of change of life time of the atom in excited state when localizing it in the photonic crystal are interesting. Applications of photon crystals for processing, storage and transfer of the information are perspective.

One-dimensional, including films, photonic structures have been used for a long time. These are dielectric mirrors with the coefficient of reflection that is close to 100%, antireflection coatings, lasers with the distributed feedback and other devices. Recently it has been shown that similar structures can also possess a total omnidirectional reflection of incident light.

Film photonic structures are usually made as alternating layers of the certain thickness of two materials with various indexes of refraction (Fig. 4).At some parameters of such layers it is possible to obtain forbidden gap for distribution of any modes in the photonic structure. Hence the total omnidirectional reflection arises for the determined spectral range of light waves.

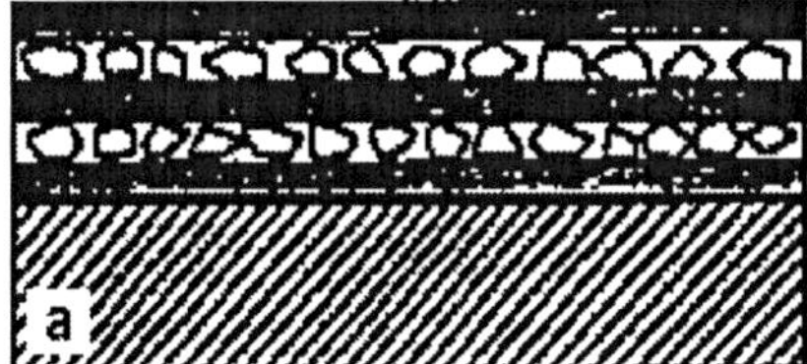

Fig. 4. Model of thin film heterostructure (a) and 48-layers of CaF_2 – LiF structure made by vacuum sedimentation (b).

Fig. 5 demonstrates such a photonic band gap for the onedimensional photonic structure of 19-layers of Na_3AlF_6 (the refractive index $\boldsymbol{n_1}$ = 1,34 in the visible range, the thickness of each layer $\boldsymbol{a_1}$ = 90nm) and ZnSe ($\boldsymbol{n_2}$ = 2,5÷2,8 in the visible range, $\boldsymbol{a_2}$ = 90 nm) .There is a forbidden band gap in the spectral range 604—638 nm (the black region in the Fig. 5). Here no propagating modes are allowed in the photonic structure at different incident angles for both *s*- and *p*-polarized waves. All waves in the spectral range of 604—638 nanometers should be reflected from this photon structure.

The corresponding calculated transmission spectra of this structure are shown in Fig. 6. They demonstrate omnidirectional reflection for s-polarized (in the region of 610—780 nm) and p-polarized (in the region of 610—640 nm) light. The experimental data display a good agreement with the calculated results.

3. Surface Phenomena in Films

3.1. SEMICONDUCTOR FILM STRUCTURES.

Films, especially consisting of small grains, possess a very much extended surface. The ratio of the surface area to the volume of a body is inverse proportional to the linear size of a body, i.e $S/V \sim 1/a$. Simultaneously with the increase of this ratio the ratio of atoms quantity

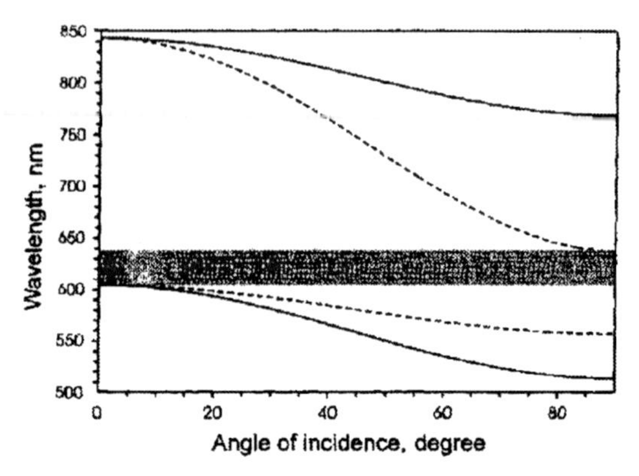

Fig. 5. Photonic band gap of periodic Na_3AlF_6 — ZnSe structure in coordinates of wavelength and incident angle. The solid (dashed) curves are for s- (p-) polarization components. The black area is the absolute omnidirectional band gap [8].

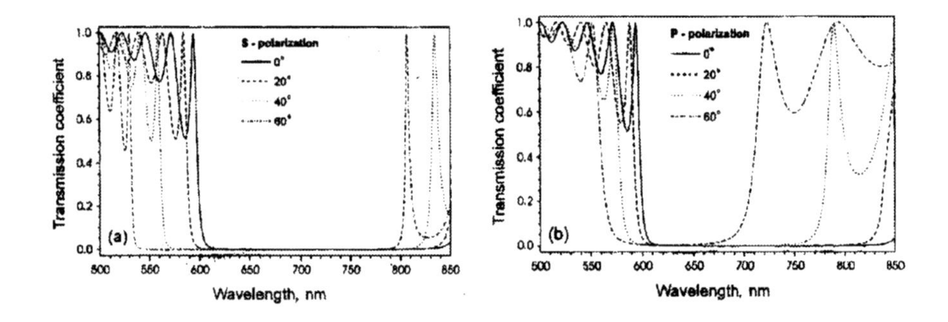

Fig. 6. Calculated transmission spectra of Na_3AlF_6 - ZnSe 19-layer struture for s-polarized (a) and p-polarized (b) light at different angles of incidence (0° — solid line, 20° — dashed line, 40° — dotted line, 60° — dash-dotted line) [8].

N_s on the surface to total quantity of atoms N in a grain also increases with the reduction of the grain size. In the table the ratios N_s/N for cubic nanoparticles with the diamond structure are presented. The big percentage of the atoms which are disposed on the surface of nanoparticles

Nanoparticle size, nm	N_s/N, %
1,7	38,7
3,39	22,0
8,48	9,5
14,1	5,8
28,3	2,9
56,5	1,5

is the strong factor of formation of nanosized media distinctive properties. In such media the phenomena connected with the special properties of surface layers, the so-called surface phenomena, should be well displayed. There is the numerous literature devoted to the researches of the surface phenomena in semiconductors. Here, the distinctive spectral properties of nanosized semiconductor media, caused by the presence of the extended surface, will be considered [9].

Luminescence spectra for CdS films with the nanocrystals size $a \approx 100$HM are presented on Fig. 7a. The sample 1 is made by chemical sedimentation. It is characterized by the high specific resistance $\sigma > 5 \cdot 10^2$ Ω·cm. The specific resistance of the sample 2, which was made by thermovacuum evaporation, is much lower and equal $\sigma = 2$ Ω·cm. The sample 2 in comparison with the sample 1 possesses more intensive luminescence in the range of zone - zone transition ($\lambda_{max} \approx 500$ nm) and less intensive one in long wave spectral range, which is typical for the defects in the volume of the semiconductor. The increase of the long-wave luminescence contribution in relation to a luminescence on zone - zone transitions was registered also for the high-resistant ZnO and ZnS films with the same size of nanocrystals.

Kinetics of absorption saturation, i.e. bleaching kinetics of CdS nanocrystals films, are presented on Fig. 7b. The change ΔD of films absorption optical density was measured after the influence on them of a radiation pulse with the wavelength $\lambda = 385$nm and the duration $\Delta\tau \approx 150$fs. The change of optical density was determined for the wavelengths of a maximum bleaching: $\lambda = 500$nm for samples 1, 2 and $\lambda = 460$nm for sample 3. Samples 1, 2 are the same samples of Fig. 7a.

For samples 1 and 2 the bleaching comes during the time $\tau \approx 2 \div 3$ps (curves 1, 2, Fig. 7b), that is typical for absorption saturation on zone - zone transitions. The relaxation of bleaching for a high-resistance sample (curve 1) occurs with two time components $\tau \approx 500 \div 600$ps (1/2 of the initial effect of bleaching) and $\tau > 1$ns. These times are typical for the excitation relaxation of excitons and bulk defects accordingly [10,11]. For the low-resistance sample at the same intensities of excitation the components with the duration $3 \div 7$ps appear in the relaxation of bleaching (curve 2). They determine the transfer of excitation energy on the surface states [12] and reduce the amplitude of the bleaching almost up to half of the initial quantity.

For the samples with the nanocrystals sizes from 60 ÷ 50nm up to 10 ÷ 15nm (curve 3, Fig. 7b) a short-wave shift of maximum bleaching wavelength is observed in comparison with the sample, where $a \approx 100$ nm. Thus, in sample 3 the quantum size effects are appeared. In the

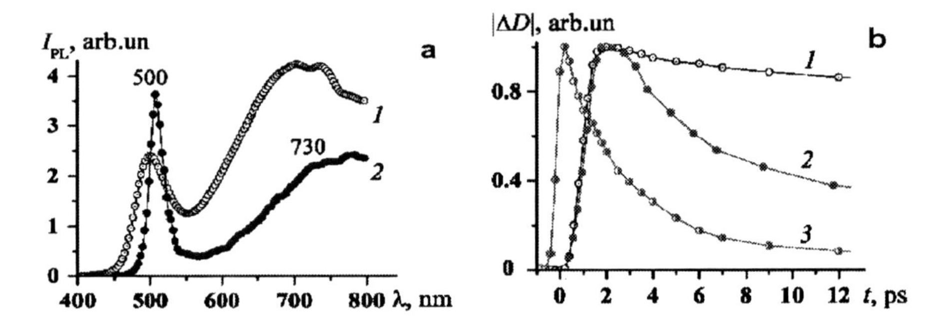

Fig. 7. Luminescence spectra (a) of high-resistant (1) and low-resistant (2) CdS structures with nanocrystals sizes 100 nm and normalized kinetics (b) of optical bleaching for wavelengths of maximum bleaching for high-resistant (1) and low-resistant (2,3) CdS structures with nanocrystals sizes 100 nm (1,2) and 10 ÷ 60 nm (3).

same sample the bleaching comes earlier, than in sample 2, and it has the larger amplitude. The last fact testifies to the greater value of optical nonlinearity for it. With the reduction of the nanoparticles sizes a relaxation component of $\tau \approx 1 \div 2$ps appears and also the contribution of the component with duration $\tau \approx 3 \div 7$ps increases.

Thus, the influence of the surface effects can cause the correlation of the optical properties and the specific conductivity of semiconductor film nanostructures, and the peculiarities of their optical bleaching and bleaching relaxation too.

3.2. FORMATION OF COLOR CENTERS (CC) IN FILMS OF ALKALI HALIDES NANOCRYSTALS

As the second example of the extended surface influence on the properties of a material we shall consider the increase in the efficiency of the intrinsic radiation defects creation, namely, the color centers in crystals. Probably, for the first time the attention to the fact of the increase in such efficiency was payed in work [13]. Before this, in work [14] it was revealed, that the ratio of the CC absorption coefficient in M-band to the CC absorption coefficient in F-band in an electron irradiated LiF crystal is in several times bigger, than such a ratio in the same crystal irradiated by γ-rays. The electron irradiation results in CC formation only in near surface layer of a crystal, while γ-rays create CC in the whole bulk crystal.

Here we shall consider the results of the absorption and luminescence characteristics research of γ-irradiated thin film structures consisting from LiF nanocrystals with size ***a*** ≈ 100nm, and also LiF bulk crystals [9]. It has been found, that the absorption coefficients in F-band (the absorption of F CC) and in M-band (the absorption F_2 and F_3^+ CC) in nanostructures are in several times digger (approximately in 3,3 and 4,7 times accordingly) than the corresponding coefficients in bulk crystals. Hence, the CC concentration in nanocrystals is bigger, than in bulk crystals at the identical dozes of irradiation, i.e. in nanocrystals the CC formation occurs more effectively.

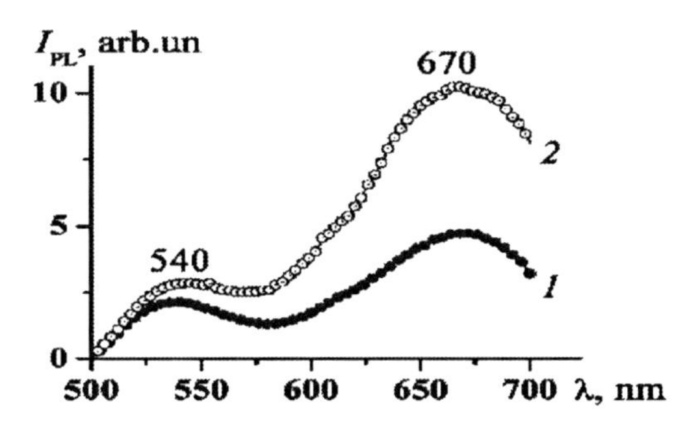

Fig. 8. Luminescence spectra of LiF film structure with nanocrystal sizes 100nm (1) and LiF bulk crystal (2).

The second observation is, that in nanostructures the ratio of the absorption coefficient in M-band to the absorption coefficient in F-band is bigger, than this ratio in bulk crystals at the identical dozes of γ-irradiation. This implies, that there are more preferable conditions of the complex CC formation (F_2 and F_3^+) in nanocrystals in comparison with bulk crystals.

In the third, the comparison of luminescence spectra of LiF nanostructures and bulk crystals (Fig. 8) at the excitation in the M-band of absorption (λ_{ex} ≈ 445nm) shows, that the ratio (0,46) of intensities of luminescence of F_3^+ CC (band with λ_{max} ≈ 540nm) and F_2 CC (λ_{max} ≈ 670nm) in nanocrystals is bigger, than the same ratio(0,28) in bulk crystals. It testifies to the best conditions for the formation of F_3 CC in nanocrystals in comparison with bulk crystals.

The interpretation of the stated results can be following. In lithium fluoride under the influence of γ-irradiation the defects appear. Among them Frenkel's defects are the most interesting for us. Frenkel's defect consists of an ion of haloid (fluorine in the case of LiF crystal), displaced from its site to the interstitial position, and a positively charged vacancy

F_1^+ corresponding to it The displaced fluorine ion is mobile when $\boldsymbol{T} > 50$K and can annihilate with a vacancy, eliminating defect. Frenkel's pair does not annihilate on the conditions of some distance between the components and of haloid ion stabilization by a defect. Further, the mobile vacancies and electrons *(e)*, created by γ-radiation, start the chain of the CC formation:

$$F_1^+ + e \rightarrow F_1, \qquad \text{(I)}$$

$$F_1 + e \rightarrow F_1^-, \qquad \text{(II)}$$

$$F_1^+ + F_1^- \rightarrow F_2, \qquad \text{(III)}$$

$$F_2 + F_1^+ \rightarrow F_3^+. \qquad \text{(IV)}$$

On the crystal surface the defects concentration is always bigger than in the crystal volume. In nanocrystals, because of the large area of a surface, the probability of the haloid ion stabilization is much bigger, than in bulk crystals and therefore the concentration of vacancies is bigger at the identical dozes of the irradiation. This fact explains the higher CC concentration in nanocrystals in comparison with bulk crystals. The concentration increase of more complex CC (F_2 and F_3^+) concerning the concentration of the elementary CC (F_1 or it is simple F) in nanocrystals in comparison with bulk crystals can follow from the competition of processes III, IV, on the one hand, and I, on the other hand. Here the ratio of these processes kinetics constants and the fact, that the CC output in processes III and IV is proportional to the concentration of the vacancies in a square and a cube accordingly, while in process I it is proportional to the first degree, are important.

Thus, the extended surface, which is typical for small-sized particles, can determine many distinctive properties of nanosized particles and thin films.

4. Conclusion

The peculiarities of spectral properties of thin films and film structures are considered. The quantum dimensional effects, limiting movement of electrons (excitons, holes) in such films and structures, result in the electron localization, so-called electron confinement, and in the essential changes of spectral characteristics of such structures: the changes of spectra, the probabilities of transitions, the nonlinear optical parameters. The change of an electromagnetic field distribution, so-called photon confinement, in the film structures as the version of photonic crystals, allows to receive high coefficients of reflection or transmission, total omnidirectional reflection and is widely used in practice.

The extended surface of films, especially consisting of nanosized grains, is the strong factor of formation of films distinctive spectral properties. The influence of the surface effects can cause the peculiarities of semiconductor luminescence spectra, optical bleaching and bleaching relaxation. The extended surface results in the increase in the efficiency of the intrinsic radiation defects creation, namely, the color centers in crystal nanosized films.

5. Acknowledgements

This work is partly supported by the Belarusian Republic Foundation for Fundamental Research.

6. References

1. M. Born, E. Wolf. Principles of Optics. Pergamon, New York (1980)
2. Ch.P. Poole, Jr., F.J. Owens. Introduction to nanotechnology. Wiley-Interscience, John Wiley and Sons, Inc. (2003)
3. V.P. Bykov. JETP, 62, 505 (1972)
4. E. Yablonovich. Phys. Rev. Lett., 58, 2059 (1987)
5. J.D. Joannopoulos, R.D. Meade, J.N. Winn. Photonic Crystals: molding the flow of light. Princeton University Press, Princeton (1995).
6. J. Pendry. J. Mod. Opt., 41, 209 (1994)
7. J. Haus. J. Mod. Opt., 41, 198 (1994)
8. D.N. Chigrin, A.V. Lavrinenko, D.A. Yarotsky, S.V. Gaponenko. Appl. Phys. A., 68, 25 (1999)
9. A.P. Voitovich, O.V. Goncharova, V.S. Kalinov. Journal Appl. Spectr. (translated from Russian), 72, 301 (2005)
10. A.I. Ekimov, I.A. Kudryavtsev, M.G. Ivanov, A.L. Efros. J. Lumin., 46, 83 (1990)
11. D. Matsuura, Y. Kanemitsu, T. Kushida, C.W. White, J.D. Budai, A. Meldrum. Appl. Phys. Lett., 77, 2289 (2000)
12. A.P. Voitovich, O.V. Goncharova. In "Physics, Chemistry and Application of Nanostructures", Eds V.E. Borisenko, A.B. Filonov, S.V. Gaponenko, V.S. Gurin, Minsk, Belarusian State University of Informatics and Radioelectronics, 215 (1995)
13. A.P. Voitovich, V.S. Kalinov, A.V. Saltanov, A.F. Zaiko, L.C. Scavarda do Carmo. In "Proceedings of the XII International Conference on Defects in Insulating Materials", Eds O. Kanert, J.-M. Spaeth, World Scientific, Singapore, v. 2, 868 (1992)
14. A.P. Voitovich, V.S. Kalinov, L.C. Scavarda do Carmo, A.V. Saltanov. Doklady AN BSSR (in Russian), 34, 21 (1990)

BASIC PHYSICS OF SEMICONDUCTOR LASERS

TOBIAS VOSS and JÜRGEN GUTOWSKI
Institute of Solid State Physics, University of Bremen
P.O. Box 330440, D-28334 Bremen, Germany
gutowski@ifp.uni-bremen.de,

Abstract

This paper is intended to provide an overview over the basic physical processes and mechanisms which are involved in the generation of laser light from semiconductor structures. After a brief look on the history of semiconductor lasers we will discuss the basic properties of the common semiconductor systems with respect to their function as active lasing media. We will give a short summary of the gain mechanisms in semiconductors and discuss design concepts of the laser structures which are intended to reduce the threshold current density and to improve the beam profile. We will then show how these concepts influence the current research activities towards semiconductor lasers operating in the green, blue, and near-UV spectral region.

1. Introduction

Semiconductor lasers are optoelectronic devices used in many areas of daily life. Data storage on CDs and DVDs relies on optical reading and writing procedures which make use of semiconductor lasers because of their small dimensions, long lifetimes and easy operation by current injection. Their drawbacks are limited to rather few aspects, as are

- a relatively small tuning range with respect to their emission wavelength (just several nm, depending on the particular device structure), often accompanied by a mode-jumping behaviour,
- the relatively low output power (normally in the order of several mW to W),
- an elliptical beam cross section together with a rather large beam divergence in the order of some tens of degrees.

These drawbacks, however, can usually rather easily be overcome, e.g., by forming arrays of semiconductor lasers which are used to increase their output power. Small external optics can be applied to correct the beam cross section, external cavities can be used to extend the wavelength tuneability without mode jumping. Therefore, today semiconductor lasers are cheap and easy-to-use sources of coherent light in the red and near-infrared spectral region. The history of semiconductor lasers dates back to the 60ies of the last century when first Basov and co-workers theoretically introduced the

B. Di Bartolo and O. Forte (eds.), Advances in Spectroscopy for Lasers and Sensing, 365–402.

concept of a semiconductor laser [1] and shortly after that, Hall, Nathan, and Quist experimentally demonstrated pulsed radiation at a wavelength of 840 nm from a forward biased GaAs pn-junction at liquid nitrogen temperatures [2-4, 5 and references therein]. Technical advances in the design of the semiconductor devices (which will be discussed in section 4) led to the first continuous-wave (cw) semiconductor laser which could be operated at room temperature, and which was developed by Hayashi and co-workers in 1970 [6]. The current research activities on semiconductor lasers with both academic and industrial background are nowadays mainly driven by three aspects:

1. Extension of the wavelength range covered by semiconductor lasers to the mid-infrared as well as to the green, blue and near-UV spectral region.
2. Improvement of the beam characteristics, especially in order to achieve an efficient coupling into optical fibres for telecom applications and to achieve minimum spot size for optical data-storage devices.
3. Improvement of the quantum efficiency of existing devices, i.e., by minimizing the loss processes for the injected carriers which finally recombine and emit the laser photons.

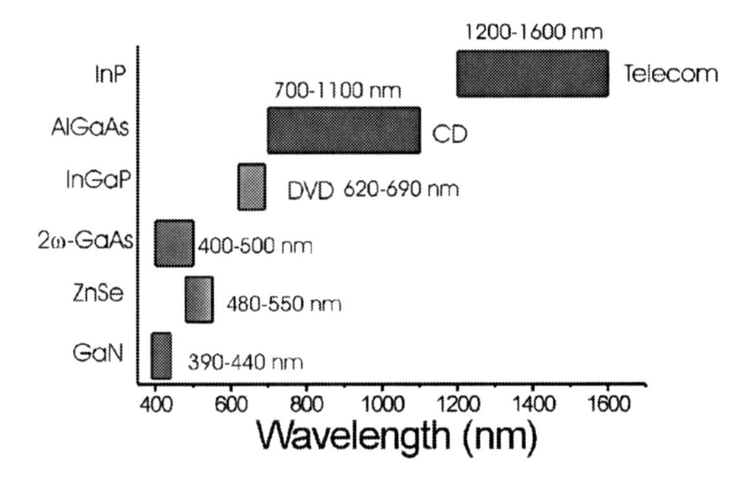

Figure 1: Semiconductor material systems, their spectral emission ranges and some of their current applications.

2. Basic Properties of Semiconductors as Laser Media

For the basic concepts of the theoretical description of solid media and especially semiconductors the reader is referred to the standard textbooks, e.g., [7-12]. Assume a semiconductor which is basically characterized by a band structure with an energy gap in the order of about 1 or a few eV between the highest completely filled and the lowest completely unoccupied band at $T = 0$ K (there is no strict energy limit distinguishing semiconductors from insulators, however, materials with more than 4 eV bandgap are usually no longer addressed as semiconductors). The highest filled band is usually referred to as the valence band, the lowest unoccupied band as the conduction band. For the following discussion we just have to consider these two bands, i.e., we will neglect all lower-lying valence and higher-lying conduction bands. To further simplify the discussion usually the concept of effective masses for electrons is used which leads to valence and conduction bands with a parabolic shape in the vicinity of the valence-band maximum and the conduction-band minimum (see Fig. 2). The isotropic effective mass m* is indirectly proportional to the second derivative of the dispersion relation E(k):

$$\frac{1}{m^*} = \frac{1}{\hbar^2}\frac{\partial^2 E}{\partial k^2} \tag{1}$$

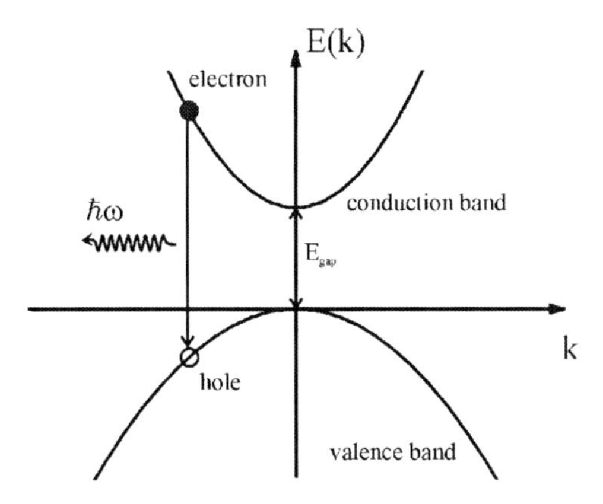

Figure 2: Band diagram of a direct-gap semiconductor around k = 0 in the effective-mass approximation.

A photon with an energy larger than the band gap of the semiconductor will be absorbed with a high probability. By this absorption process, an electron is transferred from the valence band to a free state in the conduction band, leaving behind the so-called 'hole' with positive charge in the valence band.

Those semiconductors being interesting candidates as constituents of semiconductor lasers possess a direct band gap, i.e., the minimum of the conduction band and the maximum of the valence band are located at k = 0 (the so-called Γ point of the reciprocal space). Thus, the recombination of electrons from the bottom of the conduction band (energy E_2) with holes on top of the valence band (energy E_1) takes place with respect to momentum conservation through emission of a single photon with an energy $\hbar\omega = E_2 - E_1$. Purely optical transitions (i.e., transitions without participation of other excitations such as phonons) occur almost "vertically" in k space (compare again with Fig. 2) because the momentum $p = \hbar k$ of the photon is about two orders of magnitude smaller than the typical momentum of electrons and holes in the involved bands.

For a direct-gap semiconductor a square-root function applies for the energy-dependent density of states of the electrons in the conduction band which reads for parabolic bands and a conduction-band minimum E_c

$$\rho(E-E_c)=\frac{(2m_e^*\hbar^2)^{3/2}}{2\pi^2}\sqrt{E-E_c} \tag{2}$$

For the holes in the valence band a similar relation can be easily obtained by substituting $m_e^* \rightarrow m_h^*$ and $E\text{-}E_c \rightarrow E_v\text{-}E$ (E_v: valence-band maximum). In absorption (emission) processes of photons the so-called joint density of states of the valence and conduction bands has to be used for which the theory again yields a square-root dependence [11]

$$\rho_j(\omega)=\frac{(2m_r\hbar^2)^{3/2}}{2\pi^2}\sqrt{\hbar\omega-E_g} \tag{3}$$

where the reduced mass $1/m_r = 1/m_e^* + 1/m_h^*$ of the electron-hole pair has been introduced. As a consequence, the absorption coefficient $\alpha(\omega)$ is proportional to the square-root of the photon energy:

$$\alpha(\omega)\propto\sqrt{\hbar\omega-E_g} \tag{4}$$

This expression can often be utilized as a good approximation which, however, has its limits since the Coulomb interaction between the *charged* quasi-particles electron and hole has been completely neglected. For all visible-wavelength semiconductor laser materials (the most commonly used are those on the base of GaAs and GaInP) but especially for wide-gap semiconductors such as zinc selenide (ZnSe), zinc oxide (ZnO), or gallium nitride (GaN) and their alloys, the influence of the Coulomb interaction even at room temperature is significant and must not be neglected at all.

The most prominent feature in semiconductor optics which occurs if the Coulomb interaction is taken into account are sharp emission lines below the band gap due to excitonic transitions. In a simple picture an exciton is composed of a bound electron-hole pair. Excitonic absorption and emission occurs at an energy position $E_g\text{-}E_X^B$ where E_X^B denotes the Coulomb binding energy of the electron and hole within the exciton. Theoretical calculations (see, e.g., [13,14]) show that the absorption coefficient $\alpha(\omega)$ does no longer show the simple square-root dependence but is now given by the Elliott formula,

$$\alpha(\omega)=\alpha_0^{3D}\left[\sum_{n=1}^{\infty}\frac{4\pi}{n^3}\delta(\Delta+1/n^2)+\Theta(\Delta)\frac{\pi e^{\pi\sqrt{\Delta}}}{\sinh(\pi/\sqrt{\Delta})}\right] \tag{5}$$

In this equation $\Delta = (\hbar\omega\text{-}E_g)/E_X^B$ describes a normalized and shifted energy scale. $\alpha^{3D}{}_0 = (2|M|^2)/(\hbar n_b c a_B^3)$ is the absorption coefficient of the 3D 1s-exciton (M:

inter-band dipole matrix element, a_B: Bohr radius of the exciton, n_b: refractive index of the semiconductor). $\Theta(x)$ is the heavyside step function and $\delta(x)$ the Dirac delta function.

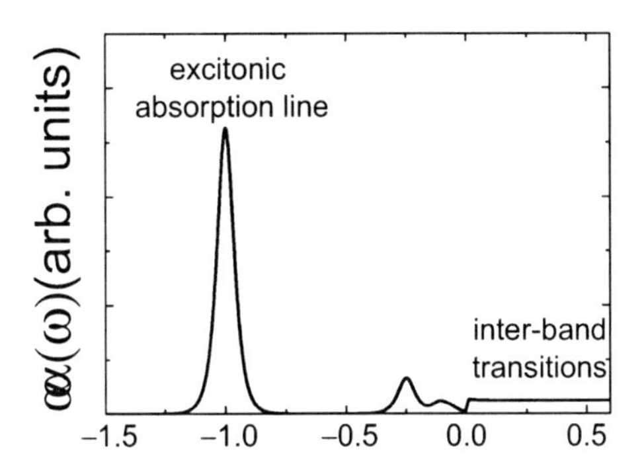

Figure 3: Simulation of the near band-edge absorption coefficient $\alpha(\omega)$ for a direct-gap semiconductor taking into account the Coulomb interaction according to the Elliott formula. $\Delta = (\hbar\omega - E_g)/E_X{}^B$ corresponds to a shifted energy scale being normalized to the exciton binding energy.

According to this formula the absorption coefficient shows a series of sharp excitonic lines below the band edge (see Fig. 3). The energy positions of these lines obey the same $1/n^2$ behaviour as the energy levels in the hydrogen atom. Their oscillator strength decreases proportional to n^{-3}. Additionally, the Coulomb interaction leads to an increase of the interband transition probability which is called the Coulomb enhancement. The Coulomb enhancement factor is defined as the second term of the Elliott formula divided by the square-root dependence of the absorption coefficient without Coulomb interaction,

$$C(\omega) = \frac{\frac{\pi}{\sqrt{\Delta}} e^{\pi/\sqrt{\Delta}}}{\sinh(\pi/\sqrt{\Delta})} \tag{6}$$

In semiconductor lasers, the material layers used as active layers, i.e. those generating the laser photons, are rather thin (near to the 2-dimensional situation), or even are composed of quantum-dot systems (see section 4). In the 2D limit, the formulas have to be modified (2D Elliott formula, see [13]). For quantum dots, the whole concept has to be changed since free motion of excitons and bands for the carriers do not longer exist. However, since this does not influence the principal tutorial understanding of the basic laser processes the reader is referred to literature (e.g., [13]) for detailed descriptions.

3. Principles of Semiconductor Lasers

3.1 GENERAL CONSIDERATIONS

The concept of every laser is based on the fundamental interaction processes of light with matter, i.e. absorption, spontaneous, and stimulated emission of a photon (compare Fig. 4).

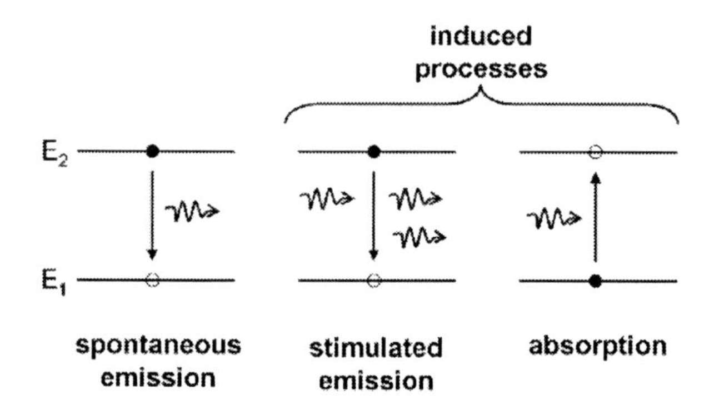

Figure 4: Fundamental interactions of light with matter.

Assume a system of N identical two-level systems with discrete energy levels E_1 and E_2 which are coupled to a resonant monochromatic field of electromagnetic radiation with a photon energy $\hbar\omega = E_2 - E_1$ and a radiation density $P(\omega)$. If N_1 particles are in the ground state the absorption rate Z_{abs} is given by $Z_{abs} = P(\omega)\, N_1\, B_{12}$ where B_{12} is the Einstein coefficient of absorption. Similarly, the rate of stimulated (induced) emission is described by $Z_{ind} = P(\omega)\, N_2\, B_{21}$ where N_2 is the number of particles in the upper energy level E_2. If the energy levels are non-degenerate the relation $B_{12} = B_{21}$ holds.
For the lasing process we need $Z_{ind} > Z_{abs}$, i.e., the rate of stimulated emission has to be larger than the absorption rate[1] which directly yields the required condition of population inversion for the lasing process,

$$N_2 > N_1 \tag{7}$$

For the two-level system it is straightforward to solve the rate equations by considering $N_1 = N - N_2$ and $d/dt\, N_2(t) = B_{12}\, N_1\, P(\omega) - B_{21}\, N_2\, P(\omega) = B_{12}\, (N-N_2)\, P(\omega)$. The solution of this equation,

$$N_2(t) = \frac{N}{2}\left(1 - \mathrm{e}^{-2B_{12}P(\omega)t}\right) \tag{8}$$

[1] Of course, Z_{ind} also has to overcome additional losses due to outcoupling of radiation energy from the laser resonator, scattering etc.

shows that for large values of t the limit $N_1 = N_2$ can be reached. However, the number of particles in the excited state N_2 cannot become larger than that in the ground state.

To achieve light amplification the use of three- or four-level systems is necessary (compare Fig. 5). The basic idea behind these concepts is the decoupling of the pump photons from the laser transitions by pumping electrons into a higher lying energy state.

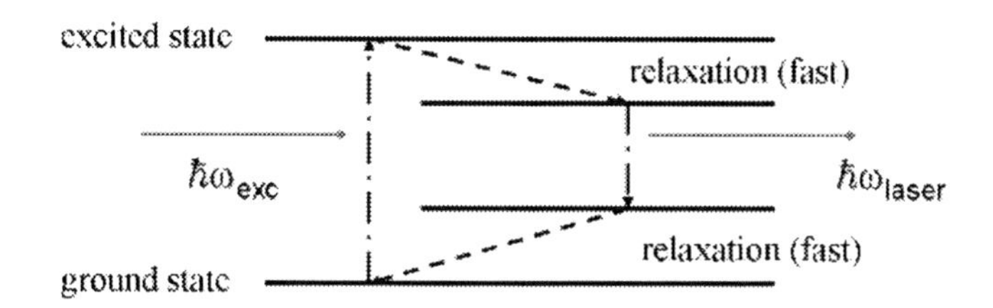

Figure 5: Four-level scheme of a laser.

The electrons in this excited state then undergo a fast relaxation into the upper laser level. After stimulated emission the electrons are in the lower laser level and (for the four-level system) again undergo a fast relaxation into the ground state. In these schemes part of the energy of the pump photons gets lost in the relaxation processes. Sufficiently fast relaxation into the upper and out of the lower laser level provided in combination with a long life time of the upper laser level, the required condition of population inversion between the two laser levels can be easily reached under moderate pumping conditions.

3.2 SEMICONDUCTOR LASER CONDITIONS

In the semiconductor band structure, the four-level scheme idea is applied by simply replacing the sharp electronic levels by the state distribution in the bands as sketched in Fig. 6.

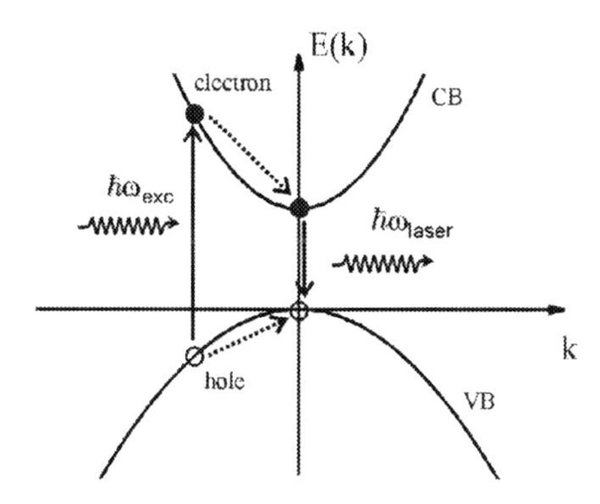

Figure 6: The semiconductor system as a four-level laser.

A pump photon with the energy $\hbar\omega_{exc} > E_{gap}$ creates an electron-hole pair upon its absorption. Being excited with excess kinetic energy the electron (hole) will relax to the lowest (highest) unoccupied energy levels near k = 0. The relaxation and thermalization processes involve electron-electron scattering and the emission of acoustic and optical phonons (lattice vibrations) and occur on typical time scales of several to several tens of picoseconds. It is important to remember that the emission of a longitudinal optical (LO) phonon which is the most important process for intraband relaxation requires a minimum excess energy (typically referred to as LO-phonon energy) of the electron or exciton whereas acoustic phonons can be created with arbitrarily small energies. As a consequence, in lower-dimensional semiconductor systems where the density of states changes from the square-root dependence of the 3D material to a step-function-like dependence for the 2D case (quantum well) or to discrete (δ-function like) states for the 0D material (quantum dot) (compare Fig. 7), the emission of LO phonons might be prevented or at least strongly suppressed if no final state with an energy difference $\Delta E = E_{LO}$ to the initial state is available. After relaxation the electron and hole with the same k value will finally recombine under (spontaneous or stimulated) emission of a photon of energy $\hbar\omega_{laser}$.

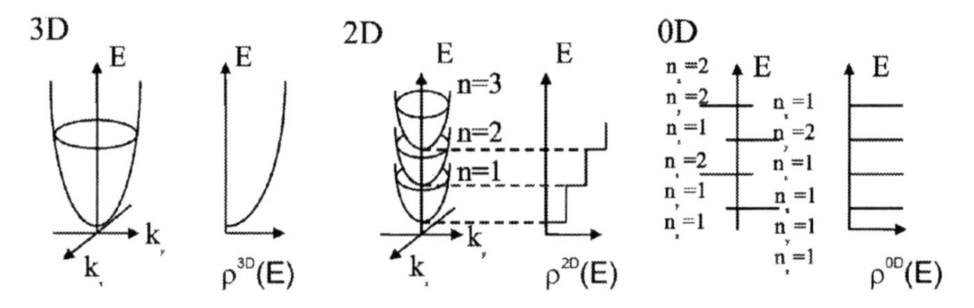

Figure 7: Dispersion relation E(k) and corresponding densities of states ρ(E) for 3D (bulk), 2D (quantum well) and 0D (quantum dot) semiconductor structure. n labels the sub-bands which occur due to quantization. See also textbooks, e.g. [9]

For the description of the lasing and laser conditions valid under these circumstances, we will present a simple phenomenological approach using rate equations based on the so-called quasi-Fermi distribution and relaxation of carriers in the bands involved. However, the full and adequate theoretical description of semiconductor lasers is necessary on a much more sophisticated level. For this, the reader is referred to the specialist's literature, e.g., [14].

For the basic description intended here, it is now straightforward to rewrite the rate equations of the last section in order to end up with some information about the energy region in which optical gain for the lasing process can be expected. The only difference is that we now deal with continua of energy states in the valence and conduction band. Thus, the product of the density of states ρ(E) and the Fermi distribution function f(E) (for holes 1-f(E)) yields the density of electrons (holes) in the conduction (valence) band n(E) = ρ(E) f(E) (n'(E) = ρ(E) [1-f(E)]). The equations for the absorption and emission rates now read

$$Z_{abs} = P(\omega_{12}) \cdot B_{12} \cdot n(E_1) \cdot n'(E_2) \tag{9}$$

$$Z_{ind} = P(\omega_{21}) \cdot B_{21} \cdot n(E_2) \cdot n'(E_1) \tag{10}$$

At this point a new problem within the theoretical description arises because we have to model the semiconductor material in a non-equilibrium situation, i.e., with a pump field applied which transfers electrons from the valence into the conduction band. This prevents the use of the Fermi distribution function which describes the distribution of fermionic particles in equilibrium conditions only. A solution of this problem can be found by making use of the so-called quasi-Fermi functions for the valence and the conduction bands:

$$f_c(E) = \left(1 + e^{(E-F_c)/(k_B T)}\right)^{-1} \tag{11}$$

$$f_v(E) = \left(1 + e^{(E-F_v)/(k_B T)}\right)^{-1} \tag{12}$$

In these functions we have substituted the Fermi level which in an equilibrium condition describes the whole semiconductor system by two separate Fermi levels F_c and F_v for conduction and valence band carriers, respectively. The idea behind this solution is that relaxation and thermalization processes within each of the two bands occur on time scales being at least 1-2 orders of magnitude faster than the inter-band recombination processes. Consequently, electrons in the conduction band and holes in the valence band separately reach a quasi-equilibrium in their respective bands described by two separate Fermi functions with the quasi-Fermi levels involved.

We replace n(E) and n'(E) in equations (9) and (10) by the product of the density of states and the quasi-Fermi distribution of the respective band. Again, the rate of induced emission has to be larger than the rate of absorption, $Z_{ind}(\hbar\omega_{21}) > Z_{abs}(\hbar\omega_{12})$. We get as a requirement for the quasi-Fermi distributions

$$f_c(E_2) > f_v(E_1) \tag{13}$$

Thus, if the occupation probability described by f_c at the energy E_2 in the conduction band is larger than the occupation probability f_v at the energy E_1 in the valence band the rate of stimulated emission will be larger than the absorption rate for the corresponding photon energy $\hbar\omega_{12}$. By inspecting the definition of the quasi-Fermi distributions (equations (11) and (12)) one notices that a semiconductor structure shows stimulated emission (optical gain) for all photon energies $\hbar\omega_{laser}$ for which the condition holds:

$$F_c - F_v > \hbar\omega_{laser} > E_g \tag{14}$$

Consequently, laser gain can be expected only if at least one quasi-Fermi level is pushed into the conduction or valence band by optical or electrical pumping. Equation (14) is sometimes referred to as the *first semiconductor laser condition.*

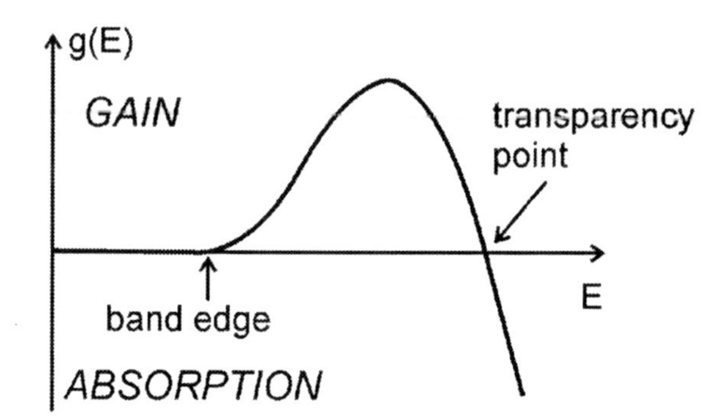

Figure 8: Typical gain profile of a semiconductor laser.

Figure 8 shows a schematic drawing of a typical gain profile of a semiconductor laser being pumped so that inversion is reached. The high-energy limit of the gain region is called "transparency point" with an energy being the difference of the quasi-Fermi levels F_c-F_v. Electromagnetic waves with photon energies between the band-gap energy on the low-energy side and the transparency point on the high-energy side will experience gain. Photons with an energy being larger than that of the transparency point will be absorbed. An idealized intrinsic semiconductor system will not show any interaction with photons having energies lower than that of the band edge, e.g., we expect neither gain nor absorption. In reality, however, due to a background doping level, impurities, or crystalline defects forming states within the gap of the semiconductor, always a slight absorption is measured for photon energies lower than that of the band edge.

We put the semiconductor as the active material of the laser into a resonator. In case of a semiconductor laser the simplest resonator is formed by the cleaved end facets of the active layer itself, i.e., oriented perpendicularly to this layer (edge-emitting lasers). Due to the different indices of refraction of the semiconductor (e.g., $n_s = n_{GaAs} \approx 3.66$ at room temperature) and the surrounding air ($n_a \approx 1.00$) a rather high reflectivity of $R = (n_s\text{-}n_a)^2/(n_s+n_a)^2 \approx 0.326$ is obtained without any further coating of the end facets. In this way, by use of two parallel end facets the semiconductor forms a Fabry-Perot resonator which imposes the condition of a standing wave on the optical modes which can be efficiently amplified inside the active material. The condition for standing waves reads (wavelength: λ, resonator length: L)

$$m\frac{\lambda_m}{2} = n_s L \rightarrow \nu_m = m\frac{c}{2n_s L} \tag{15}$$

In these equations m labels the number of half-wavelengths which fit into the resonator. The possible λ_m are called the longitudinal modes of the resonator. Only those longitudinal modes can be emitted as laser modes which spectrally fall into the gain region of the pumped semiconductor material. Thus, edge-emitting semiconductor lasers show a spacing of the longitudinal modes according to

$$\Delta\nu = \frac{c}{2L} \Leftrightarrow \Delta\lambda = \frac{\lambda^2}{2L} \tag{16}$$

Typical resonator lengths are in the order of 100 μm. For GaAs-based lasers, a $\Delta\lambda$ of about 2.4 nm can be calculated with $\lambda = 700$ nm and L = 100 μm. The intensity I(x) of a travelling wave inside the Fabry-Perot resonator will change as a function of the distance x according to the Lambert-Beer law

$$\mathrm{d}I = [\mathrm{g}(\omega) - \alpha(\omega)]\, I\, \mathrm{d}x \tag{17}$$

where $\alpha(\omega)$ summarizes all additional loss mechanisms occurring due to light scattering, free-carrier absorption etc. For laser action to occur the intensity of a travelling wave inside the resonator after one round trip (i.e., x = 2L) must fulfil I(2L) > I(0). With reflectivities R_1 and R_2 of the two cavity mirrors, I(2L) can be calculated according to:

$$I(2L) = I(0) \cdot R_1 R_2 \cdot e^{[g(\omega)-\alpha(\omega)]\cdot 2L} \tag{18}$$

These results lead to the *second semiconductor laser condition* for the threshold gain g_t:

$$g(\omega) > g_t(\omega) = \alpha(\omega) + \frac{1}{2L}\ln\left(\frac{1}{R_1 R_2}\right) \tag{19}$$

To decrease the threshold for lasing action in a semiconductor laser, obviously the losses α should be minimized, the reflectivities of the mirrors should be increased. Most interestingly, also an increase of the cavity length L will lead to a decrease of the lasing threshold.

All commercially available semiconductor lasers are pumped by an electrical current which is in most cases driven through a pn-diode layer structure (p (n): layer doped with excess holes (electrons)). In this way, excess electrons are injected into the conduction band of the active layer and excess holes into the valence band. This leads to an increase of the gain in the active layer as a function of the applied current density j = I/A, A contact or current-injection area. If we take into account the thickness d of the active layer and consider the fact that only a certain fraction η (quantum efficiency) of the electrons provided by the current really contributes to the built-up of the gain it is convenient to introduce a nominal current density by the relation $j = j_{nom} d/\eta$. As a first but already good approximation we write the gain of the active material as a linear function of the nominal current density

$$g = g_0 / j_0 \cdot (j_{\mathrm{nom}} - j_0) \tag{20}$$

Here, j_0 is the current density for which we find the intersection of the linear gain with the current-density axis, and g_0/j_0 describes the slope of the linear gain as a function of the nominal current density. By using the result of equation (19) we can now write an equation for the threshold-current density of the semiconductor laser,

$$j_{th} = \frac{j_0 d}{\eta} + \frac{j_0 d}{g_0 \eta \Gamma}\left[\alpha + \frac{1}{2L}\ln\left(\frac{1}{R_1 R_2}\right)\right] \tag{21}$$

In this formula Γ describes that part of the optical mode which is actually guided inside the active layer and is called the confinement factor (a detailed discussion will be given in the next section). As valid for g(ω) itself, a reduction of the losses α, an increase of the mirror reflectivities R_1 and R_2, and an increase of the resonator length L will also lead to a reduction of the threshold current density. Additionally, equation (21) shows that an increase of the quantum efficiency η and a reduction of the thickness d of the active layer reduce the lasing threshold. At this point one should note, however, that all these device parameters can be changed within certain ranges only. For example, a decrease of the active-layer thickness d also leads to a reduction of the confinement factor Γ which could increase j_{th} again. Similarly, an increase of the resonator length L will at some point cause additional gain saturation effects (e.g., due to an increased probability for spontaneous emission in the larger volume of the active material). In the next section we will show how common device concepts for semiconductor lasers make use of the possibilities to reduce the threshold as given by equation (21).

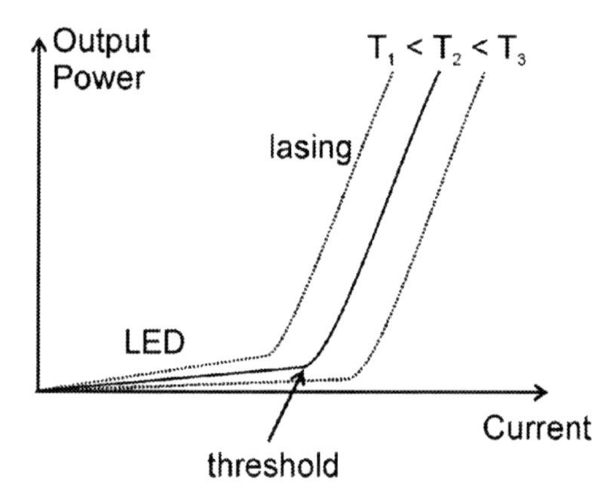

Figure 9: Output power as a function of the injection current for three different temperatures.

At the end of this section, a brief look is presented at the characteristics of the output power P of a semiconductor laser as a function of the applied current I. Figure 9 shows the typical laser characteristics for three different temperatures $T_1 < T_2 < T_3$ of the device. The value of the P(I) slope being small for low currents suddenly increases significantly at the lasing threshold current. Below threshold current the device operates in the LED mode where only spontaneous emission contributes to the output power. Above threshold, however, stimulated emission dominates, and laser emission with its characteristic features such as coherence is detected. As can be seen in Fig. 9 the threshold current is a function of the temperature T. At lower temperatures loss mechanisms in the active layer such as scattering processes are reduced leading to rather small threshold currents. With increasing temperature the lasing threshold increases. This can be phenomenologically described by the expression

$$I_{thr}(T) = I_{thr}(T_1)\exp\left(\frac{T - T_1}{T_0}\right) \tag{22}$$

In equation (22) the parameter T_0 characterizes the temperature stability of the laser device. A large value of T_0 implies large temperature stability, i.e., the threshold is rather insensitive to changes of the device temperature.

4. Common Device Structures

The first design of a semiconductor laser was based on a simple pn junction (see Fig. 10). Both the n (excess electrons) and p (excess holes) regions where doped such heavily that the Fermi levels on the n side and on the p side were pushed into the conduction and valence band, respectively. In an unbiased equilibrium condition the Fermi level in the whole semiconductor system is constant (left side of Fig. 10). If a forward bias is applied to the pn junction the bending of the bands in the contact region is reduced according to the applied voltage. The quasi-equilibrium conditions for the electrons on the n side and for the holes on the p side have now to be described in terms of two different quasi-Fermi levels (right side of Fig. 10) since the forward bias causes additional electrons and holes to flow towards the pn junction as electrical current.

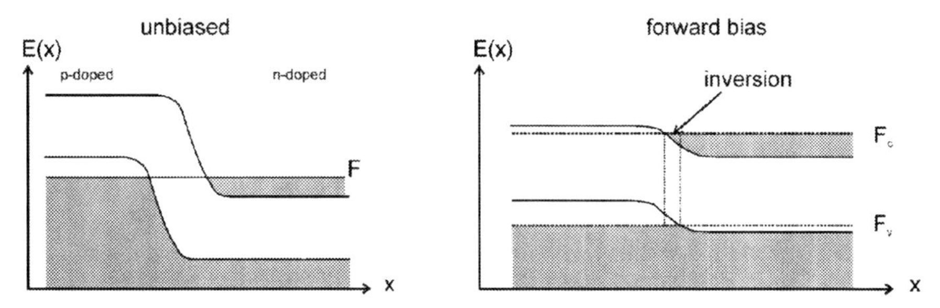

Figure 10: Unbiased (left) and forward biased (right) heavily-doped pn junctions. In the case of an applied forward bias population inversion is obtained in a certain region.

In a heavily doped pn junction the situation being necessary for fulfilling eqs. (13) and (14) can arise in which at the same point in space simultaneously excess electrons in the conduction band and excess holes in the valence band are available for a recombination process. Thus, population inversion of the carriers is generated which is the required condition for gain and lasing action to occur.

Whereas such simple semiconductor laser devices showed stimulated emission in pulsed mode at cryogenic temperatures it has not been possible to operate them at room temperature or in the continuous-wave (cw) mode. Their shortcoming can be easily understood by considering that the rate of stimulated emission is proportional to the product of the number of electrons in the upper laser level N_2 and the amplitude of the electrical field at the same point in space. On the one hand, the propagation direction of the electromagnetic laser field in a simple pn junction is, however, arbitrary and the maximum of the electrical field does not necessarily coincide with the region of population inversion. On the other hand, in regions without population inversion the wave will not only experience no amplification but will even be absorbed.

For this reason waveguide structures have been implemented in the semiconductor laser devices. These waveguide structures make use of the different refractive indices occurring in so-called heterojunctions where two semiconductors with different bandgaps are grown on top of each other. Based on the GaAs material system such a heterojunction can, for example, be realized by growing a GaAs layer sandwiched between two AlGaAs cladding layers. By alloying GaAs with aluminum the band gap of this semiconductor material is increased and its refractive index reduced. At moderately high Al concentrations the lattice constants of GaAs and AlGaAs do not significantly differ. Therefore, these two materials can be epitaxially grown on top of each other with only a small defect density introduced. A waveguide structure based on a GaAs/AlGaAs double-heterojunction is schematically shown in Fig. 11.

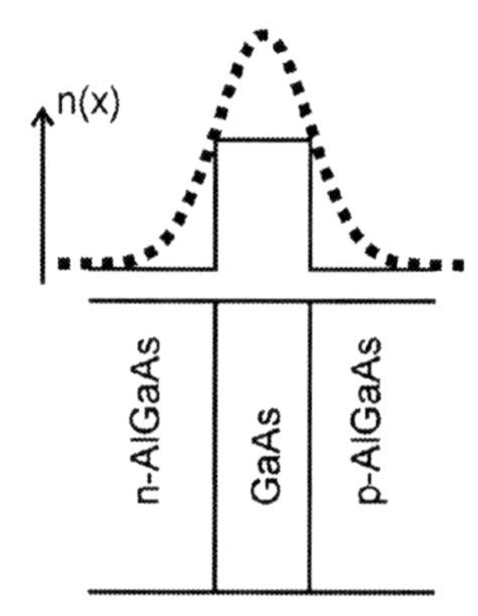

Figure 11: Refractive index profile of a waveguide p-i-n-structure of a laser diode. The dotted line shows the electrical field in the GaAs waveguide. The width of the GaAs layer is in the order of several 100 nm.

The AlGaAs cladding layers can be p- and n-type doped, respectively, so that the whole system forms a p-i-n junction (i intrinsic undoped layer) which acts similarly to the simple pn junction. However, inversion can now much more easily be obtained in the intrinsic GaAs layer. Simultaneously, due to its higher refractive index this layer acts as a waveguide for the optical modes which are thus efficiently guided in the inversion region. This concept reduces the absorption losses of the optical modes considerably. Successful cw operation of semiconductor lasers at room temperature has thus been achieved.

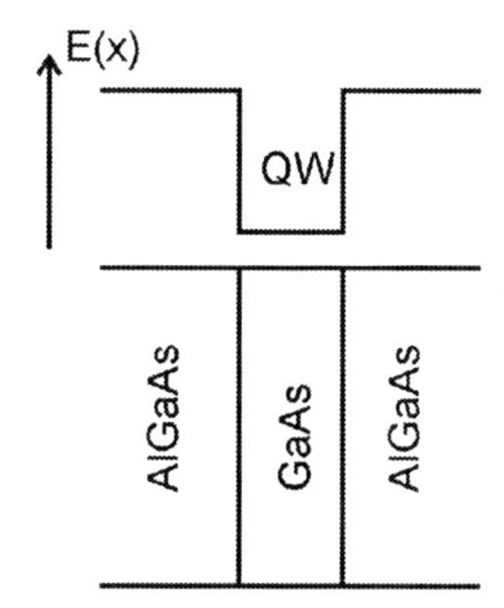

Figure 12: Energy profile of a quantum-well structure based on AlGaAs barriers and a GaAs well. Here, the width of the GaAs layer is in the order of some nm only.

Most of today's semiconductor lasers provide not only confinement of the optical modes (preferentially just one, i.e., the optical ground mode) in the waveguide but also a trapping of electrons and holes in a separate-confinement heterostructure as shown in Fig. 13. Here, the large refraction index in the very thin layer (a) represents a rather low bandgap so that carriers are efficiently collected. This layer has a thickness of at maximum a few exciton Bohr radii so that the electrons and holes are trapped in a quasi-two-dimensional layer (see also Fig. 12). This confinement of the carriers in the quantum well changes their density of states into a step function (compare Fig. 7) and additionally prevents the escape of the carriers into the surrounding material where they might be trapped and afterwards recombine radiationless at impurities or crystalline defects. For a semiconductor quantum-well laser the latter aspect is in many cases the dominant one and leads to a further reduction of the laser threshold as well as to an increase of the quantum efficiency.

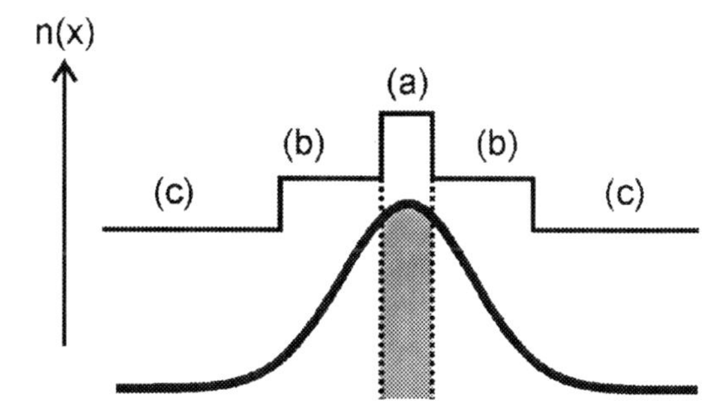

Figure 13: Refractive index profile of a separate-confinement heterostructure. The confinement factor Γ is determined by the grey-shaded area of overlapping spatial mode profile and the active layer (a) (e.g., quantum well), (b) waveguide layers, (c) cladding layers.

For the optical mode, the index step from the waveguide layers (b) to the cladding layers (c) determines the waveguiding properties. Only a small fraction of the optical mode guided between the claddings is actually overlapping with the region of the quantum well where gain can be expected. It is therefore useful to define the so-called confinement factor Γ which relates the material gain of the active layer to the gain experienced by the optical mode, $g_{mod} = \Gamma\, g_{mat}$. The confinement factor is defined as the ratio of the mode volume part in the active region to its whole volume,

$$\Gamma = \frac{\int_{-d/2}^{+d/2} J(x)\,\mathrm{d}x}{\int_{-\infty}^{+\infty} J(x)\,\mathrm{d}x} \tag{23}$$

J(x) is a function describing the profile of the optical mode, d is the thickness of the active layer which is assumed to be centred at x = 0.

In addition to the device structures defined by the sequence of the layers grown and being discussed so far, also a lateral structuring of the semiconductor laser is useful to reduce the threshold and increase the efficiency even further. On the one hand, a lateral structuring can provide a confinement and thus a waveguiding situation in a second dimension for the optical mode. On the other hand, such a process can also provide a well-defined path for the current which flows through the laser device and which provides the necessary pump energy to produce a population inversion in the active layer.

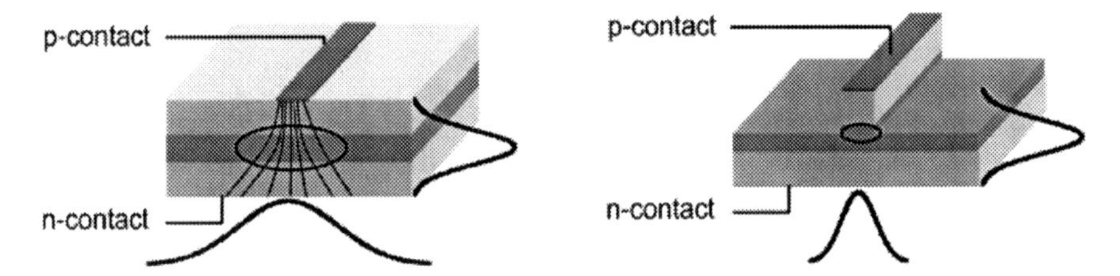

Figure 14: Gain-guided (left) and index-guided (right) semiconductor laser structures. In the gain-guided structure the p-contact defines the starting point of the current path which spreads out while travelling through the structure. In the index-guided structure the upper part of the structure has been etched so that a ridge now defines the current path into the active layer. Additionally, the lateral profile of the electric field is compressed resulting in a small aspect ratio of the field.

Figure 14 exemplarily shows two laterally structured semiconductor laser devices. In the left structure the p-side contact defines the starting point for the current path. While travelling through the structure the current spreads out so that a rather large area of the active region is pumped such strongly that inversion is reached. The only guiding mechanism for the optical mode in the lateral direction is thus the lateral extension of the effectively pumped active region (highlighted in Fig. 14, left, by the ellipse). Typically, in such gain-guided laser structures rather high overall output powers can be achieved. The profile of the optical mode, however, is described by a rather large aspect ratio, i.e., a large ratio of the lateral to the vertical width of the mode profile.

For the laser structure in Fig. 14, right, a part of the upper layers has been etched down to the active layer. In this case, the lateral width of the current path as well as of the optical mode is defined by the width of the remaining ridge. Since the lateral confinement of the optical mode is again due to the refractive-index contrast between the semiconductor and the surrounding air this structure is usually referred to as a index-guided semiconductor laser. Typically, a much better aspect ratio for the optical mode is achieved in index-guided compared to gain-guided structures. Additionally, the current spreading is prevented which leads to a higher efficiency of the devices. However, the overall output power is usually smaller than for gain-guided laser devices.

5. Green and Blue Semiconductor Lasers: Challenges for Basic and Applied Research

In this section we show how the concepts which have been introduced in this paper so far find their way into today's basic and applied research on semiconductor lasers. Several reviews available in recent literature mainly focus on the well-established "conventional" III-V material laser diodes (see, e.g., [15] and the textbooks [16,17], just to name a few). We will therefore, for the sake of keeping the contribution within a reasonable length, especially focus on the tremendously growing research activities concerning the development of semiconductor lasers operating throughout the whole UV-blue-green-yellow spectral region, since the basic features of semiconductor lasers can also well be exemplified on these modern laser materials.

5.1 GREEN ZnSe-BASED SEMICONDUCTOR LASERS

The material system Zn(Cd)Se/Zn(Mg)SSe is of considerable interest due to its possible application in optoelectronic devices emitting in the yellow (large Cd content of the active layer) to blue (binary ZnSe as active layer) spectral region [18,19]. By use of other well-known semiconductor material systems such as InAs/GaAs, InP/GaP (covering the infrared-red spectral region) and InN/GaN (blue-ultraviolet) lasing action in the green-yellow spectral region will be hardly achievable. However, the main drawback of II-VI lasers is their lifetime instability due to structural degradation under operation conditions. Although cw lasing at room temperature has already been demonstrated several years ago the lifetime of Zn(Cd)Se-based conventional edge-emitting laser diodes is still limited to about 500h [18].

Layer	Thickness
Pd/Au contact	10 nm/250 nm
Al_2O_3 insulator	80 nm
ZnSe:N/ZnTe:N multi–quantum–well contact	20 nm
ZnSe:N spacer	160 nm
ZnSSe:N spacer	120 nm
MgZnSSe:N cladding	700 nm
ZnSSe waveguide	100 nm
CdZnSSe:Cl quantum well	4 nm
ZnSSe:Cl waveguide	100 nm
MgZnSSe:Cl cladding	1000 nm
ZnSSe:Cl spacer	120 nm
ZnSe:Cl buffer	20 nm
GaAs:Si buffer	380 nm
GaAs:Si substrate	350 µm

Figure 15: CdSe/ZnSe quantum-well laser structure. The function of the layers is discussed in the text. After [19, 20].

A typical ZnCdSe based laser structure which has been realized in the Institute of Solid State Physics at the University of Bremen (Semiconductor Epitaxy Group) [19] is schematically shown in Fig. 15. We immediately recognize some fundamental device concepts of semiconductor lasers which have been introduced in the last sections:

- The active region is formed by a (quaternary) ZnCdSSe quantum well which provides confinement for the charge carriers. The use of a quaternary material enables for an engineering of the emission wavelength of the laser by adjusting the Cd and S content. Simultaneously, the sulphur partially compensates the strain introduced by the cadmium. Thus, a targeted fixing of the band gap transitions to the user's desires can be achieved.
- The active region is surrounded by ZnSSe waveguide layers. Since no Cd is added to these layers their band gap is increased compared to the active region and simultaneously their refractive index is decreased.
- The waveguide layers are embedded in MgZnSSe cladding layers. The addition of Mg to these layers increases their band gap and lowers their refractive index even further while again keeping the lattice constant. This leads to an efficient wave guiding in the ZnSSe waveguide layers and results in a refractive-index profile as has been presented in Fig. 13.

The additional buffer and spacer layers are necessary for a successful epitaxial growth and electrical contacting of the laser structure. The growth of such ZnSe-based structures is usually performed by molecular-beam epitaxy (MBE). The substrate (here an n-type doped GaAs:Si wafer) is placed in a UHV chamber which is equipped with several sources of the elements Zn, Se, Cd and S which are supplemented by a chlorine source for n-type doping and a nitrogen plasma source for p-type doping of the II-VI layers. The sources can be heated and will then emit atoms or molecules of the respective constituents. In the UHV the atomic or molecular beams hit the substrate where they are pinned and chemically react with each other thus resulting in the epitaxial growth. Since high-quality ZnSe substrates are hardly available usually a hetero-epitaxial method is used for the growth of ZnSe-based laser diodes. Due to the small lattice misfit of only 0.27% at the usual growth temperatures of about 280 to 300°C, GaAs is the mostly favoured substrate material. It first has to be cleaned and will subsequently be overgrown with an epitaxial GaAs layer. This results in a smooth and clean GaAs surface with a desired surface reconstruction. The hetero-interface is formed between this GaAs layer and a thin ZnSe layer which is grown on top of it. Since this ZnSe layer is rather thin (20 nm in the case of the structure shown in Fig. 15) it grows with the lattice constant of the GaAs material, i.e., strain is introduced into the ZnSe layer. The strained growth results in excess energy inside the ZnSe layer such that above a critical thickness of ~150-200 nm it would relax to its intrinsic lattice constant by the formation of defects (mainly dislocations). Of course, the formation of crystalline defects has to be prevented in a laser structure because they would act as non-radiative recombination centres for the electrons and holes and would additionally reduce the mobility of the carriers due to an enhanced scattering rate. This would increase the resistivity of the device and thus the generation of heat by the electrical current. The thermal energy might lead to diffusion processes of material causing mixing and defects and thus to a deterioration of the quality of the semiconductor laser structure. All the mentioned effects would cause a shortening of the lifetime of the device.

In order to avoid the formation of crystalline defects due to relaxation of strained layers the subsequent ZnSe-based layers are grown lattice matched to the GaAs substrate on top of the thin ZnSe buffer layer. This is achieved by adding either a small amount of sulphur (~ 6%) to the ZnSe since $ZnS_{0.06}Se_{0.94}$ possesses the same lattice constant as GaAs, or by use of the quaternary MgZnSSe which allows for even more flexibility concerning the achievable bandgap. Since the substrate side is intended to provide the n-type contact of the laser, the GaAs substrate is n-type doped with silicon, and the subsequent ZnSe-based layers are n-type doped during the MBE growth process with chlorine.

The layers which are grown on top of the undoped active and waveguide layers are p-type doped with nitrogen such that the whole device effectively forms a pn diode. The additional layers on top of the structure (the ZnSe/ZnTe multi-quantum well plus the Pd/Au metal layer) are necessary to achieve an ohmic contact with a low resistivity on the p-side of the device.

By varying the Cd content in the quaternary quantum wells electrically pumped CdZn(S)Se-based laser diodes have been experimentally demonstrated to operate from 500 nm (blue-green) to 560 nm (yellow) [19]. The output power of the devices amounted up to 1 W per pulse (in the pulsed mode). The long-wavelength device showed better characteristics such as lower threshold current, higher output power and higher quantum efficiencies. This superiority can be understood by considering that a longer emission wavelength corresponds to a stronger confinement of the carriers in the active quantum-well region (higher Cd content) since the barrier material has been processed to obtain the same composition for all devices. The lifetime of these devices, however, is still limited to a few hours at most.

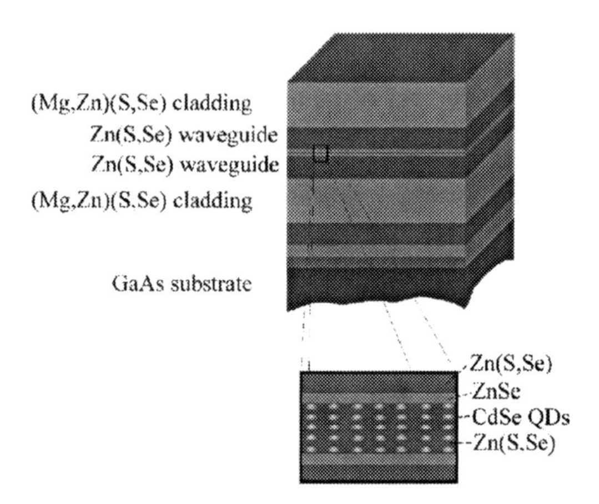

Figure 16: CdSe/ZnSe laser structure based on CdSe quantum dots as active material [24].

The use of quantum dots instead of two-dimensional quantum-well layers as active materials in semiconductor lasers has been proposed for several reasons. Besides an increase of the carrier confinement, an enhanced temperature stability of the laser threshold can be expected due to the δ–like density of states in quantum dots. Dislocations are expected to less efficiently reduce the output of dot-like active layers. For conventional III-V materials, QD lasers have been invented since several years. The structure of a respective II-VI quantum-dot laser as having been realized and investigated in Bremen [21, 23, 24] is schematically shown in Fig. 16. Whereas waveguide and cladding layers are conceptually unchanged the active region now consists of a ZnSSe matrix in which Cd-rich islands with diameters of 5-10 nm have been embedded (more exactly speaking, such islands grow in a so-called self-assembled manner on firstly evolving very thin, approximately just two atomic layers thick Cd-rich "wetting" layers if a large Cd contribution is provided by the Cd molecular beam). These Cd-rich islands provide a three-dimensional confinement and can be regarded as non-isotropic quantum dots (QDs). The height of these QDs is 1.5–2 nm, their density amounts to 10^{10}–10^{11} cm^{-2}. To obtain more efficient laser emission at such a QD density, the active layer can be designed as a multi-fold QD stack. The strain distribution in a QD layer favors an ordered QD growth in subsequent Cd-rich layers resembling the QD distribution geometry in the first QD layer. Thus, well-ordered QD stacks are fairly easily obtained in a self-assembled growth process [21].

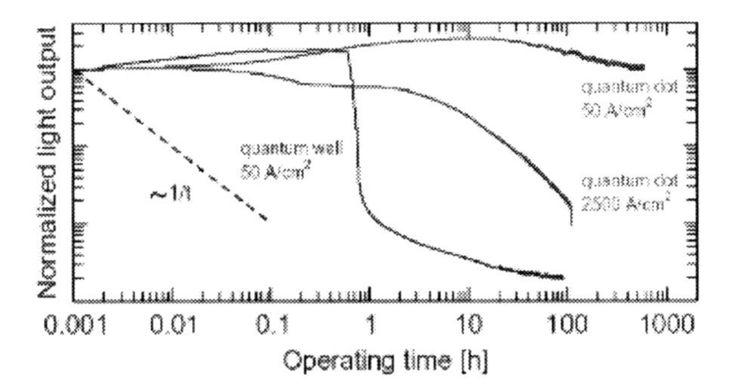

Figure 17: Lifetime of quantum-well and quantum-dot laser devices operated in LED mode (50 A/cm^2) and slightly below the laser threshold (2500 A/cm^2). Shown is the normalized light output as a function of the operating time on a double-logarithmic scale. After [21].

Figure 17 shows the normalized light output as a function of the operation time of a quantum-well and a quantum-dot laser diode on a double-logarithmic scale. Note that even under LED operating conditions fast degradation processes lead to a lifetime of 1 hour only for the quantum-well (QW) laser device (abrupt decrease of light output). Under the same operation current of 50 A/cm^2 the quantum-dot device does not show any significant decrease of the light output due to degradation effects even up to 1000 hours. The strong three-dimensional confinement of the carriers in the quantum dots prevents the escape of the carriers and their further diffusion through the laser structure which could finally lead to trapping and non-radiative recombination at structural defects. In this case, the energy of the charge carriers would be converted into thermal energy (enhanced lattice vibrations). The resulting increase of the material temperature would lead to an enhanced material diffusion in the laser structure and could also create mobile defects in the semiconductor crystal structure. In the case of a quantum-well laser the possibility for additionally generated and even mobile defects to intersect with the active layer is much larger than for a QD laser, and thus non-radiative recombination processes would be enhanced even further. For a degradation process being entirely due to recombination-enhanced defect reactions a 1/t decrease (t operating time) of the output power is expected [22]. This relation is visualized in Fig. 17 as dashed line for comparison. If the QD laser is operated just slightly below its threshold being about 7 kA/cm^2 the degradation process is faster but the lifetime of the device still amounts up to 100 hours. The results thus show that the concept of QDs as active region in a ZnSe-based semiconductor laser significantly improves the performance of the device when compared to a standard QW-based device.

Optical spectroscopy can give a deeper insight into the fundamental microscopic processes which govern the properties of QD and QW lasers. A widely used experimental technique to measure the gain of a semiconductor laser structure is the so-called variable-stripe-length (VSL) method. The basic idea of this spectroscopic technique is depicted in Fig. 18.

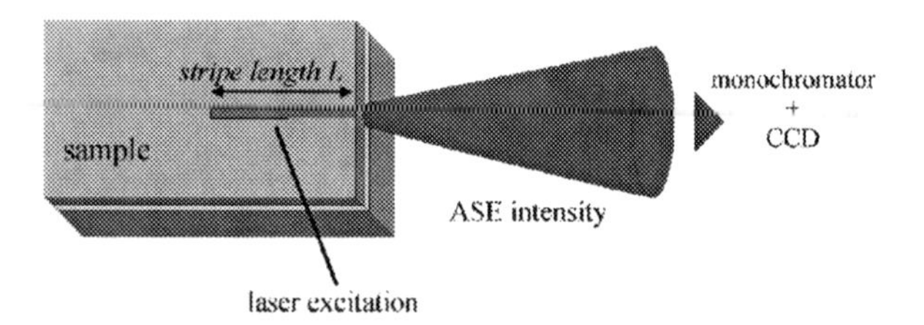

Figure 18: Gain measurements with the variable-stripe-length method.

An excitation laser is focussed down to a narrow stripe onto the sample surface such that one end of the stripe lies directly at one edge of the sample. The photon energy of the excitation laser must be chosen such that the light penetrates the cladding and waveguide layers but optically excites electron-hole pair in the active layer. The amplified spontaneous emission (ASE) emitted from the edge of the stripe is collected by a combination of lenses and imaged onto the entrance slit of a monochromator. The spectrum in the region of interest can be simultaneously detected if a charge-coupled device (CCD) is used.

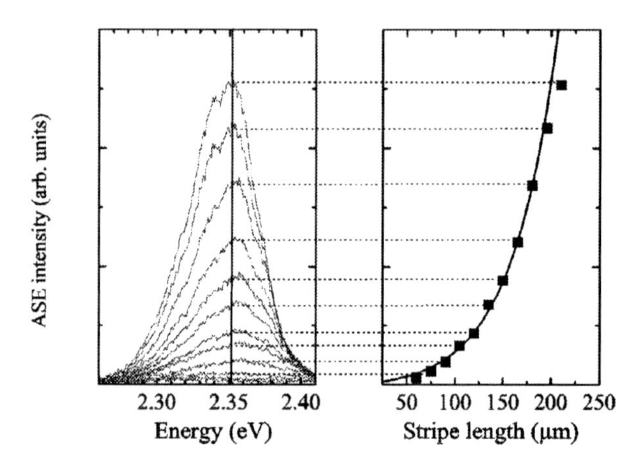

Figure 19: Determination of the gain g(E) by a non-linear fitting procedure in a VSL experiment [24].

Typical ASE spectra which have been recorded for different lengths L of the excitation stripe are shown in the left part of Fig. 19. The increase of the ASE intensity as a function of the stripe length L can be understood by considering an electromagnetic wave which is spontaneously emitted from the left end of the stripe (Fig. 18) and travels along the stripe in direction to its other end. Obeying the Lambert-Beer law the wave will be attenuated or amplified according to

$$I(x,E) = I_{sp}(E) \cdot e^{g(E)\cdot x} \tag{24}$$

The ASE intensity which is finally emitted from the edge of the sample is obtained by integrating equation (24) over the length L of the stripe

$$I_{ASE}(L,E) \propto \frac{I_{sp}(E)}{g(E)}\left(e^{g(E)\cdot L}-1\right) \tag{25}$$

Equation (25) provides a direct connection between $I_{ASE}(E)$ and the spectral gain profile g(E). However, this equation cannot be directly solved to yield g(E) as a function of all other parameters. It is therefore necessary to find an indirect way to extract the spectral gain profile from the measurements of the spectral profile of the ASE. This can be achieved by plotting the ASE intensity for a fixed energy position E_l as a function of the stripe length L (compare right part of Fig. 19). The value $g(E_l)$ is now calculated by performing a non-linear curve fit using equation (25) (solid line in Fig. 19, right). This procedure has to be repeated for all values of the photon energy E for which g has to be determined. Typical gain profiles obtained this way are shown in Fig. 20.

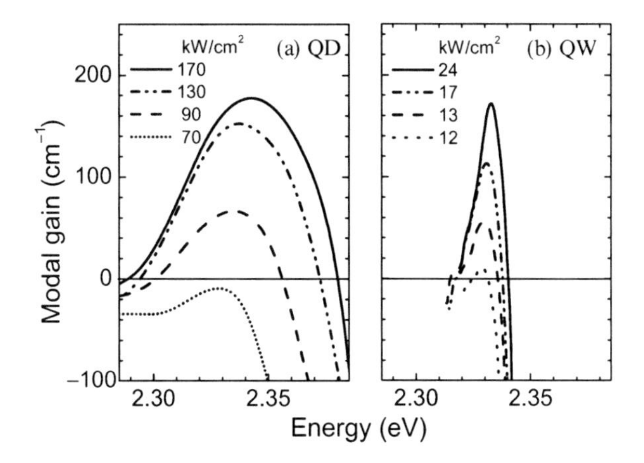

Figure 20: Comparison of the modal gain of a quantum-well (right) and a quantum-dot (left) laser structure at 10 K. The active layer of the quantum-dot laser structure consists of a five-fold quantum-dot stack [23,24].

The results presented in Fig. 20 have been experimentally obtained by use of the VSL method. The left part shows experimental data on the QD laser structure, the right part presents the corresponding results for the QW laser with, apart from the active layer, otherwise unchanged material composition and layer dimensions. In both cases the spectral gain profiles have been measured for different excitation densities. The dotted lines give the gain profiles at the lasing threshold which correspond to approximately 70 kW/cm^2 for the QD and 12 kW/cm^2 for the QW sample at 10 K. The dashed, dot-dashed and solid curves show the evolution of the spectral gain profile for pumping powers and thus densities of excited electron-hole pairs increasing above the lasing threshold. Two important differences between the characteristics of the QW and QD sample are obvious: the QD laser shows a much broader spectral gain profile and a significantly larger threshold density. The latter one can be explained by comparing the overall volume of the active material in the two laser structures: on the one hand, in the QW sample much more active material is present because in the QD laser structure the nanometer-size Cd-rich QDs can contribute to the gain process only. Consequently, the gain which the optical laser mode experiences is smaller in the QD laser. On the other hand, if the material gain is calculated by dividing the modal gain by the confinement factor Γ it is found to be more than 10 times larger in the QD laser. This is a result of the strong confinement of the carriers in the quantum dots. In each single quantum dot with discrete energy states (remember the δ–like density of states) inversion is achieved already if one electron-hole pair is injected only which results in a very efficient conversion of the pump energy into laser photons in the QD array thus increasing the material gain g_{mat}.

The difference between the spectral width of the gain profiles of the QD and QW laser structures is due to a rather large inhomogeneous broadening of the QD emission energy. This is a result of the above-mentioned self-organized growth process of the QD layer(s) which yields QDs with a certain distribution of size and Cd content. Both variations cause the confinement of the QDs to fluctuate from dot to dot thus resulting in a Gaussian-like shape of their emission energy profile.

As the excitation density is increased the gain profile of the QD laser shows a significant red shift at its low-energy side, i.e., the band edge, accompanied with a significant blue-shift at its high-energy side, i.e. the transparency point. The red shift of the band edge is mainly caused by the screening of the Coulomb interaction in the semiconductor layer by the optically excited charge carriers. The blue shift of the transparency point is due to a filling of the excitonic states in the QD layer. The occupation of the electronic ground states in the QDs is limited to two electrons and holes with opposite spin orientations due to the fermionic nature of the charge carriers. Consequently, an increase of the excitation density will subsequently fill the electronic states in the QD ensemble starting from the low-energy states. A rather strong blue shift of the transparency point such as the one observed in Fig. 20, left, is usually a strong hint that localization centres for the charge carriers are present in the investigated semiconductor system.

The change of the QW gain profile with excitation density is far less pronounced than for the QD laser structure. A detailed comparison with microscopic simulations reveals that in this case the gain results from interactions in an electron-hole plasma. Since the Coulomb interaction in the wide-bandgap semiconductor system is rather large excitonic effects still contribute to the gain mechanisms. As a consequence, the electron-hole plasma is strongly Coulomb-correlated thus yielding a different gain feature as for semiconductor lasers based on the GaAs material system [25].

Differences between the QW and QD lasers are also found if the laser threshold is analyzed as a function of the device temperature (Fig. 21). If the threshold density under optical excitation is plotted as a function of the temperature on a logarithmic scale the QW laser, although starting at a lower lasing threshold than the QD laser for lowest temperatures, immediately shows an exponential increase as described by equation (22). The temperature "stability" parameter (compare equation (22)) amounts to $T_0 = 82$ K for the investigated ZnSe based laser.

For the QD laser a much enhanced temperature stability of the threshold density is found for temperatures below T = 120K. In this range a value T_0 = 1260K is experimentally determined which again is a result of the strong confinement of the carriers and the discrete energy states in the QDs. Both properties of the QDs as gain medium prevent loss mechanisms such as the escape of the carriers from the dots, carrier-carrier scattering, or carrier-phonon scattering. Although these processes cannot be completely prevented due to the finite confinement in the QDs their strong suppression results in a superior temperature stability of the QD laser structure.

At temperatures T > 120K the thermal energy is large enough to activate some of the loss mechanisms. Nevertheless, the value T_0 = 97K is still larger than that of the QW laser. One goal of current research is to increase the confinement energy in the dots by using a higher Cd content in the dots or a barrier material with a larger bandgap, thus trying to improve the temperature stability of the QD laser even further.

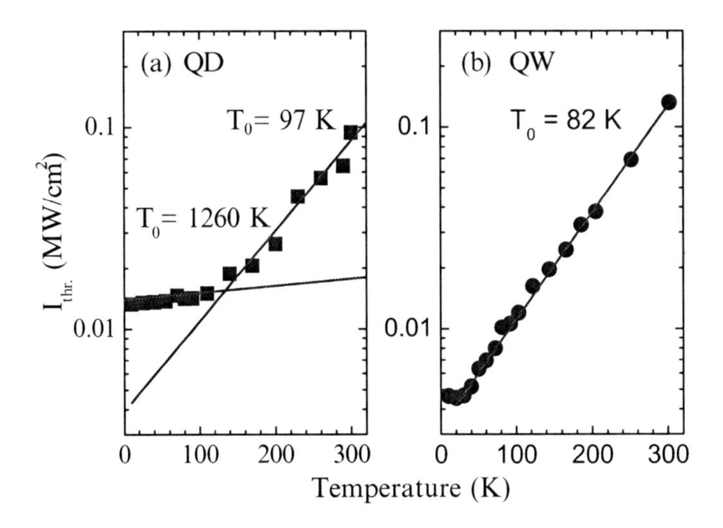

Figure 21: Temperature dependence of the lasing threshold for a quantum-well (right) and quantum-dot (left) laser structure [23,24].

5.2 InGaN/GaN-BASED SEMICONDUCTOR LASERS

Whereas ZnSe-based semiconductor lasers are not yet commercially available due to their limited lifetime, blue to UV semiconductor lasers based on the GaN material system can actually be obtained from different companies. The band gap of GaN can be increased or decreased by adding Al or In, respectively. Since the pioneering work of Nakamura et al. (see, e.g., [26] for reviews) who discovered a reproducible p-type doping of GaN much research has been devoted to the material properties of InGaN/GaN/AlGaN semiconductor lasers. In these devices, the In content is often chosen low (e.g., [27]) so that the laser emission wavelength is rather short (shorter than 365 nm) or even goes down to 340 nm if Al is added to the active layer [28]. However, it is not so easy to extend the laser wavelength into the blue by introducing a larger percentage of In. With growing In content in a ternary InGaN active layer, localization by composition fluctuations plays an increasing role in particular for the lasing process and its peculiarities [29,30].

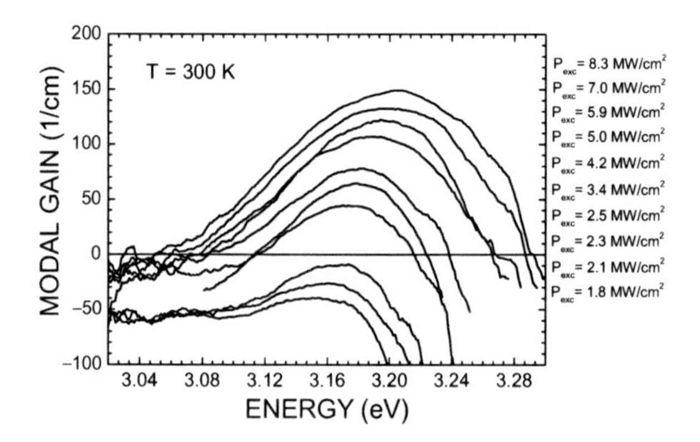

Figure 22: Gain spectra of an InGaN multiple-quantum-well laser structure at room temperature [31].

Several important aspects of InGaN based QW lasers emitting in the near-UV spectral region at about 400 nm are readily recognized if the spectral gain profile of such a laser structure is measured by, e.g., the variable-stripe-length method (Fig. 22). If the gain profile is compared to that of the ZnSe-based lasers (Fig. 20) it becomes obvious that the behavior of the InGaN quantum *well* laser resembles more that of the CdSe/ZnSSe quantum *dot* than that of the ZnCdSe quantum *well* laser. With increasing excitation density both the CdSe QD and the InGaN QW gain profiles show a significant broadening and a blue shift of the transparency point. As has been already pointed out in the previous section, these features are characteristic for systems involving strong localization centers where state-filling effects result in a strong blue shift of the transparency point with increasing excitation density. As described already there, the intentionally grown Cd-rich islands forming the quantum dots serve as strong localization centers for the CdSe/ZnSSe-based QD laser. In the InGaN QW laser, however, no QD growth was intended. Nevertheless, QD-like localization centers must have been formed in the QW to explain the experimentally determined gain profile. They are usually explained in terms of fluctuations of the In content in the QW region.

To get a deeper insight into the properties of these localization centers photo-luminescence (PL) measurements of InGaN QW structures have been performed as a function of the temperature [32,33]. As one example, in Fig. 23 the logarithmic PL intensity has been plotted as a function of the reciprocal temperature for four different heterostructures labeled SCH1-4 (thickness of the InGaN quantum well 1.0, 1.7, 3.0, 4.5 nm, respectively). The decrease of the PL intensity with temperature is due to thermally activated loss channels. The measurements clearly show that at least two loss mechanisms with significantly different activation energies are present in InGaN QWs. This can be seen in the inset where just a single activation energy has been used to describe the data points (dotted line). This ansatz clearly fails to fit the experimental results for the higher-temperature region. If two activation energies E_1 and E_2 for two different loss mechanisms are used, the fit nicely reproduces the experimental findings (solid lines).

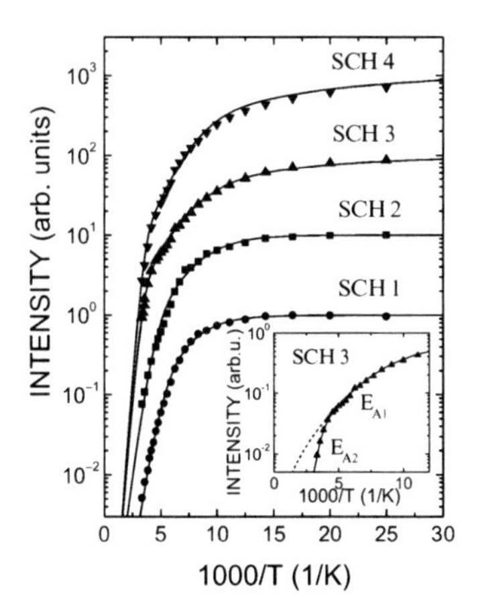

Figure 23: Temperature-dependent PL intensity of InGaN quantum-well structures [32,33].

The first activation energy being in the range of E_1 = 30-40meV for the investigated structures is ascribed to the escape of the electrons and holes from the potential fluctuations in the QW caused by the varying In content. The second activation energy is about one order of magnitude larger (E_2 = 100-250meV) and describes the activation of electrons and holes from the QW into the barrier layers.

The results clearly show that localization centers are present in InGaN QW lasers what improves the characteristics of the laser devices. Although the threshold densities for InGaN QW lasers are very high (e.g., $P_{th} = 2.1 MW/cm^2$ in Fig. 22) they can exhibit lifetimes in the order of several thousand hours because the material itself is much more stable under extreme excitation conditions than the ZnSe material system.

Of course, it would be desirable to intentionally grow InGaN quantum dots in a laser structure with a certain indium content to adjust the operating wavelength of the laser device and also to increase even further the confinement of the carriers in the In-rich localization centers (for InGaN QDs in an LED structure see, e.g., [34]). Recent results [35] confirm that it is indeed possible to fabricate such InGaN quantum dots with a ground-state emission energy of about 2.7eV (blue spectral region). Experimental results confirming the successful growth are shown in Fig. 24. The left part presents an overview of the luminescence of the InGaN/GaN structure from the red

(2.0eV) to the near UV (3.5eV) spectral region. The strong peak at about 3.5eV is related to the GaN barrier material which additionally shows typical emission bands at about 3.2 eV due to recombination associated with incorporated donors and acceptors (DAP: donor-acceptor pair recombination), and at about 2.2eV (YP: yellow PL) due to deep traps.

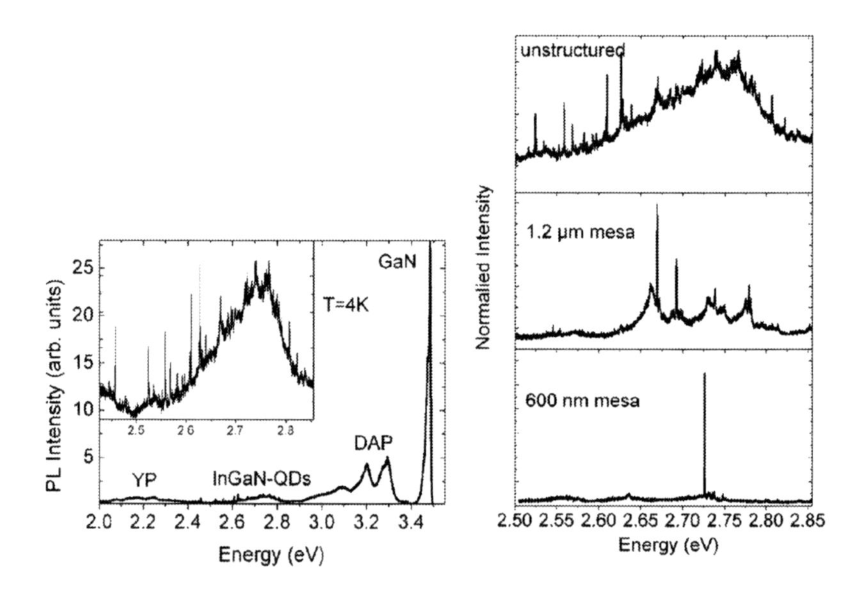

Figure 24: Photoluminescence of InGaN quantum dots [35].

In the range from approximately 2.6 to 2.8eV, a PL structure can be observed which resembles a Gaussian like distribution with several sharp emission lines superimposed to the broad band (inset of Fig. 24, left). By producing small mesa structures on the sample with diameters down to 600nm by etching with a focussed ion beam and by use of a microscope objective it is possible to selectively excite only few QDs of the ensemble (compare right part of Fig. 24). For more and more reduced mesa diameters, the emission band apparent in the unstructured sample breaks up into increasingly clearly developing and dominating single sharp emission lines. For the 600nm mesa, just one single QD line prevails the emission. Current research aims at understanding the fundamental structure of the QDs, their individual optical properties, and their excitation dynamics. The results obtained so far pave the way to a very efficient and stable InGaN-based QD laser operating in the blue spectral region around 460 nm.

The research on GaN-based laser structures is by far not limited to the investigation of In localizations and quantum dots [26,27,36]. The AlInGaN/AlGaN quaternary material combination, for example, allows to increase the emission energy of a laser beyond 3.5eV, i.e., into the UV spectral region where optically pumped room-temperature pulsed lasing has been demonstrated at 340nm [28]. Research also aims at improving the stability and performance of InGaN based QW lasers with a low In content (0.2-2.7%) which operate at about 370nm [27,36]. The main issues for a successful fabrication of such lasers are low defect densities in the active layers and a p-type doping with high concentration and mobility of the holes. At present, new companies [37] announce cw GaN-based laser diodes thus showing the intense world-wide research and development carried out on these optoelectronic devices both at universities and industry. Especially the use in next-generation DVD (blue-ray or HD-DVD) technology makes GaN an attractive material system where fundamental research achievements will often directly lead to new applications in the technology area.

5.3 ON THE WAY TO BLUE AND GREEN QUANTUM DOT VCSELS

All semiconductor lasers which have been discussed so far are edge emitters, i.e., their laser resonator is a Fabry-Perot cavity formed by the lateral end facets either uncoated or coated with a dielectric material. If no special treatment of these facets is applied the reflectivity of the cavity mirrors is determined by the index contrast between the semiconductor and the surrounding air, $R = (n_s-n_a)^2/(n_s+n_a)^2$, with $n_s(n_a)$ being the refractive index of the semiconductor material (air). As has been mentioned in section 3.2 this reflectivity is in the order of 30% for typical semiconductors which requires rather high threshold-gain values to overcome the high cavity losses due to the low R value. Additionally, for typical resonator lengths of hundreds of μm the cavity will usually support many different longitudinal lasing modes within the spectral range of the gain profile. If lasing of a single mode is required additional wavelength-selective elements have to be added to the laser device. Furthermore, the height of the active region, e.g., a several nm thick quantum well, is usually significantly smaller than the lateral width of the semiconductor laser device even if lateral structuring is applied. Thus the diffraction of the optical modes when leaving the end facets of the laser resonator leads to a large aspect ratio (ellipticity) of the far-field distribution of the beam of an edge-emitting semiconductor laser.

A vertical-cavity surface-emitting laser (VCSEL) provides a solution to decrease the threshold gain and to achieve a nearly circular beam shape together with single-mode emission. An SEM image of a VCSEL structure is shown in Fig. 25. Whereas the optical mode in edge emitters is laterally guided, the wave supported via stimulated emission in the VCSEL propagates in the vertical direction. A cavity with a vertical extension in the order of the emission wavelength (λ cavity) which includes the active region of the semiconductor laser is centred between two Bragg mirrors (distributed Bragg reflectors, DBRs). These DBRs are formed by alternating λ/4 layers of two materials with refractive indices n_1 and n_2 differing from each other as much as possible.

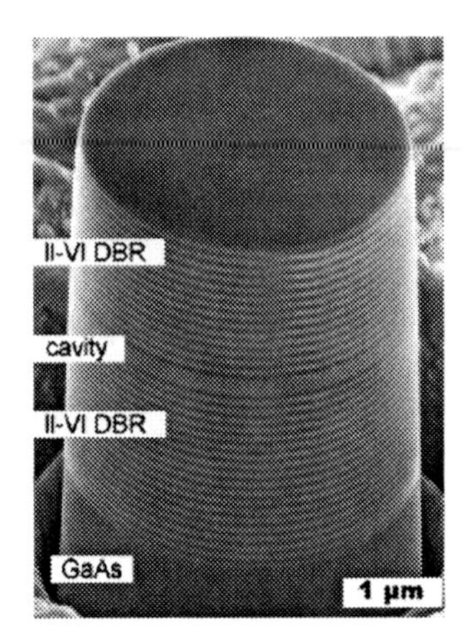

Figure 25: SEM image of a ZnSe-based VCSEL structure [38].

The thicknesses of the layers given above are adjusted such that the widths of each high-index and low-index layer exactly correspond to $\lambda/4$ with λ being the emission wavelength of the VCSEL structure inside the semiconductor, i.e., $\lambda = \lambda_{laser}/n_{layer}$, with λ_{laser} the desired emission wavelength outside the cavity. In this case the interference between the waves which are partially reflected at each boundary between the layers will lead to almost perfect destructive interference of the transmitted parts. The choice of $\lambda/4$ for each layer thickness results in a thickness of $\lambda/2$ for a double layer. By taking into account the additional phase shift which the optical mode experiences if it is reflected at the boundary to the high-index material the path difference for the reflected partial waves always corresponds to an integer multiple of λ thus providing constructive interference for back-reflection and an overall reflectivity close to 100% for $n_2 >> n_1$. Since, however, for any available solid-state material composition, the difference of the refractive indices n_1 and n_2 is somewhat limited, an overall large constructive interference and thus maximum reflectivity is only achieved if a multilayer arrangement, a so-called DBR mirror stack, is used. Any DBR stack consisting of 2m layers, i.e., m periods of each a low- and a high-index material, shows a reflectivity of

$$R = \left(\frac{(n_2/n_1)^{2m} - 1}{(n_2/n_1)^{2m} + 1} \right)^2 \tag{26}$$

In this formula R denotes the reflectivity of the intensity which is connected to the reflectivity r of the electric field by $r = R^2$ (r is used in many theoretical descriptions of DBRs). We see that for an increasing number of periods m the reflectivity of a DBR can get close to 1 even if the difference between the refractive indices n_2-n_1 is relatively small. Nevertheless, the material composition to be used is aimed at achieving index differences being as large as feasible to keep the number of mirror layers as small as possible.

For semiconductor materials suitable in laser structures, the realization of Bragg mirror material pairs with the intended large difference of the index of refraction is quite a challenge [39,40] since especially in the most desired fully epitaxial growth processes conditions like crystal structure and lattice matching as well as similar thermal expansion coefficients have to be carefully taken into account to avoid excessive strain and thus introduction, e.g., of cracks. Therefore, the material choice is often rather limited so that just relatively small differences of the indices of refraction can be realized. One solution of this problem is to use dielectric coating layers instead of epitaxially grown mirrors (for wide-gap VCSELs, see, e.g., [41-43]). However, this yields so-called hybride structures which cannot be grown in one approach without removing them from the MBE or MOVPE machine. Further, even dielectric materials do not provide index differences being large enough to achieve a high reflectivity by evaporating just a few mirror material layers. A more promising approach to this problem can be achieved by accepting the necessity of a relatively large number of alternating epitaxial layers possessing similar material parameters but, consequently, fairly small n differences only[44].

Figure 26 shows the reflection coefficient r of a DBR stack consisting of 15 periods of two different ZnSe-based materials which will be discussed in detail below. The dashed line represents a simulation of the reflectivity, the solid line shows the measurement. In the wavelength range between 510 and 560 nm a constant high reflectivity > 99% is obtained. This wavelength region is called the stop band of the DBR.

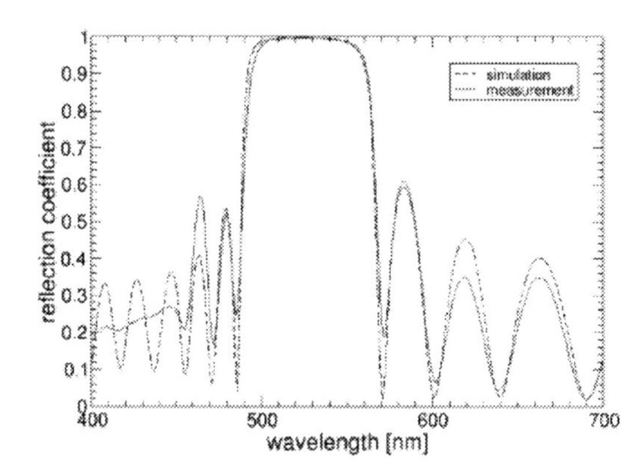

Figure 26: Measured and calculated reflectivities of a ZnSe based DBR with 15 periods. Details are given in the text. After [45].

However, for the ZnSe as well as for the GaN material system it is not straightforward to epitaxially grow VCSELs or just DBRs with high reflectivities even if materials of only small n index differences are used. Besides the above-mentioned necessary fit of lattice and lattice-expansion parameters of the mirror and cavity materials, the epitaxial growth of DBRs is complicated by two further problems.

- The final VCSEL device must be operated by current injection. Therefore, the DBR has to be grown with materials which (at least potentially) are n- and p-type dopable to a certain concentration and mobility.

- For both material systems high-quality lattice matched so-called homoepitaxial substrates (i.e., substrates being composed of constituents of the same material group) are not available at least in an industrial quantity. If substrates of other material compositions are used, strain energy builds up during the growth process and during cooling down between growth and operation temperature which finally leads to relaxation which produces crystalline defects in the DBRs.

For the ZnSe and recently also for the GaN material systems a solution to these problems has been found, and epitaxial DBRs and VCSEL structures have been successfully fabricated [44-48]. For both material systems a so-called superlattice has been introduced as the low-index material in the DBRs. A superlattice is a periodic structure consisting of thin (only a few lattice constants thick) alternating layers of two different materials (see, e.g., [9]). In the case of the ZnSe material system, $ZnS_{0.06}Se_{0.94}$ (lattice matched to the GaAs substate) is used as the high-index material (layer thickness 48nm) whereas a superlattice of 24 periods of alternating layers of 1.9nm MgS and 0.9nm ZnCdSe provides the low-index material. Since the layers within the superlattice are roughly two orders of magnitude thinner than the wavelength of the optical mode the latter one does not feel the superlattice structure but "sees" a material with an averaged effective refractive index. MgS itself cannot be used as the low-index material of the DBR because it grows in a different crystal structure (rocksalt instead of zincblende lattice) and is almost completely undopable. By adding sufficiently thin layers of the dopable zindblende material ZnCdSe both shortcomings of MgS can be overcome at the expense of a slightly reduced index contrast in the DBRs ($\Delta n = 0.6$) [45]. This way, reflectivity values beyond 0.99 have been achieved.

For the growth of GaN-based epitaxial DBR structures, the same superlattice approach is used to provide a suitable low-index material. In this case, the authors of [47,48] produced structures for which each superlattice was composed of 19.5 periods of AlN (thickness 1.6nm) and $In_{0.25}Ga_{0.75}N$ layers (thickness 0.75 nm). 42 nm thick GaN layers provided the high-index material in the nitride-DBRs. However, the reflectivity values obtained so far do not yet satisfactorily approach the desired $R = 0.99$ value. Some improved reflectivity values have been reported by [49,50] for a VCSEL system of shorter wavelength, i.e., lower In content. Further attempts to improve the R values are presently in progress for this material system.

Whereas each DBR itself shows high reflectivity as a kind of plateau over the whole range of its stop band the situation is changed in the complete VCSEL, i.e., if a λ cavity is placed between the bottom and the top DBRs. Calculations show that in this case a sharp cavity resonance occurs in the middle of the stop band at which the reflectivity strongly decreases. This cavity resonance of the VCSEL defines the output wavelength of the laser device. High reflectivities of the DBRs will result in a spectrally sharp resonance. A measure of the linewidth of the cavity resonance is the quality factor $Q = \lambda/\Delta\lambda$ with $\Delta\lambda$ the width of the resonance and λ the emission wavelength. For lasing action to occur the quality factor of a VCSEL typically has to be larger than 1000. The width of the cavity resonance is related to the width of the light emission of the structure.

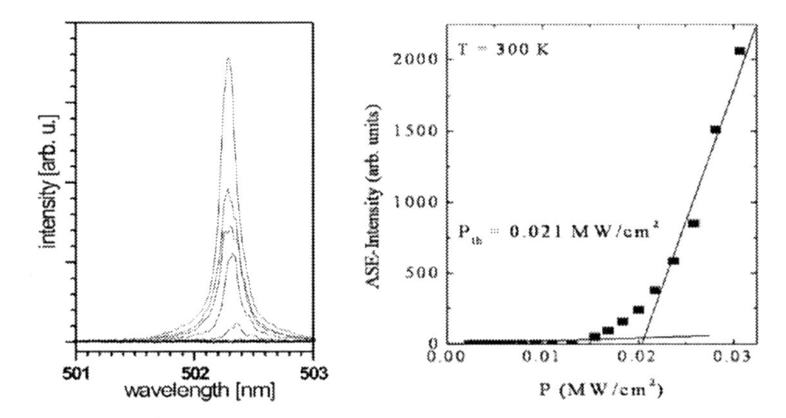

Figure 27: Left: Emission spectra of an optically pumped ZnSe-based VCSEL at room temperature for different excitation densities above lasing threshold. Right: Spectrally integrated ASE intensity of a ZnSe-based VCSEL as a function of the excitation density at room temperature. The lasing threshold was 21kW/cm^2 [51].

In Fig. 27 experimental results obtained for an epitaxially grown ZnSe-based VCSEL involving CdSe QDs as active region are depicted. All measurements presented were performed under optical excitation at room temperature. The left part shows the lasing spectra of the VCSEL structure, in the right part the output intensity is plotted as a function of the excitation density. The most important fact is that indeed a threshold behavior together with the evolution of a narrow emission line is observed. The line width (FWHM) is found to be 0.25nm which is much less than the emission band width of 8nm measured for a reference sample which has been grown under the same conditions but without the top DBR. A second important fact is the threshold density of 21 kW/cm^2 being significantly reduced compared with the edge-emitting ZnSe-based QD laser structure (70 kW/cm^2, compare Fig. 20). The quality factor of the VCSEL structure as determined by its line width amounts to Q = 2000-2500 [51].

A further necessary step towards applicable VCSEL devices is the lateral structuring of epitaxial two-dimensional VCSEL layer structures, i.e., the fabrication of pillars. This has recently very successfully been done for ZnSe- [38,46] as well as nitride-based VCSELs [47,48], compare the SEM picture Fig. 25 obtained by use of a focussed ion beam. As a consequence of lateral structuring, transverse modes of the VCSEL emission develop (Figs. 28 and 29). The modes could be measured spatially and spectrally resolved and compared with theoretical calculations [38] obtained by using a vectorial transfer-matrix method with respect to a cylindrical optical waveguide expansion for each pillar layer [52].

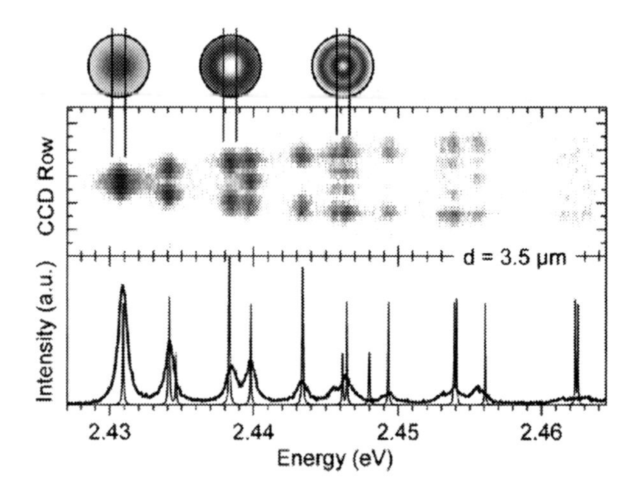

Figure 28: Spectrally and spatially (vertical axis, middle row) resolved one-dimensional CCD image of mode emission of a II-VI microcavity pillar with a diameter of 3.5µm, together with three examples of calculated transverse electric field patterns (top row). Bottom row: Measured and calculated PL spectrum of this pillar [38].

In the upper row of Fig. 28, the calculated spatial distributions of the transversal ground mode (left) and two higher transversal modes are depicted for a pillar of 3.5 µm diameter (the spatial profile of the other modes is omitted for reasons of clarity). Since the detection of the laser mode emission under optical pumping for pillar VCSELs has been performed with a CCD row detector at the output slit of a spectrometer, vertical spatial cuts through the mode profiles are recorded as shown in the middle row of Fig. 28. These findings are well correlated with the theoretically expected profiles in the upper row if one follows the features in between the black line pairs to compare the spatial mode intensities. The bottom row of Fig. 28 shows the complex spectral distribution of the PL emission of the VCSEL pillar structure resulting from the excitation of the different transversal modes. A good agreement with the theoretical mode spectrum is obtained with regard to the mode energy positions. In Fig. 29 the spectral distribution of the modes is depicted for five pillars with different diameters between 1.0 µm and 3.5 µm. As is clearly seen, the modes do not only shift to higher energies for decreasing pillar diameters, i.e., stronger lateral confinement, but also show an increasingly resolvable splitting for those modes being almost degenerate for large pillar diameters, i.e., small lateral confinement. For all diameters, the experimental results are well reproduced by the theoretical modelling. In the calculations, however, mode broadening phenomena due to surface roughness and the finite etching depth are not taken into account. In spite of this, the overall agreement concerning spectral maximum positions as well as relative strengths is almost convincing.

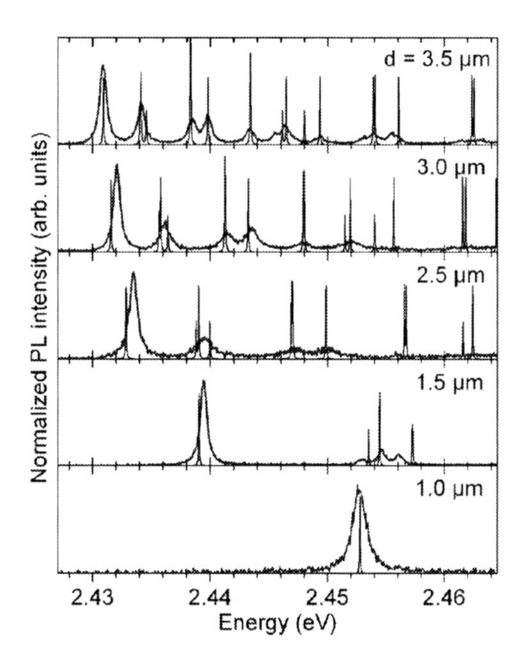

Figure 29: Measured and calculated PL spectra of II-VI microcavity pillars with indicated diameters [38,46].

The results clearly prove the applicability of the VCSEL concept to the ZnSe material system. Current research now aims at the electrical current injection into the active region. Since the DBR structure is both n- and p-type dopable no conceptual changes to the VCSEL have to be applied. However, further improvement of the VCSEL structure is still possible which will lead to an even stronger reduction of the laser threshold. The lifetime of such ZnSe-based VCSEL devices could become long enough to provide even commercially available semiconductor lasers operating in the green-yellow spectral region. Similar experiments with the nitride material system are carried out at present, and a first epitaxial VCSEL structure with InGaN QWs as active layers and a quality factor of $Q = 250$ has been demonstrated [48]. Recently, lateral structuring of such InGaN based VCSELs by use of a focussed ion beam has been demonstrated for the first time by showing pillar emission and even lateral mode profiles [47,48]. If it becomes possible to include InGaN QDs as the active material in a nitride VCSEL the wavelength range of semiconductor lasers based on this material system would be extended into the blue spectral range up to about 450 nm.

Acknowledgements

The work on the ZnSe- and InGaN-based quantum-well and quantum-dot laser structures which is discussed in this article has been mainly performed at the Institute of Solid State Physics at the University of Bremen. The authors acknowledge the experimental contributions of K. Sebald, H. Lohmeyer, M. Röwe, M. Vehse, P. Michler, S.M. Ulrich, and B. Brendemühl (Semiconductor Optics Group) and C. Kruse, T. Yamaguchi, M. Klude, T. Passow, S. Einfeldt, R. Kröger and D. Hommel (Semiconductor Epitaxy Group). TEM investigations have been performed by R. Kröger (TEM Group). InGaN quantum well samples have also been kindly provided by OSRAM Opto Semiconductors (V. Härle and coworkers). Financial support of the Bremen groups by several individual grants and the Research Group "Physics of nitride-based nano-structured light-emitting devices" provided by the Deutsche Forschungsgemeinschaft and by the Volkswagen Foundation (II-VI VCSELs) is gratefully acknowledged.

References

1. N. G. Basov, Nobel Lecture, December 11, (1964).
2. M. I. Nathan, W. P. Dumke, G. Burns, F. H. Dill, Jr., G. Laser, Appl. Phys. Lett. **1**, 62 (1962).
3. R. N. Hall, G. E. Fenner, J. D. Kingsley, T. J. Soltys, and R. O. Carlson, Phys. Rev. Lett. **9**, 366 (1962).
4. T. M. Quist, R. H. Rediker, R. J. Keyes, W. E. Krag, B. Lax, A. L. McWorther, and H. J. Zeiger, Appl. Phys. Lett. **1**, 91 (1962).
5. R. D. Dupuis, Optics & Photonics News, April 2004, p. 30.
6. I. Hayashi, M. B. Panish, P. W. Foy, and S. Sumski, Appl. Phys. Lett. **17**, 109 (1970).
7. C. Kittel, Introduction to Solid State Physics, John Wiley and Sons, 8th ed. (2004). N. W. Ashcroft and N. D. Mermin, Solid State Physics, Thomson Learning (1976).
8. K. F. Brennan, The Physics of Semiconductors – with Applications to Optoelectronic Devices, Cambridge University Press, Cambridge, U.K. (1999/2003)
9. C. F. Klingshirn, Semiconductor Optics, Springer Berlin Heidelberg New York, 2nd ed. (2005). J. H. Davies, The Physics of Low-Dimensional Semiconductors, Cambridge University Press, Cambridge, U.K. (1998/2000)
10. H. Kalt and M. Hetterich (Eds.), Optics of Semiconductors and Their Nanostructures, Springer Berlin Heidelberg New York (2004).
11. P. Y. Yu and M. Cardona, Fundamentals of Semiconductors, Springer Berlin Heidelberg New York, 2nd ed. (1999).
12. K. Seeger, Semiconductor Physics: an Introduction, Springer Berlin Heidelberg New York, 7th ed. (1999).
13. H. Haug and S. W. Koch, Quantum Theory of the Optical and Electronic Properties of Semiconductors, World Scientific Singapore (1990).

14. W. W. Chow and S. W. Koch, Semiconductor Laser Fundamentals, Springer Berlin Heidelberg New York (1999). W. W. Chow, S. W. Koch, and M. Sargent III, Semiconductor-Laser Physics, Springer Berlin Heidelberg New York (1994).
15. P. Unger, Introduction to Power Diode Lasers, in: R. Diehl (Ed.), High-Power Diode Lasers, Topics Appl. Phys. **78**, 1-54 (2000).
16. D. Meschede, Optics, Light and Lasers, Wiley-VCH Weinheim (2004).
17. W. Demtröder, Laser Spectroscopy, Springer Berlin Heidelberg New York, 3rd ed. (2003).
18. E. Kato, H. Noguchi, M. Nagai, H. Okuyama, S. Kijima, and A. Ishibashi, Electron. Lett. **34**, 282 (1998).
19. M. Klude, G. Alexe, C. Kruse, T. Passow, H. Heinke, and D. Hommel, phys. stat. sol. (b) **229**, 935 (2002).
20. M. Klude, T. Passow, R. Kröger, and D. Hommel, Electron. Lett. **37**, 119 (2001).
21. T. Passow, M. Klude, C. Kruse, K. Leonardi, R. Kröger, G. Alexe, K. Sebald, S. Ulrich, P. Michler, J. Gutowski, H. Heinke, and D. Hommel, Adv. in Solid State Phys. **42**, 13 (2002). A. Gust, C. Kruse, M. Klude, E. Roventa, R. Kröger, K. Sebald, H. Lohmeyer, B. Brendemühl, J. Gutowski, and D. Hommel, phys. stat. sol. (c) **2**, 1098 (2005).
22. S.-L. Chuang, A. Ishibashi, S. Kijima, N. Nakayama, M. Ukita, and S. Taniguchi, IEEE J. Quantum Electr. **33**, 970 (1997).
23. K. Sebald, P. Michler, J. Gutowski, R. Kröger, T. Passow, M. Klude, and D. Hommel, phys. stat. sol (a) **190**, 593 (2002).
24. K. Sebald, Ph.D. Thesis, University of Bremen, 2003.
25. P. Michler, M. Vehse, J. Gutowski, M. Behringer, D. Hommel, M.F. Pereira Jr., and K. Henneberger, Phys. Rev. B **58**, 2055 (1998) .
26. S. Nakamura, Science **281**, 956 (1998). B. Gil (Ed.), Low-Dimensional Nitride Semiconductors, Oxford University Press (2002). H. Morkoc, Nitride Semiconductors and Devices, Springer Berlin Heidelberg New York, (1999). B. Gil (Ed.), Group III Semiconductor Compounds, Oxford University Press (1998).
27. T. Mukai, S. Nagahama, T. Kozaki, M. Sano, D. Morita, T. Yanamoto, M. Yamamoto, K. Akashi, and S. Masui, phys. stat. sol. (a) **201**, 2712 (2004).
28. Y. He, Y.-K. Song, A. V. Nurmikko, J. Su, M. Gherasimova, G. Cui, and J. Han, Appl. Phys. Lett. **84**, 463 (2004).
29. P. Lefebvre, T. Taliercio, S. Kalliakos, A. Morel, X.B. Zhang, M. Gallart, T. Bretagnon, B. Gil, N. Grandjean, B. Damilano, and J. Massies, phys. stat. sol. (b) **228**, 65 (2001).
30. M. Röwe, M. Vehse, P. Michler, J. Gutowski, S. Heppel, and A. Hangleiter, phys. stat. sol. (c) **0**, 1860 (2003).
31. M. Röwe, P. Michler, J. Gutowski, V. Kümmler, A. Lell, V. Härle, phys. stat. sol. (a) **200**, 135 (2003).
32. M. Vehse, P. Michler, I. Gösling, M. Röwe, J. Gutowski, S. Bader, A. Lell, G. Brüderl, and V. Härle, phys. stat. sol. (a) **188**, 109 (2001).
33. M. Vehse, P. Michler, I. Gösling, M. Röwe, J. Gutowski, S. Bader, A. Lell, G. Brüderl, and V. Härle, Appl. Phys. Lett. **80**, 755 (2002).
34. L. W. Ji, Y. K. Su, and S. J. Chang, phys. stat. sol. (c) **1**, 2405 (2004).

35. K. Sebald, H. Lohmeyer, J. Gutowski, T. Yamaguchi, and D. Hommel, phys., stat., sol (c), in print (Proceedings of the 6th International Conference on Nitride Semiconductors, Bremen, Germany 2005).
36. M. Kneissl, D. W. Treat, M. Teepe, N. Miyashita, and N. M. Johnson, Appl. Phys. Lett. **82**, 2386 (2003).
37. Photonics Technology World, September 2005 Edition, CW GaN Laser Diodes Grown by Molecular Beam Epitaxy, Laurin Publishing.
38. H. Lohmeyer, K. Sebald, C. Kruse, R. Kröger, J. Gutowski, D. Hommel, J. Wiersig, N. Bear, and F. Jahnke, Appl. Phys. Lett., submitted.
39. F. C. Peiris, S. Lee, U. Bindley, and J. K. Furdyna, Semicond. Sci. Technol. **14**, 878 (1999).
40. P. I. Kuznetsov, V. A. Jitov, L. Y. Zakjarov, G. G. Yakushcheva, Y. V. Korostelin, and V. I. Kozlovsky, phys. stat. sol. (b) **229**, 171 (2002).
41. M. Diagne, Y. He, H. Zhou, E. Makarona, A. V. Nurmikko, J. Han, K. E. Waldrip, J. J. Figiel, T. Takeuchi, and M. Krames, Appl. Phys. Lett. **79**, 3720 (2001).
42. Y.-K. Song, H. Zhou, M. Diagne, A. V. Nurmikko, R. P. Schneider, Jr. , C. P. Kuo, M. R. Krames, R. S. Kern, C. Carter-Coman, and F. A. Kish, Appl. Phys. Lett. **76**, 1662 (2000).
43. M. Arita, N. Nishioka, and Y. Arakawa, phys. stat. sol. (a) **194**, 403 (2002).
44. C. Kruse, Ph.D. Thesis, University of Bremen, 2004.
45. C. Kruse, S. M. Ulrich, G. Alexe, E. Roventa, R. Kröger, B. Brendemühl, P. Michler, J. Gutowski, and D. Hommel, phys. stat. sol. (b) **241**, 731 (2004).
46. H. Lohmeyer, K. Sebald, J. Gutowski, C. Kruse, D. Hommel, J. Wiersig, N. Bear, and F. Jahnke, phys. stat. sol. (c), in print (Proceedings of the 12th International Conference on II-VI Compounds, Warsaw, Poland 2005).
47. H. Lohmeyer, K. Sebald, C. Kruse, R. Kröger, J. Gutowski, D. Hommel, J. Wiersig, and F. Jahnke, phys., stat. sol. (c), in print (Proceedings of the 6th International Conference on Nitride Semiconductors, Bremen, Germany 2005).
48. H. Lohmeyer, K. Sebald, J. Gutowski, R. Kröger, C. Kruse, D. Hommel, J. Wiersig, and F. Jahnke, Eur. Phys. J. B, submitted.
49. J.-F. Carlin, J. Dorsaz, E. Feltin, R. Butté, N. Grandjean, M. Ilegems, and M. Laügt, Appl. Phys. Lett. **86**, 031107 (2005).
50. J.-F. Carlin, C. Zellweger, J. Dorsaz, S. Nicolay, G. Christmann, E. Feltin, R. Butté, and N. Grandjean, phys. stat. sol. (b) **242**, 2326 (2005).
51. C. Kruse, K. Sebald, H. Lohmeyer, B. Brendemühl, R. Kröger, J. Gutowski, and D. Hommel, AIP Conference Proceedings **CP772**, Physics of Semiconductors, 27th International Conference on the Physics of Semiconductors, edited by J. Menéndez and C. G. van de Walle, p. 1521.
52. M. Benyoucef, S. M. Ulrich, P. Michler, J. Wiersig, F. Jahnke, and A. Forchel, New Journal of Physics **6**, 91 (2004).

JUDD-OFELT THEORY: PRINCIPLES AND PRACTICES

BRIAN M. WALSH
NASA Langley Research Center
Hampton, VA 23681 USA
b.m.walsh@larc.nasa.gov,

1. Introduction

With few exceptions, no single idea is conceived without bringing together the work of previous ideas and forming a complete picture. So it was in August 1962 when there appeared in the literature, simultaneously and independently, two identical formulations of a theory. One by Brian R. Judd at the University of California at Berkeley and the other by a Ph.D. student, George S. Ofelt, at the Johns Hopkins University in Baltimore. Judd and Ofelt had never met personally, and were not aware of each other's interest in the intensities of rare earth ions in solids. While there are some differences in the two formulations, the approach and the assumptions used to arrive at the final result are remarkably similar. The titles of the two articles reflect a thought along similar lines. Judd referred to *Optical Absorption*, while Ofelt referred to *Crystal Spectra*, and each to Intensities of *Rare-Earth Ions*. The formulations as originally published by Judd and Ofelt came to be known as the Judd-Ofelt theory of the intensities of rare earth ions. Regarding these publications, the late Professor Brian G. Wybourne has said, "*I suggest that the coincidence of discovery was indicative that the time was right for the solution of the problem*" [1]. In the light of the advancement in the understanding of complex atomic spectra in the quarter century preceding the 1962 publications of Judd and Ofelt, this suggestion of Wybourne is based on very sound reasoning.

The rare earths comprise Y, Sc, La, Ce, Pr, Nd, Pm, Sm, Eu, Gd, Tb, Dy, Ho, Er, Tm, Yb and Lu. The last 15 make up the lanthanide series. Most of these elements were discovered over a period of time stretching from the late 18th century to the early 20th century. Promethium (Pm) was the last to be discovered in 1947 at Oak Ridge National Laboratory. So, with the exception of Promethium, all the rare earths were discovered in the span of a little more than a century. Part of the reason why they are called rare earths is two-fold. First, they are very difficult to chemically extract from the earth. Second, they do not exist in nature in high abundance. In the universe, the rare earths are approximately 10^6 times less abundant than the more common element silicon. In spite of their scarcity and difficulty in obtaining, the rare earths are highly valued for their unique properties, especially as optically active elements in their ionized state for lasers.

B. Di Bartolo and O. Forte (eds.), Advances in Spectroscopy for Lasers and Sensing, 403–433.

Even before the advent of lasers, the rare earths presented a *puzzle* in trying to understand their spectral properties in the context of the quantum theory that blossomed in 1920's and 1930's. In 1937 J.H. Van Vleck published an article titled *"The Puzzle of Rare-Earth Spectra in Solids"* [2]. He called it a puzzle because it was well known that rare earths exhibited sharp spectral lines, which would be expected if the transitions occurred between levels inside the 4*f* electronic shell. Such transitions were known to be *forbidden* by the Laporte selection rule, which says that states with even parity can be connected by electric dipole (E1) transitions only with states of odd parity, and odd states only with even ones. Another way of saying this is that the algebraic sum of the angular momenta of the electrons in the initial and final state must change by an odd integer. For transitions within the 4*f* shell, ED transitions are *forbidden*, but *allowed* for magnetic dipole (M1) or electric quadrupole (E2) radiation. The terms *forbidden* and *allowed* are not strictly accurate. The term *forbidden* means a transition may occur in principle, but with low probability. Given the relatively strong intensities and sharp spectral features of rare earth spectra, this picture presented a *puzzle* with the following possibilities:

1. 4*f* to 5d transitions.
2. Magnetic dipole or electric quadrupole radiation.
3. Electric dipole radiation.

The operators for E1, M1 and E2 transitions are shown in figure 1 along with their selection rules. It will be seen later that these selection rules will need to be modified under the Judd Ofelt theory.

$$\vec{P} = -e\sum_i \vec{r}_i$$ Electric dipole operator (E1) (odd operator)

$$\vec{M} = -\frac{e\hbar}{2mc}\sum_i \vec{\ell}_i + 2\vec{s}_i$$ Magnetic dipole operator (M1) (even operator)

$$\vec{Q} = \frac{1}{2}\sum_i (\vec{k}\cdot\vec{r}_i)\vec{r}_i$$ Quadrupole operator (E2) (even operator)

	S	L	J (No 0 ↔ 0)	Parity
Electric Dipole	$\Delta S = 0$	$\Delta L = 0, \pm 1$	$\Delta J = 0, \pm 1$	opposite
Magnetic dipole	$\Delta S = 0$	$\Delta L = 0$	$\Delta J = 0, \pm 1$	same
Electric quadrupole	$\Delta S = 0$	$\Delta L = 0, \pm 1, \pm 2$	$\Delta J = 0, \pm 1, \pm 2$	same

Figure 1. Multipole operators and selection rules.

The solution to this *puzzle* was proposed by Van Vleck [2] in 1937 and further resolved by Broer, *et al.* [3] in 1945. The first possibility would be indicative of broad spectral lines in contrast to the sharp lines that were observed. Magnetic dipole radiation could account for some transitions, but not all transitions, and represents a special case. Quadrupole radiation could account for all the transitions, but was too weak to account for the observed intensities. Only ED radiation was a reasonable solution, but it is forbidden by the Laporte selection rule. The solution considered a distortion of the electronic motion by crystalline fields in solids, so that the selection rules for free atoms no longer applied.

Not all crystalline fields are capable of producing this effect. The crystal field must be noncentrosymmetric, that is, lacking a center of symmetry at the equilibrium position, otherwise the wavefunctions would retain an even or odd parity with regards to reflection about the origin. If the wavefunctions retained their even or odd parity, then Laporte's rule (even states can connect to odd states and odd to even) would remain rigorous, and electric dipole radiation would be strictly forbidden. In other words, In order to have a change of parity, the odd-order terms of the crystalline field, expressed as a power series in the displacement from equilibrium, must be present. These terms vanish for a central field, and no change of parity can occur. The odd-order terms of a noncentral crystalline field can force a coupling between odd and even states, resulting in mixed parity states that mitigates Laporte's rule.

It would take another 25 years after Van Vleck's paper before the appearance of the papers by Judd and Ofelt. What is the reason for the quarter century gap? The main reason is that the techniques of group theory were not applied to lanthanides until 1949 and, even then, it took some years for these ideas to be absorbed, incorporated and accepted. The seminal papers by Guilio Racah in the 1940's would revolutionize the entire subject of complex spectra. In a series of papers [4,5,6], culminating in his 1949 paper [7] regarding $4f$ electrons, he created a powerful set of tools that made possible many complex spectroscopic calculations. The work of Condon and Shortly [8] in *"The Theory of Atomic Spectra"* published in 1935 is a classic and seminal work in itself, but the tools are much more cumbersome than the newer group theory. The ideas of Racah took some time to root and proved to be a key ingredient in the solution to a problem regarding the calculation of forced electric dipole transitions. The concept of representing operators as irreducible tensors and their subsequent manipulation would play a pivotal role in the solution of the problem. Another reason for the delay between Van Vleck's paper and those of Judd and Ofelt concerns the dawn of computers. The computer made possible the tabulation of all angular momentum coupling coefficients [9]. This did not occur until 1959, but by then the stage was set for a revolution in understanding the complex spectra of lanthanide ions. The decade of the 1960's can be viewed as a revolution in spectroscopic research on rare earth ions. This decade saw the invention of the laser go from a device in search of an application to applications in search of laser devices. The theory of Judd and Ofelt remains in history at the forefront of this revolution.

2. Ions in Solids

An ion in a solid can be considered as an impurity embedded in a solid host material, usually in small quantities. These impurities replace the host ions substitutionally and form optically active centers which exhibit luminescence when excited by an appropriate excitation source. When speaking of solids in general, a glass or crystal is implied. The former is amorphous over a long range, but may contain local order. The later has definite long-range order in a lattice structure. Both glasses and crystals are insulators, distinguishing them from semiconductors, and have bandgaps greater than 5 eV, which is about the wavelength of photons in the deep

UV. The host material plays a fundamental role in determining the nature of the observed spectra of the optically active impurity ions. This is known as Ligand field theory in general and crystal field theory in the case of an ordered periodic lattice.

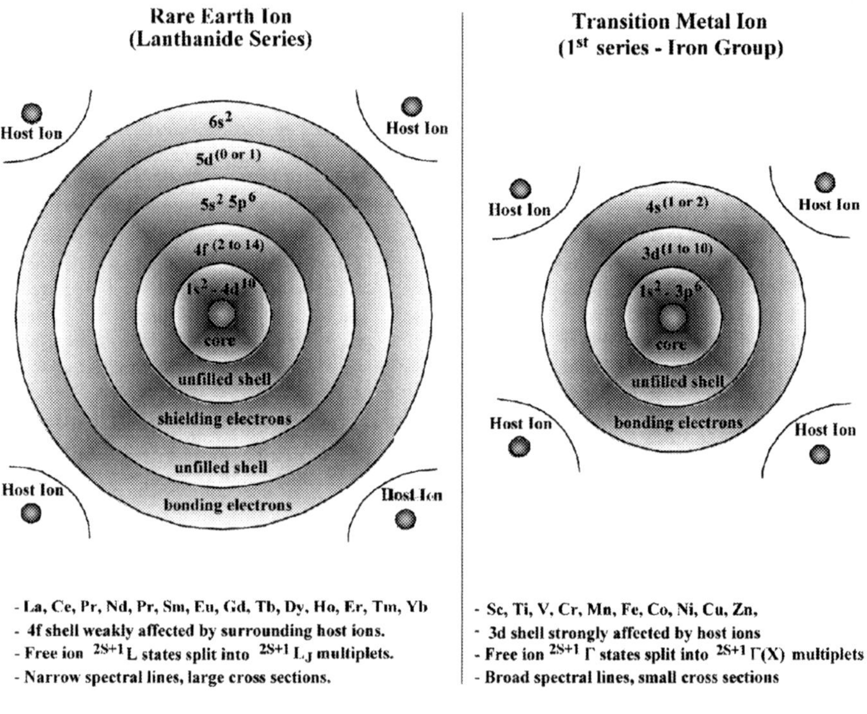

Figure 2. Rare Earth and Transition metal ion atomic structure.

Now, the impurity ions, usually called dopant ions, are the optically active centers. The host is generally transparent. The impurities, dopant ions, are transition metal or lanthanide series ions characterized by unfilled shells in the interior of the ion. The atomic structure of rare earth ions in the lanthanide series and transition metal ions of the iron group are shown in figure 2. These representations are not drawn to scale and are shown simply to give an overall visual representation of their structure. All lanthanide ions are characterized by a Xe core, an unfilled $4f$ shell, and some outer shells that screen the $4f$ shell from outside perturbing influences. This screening effect protects the optically active electrons to some extent from the influence of the crystal field, giving the lanthanides their characteristic sharp and well defined spectral features. In other words they are very similar to free ion spectra. This is in contrast to transition metals where the unfilled 3d shell is not as well screened due to only a single outer shell. The transition metals are characterized by broad, undefined features, although some sharp lines are observed. The R_1 and R_2 lines of $Cr{:}Al_2O_3$, ruby, for example, are sharp lines. Transition metals are, therefore, strongly coupled to the lattice and susceptible to the vibrational motions of the host lattice ions. The energy levels for transition metals are more vibrational-electronic, or vibronic, whereas the lanthanides are more electronic. These transitions occur within the bandgap of the material and have a range out to about 5 eV. In terms of wavelength, transitions are observed from about 0.2 to 5.0 micrometers.

As just stated, the crystal field plays an important role in influencing the features of optical spectra. It also plays an obvious role in the Stark effect regarding the splitting of energy levels of ions in solids. It is the influence it has on the ionic states regarding selection rules that makes such transitions possible, as will be discussed. The ligand, or crystalline field is totally external to the optically active dopant ion and has the symmetry determined by the chemical composition of the host. In an ionic crystal, the optically active dopant ions feel the influence of electrons, belonging to crystal host ions, as a repulsion, and of the nuclei, belonging to the crystal host, as an attraction. The accumulation of these influences can be considered as a net electric field, known as the crystalline field. In this context, lanthanide ions exist in the weak field scheme, while transition metals of the iron group such as chromium, Cr, exist in the medium field scheme. In the former the crystal field is small in comparison to spin-orbit interactions, whereas in the later the crystal field is large compared to spin-orbit interactions. The crystal field plays a fundamental role in making many laser transitions possible.

An obvious question is to what extent the crystal field splits the energy levels and how many levels are obtained for a given crystal field? To begin to answer this question, consider the free ion Hamiltonian,

$$H_F = -\frac{\hbar^2}{2m}\sum_{i=1}^{N}\nabla_i^2 - \sum_{i=1}^{N}\frac{Ze^2}{r_i} + \sum_{i<j}^{N}\frac{Ze^2}{r_{ij}} + \sum_{i=1}^{N}\xi(r_i)(\mathbf{s}_i \cdot \mathbf{l}_i) \quad (1)$$

The first term is the sum of the kinetic energies of all the electrons of a 4f ion, the second term is the potential energy of all the electrons in the field of the nucleus. The third term is the repulsive Coulomb potential of the interactions between pairs of electrons, and the last term is the spin-orbit interaction, which accounts for coupling between the spin angular momentum and the orbital angular momentum. In terms of the central field approximation, each electron can be considered to be moving independently in the field of the nucleus and a spherically averaged potential of all the other electrons. The Coulomb interaction produces different SL terms with different energies, but is independent of the total angular momentum J of the electrons. The spin-orbit interaction allows coupling between states of different SL and thus is dependent on the total angular momentum. In the language of quantum mechanics, the spin orbit operator does not commute with $\mathbf{L}^2$ and $\mathbf{S}^2$, but it does commute with $\mathbf{J}^2$ and J_z. In simple terms this means that the Coulomb interaction removes degeneracy in S and L, while the spin orbit interaction removes degeneracy in J. The M_J degeneracy remains. This is only removed by the crystal field. The atomic interactions and energy level splittings are depicted in figure 3.

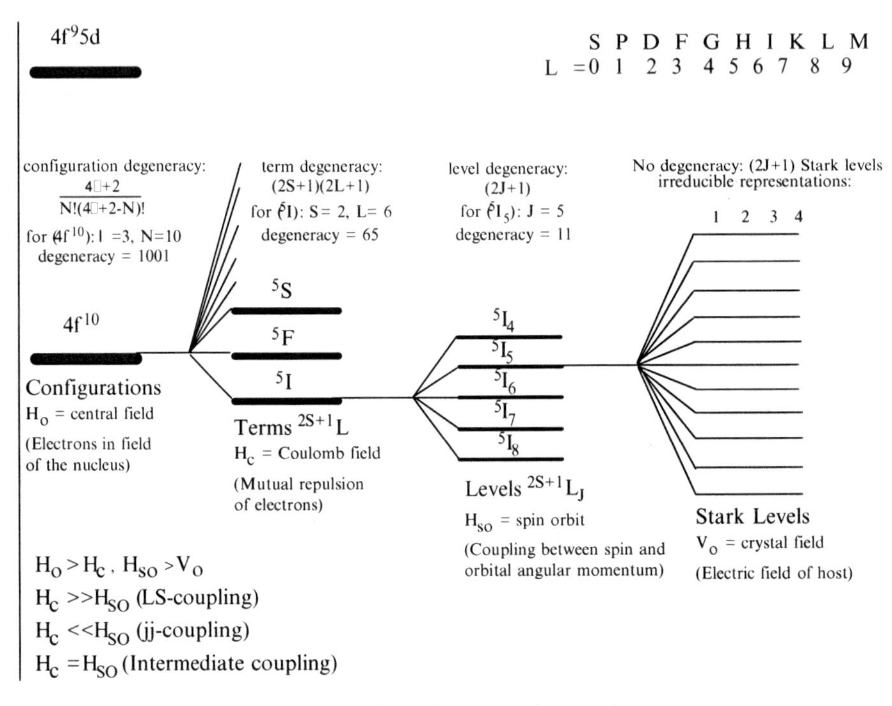

Figure 3. Rare earth and transition metal ion atomic structure.

So, in the free atom there is a spherical symmetry and each level is reduced to 2J+1 degeneracy. When the ion is placed in a crystal environment the spherical symmetry is destroyed and each level splits under the influence of the crystal field. In fact, the spherical symmetry is reduced to the point symmetry at the ion site. The degree to which the 2J+1 degeneracy is removed will depend on the point symmetry surrounding the ion. This aspect will become clear shortly. The perturbed free ion Hamiltonian for an ion in a crystal is written as,

$$H = H_F + V_{CF} \tag{2}$$

where V_{CF}, the perturbation Hamiltonian, is due to the potential provided by the crystal environment around the ion. Since the eigenfunctions of the free Hamiltonian possess complete spherical symmetry and are expressible in terms of spherical harmonics, it is natural to expand V in terms of spherical harmonics,

$$V_{CF} = \sum_{kq} A_{kq} \sum_i r_i^k Y_{kq}(\vartheta_i, \varphi_i) \quad (3)$$

where the summation over i involves all electrons of the ion of interest. The A_{kq} are structural parameters in the static crystal field expansion. They depend only on the crystal host and can be calculated in a point charge lattice sum using crystallographic data and charges of the host lattice. The point charge model assumes that the charges of the host lattice are all point charges. The A_{kq} are then given by,

$$A_{kq} = -q_e \sum_i \frac{Z_i Y_{kq}(\vartheta_i, \varphi_i)}{R_i^{k+1}} \quad (4)$$

where q_e is the electronic charge, Z_i is the size of the charge at position R_i corresponding to the surrounding atoms composing the crystal. In the calculation of matrix elements, $\langle\alpha|V|\beta\rangle$, there results matrix elements of the form $\langle r^k \rangle = \langle n\ell|r^k|n\ell\rangle$, that represent the average value of r^k. This leads to terms of the form,

$$B_{kq} = A_{kq}\left\langle r^k \right\rangle \quad (5)$$

These terms enter prominently in the calculation of energy levels. They are generally determined empirically from experimental data, that is, by fitting the measured energy levels to the theory by a least squares iterative fitting procedure. Once the point symmetry and the appropriate form of the crystal field are known, it is possible to construct the crystal field energy matrix. This matrix is then diagonalized using an estimated set of B_{kq} starting parameters. The resulting set of theoretical energy levels are compared to the set of experimental levels and, by an iterative fitting procedure, the B_{kq} parameters are adjusted to obtain the best overall fit to experiment. The thirty-two crystallographic point groups can be divided into four general symmetry classes as follows [10]:

1. Cubic: O_h , O , T_d ,T_h , T
2. Hexagonal: D_{6h} , D_6 ,C_{6v} , C_{6h} , C_6 , D_{3h} , C_{3h} , D_{3d} , D_3 , C_{3v} , S_6 , C_3
3. Tetragonal: D_{4h} , D_4 , C_{4v} , C_4 , D_{2d} , S_4
4. Lower symmetry: D_{2h} , D_2 , C_{2v} , C_{2h} , C_2 , C_s , S_2 , C_1

Now, if the initial and final states have the same parity, then k must be even. If the initial and final states have opposite parity then k must be odd. Otherwise, the matrix elements of V_{CF} are zero. So, if a matrix element of an operator of rank k connects angular momenta ℓ and ℓ' then the triangle condition, $\ell + \ell' \geq k \geq |\ell - \ell'|$, must hold. For 4f electrons, $\ell = \ell' = 3$, and k must, therefore, be even for transitions within the f^n configuration, and is limited to values k = 0, 2, 4, 6. However, if states of the f^n configuration are coupled to states of opposite parity in higher lying configurations, such as $4f^{(n-1)}$ 5d, then $\ell = 3$ and $\ell' = 2$. In this case k is odd and is limited to values k = 1, 3, 5. These odd-order terms play a key role in the Judd-Ofelt theory for forced electric dipole transitions in lanthanide and actinide ions in solids. The values for k and q are also limited by the point symmetry of the ion. That is, the number of nonzero terms is dependent on the point symmetry. This arises from the fact that the Hamiltonian must be invariant under operations of the point symmetry group. Thus, the crystal field must also exhibit the same symmetry as the point symmetry of the ion, since it is part of the total Hamiltonian. Equating the crystal field expansion with the expansion that has been transformed through operations of the point symmetry group gives the allowed crystal field parameters for a particular ionic point symmetry. Thus, the spherical symmetry of an ion in a crystal is reduced to the point symmetry at the site of the ion. It is noted that terms with k = 0 and q = 0 are spherically symmetric and affect all energy levels in the same way, resulting in only a uniform shift of all levels in the configuration. The cases where q = 1 and q = 5 occur only when there is no symmetry, as for C_1. In general, the values of q are restricted to $q \leq k$, but the point symmetry introduces further restrictions, and this determines the allowed values of q.

It can be shown that knowledge of the symmetry class at the site of the ion can be used to predict the number of levels a given J state splits into. The number of levels given J state splits into for a given point symmetry is presented in table I [10]. Alternatively, rare-earth ions can be used to probe the crystal symmetry if the number of levels of the ion can be determined.

TABLE I. Number of levels for integral J

J =	0	1	2	3	4	5	6	7	8
Cubic	1	1	2	3	4	4	6	6	7
Hexagonal	1	2	3	5	6	7	9	10	11
Tetragonal	1	2	4	5	7	8	10	11	13
Lower symmetry	1	3	5	7	9	11	13	15	17

As has already been stated, the positions of the levels arise from a combination of the Coulomb, spin-orbit, and crystal field interactions. The electrostatic interaction leads to ^{2S+1}L splitting on the order of 10^4 cm^{-1}. The spin-orbit interaction splits the levels further into $^{2S+1}L_J$, separating the J states by 10^3 cm^{-1}. Finally, the crystal field removes or partially removes the degeneracy in J yielding an energy level separation on the order of 10^2 cm^{-1}. The extent to which the Stark split sublevels spread is dependent on the strength of the crystal field. The larger the crystal field, the larger will be the spread of the J sublevels.

Consider the lanthanide ions as an example. The lanthanide ions are characterized by a shielded 4*f* shell where the atomic like transitions take place. The 4*f* states all have the same parity, that is $(-1)^{\Sigma\ell_i}$, where $\ell = 3$ for lanthanides. The question then arises, where do opposite parity states that are *mixed* in come from? They come from shells above the 4*f* shell, such as 5d. The d electrons have $\ell = 2$ and so have opposite parity to f electrons. The wavefunctions for some 1s to 6*f* states are pictured in figure 4. This figure provides a nice pictorial representation of the parity of the wavefunctions for various orbitals. For instance, the *f* orbitals clearly have odd parity since there is a change in sign on reflection about the origin. Similarly, the d orbitals have even parity as the sign is preserved on reflection about the origin.

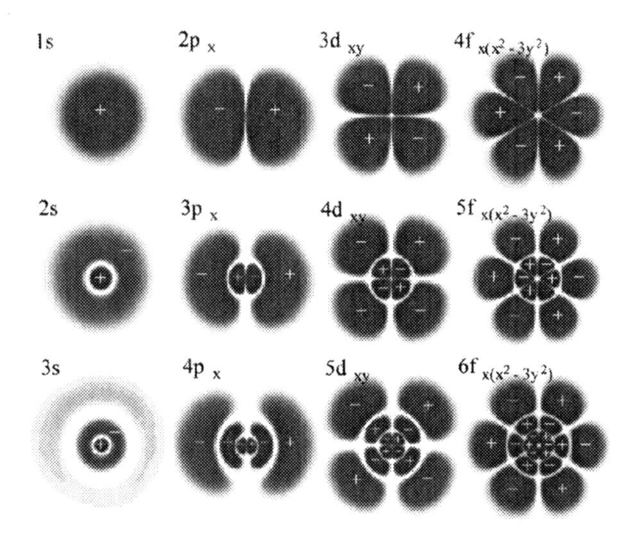

Figure 4. Wavefunctions of some s, p, d and f orbitals.

A typical wavefunction for an ion in a crystal can be expressed as a linear combination of states in the free ion, which are composed of sums of both even and odd parity wavefunctions, and form a complete orthonormal set of basis functions. This allows for mixed parity states via the odd-order terms of the crystal field mentioned earlier. As a result, electric dipole transitions can be forced through parity mixing and result from the perturbation caused by the odd-order terms of the crystal field. Nature finds a way around Laporte's rule in this way. Since these transitions come about as a result of a perturbation, they are orders of magnitude smaller than in free ions, but strong enough to produce a plethora of transitions from which many lasers can be realized. This, in itself, is a remarkable circumstance of opportunity. The theory that predicts the intensities of such transitions, the Judd-Ofelt theory, is discussed in the next section.

3. Judd-Ofelt Theory: Principles

The Judd-Ofelt theory [11,12] is based on the static, free-ion and single configuration approximations. In the static model, the central ion is affected by the surrounding host ions via a static electric field, referred to as the ligand or crystal field. In the free-ion model, the host environment produces the static crystal field, and is treated as a perturbation on the free-ion Hamiltonian. In the single configuration model, the interaction of electrons between configurations is neglected. Simply stated, the Judd-Ofelt theory describes the intensities of lanthanide and actinide transitions in solids and solutions. The utility of the Judd-Ofelt theory is that it provides a theoretical expression for the line strength, given by,

$$S_{ED}(J;J') = \sum_{\lambda=2,4,6} \Omega_\lambda \left| \left\langle f^n[SL]J \left\| \mathbf{U}^{(\lambda)} \right\| f^n[S'L']J' \right\rangle \right|^2 \tag{6}$$

where Ω_λ are the *Judd-Ofelt parameters*. The terms in brackets are doubly reduced matrix elements for intermediate coupling. Intermediate coupling refers to a situation where the mutual repulsion interaction between 4f electrons is of the same order of magnitude as the spin-orbit coupling. This effect can be incorporated by expanding the wavefunctions of the $4f$ states in a linear combination of *Russel-Saunders*, or LS-coupled states. The coupling coefficients are found by diagonalizing the combined electrostatic, spin orbit and configuration interaction energy matrices to obtain the full intermediate coupled wavefunctions, $|f^n[SL]J\rangle$. A substantial portion of the book *"Spectroscopic coefficients of the* p^n, d^n, and f^n *configurations"* by Nielson and Koster [13] is devoted to tabulating matrix elements in LS coupling. Further efforts must be devoted to converting these wavefunctions to the intermediate coupling case applicable to lanthanide ions. Fortunately, many references tabulate intermediate coupled matrix elements based on Nielson and Koster's work. Because the electric dipole transitions arise from a small crystal field perturbation, the matrix elements are not highly dependent on the host material.

These matrix elements are integrals of the dipole operator between the upper and lower wave functions of the transition, where the integration takes place over the volume of the atom. The $U^{(\lambda)}$ in Eq. (6) are the irreducible tensor forms of the dipole operator. Basically, during the transition the atom can be considered an electric dipole oscillating at some frequency whose amplitude is proportional to the value of this matrix element. It is the interaction of this dipole moment with the electric field of the electromagnetic wave that induces the transition. It is analogous to a classical oscillating dipole driven by an external electric field. Quantum mechanically, the situation is more complicated because the parity between the upper and lower electronic states must be considered. In quantum mechanics, electric dipole transitions between electronic states of the same parity are forbidden. This is basically a result of the fact that the expectation value of the position operator, **r**, is odd under reflection, and vanishes for definite parity. The electronic states have wavefunctions described by spherical harmonics, and as such, have the parity of the angular quantum number ℓ. Considering an electronic shell as a whole, the total parity for an n electron system is $\wp = (-1)^{\ell_1 + \ell_2 + \ldots + \ell_n}$, therefore, an even number of electrons has parity of $\wp = 1$ and an odd number of electrons has parity $\wp = -1$. This means that, regardless of the number of electrons, all states in the $4f$ shell always has definite parity. In free ions, this means that ED transitions within the $4f$ shell of lanthanide ions are forbidden. However, electric dipole transitions can be forced if opposite parity states from higher lying configurations outside the $4f$ shell are mixed into the upper state. This is possible when the atom is placed in a noncentrosymmetric perturbing field such as the crystal field of a lattice in which the atom is embedded. This does not happen in a central field because the Hamiltonian is invariant under coordinate inversion, and the states retain definite parity. The odd-order parts of the crystal field, expanded in a series of spherical harmonics, perturb the system and produce mixed parity states between which electric dipole transitions are allowed. This is, in fact, the starting point from which the Judd Ofelt theory is based.

If the crystal field is taken as a first-order perturbation, then the initial and final mixed parity states, $\langle\psi_a|$ and $|\psi_b\rangle$, can be expanded as:

$$\langle\psi_a| = \langle\varphi_a| + \sum_\beta \frac{\langle\varphi_a|V|\varphi_\beta\rangle}{E_a - E_\beta}\langle\varphi_\beta| \tag{7}$$

$$|\psi_b\rangle = |\varphi_b\rangle + \sum_\beta \frac{\langle\varphi_\beta|V|\varphi_b\rangle}{E_b - E_\beta}|\varphi_\beta\rangle \tag{8}$$

where $\langle\varphi_a|$ and $|\varphi_b\rangle$ are the initial and final states of single parity. The φ_β are states of higher energy, opposite parity configurations. The electric dipole matrix elements, $D = \langle\psi_a|\,\mathbf{P}\,|\psi_b\rangle$, can now be found, where **P** represents the electric dipole (ED) operator given by,

$$\mathbf{P} = -e\sum_i \mathbf{r}_i \tag{9}$$

Combining the terms from Eq. (7) and Eq. (8) about the electric dipole operator, it is clear that $\langle\varphi_a|\,\mathbf{P}\,|\varphi_b\rangle = 0$ and $\langle\varphi_\beta|\,\mathbf{P}\,|\varphi_\beta\rangle = 0$ since dipole transitions are forbidden between states of the same parity. The expression for the electric dipole matrix element, D, reduces to,

$$D = \langle\psi_a|\mathbf{P}|\psi_b\rangle = \sum_\beta \left\{ \frac{\langle\varphi_a|V|\varphi_\beta\rangle\langle\varphi_\beta|\mathbf{P}|\varphi_b\rangle}{E_a - E_\beta} + \frac{\langle\varphi_a|\mathbf{P}|\varphi_\beta\rangle\langle\varphi_\beta|V|\varphi_b\rangle}{E_b - E_\beta} \right\} \tag{10}$$

It is useful to introduce tensor forms of the crystal field and ED operator. Tensor forms are useful because they can easily be combined into a single effective tensor operator. The crystal field and ED operators, Eq. (3) and Eq. (9), respectively, have the tensor operator forms:

$$\mathbf{D}_q^{(1)} = -e\sum_i r_i \left[\mathbf{C}_q^{(1)}\right]_i \tag{11}$$

$$\mathbf{D}_p^{(t)} = \sum_{tp} A_{tp} \sum_i r_i^t \left[\mathbf{C}_p^{(t)}\right]_i \tag{12}$$

The $\mathbf{C}_q^{(k)}$ are tensor operators that transform like spherical harmonics,

$$\mathbf{C}_q^{(k)} = \left(\frac{4\pi}{2k+1}\right)^{1/2} Y_{kq} \tag{13}$$

For instance, the position operator, **r**, is a tensor of rank one and is defined as $\mathbf{r} = r\mathbf{C}^{(1)}$. This gives the spherical harmonics as $Y_{10} = -(3/4\pi)^{1/2}\cos\theta$ and $Y_{1,\pm1} = \mp(3/8\,\pi)^{1/2}\sin\theta\, e^{\pm i\varphi}$ for k = 1, q = 0, ±1. The resulting components are then written as $z = r\cos\theta$ and $\mp(x \pm iy)/\sqrt{2}$, where $x = r\sin\theta\cos\theta$ and $y = r\sin\theta\ \sin\theta$, such that $r = (x^2 + y^2 + z^2)^{1/2}$.

A couple of assumptions can now be made to simplify the problem. The first is to assume an average energy for the excited configurations above the $4f^n$. The second is to assume that the difference of the average energies, $\Delta E(4f\text{-}5d)$, is the same as the difference between the average energy of the $4f^{(n-1)}5d$ and the energy of the initial and final state of the $4f^n$ configuration. These assumptions can be summarized as follows:

1. The states of $|\varphi_\beta\rangle$ are completely degenerate in J.
2. The energy denominators in Eq. (10) are equal ($E_a - E_\beta = E_b - E_\beta$).

The relative positions of the $4f^n$ and $4f^{(n-1)}5d$ configurations are shown in figure 5 for comparison. So, for most of the rare earth ions these assumptions are only moderately met, but offer a great simplification. Otherwise, the many fold sum of perturbation expansions would not be suitable for numerical applications. The advantage of these assumptions allows the energy denominators to be removed from the summations and closure, $|\varphi_\beta\rangle\langle\varphi_\beta| = 1$, to be used. What does it mean to use closure? From basic quantum mechanics, any wavefunction can be expanded in a suitable set of orthonormal basis functions and written as $|\psi\rangle = \Sigma_n\, a_n|\varphi_n\rangle$, such that $\langle\varphi_m|\varphi_n\rangle = \delta_{mn}$ and, therefore, $a_n = \langle\varphi_m|\psi\rangle$. Hence, it follows that $|\psi\rangle = \Sigma_n\, a_n|\varphi_n\rangle\langle\varphi_n|\psi\rangle$, and so, $|\varphi_\beta\rangle\langle\varphi_\beta| = 1$. This is closure and the meaning, as it applies here, is that the states of the excited configuration form a complete orthonormal set of wavefunctions. Once closure is invoked, the angular parts of the crystal field, $\mathbf{C}^{(1)}_q = \langle\ell\|\mathbf{C}^{(1)}\|\ell'\rangle\,\mathbf{U}^{(1)}_q$ and electric dipole operator, $\mathbf{C}^{(t)}_p = \langle\ell\|\mathbf{C}^{(t)}\|\ell'\rangle\,\mathbf{U}^{(\lambda)}_Q$, can be combined into an effective tensor operator, that is,

$$\mathbf{U}^{(1)}_q\mathbf{U}^{(t)}_p = \sum_{\lambda Q}(-1)^{1+t+\lambda+Q}(2\lambda+1)\begin{Bmatrix}1 & t & \lambda\\ l & l & l'\end{Bmatrix}\begin{pmatrix}1 & t & \lambda\\ q & p & Q\end{pmatrix}\mathbf{U}^{(\lambda)}_Q \tag{14}$$

where $Q = -(q + p)$, and $\lambda = 1 + t$. The 3j symbol ( ) is related to the coupling probability for two angular momenta. The 6j symbol { } is related to the coupling probability for 3 angular momenta. These are known as Wigner symbols. The Wigner 3j and 6j symbols are related to Clebsch-Gordon coefficients. In addition, the effective tensor operator, can be further simplified by utilizing the Wigner-Eckart theorem,

$$\left\langle JM\middle|\mathbf{U}^{(\lambda)}_Q\middle|J'M'\right\rangle = (-1)^{J-M}\begin{pmatrix}J & \lambda & J'\\ -M & Q & M'\end{pmatrix}\left\langle J\middle\|\mathbf{U}^{(\lambda)}\middle\|J'\right\rangle \tag{15}$$

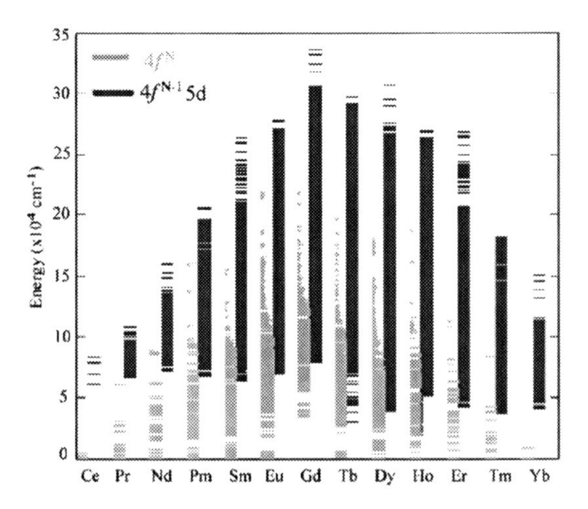

Figure 5. Energy levels of the $4f^n$ and $4f^{(n-1)}5d$ lanthanides.

The Wigner-Eckart theorem is a remarkable equation because it separates the geometry from the physics. The geometry of the angular momentum transformations are in the 3j symbol, while the physics of the dynamics is in the reduced matrix element of $\mathbf{U}^{(\lambda)}$. The full solution is,

$$D = -e\sum_{tp}\sum_{\lambda Q}(-1)^{J-M-Q}(2\lambda+1)A_{tp}Y(t,\lambda)\begin{pmatrix} 1 & t & \lambda \\ q & p & Q \end{pmatrix}\begin{pmatrix} J & \lambda & J' \\ -M & Q & M' \end{pmatrix}\left\langle \varphi_a \left\| \mathbf{U}^{(\lambda)} \right\| \varphi_b \right\rangle \quad (16)$$

where, $Y(t, \lambda)$ is given by,

$$Y(t,\lambda) = 2\sum_{n\mathrm{l}} \frac{\langle 4f|r|n\mathrm{l}\rangle\langle n\mathrm{l}\,|r^t|4f\rangle}{\Delta E_{n\mathrm{l}}}\left\langle f\left\|\mathbf{C}^{(1)}\right\|\mathrm{l}\right\rangle\left\langle \mathrm{l}\left\|\mathbf{C}^{(t)}\right\|f\right\rangle\begin{Bmatrix} 1 & t & \lambda \\ \mathrm{l} & \mathrm{l} & \mathrm{l}' \end{Bmatrix} \quad (17)$$

and A_{tp} are parameters of the static crystal field expansion, Eq. (4), with t → k and p → q. This is the full solution of the Judd-Ofelt theory. This form can be used to find electric dipole matrix elements between mixed parity states for individual Stark level to Stark level transitions [14]. In the above equation for $Y(t,\lambda)$, $\Delta E_{n\ell}$ is the energy difference between the $4f$ and opposite parity, $n\ell$ configuration. The first two terms are the interconfiguration radial integrals:

$$\langle 4f|r|n\mathrm{l}\rangle = \int_0^\infty R(4f)rR(n\mathrm{l}) \quad (18)$$

$$\langle n\mathrm{l}\,|r^t|4f\rangle = \int_0^\infty R(n\mathrm{l})r^t R(4f) \quad (19)$$

while the second two terms are reduced matrix elements of the tensor operator $\mathbf{C}_q^{(k)}$, which represents the angular part of the crystal field, and can be calculated from the angular momentum quantum numbers of the configurations according to the equations:

$$\left\langle f \left\| \mathbf{C}^{(1)} \right\| l \right\rangle = (-1)^{l} \, (2l+1)^{1/2} \, (2f+1)^{1/2} \begin{pmatrix} f & 1 & l \\ 0 & 0 & 0 \end{pmatrix} \tag{20}$$

$$\left\langle \ell \left\| \mathbf{C}^{(t)} \right\| f \right\rangle = (-1)^{f} \, (2l+1)^{1/2} \, (2f+1)^{1/2} \begin{pmatrix} l & t & f \\ 0 & 0 & 0 \end{pmatrix} \tag{21}$$

Finally, the last term is a 6j symbol, associated with transformations between angular momentum coupling schemes. In this case, it arises through the coupling of the electric dipole and crystal field operators. It is noted that t takes on only odd values in the summations above, since only the odd-order terms in the multipole expansion of the crystal field contribute to parity mixing. The even-order terms are only responsible for shifting and splitting of the energy levels. It is possible to say something about the selection rules in the Judd-Ofelt theory at this point. The selection rules can be derived from the properties of the 3j and 6j symbols, illustrated in figure 6, and are an extension from what was shown in figure 1.

6j symbols

$\begin{Bmatrix} j_1 & j_2 & j_3 \\ \ell_1 & \ell_2 & \ell_3 \end{Bmatrix} = 0$ Unless:

$j_i \geq 0$
$\ell_i \geq 0$
$|j_1 - j_2| \leq j_3 \leq j_1 + j_2$
$|\ell_2 - \ell_3| \leq j_1 \leq \ell_2 + \ell_3$
$|\ell_1 - \ell_3| \leq j_2 \leq \ell_1 + \ell_3$
$|\ell_1 - \ell_2| \leq j_3$

$\begin{Bmatrix} l & t & \lambda \\ \ell & \ell & \ell' \end{Bmatrix}$

$\lambda = 2, 4, 6$
$t = 1, 3, 5$
$\lambda \leq l + t$
$|\ell' - \ell| \leq 1$
Only d or g orbitals can mix parity

3j symbols

$\begin{pmatrix} j_1 & j_2 & j_3 \\ m_1 & m_2 & m_3 \end{pmatrix} = 0$ Unless:

$j_i \geq 0$
$m_i \leq j_i$
$m_1 + m_2 + m_3 = 0$
j_i, m_i (integer or half integer)
$|j_1 - j_2| \leq j_3 \leq j_1 + j_2$

$\begin{pmatrix} J & \lambda & J' \\ -M & Q & M' \end{pmatrix}$

$|J' - J| \leq \lambda$
$\Delta J \leq 6$
$\Delta L \leq 6$
$\Delta S = 0$
$J = 0 : J' \rightarrow$ even
$J' = 0 : J \rightarrow$ even

Selection Rules

	S	L	J (No 0 ↔ 0)	Parity
Electric Dipole	$\Delta S = 0$	$\Delta L \leq 6$	$\Delta J \leq 6$ $\Delta J = 2,4,6$ (J or J' = 0)	opposite
Magnetic dipole	$\Delta S = 0$	$\Delta L = 0$	$\Delta J = 0, \pm 1$	same
Electric quadrupole	$\Delta S = 0$	$\Delta L = 0, \pm 1, \pm 2$	$\Delta J = 0, \pm 1, \pm 2$	same

Figure 6. Selection Rules in the Judd-Ofelt Theory.

In the form of Eq. (16) the theory is suitable for calculating transitions between individual Stark levels. However, Judd and Ofelt were interested in ions in solution where individual Stark transitions cannot be distinguished. So, they simplified the problem for manifold-to-manifold transitions, which is valid for ions in glasses and, of course, crystals. Additional assumptions are needed to simplify the problem further, which are referred to as three and four to distinguish them from the first two assumptions made earlier.

3. All Stark levels within the ground manifold are equally populated.
4. The material is optically isotropic.

The third assumption is only reasonably met in most cases. There is actually a Boltzmann distribution of the population among the Stark levels. It becomes more valid the lower in energy the Stark splitting is. It also becomes more valid at higher temperatures. It is certainly a very bad assumption at low temperatures, and should not be applied to low temperature studies. The forth assumption is, of course, not valid for uniaxial or biaxial crystals. Nevertheless, polarization averaged studies can alleviate this restriction. These additional assumptions, while clear qualitatively, are often not made very clear in how they are carried out quantitatively, even in Judd's original article. To see how these assumptions are carried out in a quantitative way it is necessary to first define a quantity called the *oscillator strength*. The term oscillator strength is a historical term that has its origins relating to quantum mechanical resonance scattering in the Rutherford atomic model of a classically damped oscillator, and is directly related to a term called the *cross section*, for reasons originating in its introduction in atomic scattering theory. Nevertheless, these terms were historically adopted to describe transition intensities in the early days of studies of atomic transitions for their obvious connection to multipole transitions and their analogy to classical damped oscillators. The oscillator strength (f-number) for an electric dipole transition is defined as [3],

$$f = \frac{8\pi^2 mc}{h\bar{\lambda}e^2} n \left(\frac{n^2+2}{3n}\right)^2 \sum_{MM'} \left|\langle \alpha JM | \mathbf{P} | \alpha' J'M' \rangle\right|^2 \tag{22}$$

Squaring Eq. (16), assumption three allows the sum to be carried out a sum over M and M′, the Stark split levels. The orthogonality condition the for 3j symbols gives,

$$\sum_{MM'} \begin{pmatrix} J & \lambda & J' \\ -M & Q & M' \end{pmatrix} \begin{pmatrix} J & \lambda' & J' \\ -M & Q' & M' \end{pmatrix} = \frac{1}{2\lambda+1} \delta_{\lambda\lambda'} \delta_{QQ'} \tag{23}$$

and a factor 1/(2J+1) is introduced since M = -J, -(J-1), ..., 0, ..., (J-1), J. Assumption four allows the sum over q, the polarization. The orthogonality condition for the 3j symbols gives,

$$\sum_{q} \begin{pmatrix} 1 & t & \lambda \\ q & p & Q \end{pmatrix} \begin{pmatrix} 1 & t' & \lambda \\ q & p' & Q \end{pmatrix} = \frac{1}{2t+1} \delta_{tt'} \delta_{pp'} \tag{24}$$

and a factor of 1/3 is introduced since q = 0 for π polarization (E ⊥ c) and q = ±1 for σ polarization (E || c). In other words there are three values of q, one for π polarization and two for σ polarization. So, the summations over M, M′ and q are carried out, leaving only the summations over λ, p and t. The result for the oscillator strength is,

$$f = \frac{8\pi^2 mc}{3h\bar{\lambda}(2J+1)} n\left(\frac{n^2+2}{3n}\right)^2 \sum_{\lambda=2,4,6} \sum_{p} \sum_{t=1,3,5} (2\lambda+1)\frac{|A_{tp}|^2}{(2t+1)} Y^2(t,\lambda) \left|\left\langle \varphi_a \left\| \mathbf{U}^{(\lambda)} \right\| \varphi_b \right\rangle\right|^2 \tag{25}$$

The sum is over λ = 2, 4, 6 and t = 1, 3, 5. This result can now be condensed into its final form. The Judd-Ofelt parameters can be defined as,

$$\Omega_\lambda = (2\lambda+1)\sum_{p} \sum_{t=1,3,5} \frac{|A_{tp}|^2}{(2t+1)} Y^2(t,\lambda) \tag{26}$$

The Ω_λ, therefore, consist of odd-order parameters of the crystal field, radial integrals over wavefunctions of the $4f^n$ and perturbing, opposite parity wavefunctions of higher energy, and energies separating these states in terms of perturbation energy denominators. $Y(t,\lambda)$ was given previously in Eq. (17). With this substitution, the oscillator strength can be written as,

$$f = \frac{8\pi^2 mc}{3h\bar{\lambda}(2J+1)} n\left(\frac{n^2+2}{3n}\right)^2 \sum_{\lambda=2,4,6} \Omega_\lambda \left|\left\langle \varphi_a \left\| \mathbf{U}^{(\lambda)} \right\| \varphi_b \right\rangle\right|^2 \tag{27}$$

The summation over λ is known as the *linestrength*. It is Eq. (6) given at the beginning of this section. This is the approximate solution of the Judd-Ofelt theory. It is often used to find ED matrix elements between mixed parity states for manifold-to-manifold transitions.

In principle it is possible to calculate the Judd-Ofelt parameters *ab-initio*, but this requires accurate values for the radial integrals and odd-order crystal field components, which are not known to a high enough degree of precision. What is usually done is to treat the Judd-Ofelt parameters as a set of phenomenological parameters to be determined from fitting experimental absorption measurements determined in Eq. (28) with the theoretical Judd-Ofelt expression in Eq. (6). This is the subject of the next section. The Judd-Ofelt theory in practice.

4. Judd-Ofelt Theory : Practices

The Judd-Ofelt theory [11,12] allows for the calculation of manifold to manifold transition probabilities, from which the radiative lifetimes and branching ratios of emission can be determined. A Judd-Ofelt analysis relies on accurate absorption measurements, specifically the integrated absorption cross section over the wavelength range of a number of manifolds. From the integrated absorption cross section, the so-called linestrength, S_m, can be found,

$$S_m = \frac{3ch(2J+1)}{8\pi^3 e^2 \bar{\lambda}} n\left(\frac{3}{n^2+2}\right)^2 \int_{manifold} \sigma(\lambda)d\lambda \tag{28}$$

where J is the total angular momentum of the initial ground manifold, found from the $^{2S+1}L_J$ designation. $\sigma(\lambda)$ is the absorption cross section as a function of wavelength. The mean wavelength, $\bar{\lambda}$, can be found from the first moment of the absorption cross section data,

$$\bar{\lambda} = \frac{\sum \sigma(\lambda)}{\sum \lambda\sigma(\lambda)} \tag{29}$$

In practice, the Judd-Ofelt theory is used to determine a set of phenomenological parameters, Ω_λ (λ = 2, 4, 6), by fitting the experimental absorption, Eq. (28), or emission measurements, in a least squares difference sum, with the Judd-Ofelt expression, Eq. (6). This is most efficiently done in the following way. First, the linestrength in Eq. (28) is written as a 1xN column matrix, S_j^m, and Eq. (6) is also written in matrix form as,

$$S_j' = \sum_{i=1}^{3} M_{ij}\Omega_i \tag{30}$$

where M_{ij} are components of a N x 3 matrix for the square matrix elements of $U^{(2)}$, $U^{(4)}$ and $U^{(6)}$. The Ω_κ are components of a 1 x 3 matrix for the Judd-Ofelt parameters Ω_2, Ω_4 and Ω_6. Note that N represents the number of transitions to fit, which depends on the number of absorption manifolds actually measured. Obviously, since there are only three Judd-Ofelt parameters, N > 3. For example, since Ytterbium (Yb) has only has one absorption manifold, the Judd-Ofelt theory cannot be applied to Yb. Next, the sum of the squared difference is formed,

$$\sigma^2 = \sum_{j=1}^{N}\left(S_j^m - \sum_{i=1}^{3} M_{ij}\Omega_i\right)^2 \tag{31}$$

and minimized by taking the derivative with respect to Ω and setting the result equal to zero,

$$\frac{\partial(\sigma^2)}{\partial\Omega_k} = -2\sum_{j=1}^{N} M_{jk}\left(S_j^m - \sum_{i=1}^{3} M_{ij}\Omega_i\right) = 0 \tag{32}$$

The set of Judd-Ofelt parameters that minimizes the sum of the squared difference of measured and theoretical linestrength is written in matrix form, $\mathbf{\Omega}^{(0)} = (\mathbf{M}^\dagger\mathbf{M})^{-1}\mathbf{M}$, where $\mathbf{M}^\dagger$ is the adjoint of $\mathbf{M}$. Due to the large number of calculations to be made, matrices are suitable for computer based calculations. Once the Judd-Ofelt parameters are determined, they can be used to calculate transition probabilities, A(J;J′), of all excited states from the equation,

$$A(J';J)=\frac{64\pi^4 e^2}{3h(2J'+1)\bar{\lambda}^3}\left[n\left(\frac{n^2+2}{3}\right)^2 S_{ED}+n^2 S_{MD}\right] \quad (33)$$

where n is the refractive index of the solid, S_{ED} and S_{MD} are the electric and magnetic dipole line strengths, respectively. In this equation J′ is the total angular momentum of the upper excited state. Electric dipole linestrengths, S_{ED}, are calculated from each excited manifold to all lower lying manifolds from Eq. (6) using the matrix elements, $U^{(\lambda)}$, and Judd-Ofelt parameters.

Magnetic dipole linestrengths are calculated in a straightforward way from angular momentum considerations [15]. These values are then converted to intermediate coupling using free ion wavefunctions. The calculation of magnetic dipole linestrengths, S_{MD}, in intermediate coupling has been discussed in a previous article by the author [16]. The matrix elements for MD transitions are nonzero only if S = S′, and L = L′. Additional selection rules exist for the total angular momentum J. They are J = J′, J = J′ + 1, and J = J′ – 1. These selection rules were shown in figure 6. Calculation of magnetic dipole matrix elements is fairly straightforward. For a given S, L and J of the initial state, the matrix elements in LS-coupling are calculated from relations derived from angular momentum considerations. In the LS-coupling scheme, the matrix elements for the angular momentum operator **L** and the spin operator **S** are:

$$\left\langle f^n SLJ\|\mathbf{L}\|f^n S'L'J'\right\rangle=(-1)^{S+L+J+1}\begin{Bmatrix} S & L & J \\ 1 & J' & L \end{Bmatrix}\sqrt{L(L+1)(2L+1)(2J+1)(2J'+1)} \quad (34)$$

$$\left\langle f^n SLJ\|\mathbf{S}\|f^n S'L'J'\right\rangle=(-1)^{S+L+J+1}\begin{Bmatrix} S & L & J \\ J' & 1 & S \end{Bmatrix}\sqrt{S(S+1)(2S+1)(2J+1)(2J'+1)} \quad (35)$$

Expansion of the 6j symbols and combining Eq. (34) and Eq. (35) leads to the following equations for the MD matrix elements between LS-coupled states states:

$$\left\langle f^n SLJ\|\mathbf{L}+2\mathbf{S}\|f^n S'L'J\right\rangle=\left[S(S+1)-L(L+1)+3J(J+1)\right]\left[\frac{2J+1}{4J(J+1)}\right]^{1/2} \quad (36)$$

$$\left\langle f^n SLJ\|\mathbf{L}+2\mathbf{S}\|f^n S'L'J-1\right\rangle=\left\{\left[(S+L+1)^2-J^2\right]\left[\frac{J^2-(L-S)^2}{4J}\right]\right\}^{1/2} \quad (37)$$

$$\left\langle f^n SLJ\|\mathbf{L}+2\mathbf{S}\|f^n S'L'J+1\right\rangle=\left\{\left[(S+L+1)^2-(J+1)^2\right]\left[\frac{(J+1)^2-(L-S)^2}{4(J+1)}\right]\right\}^{1/2} \quad (38)$$

These matrix elements cannot be used directly for rare earth ions. The reason is that the LS-coupling scheme is inapplicable to rare earth ions because the interaction between electrons in the $4f$ shell and the spin orbit interaction are of the same order of magnitude. It is necessary, therefore, to find the matrix elements in another coupling scheme. Fortunately, this is easily achieved by choosing the basis states as a linear combination of LS coupling states. In this intermediate coupling scheme, which treats the electron interactions and spin-orbit interactions as being approximately the same order of magnitude, the new wavefunctions are,

$$\left|f^n[SL]J\right\rangle = \sum_{SL} C(S,L)\left|f^n SLJ\right\rangle \tag{39}$$

The coefficients C(S,L) are found by diagonalizing the combined electrostatic, spin-orbit and configuration-interaction matrices to obtain intermediate-coupled wavefunctions. MD transitions in intermediate coupling are possible if pairs of LS-coupled states are formed that satisfy MD selection rules. The MD matrix elements in intermediate coupling now follow from,

$$\left\langle f^n[SL]J\left\|\mathbf{L}+2\mathbf{S}\right\|f^n[S'L']J'\right\rangle = \sum_{SL,S'L'} C(S,L)C(S',L')\left\langle f^n SLJ\left\|\mathbf{L}+2\mathbf{S}\right\|f^n S'L'J'\right\rangle \tag{40}$$

If the interest is only in manifold-to-manifold MD transition probabilities, then it is sufficient to use free ion intermediate-coupled wavefunctions to find coupling coefficients. The manifold-to-manifold wavefunctions of ions in crystals generally show only small departures from the free ion values since the crystal field is only a small perturbation. It is known that MD transitions are allowed between states of the same parity. They are generally orders of magnitude smaller than ED transitions in free ions, but since the ED radiation for ions in solids occurs as a result of a perturbation, the ED intensities are much smaller than in free ions. As a result, some magnetic dipole transitions will make significant contributions to the observed intensities. The strongest MD transitions will usually be in the infrared, but exceptions can be expected. All the tools have been developed up to this point, and it is now a simple matter to find the radiative lifetimes, τ_r, and the branching ratio, β,

$$\frac{1}{\tau_r} = \sum_J A(J';J) \tag{41}$$

$$\beta_{J'J} = \frac{A(J';J)}{\sum_J A(J';J)} \tag{42}$$

The procedure of a Judd-Ofelt analysis is represented in a simple flow chart shown in figure 7 below. This figure shows the major steps in the process, culminating in calculation of the transition probabilities and branching ratios. The details of all these steps have been covered and form, a hopefully clear and coherent, recipe for what is called a Judd-Ofelt analysis.

It is useful at this point to show an example illustrating a real Judd-Ofelt analysis and a test of its validity. Ho^{3+} ions in $LiYF_4$ (YLF) is chosen as an example as it represents a case in extremes. YLF is a uniaxial host material exhibiting absorption that depends on polarization. Holmium also possesses many manifolds in its absorption spectra, some of which overlap with each other. This is a particularly good example because it illustrates that, despite these challenges, the Judd-Ofelt theory is remarkably accurate in most cases.

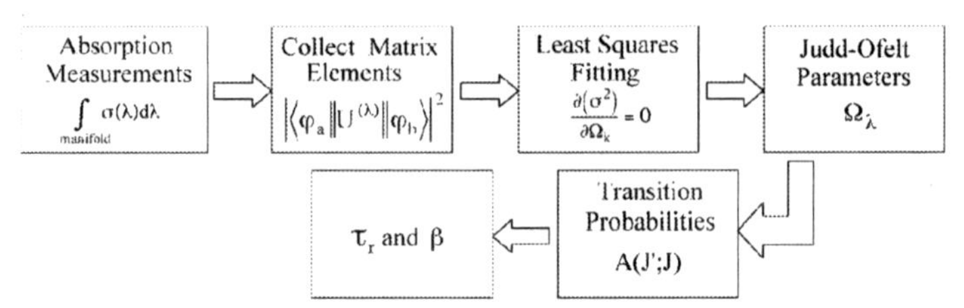

Figure 7. Procedure of Judd-Ofelt analysis.

The visible absorption spectrum of Ho:YLF is shown in figure 8. The π and σ polarized absorption cross sections of the manifolds shown in this figure were summed over wavelength and polarization averaged to obtain the integral in Eq. (28). The linestrengths were then calculated and these values were used in the Judd-Ofelt fitting. The results of the Judd-Ofelt fit are shown in table II. The measured and calculated linestrength values are shown in the right column of the table. The matrix elements and mean wavelength are also shown for convenience. The Judd-Ofelt parameters were found to be Ω_2 = 1.03 x 10^{-20} cm^2, Ω_4 = 2.32 x 10^{-20} cm^2, Ω_4 = 1.93 x 10^{-20} cm^2. The root mean square (RMS) deviation, which is a measure of the overall quality of the fit, was found to be δ = 0.13 x 10^{-20} cm^2.

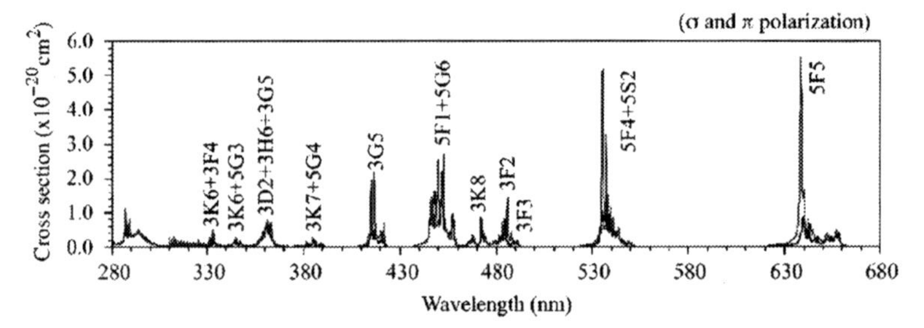

Figure 8. Visible absorption spectrum of Ho:YLF.

The Judd-Ofelt parameters can now be used to calculate the electric dipole transition probabilities between any excited manifold and any lower lying manifold. The magnetic dipole transition probabilities are calculated separately as discussed previously. The transition probabilities, ED and MD, then allow for calculation of branching ratios, β, and radiative lifetimes, τ_r. Table III collects the results for the Ho:YLF case being examined here.

Table II. Judd-Ofelt fitting in Ho^{3+} YLF.

Transition (from $^5I_8$)	$\|U^{(2)}\|^2$	$\|U^{(4)}\|^2$	$\|U^{(6)}\|^2$	λ(nm)	Linestrength(10^{-20} cm^2)	
					measured	calculated
$^3K_6 + ^3F_4$	0.0026	0.1263	0.0073	334	0.2639	0.3097
$^3L_9 + ^5G_3$	0.0185	0.0052	0.1169	345	0.2920	0.3533
$^3D_2 + ^3H_6 + ^5G_5$	0.2155	0.1969	0.1679	360	1.2803	1.0023
$^3K_7 + ^5G_4$	0.0058	0.0361	0.0697	385	0.2819	0.2243
$^3G_5$	0.0000	0.5338	0.0002	418	1.1086	1.2396
$^5F_1 + ^5G_6$	1.5201	0.8410	0.1411	448	3.7492	3.7839
$^3K_8$	0.0208	0.0334	0.1535	467	0.2282	0.3952
$^3F_2$	0.0000	0.0000	0.2041	473	0.3488	0.3941
$^3F_3$	0.0000	0.0000	0.3464	486	0.7587	0.6689
$^5F_4 + ^5S_2$	0.0000	0.2392	0.9339	540	2.5359	2.3587
$^5F_5$	0.0000	0.4250	0.5687	645	2.1426	2.0848
$^5I_5$	0.0000	0.0100	0.0936	886	0.1741	0.2039
$^5I_6$	0.0084	0.0386	0.6921	1175	1.1067	1.4347

Table III. Judd-Ofelt transition probabilities in Ho^{3+} YLF.

Transition	$\|U^{(2)}\|^2$	$\|U^{(4)}\|^2$	$\|U^{(6)}\|^2$	λ (nm)	S_{ED}[1]	A_{MD} (s^{-1})	A_{ED} (s^{-1})	β	τ_r (μs)
$^5F_4 \rightarrow ^5S_2$	0.0000	0.0159	0.0033	67656	0.043		0.00	0.000	
$^5F_4 \rightarrow ^5F_5$	0.1944	0.0923	0.0080	3173	0.429	2.83	2.82	0.002	
$^5F_4 \rightarrow ^5I_4$	0.0001	0.0234	0.2587	1887	0.554		17.67	0.005	
$^5F_4 \rightarrow ^5I_5$	0.0018	0.1314	0.4655	1327	1.206		111.33	0.033	
$^5F_4 \rightarrow ^5I_6$	0.0012	0.2580	0.1697	986	0.930		207.79	0.062	
$^5F_4 \rightarrow ^5I_7$	0.0000	0.1988	0.0324	738	0.524		284.43	0.084	
$^5F_4 \rightarrow ^5I_8$	0.0000	0.2402	0.7079	536	1.925		2748.90	0.814	295
$^5S_2 \rightarrow ^5F_5$	0.0000	0.0110	0.0036	3330	0.033		0.33	0.000	
$^5S_2 \rightarrow ^5I_4$	0.0013	0.0279	0.2795	1942	0.606		31.92	0.016	
$^5S_2 \rightarrow ^5I_5$	0.0000	0.0043	0.1062	1354	0.215		33.67	0.016	
$^5S_2 \rightarrow ^5I_6$	0.0000	0.0206	0.1541	1000	0.345		135.16	0.065	
$^5S_2 \rightarrow ^5I_7$	0.0000	0.0000	0.4096	746	0.791		747.45	0.366	
$^5S_2 \rightarrow ^5I_8$	0.0000	0.0000	0.2270	540	0.438		1100.01	0.538	489
$^5F_5 \rightarrow ^5I_4$	0.0001	0.0059	0.0040	4658	0.015		0.03	0.000	
$^5F_5 \rightarrow ^5I_5$	0.0068	0.0271	0.1649	2282	0.252		5.70	0.003	
$^5F_5 \rightarrow ^5I_6$	0.0102	0.1213	0.4995	1430	0.818		74.62	0.041	
$^5F_5 \rightarrow ^5I_7$	0.0177	0.3298	0.4340	961	1.080		324.00	0.180	
$^5F_5 \rightarrow ^5I_8$	0.0000	0.4277	0.5686	645	1.390		1394.42	0.776	556
$^5I_4 \rightarrow ^5I_5$	0.0312	0.1237	0.9099	4472	2.076	1.94	4.69	0.079	
$^5I_4 \rightarrow ^5I_6$	0.0022	0.0281	0.6640	2064	1.350		32.11	0.381	
$^5I_4 \rightarrow ^5I_7$	0.0000	0.0033	0.1568	1211	0.310		37.75	0.448	
$^5I_4 \rightarrow ^5I_8$	0.0000	0.0000	0.0077	749	0.015		7.72	0.092	11875
$^5I_5 \rightarrow ^5I_6$	0.0438	0.1705	0.5729	3831	1.547	4.07	4.44	0.067	
$^5I_5 \rightarrow ^5I_7$	0.0027	0.0226	0.8887	1662	1.771		67.93	0.538	
$^5I_5 \rightarrow ^5I_8$	0.0000	0.0099	0.0936	899	0.204		49.79	0.394	7922
$^5I_6 \rightarrow ^5I_7$	0.0319	0.1336	0.9308	2934	2.140	8.11	12.78	0.135	
$^5I_6 \rightarrow ^5I_8$	0.0083	0.0383	0.6918	1175	1.433		134.03	0.865	6455
$^5I_7 \rightarrow ^5I_8$	0.0249	0.1344	1.5217	1960	3.276	15.91	55.92	1.000	13921

[1] Line strength values are in units 10^{-20} cm^2.

It is now prudent to ask how good the results are, that is, does the Judd-Ofelt theory accurately predict the branching ratios and radiative lifetimes? A method for testing the Judd-Ofelt theory is therefore needed. The simplest way is to just measure the branching ratios and radiative lifetimes. Branching ratios can be directly measured from emission spectra, as they are just the fraction of the total photon flux from an upper to lower manifold. Selected manifold to manifold branching ratios in Ho:YLF are shown in table IV. The agreement between measurement and the Judd-Ofelt theory is quite good, the error being less than 15%.

Table IV. Measured and calculated branching ratios in Ho^{3+} YLF.

Ion	Transition	Wavelength range (nm)	$\beta_{measured}$	$\beta_{Judd-Ofelt}$	Percent difference
Ho	$^5S_2 \rightarrow {}^5I_6$	1014 – 1052	0.0548	0.0638	14.2
Ho	$^5S_2 \rightarrow {}^5I_7$	747 – 758	0.2540	0.2247	11.5
Ho	$^5S_2 \rightarrow {}^5I_8$	539 – 550	0.6916	0.6759	2.3
Ho	$^5F_5 \rightarrow {}^5I_7$	952 – 981	0.1541	0.1801	14.4
Ho	$^5F_5 \rightarrow {}^5I_8$	638 – 659	0.8458	0.7752	8.3
Ho	$^5I_5 \rightarrow {}^5I_7$	1618 – 1681	0.6174	0.5381	12.8
Ho	$^5I_5 \rightarrow {}^5I_8$	882 – 915	0.3822	0.3944	3.1

The radiative lifetimes are a bit more problematic because they are not directly measurable in most cases. The reason for this is that other processes besides spontaneous emission are usually present. These processes include energy transfer, nonradiative relaxation and radiative trapping, for example. So, directly measuring the lifetime will not always yield the radiative lifetime, even at low temperature. An alternate method is needed. Utilizing the reciprocity of absorption and emission, however, the radiative lifetimes can be indirectly derived. The emission cross section is related to the absorption cross section by the following equation originally derived by McCumber [17] and reformulated by Payne et al. [18],

$$\sigma_{em}(\lambda) = \sigma_{ab}(\lambda)\frac{Z_l}{Z_u}\exp\left[\left(E_{ZL} - \frac{hc}{\lambda}\right)\Big/kT\right] \tag{43}$$

where Z_l and Z_u are the partition functions of the lower and upper manifolds, respectively. E_{ZL} is the zero-line energy, and is defined as the energy difference between the lowest Stark level of the upper manifold and the lowest Stark level of the lower manifold. An emission cross section derived from the absorption cross section in this way can be compared to an actual measured emission cross section. The measured emission cross section is given by the equation [19],

$$\sigma(\lambda) = \frac{\lambda^5}{8\pi cn^2(\tau_r/\beta)}\frac{3I_\alpha(\lambda)}{\int[2I_\sigma(\lambda)+I_\pi(\lambda)]\lambda d\lambda} \tag{44}$$

where $I_\alpha(\lambda)$ is the emission intensity for α (π or σ) polarization, τ_r is the radiative lifetime, β is the branching ratio, and n is the index of refraction. Note that $I_\sigma(\lambda) = I_\pi(\lambda)$ for isotropic crystals.

By scaling the factor τ_r/β such that equations (43) and (44) agree, the radiative lifetime can be extracted and then compared to the values calculated in the Judd-Ofelt theory. A pictorial representation of this process is shown in figure 9 for the lowest excited state in Ho, the $^5I_7$ manifold. The top set of spectra, a and b, show the measured π and σ absorption spectra. The middle set, c and d, show the π and σ emission derived from absorption using Eq. (43). The bottom set, e and f, show the measured π and σ emission from Eq. (44) scaled to match the derived emission with a value of $\tau_r/\beta = 14000$ μs. The branching ratio in this case is one, so the radiative lifetime is 14 ms. For other manifolds, a measurement of the branching ratio must be made. Alternatively, a Judd-Ofelt branching ratio can also be used if the measurement cannot be done, but since this is a test of the Judd-Ofelt theory, this defeats the purpose of the test somewhat. However, if there is good agreement between the β's that can be measured and those determined in the Judd-Ofelt theory, then this improves confidence in using Judd-Ofelt β's for those that cannot be measured. The radiative lifetimes derived from reciprocity and those calculated from the Judd-Ofelt theory are shown in table V. The result is that the error is less than 30% for the radiative lifetimes. So, the Judd-Ofelt theory is valid within 15 to 30% as a liberal estimate. A more conservative estimate puts the error in the range of 10 to 20%.

Table V. Measured and calculated radiative lifetimes in Ho^{3+} YLF.

Ion	Transition	β	$\tau_{measured}$ (μs)	$\tau_{Judd-Ofelt}$ (μs)	percent difference
Ho	$^5F_4 + {}^5S_2 \rightarrow {}^5I_8$	0.6916	258 – 530	295 – 542	2.2 – 12.5
Ho	$^5F_5 \rightarrow {}^5I_8$	0.8458	677	556	17.8
Ho	$^5I_5 \rightarrow {}^5I_8$	0.3822	5790	7922	26.9
Ho	$^5I_6 \rightarrow {}^5I_8$	0.8651	6592	6455	2.0
Ho	$^5I_7 \rightarrow {}^5I_8$	1.0000	14000	13921	0.5

The results that have been shown here are consistent with the general consensus of the accuracy of the Judd-Ofelt theory. Usually such errors have been only approximations in the past. This analysis here shows a concrete example with definative numbers to assess the accuracy. In considering the starting point of the theory, the approximations made and the simplifications used to reduce the problem to something that can be straightforwardly calculated, this is really quite remarkable. In a sense it is a testament to simple and reasonable assumptions and approximations made with a physical basis for their implementation.

In the remaining section, some special cases where the Judd-Ofelt theory encounters problems are discussed. The lanthanide ions praseodymium and europium are discussed. These problems with the standard Judd-Ofelt theory, as discussed in the next section, lead to extensions of the Judd-Ofelt theory. These extensions are briefly outlined and indications are made, where appropriate, how these extensions may explain some problems with the standard theory, especially in the case of some europium transitions. Finally, some remaining problems towards future direction are also outlined. These are problems not yet resolved and indicate a further need for future work on this very fascinating topic.

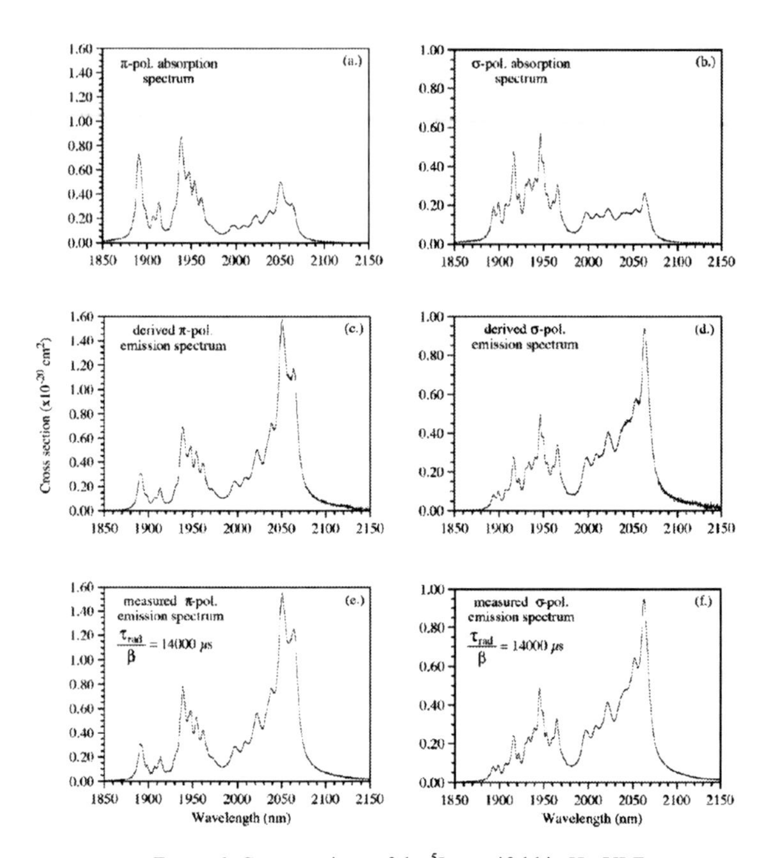

Figure 9. Cross sections of the $^5I_7$ manifold in Ho:YLF.

5. Beyond the Standard Judd-Ofelt Theory

5.1. THE CASE OF PRASEODYMIUM

Praseodymium ions suffer from several problems in applying the Judd-Ofelt theory. There are often large deviations between calculations and experimental observations, exhibited by the large RMS values obtain in Judd-Ofelt fitting. In addition, the Judd-Ofelt parameters obtained for Pr doped materials tend to be very dependent on the transitions used in the fit. Sometimes negative values are obtained for Ω_2, in opposition to its definition as given in Eq. (26). These inconsistencies are usually explained by the small difference in the average energies, $\approx$ 50,000 cm^{-1}, between the $4f^{n}$ and $4f^{(n-1)}5d$ configurations in Pr^{3+} ions. Remember that the first two assumptions of the Judd-Ofelt theory took the perturbing configurations to be degenerate and well separated from the f-f transitions of the $4f^{n}$ configuration. There have been several approaches to these problems. Dunnina, *et al.* [20] modified the standard Judd-Ofelt theory to take into account the finite 4f5d energy,

$$\Omega'_{\lambda} = \Omega_{\lambda}\left[1 + \left(\Delta E_{ij} - 2\bar{E}_{4f}\right) / \left(E^{0}_{5d} - \bar{E}_{4f}\right)\right] \tag{45}$$

where E_i and E_j are the initial and final states of the $4f$ transition, $\bar{E}_{4f}$ is the average energy of the $4f$ states, and E^{0}_{5d} is the lowest energy of the $4f5d$ state. This modified theory has been used with some success to eliminate the problem of negative Ω_2. Quimby *et al.* [21] has applied a modified Judd-Ofelt theory by introducing fluorescence branching ratios. This allows the Judd-Ofelt parameters to be reliably calculated with fewer ground state absorption measurements than would otherwise be necessary, but requires additional fluorescence measurements. These methods were applied to Pr:ZBLAN glass and produced fairly good agreement between the calculated and measured oscillator strengths. However, Goldner *et al.* [22] has argued that such modified theories are not really necessary and, in fact, do not improve the fit substantially if reasonable values for the $4f5d$ energies are used. Goldner *et al.* showed that by using a normalized least squares fitting, the measurements are properly scaled according to the error associated with each one, and positive, stable parameters are obtained with improved RMS deviations. In this new fitting procedure, Eq. (31) is replaced with,

$$\sigma^{2} = \sum_{j=1}^{N}\left[\left(S^{m}_{j} - \sum_{i} M_{ij}\Omega_{i}\right) / \sigma_{i}\right]^{2} \tag{46}$$

where σ_i are the standard deviations on the absorption measurements. This modified fitting procedure has the advantage that transitions measured with the same accuracy have the same weight in the fit. This prohibits stronger transitions from dominating the fit. In addition, there is no need to modify the Judd-Ofelt theory itself to obtain better fits with stable, positive parameters in Pr^{3+} doped materials.

5.2. THE CASE OF EUROPIUM

Europium ions provide another challenge to the standard Judd-Ofelt theory. In particular, the transitions $^{7}F_{0} \leftrightarrow {}^{5}D_{Jodd}$, $^{7}F_{Jodd} \leftrightarrow {}^{5}D_{0}$ and $^{7}F_{0} \leftrightarrow {}^{5}D_{0}$ are forbidden transitions in the Judd-Ofelt theory. They violate the selection rules derived earlier and illustrated in figure 6. Some of these transitions are primarily magnetic dipole in nature, but do occur as electric dipole transitions with low intensity in some materials. In general, for the other lanthanide ions as well, all $J = 0$ to $J' = 0$ transitions are forbidden, as well as transitions with $\Delta S = 1$. Although the later are spin-forbidden in the Judd-Ofelt theory, they account for a substantial number of observed transitions of lanthanide ions in solids. Clearly, this implies that the Judd-Ofelt theory is incomplete in its standard form. The transitions in the Eu^{3+} ion, therefore, provide an ideal testing ground for possible extensions to the standard theory. Spin-forbidden transition intensities have been treated by Burdick and Downer [23,24], where they show that spin forbidden transitions acquire a major fraction of their intensities from previously neglected spin-orbit linkages with excited configurations. In the intermediate coupling approximation, these linkages can contribute significantly when the spin-orbit interaction is comparable to the electrostatic interaction. This excited state spin-orbit interaction is known as the Wybourne-Downer mechanism. Tanaka and Kushida have covered the 0-0, 0-even, and 0-odd transitions for Eu^{3+} ions in a number of papers [25,26,27,28]. Some of these discuss the interference of J-mixing and spin-orbit effects. The europium ion continues to lend itself to testing new assumptions regarding the prevalent interactions of ions in solids and how they affect the intensities in the context of modifying the Judd-Ofelt theory towards a more complete theoretical description.

5.3. EXTENSIONS TO THE JUDD-OFELT THEORY

Since the original formulation of the Judd-Ofelt theory in 1962, there have been many investigations into extending the theory to explain its shortcomings. Some of these shortcomings have been discussed in the last two subsections. These shortcomings extend to other lanthanides besides the ones discussed here. The shortcomings have more to do with the peculiar nature of certain transitions. This topic cannot be discussed in all its details and subtleties here, as it is quite extensive. What follows is an outline of the basic mechanisms that have been considered to explain the shortcomings of the standard Judd-Ofelt theory.

1. *J-Mixing*: The wavefunctions of the $J \neq 0$ state are mixed into the $J = 0$ state by the crystal field potential. This mechanism goes beyond the free-ion model and has been used, for example, to explain the $^5D_0 \rightarrow {}^7F_0$ and $^7F_{Jodd}$ transitions in Eu^{3+} [25,29,30].

2. *Electron Correlation*: The electrostatic interaction between electrons is taken into account. This goes beyond the single configuration approximation in the standard Judd-Ofelt theory. Electron correlation within the 4f shell is used to set a theoretical description of 0-0 and 0-1 transitions in Eu^{3+} ions in hosts with C_{2v} symmetry [31].

3. *Dynamic Coupling*: The mutual interaction of the lanthanide ion and the crystal environment is taken into account. This mechanism goes beyond the static coupling model in the standard Judd-Ofelt theory and is explains hypersensitive transitions, that is, transitions which are highly dependent on changes in host environment [32,33,34].

4. *Wybourne-Downer Mechanism*: This mechanism involves spin-orbit interaction among states of the excited configurations, leading to an admixing of spin states into the $4f^n$ configuration. Originally proposed by Wybourne [35], it has been used to explain $\Delta S = 1$ spin forbidden transitions [23,24].

5. *Relativistic Contributions*: Involves a relativistic treatment of f-f transitions in crystal fields. This reformulation of the crystal field and operators in relativistic terms is a recent advancement and future calculations will indicate their significance, especially for the heavier actinides [36,37,38].

Two extensive review articles covering the intensities of f-f transitions are available in the literature. The early development of the Judd-Ofelt theory can be found in an article by Peacock [39]. More recent developments can be found in a review article by Smentek [40]. In addition, the reader is encouraged to look at the 2003 special issue of *Molecular Physics* [41] commemorating the 40th anniversary of the Judd-Ofelt theory. This volume contains many interesting articles on the state of knowledge of Judd-Ofelt theories and their applications.

5.4. FUTURE DIRECTIONS

The Judd-Ofelt theory as originally formulated is a second-order approach that treats the odd-order terms of the crystal field as a perturbation. This perturbing influence allows mixing of opposite parity states from higher lying configurations to be mixed into $4f$ states, producing mixed parity states between which electric dipole transitions become allowed. This formulation is based on the static, free-ion and single configuration approximations as discussed in section 3. In the last few subsections, it was seen that there is a need to extend the Judd-Ofelt theory to account for anomalous transitions. The various operators used to represent any additional physical mechanisms are considered in the perturbation expansion not just to second-order, but in third-order as well. It is difficult, if not impossible, to say what the relative importance of each mechanism is. In fact, the appearance of a term in third-order may have more impact than a term occurring in second-order. This remains an outstanding problem for future development. It results in a very complicated situation to disentangle which mechanisms are relevant and physically significant. The multitude of mechanisms that can be realized produces a situation that departs markedly from the simple approximations and assumptions of the standard Judd-Ofelt theory and will continue to be a challenge in the future. In fact, by extending the Judd-Ofelt theory, the simple linear parametric fitting is lost and physically meaningful descriptions can easily be obscured.

Another area of future direction is in the area of *ab-initio* calculations. Such calculations are still not entirely successful in explaining all aspects of the intensities of f-f transitions. The availability of accurate Hartree-Fock relativistic radial integrals has helped the situation, but there remains a lack of accurate knowledge of the odd-order crystal field parameters. This has hampered the development of *ab-initio* theories to some extent. Further, there is the issue of the vibrational-electronic coupling which complicates the spectral dynamics. It is clear that the theory of f-f transitions is not yet complete. As Professor Brian G. Wybourne has said, *"The Judd-Ofelt theory marked a turning point in our understanding of the fascinating spectroscopic properties of the rare earths. It has been in a very real sense the first step in the journey to an understanding of the rare-earths and their heavier cousins, the actinides, but like many journeys into the unknown, the end is not in sight"* [1].

6. Appendix

There are many interesting sidelines, or further journeys, that could be taken in learning this very fascinating subject. It is not possible to cover everything here in this article, but it is worthwhile to point out a few extra sidelines that might be of interest to the reader here in the appendix. The articles of Kushida [42,43,44] provide an interesting perspective on energy transfer theory in the context of Judd-Ofelt theory. The articles of Peacock [45,46,47,48] cover a variety of topics including vibronic transitions, Judd-Ofelt parameter variation over isostructural series and sensitivity of Judd-Ofelt parameters to transitions used. Peacock's review article [39] discusses a wide variety of topics on the Judd-Ofelt theory. Jorgensen and Reisfeld [49,50] have discussed the Judd-Ofelt parameters and chemical bonding of lanthanide ions to the solid host material. The Judd Ofelt parameter, Ω_2, is affected by covalent bonding, while Ω_6 is related to the rigidity of the solid host in which the lanthanide ions are situated. The work of Edvardsson and co-worker's [51,52,53,54,55] should not escape attention here. There are some very interesting results discussed in the context of their Molecular Dynamics Simulations. Finally, there are many articles by L. Smentek, too numerous to mention here. She has written extensively on the Judd-Ofelt theory. Her review article [40] is especially thorough. Many references to earlier works can be found in this review article. See also the 2003 special issue of *Molecular Physics* [41], celebrating the 40th anniversary of the Judd-Ofelt theory.

In the remainder of this appendix, a computer program is given. This program, written in BASIC, provides the core routines for creating a Judd-Ofelt analysis program. A full version of the program with input files can be provided on request (Brian.M.Walsh@nasa.gov).

```
REM: BEGIN JUDD-OFELT FITTING ROUTINE
REM: WRITTEN BY BRIAN M. WALSH
REM: GET INITIAL INPUT
FOR I=0 TO NI: REM NI=NUMBER OF TRANSITIONS MEASURED -1
     INPUT TAGS :  REM (2J+1) OF TERMINAL LEVEL (GROUND STATE)
     INPUT IA(I) : REM INTEGRATED ABSORPTION IN UNITS 10^-20 CM^2-NM
     INPUT W(I) :  REM AVERAGE WAVELENGTH OF TRANSITION IN NM
     INPUT NS(I) : REM INDEX OF REFRACTION AT WAVELENGTH W(I)
     FOR COL= 0 TO 2 : REM ONLY 3 COLUMNS FOR U(2), U(4), U(6)
          INPUT M(NI , COL) : REM REDUCED MATRIX ELEMENTS
     NEXT COL
     REM: FIND THE LINESTRENGTH SE(I)
     SE(I) = IA(I) * 10.41 * NS(I) * (3 / (NS(I)^2 + 2))^2 * TAGS / W(I)
     REM NOTE THAT 3hc/8π^3e^2 = 10.41
NEXT I
"100" REM JUDD-OFELT LEAST SQUARES FIT FOR ABSORPTION LINE STRENGTHS
"200" REM GET TRANSPOSE OF MATRIX COMPOSED OF REDUCED MATRIX ELEMENT'S [ME]
FOR I = 0 TO NJ: REM NJ = 2, ONLY 3 ARRAY ELEMENTS FOR U(2), U(4), U(6)
     FOR J = 0 TO NI
          A(I , J) = ME(J , I)
     NEXT J
NEXT I
"300" REM MULTIPLY TRANSPOSE [A] BY ORIGINAL MATRIX [ME]
FOR I = 0 TO NJ
     FOR J = 0 TO NJ
          FOR K = 0 TO NI
               AN(I , J) = AN(I , J) + A(I , K) * ME(K , J)
          NEXT K
     NEXT J
NEXT I
"400" REM INVERT MATRIX [AN]
N=NJ + 1
FOR K = 0 TO N - 1
     FOR I = 0 TO N - 1
          AN(I , N) = ABS(I = K)
```

```
        NEXT I
        FOR J = 1 TO N
              AN(K , J) = AN(K , J) / AN(K , 0)
              FOR I = 0 TO N-1
                    IF I = K THEN GOTO "410"
                    AN(I , J) = AN(I , J) - AN(K , J) * AN(I , 0)
"410"         NEXT I
        NEXT J
        FOR I = 0 TO N - 1
              FOR J = 0 TO N - 1
                    AN(I , J) = AN(I , J+1)
              NEXT J
        NEXT I
NEXT K
"500" REM MULTIPLY INVERSE [AN] BY TRANSPOSE [A]
FOR I = 0 TO NJ
        FOR J = 0 TO NI
              FOR K = 0 TO NJ
                    AF(I , J) = AF(I , J) + AN(I , K) * A(K , J)
              NEXT K
        NEXT J
NEXT I
"600" REM FIND JUDD-OFELT PARAMETERS [JO]
FOR I = 0 TO NJ
        FOR J = 0 TO NI
              JO(I) = JO(I) + AF(I , J) * SE(J)
        NEXT J
NEXT I
"700" REM FIND THEORETICAL LINE STRENGTHS [ST]
FOR I = 0 TO NI
        FOR J = 0 TO NJ
              ST(I) = ST(I) + ME(I , J) * JO(J)
        NEXT J
NEXT I
"800" REM FIND RMS DEVIATION
FOR I = 0 TO NI
        SSUM = SSUM + (SE(I) - ST(I))^2
NEXT I
SSUM = SSUM / (NI - 2) : REM NUMBER OF TRANSITIONS (NI + 1) > 3
RMS = SQR(SSUM) : ROOT MEAN SQUARE ERROR
"900" CALCULATE EXCITED STATE TRANSITION PROBABILITIES
FOR I = 0 TO NI - 1
        FOR K = I + 1 TO NI
              WT(K) = ABS(WL(I) - WL(K)) : REM TRANSITION WAVELENGTH IN NM
              INPUT U(0 , I),U(1 , I),U(2 , I) : REM  GET MATRIX ELEMENTS FOR EACH TRANSITION
              INPUT TAES(I) : REM (2J + 1) OF EXCITED LEVEL
              FOR J = 0 TO NJ
                    TS(K) = TS(K) + U(J , I) * JO(J) : TRANSITION STRENGTH
              NEXT J
              AED(K)=7.23E10 * NS(K) * ((NS(K)^2 +2 )/3)^2 * (TS(K) * 1E-20)/(TAES(I) * (WT(K)*1E-7)^3)
              REM NOTE THAT 64π^4e^2/3hc = 7.23E10 CM/SEC.
              REM NOTE ALSO THAT LINESTRENGTHS HAVE UNITS OF 10^-20 CM^2
              REM HENCE THE 1E-20 FACTOR MULTIPLYING TS(K).
              RS = RS + AED(K) : REM RUNNING SUM OF TRANSITION PROBABILITIES
        NEXT K
        REM CALCULATE BRANCHING RATIOS
        FOR J = I + 1 TO K - 1
              BRATIO = AED(J)/RS: REM BRANCHING RATIO
        NEXT J
        REM CALCULATE LIFETIMES (INVERSE OF SUM OVER AED FOR ALL LOWER MANIFOLDS)
        FOR KK = 1 TO NI
              AEDSUM = 0
              FOR JJ = 1 TO KK
                    AEDSUM=AED(JJ)+AEDSUM: REM SUM  TRANSITION PROBABILITIES
              NEXT JJ
              AEDSUM(KK) = AEDSUM
              IF AEDSUM = 0 THEN GOTO "910"
              LIF(KK) = (1/AEDSUM(KK)) * 1000: REM RADIATIVE LIFETIME IN MILLISECONDS
"910" NEXT KK
REM ALL LINESTRENGTHS AND JUDD-OFELT PARAMETERS ARE IN UNITS OF 10^-20 CM^2
NEXT I
```

Acknowledgements

I wish to extend my thanks to Prof. Baldassare Di Bartolo, his staff, Prof. John Collins and Mr. Ottavio Forte, as well as the Majorana Center, for my participation in this year's course. I also wish to extend my sincere thanks to Ms. Regine DeWees for many thoughtful discussions and for her support during the preparation of my lectures.

References

1. Wybourne, B.G. (2004) The fascination of Rare earths-then, now and in the future, *J. Alloys and Compounds* **380**, 96-100.
2. Van Vleck, J. H. (1937) The puzzle of rare earth spectra in solids, *J. Phys. Chem.* **41**, 67-80.
3. Broer, L.F.J., Gorter, C.J, and Hoogschagen, J. (1945) On the intensities and the multipole character in the spectra of rare earth ions, *Physica* **XI**, 231-249.
4. Racah, G. (1941) Theory of complex spectra. I., *Phys. Rev.* **61**, 186-197.
5. Racah, G. (1942) Theory of complex spectra. II., *Phys. Rev.* **62**, 438-462.
6. Racah, G. (1943) Theory of complex spectra. III., *Phys. Rev.* **63**, 367-382.
7. Racah, G. (1949) Theory of complex spectra. IV., *Phys. Rev.* **76**, 1352-1365.
8. Condon, E.U and Shortley, G.H. (1935) *The Theory of Atomic Spectra*, Cambridge University Press, Cambridge.
9. Rotenberg, M., Bivens, R., Metropolis, N., and Wooten Jr., J.K. (1959) *The 3-j and 6-j Symbols*, M.I.T Press, Cambridge.
10. Runciman, W.A. (1956) Stark-splitting in crystals, *Phil. Mag.* **1**, 1075-1077.
11. Judd, B.R. (1962) Optical absorption intensities of rare-earth ions, *Phys. Rev.* **127**, 750-761.
12. Ofelt, G.S. (1962) Intensities of crystal spectra of rare-earth ions, *J. Chem. Phys.* **37**, 511-520.
13. Nielson, C.W. and Koster, G.F. (1963) *Spectroscopic coefficients of the* p^n, d^n, *and* f^n *configurations*, The M.I.T. Press, Cambridge.
14. Leavitt, R.P. and Morrison, C.A. (1980) Crystal-field analysis of triply ionized rare earth ions in lanthanum fluoride. II. Intensity calculations, *J. Chem. Phys.* **73**, 749-757.
15. Shortley, G.H. (1940) The computation of quadrupole and magnetic dipole transitions, *Phys. Rev.* **57**, 225-237.
16. Walsh, B.M., Barnes, N.P., and Di Bartolo, B. (1998) Branching ratios, cross sections, and radiative lifetimes of rare earth ions in solids: Application to Tm^{3+} and Ho^{3+} ions in $LiYF_4$, *J. Appl. Phys.* **83**, 2772-2787.
17. McCumber, D.E. (1964) Einstein relations connecting broadband emission and absorption spectra, *Phys. Rev.* **136**, A954-A957.
18. Payne, S.A., Chase, L.L., Smith, L.K., Kway, W.L., and Krupke, W.F. (1992) Infrared cross-section measurements for crystals doped with Er^{3+}, Tm^{3+}, and Ho^{3+}, *IEEE J. Quant. Elec.* **28**, 2619-2629.
19. Moulton, P.F. (1986) Spectroscopic and laser characteristics of $Ti:Al_2O_3$, *J. Opt. Soc. Am. B* **3**, 125-133.
20. Dunnina, E.B, Kaminskii, A.A., Kornienko, A.A., Kurbanov K., and Pukhov K.K. (1990) Dependence of line strength of electrical dipole f-f transitions on the multiplet energy of Pr^{3+} ion in $YAlO_3$, *Sov. Phys. Solid State* **32**, 920-922.
21. Quimby, R.S. and Miniscalco, W.J. (1994) Modified Judd-Ofelt technique and application to optical transitions in Pr^{3+}-doped glass, *J. Appl. Phys.* **75**, 613-615.
22. Goldner, P. and Auzel, F. (1996) Application of standard and modified Judd-Ofelt theories to a praseodymium fluorozirconate glass, *J. Appl. Phys.* **79**, 7972-7977.
23. Downer, M.C., Burdick, G.W., and Sardar, D.K. (1988) A new contribution to spin-forbidden rare-earth optical transition intensities: Gd^{3+} and Eu^{3+}, *J. Chem. Phys.* **89**, 1787-1797.
24. Burdick, G.W., Downer, M.C., and Sardar, D.K. (1989) A new contribution to spin-forbidden rare earth optical transition intensities: Analysis of all trivalent lanthanides, *J. Chem. Phys.* **91**, 1511-1520.
25. Tanaka, M., Nishimura, G., and Kushida, T. (1994) Contribution of J mixing to the $^5D_0$-$^7F_0$ transition of Eu^{3+} ions in several host matrices, *Phys. Rev. B* **49**, 16917-16925.
26. Tanaka, M. and Kushida, T. (1996) Interference between Judd-Ofelt and Wybourne-Downer mechanisms in the $^5D_0$-$^7F_0$ (J = 2,4) transitions of Sm^{3+} in solids, *Phys. Rev. B* **53**, 588-593.
27. Kushida, T. and Tanaka, M. (2002) Transition mechanisms and spectral shapes of the $^5D_0$-$^7F_0$ line of Eu^{3+} and Sm^{3+} in solids, *Phys. Rev. B* **65**, 195118(1-6).

28. Kushida, T., Kurita, A., and Tanaka, M. (2003) Spectral shape of the $^5D_0$-$^7F_0$ line of Eu^{3+} and Sm^{3+} in glass, *J. Lumin.* **102-103**, 301-306.
29. Lowther, J.E. (1974) Spectroscopic transition probabilities of rare earth ions, *J. Phys. C* **7**, 4393-4402.
30. Xia, S. and Chen, Y. (1985) Effect of J-mixing on the intensities of f-f transitions of rare earth ions, *J. Lumin.* **33**, 228-230.
31. Smentek, L. and Hess Jr., B.A. (1997) Theoretical description of 0↔0 and 0↔1 transitions in the Eu^{3+} ion in hosts with C_{2v} symmetry, *Mol. Phys.* **92**, 835-845.
32. Reid, M.F. and Richardson, F.S. (1983) Electric dipole intensity parameters for lanthanide 4f→4f transitions, *J. Chem. Phys.* **79**, 5735-5742.
33. Reid, M.F., Dallara, J.J., and Richardson, F.S. (1983) Comparison of calculated and experimental 4f→4f intensity parameters for lanthanide complexes with isotropic ligands, *J. Chem. Phys.* **79**, 5743-5751.
34. Malta, O.L. and Carlos, L.D. (2003) Intensities of 4f-4f transitions in glasses, *Quim. Nova* **26**, 889-895.
35. Wybourne, B.G. (1968) Effective operators and spectroscopic properties, *J. Chem. Phys.* **48**, 2596-2611.
36. Smentek, L. and Wybourne, B.G. (2000) Relativistic f↔f transitions in crystal fields, *J. Phys. B: At. Mol. Opt. Phys.* **33**, 3647-3651.
37. Smentek, L. and Wybourne, B.G. (2001) Relativistic f↔f transitions in crystal fields: II. Beyond the single-configuration approximation, *J. Phys. B: At. Mol. Opt. Phys.* **34**, 625-630.
38. Wybourne, B.G. and Smentek, L. (2002) Relativistic effects in lanthanides and actinides, *J. Alloys and Compounds* **341**, 71-75.
39. Peacock, R.D. (1975) The intensities of f↔f transitions, *Structure and Bonding* **22**, 83-122.
40. Smentek, L. (1998) Theoretical description of the spectroscopic properties or rare earth ions in crystals, *Physics Reports* **297**, 155--237.
41. Smentek, L. and Hess Jr., A. (2003) 40th Anniversary of the Judd-Ofelt Theory (special issue), *Mol. Phys.* **101** (7).
42. Kushida, T. (1972) Energy transfer and cooperative optical transitions in rare-earth doped inorganic materials: I. Transition probability calculation, *J. Phys. Soc. Japan* **34**, 1318-1326.
43. Kushida, T. (1972) Energy transfer and cooperative optical transitions in rare-earth doped inorganic materials: II. Comparison with experiment, *J. Phys. Soc. Japan* **34**, 1327-1333.
44. Kushida, T. (1972) Energy transfer and cooperative optical transitions in rare-earth doped inorganic materials: III. Dominant transfer mechanism, *J. Phys. Soc. Japan* **34**, 1334-1337.
45. Peacock, R.D. (1972) The sensitivity of computed Judd-Ofelt parameters to the particular transitions used, *Chem. Phys. Lett.* **16**, 590-592.
46. Peacock, R.D. (1971) Spectral intensities of the trivalent lanthanides. Part I. Solution spectra of heteropoly-complexes of Pr^{3+} and Ho^{3+}, *J. Chem. Soc.* (A), 2028-2031.
47. Peacock, R.D. (1972) Spectral intensities of the trivalent lanthanides. Part II. An assessment of the vibronic contribution, *J.C.S. Faraday II* **68**, 169-173.
48. Peacock, R.D. (1973) Spectral intensities of the trivalent lanthanides. Part III. The variation of the Judd-Ofelt parameters across an isostructural series, *Mol. Phys.* **25**, 817-823.
49. Jorgensen, C.K. and Reisfeld (1983) R. Judd-Ofelt parameters and chemical bonding *J. Less-Common Metals* **93**, 107-112.
50. Reisfeld, R. and Jorgensen, C.K. (1987) Excited state phenomena in vitreous materials, Chapter 58 *Handbook on the Physics and Chemistry of Rare Earths*, Elsevier Science Publishers B.V.
51. Edvardsson, S., Wolf, M., Thomas, J.O. (1992) Sensitivity of optical-absorption intensities for rare-earth ions, *Phys. Rev. B* **45** 10918-10923.
52. Wolf, M., Edvardsson, S., Zendejas, M.A., Thomas, J.O. (1993) Molecular-dynamics-based analysis of the absorption spectra of Nd^{3+}-doped Na^+ β″-alumina, *Phys. Rev. B* **48** 10129-10136.
53. Edvardsson, S., Ojamae, L., Thomas, J.O. (1994) A study of vibrational modes in Na^+ β -alumina by molecular dynamics simulation, *J. Phys. Condens. Matter* **6** 1319-1332.
54. Edvardsson, S., Klintenberg, M., Thomas, J.O. (1996) Use of polarized optical absorption to obtain structural information for Na^+/Nd^{3+} β″-alumina, *Phys. Rev. B* **54** 17476-17485.
55. Klintenberg, M., Edvardsson, S., Thomas, J.O. (1997) Calculation of energy levels and polarized oscillator strengths for Nd^{3+}:YAG, *Phys. Rev. B* **55** 10369-10375.

PERIODIC DIELECTRIC AND METALLIC PHOTONIC STRUCTURES

MARTIN WEGENER
Institut für Angewandte Physik, Universität Karlsruhe (TH), Wolfgang-Gaede-Straße 1, D-76131 Karlsruhe, Germany and Institut für Nanotechnologie, Forschungszentrum Karlsruhe in der Helmholtz-Gemeinschaft, D-76021 Karlsruhe, Germany, and DFG-Center for Functional Nanostructures, Universität Karlsruhe (TH), D-76131 Karlsruhe, Germany
martin.wegener@physik.uni-karlsruhe.de,

1. Introduction

In a *usual crystal*, the atoms are arranged in a periodic fashion with lattice constants on the order of 0.5 nm. This is orders of magnitude smaller than the wavelength of light. For example, green light has a wavelength of 0.5 μm. Thus, the light field experiences an effective homogeneous material; it does not "see" the underlying periodicity. In such materials, the phase velocity of light, c, is generally different from the vacuum speed of light, c_0, by a factor called the refractive index $n = c_o/c > 0$. The physical origins are microscopic electric dipoles that are excited by the electric field component of the incoming light and that radiate with a certain retardation. In contrast to this, magnetic dipoles play practically no role at optical frequencies in natural materials. In other words: The magnetic susceptibility is zero, the magnetic permeability is unity.

Metamaterials are artificial periodic structures with "lattice constants" that are still smaller than the wavelength of light. Again, the light field "sees" an effective homogeneous material. The "atoms", however, are not real atoms but are rather made of many actual atoms. This allows tailoring their properties in a way not possible with normal atoms, hence allowing for realizing electric as well as magnetic dipoles at optical frequencies. This can lead to highly unusual behavior. If, for example, the electric permittivity, ε, and the magnetic permeability, μ, are both negative, it turns out that the refractive index is negative as well, i.e., $n < 0$. Such materials are called left-handed materials [1]. They can be used to realize "perfect lenses" [2].

In a *Photonic Crystal*, the artificial lattice constant is comparable to the relevant wavelength of light. The resulting physics is partly analogous to that of an electron wave moving in the periodic potential of the nuclei in a normal crystal. This spatially modulates the electrons kinetic energy and its velocity. As the corresponding lattice constant is comparable to the electron de Broglie wavelength, this leads to Bloch waves with a special dispersion relation – the band structure of crystal electrons. For the case of a semiconductor crystal such as, e.g., silicon, one furthermore gets an energy gap. This so-called band gap is *the* key physical feature of semiconductors that has led to the "silicon age" and the information revolution throughout the last decades. In analogy, for a Photonic Crystal, i.e., for visible light, a periodic modulation of the velocity of light – hence of the refractive index – is required with lattice constants on the order of a few

B. Di Bartolo and O. Forte (eds.), Advances in Spectroscopy for Lasers and Sensing, 435–458.

hundred nanometers. If a Photonic Crystal reveals a band gap, i.e., if it has no states in a certain energy or frequency interval for whatever polarization of light and for all directions of propagation, it is called a *photonic band gap (PBG) material.* Such materials can be viewed as "semiconductors for light" [3,4,5,6].

In this article, we review some of the basics of Photonic Crystals and metamaterials starting from a rather elementary level. Furthermore, we give an overview on structures that have been fabricated in my group in Karlsruhe. After reminding ourselves on some basics in the remainder of this section, we group the material into periodic dielectric structures on the one hand (section 2) and periodic metallic structures on the other hand (section 3). We briefly conclude in section 4.

1.1 MAXWELL EQUATIONS

In Photonic Crystals and in metamaterials, the light waves follow the well-known Maxwell equations of electrodynamics. In S.I. units, they are given by

$$\vec{\nabla} \cdot \vec{D} = \rho \tag{1}$$

$$\vec{\nabla} \times \vec{E} = -\frac{\partial \vec{B}}{\partial t} \tag{2}$$

$$\vec{\nabla} \cdot \vec{B} = 0 \tag{3}$$

$$\vec{\nabla} \times \vec{H} = +\frac{\partial \vec{D}}{\partial t} + \vec{j} \tag{4}$$

ρ is the electric charge and $\vec{j}$ the electric current density. In macroscopic and homogeneous media, the relation between the $\vec{E}$-field and the $\vec{D}$-field is given by[1]

$$\vec{E} = \frac{1}{\varepsilon_0}(\vec{D} - \vec{P}), \tag{5}$$

with the macroscopic polarization $\vec{P}$ (the electric dipole density), and, similarly, for the $\vec{B}$-field and the $\vec{H}$-field the relation

$$\vec{B} = \mu_0 (\vec{H} + \vec{M}), \tag{6}$$

with the magnetization $\vec{M}$. In linear optics, one has

$$\vec{P} = \varepsilon_0 \chi \vec{E} \tag{7}$$

with the linear optical susceptibility χ. In this case, (5) simplifies to

$$\vec{D} = \varepsilon_0 \varepsilon \vec{E} \tag{8}$$

with the permittivity or dielectric function $\varepsilon = 1 + \chi$. Similarly, for the magnetization (the magnetic dipole density), one has

$$\vec{M} = \chi \vec{H}, \tag{9}$$

hence,

$$\vec{B} = \mu_0 \mu \vec{H}, \tag{10}$$

---

[1] $\varepsilon_0 = 8.8542 \times 10^{-12}\ \mathrm{AsV^{-1}m^{-1}}$ and $\mu_0 = 4\pi \times 10^{-7}\mathrm{VsA^{-1}m^{-1}}$

with the magnetic permeability $\mu = 1 + \chi$. In most textbooks on optics, the magnetic permeability μ is set to unity right away, which actually applies to most of not all naturally occurring optical materials. We will see later that a pronounced magnetic response at optical frequencies can, however, occur in metamaterials.
The Maxwell equations for $\vec{j} = 0$ and $\rho = 0$ can be rewritten into the known wave equation[2] for the $\vec{E}$ -field

$$\Delta\vec{E} - \frac{1}{c_0^2}\frac{\partial^2\vec{E}}{\partial t^2} = +\mu_0\frac{\partial^2\vec{P}}{\partial t^2} + \mu_0\frac{\partial}{\partial t}(\vec{\nabla}\times\vec{M}), \tag{11}$$

or, alternatively, using (7) and (9) and assuming constant (i.e. frequently-independent) electric and magnetic susceptibilities χ, respectively, into

$$\Delta\vec{E} - \frac{1}{c^2}\frac{\partial^2\vec{E}}{\partial t^2} = 0. \tag{12}$$

Here, $c = c_0/n$ is the medium velocity of light, which is slower than the vacuum velocity of light $c_0 = 1/\sqrt{\varepsilon_0\mu_0} = 2.998 \times 10^8$ m/s by a factor identical to the (generally complex) refractive index n with

$$n^2 = \varepsilon\mu \tag{13}$$

thus

$$n = \pm\sqrt{\varepsilon\mu} \tag{14}$$

Whether the plus or the minus sign is physically relevant, depends on the situation and remains to be seen. In most textbooks on optics, only the special case of nonmagnetic materials with $\mu = 1$ is discussed, which immediately leads to a positive refractive index, i.e., to $n = +\sqrt{\varepsilon}$ provided that the permittivity ε is real and positive.

1.2. POYNTING VECTOR, LIGHT INTENSITY, AND IMPEDANCE

Our eyes and most detectors are not sensitive to the electric field itself but to the number of photons which hit the detector per unit time. In other words, classically speaking: They are sensitive to the cycle-average of the modulus of the Poynting vector $\vec{S} = \vec{E}\times\vec{H}$. For plane waves in vacuum one has $|\vec{B}| = |\vec{E}|/c_0$ or equivalently $|\vec{E}| = |\vec{H}|\sqrt{\frac{\mu_0}{\varepsilon_0}}$, with the vacuum impedance

$$Z_0 = \sqrt{\frac{\mu_0}{\varepsilon_0}} = 376.7301\ \Omega, \tag{15}$$

[2] Coming from Karlsruhe, we just have to remind you that it was Karlsruhe where Heinrich Hertz found the first experimental evidence for electromagnetic waves in the year 1887.

with the vacuum impedance Z_0, leading to

$$S = \left|\vec{S}\right| = \sqrt{\frac{\varepsilon_0}{\mu_0}} \left|\vec{E}\right|^2 , \tag{16}$$

which generally varies with time. For an electric field according to, e.g.

$$\left|\vec{E}(t)\right|^2 = \widetilde{E}_0^2 \cos^2(\omega_0 t + \phi) , \tag{17}$$

the light intensity I, which is defined as the cycle-average[3] of the modulus of the Poynting vector, becomes

$$I = \langle S \rangle = \frac{1}{2} \sqrt{\frac{\varepsilon_0}{\mu_0}} \widetilde{E}_0^2 . \tag{18}$$

In a medium rather than in vacuum, the impedance Z of the electromagnetic wave is defined as

$$Z = +\sqrt{\frac{\mu_0 \mu}{\varepsilon_0 \varepsilon}} = +Z_0 \sqrt{\frac{\mu}{\varepsilon}} . \tag{19}$$

If a wave propagates from one medium with impedance Z_{in} into another medium with impedance $Z_{\text{out}} \neq Z_{\text{in}}$, a part of the wave is generally reflected from the interface. For the special case of $\mu = 1$ used in most optics textbooks, this is equivalent to saying that reflection occurs if the refractive indicies of the two materials are different. We already note at this point that the refractive indices of two materials can be different even if the impedances are the same. In this case, *no* reflection would occur. An example for this situation is vacuum as one material (with $Z = Z_0$ and $n = 1$) and $\varepsilon = 2$ and $\mu = 2$ as the other (with $Z = Z_0$ and $n = 2$).

2. Periodic Dielectric Photonic Structures

In this section we get started by considering the analogy between light and electron waves in one dimension in 2.1., we discuss negative refraction in two-dimensional Photonic Crystals in 2.2., and describe our efforts to fabricate high-quality three-dimensional Photonic Crystals and photonic band gap materials in subsection 2.3.

2.1. ONE-DIMENSIONAL PHOTONIC CRYSTALS

Some readers might already be familiar with crystals for electrons, i.e., with solutions of the stationary Schrödinger equation for an electron in a periodic potential. In one dimension it reads

$$-\frac{\hbar^2}{2m_e} \frac{\partial^2}{\partial x^2} \psi(x) + V(x)\psi(x) = E\psi(x) \tag{20}$$

equivalent to

---

[3] Remember that $\left\langle \cos^2(\omega_0 t + \phi) \right\rangle = 1/2$.

$$\frac{\partial^2}{\partial x^2}\psi(x)+\underbrace{\frac{2m_e}{\hbar^2}(E-V(x))}_{=a(x)}\psi(x)=0\,, \tag{21}$$

with the electron mass m_e, the potential energy $V(x)$, the eigenenergy E and the stationary wave function $\psi(x)$. At first sight, the wave equation for light (12) seems quite different. Its one-dimensional version is

$$\frac{\partial^2}{\partial x^2}E(x,t)-\frac{1}{c^2(x)}\frac{\partial^2}{\partial t^2}E(x,t)=0 \tag{22}$$

Note that we have tacitly assumed that the medium velocity of light c is constant with respect to the coordinate x or at least piecewise constant. Upon inserting the ansatz

$$E(x,t) = E(x)e^{-i\omega t} + \text{c.c.} \tag{23}$$

into the wave equation, we obtain

$$\frac{\partial^2}{\partial x^2}E(x)+\underbrace{\frac{\omega^2}{c_0^2}(x)}_{=a(x)}E(x)=0\,, \tag{24}$$

which is mathematically equivalent to (21) with $\psi(x) \leftrightarrow E(x)$ and $a(x)$ as indicated. Obviously, apart from a shift and a prefactor, the negative square of the refractive index, i.e., $-n^2(x)$, plays the role of the potential energy $V(x)$ for the light waves.

For example, a periodic stack of two alternating dielectric materials – just a dielectric mirror – is strictly equivalent to an electron moving in a periodic arrangement of finite depth potential wells. This problem is known as the *Kronig-Penney model* and is discussed in most elementary textbooks on solid-state physics. The free-electron model of solid-state physics can similarly be translated to photonics.

While this strict analogy between light and electron waves is restricted to the above assumptions (piecewise constant $n(x)$), we suspect that this analogy also holds qualitatively in a much broader sense. There are, however, obvious differences between light waves and matter waves, one of which is that the electric field is generally a vector whereas the electron wave function is a scalar. This becomes especially important in two and three dimensions.

2.2. TWO-DIMENSIONAL PHOTONIC CRYSTALS

Indeed, the general form of the wave equation suitable for treating two- and three-dimensional Photonic Crystals is more complex than (12) [7]. Using the ansatz

$$\vec{H}(\vec{r},t)=\vec{H}(\vec{r})e^{-i\omega t}+c.c., \tag{25}$$

we derive the general form of the eigenvalue equation from the Maxwell equations

$$\frac{1}{\mu(\vec{r})}\vec{\nabla}\times\left(\frac{1}{\varepsilon(\vec{r})}\vec{\nabla}\times\vec{H}(\vec{r},t)\right)=\frac{\omega^2}{c_0^2}\vec{H}(\vec{r},t)\,, \tag{26}$$

This equation is of the form

$$O(\vec{r})\vec{H}(\vec{r},t)=\frac{\omega^2}{c_0^2}\vec{H}(\vec{r},t)\,, \tag{27}$$

where $O(\vec{r})$ is a complicated but linear and hermitian operator. For periodic structures with

$$\varepsilon(\vec{r}+\vec{T})=\varepsilon(\vec{r}) \tag{28}$$

and

$$\mu(\vec{r}+\vec{T})=\mu(\vec{r}) \tag{29}$$

for any lattice translation vector $\vec{T}$, the operator is also lattice translational invariant, i.e.

$$O(\vec{r}+\vec{T})=O(\vec{r})\,. \tag{30}$$

In perfect analogy to solid-state physics, this leads to the Bloch theorem, i.e., the solutions for the magnetic component of the light field can be expressed as

$$\vec{H}(\vec{r})=h_{\vec{K}}(\vec{r})e^{i\vec{K}\vec{r}}\neq\vec{H}(\vec{r}+\vec{T})\,, \tag{31}$$

with the periodic function

$$h_{\vec{K}}(\vec{r}+\vec{T})=h_{\vec{K}}(\vec{r})\,. \tag{32}$$

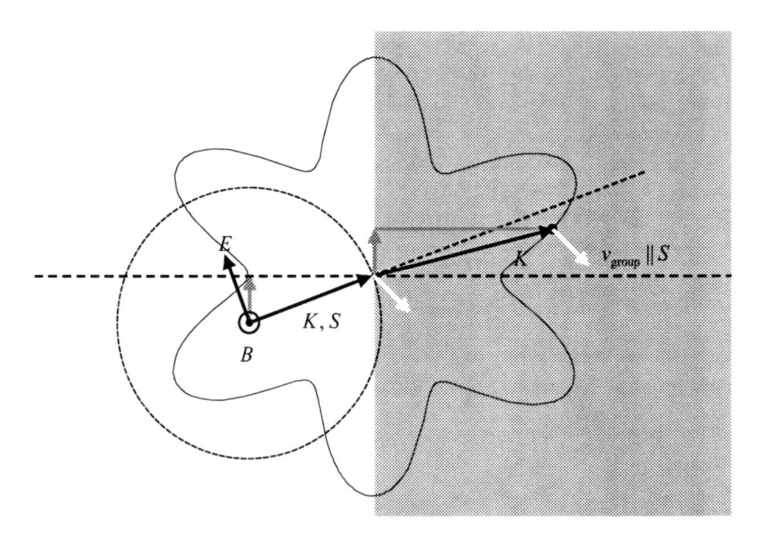

Figure 1: Illustration of negative refraction at the interface between a vacuum or air halfspace on the LHS and a Photonic Crystal halfspace on the RHS for p-polarization of the incident light.

Using numerical methods translated from solid-state physics, the eigenvalue problem (27) can be solved, leading to the dispersion relation $\omega(\vec{K})$, or using modern jargon, to the photonic band structure.

In analogy to electron band structures for semiconductors, photonic band structures can exhibit a band gap. As already pointed out in the introduction, such structures are called photonic band gap (PBG) materials. Much research effort has been focused onto designing and fabricating such structures. We will come back to this aspect in section 2.3.

Negative refraction. There are, however, other interesting effects of Photonic Crystals that are not related to band gaps. One of them is the phenomenon of *negative refraction*, e.g., discussed in [8]. Consider light impinging onto the surface of a Photonic Crystal under an angle with respect to the surface normal (see Fig. 1). For normal dielectrics, Snell's law of refraction tells us that light rays (i.e., the Poynting vectors of plane waves) are refracted towards the surface normal if the materials refractive index is larger than unity. What happens for a Photonic Crystal? It is clear that Snell's law cannot be applied directly because it assumes a homogeneous dielectric. For the fundamental beam, two quantities are conserved: The frequency of light ω and the tangential component of the wave vector of light. (For higher-order diffracted beams, the tangential component of the wavevector is only conserved modulo reciprocal lattice vectors.) To construct the wavevector of light inside the Photonic Crystal, one furthermore needs to know the iso-frequency curves (or surfaces), i.e., the relation $\omega(\vec{K}) = \text{const}$. These iso-frequency surfaces are the analogue of Fermi surfaces in usual solids. Generally, they can have a rather complex shape (see Fig. 1.). From these two conditions one gets a point in wavevector space on the iso-frequency surface. Importantly, the wavevector of light is generally not parallel to the direction of the flow of energy, the Poynting vector or the group velocity. The group velocity vector is rather given by

$$\vec{v}_{group} = \vec{\nabla}_{\vec{K}}\,\omega(\vec{K}) \parallel \vec{S}\,, \tag{33}$$

i.e., it is given by the gradient of the frequency, which is normal to the iso-frequency surfaces in wavevector space (RHS of Fig. 1). Depending on the parameters, this normal direction can point in about any direction. In particular, it is generally not parallel to $\vec{K}$. Furthermore, it can point into a direction corresponding to a negative material angle in Snell's law. This can be interpreted as an effective refractive index that is negative. Note that this effective refractive index not only depends on the material but also on the incident angle. Furthermore, note that this type of "refraction" is not refraction in the usual sense, it is rather Bragg diffraction of light by the Photonic Crystal into the lowest order mode. In section 3.3, we will encounter another distinct mechanism leading *to negative refraction*, which is not due to Bragg diffraction.

2.3. THREE-DIMENSIONAL PHOTONIC CRYSTALS

Our approach for the fabrication of three-dimensional Photonic Crystals and photonic band gap materials consists of three "ingredients": Holographic lithography of photoresist templates, direct laser writing of photoresist templates, and subsequent infiltration or double-infiltration with high-index materials (e.g., silicon).

Holographic lithography. The main idea of holographic lithography [9,10] is to expose a thick-film photoresist by a three-dimensional standing wave pattern originating from the interference of at least four laser beams.

We use the commercially available photoresist SU-8 from MicroChem. It consists of the epoxy EPON SU-8 and a photo initiator both dissolved in gamma-butyrolactone (GBL). Upon irradiation by near-UV light (350-400 nm), the photo initiator generates an acid. The spatial acid concentration is an image of the irradiation dose. In a post-exposure bake the latent picture is converted into a cross-linking density by cationic polymerization during this thermal treatment. The cross-linking degree determines the solubility in the "developing" solvent. GBL or another appropriate solvent is used in this step. Thus, sufficiently illuminated resin remains ("negative" photoresist) whereas underexposed resin is washed away. The remaining material therefore has a shape that follows the surface for the threshold irradiation which is the boundary between over- and under-exposed regions. Effectively, the interference pattern is stored in a digitized form: It is a porous air-polymer-structure showing the iso-dose surface for the threshold dose value.

The interference pattern with local intensity $I(\vec{r})$, resulting from the superposition of four light beams (plane waves), can be expressed as

$$I(\vec{r}) \propto \left| \sum_{n=1}^{4} \vec{E}_n e^{i(\vec{K}_n \vec{r} - \omega t)} \right|^2 = \left| \sum_{n=1}^{4} \sum_{m=1}^{4} a_{nm} e^{i\vec{G}_{nm}\vec{r}} \right|^2 \tag{34}$$

with the reciprocal lattice vectors

$$\vec{G}_{nm} = \vec{K}_n - \vec{K}_m \tag{35}$$

and the form factors

$$a_{nm} = \vec{E}_n \cdot \vec{E}^*_m \,. \tag{36}$$

The latter result from the relative amplitudes and polarizations of the four beams. It is clear that the reciprocal lattice vectors determine the translational symmetry of the crystal. Four wave vectors of light allow for three linearly independent reciprocal lattice vectors, hence allowing for three-dimensional crystals. The shape of the motif (in the language of crystallography) results form the form factors. In general, lattice and motif have different point symmetries, resulting in some overall crystal symmetry common to both. At present, there are mainly two four-beam configurations under discussion: The "two-planes" geometry has two pairs of beams traveling in planes perpendicular to each other (Fig. 2 (b)). In this geometry, the substrate has to be passed by two beams. Moreover, rather small lattice constants result since beams are counter-propagating. Therefore, we use the experimentally more convenient "umbrella-like" configuration in which three beams (#2, #3, #4) are equally distributed on a cone with a certain apex angle γ. The fourth beam (#1) is directed along the axis of the cone (Fig. 2 (a)).

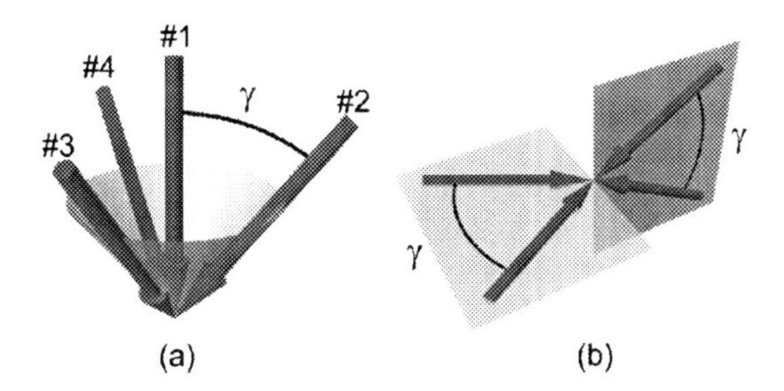

Figure 2: (a) The umbrella-like configuration: The central beam (#1) is along the axis of a cone with apex angle γ, three side beams (#2, #3, #4) are equally distributed on the cone. (b) The two-planes geometry: Two pairs of counter-propagating beams in two planes perpendicular to each other. Taken from [14].

While the above equations suggest that any of the 14 Bravais lattices in three dimensions should easily be accessible via holographic lithography, refraction of the incident laser beams at the air/photoresist interface inhibits that: If the angles have, for example, been properly adjusted to give fcc translational symmetry in air, an undesired trigonal translational symmetry results within the photoresist. Even for grazing incidence, desired symmetries such as fcc can *not* be achieved [11]. However, by adding an appropriately shaped dielectric object on the air/photoresist interface, refraction at this interface can be suppressed and, e.g., fcc translational symmetry becomes possible. Our corresponding results have been published in 2003 [11]. In this work, the desired fcc translational symmetry has been realized. Moreover, for the first time, optical spectra of Photonic Crystals made via holographic lithography have been published. The measured transmission spectra reveal a pronounced stop band around 700 nm wavelength, which agrees well with band structure calculations [11].

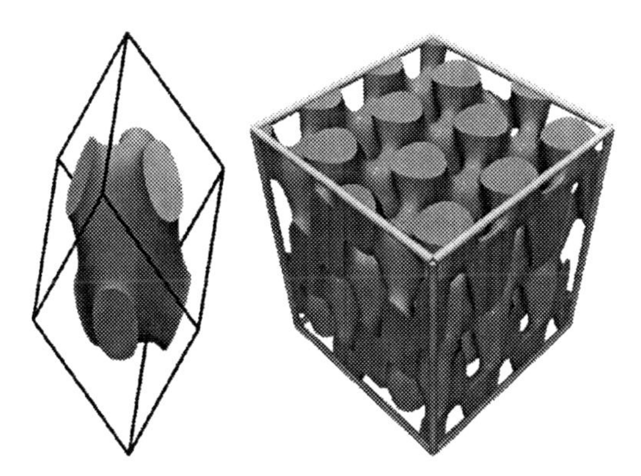

Figure 3: LHS: Motif of the rhombohedra structure within the rhombohedra unit cell for a silicon volume filling factor of 37%. RHS: Fragment of the corresponding crystal structure. Taken from [14].

The total number of free parameters in the holographic lithography of three-dimensional Photonic Crystals with four beams is 20: For each of the four beams one has 2 parameters for the orientation of the beam for fixed exposure wavelength, 1 parameter for the beam amplitude, and 2 additional parameters for the generally elliptical polarization. On the one hand, this offers an enormous degree of flexibility. On the other hand, this twenty-dimensional parameter space makes it truly difficult to actually identify parameter sets that eventually lead to complete three-dimensional band gaps and, at the same time, correspond to a sufficiently large contrast of the corresponding interference pattern. As already mentioned above, the umbrella geometry is to be preferred from an experimental viewpoint. Unfortunately, the first theoretical publications identifying complete gaps took advantage of the two-planes geometry [12, 13]. For the refractive index of silicon, they found gap/midgap ratios as large 23%. To date, none of these proposals has been realized experimentally. For the umbrella geometry, parameters were independently suggested for the first time by us [14] and by a collaboration of groups from MIT and from Bell Laboratories [15] in 2004. Our approach takes advantage of the fact that functions (essentially sums consisting of trigonometric functions) of known crystal symmetry are available from crystallography for all of the 230 space groups in three dimensions. These functions are given in the crystal coordinate system (generally not a Cartesian system), whereas the above interference pattern is quoted in the Cartesian laboratory frame. Thus, in a first step, the interference pattern has to be represented in the corresponding crystal system. In the second step, a comparison of the coefficients delivers the form factors (if possible), required for a certain space group. Band structure calculations with variable filling fraction follow. It turns out that an infinite number of combinations of parameters deliver exactly the same form factors. This freedom can be used to independently optimize another very important aspect for the practical realization, namely the interference contrast. The resulting algebraic equation can be solved analytically [14] and interference contrasts as large as ten are possible indeed. One of the novel solutions found by us is a crystal with fcc translational symmetry and rhombohedra crystal symmetry, which delivers a moderate gap/midgap ratio of 5.7% for infiltration with silicon at a volume filling fraction of 37% (see Fig. 3). This complete gap between the second and the third band (see Fig. 4) is expected to be quite robust. Interestingly, this structure – which roughly resembles the famous "Yablonovite" – is balanced. Fore example, for 50% filling fraction, this means that the structure is simply identical to its inverse (apart from a shift in space). Thus, complete band gaps can be achieved along these lines for inversion of the templates as well as for double-inversion – leaving both options available. In addition, a somewhat less attractive solution with cubic translational and crystal symmetry has been found as well [14].

Direct laser writing. Holographic lithography is capable of producing samples with defect-free areas of several square millimeters and offers a high flexibility in tailoring the interior of the unit cell. However, due to the fabrication principle, these structures are strictly periodic, i.e., defect and waveguide structures can not easily be included in a controlled fashion. Therefore, a second complementary technique is required to inscribe functional elements into Photonic Crystals provided by holographic lithography. An excellent candidate for this task is direct laser writing (DLW) through multi-photon polymerization [16, 17, 18].

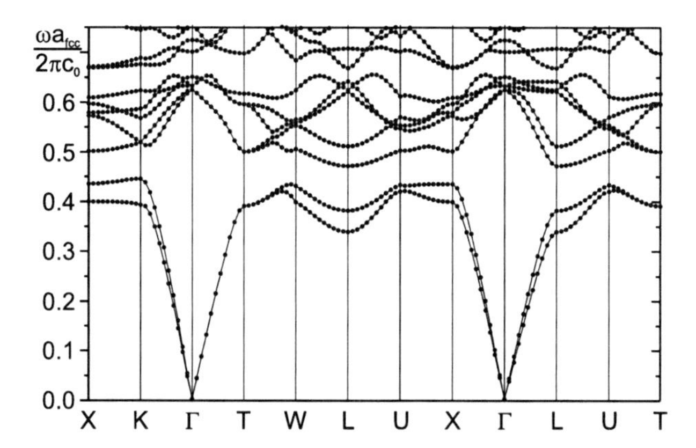

Figure 4: Band structure of the Photonic Crystal shown in Fig. 3. Note that there is a complete photonic band gap between the 2nd and the 3rd band as well as another complete gap between the 7th and the 8th band. Taken from [14].

To guarantee compatibility with holographic lithography, we again use the SU-8 negative photoresist [19,20]. DLW then simply may be added as a second exposure step to the holographic fabrication process prior to development of the photoresist. Additionally, the cationic polymerization mechanism of SU-8 is advantageous, since the monomers do not polymerize directly after exposure, leading to negligible refractive index contrast of exposed and unexposed photoresist. Therefore, structures that have already been exposed will not affect additional exposure steps. A further advantage is that SU-8 is solid during the writing process. This does not only make the sample handling much more comfortable, but also offers a higher degree of freedom in the scanning pattern, since successively written structures do not have to be interconnected immediately.

In multi-photon DLW, a photoresist exhibiting an intensity threshold for exposure is illuminated by laser light whose photon energy is insufficient to expose the photoresist by a one-photon absorption process. If this laser light is tightly focused into the resist, however, the light intensity inside a small volume element ("voxel") inclosing the focus may become sufficiently high to exceed the exposure threshold by multi-photon processes. By scanning the focus relative to the photoresist, in principle, any three-dimensional connected structure consisting of these voxels may be written directly into the photoresist. The size and shape of the exposed voxels depend on the isophotes of the microscope lens and the multi-photon exposure threshold of the photosensitive medium. The isophotes in the vicinity of the geometrical focus typically exhibit a near-ellipsoidal shape with a lateral diameter as small as 100 nm and an aspect ratio of about 2.7 for the parameters used by us.

Prior to exposure, a 20 μm thick SU-8 photoresist film is spin-coated on a microscope cover slide and solidified by a soft-bake process. The laser, a regeneratively amplified Ti:sapphire laser system (Spectra Physics Hurricane, pulse duration 120fs) is tuned to a central wavelength of 800 nm, where the one-photon absorption of SU-8 is negligible. The output beam is attenuated by a halfware-plate/polarizer combination and after beam expansion, typically a few ten nJ of single pulse energy are focused into the photoresist sample is placed on a three-axis piezoelectric scanning stage (Physic Instrumente) that provides a resolution of 5 nm at a full scanning range of 200 μm by 200 μm in the plane and 20 μm in the normal direction. A personal computer controls the scanning operation of the piezo stage and synchronizes its movement with the pulsing of the laser system via a laser controlling interface. After DKW of a pre-programmed pattern, the exposed sample is post baked and developed.

We have fabricated a large variety of different structures via DLW [19,20]. Here, we briefly discuss our work on woodpile structures [19]. Owing to their simple geometrical structure, the so-called layer-by-layer [6] or woodpile Photonic Crystals are ideal structures for DLW. Layers of straight parallel rods with a center-to-center distance are stacked to a three-dimensional lattice. Adjacent layers have the orientation of the rods rotated by 90 degrees and second-nearest neighboring layers are shifted by a distance of $a/2$ perpendicular to the rod axes. This stacking sequence repeats itself after every four layers with a lattice constant c. For $c/a = \sqrt{2}$, this lattice exhibits a fcc unit cell with a two-rod basis and can be derived from a diamond lattice by replacing the (110) chains of lattice points with rods. With our fabrication technique, these rods are built up of joining individual voxels. Fig. 5 shows electron micrographs of corresponding structures. It should be noted that these samples have lattice constants which are sufficiently small for telecommunication wavelengths and, at the same time, are sufficiently large in area for measurements as well as for potential devices. With the optimized writing procedure used by us, the exposure time for the Photonic Crystal shown in Fig. 5 (a) is about 25 minutes.

The optical spectra (not shown, see [19]) reveal pronounced stop bands. For the first time in DLW, we have achieved lattice constants sufficiently small for telecommunication wavelengths. The optical transmission is close to 100% for wavelengths longer than the stop band, already indicating excellent quality. This assessment is further confirmed by a direct comparison with calculated transmission spectra from the groups of Kurt Busch and Costas M. Soukoulis [19], using a scattering matrix formalism.

More recently, we have also fabricated Photonic Crystals from the "Slanted Pore" family for the first time [20].

Yet more recently, we have succeeded in replicating such photoresist templates by silicon with high fidelity using chemical-vapour-deposition (or atomic-layer deposition) techniques for the first time [21].

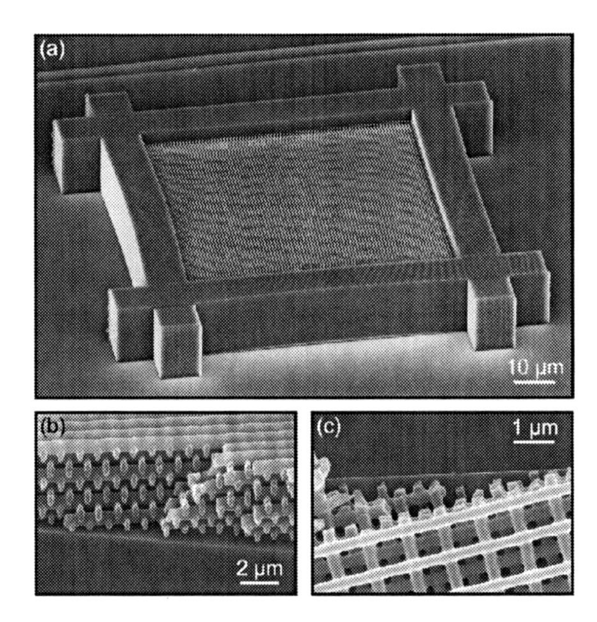

Figure 5: Three-dimensional Photonic Crystals fabricated by direct laser writing (DLW). (a) Layer-by-layer structure with 40 layers and a massive wall that prevents bending and reduces distortions due to polymer shrinkage during polymerisation, completely fabricated by DLW. (b) Side and (c) top view of a different broken sample with 12 layers, illustrating the sample quality obtained with the DLW process. Taken from [19].

3. Periodic Metallic Photonic Structures

In this section we address two different types of periodic metallic structures for photonics: metallic Photonic Crystal slabs in 3.1. and left-handed metamaterials in 3.2.

3.1. METALLIC PHOTONIC CRYSTAL SLABS

The metallic photonic crystal slabs as well as the metamaterials to be discussed below contain nanoscopic metallic structures. The simplest optical excitation of a sub-wavelength metallic structure is the *particle plasmon resonance* (or Mie resonance). This Lorentz oscillator-like resonance on the nanoscale replaces the Drude response of a bulk metal on the macroscopic scale. We briefly review this well-known phenomenon before proceeding.

Particle plasmon resonances. Let us consider the limit of a small dielectric or metallic particle in vacuum [24]. Its dimension a shall be much smaller than the wavelength of light, λ, i.e., $a << \lambda$. This is equivalent to the limit $\lambda \rightarrow \infty$. With the vacuum dispersion relation of light, $\lambda f = c_0$, this is further equivalent to the limit of vanishing frequency f

of light, $f \to 0$. In other words: A small particle exposed to an electromagnetic plane wave is equivalent to a particle between the plates of a plate capacitor in electrostatics. Here, the external electric field of the plate capacitor, $\vec{E}_{\rm ext}$, induces a polarization $\vec{P}$ in the particle, which gives rise to surface charges leading to the well-known depolarization field, i.e.,

$$\varepsilon_0 \vec{E}_{dp} = -\hat{N}\vec{P}\,, \tag{37}$$

with the depolarization matrix $\hat{N}$. For the special case of a sphere, the depolarization matrix reduces to a number and we have $\hat{N} = 1/3$. This allows us to write

$$\vec{P} = \varepsilon_0 \chi \vec{E} = \varepsilon_0 \chi (\vec{E}_{ext} + \vec{E}_{dp}) = \varepsilon_0 \chi \vec{E}_{ext} - \frac{1}{3}\chi \vec{P} \tag{38}$$

Solving for the polarization with respect to external electric field and introducing $\varepsilon = 1 + \chi$ leads to

$$\vec{P} = 3\varepsilon_0 \frac{\varepsilon - 1}{\varepsilon + 2}\vec{E}_{ext} \tag{39}$$

This expression obviously has a resonance for $\varepsilon(\omega) = -2$. For a metal, the Drude dielectric function (neglecting losses) is given by

$$\varepsilon(\omega) = 1 - \frac{\omega_{pl}^2}{\omega^2}\,, \tag{40}$$

hence the resonance condition is met for the frequency

$$\omega = \frac{1}{\sqrt{3}}\omega_{pl}\,, \tag{41}$$

which is lower than the metal plasma frequency $\omega_{\rm pl}$. For typical material parameters, the resonance frequency ω lies in or near the visible part of the spectrum.

If the particle is embedded in a dielectric medium with permittivity $\varepsilon_{\rm med}$, the particle plasmon resonance condition becomes $\varepsilon(\omega) = -2\varepsilon_{\rm med}$. For particles outside the electrostatic approximation, the overall qualitative behaviour remains the same, but the resonance frequency is lowered with increasing particle size. Particle shapes other than spheres lead to modified resonance via their respective depolarization factors.

The waveguide plasmon-polariton. In section 2, we have discussed Photonic Crystals made by a periodic arrangement of a dielectric material with frequency-independent refractive index and air. We have seen that, for frequencies near the Bragg-frequency, a stop band or even a band gap can evolve. What happens if we rather have a dielectric with a *frequency-dependent* refractive index $n(\omega)$? A strong frequency-dependence can, e.g., arise for a phonon resonance or for a particle plasmon resonance. In both cases, the dielectric function can be approximated by a Lorentz oscillator model to lowest order. If the frequency of light, the oscillator eigenfrequency and the Bragg frequency all coincide, we expect new and interesting features [23-26].

Figure 6 schematically shows the structure. Metal nanowires are arranged into a one-dimensional lattice with lattice constant a. Their coupling is mediated by a slab waveguide located underneath the nanowires. Using nanowires rather than spheres or dots offers the interesting possibility to conveniently control the coupling to light via the polarization of the incident light: If the electric field vector is oriented perpendicular to the nanowires (TM-polarization), a pronounced depolarization field arises, giving rise to a strong optical resonance. In contrast, if the electric field vector is along the wire axis (TE-polarization), the depolarization factor is zero and one rather gets a Drude-type response of the metal.

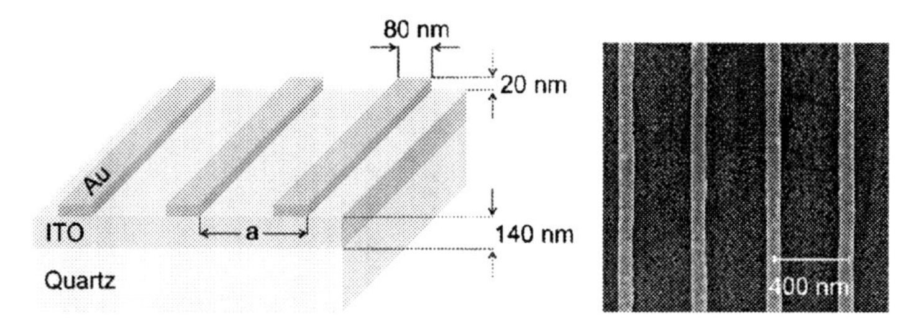

Figure 6: LHS: Scheme of a one-dimensional metallic Photonic Crystal slab, RHS: electron micrograph of an actual structure with lattice constant $a = 400$ nm. After [25].

In Fig. 7 we depict optical spectra (the negative logarithm of the optical transmission) for both TE and TM polarization. For TE polarization we observe an extinction peak that simply shifts with lattice constant. This peak is due to Bragg diffraction of the incident light into the waveguide mode, leading to reduced transmission in the forward direction. The behavior for the TM polarization is much more complex. Here two resonances, the particle plasmon and the Bragg resonance interact and lead to a pronounced avoided crossing. While the avoided crossing is expected for any two interacting harmonic oscillators (Lorentz oscillators), the actual lineshape of the extinction spectra reminds one of Fano lineshapes in quantum mechanics. These lineshapes have been explained by numerical calculations of the extinction spectra using a scattering matrix approach [24]. In the following we show that these features can be explained on a much simpler footing as well [26]. Broadly speaking, they are an interference phenomenon occurring for any two interacting oscillators, one of which is much narrower than the other one.

Consider two coupled oscillating particles of equal mass m. Newton's second law can easily be arranged into the form:

$$\ddot{x}_{pl} + 2\gamma_{pl}\dot{x} + \Omega_{pl}^2 x_{pl} - \Omega_c^2 x_{wg} = \frac{q_{pl}}{m} E(t)\,, \tag{42}$$

$$\ddot{x}_{wg} + 2\gamma_{wg}\dot{x}_{wg} + \Omega_{wg}^2 x_{wg} - \Omega_c^2 x_{pl} = \frac{q_{wg}}{m} E(t)\,. \tag{43}$$

Here, x_{pl} (t) and x_{wg} (t) are the displacements representing the plasmon and waveguide oscillations, respectively. The resonance frequencies, half-widths at half maximum, and charges (oscillator strength) of the uncoupled system are denoted by Ω_j, γ_j, and q_j (j = pl, wg), respectively. Ω^2_c represents the coupling strength between the oscillators.

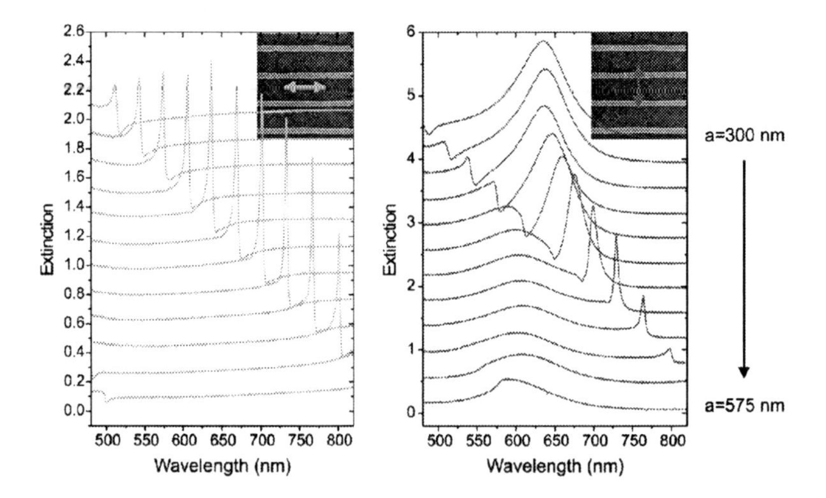

Figure 7: Normal-incidence, far-field extinction spectra for two orthogonal linear polarizations and for different lattice constants a from 300 nm to 575 nm in steps of 25 nm. The curves for different values of a are shifted for clarity. LHS: TE-polarization, i.e. incident electric field vector along the metal nanowires, RHS: TM-polarization, i.e. electric field vector perpendicular to the wires. After [25].

We consider the limit $\gamma_{\mathrm{wg}} << \gamma_{\mathrm{pl}}$ or $\gamma_{\mathrm{wg}} = 0$, which is crucial for the occurrence of Fano-like resonances. In order to make the resulting formulae transparent, we also chose $q_{\mathrm{wg}} = 0$, which means that the oscillator strength of the waveguide mode is much smaller than that of the particle plasmon. This is justified for the above experiments (see Fig. 7). Following along the usual lines of Lorentz oscillators, it is straightforward to derive the absorption coefficient

$$\omega^2 = \frac{N_{pl} q_{pl}^2}{V 2m\varepsilon_0 c_0} \frac{4\gamma_{pl}\omega^2(\omega^2 - \Omega_{wg}^2 - \frac{q_{wg}}{q_{pl}}\Omega_c^2)^2}{((\omega^2 - \Omega_{pl}^2)(\omega^2 - \Omega_{wg}^2) - \Omega_c^4)^2 + 4\gamma_{pl}^2\omega^2(\omega^2 - \Omega_{wg}^2)^2}, \tag{44}$$

which mimics the experimental extinction spectra. ε_0 is the vacuum permittivity, c_0 the vacuum speed of light, and N_{pl} the number of particle plasmon oscillators in volume V. Examples of absorption spectra are shown in Ref. [26]. One obtains the anticipated avoided crossing. For not too large damping, absorption maxima appear at the spectral positions ω given by

$$\omega^2 = \frac{\Omega_{pl}^2 + \Omega_{wg}^2}{2} \pm \sqrt{\frac{(\Omega_{pl}^2 - \Omega_{wg}^2)^2}{4} + \Omega_c^4}\ . \tag{45}$$

These positions coincide with the normal mode frequencies of the coupled, but undamped system. In contrast to frequent believe, however, the lineshape does not correspond to the sum of two effective Lorentz oscillators. One rather gets a highly asymmetric, Fano-like lineshape. Usually, a Fano resonance results from the coherent interaction of a discrete quantum mechanical state with a continuum of states. In our purely classical model, a single sharp oscillator coherently interacts with a strongly broadened second oscillator. The latter replaces the continuum. One result of the Fano-like interaction is that one obtains zero absorption in between the two absorption maxima. For $q_{\mathrm{wg}} \approx 0$, the position of this zero appears at the root of the numerator of (44), i.e. at the spectral position of the (uncoupled) waveguide mode, Ω_{wg}. Intuitively, this minimum is a result of destructive interference, which effectively suppresses the response of the two absorption "channels", because the plasmon and waveguide polarizations have a phase difference near π.

3.2. LEFT-HANDED METAMATERIALS

The second Maxwell equation tells us that the vectors $\vec{K}$, $\vec{E}$, and $\vec{B}$ (in that sequence or in any cyclic permutation) form a right-handed system, i.e., $(\vec{K} \times \vec{E}) \parallel \vec{B}$. For a plane wave in vacuum, we furthermore have that the wave vector of light $\vec{K}$ and the Poynting vector $\vec{S}$ point in the same direction. Thus, the vectors $\vec{S}$, $\vec{E}$, and $\vec{B}$ also form a right-handed system. In most optical materials, this statement is valid as well. However, in so-called left-handed materials, where the refractive index is negative, $\vec{S}$, $\vec{E}$, and $\vec{B}$ form a left-handed system. As Veselago [1] predicted materials like that many years ago, they are sometimes also called Veselago materials.

We start our discussion with the refraction of light at the interface between two materials, for example between vacuum and a medium with refractive index n. From the first and the third Maxwell equation, respectively, we get that the components of $\vec{D}$ and $\vec{B}$ normal to the surface are continuous (i.e., they make no jumps). From the second and the fourth Maxwell equation we get that the tangential components of $\vec{E}$ and $\vec{H}$ are also continuous. All other components generally do make jumps at the interface. Evaluating the details leads to the well-known Snell's law of refraction

$$\frac{\sin(\alpha_{vac})}{\sin(\alpha_{med})} = n\,, \tag{46}$$

with the angles between the Poynting vector and the surface normal in vacuum α_{vac} and in the medium α_{med}, respectively (as described in detail in any good optics textbook). While the first two cases are well-known, one is tempted to think that a negative index of refraction can not simply be inserted into Snell's law. Actually, what does a negative n mean? Can it occur in principle? How could a negative-index material be made?

In order to answer these fairly fundamental questions in some detail, we consider a simple but very instructive case, namely a fictitious medium with constant permittivity $\varepsilon = -1$ and constant magnetic permeability $\mu = -1$. We do not yet ask how it could be made, but we will see later that it can be made indeed. The impedance of this medium is $Z = Z_0$. Thus, a light wave impinging from vacuum onto this medium would be completely transmitted, no reflected wave occurs for any angle of incidence. This aspect makes the detailed discussion of the different components of the electromagnetic vectors very transparent. Let us first consider p-polarization of light, i.e., the electric-field vector lies in the plane of incidence. In this case, we have for the tangential (t) and normal (n) components of $\vec{E}$ and $\vec{B}$, respectively: $E_t \rightarrow E_t$, $E_n \rightarrow -E_n$, $B_t = 0$, and $B_n \rightarrow B_n$. This situation is visualized in Fig. 8. As always, $\vec{K}$, $\vec{E}$, and $\vec{B}$ form a right-handed system but $\vec{S}$, $\vec{E}$, and $\vec{B}$ form a left-handed system. In other words: the wave vector of light $\vec{K}$ and the Poynting vector $\vec{S}$ point in the opposite direction. In this sense, the phase velocity of the wave inside the medium is indeed negative as expected for a negative index of refraction. Also, the light wave is refracted towards the "wrong" side of the surface normal as expected for a negative index of refraction from Snellius law.

Hence, in the special case considered, we have to take the minus sign in $n = \pm\sqrt{\varepsilon\mu}$ (see section 1.1.) if both ε and μ are negative[4]. This statement turns out to be true beyond this special case. The discussion for the s-polarization, where the electric-field vector is normal to the plane of incidence, is analogous: $E_t \rightarrow E_t$, $E_n = 0$, $B_t \rightarrow -B_t$, and $B_n \rightarrow B_n$. This leads to the same conclusions as for the p-polarization.

Note that the negative refraction in a Photonic Crystal metamaterial with negative permittivity and negative permeability is distinct from the negative refraction in a

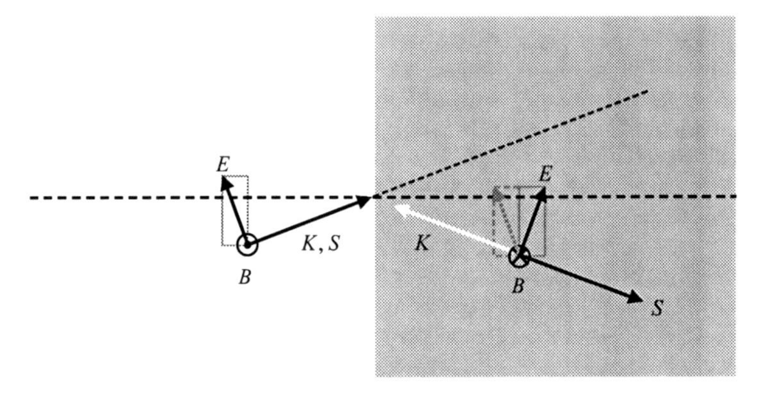

Figure 8: Illustration of negative refraction at the interface between a vacuum or air halfspace on the LHS and a left-handed material with $\varepsilon = \mu = -1$ on the RHS for p-polarization of the incident light. The resulting negative refraction leads to a refractive index of $n = -\sqrt{\varepsilon\mu} = -1$. Also note that the wavevector (white) and the Poynting vector (black) of light inside the left-handed material are opposite to one another. In this sense, the phase velocity of light is negative indeed.

---

[4] Note that the parameter combination that we have discussed, i.e. $\mathrm{Re}(\varepsilon) < 0$, $\mathrm{Re}(\mu) < 0$, and $\mathrm{Im}(\varepsilon) = \mathrm{Im}(\mu) = 0$, is a sufficient but not a necessary condition for $\mathrm{Re}(n) < 0$. For the case of substantial imaginary parts $\mathrm{Im}(\varepsilon) \neq 0$, and $\mathrm{Im}(\mu) \neq 0$, the condition $\mathrm{Re}(n) < 0$ (inherently accompanied by $\mathrm{Im}(n) > 0$, i.e., attenuation of the wave) can also be achieved for, e.g., $\mathrm{Re}(\mu) > 0$. As large imaginary parts are, however, rather undesired, we will not further get into this discussion here.

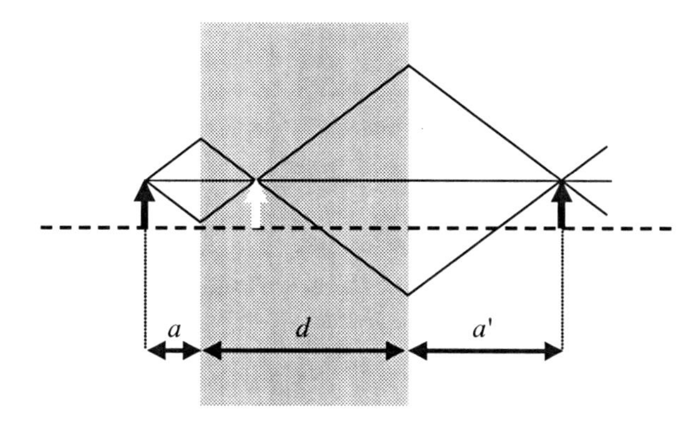

Figure 9: Illustration of a "perfect lens". The grey plate with thickness d in the center is a left-handed material with $\varepsilon = \mu = -1$ (see Fig. 8), the outer parts are vacuum or air with $\varepsilon = \mu = n = +1$. Ref. [2] showed that the sharpness of the image (RHS black arrow) of the object (LHS black arrow) is not restricted by the usual diffraction limit.

Photonic Crystal as discussed in section 2.2. The latter is actually due to diffraction, or in other words, it arises from interference of partial waves from different lattice points, whereas the light effectively averages over the different photonic "atoms" in a metamaterial.

"Perfect lenses". What are the consequences of negative refraction in a metamaterial? Let us consider an infinite parallel plate of thickness d made of the above medium with $\varepsilon = \mu = n = -1$, embedded in vacuum (or air) [2]. Fig. 9 shows selected rays from an object on the left-hand side of this plate at a distance of a from the LHS of the plate. This leads to a real image of the object on the RHS of the plate with a distance a' given by

$$d = a + a' \tag{47}$$

as indicated in Fig. 9, provided that $d > a$. For $d \leq a$, the image is virtual, e.g., $a' < 0$. This behaviour is closely similar to the equation for a usual "thin" spherical lens with focal length f

$$\frac{1}{f} = \frac{1}{a} + \frac{1}{a'}, \tag{48}$$

where a is the distance between the object and the lens and a' the distance between lens and the image of the object. For $a > f$, the image is real, for $a \leq f$, it is virtual, i.e., a' is negative. Obviously, the parallel plate with $n = -1$ serves as a lens. It is actually a "perfect lens" in the sense that there are no aberrations associated with imperfections of the geometrical shape (such as spherical aberrations for a usual lens) provided that a parallel plate can be fabricated without tolerances. More importantly, a more detailed analysis [2] shows that such a perfect lens can indeed beat the usual diffraction limit, i.e., the image can have sub-wavelength resolution. Chromatic aberrations, on the other hand, can still arise and would actually be horrible as will become obvious in just a moment. Still, such a "perfect lens" could possibly be useful for imaging a mask onto a photoresist in high-resolution photolithography at a fixed wavelength.

Nanoscopic split-ring resonators. We have seen that the key for a negative refractive index is to have a magnetic response with $\mu < 0$, while $\varepsilon < 0$ occurs in any metal for frequencies below the plasma frequency. Actually, getting any magnetic response at all is unusual as practically all naturally occurring materials have $\mu = 1$ at optical frequencies. Asking for a magnetic response means that some circulating current has to flow, which generates a magnetic field opposing the magnetic component of the light field. One possibility is to mimic a usual *LC*-circuit, consisting of a plate capacitor C and a magnetic coil with inductance *L*, on a scale much smaller than the relevant wavelength of light. Fig. 10 shows the analogy of a conventional *LC* circuit and a metallic split-ring resonator (SRR) on a dielectric surface. The RHS shows an electron micrograph of a single gold SRR fabricated by standard electron-beam lithography.

The anticipated LC-resonance frequency – where μ is expected to change sign – can be roughly estimated by the following extremely crude approach: Assume that we can describe the capacitance by the usual textbook formula for a large capacitor with nearby plates and the inductance by the formula for a "long" coil with N windings for $N = 1$. Using the nomenclature of Fig. 10B, i.e., the width of the metal w, the gap of the capacitor d, the metal thickness t, and the width of the coil l, we have

$$C = \varepsilon_0 \varepsilon_C \frac{wt}{d}, \tag{49}$$

with the effective permittivity of the material in between the plates ε_C and

$$L = \mu_0 \frac{l^2}{t}. \tag{50}$$

This leads to the eigenfrequency

$$\omega_{LC} = \frac{1}{\sqrt{LC}}, \tag{51}$$

and to the *LC*-resonance wavelength

$$\lambda_{LC} = \frac{2\pi c_0}{\omega_{LC}} = l 2\pi \sqrt{\varepsilon_C} \sqrt{\frac{w}{d}} \propto \text{size.} \tag{52}$$

Note that the thickness t of the metal film has dropped out. For relevant parameters ($\varepsilon_C = 1$ and $w \approx d$), this formula tells us that the *LC*-resonance wavelength is about an order of magnitude larger than the linear dimension of the coil l. Thus, it is possible to arrange these split-ring resonators in the form of an array such that the lattice constant a is much smaller than the resonance wavelength, i.e. $a \ll \lambda_{LC}$. Due to the close proximity of adjacent split-ring resonators, the degeneracy of the *LC* resonances is lifted, which leads to a certain broadening of the resonance that depends on the lattice constant a.

A second source of broadening of the *LC* resonance is the finite Ohmic resistance *R* in the *LC* circuit. It is well known that the differential equation for the electrical current I in an *LRC* circuit closely resembles that of a damped harmonic oscillator, i.e.

$$\ddot{I}+\frac{R}{L}\dot{I}+\frac{1}{LC}I=0, \tag{53}$$

where the ratio R/L plays the role of the Stokes damping. The total resistance R of the SRR shown in Fig. 10 is given by

$$R=\frac{4(l-w)-d}{wt\sigma}, \tag{54}$$

with the metals conductivity σ. Obviously, the resistance goes up when decreasing the size of the SRR, while L and C decrease at the same time. Hence, the ratio of damping and (undamped) resonance frequency (which is proportional to the quality of the resonance) is given by

$$\frac{R/L}{\omega_{LC}}\propto\frac{1}{size}. \tag{55}$$

This is true if all dimensions (*w, t, d, l*) are scaled sown by the same factor as well as if the metal film thickness t is kept constant. For the latter case, the resistance *R* would be size-independent. The bottom line of this fairly naive reasoning is that we expect Ohmic damping to become a problem for very small *SRR*, equivalent to large *LC* frequencies. This trend is further enhanced by the fact that the losses of metals tend to increase as the frequency approaches the plasma frequency (typically of order 1000 THz). A third contribution to the width of the magnetic resonance is radiation damping, which also increases with increasing frequency. Finally, a fourth broadening mechanism is inhomogeneous broadening due to fabrication tolerances of the SRR: While there is no intrinsic frequency dependence, one has to push fabrication technology to the limits to achieve sufficiently small structures. This leads to an effective (extrinsic) increase of the influence of inhomogeneous broadening with increasing resonance frequency.

Historically, the first demonstration of left-handed materials was in 2001 at around 10 GHz frequency [27], where the split-ring resonators can easily be fabricated on electronic circuit boards. The negative permittivity was achieved by additional metal stripes. In 2004 [28], $\mu < 0$ has been demonstrated at about 1 THz frequency using standard micro fabrication techniques for the SRR.

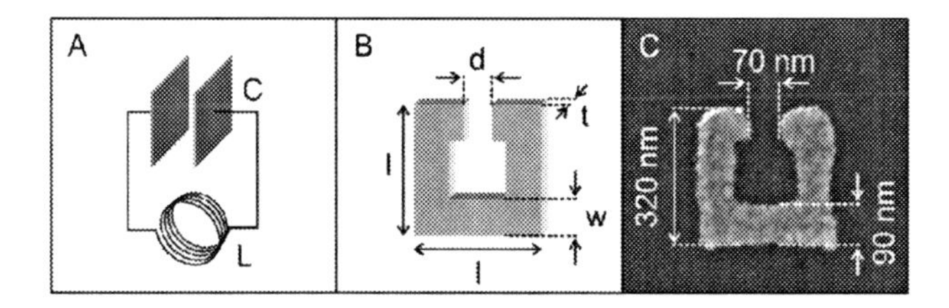

Figure 10: Illustration of the analogy between a usual *LC* circuit, A, and split-ring resonators (SRR), B. The electron micrograph in C shows an actually fabricated structure with gold SRR ($t = 20$ nm) on a glass substrate. Taken from [29].

More recently, we have fabricated nanoscopic split-ring resonators [29] (see Fig. 10). Using the above formulae, we estimate *LC* resonance frequencies of about 3 μm wavelength. The spectroscopic results are discussed in Ref. [29]. All features of the measured spectra are reproduced by numerical calculations using a three-dimensional finite-difference time-domain approach [29].

4. Summary

We have seen that artificial periodic dielectric and/or metallic structures offer fascinating opportunities for tailoring the optical properties of materials by taking advantage of nanofabrication tools. This allows for achieving properties that simply do not occur in natural materials (e.g., photonic band gaps, negative refraction, or left-handed behaviour). In this brief review we have attempted to give a first flavour by discussing selected examples rather than going into very much detail for any particular system. It seems more than likely that many more scientifically interesting periodic photonic structures will be discussed in the next decade. Also, the nonlinear optical and quantum optical properties of many of these structures are not well explored yet. In parallel, those structures which do appear to be attractive for applications have to make the transition from science into the "real world".

5. Acknowledgements

Thanks to Prof. Dr. B. Di Bartolo and all the participants of the school for another wonderful time in Erice. The scientific results (both experimental and theoretical) described in this article have mainly been obtained by the young scientists Georg von Freymann and Stefan Linden as well as by the graduate students Christiane Becker, Markus Deubel, Christian Enkrich, Martin Hermatschweiler, Matthias Klein, Daniel Meisel, and Sean Wong. We acknowledge cooperations with the theory groups of Sajeev John (Toronto), Costas Soukoulis (Iowa State), Kurt Busch (Karlsruhe and CREOL, Florida), and Sven Burger (Matheon, Berlin), with the chemistry group of Geoffrey Ozin (Toronto) and with the experimental physics groups of Dagmar Gerthsen (Karlsruhe), Jürgen Kuhl (MPI Stuttgart) and Harald Giessen (Bonn). We acknowledge support by the Deutsche Forschungsgemeinschaft (DFG) through subprojects A1.4 and A1.5 of the DFG-Forschungszentrum "Functional Nanostructures" (DFG-CFN) and via the DFG Leibniz-award 2000 for M.W.

6. References

1. Veselago, V.G. (1968) The electrodynamics of substances with simultaneously negative values of ε and μ, *Sov. Phys. Usp.* **10**, 509
2. Pendry, J.B. (2000) Negative refraction makes a perfect lens, *Phys. Rev. Lett.* **85**, 3966
3. John, S. (1987) Strong localization of photons in certain disordered dielectric superlattices, *Phys. Rev. Lett.* **58**, 2486

4. Yablonovitch, E. (1987) Inhibited spontaneous emission in solid-state physics and electronics, *Phys. Rev. Lett.* **58**, 2059
5. Ho, K.-M., Chan, C.T., and Soukoulis, C.M. (1990) Existence of a photonic gap in periodic dielectric structures, *Phys. Rev. Lett.* **65**, 3152
6. Ho, K.-M., Chan, C.T., Soukoulis, C.M., Biswas, R., and Sigalas, M. (1994), Photonic band gaps in three dimensions: new layer-by-layer periodic structures, *Solid State Comm.* **89**, 413
7. Joannopoulos, J.D., Meade, R.D., and Winn, J.N. *Photonic Crystals*, Princeton University Press
8. Kosaka, H., Kawashima, T., Tomita, A., Notomi, M., Tamamura, T., Sato, T., and Kawakami, S. (1998) Superprism phenomena in photonic crystals, *Rhys. Rev. B.* **58**, R10096
9. Campbell, F., Sharp, D.N., Harrison, M.T., Denning, R.G., and Turberfield, A.J. (2000) Fabrication of photonic crystals for the visible spectrum by holographic lithography, *Nature* **404**, 53
10. Shoji, S. and Kawata, S. (2000) Photofabrication of three-dimensional photonic crystals by multibeam laser interference into a photopolymerizable resin, *Appl. Phys. Lett.* **76**, 2668
11. Miklyaev, Y.V., Meisel, D.C., Blanco, A., von Freymann, G., Busch, K., Koch, W., Enkrich, C., Deubel, M., and Wegener, M. (2003) Three-dimensional face-centered-cubic photonic crystal templates by laser holography: fabrication, optical characterization, and band-structure calculations, *Appl. Phys. Lett.* **82**, 1284
12. Maldovan, M., Urbas, A.M., Yufa, N., Carter, W.C., and Thomas, E.L. (2002) Photonic properties of bicontinuous cubic microphases *Phys. Rev. B.* **65**, 165123
13. Toader, O., Chan, T.Y.M., and John, S. (2004) Photonic band gap architectures for holographic lithography, *Phys. Rev. Lett.* **92**, 043905
14. Meisel, D.C., Wegener, M., and Busch, K. (2004) Three-dimensional photonic crystals by holographic lithography using the umbrella configuration: Symmetries and complete photonic band gaps, *Phys. Rev. B* **70**, 165104
15. Ullal, C.K., Maldovan, M., Thomas, E.L., Chen, G., Han, Y.-J., and Yang, S. (2004) Photonic crystals through holographic lithography: Simple cubic, diamond-like, and gyroid-like structures, *Appl. Phys. Lett.* **84**, 5434
16. Sun, H.-B., Matsuo, S., and Misawa, H. (1999) Three-dimensional photonic crystal structures achieved with two-photon-absorption photopolymerization of resin, *Appl. Phys. Lett.* **74**, 786
17. Kawata, S., Sun, H.-B., Tanaka, T., Takada, K. (2001) Finer features for functional microdevices, *Nature* **412**, 697
18. Straub, M. and Gu, M. (2002) Near-infrared photonic crystals with higher-order bandgaps generated by two-photon photopolymerization, *Opt. Lett.* **27**, 1824
19. Deubel, M., v. Freymann, G., Wegener, M., Pereira, S., Busch, K., and Soukoulis, C.M. (2004) Direct laser writing of three-dimensional photonic-crystal templates for telecommunications, *Nature Materials* **3**, 444
20. Deubel, M., Wegener, M., Kaso, A., and John, S. (2004) Direct laser writing and characterization of "Slanted Pore" photonic crystals, *Appl. Phys. Lett.* **85**, 1895
21. Tétreault, N., von Freymann, G., Deubel, M., Hermatschweiler, M., Pérez-Willard, F., John, S., Wegener, M., and Ozin, G.A. (2005) New route towards three-dimensional photonic bandgap materials: silicon double inversion of polymeric templates, submitted
22. Kreibig, U. and Vollmer, M. (1995) *Optical Properties of Metal Clusters*, Springer-Verlag, Berlin
23. Linden, S., Kuhl, J., and Giessen, H. (2001) Controlling the interaction between light and gold nanoparticles: selective suppression of extinction, *Phys. Rev. Lett.* **86**, 4688
24. Christ, A., Tikhodeev, S.G., Gippius, N.A., Kuhl, J., and Giessen, H. (2003), Waveguide-plasmon polaritons: strong coupling of photonic and electronic resonances in a metallic photonic crystal slab, *Phys. Rev. Lett.* 91, 183901

25. Linden, S., Rau, N., Neuberth, U., Naber, A., Wegener, M., Pereira, S., Busch, K., Christ, A., and Kuhl, J. (2005), Near-field optical experiments on one-dimensional metallic photonic crystal slabs, *Phys. Rev. B.*, in press
26. Klein, M., Tritschler, T., Wegener, M., and Linden, S. (2005) Lineshape of harmonic generation on metal nanoparticles and metallic photonic crystal slabs, *Phys. Rev. B*, in press
27. Shelby, R.A., Smith, D.R., Schultz, R. (2001) Experimental verification of a negative index of refraction, *Science* **292,** 77
28. Yen, T.J., Padilla, W.J., Fang, N., Vier, D.C., Smith, D.R., Pendry, J.B., Basov, D.N., Zhang, X. (2004) Terahertz magnetic response from artificial materials, *Science* **303**, 1494
29. Linden, S., Enkrich, C., Wegener, M., Zhou, J., Koschny, T., and Soukoulis, C.M. (2004) Magnetic response of metamaterials at 100 Terahertz, *Science* **306**, 1351

COOLING AND TRAPPING OF ATOMS

CARL E. WIEMAN
University of Colorado
Boulder, Colorado USA
cwieman@Jila.colorado.edu,

Cooling and trapping of atoms produces samples of atoms with greatly increased phase space density (i.e. more atoms within a given velocity range and volume). Cold, trapped atoms have a variety applications in precision spectroscopy and measurement. For example, they have been used for ultra-sensitive, ultra-selective detection of rare isotopes (even in the presence of large backgrounds) and to produce significantly better atomic clocks. A new type of matter, a macroscopic quantum wave function call a Bose-Einstein condensate, was produced when atoms were cooled to well below 1 microkelvin.

This series of lectures described the physics of laser cooling and trapping of atoms, practical techniques for accomplishing laser cooling and trapping, and limitations of different approaches. It then described the technique of evaporative cooling for producing ultracold samples and the creation of Bose-Einstein condensates.

The two Resource Letters listed below provide an excellent set of references on these subjects, including a number of books and review articles.

N. R. Newbury and C. Wieman, "Resource letter TNA-1: Trapping of neutral atoms," Am. J. Phys. **64**, pp. 18-20 (1996).

D. S. Hall, "Resource Letter: BEC-1: Bose–Einstein condensates in trapped dilute gases," Am. J. Phys. **71**, pp. 649-660 (2003).

B. Di Bartolo and O. Forte (eds.), Advances in Spectroscopy for Lasers and Sensing, 459.

NON-LINEAR PROPAGATION OF FEMTOSECOND TERAWATT LASER PULSES IN AIR AND APPLICATIONS

JEAN-PIERRE WOLF
LASIM (UMR CNRS 5579), University Claude Bernard Lyon 1
43, Bd du 11 Novembre 1918
69622 Villeurbanne Cedex (France)
wolf@lasim.univ-lyon1.fr
&
GAP-Biophotonics, University of Geneva
20, rue de l'Ecole de Médecine
1211 Geneva 4 (Switzerland)

1. Introduction

The advent of ultra-short and ultra-intense lasers has recently allowed experiments that involve highly non-linear effects even in weakly non-linear optical media. A particular case of such interactions is the non-linear propagation of terawatt (TW) laser pulses in the atmosphere and the related phenomena, such as filamentation and supercontinuum generation. Filamentation is a self-guided propagation mode characterized by tiny localized structures (typ. 100 μm in diameter) that propagate over hundreds of meters and transport intensities as high as 10^{14} W/cm^2. While propagating, the filamented laser light generates a coherent supercontinuum analogous to a "white light laser" that can be used for atmospheric analysis. In this chapter, we describe the basic phenomena leading to filamentation and supercontinuum generation, as well as atmospheric applications such as multi-pollutants Lidar (Light Detection and Ranging) measurements, detection of bacteria in air, and aerosols identification using quantum control schemes.

Air pollution is an extremely dynamic phenomenon, which makes difficult its understanding and therefore its control. This dynamic behavior appears in physical terms by the spread and transport of emitted pollutants as well as chemically through the many reactions that take place in the atmosphere. It is thus of outstanding importance to characterize the impacts of different pollution sources (industries, vehicles, but also sources of bacteria) on the environment and on public health. Recent public safety issues linked to bioterrorism and epidemic spread (e.g; SARS, legionnellosis) are acute examples of this need. Presently existing devices, although very sensitive, only provide spot measurements at the ground level. Three-dimensional information, reflecting the dynamic character of pollution, has been for a long time sorely lacking.

B. Di Bartolo and O. Forte (eds.), Advances in Spectroscopy for Lasers and Sensing, 461–482.

These considerations have induced the development of the Lidar technique [1-4], and in particular of Differential Absorption Lidar (DIAL). Lidar devices allow for selective measurements of pollutants concentration over several kilometers, range resolved like a Radar (Radiowaves Detection And Ranging), and in an interactive manner (i.e. without the usual need for samples). It then became possible to get "3D-maps" of concentrations, which display the propagation, spread and chemical evolution of the emitted pollution.

In spite of the great success of the Lidar technique for monitoring the atmosphere [1-4], it is still limited by the following drawbacks: (1) Since the principle of operation of conventional Lidar systems is based on quasi-isotropically distributed scatter processes like Rayleigh, Mie, Raman and fluorescence, only a very small fraction of the signal is captured by the receiver, limiting the detectability of trace gases from large distance (2) The pollutants to be detected are chosen *a piori* by selecting the laser wavelengths sent into the atmosphere (which prevents an actual chemical *analysis*), (3) Lidar soundings in the IR are usually restricted to aerosol containing atmospheres, because of the strongly decreasing Rayleigh scattering cross-section with wavelength (λ^{-4}) and (4) The characterization of aerosol properties with Lidar is limited to estimations of concentration and size distribution [5-7], with strong *a priori* assumptions; Almost no information about its composition can be retrieved.

In 1998, a pioneering experiment showed that most of the above mentioned difficulties might well be overcome by employing atmospheric white light filamentation [8,9]. They shined intense ultra-short pulses from a high-power femtosecond laser (220 mJ 100 fs) vertically into the sky and observed for the first time the phenomenon of white light generation in an extended plasma channel in the atmosphere. Time-resolved measurements of the backscattered white light using a Lidar set-up showed atmospheric features at altitudes as high as 12 km. This experiment opened a new field for Lidar investigations: the white light femtosecond Lidar technique based on the non-linear propagation of ultra-short and ultra intense laser pulses in the atmosphere.

2. Non-linear Propagation of Ultra-Intense Laser Pulses

High power laser pulses propagating in transparent media undergo nonlinear propagation. Nonlinear self-action leads to strong evolutions of the spatial (self focusing [10-11], self guiding [12], self reflection [13]), spectral (four wave mixing [14], self phase modulation [15-17]) as well as temporal (self steepening [18], pulse splitting [19]) characteristics of the pulse. The propagation medium is also affected, as it is partially ionized by the propagating laser beam [20-23]. Those phenomena have been extensively studied since the early 1970's, from the theoretical as well as from the experimental point of view. However, it was only in 1985 that the development of the chirped pulse amplification (CPA) technique [24,25] permitted to produce ultrafast laser pulses that reach intensities as high as 10^{20} W/cm^2 and hence to observe highly nonlinear propagation even in only slightly nonlinear media such as atmospheric pressure gases. We will focus here on non-linear propagation in air and processes related to coherent white-light generation and filamentation.

2.1. KERR EFFECT INDUCED SELF-FOCUSING

For high intensities *I*, the refractive index *n* of the air is modified by the Kerr effect [10,11]:

$$n(I) = n_0 + n_2 \cdot I \quad (1)$$

where n_2 is the non-linear refractive index of air ($n_2 = 3.10^{-19}$ cm^2/W). As the intensity in a cross-section of the laser beam is not uniform, the refractive index increase in the center of the beam is larger than on the edge (Fig. 1a). This induces a radial refractive index gradient equivalent to a lens (called 'Kerr lens') of focal length *f(I)*. The beam is focused by this lens, which leads to an intensity increase, and thus to a shorter focal length lens, a.s.o. until the whole beam collapses. Kerr self-focusing should therefore prevent propagation of high power lasers in air.

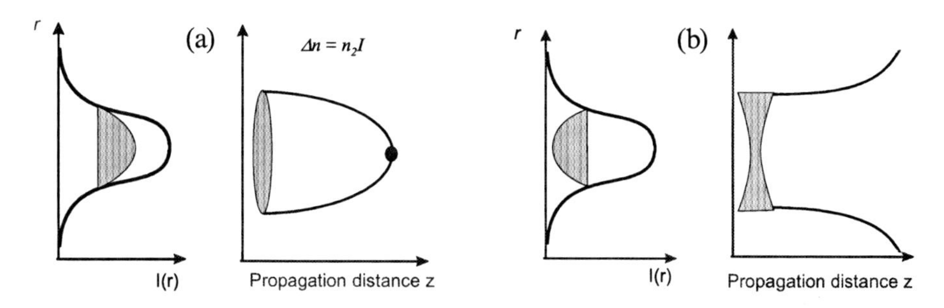

Figure 1: Kerr self-focusing (a) and plasma defocusing (b) of a high power laser beam.

The Kerr effect becomes significant when self-focusing is larger than natural diffraction, i.e. above a critical power *Pcrit*:

$$P_{crit} = \frac{\lambda^2}{4\pi \cdot n_2} \quad (2)$$

It should be pointed out that this is a critical *power* rather than a critical intensity. For a titanium-sapphire laser ($\lambda = 800$ nm) in air, $P_{crit} \sim 2$ GW (a more detailed treatment shows that in case of pulses shorter than 100 fs, $P_{crit} \sim 6$ GW). However, the distance at which the beam is focused is related to the successive focal lengths *f(I)*, and is a function of the initial *intensity*.

2.2. MULTIPHOTON IONIZATION AND PLASMA GENERATION

If the laser pulse intensity reaches 10^{13}–10^{14} W/cm^2, higher order non-linear processes occur such as multiphoton ionization (MPI). At 800 nm, typically 10 photons are needed to ionize N_2 and O_2 molecules and give rise to plasma [26]. The ionization process can involve tunneling as well, because of the very high electric field carried by the laser pulse. Following Keldysh's theory [27], MPI dominates for intensities lower than 10^{14} W/cm^2. In contrast to longer pulses, fs pulses combine high ionization efficiency due to their very high intensity, and a moderate overall energy, so that the generated electron densities (10^{16}–10^{17} cm^{-3}) are far from saturation. Losses by inverse Bremsstrahlung are therefore negligible, in contrast with nanosecond (or longer pulse)-laser generated plasma. However, the electron density ρ induces a negative variation of the refractive index, and because of the radial intensity profile of the laser beam, a negative refractive index gradient. This acts as a negative lens, which defocuses the laser beam, as schematically shown in figure 1b.

2.3 FILAMENTATION OF HIGH POWER LASER BEAMS

Kerr self-focusing and plasma defocusing should prevent long distance propagation of high power laser beams. However, a remarkable behavior occurs in air, where both effects exactly compensate and give rise to a self-guided quasi-solitonic [28] propagation. The laser beam is first self-focused by Kerr effect. This focusing then increases the beam intensity and generates a plasma by MPI, which in turns defocuses the beam. The intensity then decreases and the plasma generation stops, which allows Kerr re-focusing to take over again. This dynamic balance between Kerr effect and plasma generation leads to the formation of stable structures called "filaments" (Fig. 2). Light filaments were first observed by A. Braun *et al.* [29], who discovered that mirrors could be damaged by high-power ultrashort laser pulses even at large distance from the laser source. These light filaments have remarkable properties. In particular, they can propagate over several hundreds of meters, although their diameter is only 100-200 μm (thus widely beating the usual diffraction limits), and have almost constant values of intensity (typ. 10^{14} W/cm^2), energy (some mJ), and diameter.

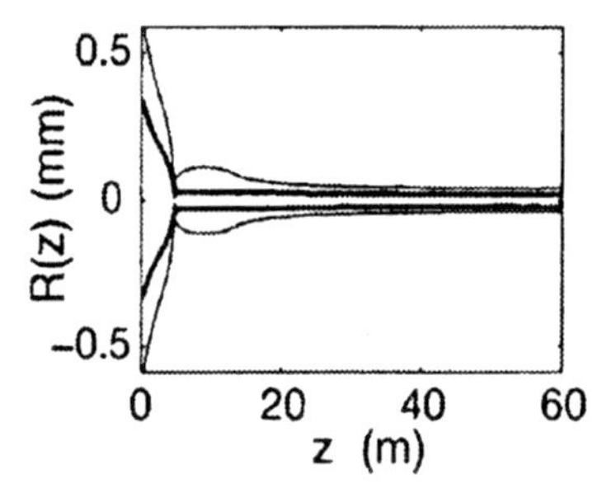

Figure 2: Filamentation of a high power laser beam as it propagates in air [28].

The laser pulse propagation is governed by the Maxwell wave equation:

$$\nabla^2 E - \frac{1}{c^2} \cdot \frac{\partial^2 E}{\partial t^2} = \mu_0 \cdot \sigma \cdot \frac{\partial E}{\partial t} + \mu_0 \cdot \frac{\partial^2 P}{\partial t^2} \tag{3}$$

where σ is the conductivity that accounts for losses, and P the polarization of the medium. In contrast with the linear wave propagation equation, P contains a self-induced non-linear contribution corresponding to Kerr focusing and plasma defocusing:

$$P = P_L + P_{NL} = \varepsilon_0 \cdot (\chi_L + \chi_{NL}) \cdot E \tag{4}$$

where χ_L and χ_{NL} are the linear and non-linear susceptibilities, respectively. Considering a radially symmetric pulse propagating along the z-axis in a reference frame moving at the group velocity v_g yields the following non-linear Schrödinger equation (NLSE) [28]:

$$\nabla_{\perp}^2 \varepsilon + 2i\left(k\frac{\partial \varepsilon}{\partial z}\right) + 2k^2 n_2 \cdot |\varepsilon|^2 \cdot \varepsilon - k^2 \frac{\rho}{\rho_c} \cdot \varepsilon = 0 \tag{5}$$

where $\varepsilon = \varepsilon\,(r, z, t)$ is the envelope of the electric field and ρ_c the critical electron density (1.8 10^{21} cm^{-3} at 800 nm [28]). ε is assumed to vary slowly as compared to the carrier oscillation and to have a smooth radial decrease. In this first order treatment, group velocity dispersion (GVD) and losses due to multiphoton and plasma absorption are neglected ($\sigma = 0$). In (5), the Laplacian models wave diffraction in the transverse plane, while the two last terms are the non-linear contributions: Kerr focusing and plasma defocusing (notice the opposite signs). The electronic density ρ (r, z, t) is computed using the rate equation (6) in a self-consistent way with (5):

$$\frac{\partial \rho}{\partial t} - \gamma |\varepsilon|^{2\alpha} (\rho_n - \rho) = 0 \tag{6}$$

where ρ_n is the neutral molecular concentration in air, γ the MPI efficiency and α the number of photons needed to ionize an air molecule (typ. $\alpha = 10$ [26]). Numerically solving the NLSE equation [28] leads to the evolution of the pulse intensity $I = |\varepsilon|^2$ as a function of propagation distance, as shown in Fig. 2. Initial Kerr lens self-focusing and subsequent stabilization by the MPI-generated plasma are well reproduced by these simulations. Notice that the filamentary structure of the beam, although only 100 µm in diameter, is sustained over 60 m. Numerical instability related to the high non-linearity of the NLSE prevents simulations over longer distances.

For higher laser powers $P >> P_{crit}$ laser beams break up into several localized filaments. The intensity in each filament is clamped to 10^{13} -10^{14} W/cm^2 corresponding to a few mJ, so that an increase in power leads to the formation of more filaments. Figure 3 shows a cross-section of laser beams undergoing mono-filamentation (Fig 3a, 5 mJ) and multi-filamentation (Fig. 3b, 400 mJ). The stability of these quasi-solitonic structures are remarkable: filaments have been observed to propagate horizontally over kilometric distances [41].

Many theoretical studies have been carried out to simulate the non-linear propagation of high power laser beams, both in the mono-filamentation [30-38] and in the multi-filamentation [39-41] regimes. The reader is encouraged to refer to these studies to get further insight in this subject.

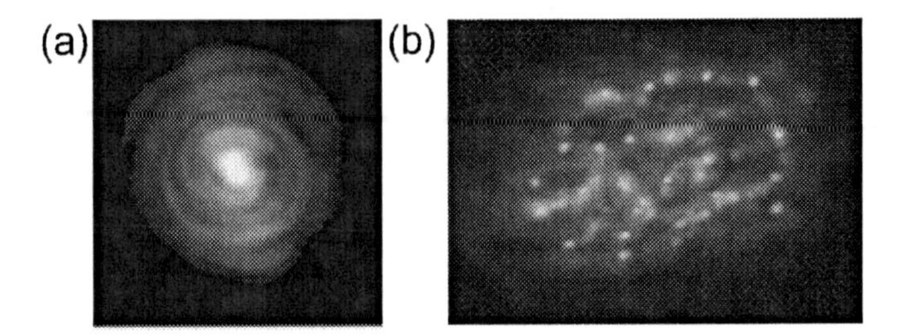

Figure 3: (a) Monofilamentation (5mJ) and (b) multifilamentation (400 mJ) of high power laser beams. Scale: The size of the filament in Fig. 3(a) (200 μm) is similar to one of the many filaments on Fig. 3b. The rings around the filaments are colourful and correspond to conical emission.

2.4 SUPERCONTINUUM GENERATION AND SELF-PHASE MODULATION

The spectral content of the emitted light is of particular importance for Lidar applications. Non-linear propagation of high intensity laser pulses not only provides self-guiding of the light but also an extraordinary broad continuum spanning from the UV to the IR. This supercontinuum is generated by self-phase modulation as the high intensity pulse propagates. As depicted above, Kerr effect leads, because of the spatial intensity gradient, to self-focusing of the laser beam. However, the intensity also varies with time, and the instantaneous refractive index of the air is modified as:

$$n(t) = n_0 + n_2 I(t) \tag{7}$$

This gives rise to a time dependent phase shift $\phi = -n_2 I(t)\, \omega_0\, z/c$ (where ω_0 is the carrier frequency), which generates new frequencies ω in the spectrum $\omega = \omega_0 + d\phi/dt$. The smooth temporal envelope of the pulse induces thus a strong spectral broadening about ω_0. Fig. 4 shows the spectrum emitted by filaments that were created by the propagation of a 2 TW pulse in the laboratory. The supercontinuum spans from 400 nm to over 4 μm, which covers absorption bands of many trace gases in the atmosphere (methane, VOCs, CO_2, NO_2, H_2O, etc..). Recent measurements showed that the supercontinuum extends in the UV down to 230 nm (see below), due to efficient third harmonic generation (THG) and frequency mixing [42,43]. These results open further multispectral Lidar applications to aromatics, SO_2, and ozone.

2.5 ANGULAR DISTRIBUTION OF THE SUPERCONTINUUM

Most of the filamentation studies showed that white light was generated in the filamentary structure, and was leaking due to coupling with the plasma in form of a narrow cone in the forward direction (called "conical emission", see fig 3a) [44,45]. This cone spans from the longer wavelengths in the center to the shorter wavelengths at the edge, with a typical half-angle of 0.12°.

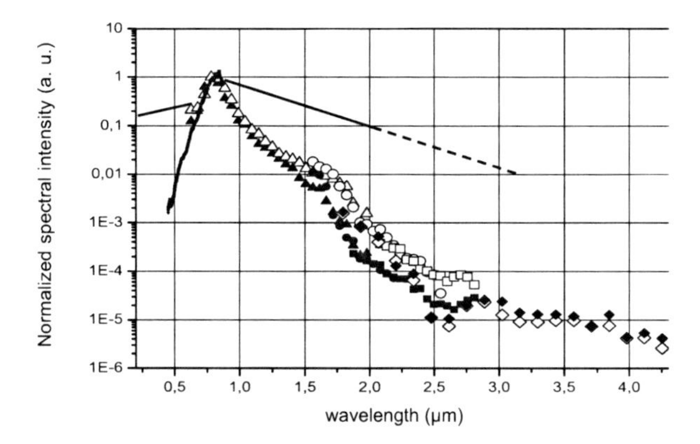

Figure 4: Supercontinuum generation in air; Dots are measurements in the laboratory with a few meters propagation while the lines correspond to further broadening when the laser propagates over kilometric distances [51]

An important aspect for Lidar applications is the angular distribution of the white light continuum in the near backward direction. Already in the first fs Lidar experiments, a pronounced backscattering component of the emitted white light was observed. For this reason angular resolved scattering experiments have been performed. The emission close to the backward direction of the supercontinuum from light filaments was found to be significantly enhanced as compared to linear Rayleigh-Mie scattering [46]. The enhancement may qualitatively be attributed to a co-propagating, self-generated longitudinal index gradient due to plasma generation, inducing a backreflection. Combined with self guiding that drastically reduces beam divergence, this aspect is extremely important for Lidar experiments: most of the white light emitted from the distance R is then collected by the Lidar receiver, unlike conventional elastic scattering of white light sources (e.g; flashlamps based Lidars, [47]).

To summarize, non-linear propagation of TW-laser pulses exhibits unique properties for multispectral Lidar measurements: extremely broadband coherent light emission ("white light laser"), confined in a self-guided beam, and back-reflection to the emitter as the laser pulse propagates.

3. The "TERAMOBILE" Project

On the basis of the first experiments [8,9] using femtosecond white light in a Lidar arrangement, a large frame french-german project, called "Teramobile" (for "Terawatt laser in a mobile system") was launched in 1999. This project, which is jointly funded by the CNRS and the DFG, federates four laboratories: The Unversity Jena (Prof. R. Sauerbrey), University Berlin (Prof. L. Woeste), University Lyon (Prof. J.P. Wolf) and the ENSTA (A. Mysyrowicz). The aim of the project was to construct the first mobile Terawatt-Laser based Lidar system, investigate fundamental processes like long range propagation and filamentation, and new possibilities of sounding the atmosphere.

The Teramobile system [48,49] consists of a femtosecond-terawatt laser and a multispectral Lidar detection system integrated in an air conditioned container. Since mobility is a crucial aspect as well as a strong constraint, particular care has been dedicated to the design of the ISO 20 mobile container-laboratory. The system only needs cooling water and power as external supplies (Diesel power generator and cooling system are separated), which allows full stand-alone operation.

The general structure of the container is divided in two compartments, a laser room and a control room. The container is equipped with 4 windows: two emission windows, in the roof and in the side wall, located in the laser room, and two detection windows, of the same diameter as the receiving telescope, located in the control room. They permit horizontal as well as vertical emission of the laser beam.

The heart of the system is a femtosecond-terawatt CPA laser system provided by Thales Laser (formerly BMI-Thomson). The femtosecond pulses (few nJ, 60 fs) are generated by a Kerr-lens mode locked Ti:Sapphire oscillator, pumped by a cw Nd:YLF laser. The pulses are then stretched using a grating arrangement to about 500 ps, in order to prevent damages in the amplifier chain. Amplification is performed by a regenerative amplifier, a multipass preamplifier, and a final multipass amplifier, pumped by two Nd:Yag laser of 1 J energy each at 10 Hz. The amplified pulses are then recompressed by a second pair of gratings to obtain 60 fs pulses of very high intensity (330 mJ, 5.5 TW). A key feature of the grating compressor is that a chirp can be set to the output laser pulses: the spectral components of the pulse can be ordered in time (for example the shorter wavelength components precede the longer wavelength components, called "negative chirp"). The pulse duration, and thus the peak intensity, can also be modified in this way.

A laser emitting telescope is placed over the laser table. It expands the laser beam diameter from 5 cm to 15 cm in order to reduce the fluence at the exit window. The focal length of the emitting telescope is adjustable from 10 m to infinity, permitting to vary the initial laser divergence which is important for Lidar experiments and long-range filamentation studies.

The Lidar receiver includes a double telescope, composed of two primary mirrors (one vertical and one horizontal) of 400 mm diameter and a switchable secondary mirror, which permits to either select the vertical or the horizontal detection direction. The Lidar detection set-up uses a set of fast detectors (ICCD for the visible and fast InSb photodiode for the infrared up to 5 μm) and a VIS-IR spectrometer for multi-spectral Lidar measurements. The whole detection system is integrated in a compact unit held on a vertical breadboard, which is rigidly attached to the laser table on mounts crossing the separation wall.

For eye-safety considerations, the femtosecond Lidar measurements are performed vertically. Moreover, a stand alone low power eye-safe Lidar (from Elight Laser Systems) with a wide field of view is attached to the Teramobile, in order to control that no aircraft enters the measurement region.

4. White Light Femtosecond Lidar Measurements

As filamentation counteracts diffraction over long distances, it allows to deliver high laser intensities at high altitudes and over long ranges. This behavior contrasts with linear propagation, for which the laser intensity decreases monotonely as the pulse propagates away from the source.

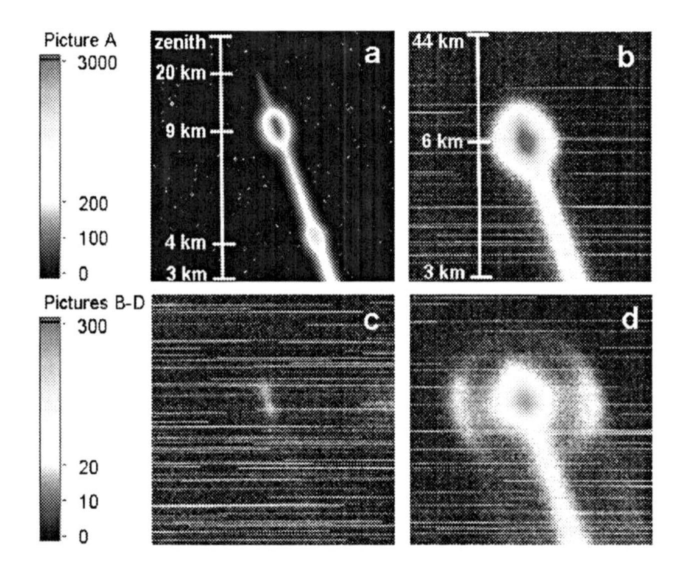

Figure 5: Long-distance filamentation and control of non-linear optical processes in the atmosphere [56]. Pictures of the beam propagating vertically, imaged by the CCD camera of a 2-m telescope. a): Fundamental wavelength, visible up to 25 km through 2 aerosol layers. b) – d): Supercontinuum (390-490 nm band) generated by 600-fs pulses with respectively negative (GVD precompensating), positive, and slightly negative chirp. On picture d), conical emission appears as a ring on the high-altitude haze layer.

The distance R0 at which high powers are reached and thus where filamentation starts, can be controlled by the following laser parameters: the initial laser diameter and divergence, and the pulse duration and chirp. The geometrical parameters are set by the transmitting telescope while the temporal parameters, leading to "temporal focusing" are determined by the grating compressor. A particular aspect is temporal focusing using an initial chirp, as it can be used together with the air GVD to obtain the shortest pulse duration (and thus the onset of filamentation) at the desired location R0. The compressor is then aligned in such a way that a negatively chirped pulse is launched into the atmosphere, i.e. the blue component of the broad laser spectrum precedes its red component. In the near infrared, air is normally dispersive, and the red components of the laser spectrum propagate faster than the blue ones. Therefore, while propagating, the pulse shortens temporally and its intensity increases. At the preselected altitude R0, the filamentation process starts and white light is generated.

Chirp-based control of the white-light supercontinuum generation has been demonstrated using the high-resolution imaging mode of the 2 m diameter telescope of the Thüringer Landessternwarte Tautenburg (Germany). For these experiments, the Teramobile laser was placed next to the astronomical telescope. The laser beam was launched into the atmosphere and the backscattered light was imaged through the telescope. Figure 5a shows a typical image at the fundamental wavelength of the laser (λ = 800 nm), over an altitude range from 3 to 20 km. In this picture, strong scattering is observed from a haze layer at an altitude of 9 km (and weaker around 4 km). In some cases, scattered signal could be detected from distances up to 20 km. Turning the same observation to the blue-green band (390 to 490 nm), i.e. observing the white-light supercontinuum, leads to the images shown in Fig 5b and 5c. As observed, filamentation and white light generation strongly depends on the initial chirp of the laser pulse, i.e. white-light signal can only be observed for adequate GVD precompensation (Fig. 5b). With optimal chirp parameters, the white light channel could be imaged over more than 9 km. It should also be pointed out that, as presented above, the angular distribution of the emitted white light from filaments is strongly peaked in the backward direction, and that most of the light is not collected in this imaging configuration.

Under some initial laser parameter settings, conical emission due to leakage out of the plasma channel could also be imaged on a haze layer, as shown in Figure 5d. Since conical emission is emitted sidewards over the whole channel length, the visible rings indicate that under these experimental conditions, the channel was restricted to a shorter length at low altitude.

This femtosecond white-light laser is an ideal source for Lidar applications. Recall that linear processes like Rayleigh-, fluorescence or Mie-scattering return only a small fraction of the emitted light back to the observer. This necessarily leads to an unfavorable $1/R^2$ – dependency of the received light, where R is the distance from the scatter location to the observer. When spectrally dispersed, this usually leaves too small signals on the receiver, as arc-lamp-based Lidar experiments have shown in the past [47]. Unlike these linear processes, the strong backward emission from white light channels, as described above, allows high spectral resolution of the observed signals, even at large distances. As a result, spectral fingerprints of atmospheric absorbers along the light path can be retrieved.

The schematic principle of a corresponding white light Lidar experiment is depicted in figure 6. It shows the TW fs-laser pulse, which – after passing the compressor set as a chirp generator – is vertically sent into the atmosphere.

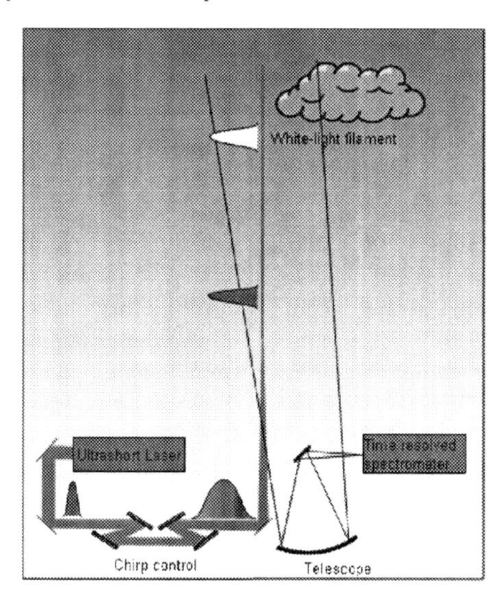

Figure 6: Principle of operation of a "white light" femtosecond Lidar.

Depending on the chirp-preset, filamentation starts at a distance R0 as mentioned before, and spans over several hundreds meters distance. The resulting white light, which co-propagates up to very high altitudes, is detected using a telescope and a multichannel spectrometer with a gated ICCD camera. While propagating in the atmosphere, the supercontinuum is partially absorbed by the atmospheric gases, and measuring the spectrum as a function of altitude provides information on the concentration of the absorbing species. The resolution in altitude is obtained by measuring the spectrum as a function of time, i.e. the flight time of the laser pulse, as for other Lidar systems.

Figure 7 shows examples of spectrally filtered white light Lidar returns, in three different spectral regions (visible at 600 nm and UV around 300 and 270 nm, 1000 shots average). These profiles of in-situ generated white-light show scatter features of the planetary boundary layer. The much faster decrease of the UV signal is due to the stronger Rayleigh scattering at shorter wavelengths and absorption at 270 nm (compared to 300 nm) due to high ozone concentration.

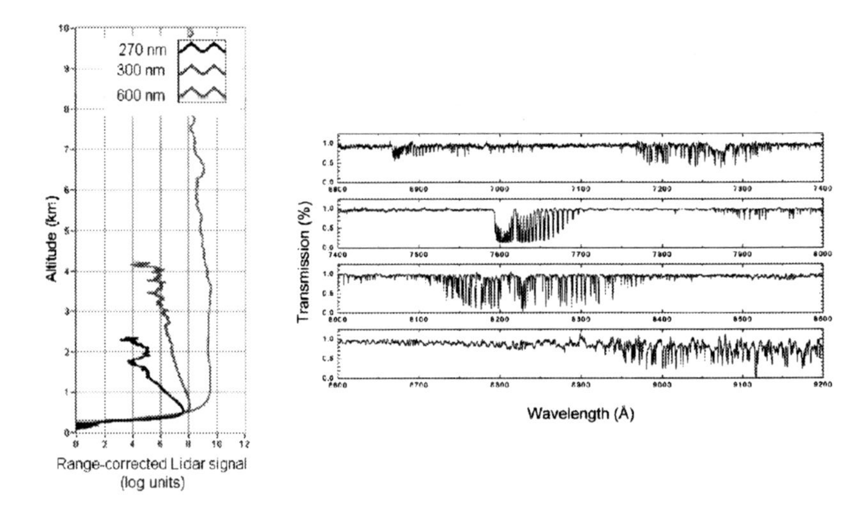

Figure 7: White light Lidar returns [1] O_3 absorption at 270 nm is clearly visible.

Figure 8: White light Lidar resoved spectrum returned from 4.5 km altitude [1].

The white light spectrum generated over long distances in the atmosphere shows significant differences (Fig. 4, [51]) with the one previously recorded in the laboratory [50]: The infrared part of the spectrum (recorded with filters) is significantly stronger (full line, typ. 2 orders of magnitude higher in Fig. 4) than in the laboratory, which is very encouraging for future multi-VOCs detection. A quantitative explanation of this IR-enhancement requires the precise knowledge of the nonlinear propagation of the terawatt laser pulse, which can not be simulated with the present numerical codes. However, it qualitatively indicates that the pulse shortens and/or splits while propagating, introducing wider frequency components in the spectrum.

On the other end of the spectrum (not shown), it was observed that the supercontinuum extends continuously down to 230 nm (limited by the spectrometer). This UV-part of the supercontinuum is the result of efficient third harmonic generation in air [42,43] and mixing with different components of the Vis-IR part of the spectrum. This opens very attractive applications, such as multi-aromatics (Benzene, Toluene, Xylene,..) detection (without interference), NOx and SO_2 multi-DIAL detection, and O_3 measurements, in which the aerosol interference can be subtracted (broadband UV detection) [52].

Very rich features arise from the white light backscattered signal, when it is recorded across a high-resolution spectrometer, as shown in Figure 8. The spectrum, which was detected from an altitude of 4.5 km with an ICCD, shows a wealth of atmospheric absorption lines at high resolution (0.01 cm^{-1}). The excellent signal to noise ratio (2000 shots average) demonstrates the advantages of using a high brightness white light channel for multi-component Lidar detection.

The well-known water vapor bands around 720 nm, 830 nm and 930 nm are observed simultaneously. Depending on the altitude (i.e. the water vapor concentration), the use of stronger or weaker absorption bands can be used. A fit of the spectrum around 815 nm of water vapor using the Hitran database leads to a mean water vapor concentration of 0.4%. Excellent agreement was found with the database spectrum, demonstrating that no non-linear effects or saturation are perturbing the absorption spectrum. This is explained by the fact that the white light returned to the Lidar receiver is not intense enough to induce saturation, and that the volume occupied by the filaments (the white light sources) is very small compared to the investigated volume.

Information about atmospheric temperature (and/or pressure) could be retrieved from the lineshapes of the absorption lines. Another possibility is to measure the intensities of the hot bands, in order to address the ground state population. As the molecular oxygen spectrum is very well known, O_2 is particularly well suited for this purpose. The access of the whole spectrum should again allow to obtain significantly better precision than in former DIAL investigations [53,54]. The spectrum covered by the white light gives access to many bands to measure the H_2O concentration and 2 bands of O_2 ((0)"->(0)' and (0)"->(1)' sequences of the X–>A transition, around 760 nm and 690 nm respectively) to determine the atmospheric temperature. The combination of both information with good precision could give rise to an efficient "relative humidity Lidar profiler" in the future.

5. Non-Linear Interactions with Aerosols

Particles are present in the atmosphere as a broad distribution of size (from 10 nm to 100 μm), shape (spherical, fractal, crystals, aggregates, etc.), and composition (water, soot, mineral, bioagents, e.g., bacteria or viruses, etc.). The Lidar technique has shown remarkable capabilities in fast 3D-mapping of aerosols, but mainly only qualitatively through the measurement of statistical average backscattering and extinction coefficients. The most advanced methods use several wavelengths (typ. 2-5), usually provided by the fixed outputs of standard lasers (Nd:Yag, Ti:Sapphire, Excimers, CO_2). The set of Lidar equations derived from the obtained multiwavelength Lidar data is subsequently inverted using sophisticated algorithms or multiparametric fits of pre-defined size distributions with some assumptions about the size range and complex refractive indices. In order to obtain quantitative mappings of aerosols, complementary local data (obtained with, e.g., laser particle counters, or multi-stage impactors to identify the sizes and composition) are often used together with the Lidar measurements. The determination of the size distribution and composition using standard methods must, however, be taken cautiously as complementary data, because of its local character in both time and space.

Remote range resolved "all optical" measurement of aerosols without *a priori* assumptions remains therefore an important challenge, for both environmental and strategic reasons. It is of particular interest to remotely detect and identify aerosol populations such as traffic related soot particles and/or biological agents in a background of harmless aerosols.

5.1 MULTI-PHOTON-EXCITED FLUORESCENCE (MPEF) AND MULTI-PHOTON-IONIZATION (MPI) IN AEROSOL MICROPARTICLES

Femtosecond laser pulses provide very high pulse intensity at low energy, which allows inducing non-linear processes in particles without deformation due to electrostrictive and thermal expansion effects.

The most prominent feature of non-linear processes in aerosol particles is strong localization of the emitting molecules within the particle, and subsequent backward enhancement of the emitted light [55,56]. This unexpected behavior is extremely attractive for remote detection schemes, such as Lidar applications. Localization is achieved by the non-linear processes, which typically involve the n-th power of the internal intensity $I^n(\mathbf{r})$ (**r** for position inside the droplet). The backward enhancement can be explained by the reciprocity (or "time reversal") principle: Re-emission from regions with high $I^n(\mathbf{r})$ tends to return toward the illuminating source by essentially retracing the direction of the incident beam that gave rise to the focal points. Backward enhancement has been observed for both spherical and non spherical [57] microparticles.

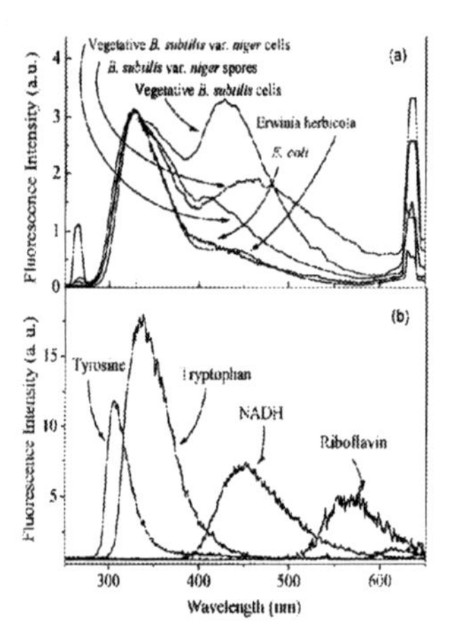

Figure 9: Fluorescence spectrum of bacteria and respective contributions of amino acids, NADH and flavins [61].

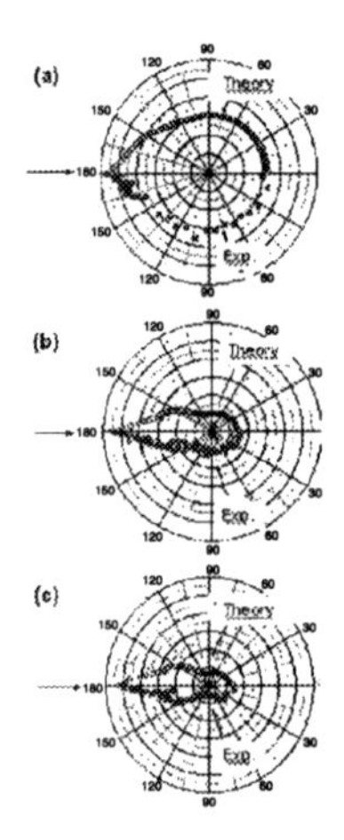

Figure 10: Backward enhancement of the fluorescence from aerosol microparticles in the case of non linear excitation [55].

We investigated, both theoretically and experimentally, incoherent multiphoton processes involving n = 1 to 5 photons [56]. For n = 1, 2, 3, MPEF occurs in bioaerosols because of natural fluorophors such as amino acids (tryptophan, tyrosin), NADH (nicotinamide adenine dinucleotide), and flavins. Figure 9 shows the LIF spectra of various bacteria, and the contribution of each fluorophor in the fluorescence spectrum under 266 nm excitation. The strongly anisotropic MPEF emission was demonstrated on individual microdroplets containing tryptophan, riboflavin, or other synthetic fluorophors [55-57]. The experiment was performed such that each individual microparticle was hit by a single laser shot. The aerosol source was based on a piezo driven nozzle, which precisely controlled the time of ejection of the microparticles. Figure 10 shows the MPEF angular distribution and the comparison between experimental and theoretical (Lorentz-Mie calculations) results for the one- (400 nm) (Fig. 10a), two- (800 nm) (Fig. 10b) and three-photon (1,2 μm) (Fig. 10c) excitation process. They show that the fluorescence emission is maximum in the direction towards the exciting source. The directionality of the emission is dependent on n, because the excitation process involves the nth power of the intensity $I^n(\mathbf{r})$. The ratio Rf = P(180°)/P(90°) increases from 1.8 to 9 when n changes from 1 to 3 (P is the emitted light power). For 3PEF, fluorescence from aerosol microparticles is therefore mainly backwards emitted, which is ideal for Lidar experiments.

For n = 5 photons we investigated laser induced breakdown (LIBS) in water microdroplets, initiated by multiphoton ionization. The ionization potential of water molecules is Eion = 6.5 eV, so that 5 photons are required at a laser wavelength of 800 nm to initiate the process of plasma formation. The growth of the plasma is also a nonlinear function of $I^n(\mathbf{r})$. We showed that both localization and backward enhancement strongly increases with the order n of the multiphoton process, exceeding Rf = U(180°)/U(90°) = 35 for n = 5 [58]. As for MPEF, LIBS has the potential of providing information about the aerosols composition. In particular, we showed that in the case of saline droplets the spectrum clearly exhibited Na D-lines, identifying the salt content within the droplet. Also, the blackbody emission spectrum was used to determine the plasma temperature (5000 to 7000 K).

5.2. PUMP-PROBE MEASUREMENT OF PARTICLE SIZE: BALLISTIC TRAJECTORIES IN MICROPARTICLES

The very small spatial extension of femtosecond pulses (30 fs corresponds to 6.8 μm in water, i.e. the equator length of a 1.1 μm droplet) has recently been used to measure the size of microparticles [59,60]. Moreover, the ballistic trajectories used by the wavepackets within the microparticle could be clearly observed for the first time.

The experimental scheme used 2PEF (in dye containing microdroplets) to create an optical correlator between a wavepacket at wavelength λ_1= 1200 nm, and a wavepacket at a different wavelength (λ_2 = 600 nm), each circulating on ballistic orbits. Fluorescence was then recorded as a function of the time delay between the two wavepackets in order to quantify the path length traveled within the particle. With this method, the size of droplets up to 700 μm could precisely be measured [59].

In order to address applications of these pump-probe results for Lidar measurements of both size and composition of atmospheric aerosols, time resolved Lorentz-Mie calculations have been performed using plane wave excitation (50 femtosecond pulses) [60]. In pump-probe 2PEF Lidar experiments, the composition would be addressed by the excitation/fluorescence signatures and the size by the time delay between the 2 exciting pulses. The high peak contrast obtained shows that measurements should be feasible for microparticles even of sub-micronic size.

5.3. MPEF-LIDAR DETECTION OF BIOAEROSOLS

We performed the first multiphoton excited fluorescence Lidar detection of biological aerosols. The particles, consisting of 1 μm size water droplets containing 0.03 g/l Riboflavin (typical for bacteria), were generated at a distance of 50 m from the "Teramobile" system.

Riboflavin was excited with two photons at 800 nm and emitted a broad fluorescence around 540 nm. This experiment [49,62] is the first demonstration of the remote detection of bioaerosols using a 2PEF-femosecond Lidar. The broad fluorescence signature is clearly observed from the particle cloud (typ. 10^4 p/cm^3), with a range resolution of a few meters (Fig. 11). As a comparison, droplets of pure water did not exhibit any parasitic fluorescence in this spectral range. However, a background is observed for both types of particles, arising from the scattering of white light generated by the filaments in air.

MPEF might be advantageous as compared to linear LIF for the following reasons: (1) MPEF is enhanced in the backward direction and (2) the transmission of the atmosphere is much higher for longer wavelengths. For example, if we consider the detection of tryptophan (which can be excited with 3 photons of 810 nm), the transmission of the atmosphere is typically 0.6 km^{-1} at 270 nm, whereas it is 3 x 10^{-3}km^{-1} at 810 nm (for a clear atmosphere, depending on the background ozone concentration). This compensates the lower 3-PEF cross-section compared to the 1-PEF cross-section for distances larger than a couple of kilometers [62]. The most attractive feature is however the possibility of using pump-probe techniques, as described hereafter in order to discriminate bioaerosols from background interferents such as traffic related soot or PAHs.

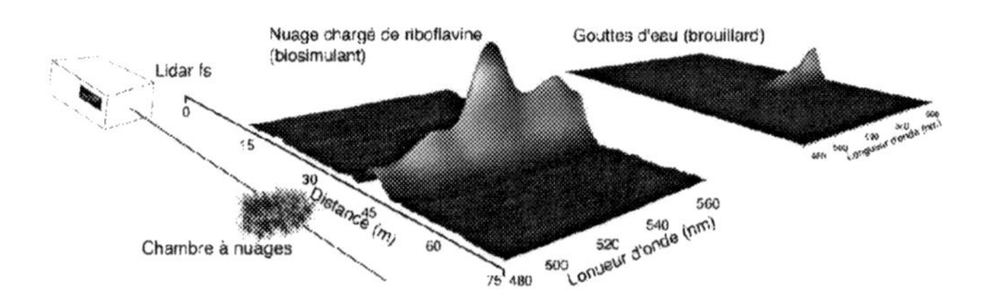

Figure 11: 2-photon excited fluorescence (2PEF)-Lidar detection of bioaerosols [62].

5.4 PUMP-PROBE SPECTROSCOPY TO DISTINGUISH BIOAEROSOLS FROM BACKGROUND PARTICLES IN AIR

A major drawback inherent in LIF instruments is the frequency of false identification because UV-Vis fluorescence is incapable of discriminating different molecules with similar absorption and fluorescence signatures. While mineral and carbon black particles do not have strong fluorescence signals, aromatics and polycyclic aromatic hydrocarbons (PAH) from organic particles and Diesel soot strongly interfere with biological fluorophors such as amino acids [61]. The similarity between the spectral signatures of organic and biological molecules under UV-Vis excitation lies in the fact that similar π-electrons from carbonic rings are involved. Therefore, aromatics or PAHs (such as naphtalene) exhibit absorption and emission bands similar to those of amino acids like tyrosine or tryptophan (Trp). Some shifts are present because of differences in specific bonds and the number of aromatic rings, but the broad featureless nature of the bands renders them almost indistinguishable. Moreover, the different environments of Trp in bacteria (e.g., they are contained in many proteins) and the mixtures of PAHs in transportation generated particles, blur their signatures. For example, Figure 12 compares the fluorescence spectra (resulting from excitation at 270 nm) of Trp in water and Trp-containing bacteria (Bacillus subtilis) to those of Diesel fuel and soot emitted by Diesel engines.

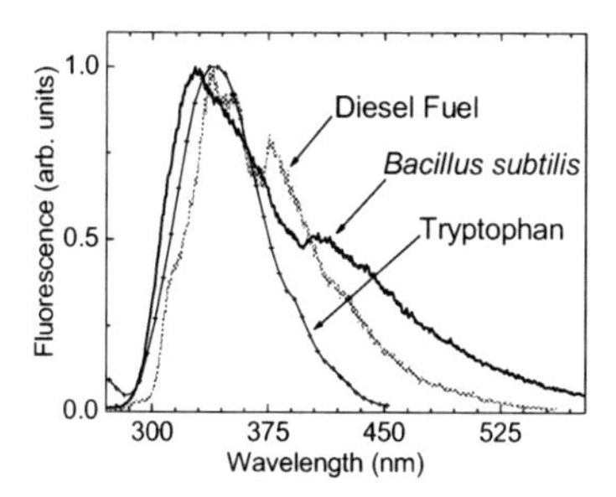

Figure 12: Comparison of the fluorescence spectra of Tryptophan, *B. Subtilis* and Diesel fuel.

We recently developed a novel femtosecond pump-probe depletion (PPD) concept [63], based on the time-resolved observation of the competition between excited state absorption (ESA) into a higher lying excited state and fluorescence into the ground state. This approach makes use of two physical processes beyond that available in the usual fluorescence spectrum: (1) the dynamics in the intermediate pumped state and (2) the coupling efficiency to a higher lying excited state. More precisely, as shown in Fig. 13a a femtosecond pump pulse (τ = 120fs) at 270 nm transfers a portion of the ground state S_0 population of Trp to its $S_1(\{v'\})$ excited state (corresponding to a set of vibronic levels, i.e., $\{v'\}$). The vibronic excitation relaxes by internal energy redistribution to lower $\{v'\}$ modes, associated with charge transfer processes (CT), conformational relaxation, and intersystem crossing with repulsive $\pi\sigma^*$ states. After vibronic energy redistribution, fluorescence is emitted from $S_1(\{v'\})$ within a lifetime of 2.6 ns. By illuminating the amino acid with a second pulse (probe) at 800 nm after a certain time delay from the pump pulse, the S_1 population is decreased and the fluorescence depleted. The 800 nm femtosecond laser pulse induces a transition from S_1 to an ensemble of higher lying Sn states, which are likely to be autoionizing [64] but also undergo radiationless relaxation into S_0 [65]. By varying the temporal delay between the pump and probe, the dynamics of the internal energy redistribution within the intermediate excited potential surface S_1 is explored. In principle, as different species have distinct S_1 surfaces, discriminating signals can be enhanced in this fashion.

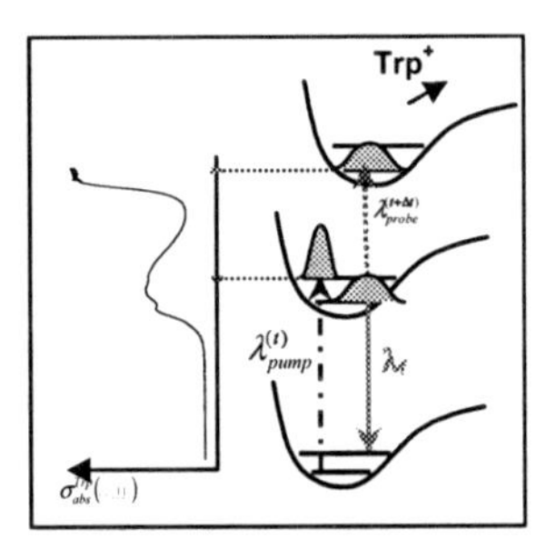

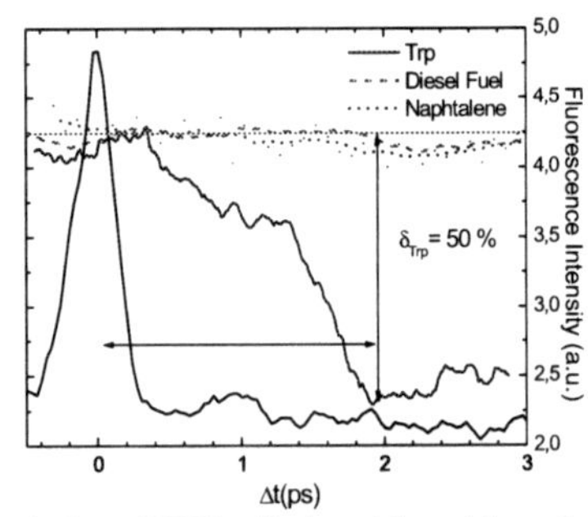

Figure 13: Femtosecond pump-probe depeletion (PPD) distinguishes biomolecules from organic interferents [63].

Figure 13b shows the pump-probe depletion dynamics of the S_1 state in Trp as compared to a circulation of Diesel fuel and Naphtalene in cyclohexane, one of the most abundant fluorescing PAHs in Diesel. While depletion reaches as much as 50% in Trp for an optimum delay of $\delta t = 2$ ps, Diesel fuel and Naphtalene appear almost unaffected (within a few percent), at least on these timescales. This remarkable difference allows for efficient discrimination between Trp and organic species, although they exhibit very similar linear excitation/fluorescence spectra (Fig. 12). Two reasons can be invoked to understand this difference: (1) the intermediate state dynamics is predominantly influenced by the NH- and CO- groups of the amino acid backbone and (2) the ionization potential and other excited states are higher for the PAHs contained in Diesel than for Trp by about 1 eV so that excitation induced by the probe laser is much less likely in the organic compounds. The particular dynamics of the internal energy redistribution in the S_1 state of Trp (fig. 13b), and in particular the time needed to start efficient depletion, is not fully interpreted as yet. Further electronic structure calculations are required to better understand the process, especially on the higher lying Sn potential surfaces.

In order to more closely approach the application of detecting and discriminating bioagents from organic particles in air, we repeated the experiment with bacteria, including Escherichia coli, Enterococcus and Bacillus subtilis (BG). Artefacts due to preparation methods have been avoided by using a variety of samples, i.e. lyophilized cells and spores, suspended either in pure or in biologically buffered water (i.e. typically 10^7-10^9 bacteria per cc,). The observed pump-probe depletion results are remarkably robust, with similar depletion values for all the considered bacteria (results for Enterococcus, not shown in the figure, are identical), although the Trp microenvironment within the bacteria proteins is very different from water. These unique features can be used for a novel selective bioaerosol detection technique that avoids interference from background traffic related organic particles in the air: The excitation shall consist of a pump-probe sequence with the optimum delay $\delta t = 2$ ps, and the fluorescence emitted by the mixture will be measured as the probe laser is alternately switched on and off. This pump-probe two-photon differential fluorescence method will be especially attractive for an active remote detection technique such as MPEF-Lidar, where the lack of discrimination between bioaerosols and transportation related organics is currently most acute.

These results, based on very simple pump-probe schemes, are very encouraging and open new perspectives in the discrimination capability of bioaerosols in air. We intend to extend the technique by applying more sophisticated excitation schemes (e.g., optimally shaped pulses), related to coherent control, in order to better distinguish bioaerosols from non-bioaerosols, but also to gain selectivity among the bacteria themselves. We recently performed theoretical calculations with H. Rabitz at Princeton, which show that under some conditions, optimal dynamic discrimination (ODD) can lead to efficient distinction between 3 species that exhibit almost the same spectral characteristics [66].

6. Conclusion

The non-linear propagation of ultrashort ultra-intense laser pulses provides unique features for Lidar applications: a coherent supercontinuum, self-guided, and back-reflected towards the source. Backward enhancement also occurs, for other reasons, for MPEF and LIB processes in aerosol particles. These characteristics open new perspectives for Lidar measurements in the atmosphere: multi-component detection, reduced spectral interference, better precision (more absorption lines), improved IR-Lidar measurements in aerosol-free atmospheres, and remote measurement of aerosols size distribution and composition.

Atmospheric applications of filamentation other than Lidar are also investigated. The first application concerns lightning control. Filaments are indeed conductive because of the generated plasma and might be used as a laser lightning rod. Promising experiments have been performed at a HV-facility in Berlin, which showed that megavolt discharges could be triggered and guided over distances of several metres [67-68]. The second application concerns triggered nucleation of water droplets. The plasma charges can act as condensation nuclei in saturated atmospheres, in a similar way as in cloud chambers for detecting ionizing particles. The demonstration of droplet nucleation induced by femtosecond laser filaments was recently performed in the laboratory [49].

The potential applications of filamentation are numerous and wide spread. Ultrashort and ultra intense lasers not only opened new perspectives in atmospheric sensing, they opened a new field in physics and chemistry.

7. Acknowledgements

The Teramobile project (www.teramobile.org) is funded by the Centre National de la Recherche Scientifique (CNRS) and the Deutsche Forschungsgemeinschaft (DFG).

The authors gratefully acknowledge the members of the Teramobile consortium, formed by the groups of L. Woeste in Berlin, J.P. Wolf in Lyon, R. Sauerbrey at the University Jena and A. Mysyrowicz at the ENSTA (Palaiseau). In particular, they wish to thank J. Kasparian, V. Boutou, J. Yu, E. Salmon, D. Mondelain, G. Méjean, M. Rodriguez, H. Wille, S. Frey, R. Bourayou, Y. B. Andre. The laboratory measurements on aerosols were performed in collaboration with the groups of R.K. Chang (Yale University), S.C. Hill (US Army Research Laboratories) and H. Rabitz (Princeton University). For these studies, RKC and JPW acknowledge NATO support SST-CLG977928.

The authors also acknowledge strong support by the technical staffs in Berlin, Jena and Lyon, in particular M. Barbaire, M. Kerleroux, M. Kregielski, M. Neri, F. Ronneberger, and W. Ziegler.

8. References

1. Weitkamp, C. (2005) *LIDAR*, Springer Verlag, New York
2. Measures, R.M. (1984) *Laser Remote Sensing*, J. Wiley & Sons, New York
3. Meyers, R.A. (2000) *Encyclopedia of Analytical Chemistry Vol 3*, J. Wiley & Sons, New York
4. Measures, R.M. (1988) *Laser Remote Chemical Analysis*, J.Wiley & Sons, New-York
5. Kasparian, J., Fréjafon, E., Rambaldi, P., Yu, J., Ritter, P., Viscardi, P. and Wolf, J.P. (1998) *Atmos. Environ.* **32(17)**, 2957-2967
6. del Guasta, M., Morandi, M., Stefanutti, L., Stein, B. and Wolf, J.P. (1994) *App. Optics* **33(24)**, 5690
7. Stein, B., del Guasta, M., Kolenda, J., Morandi, M., Rairoux, P., Stefanutti, L., Wolf, J.P. and Wöste, L. (1994) *Geo.Res.Lett.* **21(13)**, 1311
8. Woeste, L., Wedekind, C., Wille, H., Rairoux, P., Stein, B., Nikolov, S., Werner, C., Niedermeier, S., Ronneberger, F., Schillinger, H. Sauerbrey, R. (1997) *Laser und Optoelektronik* **E 2688**
9. Rairoux, P., Schillinger, H., Niedermeier, S., Rodriguez, M., Ronneberger, F., Sauerbrey, R, Stein, B., Waite, D.,Wedekind, C., Wille, H., Wöste, L. and Ziener, C. (2000) *App. Phys. B* **71**, 573
10. Kelley, P.L. (1965) *Phys. Rev. Lett.* **15**, 1005
11. Ashkaryan, G.A. (1974) *Sov. Phys. J.* **16(5)** 680
12. Braun, A., Korn, G., Liu, Du, D., Squier, J. and G. Mourou (1995) *Opt. Lett.* **20**, 73
13. Roso-Franco, L. (1995) *Phys. Rev. Lett.* **55**, 2149
14. Alfano, R.R. and Shapiro, S.L. (1970) *Phys. Rev. Lett.* **24**, 584
15. Alfano, R.R. and Shapiro, S.L. (1970) *Phys. Rev. Lett.* **24**, 592
16. Alfano, R.R. and Shapiro, S.L. (1970) *Phys. Rev. Lett.* **24**, 1217
17. Brodeur, A. and Chin, S.L. (1999) *J. Opt. Soc. Am. B* **16**, 637
18. Shen, Y. (1984) *The principles of nonlinear optics,* John Wiley & Sons, New York
19. Ranka, J.K., Schirmer, R.W.and Gaeta, A.L. (1996) *Phys. Rev. Lett.* **77**, 3783
20. Proulx, A., Talebpour, A., Petit, S. and Chin, S. L. (2000) *Optics Comm.* **174**, 305
21. Tzortzakis, S., Franco, M.A., André, Y.-B., Chiron, A., Lamouroux, B., Prade, B.S. and Mysyrowicz, A. (1999) *Phys. Rev.* E **60**, R3505
22. Tzortzakis, S., Prade, B., Franco, M. and Mysyrowicz, A. (2000) *Optics Comm.* **181**, 123
23. Schillinger, H. and Sauerbrey, R. (1999) *Appl. Phys. B* **68**, 753
24. Strickland, D. and Mourou, G. (1985) *Opt. Comm.* **56**, 219
25. Maine, P., Strickland, D., Bado, P., Pessot, M. and Mourou, G. (1998) *IEEE J. Quantum Electron.* **24**, 398
26. Talebpour, A., Young, J. and Chin, S.L. (1999) *Optics Comm.* **163**, 29
27. Keldysh, L.V. (1965) *Sov. Phys. JETP* **20**, 1307

28. Berge, L. and Couairon, A. (2001) *Phys. Rev. Lett.* **86**, 1003-1006
29. Braun, A., Korn, G., Liu, X., Du, D., Squier J. and Mourou, G. (1995) *Opt. Lett.* **20**, 73
30. Wagner, W.G., Haus H.A. and Marburger, J.H. (1971) *Phys. Rev. A* **3**, 2150
31. Wagner, W.G., Haus H.A. and Marburger, J.H. (1968) *Phys. Rev.* **175**, 256
32. Goorjian, P.M., Taflove, A., Joseph, R.M. and Hagness, S.C. (1992) *IEEE j. of quantum electronics* **28**, 2416
33. Ranka, J.K. and Gaeta, A.L. (1998) *Opt. Lett.* **23**, 534
34. Mlejnek, M., Wright, E.M. and Moloney, J.V. (1998) *Opt. Lett.* **23**, 382
35. Chiron, A., Lamouroux, B., Lange, R., Ripoche, J.-F., Franco, M., Prade, B., Bonnaud, G.,Riazuelo, G. and Mysyrowicz, A. (1999) *European Physical J.* D **6**, 383.
36. Akšzbek, N., Bowden, C. M., Talepbour, A. and Chin, S.L. (2000) *Phys. Rev. E* **61**, 4540
37. Hovhannisyan, D.L. (2001) *Optics Comm.* **196**, 103
38. Sprangle, P., Pe–ano, J. R. and Hafizi, B. (2002) *Phys. Rev. E* **66**, 046418
39. Mlejnek, M., Kolesik, M., Moloney, J.V. and Wright, E.M. (1999) *Phys. Rev. Lett.* **83**, 2938
40. Ren, C., Hemker, R.G., Fonseca, R.A., Duda, B.J. and Mori, W.B. (2000) *Phys. Rev. Lett.* **85**, 2124
41. Mechain G, Couairon A, Andre YB, D'Amico C, Franco M, Prade B, Tzortzakis S, Mysyrowicz A, Sauerbrey R (2004) *App.Phys. B.* **79 (3)** 379-382
42. Akšzbek, N., Iwasaki, A., Becker, A., Scalora, M., Chin, S.L. and Bowden, C.M. (2002) *Phys. Rev. Lett.* **89**, 143901.
43. Yang, H., Zhang, J., Zhao, L.Z., Li, Y.J., Teng, H., Li, Y.T., Wang, Z.H., Chen, Z.L., Wei, Z.Y., Ma, J.X., Yu, W. and Sheng, Z.M. (2003) *Phys. Rev. E* **67**, 015401
44. Kosareva, O.G., Kandidov, V.P., Brodeur, A., Chen, C.Y. and Chin, S.L. (1997) *Opt. Lett.* **22**, 1332
45. Nibbering, E.T.J., Curley, P.F., Grillon, G., Prade, B.S., Franco, M.A., Salin F. and Mysyrowicz, A. (1996) Opt. Lett. **21**, 62
46. Yu, J., Mondelain, D., Ange, G., Volk, R., Niedermeier, S., Wolf, J.-P., Kasparian, J. and Sauerbrey, R. (2001) *Opt. Lett.* **26**, 533
47. Strong, K. and Jones, R.L. (1995) *Applied Optics* **34(27)**, 6223
48. Wille, H., Rodriguez, M., Kasparian,J., Mondelain, D., Yu, J., Mysyrowicz, A, Sauerbrey, R., Wolf, J.P. and Woeste, L. (2002) *European Physical J. D* **20(3)** 183-189
49. Kasparian, J., Rodriguez, M., Mejean, G., Yu, J., Salmon, E. Wille, H., Bourayou, R., Frey, S., Andre, Y.-B., Mysyrowicz, A., Sauerbrey, R., Wolf, J.-P.and Woeste, L. (2003) *Science* **301(5629)**, 61-64
50. Kasparian, J., Sauerbrey, R., Mondelain, D., Niedermeier, S., Yu, J., Wolf, J.-P., Andre, Y.-B., Franco, M., Prade, B., Mysyrowicz, A., Tzortzakis, S., Rodriguez, M. Wille, H. and Woeste, L. (2000) *Opt. Lett.* **25**, 1397
51. Méjean, G., Kasparian, J., Salmon, E., Yu, J., Wolf, J.P., Bourayou, R., Sauerbrey, R., Rodriguez,M., Woeste, L., Lehmann, H., Stecklum, B., Laux, U., Eilfoeffel, J., Scholtz, A., Hatzes, A.P.(2003) *Appl.Phys. B* **76(7)**, 357-359
52. Yu, J., Bourayou, R., Frey, S., Kasparian, J., Salmon, E., Mejean, G., Rodriguez, M., Sauerbrey, R., Wolf J.-P. and Woeste, L. (2005) "Simultaneous measurements of ozone and aerosols using a white light femtosecond laser", in preparation.
53. Megie, G. (1980) *Applied Optics* **19**, 34
54. Schwemmer, G. and Wilkerson, T.D. (1979) *Applied Optics* **18**, 3539
55. Hill, S.C., Boutou, V., Yu, J., Ramstein, S., Wolf, J.P., Pan, Y., Holler, S. and Chang, R.K. (2000) *Phys.Rev.Lett.* **85(1)**, 54-57
56. Boutou, V., Favre, C., Hill, S. C., Pan, Y., Chang, R. K. and Wolf, J.P. (2002) *App. Phys .B* **75**, 145-153
57. Pan, Y., Hill, S. C., Wolf, J.P. , Holler, S., Chang, R.K. and Bottiger, J.R. (2002) *App. Optics* **41(15)**, 2994-2999
58. Favre, C., Boutou, V., Hill, S.C., Zimmer, W., Krenz, M., Lambrecht, H., Yu, J., Chang, R.K., Woeste, L. and Wolf, J.P. (2002) *Phys.Rev.Lett.* **89(3)**, 035002
59. Wolf, J.P., Pan, Y., Holler, S., Turner, G.M., Beard, M.C., Chang, R.K. and Schmuttenmaer, C. (2001)*Phys.Rev. A* **64**, 023808-1-5
60. Mées, L., Wolf, J.P., Gouesbet, G. and Gréhan, G. (2002) *Optics Comm.* **208**, 371-375 (2002)
61. Hill, S.C, Pinnick, R.P., Niles, S., Pan, Y.L, Holler, S., Chang, R.K., Bottiger, J., Chen, B.T., Orr, C.S. and Feather, G (1999) *Field Anal Chem. Tech.* **5**, 221-229 (1999)
62. Méjean, G., Kasparian, J., Yu, J., Frey, S., Salmon, E. and Wolf, J.P. (2004)*Applied Physics B* **78(5)**, 535-537
63. Courvoisier, F., Boutou, V., Wood, V., Wolf, J.-P., Bartelt, A., Roth, M. and Rabitz H. (2005) *App. Phys. Lett.* **87(6)**, 063901
64. Teraoka, J., Harmon, P.A. and Asher, S. (1989)*J. Am. Chem. Soc.*, **112**, 2892
65. Steen, H.B. (1974) *J. Chem. Phys.*, **61(10)**, 3997

66. Li, B., Rabitz, H. and Wolf, J.P. (2005) *J. Chem. Phys.* **122,** 154103-1-8
67. Rodriguez, M., Sauerbrey, R., Wille, H., Woeste, L., Fuji, T., Andre, Y.B., Mysyrowicz, A . Klingbeil, L., Rethmeier, K., Kalkner, W., Kasparian, J., Yu, J. and Wolf, J.P. (2002) *Opt. Lett.* **27(9)** 772-775
68. Ackermann, R., Stelmasczyck, K., Rohwetter, P., Méjean, G., Salmon, E., Yu, J., Kasparian, J., Bergmann, V. Schaper, S., Weise, B., Kumm, T., Rethmeier, K., Kalkner, W., Wolf, J.P. and Woeste, L. (2004) *Appl. Phys. Lett.* **85(23)** 5781-5783

SPECTROSCOPY OF THE GAP STATES IN *Ge* BASED ON ITS NEUTRON TRANSMUTATION DOPING KINETICS

A.G. ZABRODSKII
Ioffe Physical Technical Institute,
Polytechnicheskaya, 26, 194021 Saint Petersburg, Russia
Andrei.Zabrodskii@mail.ioffe.ru,

1. Introduction

An energetic diagram of doped semiconductor contains a set of impurity levels in the band gap presented in the *Figure 1*. At the low temperature the Fermi level separating occupied states from unoccupied ones is fixed in the upper (n–type) or lower (p–type) part of the gap.

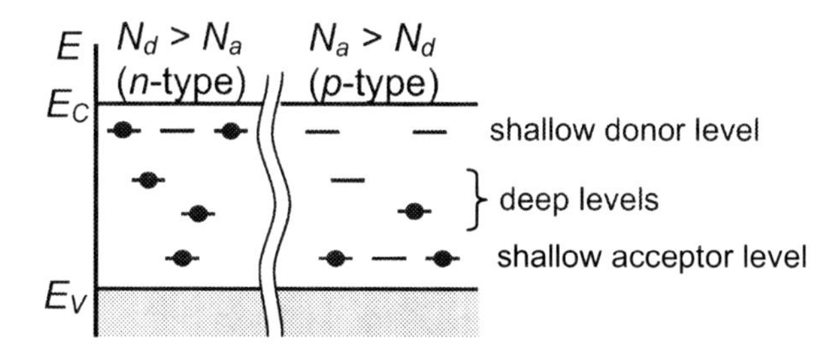

Figure 1. Energetic diagram of a compensated semiconductor.

There are deep levels in the forbidden gap apart from shallow (generally hydrogen–like) donor levels. For the IV group semiconductor in particular shallow acceptor and donor levels are relevant to III or V group impurities. In contrast to this, deep multicharged nonhydrogen–like centers are caused by the presence of other group impurities. When donor and acceptor states are in existence simultaneously, one speaks about compensated semiconductor characterized by a compensation degree which is defined as the ratio of the minority impurity density to the majority one.

Impurity level spectroscopy in semiconductors measures their position in the gap (from optical or photo conductivity edges, activation energies for conductivity or charge concentration, deep level transient spectroscopy) and density. In the methods above the Fermi level position is fixed!

We will speak about a different and very sensitive method of the impurity (and any defect) level spectroscopy in the semiconductor (*Ge*, *Si*:*Ge*) gap when the Fermi level is moving across the gap between shallow donor and acceptor states with a fixed velocity.

B. Di Bartolo and O. Forte (eds.), Advances in Spectroscopy for Lasers and Sensing, 483–497.

The method is based on our previous investigations of the neutron transmutation doping kinetics in *Ge*.

Neutron transmutation doping of *Si* and *Ge* is based on the following β–decay reactions [1]:

$$^{30}Si(n,\gamma)^{31}Si \xrightarrow[2.6\,h]{\beta} {}^{31}P + \tilde{\nu} \tag{1}$$

$$^{74}Ge(n,\gamma)^{75}Ge \xrightarrow[82\text{ min}]{\beta} {}^{75}As + \tilde{\nu} \tag{2}$$

$$^{76}Ge(n,\gamma)^{77}Ge \xrightarrow[11.3\text{ h}]{\beta} {}^{77}As + \tilde{\nu} \tag{3a}$$

$$^{77}As \xrightarrow[38.3\text{ h}]{\beta} {}^{77}Se + \tilde{\nu} \tag{3b}$$

$$^{70}Ge(n,\gamma)^{71}Ge \xrightarrow[11.4\text{ day}]{\varepsilon} {}^{77}Ga + \nu \tag{4}$$

The reaction (4) proceeds with trapping of an orbital electron as distinct from the conventional reactions of β–decay (1) and (3). The reaction (4) known as "reaction of the internal conversion" is symbolized by ε. In *Ge* the yield of each chain (*As*, *Se* and *Ga*) after the β–decay is

$$N_i = \overline{\Sigma_i}F = N_{Ge}p_i\overline{\sigma_i}F \tag{5}$$

N_{Ge}=4.42·10^{22} cm^{-3}, p_i is the i–*th* isotope relative constant, $\overline{\sigma_i}$ is the averaged neutron capture cross section of the atom, F is the neutron flux, and $\overline{\Sigma_i} = N_{Ge}p_i\overline{\sigma_i}$ is the averaged macroscopic neutron–capture cross section.

- Which neutrons are important in the averaging of $\overline{\sigma_i}$?
- Thermal (with the Maxwellian spectrum) and resonance.
  This fact is illustrated by *Figure 2*.

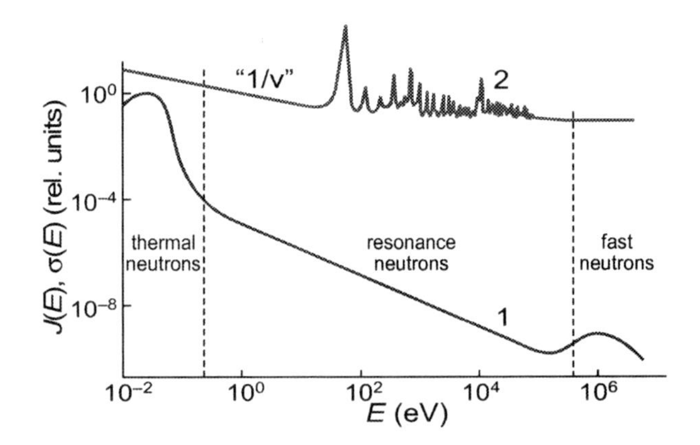

Figure 2. Spectra of the reactor neutrons and their absorption.

Particularly for the Maxwellian spectrum

$$\sigma_i = \frac{1}{2}\pi^{1/2}\sigma_i^{th} \tag{6}$$

TABLE 1. Some data on the "doping" *Ge* isotopes [2].

Isotope	p,%	σ^{th}, barn	$\sum^{th}$, cm^{-1}
^{70}Ge	20.5	3.43	$2.75\cdot10^{-2}$
^{74}Ge	36.5	0.51	$7.29\cdot10^{-3}$
^{76}Ge	7.8	0.15	$4.58\cdot10^{-4}$

Thus the neutron transmutation doped (NTD) *Ge* is p–*Ge*:*Ga* with moderate compensation

$$K = N_d / N_a = (N_{^{75}As} + 2N_{^{77}Se}) / N_{^{71}Ga} \tag{7}$$

which doping level depends lineary versus fluence of the neutrons. Factor "2" here reflects that the deep VI group *Se* donor has to be so called "double" donor: being neutral it must containe two electrons not involved into valence bond.

Application of the NTD *Ge*: series of the homogeneously doped and compensated samples, for the metal–insulator transition and hopping (tunnel) electron transport investigation at the low temperatures manufacturing of the cooled deeply bolometers and thermoresistors.

For the theoretical analyses of the low temperatures properties one needs strongly in the exact values of the main parameters: *N* and *K*. Unfortunately the corresponding nuclear data varied strongly from one source to another one (see *Figure 3*).

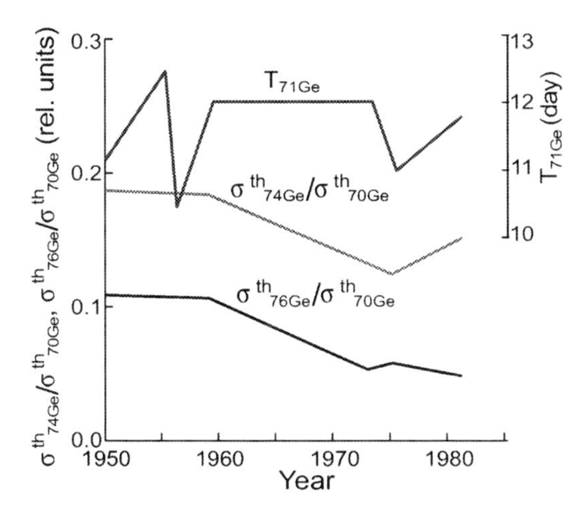

Figure 3. Divergence of the nuclear physical data on the *Ge* isotopes until to early of the 1980s.

Just to measure the exact value of K for the NTD *Ge*:*Ga* the first investigation of the neutron transmutation doping kinetics was used [3].

2. What one can Extract Directly from the Investigations of the Neutron Transmutation Doping Kinetics of *Ge*?

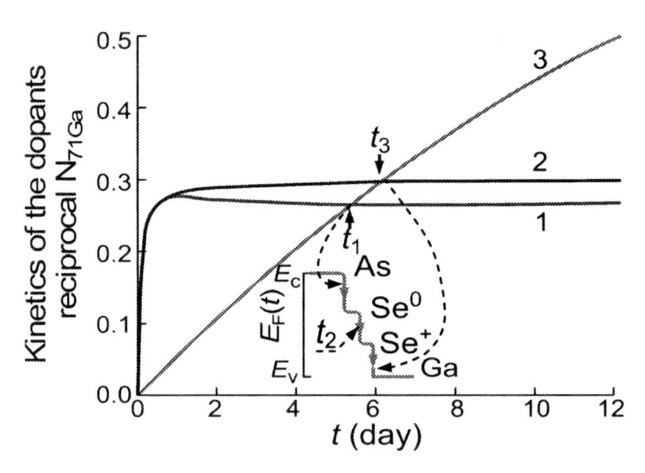

Figure 4. Calculated kinetics of the relative yield of the transmutation impurities in *Ge* for irradiation by the thermal neutrons [4]:1—$N_{^{75}As}(t) + N_{^{77}As}(t)$; 2—$N_{^{75}As}(t) + N_{^{77}As}(t) + 2N_{^{77}Se}(t)$; 3—$N_{^{71}Ga}(t)$.

The insert shows a three drop moving of the Fermi level. The characteristic times t_1, t_2 and t_3 correspond to a moments of the exact compensation of

- shallow donors (t_1);
- shallow donors and upper level of the *Se* donor (t_2);
- all donor levels (t_3).

Measured by the Hall effect the electron (n) and hole (p) density kinetics (after annealing) are presented in the *Figure* 5 in the semilogarithmical scale. The scale in this figure is not suit for the analysis.

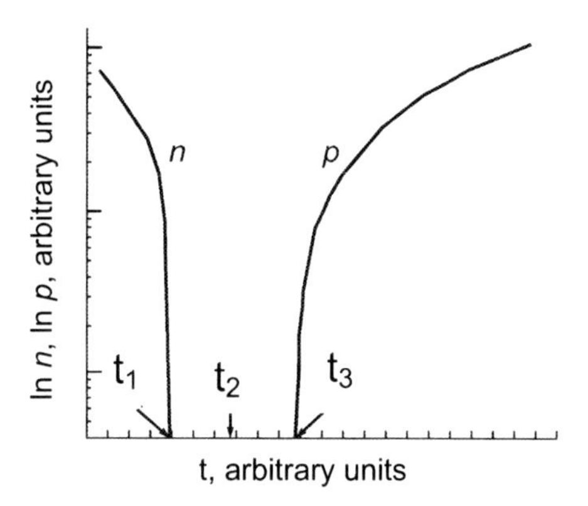

Figure 5. The electron (n) and hole (p) density kinetics after annealing.

Important simplification: several days after neutron irradiation charge carrier kinetics are linearized in the scale of the dimensionless parameter "τ":

$$\tau = 1 - \exp[-\lambda_{71_{Ge}} t], \quad \text{where} \quad \lambda_{71_{Ge}} = \ln 2 / T_{71_{Ge}} \tag{8}$$

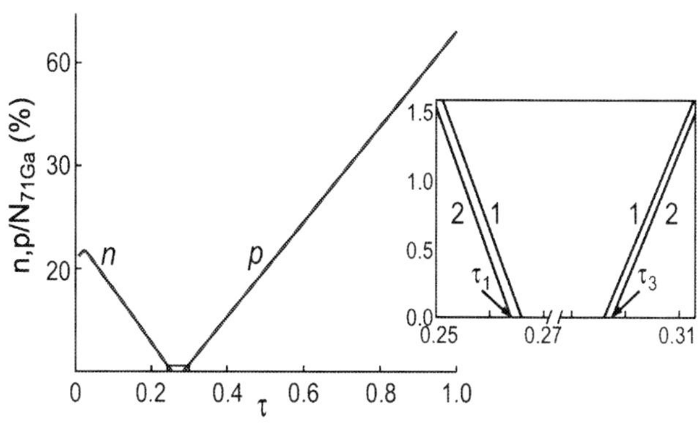

Figure 6. Linearized kinetics of the electron (n) and hole (p) densities reduced to the steady–state concentration of the ^{71}Ga isotope. The insert shows regions near the characteristic times τ_1 and τ_3: (1) fore the real rates, (2) calculated on the assumption that all ^{77}Se impurity is created.

$$p(\tau) = N_{71_{Ga}} \tau - (N_{75_{As}} + 2N_{77_{Se}}) = \alpha_p \tau - \beta_p \tag{9}$$

$$p(\tau_3) = 0 \rightarrow K = N_d / N_a = \tau_3 = \beta_p / \alpha_p \tag{10}$$

$$n(\tau_1) = \beta_n - \alpha_n \tau \tag{11}$$

$$n(\tau_1) = 0 \rightarrow N_{As} / N_a = \tau_1 = \beta_n / \alpha_n \tag{12}$$

- What one can extract from an experiment ?
- $T_{^{71}Ge}$, $N_{^{71}Ga}$ (from "p"–branch of *Figure 6* and Eqns. (9,10)),
  $N_{^{77}Se} : N_{^{75}As} : N_{^{71}Ga} = ((\tau_3 - \tau_1)/2) : \tau_1 : 1$ (Eqns. (9–12))

3. Experiment. Neutron Transmutation Doping by the Strongly Moderated Neutrons

The experimental procedure was as follows:

= High pure ($N \approx 10^{10}\ cm^{-3}$) starting *Ge*
= Irradiation in WWR–M reactor in the channel with thermal to fast neutron ratio 30–50
= Radiation damage annealing at 450°C in 3 – 4th day after irradiation
= 6–armed samples in the Hall geometry
= Measurement of the densities $n(t)$, $p(t)$, and conductivity $\sigma(t)$ at 77 K until to 3 months
= 3 experiments: $N_{^{71}Ge} = 10^{14} - 1.5 \cdot 10^{15}\ cm^{-3}$

The representative experimental kinetics for carrier densities is shown in the *Figure 7*.

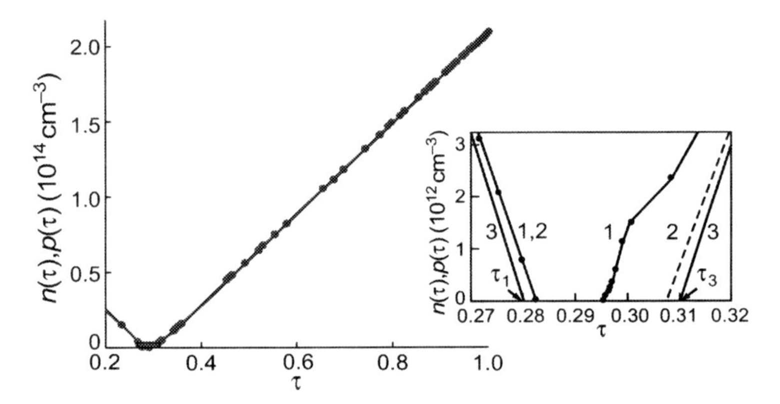

Figure 7. Linerized kinetics for n– and p–branches, $N_{^{71}Ge} = 3.1 \cdot 10^{14}\ cm^{-3}$. The insert shows a range in the vicinity of the τ_1 and τ_3 characteristic times: 1 – experiment, 2 – calculated for the real rates, 3 – calculated on the assumption all ^{77}Se was created.

The question arises about possible attendant defect generation by the β–decay. To receive an answer we did two special experiments. In one of them pure *Ge* detector sample was used to detect the radiation defects, that could be created by the sample investigated. In the another one an annealing procedure was carried much later (one month) after neutron irradiation. No defects was measured. As the first result the exact value of $T_{^{71}Ge}$ was measured: $T_{^{71}Ge} = 11.36 \pm 0.04$ day. One can receive an imagination about a precision of our data from *Figure 8*, where we compare them with the nuclear data available.

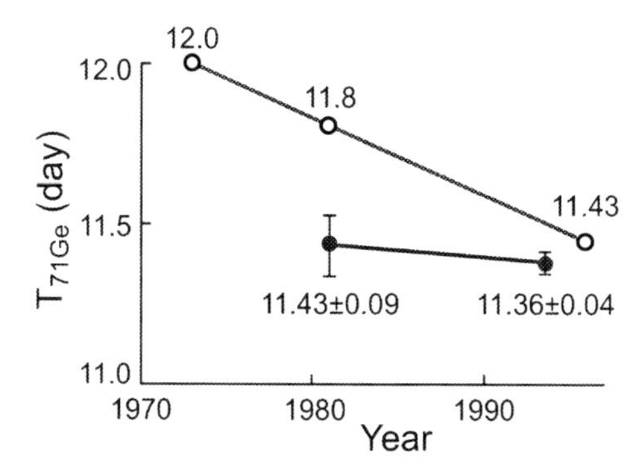

Figure 8. Data of the BNL (USA) (o) and Ioffe Physical Technical Institute (PTI) (•) on $T_{^{71}Ge}$.

As the second result the exact ratio $N_{^{77}Se} : N_{^{75}As} : N_{^{71}Ga} \rightarrow \bar{\sigma}_{^{76}Ge} : \bar{\sigma}_{^{74}Ge} : \bar{\sigma}_{^{70}Ge}$ was obtained.

It is very important for lots of applications to divide a contribution of the thermal neutrons and of the resonance ones in the transmutation process.

4. Separation of the Contributions of Thermal and Resonance Neutrons

In the neutron physics the contributions of thermal ($E \leq 0.2\ eV$) and resonance neutrons into the target activation are separated by the cadmium difference method.

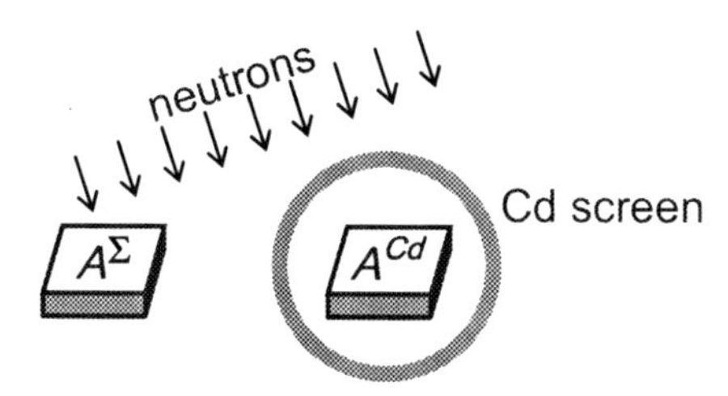

Figure 9. The experimental schema of a measurement of the thermal neutrons contribution into the target activation.

The reason is that the cadmium absorbs neutrons with the energies less than some cut off energy: $E \leq\ E_c(d)$. This is illustrated in the *Figure 10.*

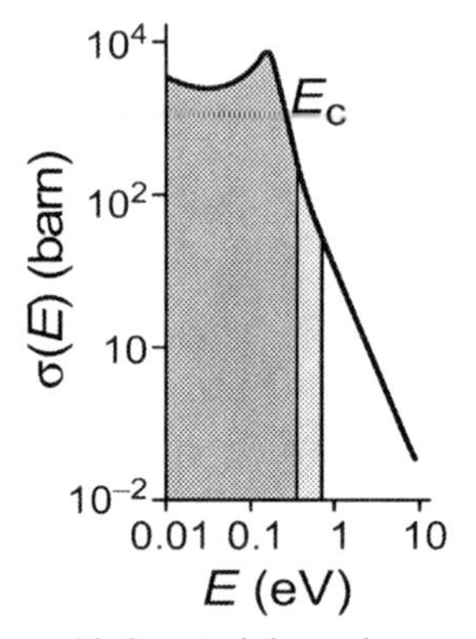

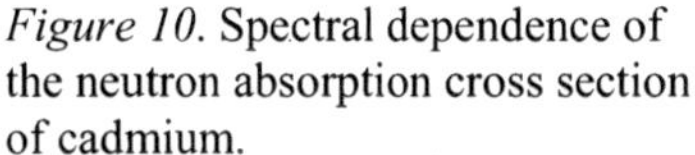

Figure 10. Spectral dependence of the neutron absorption cross section of cadmium.

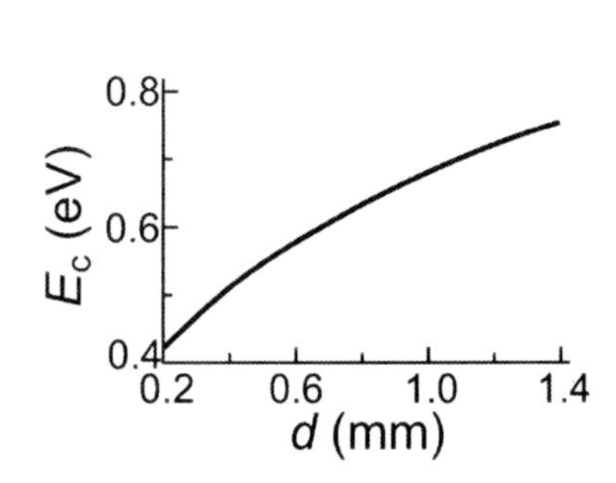

Figure 11. Dependence of the cadmium transmission edge on its thickness.

Of course the cadmium transmission edge depends versus a thickness of cadmium (see *Figure 11*). Taking this into account a following simple formula was used to measure of thermal neutrons contribution into the target activation:

$$A^{th} = A^{\Sigma} - A^{Cd} F, \tag{13}$$

where $F = F(d)$ (see *Figure 12*) is a coefficient of the cadmium correction corresponding to the fact that the cadmium cut off energy is higher than a boundary energy for thermal neutrons equal to 0.2 eV: $E_c > 0.2\ eV$.

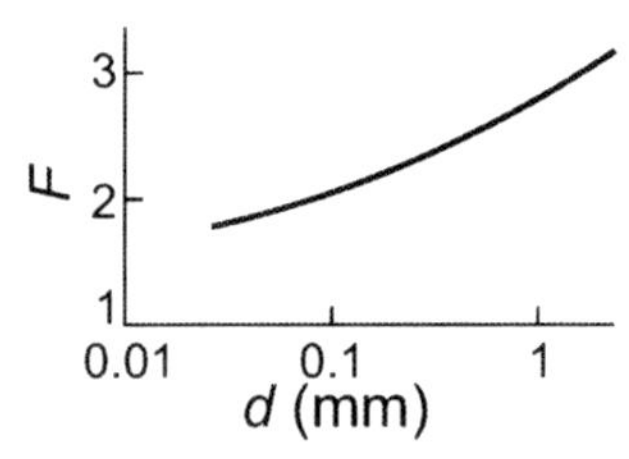

Figure 12. Coefficient of the cadmium correction versus its thickness.

We are interested in the doping density, not activation and have to change in Eqn. (13) In our case: $A \leftrightarrow N_i$ and thus receive

$$N^{th}_{^{71}Ga} = N^{\Sigma}_{^{71}Ga} - N^{Cd}_{^{71}Ga} \cdot F \tag{14}$$

An idea of our experiment was to combine method of the cadmium difference with the investigation of the neutron transmutation doping kinetics [4]. The experimental results obtained are shown in *Figure 13*.

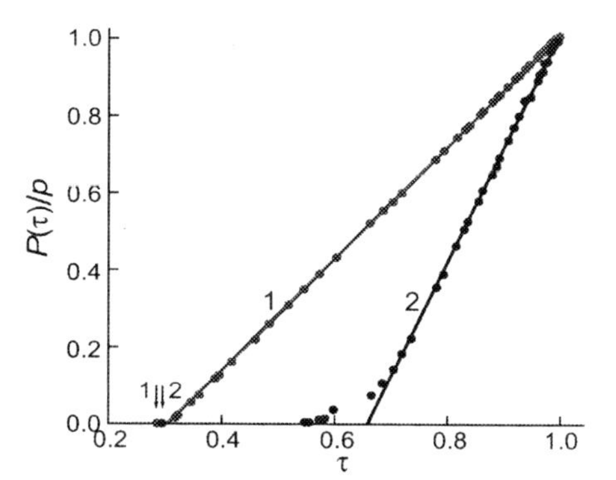

Figure 13. NTD kinetics in the cadmium difference τ method, irradiations without *Cd* (1) and in cadmium screen (2) arrows indicate τ_1 time for the short n–branches.

As the first result a high precision ratio was obtained:

$$N^{th}_{^{77}Se} : N^{th}_{^{75}As} : N^{th}_{^{71}Ga} \rightarrow \sigma^{th}_{^{76}Ge} : \sigma^{th}_{^{74}Ge} : \sigma^{th}_{^{70}Ge} \tag{15}$$

A precision of the result obtained (15) can be inferred from scheme below.

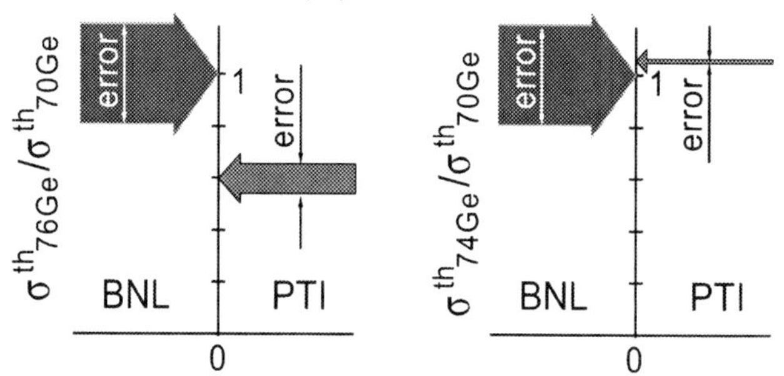

Figure 14. Comparison of our PTI results on the thermal neutron capture cross section with the BNL (USA) data that time.

The second important result was the next high precision ratio:

$$N^{Cd}_{^{77}Se} : N^{Cd}_{^{75}As} : N^{Cd}_{^{71}Ga} \rightarrow I_{\gamma}\,{}^{76}Ge : I_{\gamma}\,{}^{74}Ge : I_{\gamma}\,{}^{70}Ge, \tag{16}$$

where I_{γ} is a corresponding resonance integral related to the resonance neutrons absorbtion.

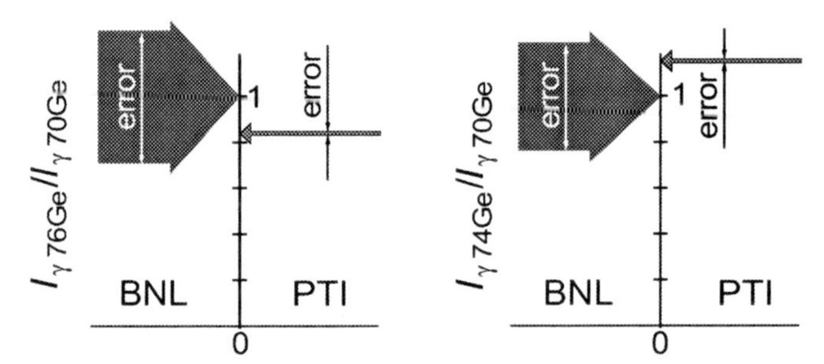

Figure 15. Comparison of our PTI results on the resonance integrals with the BNL data that time.

- What are the reasons for the precision of our data?
- Neutron transmutation of the germanium isotopes is the unique internal detector characterized by the huge dynamical range, the absence of losses and background and by the fact that only relative changes of the carrier concentrations are essential in both "n–" and "p–" branch registration.
- What is resulted for the physics of the semiconductors?
- An answer is presented in the sections 5–7.

5. Effect of the Spectrum of Fusion Reactor Neutrons on the Yield of the Transmutation Impurities in Germanium [5]

Experiments have been performed in the WWR–M reactor channel with a ratio of the thermal to the fast neutron flux $F^{th}/F^{f} = 8 \div 10$. Results of the corresponding measurement of the NTD kinetics are shown below.

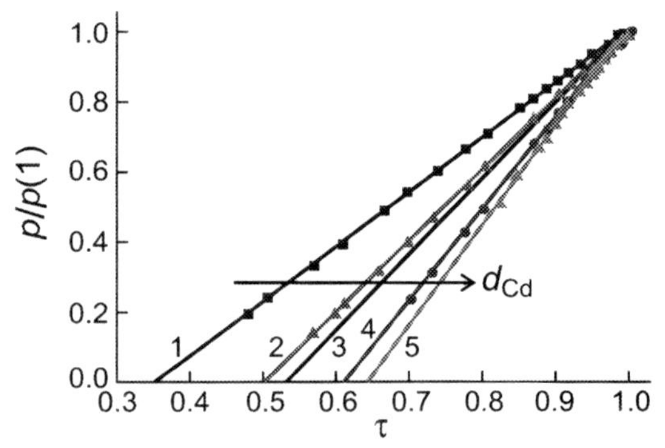

Figure 16. Effect of the neutron spectrum on the NTD kinetics; 1–5: d_{Cd} = 0; 0.17; 0.5; 0.9 mm, and composed screen (0.5 mm *Cd* and 2.7 mm *In*).

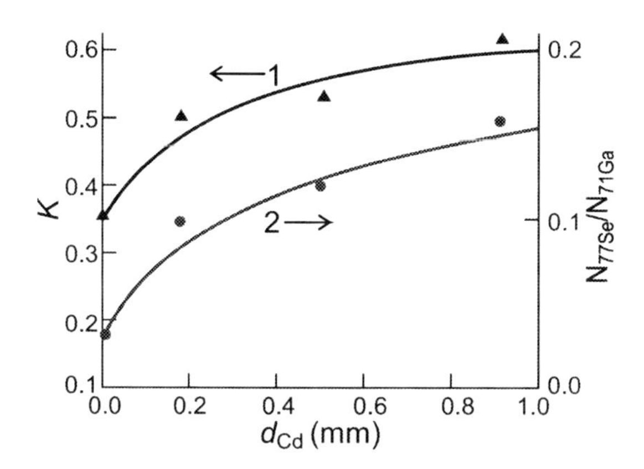

Figure 17. Effect of the cadmium thickness on the compensation degree and N_{Se} / N_{Ga} ratio.

As the result it was established that the ratio of the transmutation dopant densities one can easy vary by the neutron spectrum using absorbing screen. This is illustrated in the *Figure 17.*

6. Spectroscopy of the Gap States in *Ge* Based on its Neutron Transmutation Doping Kinetics: The Idea

A unique peculiarity of the NTD kinetics in *Ge* is that the minority donor states are created first while the majority *Ga* states appear mach later through $^{71}Ge \rightarrow ^{71}Ga$ (T_{71} = 11.36 day) electron capture reaction what is not follow by any defect generation. Thus after annealing of the radiation damage neutron transmutation of *Ge* is followed by n→p conversion or by the Fermi level sharp shift through the forbidden gap in *Ge* at low enough temperature. Kinetics of the Fermi level being measured allows to determine the energy of the levels in the gap and their densities.

That was the idea of the so called "Fermi level spectroscopy" (FLS) suggested by J. Cleland in 1960, who was failed in his attempt to realize it experimentally. He observed only very smooth n → p conversion and explained it by the inhomogeneities of the sample. They could be origin from the start doping or from the radiation damage. We moved the FLS realization from our own experience of the NTD kinetics investigation and know nothing about Cleland's idea and his unsuccessful attempt in the beginning.

7. Spectroscopy of the Gap States in *Ge* Based on its Neutron Transmutation Doping Kinetics: Application to the Old Problem of the Deep *Se* States in *Ge* [6,7]

A problem was the following. While a position of one of the two *Se* states in the gap of *Ge* was known well ($Se^0: E_c - 0.3\ eV$), a position of the Se^+ ion was unknown. This raises the question of whether *Se* (VI group element) is a double donor in *Ge* or not. To receive an answer we combined NTD kinetics studying method with the FLS idea. In the *Figure 18* we shown what one could expect for the NTD and of the Fermi level kinetics for 1 and 2 *Se* levels in the gap.

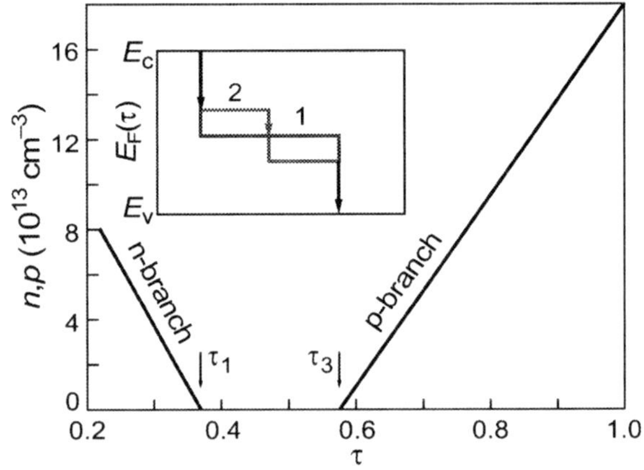

Figure 18. The kinetics of NTD and Fermi level (insert) for 1 (curve 1) and 2 (curve 2) *Se* levels in the gap.

The experimental procedure was the following:

- starting material was high purity *Ge*, manufactured by E. Haller;
- to increase a yield of *Se* atoms the irradiation in *Cd* was used;
- $R_H(t)$, $\sigma(t)$ and photoconductivity edge were measured in the $t_1 - t_3$ interval;
- in parallel 2 irradiated samples were measure: 1 – $R_H(t)$ and $\sigma(t)$;
  2 – photoconductivity spectra.

Measured kinetics of the electron density versus reverse temperature ($t \le t_2$ and $t \ge t_2$), are presented in the *Figure 19 a,b*.

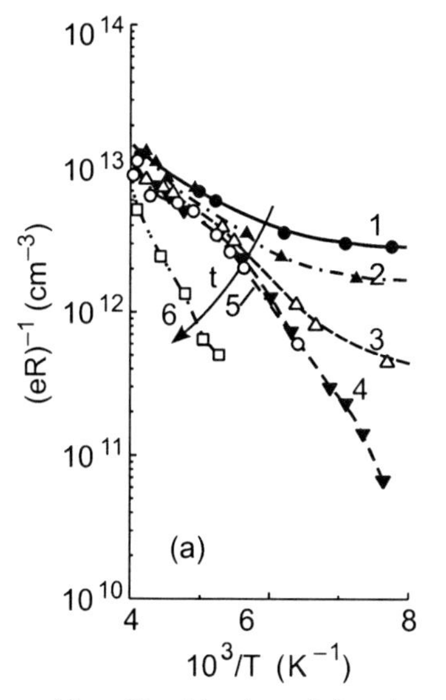

Figure 19 a. The kinetics of the electron density versus reverse temperature, $t \le t_2$: 1–6.67; 2–7.1; 3–7.67; 4–8.11; 5–8.67; 6–10.67.

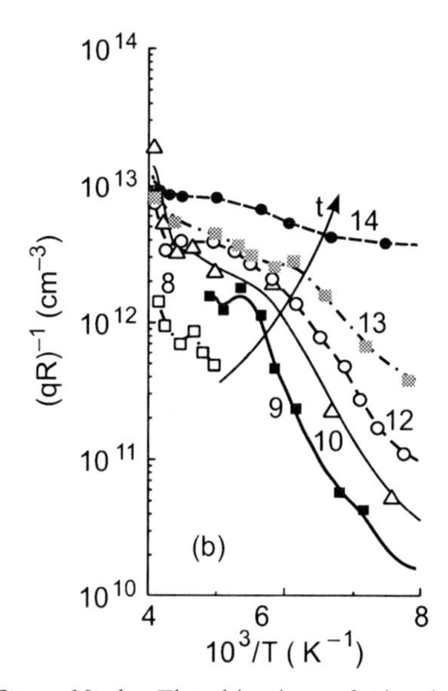

Figure19 b. The kinetics of the hole density versus reverse temperature, $t \ge t_2$: 8–11.12; 9–11.69; 10–12.09; 11–12.74; 12–13.73; 13a–14.69; 13b–14.92;14–18.08.

The data in the *Figures 19a* and *19b* resulted to the *Se* levels:

$$E_{Se^0} = E_c - 0.27\ eV \tag{17}$$

$$E_{Se^+} = E_V + 0.22\ eV \tag{18}$$

Both of these results (17) and (18) were supported by analysis of the conductivity $\sigma(t)$ and photoconductivity edge kinetics.

The final results obtained for the FLS of the deep *Se* states in *Ge* are illustrated in the *Figure 20*.

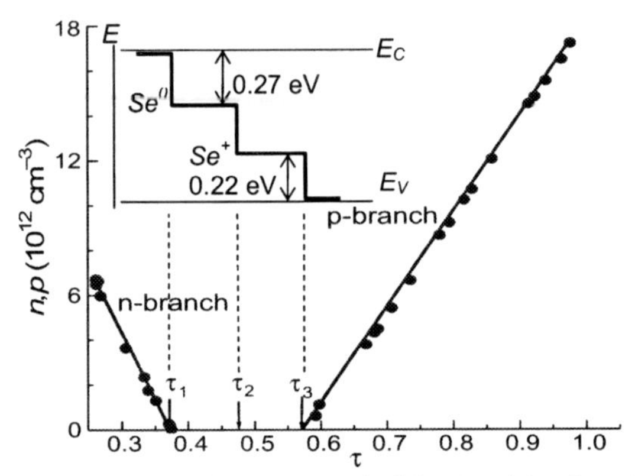

Figure 20. Experimental kinetics of the NTD and of the Fermi level position in the gap at low temperature (insert).

8. Conclusions

- The method of the Fermi level spectroscopy of the deep levels in *Ge* was created. This method is based on the reaction of orbital electron capture: $^{71}Ge \rightarrow ^{71}Ga$.
- This reaction fulfils two main functions:
  Fermi level scan of the electronic states in the gap of *Ge* and
  determination of the time and concentration scales.
- Once all transmutation states has been much studied, the application range of the method can be extended to the different deep states in *Ge* and to *Si* doped by *Ge* to the density no less than 1% (because of a macroscopic cross section of thermal neutron absorbtion for the main *Ge* isotope (^{70}Ge) is two order of magnitude higher than this one for *Si* doping isotope (^{30}Si)).
- The skill and the experience are desirable to use this method in a practice.

Acknowledgements

The author thanks Prof. Dr. B. Di Bartolo and his team for the invitation to come and to lecture at the School, Dr. M. Alekseenko for the assistance in performance of measurements, Prof. Dr. E.E. Haller for making high purity *Ge* samples and Dr. O. Proshina for the help in preparing the lecture materials for printing.

References

1. Cleland, J.W., Lark–Horovitz, K., and Pigg, C. (1950) Transmutation–produced germanium semiconductor, *Phys. Rev.* **78**, 814.
2. Mughaghab, S.F., Divadeenam, M., and Holden, N.E. (1981) Neutron cross section, *4th ed., Reports BNL*–**325**, 32–1–2–6, Brookhaven National Laboratory, Upton, NY.
3. Zabrodskii, A.G. (1981) Experimental determination of the compensation of neutron–doped germanium, *JETP Lett.* **33,** 243.
4. Zabrodskii, A.G. and Alekseenko, M.V. (1993) The kinetics of neutron–transmutation doping of germanium: characterization of the material and determination of the nuclear physical constants, *Semiconductors* **27,** 1116.
5. Zabrodskii, A.G. and Alekseenko, M.V. (1994) Effect of the spectrum of fission–reactor neutrons on the kinetics of neutron transmutation doping and on the dopant yield in germanium, *Semiconductors* **28,** 101.
6. Zabrodskii, A.G. and Alekseenko, M.V. (1996) *In 23rd Intern. Conf. Phys. Semicond.* **4,** 2681.
7. Zabrodskii, A.G. and Alekseenko, M.V. (2005) Fermi level scan spectroscopy of gap states in Ge and Si–Ge alloys based on the kinetics of neutron transmutation doping, *Proc. of the 23th International Conference on Defects in Semiconductors, Awaji Island, Hyogo, Japan,* to be published.

THE STATUS OF UNIFIED THEORIES OF FUNDAMENTAL INTERACTIONS

GIOVANNI COSTA
Dipartimento di Fisica "G.Galilei", Università di Padova
Istituto Nazionale di Fisica Nucleare, Sezione di Padova
Via F.Marzolo 8, 35131 Padova, Italy
giovanni.costa@pd.infn.it,

Abstract

The electromagnetic, weak and strong interactions are well described in terms of quantum field theories; they are included in the "Standard Model" of particle physics, and their complete unification can be accomplished in the frame of Grand Unified Theories. Unfortunately, gravity cannot be included in such theories, because of the difficulty of combining Einstein's general relativity with quantum mechanics. A possible solution is offered by string theories, where the basic point-like constituents of field theories are replaced by extended objects. They unify all the fundamental interactions; however, for their consistency, they require very stringent conditions. In particular, while the usual field theories are formulated in d = 3+1 space-time dimensions, the string theories require d = 10. The presence of extra dimensions would imply a modification of Newton's law at short distances, which is not excluded by the present experimental tests.

1. Introduction

The first attempts in the unification of different interactions goes back to the work of Einstein who, after the formulation of the theory of general relativity in 1916, was looking for a unified picture of the two kinds of interactions which were known at that time: gravitational and electromagnetic. However, the first theoretical model was published by Kaluza in 1921, and generalized by Klein in 1926.

After this work, there was a loss of interest in the problem of unification, because most physicists were involved, on one hand, in the development of quantum mechanics and in the construction of quantum field theory and, on the other hand, in the study of the other two kinds of interactions discovered in the 30's: namely the "strong" and "weak" interactions.

B. Di Bartolo and O. Forte (eds.), Advances in Spectroscopy for Lasers and Sensing, 499–512.

Only at the end of the 60's unification theories were considered again and specific models were proposed: first for the unification of the electromagnetic and weak interactions, and later on for the inclusion also of the strong interactions. These unifications were formulated in terms of quantum field theories.

Gravity was excluded in these unifications, essentially because nobody succeeded in building a satisfactory quantum field theory for the gravitational interactions. Only more recently, the task of unifying all forces, including gravity, was performed outside the domain of field theory, namely in the frame of string theories.

In the following we shall give a brief historical account of these developments. Since the gravitational interaction need a special description with respect to the other forces, we shall start with a short introduction to general relativity, which is the modern theory of gravity.

We shall see that the string theories imply the extension of space, specifically the introduction of six extra dimensions. This feature can give rise to modifications of the Newton's law of gravity at very small distances. In the last section, we shall briefly discuss these modifications.

2. From Special to General Relativity

The year 1905 was the "annus mirabilis", in which Albert Einstein developed three fundamental ideas, which opened the road to new important developments: the photoelectric effect, the brownian motion and the special theory of relativity.

I shall mention only the last contribution, for its relevance with the later discussion (For detailed information about the work of Einstein and the topics discussed in sections 2 and 3, see [1,2]. This theory is based on the two assumptions:

a) the speed of light has the same constant value in all inertial frames; it is the maximum value allowed for the transmission of physical signals;

b) the laws of physics are the same in all inertial frames.

The main implication of these assumptions is the fact the Newton's concept of absolute space and time has to be abandoned: the measurements of space distances and time intervals depend on the relative speeds of the observers.

The relation between two different reference frames are given by the Lorentz transformations, from which one derives the specific properties of length contraction and time dilation. What is invariant is the square distance in the four-dimensional Minkowsky space-time:

$$ds^2 = \eta_{\mu\nu} x^\mu \; x^\nu = x_0^2 - x_1^2 - x_2^2 - x_3^2 \qquad (1)$$

where $x_0 = ct$ and $\eta_{\mu\nu}$ ($\mu,\nu = 0, 1,2, 3$) is the metric tensor

$$\eta_{\mu\nu} = (1,-1,-1,-1). \tag{2}$$

Another very important result is, of course, the equivalence of mass and energy, represented by the famous formula $E = mc^2$.

Soon after the formulation of the theory of special relativity, Einstein was trying to include gravity in the theory. Newton's law

$$F = G_N m_1 m_2 / r^2, \tag{3}$$

which specifies the gravitational force between two masses, does not depend on time; it assumes implicitly that there is an *instantaneous action* between the two masses even if they are very far apart. In fact, Newton himself was not satisfied with the idea of an action at a distance, but he did not try to solve this difficulty.

This was a crucial point for Einstein, since this law is in contrast with the postulate of special relativity, i.e. that no physical signal can travel faster than light.

The starting point of his research was the *equivalence principle*, i.e. the equivalence between inertial and gravitational masses. This idea is implicit in the results of the experiments carried out by Galileo on the free falling bodies. He deduced that they fall with the same acceleration (in vacuum, i.e. in the absence of any friction). Since then this equivalence was verified with great accuracy, in particular by Lorand Eötvös and by Robert H.Dicke.

The relevant equations are the second principle of dynamics

$$F = m_I a \tag{4}$$

and Newton's law of gravitation for a body on the Earth

$$F = G_N m_G M / R^2, \tag{5}$$

where M and R are, respectively, the mass and the radius of the Earth. In the above equations, we make distinction between the inertial mass m_I (which measures the *passive* resistance to the change of motion) and the gravitational mass m_G (which represents an *active* element of the gravitational force). In the case of a falling body, they become

$$P = m_I g$$
$$P = G_N m_G M / R^2, \qquad (6)$$

from which we get for the gravitational acceleration

$$g = (m_G / m_I) G_N M / R^2, \qquad (7)$$

and the experiments indicate that

$$m_I = m_G . \qquad (8)$$

This equality lead Einstein to formulate a stronger version of the equivalence principle, as the equivalence between *gravity* and *acceleration*.

This idea was considered by himself as the happiest thought in his life; he wrote:

...the gravitational field has only a relative existence. In fact, for an observer who is falling from the roof of his house there is [at least in his closed vicinity] *no gravitational fiel.*

A reference system in acceleration is locally equivalent to a system in presence of gravity. Inertial forces and gravity can compensate each other (locally), and the system behaves (for an observer located in it) as an inertial frame. For instance, in a spacecraft or in a satellite in stationary orbit around the earth, the gravitational force is compensated by the centrifugal forces and the astronauts do not experience any force.

In general, when gravity is present, there is no longer distinction between inertial and non-inertial systems. For this reason, the program of Einstein was to extend the requirement of invariance of the laws of physics under the Lorentz transformations to the invariance of under general transformations between two reference systems in arbitrary relative motion.

In the presence of gravity, a free falling body describes, in general, a curved trajectory, so that it is convenient to use a curvilinear system of coordinates. The transformations are no longer linear as in the case of the Lorentz ones, and this fact requires a new mathematical formalism, which is similar to that employed in the treatment of non-Euclidean geometries.

Let us consider a simple example, i.e. a system which consists in a disk rotating with uniform speed about an axis perpendicular to it. While the length of the segments along the radius of the disk are not modified since there is no radial movement, there is a modification in the length of the segments along the circumference: as a consequence, the ratio of the circumference to the radius is different from 2π. For the equivalence between gravity and acceleration, the same occurs, in general, in the case of pure gravity. This shows that the geometry is no longer Euclidean (e.g. the ratio between the circumference and the radius is different from 2π; the sum of the internal angles of a triangle is different from 180°, etc.).

In 1828 the great mathematician Carl Friedrich Gauss introduced the idea of non-Euclidean geometries, for which the fifth postulate by Euclides (parallel axiom) is no longer valid. Few years later, also Nikolaij Lobacevskij and Janos Bolyai, independently, proposed non-Euclidean geometries. We should quote also Elwin Christoffel who generalized some methods introduced by Gauss, but it was Georg Berhard Riemann who extended the differential geometry to spaces (manifolds) with arbitrary curvature and dimensions.

It is easy to visualize non-Euclidean geometries in the case two-dimensional spaces. In this case one defines a coefficient of curvature

$$K = (\alpha + \beta + \gamma - \pi) / A, \tag{9}$$

where α, β, γ are the internal angles of a triangle and A is its area. One makes the distinction: elliptic geometry for $K > 0$; hyperbolic *for* $K < 0$, and Euclidean for $K = 0$.

In fact, we are dealing with a four-dimensional space-time, and in this case one need many more parameters to specify the curvature or, in general, the local properties of the space. The metric tensor $\eta_{\mu\nu}$ of the *flat* (pseudo-Euclidean) Minkowsky space of special relativity is replaced by the symmetric tensor $g_{\mu\nu}$ which is a function of the coordinate x^μ:

$$ds^2 = g_{\mu\nu}(x) x^\mu x^\nu \tag{10}$$

Einstein worked very hardly to obtain the right equations of general relativity. His friend, the mathematician Marcel Grossman, informed him that what he was looking for was already available. It was the absolute differential calculus developed by Gregorio Ricci-Curbastro and Levi-Civita in Padua. In fact, as Einstein himself wrote:

The mathematical instruments necessary for the theory were already available in the "absolute differential calculus", which is based on the researches by Gauss, Riemann and Christoffel on non-Euclideal manifolds, and which was raised to system by Ricci-Curbastro and Levi-Civita and applied by them to problems of theoretical physics.

With the help of this formalism and the use of the *tensors* invented by Ricci-Curbastro, Einstein was able to obtain the famous formula ($k = 8\pi G_N / c^2$):

$$R_{\mu\nu} - g_{\mu\nu} R = kT_{\mu\nu}, \tag{11}$$

where the term on the l.h.s. describes the local properties of the space-time, and $T_{\mu\nu}$ on the r.h.s. is the energy-momentum tensor. In words (again by Einstein):

...matter tells space how to curve, but the curved space also tells matter how to move.

Unlike the other equations of physics, Einstein's equation has a dual role: as a field equation and as an equation of motion.

In the above equation, $g_{\mu\nu}$ is the tensor that, in the case of special relativity, reduces to $\eta_{\mu\nu} = \eta^{\mu\nu} = (1, -1, -1, -1)$; it contains 10 independent elements, which are needed to specify the local properties of space-time and which, in general, depend on the space-time coordinates. $R_{\mu\nu}$, that is called the Ricci tensor, can be expressed in terms of the derivatives of $g_{\mu\nu}$, and $R = g^{\mu\nu} R_{\mu\nu}$.

It is important to remark that the tensor $g_{\mu\nu}$ plays the role of the *gravitational field* in the theory of general relativity.

We want to stress that the geometrical properties of space-time vary from point to point, so that one is always dealing with local properties. However, since the gravitational effects are in general very weak, they produce very small deviations from the *flat* geometry of the Minkowski space-time.

From the principle of least action one can derive the path of a particle in a gravitational field. A particle in free fall moves on a *geodesics*, which is a curved line whose length is an extremum in the 4-dimensional space-time.

Before leaving this section, it seems appropriate to devote a few lines to the experimental tests of general relativity. When it was formulated in 1916, the only prediction which could be tested successfully was the anomalous motion (precession of the perihelion) of the planet Mercury.

No experimental information was available for the other predictions, such as the light deflection by a gravitational field and the gravitational red shift. It was only on the occasion of the total solar eclipse in 1919 that the first prediction was tested by astronomers. The light from a distant star, according to the general relativity, is bent by the Sun, but the deviation could be observed only during an eclipse: the theoretical estimate agreed well with the measurements. The news spread all around the world and Einstein became the most famous scientist.

Nowadays general relativity has been verified accurately, but still an important prediction has not been tested yet, i.e. that of the gravitation waves.

The Einstein's equation predicts that in the presence of gravitational perturbations (such as in the case of a supernova explosion or of a rotating binary system) there is emission of gravitational waves. In the case of small perturbations, the gravitational field $g_{\mu\nu}$ can be written in terms of small deviations with respect the Minkowksky tensor:

$$g_{\mu\nu} = \eta_{\mu\nu} + h_{\mu\nu} \tag{12}$$

The quantity $h_{\mu\nu}$ satisfies a wave equation, which shows that the gravitational perturbations are waves which propagate with the speed of light. The instantaneous gravitational action of Newton is then replaced by the propagation of a field, in analogy with the electromagnetic case, and this solves the problem of Newton action at a distance.

There is an indirect evidence due to two astronomers, Russel Hulse and Joseph Taylor, who discovered in 1982 a special pulsar. Pulsars are believed to be rotating neutron stars which emit pulses of radio waves at very regular intervals. The detected pulsar was in orbit around another compact object; the two astronomers observed a decrease of the orbital period and they explained it as a consequence of the energy loss due to the emission of gravitational waves. The theoretical estimate agrees well with the experimental observation.

Several experiments are running for a direct detection of gravitational waves. However, due to the extreme weakness of the gravitational interaction, it is extremely hard to detect such waves. These experiments make use of large detectors; the largest of these are based on laser interferometers; also a space mission (LISA = Laser Interferometer Space Antenna) is planned by NASA and the European Space Agency.

The discovery of the gravitational waves will provide a further important test of general relativity but also a new mean to explore the Universe.

3. First Attempts of Unification

As already pointed out, before 1930 only two kinds of interactions were known: gravitational and electromagnetic. The analogy between gravitational waves and the electromagnetic ones induced Einstein to study the possible unification of these interactions. But it was Teodor Kaluza, an unknown mathematician in Könisberg (now Kaliningrad), who proposed an original model of unification in 1919, which was published in 1921. His idea was improved by a Swedish physicist, Oskar Klein, and now the unification theory goes under the name of Kaluza-Klein (KK theory).

Kaluza had the strange idea of introducing an extra spatial dimension. As a matter of fact, this idea was previously formulated by Gunnar Nordström in 1914; however, since the general relativity was not existing at that time, his model was not based on the correct theory of gravity.

In the KK theory, the 4-dimensional space-time of general relativity is replaced by a 5-dimensional one:

$$x_\mu \to z_M \equiv (x_\mu, x_5), \tag{13}$$

and the gravitational field $g_{\mu\nu}$ is replaced by the tensor g_{MN} (M, N =1,... 5), which satisfies a generalized Einstein's equation. Kaluza assumed that the fifth dimension is curled up into a tiny ring, so small that it could not be observed by any instruments. This should explain why we do not detect the extra dimension.

The set of the four components $g_{\mu 5}$ of the metric tensor g_{MN} behaves as a four-vector which can be identified with the electromagnetic potential $A_\mu(x)$: it is remarkable that it satisfies the Maxwell equations. The component g_{55} behaves as a scalar field ϕ(x) which satisfies the Klein-Gordon equation. The situation can be summarized as follows:

$$g_{MN} \to \begin{vmatrix} g_{\mu\nu} + \kappa^2 A_\mu A_\nu & \kappa A_\mu \\ \kappa A_\mu & \phi \end{vmatrix} \tag{14}$$

where $\kappa^2 = G_N / \hbar c.$

The *KK* idea was rather interesting but the theory, even if appreciated by Einstein, did not received much attention; perhaps because time was not yet mature for this kind of ideas, and also because new kinds of interactions (strong and weak) where disclosing in the realm of nuclear and particle physics, and physicists were more interested in understanding their properties.

Moreover, the *KK* theory put on the same footing the gravitational and the electromagnetic force, disregarding the huge difference in their strengths, and this feature was rather unsatisfactory.

4. Unification of Electroweak and Strong Interactions

Only many years later, namely in the late 60's, unification of interactions was becoming fashionable again. But it was the unification of the strong and weak interactions with the electromagnetic one, and not with gravity.

Before speaking of these unified theories, it convenient to summarize very briefly the properties of those interactions (See also [3]).

First of all, we remind that the elementary constituents of matter are the following:

a) quarks *(u* = up, with electric charge Q = 2/3; *d* = down, Q = –1/3)

b) leptons (*e* = electron, Q = –1, ν = neutrino, Q = 0) .

Each quark occurs in three different species identified by different quantum numbers conventionally called *colors*, which we shall distinguish by the numbers (1,2,3). Moreover, for each particle there is an antiparticle with opposite electric charge and additive quantum numbers $(\bar{u}, \bar{d}, e^+, \bar{\nu})$.

All this particles have half-integer spin ½ , so that they satisfy the Fermi-Dirac statistics. Each particle can have two different spin orientations or two helicity states (spin along the direction of motion) which are called left-handed and right-handed, except for the neutrino which can be only left-handed. If we consider left-handed helicity, there is a set of 15 different particle states:

$$(u_1, u_2, u_3, d_1, d_2, d_3, e^-, \nu, \bar{u}_1, \bar{u}_2, \bar{u}_3, \bar{d}_1, \bar{d}_2, \bar{d}_3, e^+) \tag{15}$$

Similar situation occurs for the right-handed states, with the exception of the neutrino, which is replaced by the antineutrino $\bar{\nu}$.

There are two other sets of 15 particles which are the replicas of this pattern at higher masses: these particles are produced in collisions at very high energies; they are all unstable and they decay in very short times into the particles specified above. Here we shall limit ourselves to the set given in eq. (15).

The interactions in which the particles participate are: strong, weak and electromagnetic, while the effect of the gravitational ones can be completely neglected. Only quarks interact strongly; they are bound in the nucleons (proton = *uud* and neutron = *udd*) and they are subject also to weak and electromagnetic interactions. Electrons interact weakly and electromagnetically and neutrinos only weakly

The different kinds of interactions are described in terms of fields; quantization of these fields gives rise to a set of quanta (photons, gluons and intermediate bosons), which are bosons of spin 1 (they satisfy the Bose-Einstein statistics). Each interaction among particles is represented by exchange of the quanta of the field.

The properties of the different interactions and the relative quanta are listed in Table I where, for the sake of completeness, gravity is also included (gravitons are the quanta of the field). They differ not only in the properties of the quanta exchanged, but also on their strengths i.e. on their couplings.

The different interactions possess specific symmetry properties (they are called "internal" because they do not take place in the usual space-time), and they are described by field Lagrangian which exhibits such symmetries.

TABLE 1. Properties of the fundamental interactions.

Type of interaction	Relevant system or process	Elementary Fermions involved	Strength ($\eta = c = 1$; m_P= proton mass)	Range (cm)	Quanta of radiation
EM	atoms	charged	$\alpha = \frac{e^2}{4\pi} \approx \frac{1}{137}$	∞	photons
Weak	β-decay	leptons and quarks	$\alpha_W = \frac{G_F m_P^2}{4\pi} \approx 10^{-5}$	2.5×10^{-16}	$W^{\pm}, Z^{o}$
Strong	nuclei	quarks	$\alpha_G = \frac{g^2}{4\pi} \approx 1$	1.4×10^{-13}	gluons
Gravity	solar system	all	$\alpha_G = \frac{G_N m_P^2}{4\pi} \approx 6x10^{-39}$	∞	gravitons

The different interactions possess specific symmetry properties (they are called "internal" because they do not take place in the usual space-time), and they are described by field Lagrangian which exhibit such symmetries.

In the years 1961-68, it was shown by Glashow, Weinberg and Salam that the electromagnetic and weak interactions can be unified in a field theory which possesses the "gauge" symmetry described by the group

$$SU(2) \times U(1), \tag{16}$$

where *gauge* means that the group transformations which leave the Lagrangian are local, i.e. they vary from point to point. Such symmetry is valid above an energy scale of the order of 100 GeV; above this energy both the photon and the weak bosons ($W^{\pm}$ and Z^0) are massless. However, below this energy the symmetry is broken (by a mechanism of spontaneous breaking) and the weak bosons acquire mass different from zero. They are very heavy with respect to the proton mass $M_P \approx 0.94\ GeV$: $M_W \approx 80\ GeV$ and $M_Z \approx 90\ GeV$. These high values of the masses explain why the weak interactions become much smaller than the electromagnetic ones, since the effective coupling is proportional to the inverse squared mass:

$$\alpha_W = g_W^2 / M_W^2 \tag{17}$$

where $g_W \approx e$ (elementary electric charge).

A complete description of the three interactions which are relevant for elementary particles requires a Lagrangian field theory invariant under the gauge transformations of the group

$$SU(3) \times SU(2) \times U(1). \tag{18}$$

With respect to this group quarks behave as SU(3)-triplet and SU(2)-doublets:

$$\begin{vmatrix} u_1 & u_2 & u_3 \\ d_1 & d_2 & d_3 \end{vmatrix} \tag{19}$$

while the leptons are SU(3)-singlets and SU(2) doublets or singlets

$$\begin{vmatrix} e^- \\ \nu \end{vmatrix} \qquad \begin{vmatrix} e^+ \end{vmatrix} \tag{20}$$

This situation represents only a partial unification, since for each simple group in the product there is a different coupling: g_S for strong, g_W for weak and e for electromagnetic. Their values are rather different at low energies (below 100 GeV). It was discovered that the couplings are not really constant, since the higher order corrections modify their values and they depend on the energy scale. Going to very high energies their values become closer; imposing also the "supersymmetry", which implies symmetry relations between fermions and bosons, the three couplings become closer and closer at increasing energies, and coincide at a certain scale called unification scale.

Beyond the unification scale the appropriate field theories are the Grand Unified Theories (GUT), which possess a higher symmetry described by a group such as *SU(5)* or *SO(10)*. The unification scale is extremely high, of the order of 10^{15} GeV; this is due to the fact that the coupling corrections vary logarithmically with the energy, so that it takes a large energy scale to get unification.

It is interesting to note that all the 15 particle states listed above can be classified in a representation (multiplet) of the group:

In the case of SU(5): 15 = 5 + 10

In the case of SO(10): 16

The structure of these bigger multiplets is just the right one to include the SU(3) and SU(2) multiplets shown above.

The case of SO(10) is very interesting, since all particles can fit into a single multiplet (irreducible representation of the group); moreover the 16-miltiplet contain an extra state which would correspond to a right-handed neutrino, i.e. the missing element for generating a neutrino mass different from zero, as discovered in the last few years.

5. Inclusion of Gravity in GUT

We said that the gravitational force can be completely neglected in the realm of elementary particles. This is true in general; however at extremely high energy also gravity becomes relevant. This occurs at the so-called Planck scale $c^2(\hbar c/G_N)^{1/2} \approx 10^{19}\ GeV$ corresponding to distances of the order of the Planck length = $(G_N\hbar/c^3)^{1/2} \approx 10^{-33}\ cm$. Even if such energies are obviously completely un-accessible, it is in principle interesting to understand the behavior of matter in such conditions; moreover this can become important in cosmological problems.

There is however a big problem: to include gravity in the world of particles it is necessary to marry gravity with quantum mechanics and this is really a difficult task. In fact, it was shown that it is not possible to formulate a quantum field theory for gravitational interactions, since such theories are not renormalizable, that is they are plagued by infinities which cannot be reabsorbed in theoretically undetermined parameters.

In order to build a consistent theory one has to abandon the usual schemes and introduce new concepts. Specifically, one has to go beyond the usual field theory in which particles are considered point-like and replace them by strings. In the last 20 years lot of theoretical physicists have developed different kinds of string theories; they possess interesting features, but no phenomenological tests are yet available.

Here I would like to discuss a specific feature of these theories, i.e. the fact that, in order to have a coherent and consistent theories, without infinities, one has to extend the space-time from 4 to 10 dimensions, i.e. 6 extra space dimensions are required. This is due to the fact that the symmetries of the theories have to be preserved at quantum level to keep all the good properties; these symmetries must not be spoiled by the higher order corrections, and this fixes the number of extra dimensions.

Superstring theories, which possess also supersymmetry, can unify all the fundamental interactions. We mentioned already the attempt of unification of gravity and electromagnetic interactions by Kaluza and Klein, who introduced one extra dimensions. If one follows the same attitude of the KK theory, one should assume that the six extra dimensions are compactified in very tiny regions. If this were the case, it would be difficult to test these theories, and one would look for indirect tests.

It was shown recently that not all the extra dimensions need to be small; some of them could be large enough to give rise to detectable effects. We would like to discuss briefly these models, since they would imply some interesting physical consequences.

In 1998, N. Arkani-Hamed, S. Dimopoulos and G.Dvali [4] suggested that all the interactions, except the gravitational one, are confined on a 4-dimensional *brane* (i.e. our space-time), while only gravity is aware of some of the extra dimensions. The electromagnetic, weak and strong forces and all the particles which are sources or carriers (quanta) of these interactions would be trapped in the three spatial dimensions, while only gravitons would be able to leave the 3-dimensional space and move into (at least some of) the extra dimensions.

In fact, since the highest energy particle accelerators extend our range of sight to include the electro-weak and strong forces down to small scales of the order of 10^{-16} cm, we are obliged to assume that either these interactions are acting in very small regions of the extra dimensions, or they are bound in the 3-dimensional space and they do not propagate in the extra dimensions. On the other hand, gravity has been tested only down to the millimeter range, so that there is no stringent limit for its extension to the extra dimensions.

Let us suppose that gravity is extended to some (say n = 1, 2,..) of the extra dimensions. Then the force is modified with respect to the Newton's law; according to Gauss theorem, one gets

$$F = G_{(3+n)} m_1 m_2 / r^{2+n} \tag{21}$$

where $G_{(3+n)}$ replaces the coupling constant G_N given in eq.(3). If this is the case, one would expects deviations from the Newton's law, which has been tested from astronomical distances to a few fractions of millimeter.

Let us suppose that the extra dimensions are compactified into a hyper-spherical region of radius r_C. Then the above relation will be valid for $r < r_C$, and it could be rewritten as

$$F = m_1 m_2 / M_F^2 r^{2+n} \tag{22}$$

with the introduction of a new mass scale which can be much smaller that the Planck mass M_P. On the other hand, for distances much larger that r_C the force can be described by the relations:

$$\begin{aligned} F &= G_{(3+n)} m_1 m_2 / r_C^n r^2 = m_1 m_2 / M_F^2 r_C^2 r^2 \\ &= m_1 m_2 / M_P^2 r^2 = G_N m_1 m_2 / r^2, \end{aligned} \tag{23}$$

where we have used the definitions $M_P^2 = r_C^n M_F^2$ and $G_N = M_P^{-2}$ and, for the sake of simplicity, the system of the natural units $\hbar = c = 1$).

As an example, by assuming that the new energy scale is of the order 10^3–10^4 GeV, and considering the case n = 2, one gets for the compactification radius r_C values in the range 0.02–2.0 mm. With n = 1 one would obtain too large values for r_C which are completely excluded by experiments, while for $n > 2$ the values of r_C become too small.

This kind of considerations, together with of the general interest in testing Newton's law for distances smaller that 1 mm, have stimulated a series of experiments. They are of the type of those carried out by Cavendish for the determination of the gravitational constant G_N, of which they represent more refined and sophisticated versions. In fact, they have to disentangle the pure gravitational force by other perturbations such as the Casimir effect due to quantum fluctuations in the region between the test masses. The present limit shows that there are no modification of Newton's law down to 0.2 mm. Future experiments are planned to lower this limit to a few microns [5].

Acknowledgements

I thank Professor Rino Di Bartolo for inviting me to participate in the stimulating and friendly atmosphere of his School.

References

1. Pais, A. (1982) *'Subtle is the Lord...' The Science and Life of Albert Einstein*, Oxford University Press, Oxford.
2. Will, C.M. (2005) Relativity at the centenary, *Physics World*, pp. 27-32.
3. Costa, G. (1997) Unification of the fundamental interactions: problems and perspectives, in B. Di Bartolo (ed.) *Spectroscopy and Dynamics of Collective Excitations in Solids*, Plenum Press, New York, pp. 581-597.
4. Arkani-Hamed, N., Dimopolous, S. and Dvali, G. (2002) Large extra dimensions: a new arena for particle physics, *Physics Today* 55N, 35-40.
5. Binétruy, P. (2002) How many dimensions to our Universe?, *Europhysics news*, 33/2, 54-57.

ANGULAR MOMENTUM OF THE HUMAN BODY

JOHN DI BARTOLO
Polytechnic University of Brooklyn
6 Metrotek Center
Brooklyn, NY 11201, USA
jdibarto@poly.edu,

Movement of the human body in the sagittal plane (the plane which slices the body between left and right) can be simulated with a simple model made up of four rods and a sphere. In a situation where the torque about the center of mass is zero (such as diving into water) the angular momentum of the body about the center of mass is conserved. In a situation where the torque about the center of mass is non-zero (such as jumping or swinging by one's hands from a bar), the angular momentum of the body about the center of mass has a rate of change equal to the torque. Based on these principles, as the configuration of the model's limbs changes, the orientation of the model changes accordingly. Using this model, a software authoring-application called Director (by Macromedia) was used to program simulations for Physics Curriculum and Instruction. The simulations demonstrate the three above-mentioned activities: diving, swinging, and jumping. A user of this software can determine how the "athlete" will move his/her joints over a period of time, and the resulting motion of the body is shown.

B. Di Bartolo and O. Forte (eds.), Advances in Spectroscopy for Lasers and Sensing, 513.

CLIMATE CHANGE AND GLOBAL SECURITY

Anthropogenic Greenhouse Gases, Fossil Fuels and the Earth's Climate: the overlooked self-imposed threat to our security. A Brief Review

GEORGE J. GOLDSMITH
Department of Physics
Boston College, Chestnut Hill, MA 02467
goldsmig@bc.edu,

1. Introduction

It is generally conceded that human activities over the past five centuries have had a measurable impact on the world environment. Industrialization, urbanization, agriculture, deforestation, particularly during the 19th century to the present, have contributed to what is now regarded as general warming of the atmosphere brought about by an increase in the concentration of anthropogenic greenhouse gases in the atmosphere. Within the past few decades improved measurements of environmental factors and computer modeling have led to predictions that this warming will lead to serious alterations in the climate causing shifting rainfall patterns, costal flooding, periodic droughts, and other undesirable changes. The models suggest that large scale world wide action must be undertaken, especially by the heavily industrialized nations, to reduce or even reverse this trend. While it is certain that warming is occurring, the magnitude of the accompanying climatic changes and their time scale is uncertain. This is because of the difficulty in developing a comprehensive model of a system so complex as the atmosphere of the Earth. Thus major questions arise such as: "To what extent are these predictions valid?" "Insofar as they are valid what actions can we, or should we, undertake to hold off or reverse the current trend?" "Are long term temperature fluctuations really natural phenomena only incrementally impacted by human activity?.

These matters had been gaining more attention with time, education, and collection of more pertinent data as a consequence of improvement in measurement techniques and of models. Now however, with the current sensitivity to homeland and world security threats of the near-term, this potential long-term environmental security threat has drifted into the background. This paper will address questions of the magnitude and indeed the validity of concern over the long term impact of global warming. Is there anything that can be done in the near term to reverse the current trend.

Now, in the US and in many other nations, security calls to mind what the US and many other nations are said to be engaged in what US president has characterized as "The War on Terror". If we examine all matters related to world wide hazards not just from terrorism but to national security in general, evidence of the impact of global warming looms large as a serious threat. We are aware that anthropogenic substances introduced into the atmosphere, are increasing the rate of rise in global average temperature. Convincing evidence points to the expectation that within the next ten decades we could encounter significant coastal flooding,

B. Di Bartolo and O. Forte (eds.), Advances in Spectroscopy for Lasers and Sensing, 515–538.

perplexing shifts in rainfall, in seasonal temperature patterns, and in extensive new regions of drought. If not anticipated and planning is not initiated now to remedy the consequences, it is likely to result in destruction of vulnerable areas of the global seacoast, serious reduction in food productivity, and consequent widespread famine. Because of oceanic thermal inertia, these events may occur even if no further greenhouse gases are added to the environment. Under such circumstances it would be prudent to initiate planning now for dealing with such eventualities(1).

The topic divides itself into two parts. First is an historical account looking back from the present to a few million years ago, and then surveying the progress of awareness of the phenol-menon of global warming from the beginning of the 19th century. We will survey the evidence of that has yielded reliable evidence of the cause of several large fluctuations of the Earth's temperature which caused "ice ages" when massive glaciers covered great portions of the Earth's surface. Further we will consider how much we now know about the duration and occurrence of those "ice ages" and how that understanding was reached[1].

In the second part we will survey the nature, composition, and origin of the greenhouse gases, why they contribute to global temperature, and the technology developed to determine the history and magnitude factors. We will find that this relatively small community of courageous, audacious, and determined climate scientists pursued successfully a very complex and daunting problem. Further we will discuss some of the anticipated consequences of phenomena caused by global warming and consider what sort of mitigating action may be warranted.

It is customary climatologists to designate time periods in the evolution of the Earth in terms of epochs. Those definitions of the geologic epochs are illustrated in Table 1 below.

First, we pause to consider the unique character of *this planet* among all the other members of our solar system, one that nurtures and protects all of humanity and all of the resident flora and other fauna. Because of its particular mass and distance from the sun, the consequent gravitational force maintains a thin layer of life–supporting gases that sustain all life. It is also particularly significant that its temperature is compatible with that of liquid water nearly everywhere and that *ice is less dense than liquid water*. We and our entire environment are *what they are* because of this. But we must also be aware that in spite of its apparent robust nature it is a very fragile and delicately balanced system. Indications of long term as well as abrupt changes in the climate throughout time have been detected since it was appropriate to consider a "climate" at all.

The early searches for the causes of climate change were inspired in part by evidence of the occurrence of *Ice Ages* over the early Earth's history, raising the concern that we might be

[1] A thorough narrative of the history of Global Warming is the publication of Spencer R. Weart, *The Discovery of Global Warming*, Harvard University Press, 2003. For a comprehensive list of the highlights of the study of Global Warming, see http://www.aip.org/history/climate/timeline.htm.

Table 1 Geological Epochs (from Encyclopedia Britannica).

GEOLOGICAL EPOCHS

Years Ago	Era	Period	Epoch	Major Events
10,000	Cenozoic	Quaternary	Holocene	Extinction of many large mammals. Humans evolve
2 million			Pleistocene	
5		Tertiary	[Pliocene]	
25			[Miocene]	
38			[Oligocene]	
55			[Eocene]	
65			[Paleocene]	
144	Mesozoic	Cretaceous		Dinosaurs reach peak, go extinct. Primitive placental mammals
213		Jurassic		Marsupial mammals
248		Triassic		Egg-laying mammals
286	Paleozoic	Permian		
360		Carboniferous		
408		Devonian		
438		Silurian		
505		Ordovician		Invertebrates dominant; first fishes
570		Cambrian		Major diversification

suddenly confronted with an abrupt return of a new Ice Age. NOAA Paleoclimatologists search for clues to this phenomenon call them "Proxy Data" because there is no direct measure of events so far into the past. They include historical records, corals, fossil pollen, tree rings, ice cores, ocean and lake sediments and physical evidence of glacial movement over the Earth's surface (Fig. 1a).

The horizontal striations in this outcrop, in New York's Central Park, were caused by scouring from boulders imbedded beneath an advancing glacier. Photo by Tina Gaud, © American Museum of Natural Historv.

Fig.1a. Glacial Tracks.

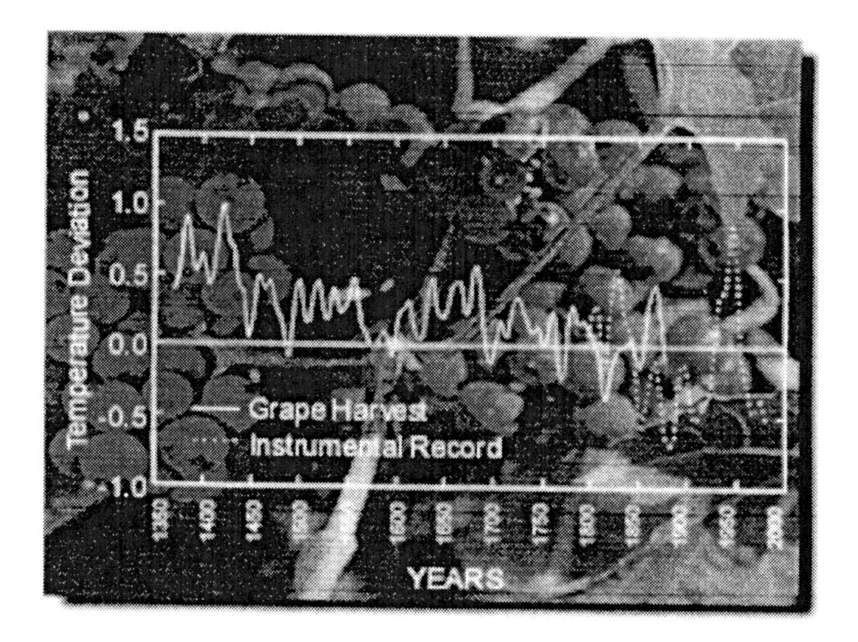

Fig. 1 Grape Harvest (Germany) vs. Temperature 1350–2000 (Ladurie and Boulant, 1981[R2]).

Figure 1 is an early historical record of the German grape harvest from the year 1350 until 1900 matched with the temperature record, but only from 1750 to 2000.

We all know that the Earth receives a net thermal input on the "day" side while radiating most of its new burden of heat into the void on the night side. How to account for the Earth's relatively uniform life-supporting temperatures was less obvious to scientists and philosophers of the late 18th and early 19th centuries.

At least two important questions persisted. The first: *Why doesn't life on Earth, exposed to a constant bath of sunlight as it rotates on its axis, simply keep warming until it burns up like a chicken rotating too long over a sp*it. The second, and perhaps more perceptive: *The heat deposited by the sun on the day side radiates away during the night hours so that it hardly warms up at all except for the solar heat stored within it during the day.* (Fig. 2).

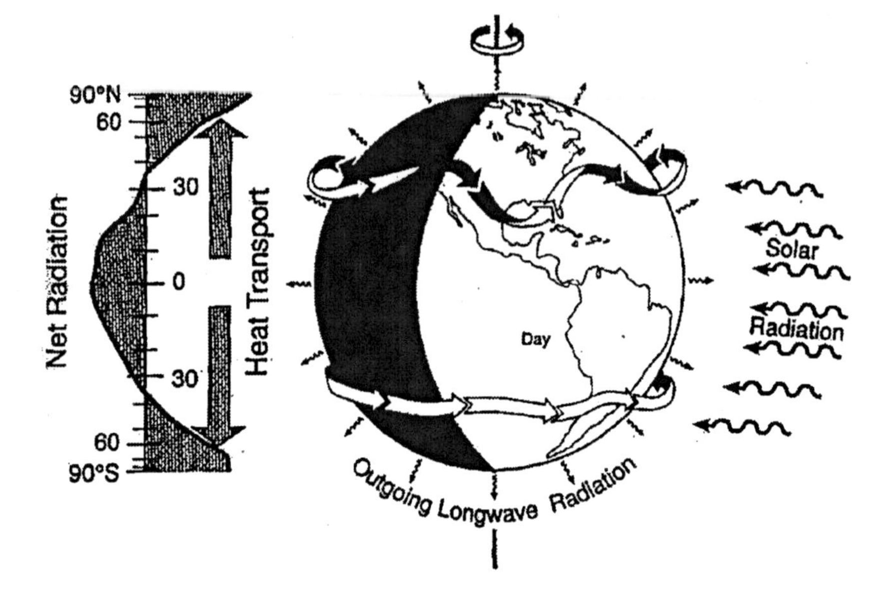

Fig. 2 Earth Radiation Input—output.

From Fig. 2 it is obvious that as the Earth turns, at any instant of time there is heat input on the sunlit side, and thermal output on the dark side. The question is "By what mechanism does the mean temperature on the Earth reach equilibrium?"

Joseph Fourier (1768-1830), best known by every science student for his "series"and his explanation of the composition of wave forms, who, in the 1820's, took up the question(R2a) *(Fourier, Joseph, Remaques Generals Sur Les Temperatures Du Globe Terrestre Et Des Espaces Planitaire)* .

Gale Christianson in her book, *Greenhouse*: *The 200 Year\ Story of Global Warming* (R3), wrote: "...Fourier began to ponder the question of how the Earth stays warm enough to support the diverse range of flora and fauna inhabiting its surface. Why is the heat generated by the sun's rays not lost after striking and bouncing off the great oceans and landmasses of the world? Taking a pen in hand, he set down a novel hypothesis. Much of the heat does in fact escape back into the void, but not all. The invisible dome that is the atmosphere absorbs some of the Sun's warmth and radiates it downward to the Earth's surface. Fourier likened this thermal envelope to a domed container made of glass, a gigantic bell jar formed out of clouds and invisible gases. In coming together, the water vapor and other gases simulate a vault that conserves heat, without which all life would surely perish" What Fourier and others among his contemporaries did not initially take into account was the more subtle differences between the daytime radiation input to the Earth and the nighttime radiation output (Fig. 3).

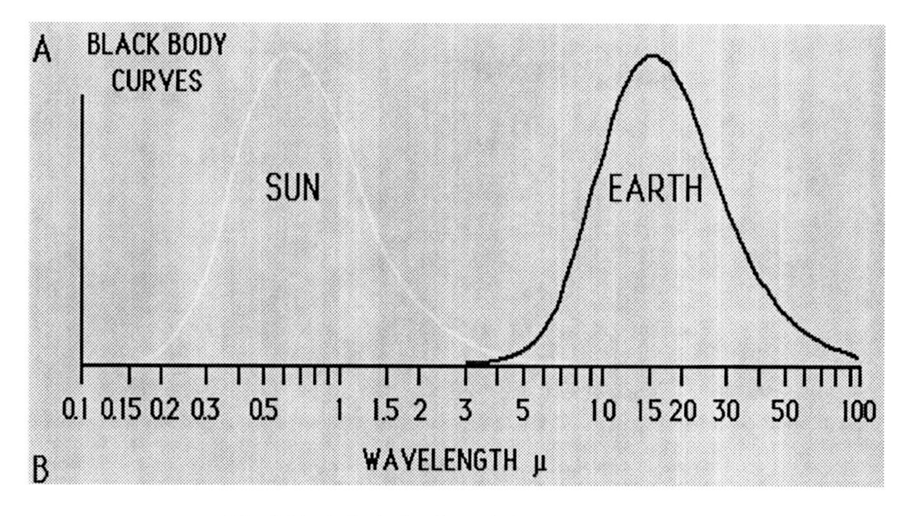

Fig. 3 Black Body Earth Radiation Input—Output.

As one would expect there is a significant difference in the spectral range between the input (uv—visible) and the output (IR).

John Tyndal(1820–1893) further pursued the matter by finding that water vapor was a strong absorber of infrared radiation as was carbon dioxide. There also are a number of other substances flushed into the air that have either heat trapping or other deleterious effects on the energy balance of the atmosphere.

(*John Tyndall's Research on Trace Gases and Climate*: "The solar heat process…the power of crossing an atmosphere; but when the heat is absorbed by the planet is so changed in quality that the rays emanating from the planet cannot get with the same freedom back into space. Thus the atmosphere admits of the entrance of the solar heat, but checks its exit; and the result is a tendency to accumulate heat at the surface of the planet (R4))

John Tyndall found that the oxygen and nitrogen displayed no significant optical absorption (see Fig. 5 for details of Tyndall apparatus). When, however, he thought to measure the absorption of coal gas, available from the jet in his laboratory, a mixture of methane and other hydrocarbons, carbon dioxide, sulfur dioxide and who knows else, he found the behavior that led him to find the role that carbon dioxide and other atmospheric gases played in the behavior of the night time atmosphere with regard to energy transport. These trace contaminants in the dominant oxygen–nitrogen mix, have become known as the "greenhouse gases".

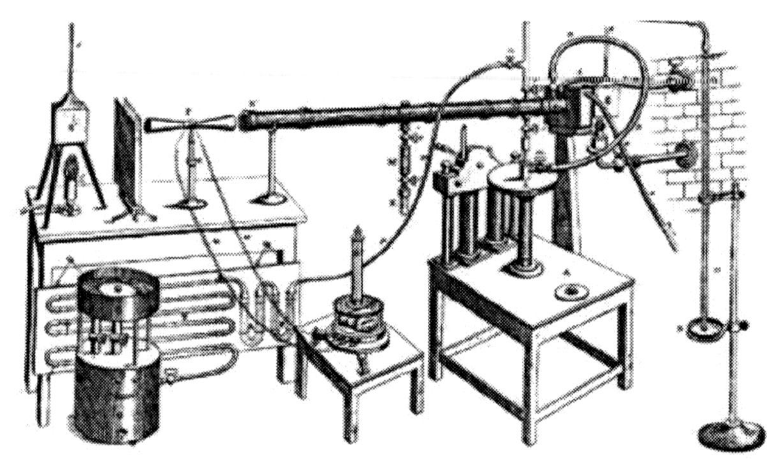

Tyndall's experimental apparatus, the first ratio spectrophotometer, consisted of a long tube that he filled with various gases. The ends of the tube were capped with slabs of rock salt crystal, a substance known to be highly transparent to heat radiation. A standard Leslie cube emitted radiation that traversed the tube and interacted with the gas before entering one cone of a differential thermopile. Radiation from a second Leslie cube passed through a screen and entered the other cone. The common apex of the two cones, containing the differential thermopile junction, was connected in series to a galvanometer that measured small voltage differences. The intensity of the two sources of radiation entering the two cones could be compared by measuring the deflection of the galvanometer, which is proportional to the temperature difference across the thermopile. Different gases in the tube would cause varying amounts of deflection of the galvanometer needle. If the intensity of the reference source of radiation was known, the intensity of the other source (and thus the absorptive power of the gas in the tube) could be calculated.

Fig. 4 Tyndall Experiment.

Table 2 (IPCC) lists the more common anthropogenic greenhouse gases. The optical absorption spectrum of a typical air sample is illustrated in Fig. 5. The influence of the greenhouse gases is termed "*radiative forcing*".

Table 2. Greenhouse Gases.[2]

	CO_2 (Carbon Dioxide)	CH_4 (Methane)	N_2O (Nitrous Oxide)	CFC-11 (Chlorofluoro-carbon-11)	HFC-23 (Hydrofluoro-carbon-23)	CF_4 (Perfluoro-methane)
Pre-industrial concentration	about 280 ppm	about 700 ppb	about 270 ppb	zero	zero	40 ppt
Concentration in 1998	365 ppm	1745 ppb	314 ppb	268 ppt	14 ppt	80 ppt
Rate of concentration change [b]	1.5 ppm/yr [a]	7.0 ppb/yr [a]	0.8 ppb/yr	-1.4 ppt/yr	0.55 ppt/yr	1 ppt/yr
Atmospheric lifetime	5 to 200 yr [c]	12 yr [d]	114 yr [d]	45 yr	260 yr	>50 000 yr

[a] Rate has fluctuated between 0.9 ppm/yr and 2.8 ppm/yr for CO_2 and between 0 and 13 ppb/yr for CH_4 over the period 1990 to 1999.

[b] Rate is calculated over the period 1990 to 1999.

[c] No single lifetime can be defined for CO_2 because of the different rates of uptake by different removal processes.

[d] This lifetime has been defined as an "adjustment time" that takes into account the indirect effect of the gas on its own residence time.

[2] It should be noted that the most significant of the Greenhouse gases, H_2O, is not listed because it is largely not of "anthropogenic origin".

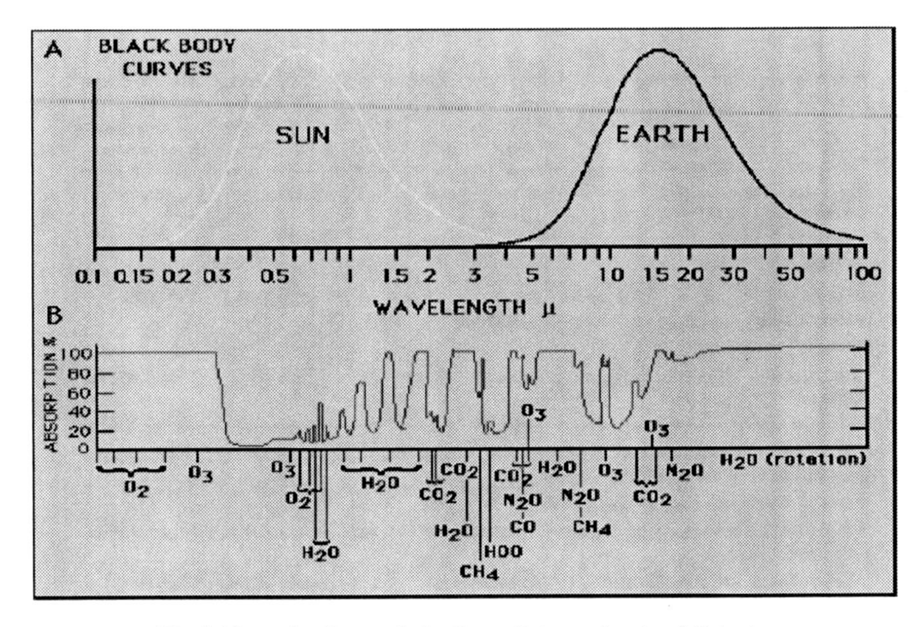

Fig. 5 Absorption Spectra in the Range Between Input and Output.

Svante Arrhenius (1859–1927) (Nobel Prize for Chemistry 1903) related the atmospheric carbon dioxide to global warming in a paper to the Stockholm Physical Society in 1895: *"On the Influence of Carbonic Acid in the Air to the Temperature on the Ground"(R*4a) He described the radiative effects of carbon dioxide and water vapor on the surface temperatures of the Earth. He calculated that the temperature of the Arctic regions would rise about 8 or 9 degrees Celsius if the carbonic acid increased 2.3 to 2.5 times its present value. In order to reach the temperature of the ice age between the 40th and 50th parallels, the carbonic acid in the air should sink to 0.62 to 0.55 of present value (Lowering the 4 to 5 degrees Celsius)". By 1904, Arrhenius became concerned with anthropogenic carbon emissions and recognized that "...the slight percentage of carbonic acid may, by the advances of industry, be changed to a noticeable degree in the course of a few centuries." He eventually suggested that an increase in atmospheric carbon dioxide due to the burning of fossil fuels could be beneficial, making the Earth's climates "more equitable", stimulating plant growth and providing more food for a larger population."

But it wasn't the warming that concerned climatologists at time, but rather the occurrence and cause of the great periods of glaciation that left their traces all over the temperate zones. How many were there over time and what was the temporal behavior of their occurrence.

On Jan 30, 1961, Walter Sullivan, the New York Times' long time science reporter, described, a week-long meeting of an international group of climate change experts in New York City under the headline,

2. Scientists Agree World is Colder

He wrote in part:

> "After a week of discussions on the causes of climate change, an assembly of specialists from several continents seems to have reached unanimous agreement on only one point: it is getting colder"
> "...Their techniques ranged from observations with Earth satellites to such methods as palyology [analysis of ancient pollens], dendrology [analysis of tree rings], and the deciphering of ancient oriental scripts.":
> "These and many other many other methods have been used to test some of the scores of theories that have been advanced to explain the ice ages. Such periods of glaciation have not occurred at regular intervals. Rather they seem to have come in batches lasting a million years or so at intervals of one or two hundred million years. We are in the midst or at the end of such a batch, which began roughly 1,000,000 years ago—the Pleistocene"
> The theories most extensively discussed this last week depended upon celestial mechanics, upon changes in transparency of the atmosphere, upon changes in the sun and upon acyclic sequence of event centering on the presence or absence of ice in the Arctic Ocean.
> "...Objections were raised to all of them. A number of the participants seemed to have come away with a suspicion that more than one factor was responsible for the long term and short term changes. Thus several of the speakers supported a modification of theory advanced during the period between the wars by M. Milankovitch (see below) ....Despite the failure of the conferees to agree, it appeared that the answer might be within their reach. As some of them noted, much remains to be done in refining dating methods with radioisotopes and in collecting data from space, from the ocean floor, and from the continents, before convincing case can be made for one of the rival theories"

3. The Search for Evidence of Global Warming

The answer to the questions raised by these many theories (often contradictory) requires determination of temporal, thermal and climatic temperature profiles over eons, a daunting prospect. By the middle of the 20th Century it was clear that not only natural phenomena, but also the acceleration of industrial activity, the increased combustion of fossil fuels, and the evolution of huge urban centers were causing changes in global weather patterns and global temperatures.

The earliest experimental studies came from, what seems to be a most unlikely source. Borings into lake beds and ocean floors revealed micron size pollen particles as well as sedimentary remains of tiny marine animals. In the 1950's better tools evolved that perhaps could point to a solution. The hardy micron–size pollen particles survive intact for many millennia and each article is characteristic of a particular plant species. From knowing the type of plant and knowledge of habits of each plant species, pollen experts, palyologists can deduce the range of temperature in a particular layer of the cores.

Throughout the last half of the 20th century several new techniques for measuring ancient temperatures profiles emerged as did other methods for dating. In the same kinds of layers that yielded pollens, there also were the shells of microscopic marine animals which are composed mostly of calcium carbonate. Dating the layer was established by measuring the relative content of the radioactive carbon isotope. C^{14} (half life: 5715 years) to the stable Carbon^{14}. Of course with a half life relatively short compared with the time scale, it was not useful for exploring very far down the geologic time line. On occasion it was possible to employ isotopes other than C^{14} such as uranium..

4. Milankovitch

One of the more popular theories for "sudden cooling" pointed to the work of a Serbian astronomer/mathematician, Milutin Milankovitch, who in the 1930's developed mathematical formulae based on the orbital variations of the sun (Fig. 6). (Rm) (Milankovitch, Milan (1930). "Mathmatische Klimilere und Astromische Theorie Der Klimaschwankungen." In *Handbuch Der Klimatologie,* W. Koppen and R. Geiger, Vol. **1, Pt. A,** pp. 1-176 Berlin :Borntraeger).

He found that the *eccentricity* of the orbit changed from a circle to a thin ellipse over a 95,000 year cycle while its *obliquity*, the angle of the axis relative to the plane of revolution around the sun , between 22.1^{o} and 24.5^{o}. Less of an angle than our current 23° causes smaller seasonal differences between the Northern and the Southern Hemispheres while larger angles produce larger seasonal differences. *Precession*, in about 12000 years will have rotated the solar axis toward Vega rather than Polaris as it is now. The confluence of the correct combinations of extreme deviations in the solar axial relation to the Earth could lead to a large reduction in the Earth's solar input .

In general the popularity of Milankovitch's astronomical explanation did not satisfy the many skeptics who believed that the Milankovitch effects were too small to create an ice age. It wasn't until the 1970's when a group, James Zachos, Justin Ravenaugh, Nicholas Shackelton, and Haiko Paliki of Cambridge University and Benjamin Flower of University of South Florida, reported an analysis of the oxygen isotope ratio O^{18} to O^{16} differences and hence the isotopic composition of ice at a particular time (—in this case as far back as the Oligocene and the Miocene epochs). It seems incredible. but the results were comparable to studies carried out years earlier from borings through sedimentary layers on the bottom of lakes and seas.

Orbital parameters

- Earth's orbit varies over time due to influence of the Sun, Jupiter, and the Moon.

- Eccentricity (ellipticity) ~100 kyr, 400 kyr

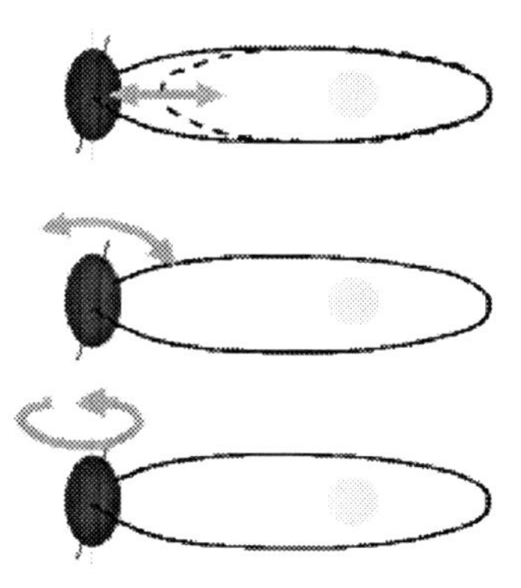

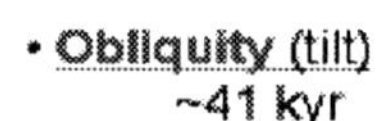

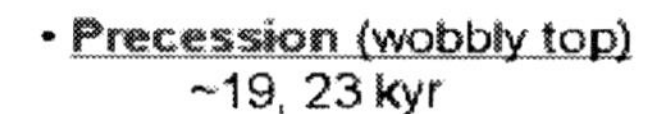

Fig. 6 Milankovitch Cycles.

The questions that arise are threefold: How and why did the periodic ice ages occur? What was the record of global temperature fluctuations like during the Pleistocene epoch (1.6 x 106 years to 1 x 10^4 years from today) and the Holocene epoch (from 1 x 10^6 years from today to today), especially for the first few hundred years of global industrialization (see Table 1 above)? And what was the mechanism of heat capture that kept the Earth warm and warming up? The Milankovitch Cycle turned out to be an acceptable match to the sedimentary data. Yet the answer to the first question is still uncertain.

Consider the mechanism of warming (Table 2, Fig. 5). In Fig. 5, we find there is strong absorption by the greenhouse gases in the spectral range 0.6 and 100 micrometers. This is the range where solar input is captured (*see Table 2 for a list of greenhouse gases*).

It is clear from the spectrum in Fig. 5 that CO_2 and H_2O are the most prominent "greenhouse gases", while methane is a strong contender as well. Both H_2O and CH_4 are largely products of natural phenomena; while H_2O, represents water and ice, and CH_4 is primarily the product of the decomposition of vegetable matter. The largest *anthropogenic compound* is CO_2 produced by the respiration of fauna and combustion of vegetable matter, geologic eruptions, and fossil fuels.

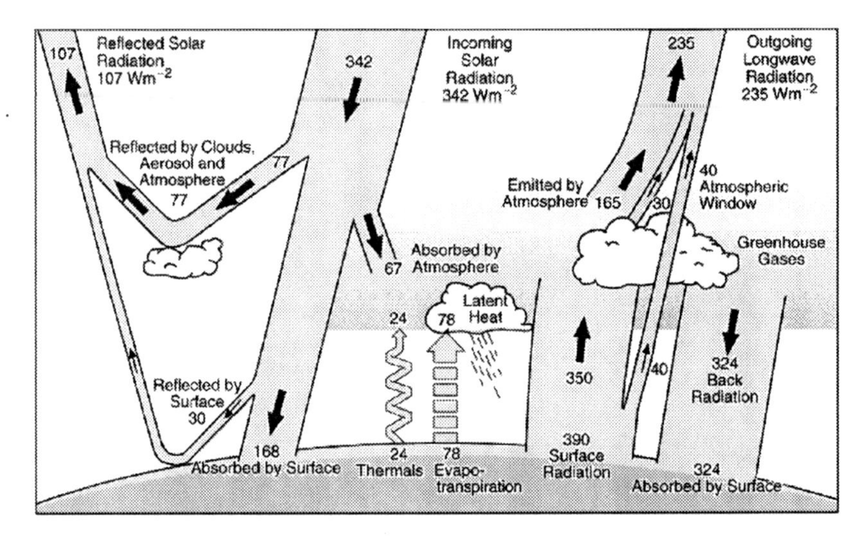

Fig. 6a Radiative budget of the Earth–sun sources.

The radiative budget is complex. Above (Fig. 6a) is a rough diagram of the tortured paths between Solar input and Earth output. The average input at noon on summer day the input is 342 W/m2 and the net output is 235 W/m2 distributed among the various elements as indicated in the diagram some into the atmosphere and some to the surface. The mean total of incident radiation incident on the Earth is 342 Wm^{-2}. Of that quantity, 107 Wm^{-2} are reflected from the Earth surface and from clouds, aerosols, and the atmosphere, while 67 Wm^{-2} are absorbed directly, by the atmosphere. Thermal radiation returns to the atmosphere from the Earth surface by thermal radiation and a 78 Wm^{-2} comes from latent heat through precipitation which brings a total of 390 Wm^{-2}, some of which is reradiated from the warmed Earth. Of this, 40 Wm^{-2} escapes through the atmospheric window and 235 Wm^{-2} as infra red radiation from the night side and 324 Wm^{-2} stored in the greenhouse gases are returned to the Earth surface.

The earliest systematic study of the temporal burden of atmospheric CO_2 has been measured by Charles Keeting et al. (RZ Atmospheric Carbon Dioxide Variations at Mauna Loa Observatory, *Tellus* **28**: 538–51 (1976) (Fig. 7).

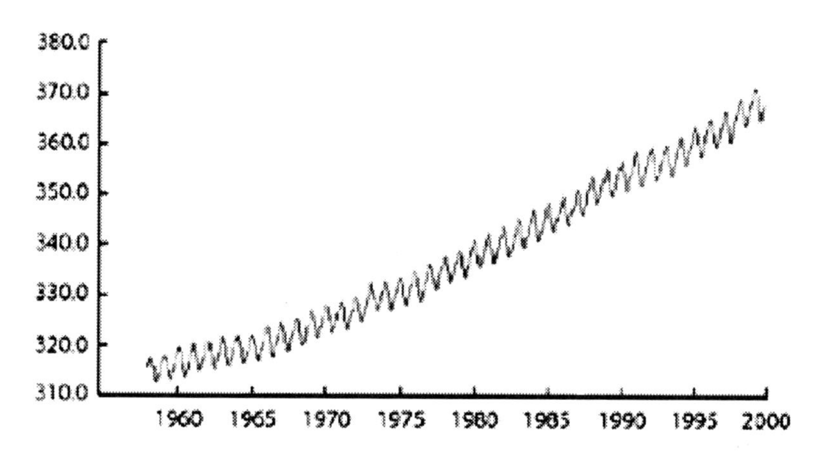

Fig. 7 (*the vertical scale is in parts per million.*)

Rising levels of "greenhouse gases" from the burning of fossil fuels and other human activities cause global warming, with potentially grave consequences for human agriculture and society. One of the clearest determinations over millennia of the content of the atmosphere are taken from analyses of ice cores measured near the Russian Vostok station in Antarctica and in Greenland. They track temperature and atmospheric levels of carbon dioxide (CO_2) and methane (CH_4) from the present back about 160,000 years. (This represents about 11,350 feet of ice accumulation.) The Vostok graphs (fig. 8) as well as the dust concentration and clearly show the systematic rise in global temperature over more than 400,000 years . Also note that, at about 360 parts per million (Table 2) the amount of CO_2 in the atmosphere today far exceeds levels at any time in the past 160,000 years–indeed, in the past few million years. For those worried about global warming, this is a sobering statistic.

The richest sources of temperature and other essential data are from two very stable ice sheets, one in the Antarctic and the other in Greenland. Ancient water is locked in this ice for eons so that cores withdrawn from deep into the ice create a temperature profile through measurements of the hydrogen/deuterium and the O^{16}/O^{18} ratios. Dust and trapped air from bubbles in the ice reveal the composition of ancient air and dust (mostly calcium carbonate). Since the vapor pressures of O^{16} and O^{18} have different thermal profiles, the ratio of the two make a reliable elative temperature determination. Similarly the H_2/D_2 ratios perform the same function. Carbon dating of calcium carbonate is useful for confirming the relative age of each layer, unfortunately over a short period of time because of the relatively short half-life of C^{14}. CO_2 is measured from bubbles in the ice core. It should be noted that the temperature profiles (Fig. 8) indicate a regular sequence of high and low temperatures, but that the profile for the most recent period is noticeably different from the earlier maxima. This modified high temperature structure can be attributed to human activity in the recent centuries

From the Vostok site in Antarctica, a Russian Station, more than three kilometer long cores representing more than one hundred thousand years into the past have yielded remarkable temporal profile as illustrated in Fig. 8.

The Fig. 8 profiles reveal a remarkable succession of warm periods and cold periods all with similar profiles save for the warm period we are now in where the CO_2 profile is more irregular as is the temperature profile.

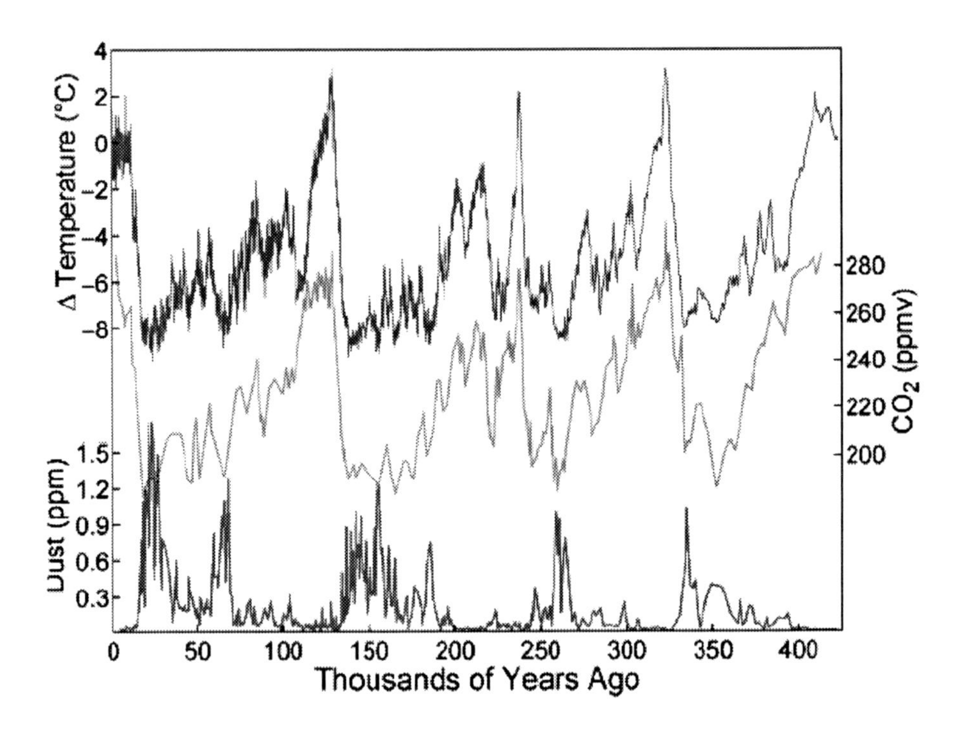

Fig. 8 (temporal intensity of ice core content).

Another element very useful for establishing the temperature at a particular layer is the ratio of O^{16} to O^{18}. Since the heavier O isotope has a slightly smaller vapor pressure, the colder the water the larger is its relative concentration. The isotopic pair, deuterium/hydrogen exhibits similar behavior (Fig. 9).

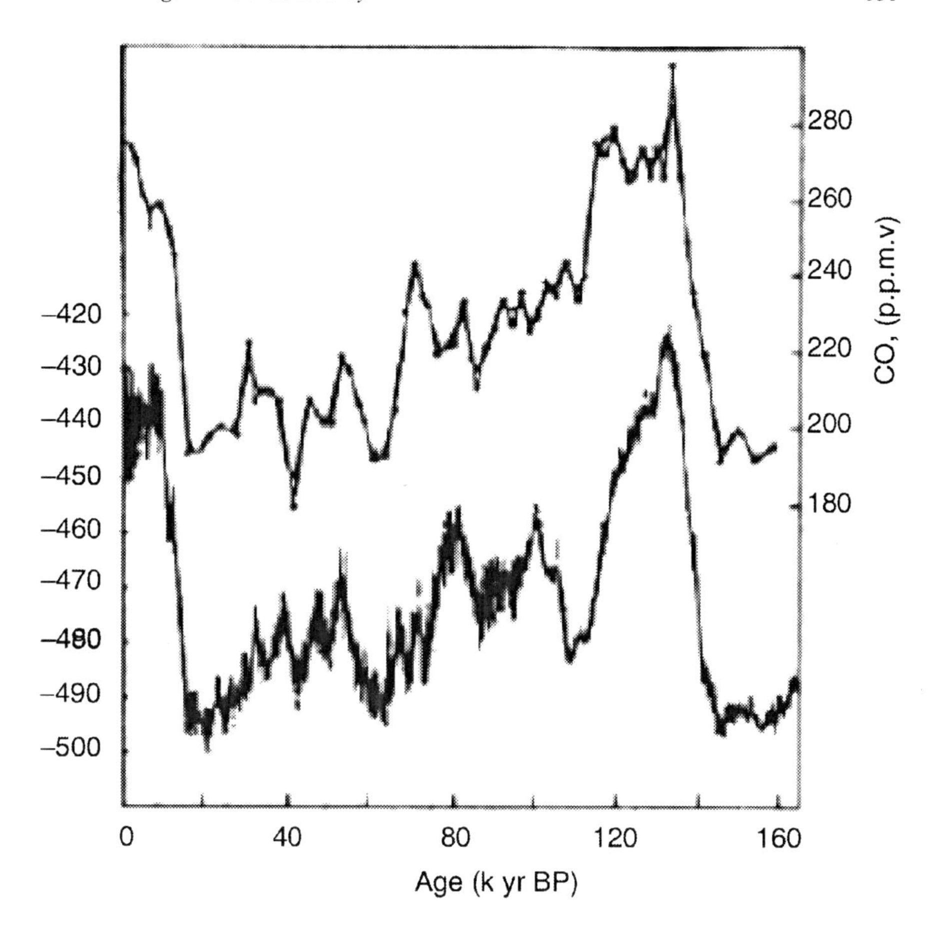

Fig. 9 Comparison of O^{16}/O^{18} to H_2/D_2 Temporal Thermal Profiles.

The validity of the thermal measurements is well authenticated by the nearly identical profiles for O^{16}/O^{18} and H_2/D_2 measurements.

Fig. 10 Cutting of a Segment of a Greenland core.

The most recent report, published earlier this year, is a study of the Greenland North GRIP Project by a group from University of Copenhagen, Alfred Wegner Institute, Bremerhaven, and from the Physics Institute at the University of Bern headed by Anders Svenson of a core from 1330 meters to bedrock at 3085 meters covering the last glacial period. They employed a "visual stratigraphy" which is a technique for enhancing the visibility and producing a digital image of the core (Fig. 12).

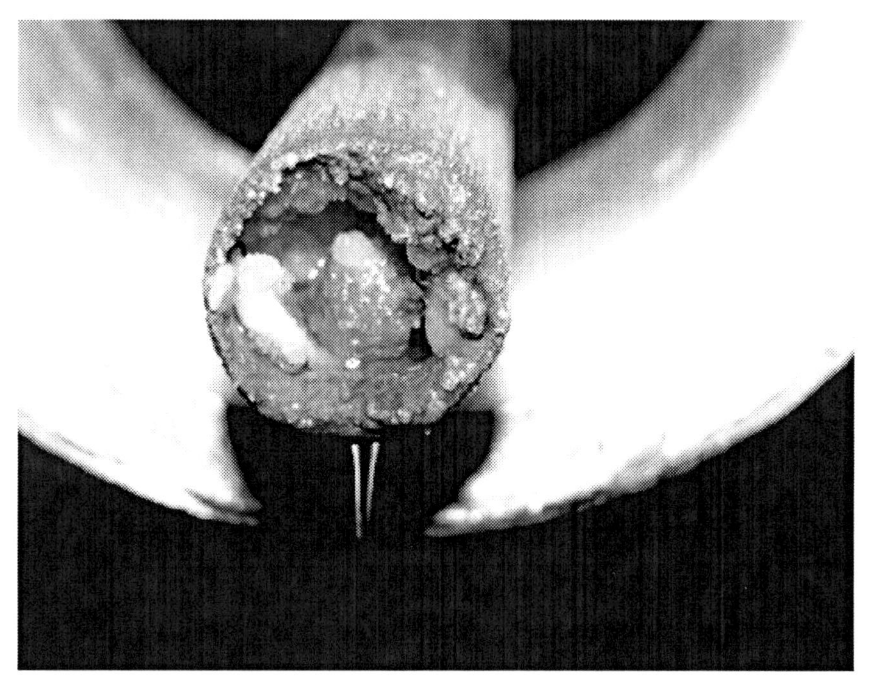

Fig. 11 The Emerging Core of a Greenland Boring.

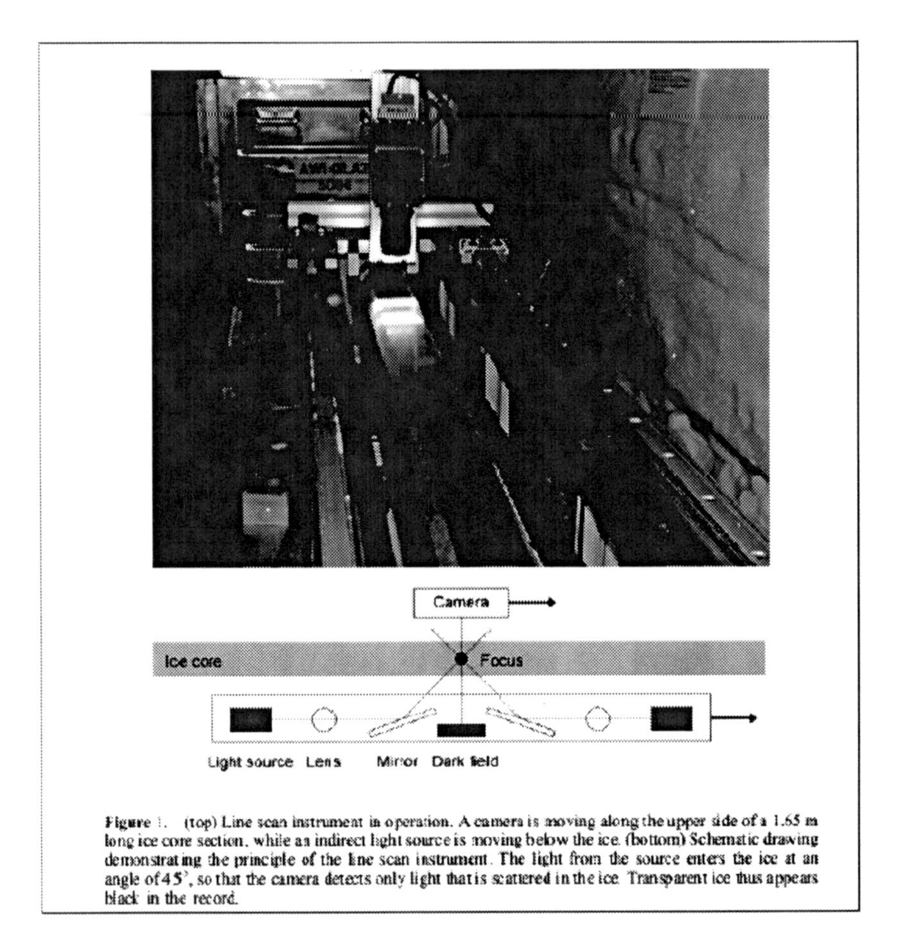

Figure 1. (top) Line scan instrument in operation. A camera is moving along the upper side of a 1.65 m long ice core section, while an indirect light source is moving below the ice. (bottom) Schematic drawing demonstrating the principle of the line scan instrument. The light from the source enters the ice at an angle of 45°, so that the camera detects only light that is scattered in the ice. Transparent ice thus appears black in the record.

Fig. 12 Line Scan Instrument in Operation.

Samples of the results of the Visual Stratigraphic (VS) scans are llustrated in Fig. 13

According to the authors[3] "The obtained VS profiles are generally of very high quality and provides a detailed visual documentation of the entire glacial part of the NorthGRIP ice core, in addition to Prehoreal ice and Eemian ice (2985-3085 meters in depth (see Table 1).

---

[3] Svensson A.S. W. *J. Geophys, Res.* 110, D062108, doi:10.1029/2064JD005134.

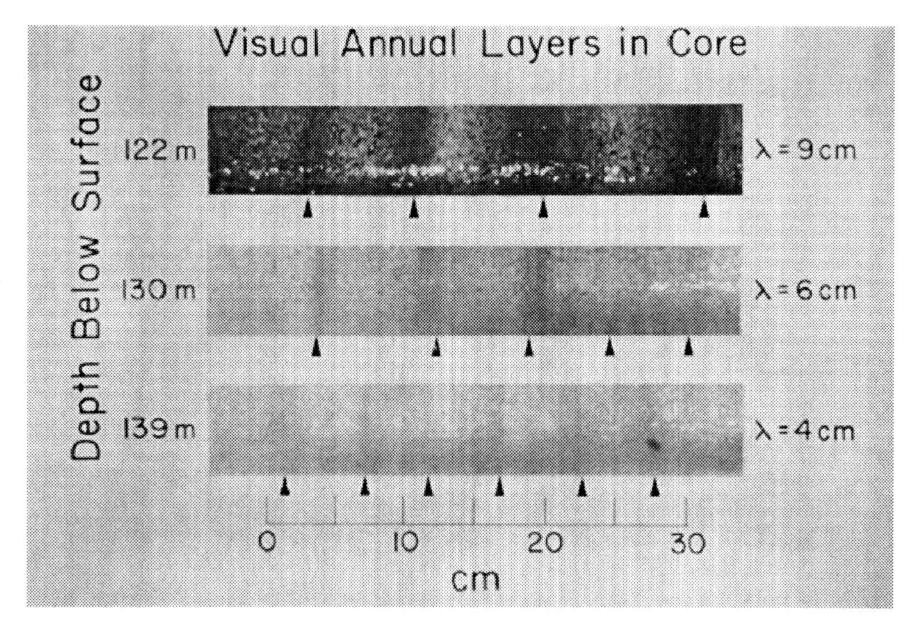

Fig. 13 Examples of the VS Output

Further confirmation of the impact of industrial activity particularly in the western hemisphere is found in the detailed short-term profiles illustrated in Fig. 14.

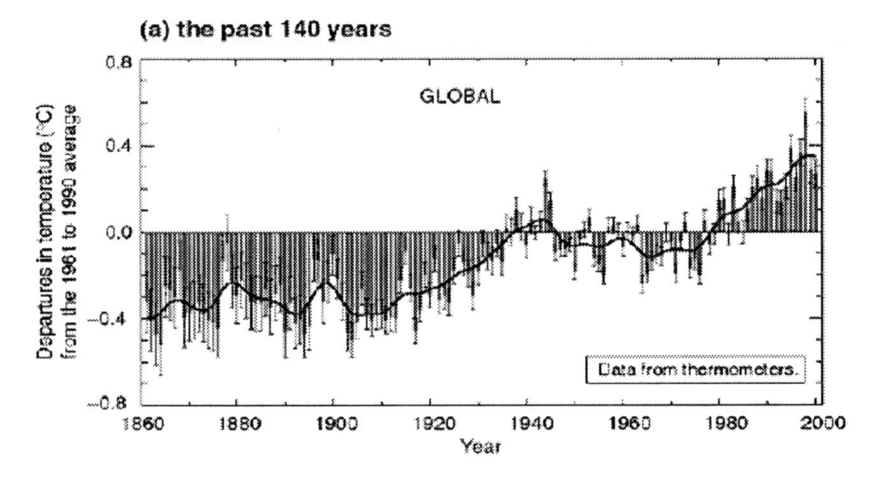

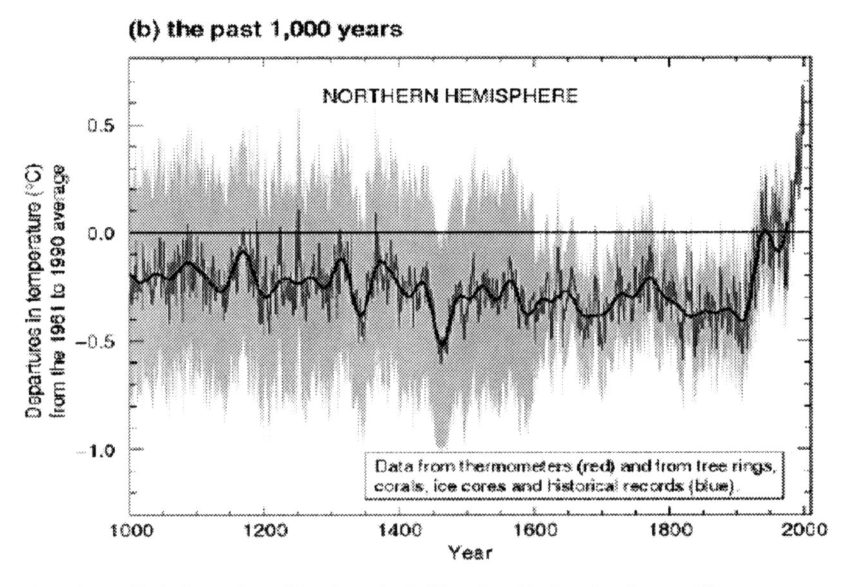

Fig. 14 Detailed Thermal Profiles Over the Millennium Before the Current Time.

5. Conclusions

The ice core evidence and other measurements of climate change leaves no question that industrial activity over the past two centuries has had considerable impact on the rate of Global Warming. The Earth's climate system is so complex that reliable models present daunting challenges for climatologists. Separating *natural* phenomena, over the distant past to the present, from *anthropogenic* influences is fraught with uncertainties. There is a reluctance within the industrial communities to confront the problem realistically, especially since, aside from the accompanying hazards of urban air pollution, evolving shortages of fossil fuel resources, the apparent increase of the number and intensity of catastrophic ocean storms, and mysterious long term droughts particularly on the African continent, the effects of Global Warming are barely visible to a large segment of the Earth's population.

There is, however, ample anecdotal and scientific evidence to raise serious concerns: glaciers are shrinking; the Arctic ice pack is also shrinking giving rise to the abandonment of coastal villages and limiting access to traditional hunting grounds necessary for the survival of the indigenous population; rise in the sea level already threatens some coastal areas. Evidence is evolving that the traditional role of the Gulf Stream circulation in maintaining the warmth of the European continent is undergoing worrisome modifications.

Rising sea level in New York City: the tide gauge at the tip of Manhattan has measured an increase of roughly a quarter meter (ten inches) since 1920. This is similar to what other sea-level gauges around the world have seen since the late 19th century. Thermal expansion of the water due to higher temperature is one main cause. The effects of melting glaciers have been hard to determine.

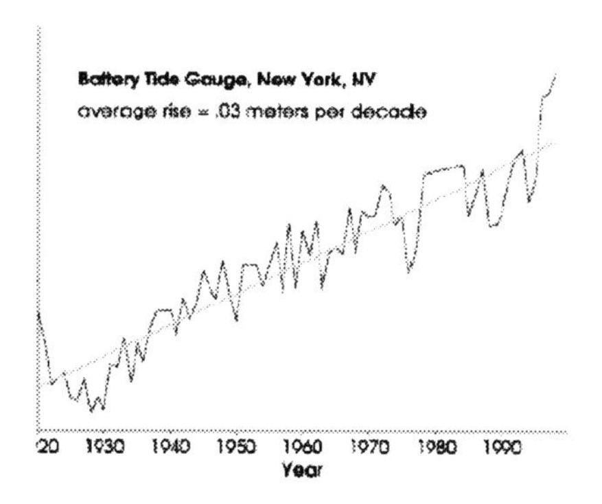

Fig. 15 New York City Sea Level 1920 to 1990.

While it is unlikely that any serious disruption of living conditions over the near term, future generations are clearly at risk. Perhaps the most vulnerable are low lying coastal regions. The Netherlands is already reinforcing the dykes that protect the Polders, large areas of land recovered from the sea. There is evidence already in a steady increase of the sea level, e.g. Fig. 15.

Even should we curtail further greenhouse gas emissions, to reduce pollution and curtail dependence fossil fuels of fossil fuels, the massive inertia of the Earth–Oceans system will continue to influence the climate (Wigley, Tom M. L, *The Climate Change Commitment,* **Science, 307,** 1766–69, (2005)). Thus, while there is urgent need to eliminate terrorism and to settle unrest in so many places in the World, it would be prudent to initiate international long range planning for coping with these threats which surely lie in the not so distant future.

SEMINARS

UV-LASER IRRADIATION OF AMORPHOUS SIO_2: GENERATION AND CONVERSION OF POINT DEFECTS AND POST-IRRADIATION PROCESSES

F. Messina, M. Cannas, and R. Boscaino
Dipartimento di Scienze Fisiche ed Astronomiche dell'Università, Palermo (Italy)

The effect of radiation on amorphous silicon dioxide (silica) is a timely research field due to the wide use of SiO_2 in many optical and electronic technologies, such as optical components in lasers, metal-oxide-semiconductor (MOS) transistors and optical fibers. Exposure of the material to radiation induces stable alterations of its macroscopic properties, often related to generation and conversion processes of point defects. In some cases, these radiation-induced processes deteriorate those properties of silica, such as UV-transparency, on which applications are founded; at variance, in other cases, the effect of radiation can be exploited to induce in pure- or doped-SiO_2 novel physical characteristics, like for example photosensitivity or optical non-linearity. Moreover, point defects in amorphous solids are still a very open research issue since their peculiar properties due to the disordered structure of the host matrix are not yet completely clarified.

We propose here a study of the effects induced by pulsed UV Nd:YAG laser (266 nm) radiation on silica obtained by fusion of natural quartz. This variety of SiO_2 naturally contains small amounts of extrinsic impurities such as germanium, so being a suitable material to study both intrinsic and extrinsic point defects in SiO_2 as well as their photo-induced transformation mechanisms. A convenient experimental approach to investigate point defects embedded in the glass network consists to correlate measurements obtained with different spectroscopic techniques, such as optical absorption (OA), Photoluminescence (PL), and electron spin resonance (ESR). As regards the OA measurements, we used an experimental approach suitable to monitor defect-related absorption bands induced by irradiation *in situ*, i.e. during laser exposure. This is an innovative technique, since most of the current knowledge on radiation-induced point defects in SiO_2 is based on *ex situ* measurements, i.e. performed after the end of exposure.

In our experiments, during laser irradiation we observe the generation of the intrinsic defect known as *E'* center, monitored by the observation of its 5.8 eV-peaked OA band. The *E'* is an important defect in SiO_2 consisting in un unpaired electron on a three-fold coordinated silicon atom (≡ Si•, where each – stands for a bond with oxygen atoms, while the • represents an electron). After the end of exposure we observe a decrease of the concentration of *E'*, accompanied by other two concurrent defect conversion processes: the growth of the extrinsic paramagnetic *H(II)* (= Ge-H•), monitored by ESR measurements, and the decrease of the photo-luminescent *GLPC* (= Ge••), pre-existing in the material prior to UV-exposure. Post-irradiation processes are due to reaction of diffusing hydrogen (H_2) of radiolytic origin with the defects. A comprehensive scheme of chemical reactions explaining the observed post-irradiation processes is proposed and tested against experimental data.

CAVITY RING-DOWN SPECTROSCOPY AS A DETECTOR IN LIQUID CHROMATOGRAPHY

Lineke van der Sneppen, Freek Ariese, Cees Gooijer, Arjan Wiskerke, Wim Ubachs

A novel method for applying cavity ring-down spectroscopy (CRDS) in the liquid phase is presented. Liquid chromatography (LC) is a separation technique that is widely used in many different research areas, e.g. analytical chemistry, pharmaceutical sciences and environmental sciences. Quite often, the scope of these studies is to find traces of the compounds of interest in complex mixtures. Hence, the development of sensitive detection methods is imperative.

Absorption detection is, due to its simplicity and versatility, often the detection method of choice. Since it is non-destructive, this detection method can be used in tandem with other detection methods such as mass spectrometry. The sensitivity of conventional absorbance detection or laser-based techniques is usually dictated by the precision by which $\Delta I/I$ can be determined. Currently, the sensitivity of absorbance detection is limited by the stability of the light source.

CRDS is based on the injection of a short light pulse into a cavity with high-reflectance mirrors, followed by the exponential decay of the energy that is stored in the resonator. Measurements of the rate of decay will be a measure of the absorbance inside the resonator. This technique has two major advantages: the light inside the resonator will only decay significantly after hundreds or thousands of passes, providing for a large path length. Furthermore, the technique is insensitive to laser power fluctuations, since the rate of decay is measured.

In our setup, a flow-cell is formed by clamping a 2-mm thick silicon rubber spacer with a 12-μl hole in it, leak-tight between two high-reflectivity mirrors. The mirrors are in direct contact with the liquid, eliminating scatter losses on surfaces inside the cavity. The eluent from a LC column is introduced inside the cavity via capillary tubing that is inserted into the spacer. Typical ring-down times τ are between 65 and 75 ns. These short ring-down times are due to the short mirror distance and the additional Rayleigh scattering of the solvent. The chromatogram as measured directly in τ can be converted to absorbance units following:

$$\alpha = 2.303\varepsilon C = (n/c)\left[\frac{1}{\tau} - \frac{1}{\tau_0}\right]$$

in which $1/\tau_0$ includes absorption and Rayleigh scatter losses introduced by the liquid and $1/\tau$ is measured when a compound passes through the cavity.

With flow-injection measurements, it is proven that the sensitivity of conventional LC absorbance detectors is surpassed. Measurements of a chromatogram using CRDS in series with a conventional absorbance detector have shown that the peak-to-peak baseline noise, and hence the detection limit, is about a factor of 30 lower for CRDS detection. The peak-to-peak baseline noise is determined to be $2.7 \cdot 10^{-7}$ A.U. as compared to 10^{-4} A.U., which is common for conventional absorbance detectors. The current system can further be improved by designing a Z-shaped flow cell to increase the sample path length or by increasing the laser repetition rate of the laser allowing for more signal averaging.

MAGNETO-OPTICAL SPECTROSCOPY AND SPIN INJECTION

Wolfgang Loeffler
(1)Institut fuer Angewandte Physik, Universitaet Karlsruhe (TH), D-76128 Karlsruhe (Germany)
(2)Lab. fuer Elektronenmikroskopie, Universitaet Karlsruhe (TH), D-76128 Karlsruhe (Germany)
(3)DFG Center for Functional Nanostructures (CFN), Universitaet Karlsruhe (TH), D-76128 Karlsruhe (Germany)

By use of optical spectroscopy in a variable magnetic field we investigate electrical spin injection into semiconductor nanostructures. The spins of the electrons are aligned in a layer of the diluted magnetic semiconductor ZnMnSe which shows giant Zeeman splitting. The carriers relax into an optically active region (InGaAs quantum wells or InAs quantum dots). As the circular polarization of the emitted photons mirrors the spin orientation of the carriers through selection rules, information about spin relaxation processes during the transfer is retrieved.

NEW CRYSTALS OF SARCOSINE COMPLEXES: STRUCTURE, VIBRATIONAL SPECTRA AND PHASE TRANSITIONS

M. Trzebiatowska-Gusowska, J. Baran, M. Śledź, M. Drozd, A. Pietraszko
Institute of Low Temperature and Structure Research, Polish Academy of Sciences
Wrocław 3, Poland

Sarcosine is the N-methyl derivative of glycine ($CH_3NH_2^+CH_2COO^-$). This amino acid belongs to a group of very important biological compounds as it is an intermediate in trans- and demethylation reactions. It forms complexes with inorganic acids and salts of which the most known due to its interesting properties is trisarcosine calcium chloride (TSCC). Since many amino acid complexes exhibit phase transitions, e.g. TSCC, we have undertaken the effort of obtaining new systems and also studying the existing ones.

The results for trisarcosine strontium chloride crystals, a complex that shows phase transition at higher temperatures (388 (cooling)–390 (heating) K), will be presented in the seminar.

Some studies on TSSC were undertaken by C. Haridass and P. Muthusubramanian. But their assumption that TSSC was isomorphus to trisarcosine calcium chloride was not correct as well as the assignment of the bands in IR spectrum.

The crystal structure (at room temperature) was determined by X-ray diffraction with the final R = 0,0294. The compound crystallizes in $P2_1/c$ space group with cell parameters: a = 17,922(4) Å, b = 10,402(2) Å, c = 8,796(2) Å, $\beta = 92{,}08(3)^{\circ}$ and Z = 4. Amino acid molecules reside as zwitterions of three types. All molecules lie in general positions. Crystal possesses pseudo-hexagonal symmetry when viewed down **c** axis. There are six hydrogen bonds of N-H⋯Cl type. Strontium ions are coordinated by seven oxygen atoms arranged in polyhedron. The crystal decomposes at *ca.* 483 K.

TSSC crystals are ortorhombic above the phase transition temperature. This fact was deduced from the data obtained both from the X-ray diffraction measurements which were carried out up to 450 K and also from spectroscopic studies.

The change of the crystal symmetry is forced by the disorder of the oxygen atoms around strontium atoms at higher temperature (quasi-octahedral coordination), which leads to the reduction of indepedent sarcosine molecules (a mirror plane appears).

SCATTERING STATE SPECTROSCOPY AS A PROBE OF CHEMICAL REACTION DYNAMICS AND NON-RADIATIVE ENERGY TRANSFER: LI(NP) + M SYSTEM (M = AR, H_2, N_2,CH_4,C_2H_4,C_2H_6,C_3H_8), (N = 2,3,4)

Solomon Bililign
Department of Physics, North Carolina A&T State University
Greensboro, NC, 27411, USA

Quenching mechanisms of the Li (3p) and Li(4p) states in collision with different molecules (M = Ar, H_2, N_2,CH_4,C_2H_4,C_2H_6,C_3H_8) are studied by far wing scattering state spectroscopy and by *ab initio* quantum chemical calculations.* The Li(3p) state is observed to be efficiently quenched to the Li(3s) state detected as intense 3s→2p emission. The Li(4p) state is efficiently quenched to the Li(4s) and Li(3d) states detected as 4s - 2p and 3d - 2p emission, respectively. Reactive products are probed using a second laser in a pump-probe arrangement by detecting laser-induced fluorescence of LiH (v", J") products. We showed that the Li (2p) + $H_2 \Rightarrow$ LiH + H reaction takes place when the kinetic energy of the reactants are high enough. We think that the Li (2p) + H_2 collisions take place preferentially in bent near-C_2 geometry and that the LiH_2 2A' intermediate plays a major role where wide-amplitude internal vibrations eventually result in losing a hydrogen atom from the LiH_2 complex. We also observed the Li (3p) + $H_2 \Rightarrow$ LiH + H reaction . We show different possibilities for the reactive and nonreactive collisions and provide an explanation of the reaction mechanism using the highly accurate *ab initio* potential energy surfaces. We also show that the harpooning model cannot be used to explain the reaction mechanism in the Li* + H_2 collision for all the atomic states from Li (2s) to Li (3d), and that no long range electron transfer occurs from the metal atom to the hydrogen molecule. Despite the measured large quenching cross sections of quenching of Li (3p) by hydrocarbons we found that only the collision of Li (3p) with C_2H_4 leads to the formation of LiH. For most of the perturbers studied we show that despite the complexity of the collision partners our results show that we can still obtain useful information on non radiative energy transfer processes by using scattering state spectroscopy techniques which serves as a probe of molecular dynamics. ignoring the their internal structure in the interpretation of the absorption profile is not valid as the shapes of the profiles are different for each perturber. For the Li-N_2 system the potential energy surfaces for the Li (2s–4p) N_2 states show a large number of conical intersections and avoided crossings resulting from the couplings between the ionic ($Li^+(N_2)^-$) and covalent configurations.

**All quantum mechanical calculations were performed by Prof. Gwang-Hi Jeung of Chimie Théorique, Université de Provence, Case 521 (CNRS UMR6517), Campus de St-Jérôme, 13397 Marseille Cedex 20, France.*

OPTICAL AND ELECTRICAL INVESTIGATION OF THE PHOTOREFRACTIVE PROPERTIES OF HAFNIUM-DOPED CONGRUENT LITHIUM NIOBATE SINGLE CRYSTALS.

F. Rossella[1], P.Galinetto, G. Samoggia
INFM-Dipartimento di Fisica "A. Volta", Università di Pavia, Pavia, Italy
L. Razzari, P. Minzioni, I. Cristiani, V. Degiorgio
INFM-Dipartimento di Ingegneria Elettronica, Università di Pavia, Pavia, Italy
E. P. Kokanyan
Institute for Physical Research National Academy of Sciences of Armenia, Armenia.

Photoinduced birefringence photoelectronic-transport measurementsin congruent $LiNbO_3$:Hf single crystals with different Hf concentration are presented. The crystalline quality of the crystals is determined using microRaman spectroscopy. Hf doping induces a progressive decrease of photorefractive (PR) damage up to a threshold concentration of about 4 mol%. Direct measurements of dark current, photoconductivity (PhC) and photovoltaic current demonstrate how such a decrease in PR-damage is correlated to a corresponding increase in PhC. Heating effects attributed to nonlinear absorption processes are taken into account to interpretate results. Hf-doping seems at least as effective as usual Mg-doping in reducing photorefraction, requiring a lower dopant concentration. The obtained results are relevant also because, among the known optical-damage-resistant impurities, Hafnium oxide is, up to now, the only additive used to create periodically-poled LN crystals by a direct-growth Czochralski technique.

Acknowledgements.
This work was partially supported by: FIRB-MIUR Project "Miniaturized Systems for Electronics and Photonics ISTC project A-1033 Italian Grant FIRB RBNE01KZ94".

EFFECT OF THE COMPOSITION ON SPECTROSCOPIC AND STRUCTURE OF TM^{3+}: TEO_2-PBF_2 GLASSES

İ. Kabalcı[1], M.L. Öveçoğlu[2], A. Sennaroğlu[3], G.Özen[1]
[1]Department of Physics, Istanbul Technical University, Maslak-İstanbul, Turkey
[2]Dept. of Metallurgical and Materials Engineering, Istanbul Technical University, Maslak-İstanbul, Turkey
[3]Laser Research Laboratory, Department of Physics, Koç University, Sariyer, İstanbul, Turkey

We present the effect of glass compositions on the spectroscopic and structural properties of Tm^{3+} doped tellurite glasses. We used the absorption, luminescence, the differential thermal analysis (DTA), X-ray diffraction (XRD), and scanning electron microscope (SEM) techniques. Judd-Ofelt intensity parameters were determined from the absorption spectra measured at room temperature. Luminescence spectra of each sample shows two emission bands centered at about 1470 nm ($^3F_4 \rightarrow {}^3H_4$) and 1800 nm ($^3H_4 \rightarrow {}^3H_6$). The full band width at half maximum intensity (FWHM) is independent of the composition but varies with the temperature of the sample. We used DTA, XRD, and SEM techniques to examine the structural changes of the undoped glasses as a function of the composition. The compositions contain $PbTe_3O_7$, $\gamma-TeO_2$, $\alpha-TeO_2$, and/or $\delta-TeO_2$, crystalline phases.

FOURIER TRANSFORM INFRARED SPECTROSCOPY ON $ZN_{1-X}MN_XSE$ EPILAYERS

K. C. Agarwal, B. Daniel, M. Hetterich, C. Klingshirn
Institut fuer Angewandte Physik, Universitaet Karlsruhe (TH),
Wolfgang-Gaede-Strasse-1, D-76131 Karlsruhe,
Germany.

The large band gap and the close matching of ZnSe lattice constant to that of GaAs make ZnSe based materials important for variety of potential applications in optoelectronics devices, e.g. blue-green laser diodes and light-emitting diodes. $Zn_{1-x}Mn_xSe$ belongs to the group of II-Mn-VI semimagnetic or diluted magnetic semiconductors (DMS) with energy gap tunable between 2.7–3.4 eV, depending on the Mn composition, temperature and the applied magnetic field. In $Zn_{1-x}Mn_xSe$ magnetic ions (Mn^{2+}) are randomly substituted with the Zn^{2+}, in the cation positions, leading to interesting magneto-optical properties, such as a giant Zeeman splitting of the band edges at low temperatures. The latter made it a possible candidate for spin aligner in spin-based optoelectronic devices. However, information regarding the influence of the Mn composition on several fundamental material parameters is still poor.

To extract the information about electron effective mass we performed room-temperature plasma edge measurements on chlorine-doped n-type $Zn_{1-x}Mn_xSe$ epilayers ($0 \leq x \leq 0.13$). By making Drude-Lorentz-type multi-oscillator fits to our data, we extracted the optical electron effective mass (m^*) in doped Zn(Mn)Se:Cl samples with different Mn content and free-electron concentrations. The doping concentration in our samples was determined using Hall measurements in the van-der-Pauw geometry.

Our results indicate that m^* in $Zn_{1-x}Mn_xSe$ is lower than that for ZnSe. In n-type chlorine-doped ZnSe samples with different free-electron concentrations, m^* varied from 0.133 m_0 to 0.152 m_0, while in $Zn_{0.87}Mn_{0.13}Se$:Cl samples, we found a variation from 0.095 m_0 to 0.115 m_0 within ±9% experimental accuracy. From theoretical calculations, we determined the extrapolated room temperature band edge electron mass in ZnSe and $Zn_{0.87}Mn_{0.13}Se$ to be about 0.132 m_0 and 0.093 m_0, respectively . Additionally, from electrical measurements on $Zn_{1-x}Mn_xSe$:Cl epilayers we found a drastic reduction in the free-electron concentration with increasing Mn composition .

Our preliminary Far Infrared (FIR) studies carried out on a new series of undoped ZnMnSe epilayers suggests a "intermediate-mode" behavior for the optical phonon modes in the composition range $0 \leq Mn \leq 0.43$.

FEMTOSECOND LASER DISSECTION OF NEURONS IN *C. ELEGANS*

Samuel Chung[1] *and Eric Mazur*[1,2]
[1]*Division of Engineering and Applied Sciences,* [2]*Department of Physics; Harvard University, Cambridge Massachussets USA*

Tightly-focused femtosecond laser pulses of a few nanojoules sever individual dendrites in the nematode worm *C. elegans*. Quantification of the resulting behavioral deficits identifies the contribution of the dissected structures. Due to nonlinear absorption of laser light, the dissection has submicrometer resolution with no collateral damage, permitting subcellular surgery on live animals. Future work include an examination of the molecular basis of neurodegeneration that has application to diseases such as Parkinsons and Alzheimers.

ULTRAFAST DYNAMICS OF METALLIC THIN FILMS

Tina Shih, Maria Kandyla, Eric Mazur
Division of Engineering and Applied Sciences,
Harvard University, 9 Oxford Street,
Cambridge, MA 02138, USA

With the coming of the computer age, there is an ever-increasing demand for faster responsivity of materials to be used in technological applications. Such an understanding of the processes that result from matter in extreme states can be obtained through observations of the interaction between matter and light.

The interaction of matter and femtosecond laser pulses generate extreme states in the material that affect the dynamic response of electrons. These dynamics not only define many basic properties of solids such as conductivity and magnetism, but also play a crucial role in adsorption and scattering. Previous research in the Mazur group involved excitation of a semiconductor substrate by an incident femtosecond pulse that created ultrafast semiconductor-to-metal phase-transitions. Using a dual-angle broadband pump-probe setup, the excitation can be probed over a large range of frequencies and the femtosecond-time-resolved dielectric function of the material can be derived. This technique will be extended to analyze the electron thermalization process in metallic thin films under laser excitation. Of the metallic thin films, gold thin films, in particular, are of interest because of its applications as ohmic contacts in the semiconductor industry. The initial experimental setup, along with metallic thin film sample fabrication and research goals will be discussed.

RAMAN SCATTERING OF MG_2SI PRODUCED BY IBS

A. Atanassov, G. Zlateva, and M. Baleva
Faculty of Physics, Sofia University, 5 J. Bouchier Blvd, 1164 Sofia, Bulgaria

Raman scattering is used as a tool for investigating the 'third generation' semiconducting material Mg_2Si. The samples under investigation are prepared at different technological conditions – doze of implantation of the $^{24}Mg^+$ ions, energy of implantation and annealing time – by Ion Beam Synthesis, followed by Rapid Thermal Annealing. They are studied at room temperature, using SPEX 1403 double spectrometer, equipped with photomultiplier, working in the phonon counting mode. The spectra are taken in the range from 100 cm^{-1} to 800 cm^{-1} with 2 cm^{-1} resolution, excited with a wavelength of the Ar^+ laserline 488 nm with power of 60 mW. The Mg_2Si phase is detected in all of the investigated samples and, depending on the technological conditions, it is either in the form of precipitates or as a continuous or quasicontinuous layer.

OPTICAL PROPERTIES OF FEW AND SINGLE ZNO NANOWIRES

Lars Wischmeier, Chegnui Bekeny, Tobias Voss
Institut für Festkörperphysik, Universität Bremen, P.O. Box 330440, D-28334 Bremen

Zincoxide (ZnO) is a direct wide band-gap semiconductor (3.37 eV at room temperature) with an exciton binding energy of 60 meV being large compared to the thermal energy at room temperature of $k_BT \approx 25$ meV. Due to these properties ZnO nanowires have recently attracted many research activities because they are considered to be promising building blocks for nanometer-scale optolectronic devices operating in the blue to UV spectral region. A detailed knowledge about the electronic processes which are involved in the generation of the photoluminescence (PL) or even stimulated emission is of fundamental importance for building and optimizing optical devices. Here, we give an overview of our research activities on the optical properties of single ZnO nanowires.

INVESTIGATIONS OF A FADOF SYSTEM AS AN EDGE-FILTER RECEIVER FOR A BRILLOUIN-LIDAR FOR REMOTE SENSING OF THE OCEAN TEMPERATURE

Alexandru Popescu, Daniel Walldorf, Kai Schorstein and Thomas Walther
Darmstadt University of Technology Institute for Applied Physics, Schlossgartenstr. 7, D-64289 Darmstadt, Germany

Ocean temperature profiles are of great importance in oceanography. An airborne remote measurement technique could provide area-wide data for climate studies and input for weather forecasts. Therefore a practical LIDAR system capable of evaluating the temperature information encoded in the Brillouin scattered light is highly desirable. The technical implementation of such a

system has to withstand a harsh environment aboard of aircrafts. On the transmitter side the system must provide single mode radiation close to the absorption minimum of water. The receiver must be able to resolve the Brillouin shift and be insensitive to vibrations, so that standard interferometric techniques, such as etalons, can not be used. This work focuses on an edge filter technique based on a **F**araday **A**nomalous **D**ispersion **O**ptical **F**ilter, which is capable to act as a sensitive spectroscopic receiver system resolving the small frequency shift caused by the Brillouin-scattering. Our theoretical and experimental studies show that the desired edge-filter characteristics can be achieved. Furthermore FADOF systems are not bound to this particular purpose. Due to the existence of several tuning parameters, such as the magnetic field, the temperature of the gas cell and eventually the pumping geometry for Excited-state FADOFs, the filter transmission spectrum can be tailored to the needs of the application. Therefore, other applications such as filtered Rayleigh scattering might benefit from FADOF
systems as well.

RAMAN STUDY OF THE EFFECT OF THE ANNEALING ON ION-BEAM SYNTHESIZED β-FESI$_2$ LAYERS

M. Marinova, G. Zlateva, and M. Baleva
Faculty of Physics, Sofia University, 5 J. Bouchier Blvd, 1164 Sofia, Bulgaria

The samples under investigation are prepared by ion beam synthesis (IBS), followed by rapid thermal annealing (RTA). A two-step implantational process is performed on n- and p-type Si wafers with two different doses of the implanted Fe^+ ions, D2 = 5×10^{16} cm^{-2} and D3 = 5×10^{17} cm^{-2}, each of them implanted with two different energies E_1 = 60 keV E_2 = 20 keV. Subsequently the implanted samples are annealed at T_a = 900°C for three different times t_a = 60 s, 90 s and 300 s. The influence of the annealing on the formation and the crystal orientation of the β-$FeSi_2$ phase is studied by Raman scattering. The dependence of the Raman peaks intensity on the annealing time indicates an orientational change of the grains, constituting the polycrystalline layers independently on the implantation iron ion dose. The behavior of the spectra in samples, implanted in n- and p-type substrates is similar.

AVOIDING THERMAL EFFECTS IN Z-SCAN MEASUREMENT BY HIGH REPETITION RATE LASERS

Andrea Gnoli[a] and Marcofabio Righini[b]
[a]CNR – Istituto di Struttura della Materia, Area di Ricerca Tor Vergata, via del Fosso del Cavaliere, 100 – 00133 Roma, Italy
[b]CNR – Istituto dei Sistemi Complessi, Sezione di Roma Tor Vergata, via del Fosso del Cavaliere, 100 – 00133 Roma, Italy

The Z-scan technique is a straightforward method to measure both sign and magnitude of the third order susceptibility of a large variety of materials. Since the technique is sensitive to the variation of the intensity dependent refractive index, unwanted effects, such as thermal ones, can hinder the pure electronic nonlinear optical response of the sample. Thermal cumulative effects are usually reported for high repetition rate lasers. In this work, we demonstrate the effectiveness of a simple method to utilize Z-scan technique managing cumulative thermal effects despite of the repetition rate. As suggested by Falconieri [J. Opt. A, **1** (1999) 662] time evolution of Z-scan signal is recorded. We use time correlation of data to extrapolate with high accuracy the instantaneous (i.e., single pulse) nonlinear optical response of the sample. The method allows us to estimate also the refractive index variations due to thermo-optical nonlinearities. Using a mode locked 76 MHz repetition rate laser with 120 fs pulse width we have been able to measure nonlinear electronic susceptibility and thermo-optical nonlinearity in CS_2 and toluene.

TOWARDS CONTROL OF PHOTODYNAMICS BY FEMTOSECOND POLARISATION SHAPING AND COINCIDENCE VELOCITY MAP IMAGING

Arno Vredenborg, Wim G. Roeterdink & Maurice H.M. Janssen*
Laser Centre and Department of Chemistry, Vrije Universiteit
De Boelelaan 1083, 1081 HV Amsterdam, The Netherlands

The use of femtosecond laser pulse shaping is currently proliferating as a tool for the research on photo dynamics of small molecules. Complex pulses can be obtained by a Fourier-domain shaping technique with the use of a Liquid-Crystal Display (LCD) We propose to use the same device for polarization shaping of femtosecond laser pulses The spectral phase modulation can be imposed independently onto two orthogonal polarization directions with a two-layer LCD, therefore giving rise to time-varying polarization states in the laser pulse. This is in contrast to most of the current experiments, femtosecond laser pulses are shaped either in phase or amplitude. These pulses will be used via multiphoton ionization to study and control photo dynamics of small molecules.

The most complete information on photo dynamics can be obtained with the use of the photo-electron/photoion coincidence imaging technique This is a technique in which both the recoiling photo-electron and the correlated ionic photofragment originating from isolated dissociation events are detected. It generates three dimensional energy- and angle resolved images, which reveal the energy distribution over the different photofragments and correlations between them. With this information it is possible to extract the photo-electron angular distribution in the molecular frame. A change in this distribution over time directly reflects the photo dynamics of the system under investigation.

LABORATORY FREQUENCY METROLOGY AND THE SEARCH FOR A TEMPORAL VARIATION OF THE FINE STRUCTURE CONSTANT α ON A COSMOLOGICAL TIME SCALE

E. Salumbides, S. Hannemann, E. Reinhold, I. Labazan, S. Witte, R. Zinkstok, K. Eikema, W. Ubachs
Laser Centre, Vrije Universiteit, Amsterdam

The issue of the possibility of a temporal variation of the fundamental constants has been put high on the agenda of modern physics, now that accurately calibrated spectra of quasars become available. Such data allow for a comparison between the physical constants underlying the spectra for today and those of 12 Billion years ago. Claims have been made that the fine structure constant α has changed outside a 5σ uncertainty. More recent observations of quasar spectra, employing the Very Large Telescope of the European Southern Observatory contradict the claims of the Sydney group, that are based on observations with the Keck telescope; note that differing observations form the Northern and Southern hemisphere might point at space-time effects on spectroscopic properties. Up till now only 23 atomic (and ionic) transitions have been used following the so-called Many-Multiplet analysis, where for each line a *q*-parameter , representing the relativistic dependence of a variation of α on the transition frequency is calculated in atomic structure calculations.

For test on variations of α more spectral lines can be used, but many important line positions of atomic transitions (mostly ionized atoms) are not known to sufficient accuracy. Our group has started a line of research to produce accurate laboratory transition frequencies of some relevant species, also observed in quasars. For this purpose we perform high-resolution spectroscopy and calibrate the transitons with a frequency comb laser, built at the Laser Centre VU. First results on the resonance lines of atomic carbon (at 94.5 nm) and of atomic magnesium (at 202 nm) will be presented. Results include accurate determination of transition isotope shifts. Such effects are of importance to include in the data analysis, since differing isotopic abundances in the quasars may mimic variation of α. Knowledge of isotopic shifts also allows to study another important problem: isotopic evolution in the Universe. This provides a very sensitive test of models of nuclear processes in stars.

The authors acknowlegde a fruitful coolaboration with Prof. V. Flambaum (University of New South Wales, Sydney, Australia)

DIPOLE STRENGTHS OF THE PHOTOSYNTHETIC PIGMENTS AND CAROTENOIDES FROM BACTERIA PROSTECOCHLORIS AESTUARI

Jacek P. Goc, Adam Bartczak and Marek Zaparucha*
Institute of Physics, Poznań University of Technology, Piotrowo, Poznań, Poland

Bacteriochlorophyll and carotenoide pigments have a lot of unique properties, and their inter- and intra-molecular interactions are still a mater of investigations. Photosynthetic pigments are also interested for the application, in a model of molecular converter of solar energy, in photodynamic therapy and diagnostic as well as for molecular bioelectronics. For Chl *a,b* and BChl *a* dipole strength is a linear function of refractive index. Knox formula was also applied to Bchl *c* dissolved in mixed two-solvent system, but linear dependency of dipole strength on refractive index, have been obtained for $Q_y(0,0)$ transition.

In this work, the spectral properties of photosynthetic pigments from anaerobic green bacteria *Prostecochloris Aestuari*: bacteriochlorophyll *c*, bacteriopheophytine *c* as well as betacarotene and hydroxychlorobactene, are characterized. Pigments were dissolved in seventeen organic solvents with wide range of refractive index. From the absorption and fluorescence measurements, basic parameters were calculated: module of squared electronic transition moment (dipole strength), fluorescence quantum yield, fluorescence natural lifetime and fluorescence quenching time. For monomeric BChl *c*, BPhe *c* and β-Carotene, linear dependence of dipole strengths on solvent refractive index, was confirmed. Only for the hydroxychlorobactene this dependence, calculated according to Knox's formula, was nonlinear.

This work was supported by the Polish BW 62/207/2005 grant.

PERSISTENT LUMINESCENCE OF LU_2O_3:TB CERAMICS

J. Trojan-Piegzal[1], *E. Zych*[1], *J. Niittykoski*[2,3], *J. Hölsa*[2]
[1] *Faculty of Chemistry, University of Wroclaw, Wroclaw, Poland*
[2] *Department of Chemistry, University of Turku,Turku, Finland*
[3] *Graduate School of Materials Science, Turku, Finland*

Persistent luminescence properties of Lu_2O_3:Tb and Lu_2O_3:Tb, M (M = Mg, Ca, Sr, Ba) vacuum-sintered pellets were investigated. The ceramics show a long lasting luminescence which can be easily observed by eye 15-20 hours after ceasing the irradiation with light of wavelength shorter than about 330 nm. It was shown that the Ca co-doping of Lu_2O_3:Tb ceramics simplifies their thermoluminescence spectra structure leaving practically only one intense band peaking at about 120°C. The decay of the persistent emission is non-exponential. The time dependence of the emission reciprocal intensity is, however, perfectly linear, which proves that the process is ruled by the second order kinetics.

SPECTROSCOPIC MONITORING OF THE FORMATION OF DISPERSED MANGANESE OXIDE ON ALUMINA SURFACE

L. Davydenko, I. Babich, Y. Plyuto, A. D. van Langeveld, J. A. Moulijn
Institute of Surface Chemistry, National Academy of Sciences of Ukraine, Naumov Str. 17, 03164 Kiev, Ukraine, Delft ChemTech, Delft University of Technology, Julianalaan 136, 2628 BL Delft, The Netherlands

Supported manganese oxides are widely used in catalysis and sorbtion. The activity of such systems correlate with dispersion of supported phase. Using of acetylacetonate precursors is known to be promising way to achieve high dispersion of supported oxide phase.

The aim of the present work is the design of dispersed manganese oxide on the alumina surface by using manganese (III) acetylacetonate ($Mn(acac)_3$) precursor. The questions which should be cleared during the work are processes which take place during the deposition of $Mn(acac)_3$ on alumina support, an influence of support surface on the thermal transformation of $Mn(acac)_3$, composition and distribution of surface species after thermal activation of alumina supported $Mn(acac)_3$.

Combination of several spectroscopic methods gives opportunity to monitor all stages of formation of alumina supported dispersed manganese oxide obtained by deposition and subsequent thermal activation of $Mn(acac)_3$ on alumina surface.

By using FTIR spectroscopy deposition mechanism of $Mn(acac)_3$ on alumina surface was cleared. At low loadings (0.35, 0.74 and 1.38 $Mn(acac)_3/nm^2$) $Mn(acac)_3$ interacts with the hydroxyl groups of alumina surface due to substitution of acetylacetonate ligand(s) with the oxygen atom(s). Acetylacetonate which is evolved, reacts with Al^{3+} sites of the alumina surface:

$$Al_{surf} - OH + Mn(acac)_3 \rightarrow (Al_{surf} - O)_x Mn(acac)_{3-x} + acacH$$

$$Al^{3+}_{surf} + x\, acacH \rightarrow \left[(Al^{3+}_{surf})\, acac_x\right]^{(3-x)+} + H_{surf}$$

At high loadings (2.38 and 3.5 $Mn(acac)_3/nm^2$) the excess of $Mn(acac)_3$ is deposited on the alumina surface in the form of bulk $Mn(acac)_3$.

By combination of TG analysis and FTIR spectroscopy the set of intermediate compounds was determined for the process of decomposition of bulk $Mn(acac)_3$. Thermal decomposition of bulk as well as alumina supported $Mn(acac)_3$ occurs by oxidative destruction of ligands and results to manganese oxocarbonate and supported manganese oxide, respectively. The temperatures of transformations were defined.

Raman mapping technique and XRD spectroscopy gave the information about composition, dispersion and distribution of manganese oxospecies over alumina support after thermal decomposition of supported $Mn(acac)_3$ in air. In the case of the sample with molecular level of dispersion of supported $Mn(acac)_3$ disperse randomly distributed Mn_2O_3 and Mn_3O_4 surface species are formed. If formation of bulk $Mn(acac)_3$ occurs on alumina surface during the deposition thermal treatment leads to highly dispersed Mn_3O_4 species covered by layer of crystalline Mn_2O_3.

TRANSIENT LOCAL STRUCTURE OF SOLVATED TRANSITION METAL COMPLEXES PROBED BY PICOSECOND TIME-RESOLVED XAFS

W. Gawelda[1,2], B. Sobanek[1], V. T. Pham[1], M. Kaiser[1], Y. Zaushitsyn[1,2], A. Tarnovsky[1,2], D. Grolimund[2], S.L. Johnson[2], R. Abela[2], M. Chergui[1] and C. Bressler[1]
[1]Laboratory of Ultrafast SpectroscopyInstitute of Chemical Sciences and Engineering
Swiss Federal Institute of Technology - EPFLCH-1015 Lausanne, Switzerland
[2]Swiss Light SourcePaul Scherrer Institut CH-5232 Villigen PSI, Switzerland

Time-Resolved X-ray Absorption Fine Structure Spectroscopy is a new method to study short-lived reaction intermediates and radicals in the condensed phase. With this technique we can access the transient local geometric structure (bond distances) of the complex via its EXAFS, while bound-bound transitions of the near-edge (XANES) region serve as a sensitive probe of the local electronic structure including the geometrical symmetry of the studied molecule. We have employed this technique to study the time-dependent structure of the triplet Metal-to-Ligand Charge Transfer (^{3}MLCT) states of aqueous ruthenium (II) tris-(2,2')bipyridine, $[Ru^{II}(bpy)_3]^{2+}$. Upon photoexcitation of $Ru(bpy)_3$ with 200 fs/400 μJ/400 nm laser light, a 4*d* electron of the Ru central atom is excited to the singlet ^{1}MLCT state, which undergoes ultrafast intersystem crossing into the manifold of lower-lying triplet ^{3}MLCT excited states, where the electron stabilizes on the bpy ligand system within < 300 fs. The excited state XANES in the L_3 and L_2 regions reveals the change of the crystal field splitting, 10 Dq, between $4d(t_{2g})$ and $4d(e_g)$ levels of the central Ru atom. The measured difference of 10 Dq by 0.15 eV between the ground (3.74 eV) and the excited states (3.89 eV) indicates a decrease in the average Ru-N bond length of about –0.02 Å. Likewise, the EXAFS analysis of the short-lived reaction intermediate shows a nearly identical bond length decrease. In addition, applying Natoli's rule to the measured shift of XAFS features, we can correlate their blue shift in the excited state with the structural contraction of the Ru-N bond ($\Delta R = -0.04$ Å). Although the excited state spectra are accurate only within the 6% of the absorption edge jump, we can derive bond length *changes* to a much greater accuracy via our measurement of transient absorption XAFS when we rely on the well-known static structure of the compound under investigation.

STRUCTURAL AND LUMINESCENCE PROPERTIES OF EUROPIUM DOPED $BA_{1-X}SR_XTIO_3$ NANOCRYSTALLITES PREPARED BY DIFFERENT METHODS

R. Pazik[1], *D. Hreniak*[1], *W. Strek*[1], *A. Speghini*[2], *M. Bettinelli*[2]
[1]*Institut of Low Temperatures and Structure Research, Polish Academy of Science, Wroclaw, Poland*
[2]*Dipartamento Scientifico e Tecnologico, Universita di Verona, Verona, Italy*

$Ba_{1-x}Sr_xTiO_3$ (BST) powder samples have been prepared using three different techniques: molten salts, sol-gel and hydrothermal method in order to compare their structural and luminescence properties. Concentration of Sr^{2+} was chosen from 0.05 to 0.25 mole percentage. Moreover, BST samples were doped with 1% of Eu^{3+} ions, taking into account Ba/Sr molar ratio, and then they were used as an optical probe for the investigation of structure changes. Sol-gel and hydrothermally obtained BST powders grain size were below 60 nm, whereas samples fabricated by molten salts method were about 500nm. It has been found that the luminescence characteristics are strongly dependent on the synthesis method. All powders demonstrate a luminescence behavior characteristic to the lack of inversion symmetry of Eu^{3+} sites. Substitution of Ba^{2+} to Sr^{2+} results in shift of the zero phonon line towards higher energies along with increase of Sr^{2+} concentration.

WAVEGUIDE STRUCTURES WRITTEN WITH FS-LASER PULSES ABOVE THE CRITICAL SELF-FOCUSING THRESHOLD IN SIO_2-PBO BASED GLASSES

V. Diez Blanco, J. Siegel and J. Solis
Instituto de Óptica, CSIC, Serrano 121 28006 Madrid, Spain

The use of ultrashort (fs) laser pulses to generate 3D-optical structures inside dielectric media has recently become an important research topic. This technique is compromised by non-linear propagation phenomena which can lead to catastrophic material damage when exceeding a certain threshold power, P_{cr}. In this work, we demonstrate that waveguide structures can be generated in SiO_2-PbO glasses presenting high values of the linear, n_0, and non-linear, n_2, refractive indices in spite of working above the corresponding P_{cr}. We also propose multiple structures as an alternative route by means of which we can control the shape and the size of the guided mode.

THE LATTICE RESPONSE OF QUANTUM SOLIDS TO AN IMPULSIVE LOCAL PERTURBATION

L. Bonacina, P. Larrégaray, F. van Mourik, M. Chergui, and V. Troncale
Ecole Polytechnique Fédérale de Lausanne
Laboratoire de Spectroscopie Ultrarapide
ISIC, FSB-BSP, EPFL, CH-1015 Lausanne, Switzerland

We address the ultrafast dynamics of a *quantum solid*: crystalline hydrogen. This is accomplished by optical excitation of a dopant molecule, Nitric Oxide (NO), to a large orbital Rydberg state, which leads to a bubble-like expansion of the species surrounding the impurity. The dynamics is directly inferred from the time-resolved data, and compared with the results of molecular dynamics simulations. We report the presence of three time-scales in the structural relaxation mechanism: the first 150 fs are associated with the ultrafast inertial expansion of the first shell of lattice neighbors of NO. During the successive 0.7 ps, as the interactions between the molecules of the first and of the successive shells increase, we observe a progressive slowing-down of the bubble expansion. The third timescale (~10 ps) is interpreted as a slow structural re-organization around the impurity center. No differences were observed between the dynamics of *normal*- and *para*-hydrogen crystals, justifying the simplified model we use to interpret the data, which ignores all internal degrees of freedom of the host molecules. The molecular dynamics simulations reproduce fairly well the static and dynamic features of the experiment. In line with the measurements, they indicate that the quantum nature of the host medium plays no role in the initial ultrafast expansion of the bubble. The overall response resembles that of a liquid.

APPLICATION OF IN-PLANE TRANSPORT IN GAAS-BASED NANOSTRUCTURES FOR BROAD BAND AND SELECTIVE THZ SENSING

D. Seliuta
Semiconductor Physics Institute,
Gostanto 11, LT-01108,
Vilnius, Lithuania

The range terahertz (THz) frequency is becoming increasingly important in a large variety of scientific and electronic applications. THz spectroscopy allows to detect and identify various chemical materials via their spectroscopic signatures; THz radiation is a powerful tool in making different type of images in medical and security applications, it has a big perspective in satellite communication, etc. One of the key issues in the evolution of the THz electronics is the development of compact solid state radiation sources and sensitive broad band detectors

Semiconductor nanostructures are essential ingredients in many modern optoelectronic devices. As their design – well width, barrier thickness, doping profile, etc – more than the material itself defines the properties of a structure, it is possible to fabricate a compound having desired features fo requested task. These facilities, within recent years in particular, gained huge attention due to the possibility to apply the to the above mentioned needs of THz electronics.

In this work we present two different concepts of THz sensing applying properties of an in-plane carrier transport in GaAs-based quantum wells. The first one – so-called "bow-tie" diode – is promising due to the broad frequency band and room temperature operation. The device is fabricated from GaAs/AlGaAs-modulation-doped structure and combines the confluence of the features of a bow-tie antenna (here the origin of the name) and non-uniform two-dimensional electron heating. More specifically, on the one hand, it, as antenna, allows coupling of the incident radiation to the structure, on the other hand, its asymmetrical shape gives rise to non-uniform electron heating at the apex and consequently inducing the electromotive force along the structure. We demonstrate that the detection sensitivity has a wide plateau, from 10 GHz up to 0.8 THz (Electr. Lett. **40**, p. 631, 2004) with the sensitivity of 0.3–0.5 V/W (Acta Physical Polonica A **107**, P 184, 2005). With the decrease of temperature down to the liquid nitrogen, the sensitivity rises proportionally to the carrier mobility and reaches a value of 20 V/W in the GHz range (Semicond. Sci. Technol. **19**, p. S436, 2004).

The second approach is attractive because of its selectivity and option to tailor the detection wavelength by controlling the GaAs/AlAas quantum well parameters. It is known that in GaAs beryllium is relatively stable acceptor, commonly used electronic devices. Its binding energy (28 meV in the bulk) can be increased substantially when incorporated into GaAs/AlAs quantum wells. Here we present optical and terahertz photoreflectance spectra and the surface photovoltage spectra allows estimate of internal electric fields, interband transition energies and broadening of the special features. In the THz, photoconductivity measured under pulsed and continuous wave THz excitation within the range 1.6–7.5 THz at low (4 K-40 K) temperatures is consistent with Be transition energies obtained from the photoluminescence data and from theoretical calculations.

MICROSTRUCTURE AND PROPERTIES OF PERIODIC MULTILAYER THIN-FILM STRUCTURES WITH NANOCRYSTALS

Olga Goncharova
Institute of Molecular and Atomic Physics,
National Academy of Sciences of Belarus,
F. Scaryna Prospect 70, 220067 Minsk, Belarus

Recently there has been an enormous interest in film nanostructures composed of nano-sized crystals and/or local regions of colour centres (CC), because they possess strong optical nonlinearity and/or luminescence, a great potential for miniature modulators, light-emitting sources, dosimeters. Among these film nanostructures, a class of nanolayer-by-nanolayer nanostructured films is one of the most attractive thin film structures. These structures composed of semiconductor A^2B^6 nanocrystals (NC) exhibit large, ultrafast third-order optical nonlinearity at visible wavelengths as well the periodic structures formed of irradiated dielectric NCs have efficient luminescent CCs. Extra feature of NC composed film structures in comparison with bulks is the big relation of NC surface area to their volume , which is in inverse proportion to their linear size. Various film nanostructures had been extensively investigated , but determination of the correlations between their distinctive properties and their microstructure (size, ordering, arranging, developed intrinsic surface of NCs) have not been completed .

The present report is devoted to the specific properties demonstration and consideration of their correlations with microstructure and intrinsic surface of NC lithium fluoride (LiF) and cadmium sulfide (CdS) films, and also of two-componential LiF/CaF_2, CdS/CaF_2, CdS/SiO_2 structures, prepared by the same method on silica substrates. The developed surface in alkali-halide nanostructures explains features of CCs formation in them under irradiation. Among such features detected in our experiments are the followed ones. In nanostructures (i) CCs formation occurs more effectively; (ii) more complex CCs (F_2 and F_3^+) are formed more effectively, than more simple CCs (F); (iii) there is more substantial improvement of conditions for F_3^+-CCs formation in comparison with ones for F_2-CCs. Some examples of distinctive properties of film structures with semiconductor NCs, related to their microstructure, are also will be discussed. Among former properties are correlations detected between the size and conductivity of NCs and their luminescent spectra, optical nonlinearity spectra, photoexcited carriers dynamics.

STRUCTURAL AND OPTICAL STUDIES OF ERBIUM DOPED NANOCRYSTALLINE SILICON THIN FILMS PRODUCED BY R.F. SPUTTERING

Cristina Gaspar de Oliveira[1], Paulo Coelho de Oliveira[1]
[1]Departamento de Física, Instituto Superior de Engenharia do Porto, Rua Doutor António B. Almeida 431, 4200-072 Porto, Portugal.

Both, the nanocrystalline silicon (nc-Si) and Er- doped silicon have been attracting enormous interest of researchers as promising candidates for the realization of Si-based visible and infrared light emitters. This last take on special significance for optical communication systems due to their emission line at 1.54 μm (originating by the intra-4f transitions of Er^{3+} ions), which passes the absorption minimum of silica-based glass optical fibers. However, bulk crystalline Si:Er operating under forward bias still present unsolved problems like the strong temperature quenching of Er emission. The situation may be considerably improved by incorporation of Er ions in nanocrystalline silicon due to the band- gap widening of nanometer size Si, which consequently has to result in reducing of the thermal quenching of Er luminescence.

In this contribution we show the ability to produce by the reactive magnetron sputtering method Er-doped nc-Si:H thin films emitting at room temperature. The advantages of these films, if compare to usually investigated nanocrystal containing SiO_2 structures, is their relative high conductivity, which makes the material attractive for device applications. Erbium doped nanocrystalline silicon thin films were produced by reactive magnetron sputtering on glass substrates under several different conditions (RF power, temperature, Er content and gas mixture composition). In this work we will discuss the PL of Er centers as a function of crystal sizes and more important as a function of crystal distributions (isolated or agglomerated) on the matrix. We have shown that the position of the Er^{3+} PL peak is a function of kind of crystal distribution on the matrix. Efficient photoluminescence was observed in these structures. Since the luminescence efficiency of Er doped nc-Si films is strongly influenced by the microstructure and impurity content of the layers we also discuss the PL characteristic as a function of film anatomy obtained by spectroscopic ellipsometry.

MICROSTRUCTURE AND THERMAL FEATURES OF A-SI:H AND NC-SI:H THIN FILMS PRODUCED BY R.F. SPUTTERING

Paulo Coelho de Oliveira [1], Cristina Gaspar de Oliveira [1]
[1]Departamento de Física, Instituto Superior de Engenharia do Porto, Rua Doutor António B. Almeida 431, 4200-072 Porto, Portugal.

We have grown undoped a-Si:H and nc-Si:H thin films using a reactive RF sputtering apparatus, with a varied Si nanocrystal fraction.

The produced samples have been studied by several techniques. For the structural characterization we used X-ray in the grazing incidence geometry and Raman spectroscopy. For the thermal properties we use thermal diffusivity measurement using the photothermal beam deflection technique.

The optical study, namely absorption coefficient, refractive index and energy gap was obtained using Tauc plot.

With the measurements of X-ray and Raman give us the ability to determine average crystal sizes and crystalline volume fraction of the samples.

We verify that more crystalline the sample less step is the transition present in the optical density graph and bigger is the Tauc energy gap.

Thermal transport properties of amorphous and nanocrystalline silicon samples have been measured.

The measurements at high frequencies allow the elimination of the substrate contribution and then the measurements of the thermal diffusivity of the film. We verify that the thermal diffusivity increases with the crystallinity.

CONFOCAL MICROSCOPY FOR INVESTIGATIONS ON LITHIUM FLUORIDE COLOUR CENTERS

Salvatore Almaviva, Giuseppe Baldacchini, Francesca Bonfigli, Rosa Maria Montereali, Maria Aurora Vincenti.
ENEA, Centro Ricerche Frascati, Via E. Fermi 45, 00044 Frascati, Italy.

Low penetrating radiations can create primary and aggregate defects in alkali halide crystals. The current search for novel light emitting materials has demonstrated the relevance of the emission features of laser active colour centres (CCs) in lithium fluoride (LiF) for optoelectronic devices. Such CCs are characterized by broad absorption bands in the visible and near UV spectral range and by efficient, broad-band emissions from the visible to the NIR. The increasing demand for low-dimensionality photonic devices imposes the use of advanced irradiation methods for producing luminescent structures with high spatial resolution. Photoluminescent patterns produced in LiF by soft X-rays have been investigated with Confocal Laser Scanning Microscope (CLSM) in fluorescence mode, in order to show the potentialities offered by this powerful microscopic technique. CLSM can be used for high spatial resolution imaging and to have details about CCs depth distribution in irradiated LiF crystals and films, using its focal plane-sectioning property. Submicrometric details of geometric structures as well as of biological samples and living cells have been observed by CLSM technique.

CAVITY RING-DOWN SPECTROSCOPY AS A DETECTOR IN LIQUID CHROMATOGRAPHY

Lineke van der Sneppen, Freek Ariese, Cees Gooijer, Arjan Wiskerke, Wim Ubachs

A novel method for applying cavity ring-down spectroscopy (CRDS) in the liquid phase is presented. Liquid chromatography (LC) is a separation technique that is widely used in many different research areas, e.g. analytical chemistry, pharmaceutical sciences and environmental sciences. Quite often, the scope of these studies is to find traces of the compounds of interest in complex mixtures. Hence, the development of sensitive detection methods is imperative.

Absorption detection is, due to its simplicity and versatility, often the detection method of choice. Since it is non-destructive, this detection method can be used in tandem with other detection methods such as mass spectrometry. The sensitivity of conventional absorbance detection or laser-based techniques is usually dictated by the precision by which $\Delta I/I$ can be determined. Currently, the sensitivity of absorbance detection is limited by the stability of the light source.

CRDS is based on the injection of a short light pulse into a cavity with high-reflectance mirrors, followed by the exponential decay of the energy that is stored in the resonator. Measurements of the rate of decay will be a measure of the absorbance inside the resonator. This technique has two major advantages: the light inside the resonator will only decay significantly after hundreds or thousands of passes, providing for a large path length. Furthermore, the technique is insensitive to laser power fluctuations, since the rate of decay is measured.

In our setup, a flow-cell is formed by clamping a 2 mm thick silicon rubber spacer with a 12 μl hole in it, leak-tight between two high-reflectivity mirrors. The mirrors are in direct contact with the liquid, eliminating scatter losses on surfaces inside the cavity. The eluent from a LC column is introduced inside the cavity via capillary tubing that is inserted into the spacer. Typical ring-down times τ are between 65 and 75 ns. These short ring-down times are due to the short mirror distance and the additional Rayleigh scattering of the solvent. The chromatogram as measured directly in τ can be converted to absorbance units following:

in which $1/\tau_0$ includes absorption and Rayleigh scatter losses introduced by the liquid and $1/\tau$ is measured when a compound passes through the cavity.

$$\alpha = 2.303\varepsilon C = (n/c)\left[\frac{1}{\tau} - \frac{1}{\tau_0}\right]$$

With flow-injection measurements, it is proven that the sensitivity of conventional LC absorbance detectors is surpassed. Measurements of a chromatogram using CRDS in series with a conventional absorbance detector have shown that the peak-to-peak baseline noise, and hence the detection limit, is about a factor of 30 lower for CRDS detection. The peak-to-peak baseline noise is determined to be $2.7 \cdot 10^{-7}$ A.U. as compared to 10^{-4} A.U., which is common for conventional absorbance detectors. The current system can further be improved by designing a Z-shaped flow cell to increase the sample path length or by increasing the laser repetition rate of the laser allowing for more signal averaging.

OPTOELECTRONIC DEVICES USING FEMTOSECOND LASER MICROSTRUCTURED SILICON

James Carey, Brian Tull, Michael Sheehy ,Eric Diebold and Eric Mazur
Division of Engineering and Applied Sciences, Harvard University
Cambridge, Massachusetts 02138 USA

Irridiating silicon surfaces with trains of ultrashort laser pulses in the presence of a sulfur containing gas drastically changes the structure and properties of silicon. The normally smooth and highly reflective surface develops a forest of sharp microscopic spikes. The microstructured surface is highly absorbing even at wavelengths beyond the bandgap of silicon and has many interesting novel applications including infrared photodetectors, solar cells, and flat panel displays.

INDEX

Printed in the United States
53987LVS00001B/52-69

9 781402 047886